3rd International Conference on Inventive Research in Material Science and Technology (ICIRMCT 2021)

IOP Conference Series: Materials Science and Engineering
Volume 1091

Coimbatore, India
22 - 23 January 2021

Editors:

S. Smys
Subarna Shakya
Joy Iong-Zong Chen
M. V. Reddy

ISBN: 978-1-7138-4592-8
ISSN: 1757-8981

Printed from e-media with permission by:

Curran Associates, Inc.
57 Morehouse Lane
Red Hook, NY 12571

Some format issues inherent in the e-media version may also appear in this print version.

Institute of Physics
Dirac House, Temple Back
Bristol BS1 6BE UK

Phone: 44 1 17 929 7481
Fax: 44 1 17 920 0979

techtracking@iop.org

Additional copies of this publication are available from:

Curran Associates, Inc.
57 Morehouse Lane
Red Hook, NY 12571 USA
Phone: 845-758-0400
Fax: 845-758-2633
Email: curran@proceedings.com
Web: www.proceedings.com

TABLE OF CONTENTS

Preface

This Conference Proceedings volume contains the written versions of most of the contributions presented during the conference of ICIRMCT 2021. The Conference has provided a platform to share and exchange the recent innovations and developments in the areas of material science and technology.

The Conference has emerged as an exclusive opportunity for the participants from all over the globe to present and discuss the reviews and results in their respective material science research areas. The goal of this conference is to provide a platform for both industrialists and academicians to promote, share, and discuss various new issues and developments in evolving areas of Material Science and Technology applications. It includes a selection of 77 papers from 272 papers submitted to the conference from universities and industries all over the world. All the accepted papers were subjected to double blind review process by 2-4 expert referees. The papers have been selected for this volume by solely considering their quality and the relevance to the conference.

ICIRMCT 2021 would like to express our sincere gratitude to all the authors for their dedicated contributions to this book. We would like to extend our thanks to all the referees for their constructive comments on all papers, especially, we would like to thank their organizing committee for their sincere and dedicated work. Finally, we would like to thank the AIP publications for producing this volume.

Dr. S. Smys
Conference Chair
ICIRMCT 2021

VIRTUAL DISCLAIMER

In order to serve the research communities better even during the global pandemic situation like COVID-19, unlike the previous year conferences, 3rd ICIRMCT 2021 was hosted as a virtual event. With the previous conference organizing resources, skills and experience, the committee has hosted the conference in the online meeting platform. Prior to the day of presentation, to avoid the technical glitches during the presentation, authors are also instructed to submit a pre-recorded 15-minutes presentation two weeks before the conference dates through google drive. On the conference day, followed by the inaugural, a keynote speech has lasted for about 1 hr 30 minutes including the Q&A session. Each and every talk has included a set of active conference participants. Then, the conference attendees are allowed to present their paper according to the conference presentation schedule, which was sent 20 days before the conference event. Amidst the crisis, the virtual conference 3rd ICIRMCT 2021 has provided the conference participants an opportunity to explain their innovative research to the audience, and make the conference more informative, valuable and engaging for other attendees.

Guest Editors:

- Dr. S. Smys, Professor, RVS Technical Campus, Coimbatore, India. Emailid: smys375@gmail.com
- Dr Subarna Shakya, Department of Electronics and Computer engineering, Pulchowk Campus, Institute of Engineering, Tribhuvan University, Kirtipur-44618, Nepal. {drss@ioe.edu.np}
- Dr Joy Iong-Zong Chen, Professor, Electrical Engineering, Dayeh University, Taiwan. {jchen@mail.dyu.edu.tw}
- Dr. M V Reddy, Institute of Research Hydro Quebec, Canada. {MogalahalliVenkatesh.VenkatashamyReddy@ireq.ca/reddymvvr@gmail.com}

Technical Program Committee:

- Dr. Ke-Lin Du Associate Professor, Department of Electrical and Computer Engineering, Concordia University, Montreal, Canada.
- Dr. Robert Bestak Professor, Czech Technical University in Prague.
- Dr. Ayoub Khan Associate Editor, IEEE Access Journal, USA, Scientific & Scholarly Editor-eContent Pro Int., USA.
- Dr. Xiao-Zhi Gao University of Eastern Finland, Finland.
- Dr. Francisco M. Gonzalez-Longatt Universities I Sørøst-Norge/University of South-Eastern Norway.
- Dr. R. DHAYA Asso.Prof. / Dept. of Computer Science, King Khalid University, Kingdom of Saudi Arabia.
- Prof. Dr. Kuan Yew Wong School of Mechanical Engineering, Universiti Teknologi Malaysia, 81310 Skudai, Johor, Malaysia.
- Prof. Dr. Engr. Mohiuddin Ahmad Khulna University of Engineering & Technology (KUET), Khulna -9203, Bangladesh.
- Prof. Tomonobu Senjyu University of the Ryukyus, Nakagami District, Nishihara, Senbaru.
- Dr. Yo-Sheng Lin Distinguished Professor, Department of Electrical Engineering, National Chi Nan University Puli, Nantou, Taiwan.
- Dr. Ing. Pavel Lafata Assistant Professor, Department of Telecommunication Engineering, Czech Technical University in Prague, Čeština.
- Prof. Dr. Subarna Shakya Professor, Electronics and Computer Engineering, Pulchowk Campus, Institute of Engineering, Tribhuvan University, Pulchowk, Lalitpur, Nepal.
- Professor Dr Hui-Ming Wee, Distinguished Professor, Department of Industrial and Systems Engineering, Chaplain, Chung Yuan Christian University (CYCU), Former

Associate Dean, College of Electrical Engineering Computer Science, Chung Yuan Christian University, Taoyuan, Taiwan.

- Dr. Hoaxing Wang Go Perception Laboratory, Cornell University, USA.
- Dr. Girmay Hailu Assistant Professor, Department of Communication Systems, Eritrea Institute of Technology, Eritrea.
- Dr. D. Vasanthi Balasubramanian Professor, CSE, IFET College of Engineering, Villupuram, India.
- Dr. Avudaiammal R, Professor/ECE, St. Joseph's College of Engineering, Chennai, India.
- Dr. L.M.Merlin Livingston Principal, Jeppiaar Institute of Technology, India.
- Dr. N.Poonguzhali, Associate Professor, Department of CSE Manakula Vinayagar Insititue of Technology, Puducherry, India.
- Dr. Latha Rajagopal Professor, CSE, KSK College of Engineering & Technology, Thanjavur, Tamil Nadu , India.
- Dr. Jawahar.P.K B.S.Abdur Rahman crescent institute of science & technology GST Road, Vandalur, Chennai, India.
- Dr. Thippeswamy Mn Nitte Meenakshi Institute of Technology, Yelahanka, Bangalore, India.
- Dr. Arti Jain, Assistant Professor, Department of Computer Science & Engineering, Jaypee Institute of Information Technology, Noida, Uttar Pradesh, India.
- Dr. V. Nandalal, Associate Professor, Department of Electronics and Communication Engineering, Sri Krishna College of Engineering and Technology, Coimbatore, India.
- Dr. Vijay Kumar Banga Professor, IKG Punjab Technical University, Kapurthala, Punjab, India.
- Dr. Anand Raghavendra Rao Professor, PESIT, Bangalore, Karnataka, India.
- Dr. Amlana Panda, Assistant Professor, School of Mechanical Engineering, Kalinga Institute of Industrial Technology, Bhubaneswar, Odisha, India.
- Dr. Nehru K Electrical and Computer Engineering, in national university of Singapore, Singapore.
- Dr. Jennifer S. Raj Professor, ECE, Gnamani College of Technology, Namakkal, India.
- Dr. Sikhinam Nagamani Assistant Professor, Department of Information Technology, Lakireddy Balireddy college of Engineering, Mylavarm, Krishna, India.
- Dr. N.Nithyavathy IEEE Branch Counsilar, Kongu Engineering College, Perundurai, Erode, Tamil Nadu, India.
- Dr. Nikhil Padhye Department of Mechanical Engineering, Indian Institute of Technology Kanpur, Kanpur-208016, U.P., India.
- Dr. Joshua Samuel Raj Professor, KVM College Road, Kokkothamangalam P.O, Cherthala, Alappuzha, India.
- Dr. K.Rajeswari, Assistant Professor, Dept. of ECE, TCE, Madurai, India.
- Dr. A. Ayyasamy, Department of Computer Engineering, Government Polytechnic College, Konam, Nagercoil, Tamilnadu, India.

Reviewer Committee:

- Engr. Prof. Abdulwahab GIWA, Chemical and Petroleum Engineering Department, College of Engineering, Afe Babalola University, Ado-Ekiti, Ekiti State, Nigeria.
- Dr. Manoj Gupta, professor, NUS, Singapore.
- Ts. Morris Ezra, Associate Professor, Deputy Director, Institute of Postgraduate Studies and Research (IPSR), Universiti Tunku Abdul Rahman, Jalan Sungai Long, Bandar Sungai Long, Cheras, Malaysia.
- Dr. Hedayat Omidvar, National Iranian Gas Company, Tehran.
- Dr. Pradeep Kumar, University of KwaZulu-Natal, South Africa.
- Dr. Badrul Hisham bin Ahmad , Assoc. Professor, Department of Aerospace Engineering, Universiti Teknikal Malaysia Melaka.
- Prof. Dr. Kuan Yew Wong, Department of Manufacturing and Industrial Engineering, Faculty of Mechanical Engineering, Universiti Teknologi Malaysia.
- Dr.Ranjit Singh Sarban Singh, Universiti Teknikal Malaysia Melaka, Faculty of Electronics and Computer Engineering, Hang Tuah Jaya, Melaka, Malaysia.
- Dr. Mohammed Mousa Mohsin, Department of Statistics & university of Basra, Iraq.
- Prof. Dr. Engr. Mohiuddin Ahmad, Dept. of Electrical and Electronic Engineering, Khulna University of Engineering & Technology (KUET), Bangladesh.
- Dr. Ali Newaz Bahar, Assistant Professor, Department of Information and Communication Technology, MawlanaBhashani Science and Technology University,Tangail, Bangladesh.
- Dr. Alexandra Queirós, Escola Superior de Saúde da Universidade de Aveiro, Campus Universitário Santiago, Portugal.
- Dr.AamirSohail, Lecturer, Mechanical Engineering Department, University of Engineering and Technology Taxila, Pakistan.
- Dr.Alexander Kalashnikov, Senior Lecturer In Embedded Systems, Department Of Engineering And Maths, Faculty ACES, Sheffield Hallam University, UK.
- Dr. S. M. ShamsulAlam, Faculty Member, Electronics and Communication Engineering Discipline, Khulna University, Khulna, Bangladesh.

- Dr. SubraSubramanya, Professor, School of Engineering and Computing, National University, San Diego, California.
- Dr.Trong-Phuoc Huynh, Department of Civil and Construction Engineering (DCE),National Taiwan University of Science and Technology, Taiwan.
- Dr. YuwaIgbinosa, School of Electrical, Electronic and Computer Engineering ,Howard College Campus , University of Kwazulu-Natal , King George V Avenue , Durban, South Africa.
- Dr.Holyheavy Balami , Department of Mathematics, Ahmadu Bello University, Zaria, Nigeria.
- Dr.MD. Al-Amin, Assistant Professor, Department of CSE, Shahjalal University of Science and. Technology. Sylhet, Bangladesh,
- Dr. AbdollahMehrpanahi, Faculty of Mechanical Engineering, ShahidRajaee Teacher Training University, Iran.
- Dr. İlkerErkan, Asst. Prof. Faculty of Architecture, SüleymanDemirel University Cunur, Isparta, Turkey.
- Dr. Vasile Avram, Bucharest University, Romania.
- Dr. KyungHi Chang, INHA University, Korea.
- Dr. Michel Plaisent, University of Quebec in Montreal, Canada.
- Dr. Rasiah Ladchumananandasivam, Professor, B.S.A.R. Crescent University, Vandalur, Chennai, India.
- Dr. S. Das, Professor (Adjunct), MME, MANIT Bhopal, Former Director CSIR AMPRI BHOPAL, INDIA & Visiting Professor IIT KANPUR, INDIA.
- Dr. Ranjana Jha, Professor& Head, Research Lab for Energy Systems, Department of Physics, Netaji Subhas Universityof Technology, India.
- Dr. VINOD SINGH YADAV, Professor, Faculty in Department of Mechanical Engineering, National Institute of Technology, Uttarkhand. E-mail :
- Dr Vinod Singh Yadav, Assistant Professor, Department of Mechanical Engineering, NIT Uttarakhand, ITI Campus, Srinagar, Pauri, Uttarakhand.
- Dr.K.V. Madhuri, Professor of Physics, Associate Dean - Research & Development Vignan's Foundation for Science, Technology & Research University, Vadlamudi, Guntur - Andhra Pradesh.

- I Sardar, Assistant Professor, Department of Mechanical Engineering, Yanbu University, Saudi Arabia.
- S Satheesh, Assistant Professor, Department of Mechanical Engineering, Eritrea Inst of Tech, Eritrea.
- Paval Lafta, Professor, Czech Technical Univeristy. paval
- Malarvizhi V, Professor, SNS College of Tech, India.
- Dr. Shahed Mohammadi, Profesor, Ayandegan University, Tonekabon, Iran.
- Dr. Francisco M. Gonzalez-Longatt, Universitetet I Sørøst-Norge/University of South-Eastern, Norway.
- Vanitha, Assistant professor, Park College of Engineering, India.
- Deepa S, Nehru Inst of Tech, India.
- P.Sivaranjani, Kongu Engineering College, India.
- Varun P., Gopi Government College of Engineering, India.
- G. Jaspher, Professor, Karunya University, India.
- Dattaguru V. Kamath, Professor, Manipal Inst of Technology, India.
- Radha N, Professor, SSN College of Engineering, India.
- Gunasekaran R, Professor, Anna University, India.
- Prof. Dr. Shashikantha Karinka, Prof. & Head of Mechanical Engineering Department. NMAM Institute of Technology, Nitte.
- Dr. Prasanna Kumar S Mural, Assistant Professor, Dept of Chemical Engineering, National Institute of Technology, Calicut.
- Dr. H S Kumar, Associate Professor, Dept. of Mechanical Engg., NMAMIT, Nitte.
- Dr. Rajib Ghosh Chaudhuri, Assistant Professor, Chemical Engineering, National Institute of Technology, Durgapur, Durgapur, West Bengal, India.
- Dr. Janita Saji, Assistant Professor of Physics, Department of Sciences and Humanities, School of Engineering & Technology, Christ (Deemed to be University), Bangalore.
- Dr. Barnali Dasgupta Ghosh, Assistant Professor, Department of Chemistry Birla Institute of Technology, Mesra, Ranchi, India.
- Dr. Pinki Chakraborty, Assistant Professor, Chemistry, SBAS, Galgotias University, U.P.

IOP Conf. Series: Materials Science and Engineering 1091 (2021) 011001 doi:10.1088/1757-899X/1091/1/011001

- Dr. J. Vijayaraghavan, Assistant Professor, Department of Civil Engineering, Anna University, University College of Engineering Ramanathapuram, Ramanathapuram, India.
- Dr. Nilamber Kumar Singh, Assistant Professor, Department of Mechanical Engineering, National Institute of Technology Patna, India.

IOP Conf. Series: Materials Science and Engineering 1091 (2021) 011002 doi:10.1088/1757-899X/1091/1/011002

Peer review declaration

All papers published in this volume of *IOP Conference Series: Materials Science and Engineering* have been peer reviewed through processes administered by the Editors. Reviews were conducted by expert referees to the professional and scientific standards expected of a proceedings journal published by IOP Publishing.

- **Type of peer review: Double-blind**

Single-anonymous: authors' identities are known to the reviewers, reviewers' identities are hidden from authors

Double-anonymous: author and reviewer identities are hidden to each other

Triple-blind: author and reviewer identities are hidden to each other, and from the Editor(s)

Open: author and reviewer identities are known to each other

- **Describe criteria used by Reviewers when accepting/declining papers. Was there the opportunity to resubmit articles after revisions?**

  Based on Experience and Qualification (professor, industrial person or Ph.D qualified) reviewers are identified. Based on the journal publication and previous reviewing experience, reviewers are selected. The review process is going to be according to:
    - **Initially, all submitted papers will be checked for plagiarism using crosscheck iThenticate;**
    - **Papers will be classified based on their subject area by the technical team;**
    - **The technical head will pass the classified papers to the subject Experts;**
    - **The subject Experts will send each paper to at least three Reviewers for review after removing the Author's information from the papers;**
    - **The papers will be accepted only if at least two Reviewers suggest the acceptation.**

- **Conference submission management system:**
  Easychair

- **Number of submissions received:**
  272
- **Number of submissions sent for review:**
  258
- **Number of submissions accepted:**
  83
- **Acceptance Rate (Number of Submissions Accepted / Number of Submissions Received X 100):**
  30%
- **Average number of reviews per paper:**
  6
- **Total number of reviewers involved:**
  120
- **Any additional info on review process (ie plagiarism check system):**

IThenticate Plagiarism tool

- **Contact person for queries:**
  Dr. S. Smys smys375@gmail.com

Battery energy storage systems (bess) state of the art

Deepa Jose [1], Jhonatan Meza[2], J S Prashanth[3]
[2]*Instituto Superior Tecnológico 17 de Julio*
Ibarra, Ecuador.
[1,3]*KCG College of Technology, Chennai-600097, TamilNadu, India.*
[2]*jmezac@est.ups.edu.ec,* [1]*deepa.ece@kcgcollege.com*

Summary: This article presents a review of the battery energy storage method or BESS for its acronym in English (Battery Energy Storage System), considering that the era of high capacity storage batteries is approaching and why historically the storage of energy did not appear in the planning or operation of electrical systems. However, this solution is currently setting a global trend for having high levels of efficiency and why it is also being driven by the need to modernize the network in order to increase the use of renewable energies.

Keywords: Batteries, Energy storage systems, Technologies, Efficiency, Renewable energies.

1. Introduction

In a report entitled The Global Market for Grid-Connected Energy Storage Battery Systems, analysts at US consultancy Frost & Sullivan say we are facing a period of “dynamic growth” where commercialization of systems is expected to of storage will skyrocket after 2017. The great 'boom' of storage systems is predicted for after 2020, when the fall in their costs will allow the facilities to be amortized in five six years.[one]

Demand for Battery Energy Storage Systems (BESS) is being driven by the need to modernize the grid for the greater use of renewable energy.

Although there are barely 430 MW of BESS operating in the current market, this market is expected to reach between 10 and 12 GW connected to the grid in 2024. This will mean that the market will exceed 460 million dollars (407 million euros) entered in 2014 to about 8,440 million dollars (7,500 million euros) in 2024.

As indicated by various other reports and projects, lithium-ion (Li-ion) technology will continue to dominate the market, "at least for the next 2-3 years," remaining the leading technology in terms of installed power.

In fact, the penetration of renewable sources in the SEP's networks has increased notably in recent years. However, the power supplied by these energy sources is not as safe and easy to adjust to changes in demand as the power supplied by traditional energy systems. As a consequence, and in order to guarantee the reliability, quality and performance of these systems, energy storage devices are integrated into the networks, simultaneously with these sources.[two]

This document presents the development of battery energy storage systems that propose their use in SEP's. The characteristic expressions of the same, specific applications and comparative data of the different technologies in technical and economic terms are shown.[one]

Fig. 1. Installation of batteries connected to the network[3]

2. STATE OF THE ART

Geographically, the US will lead the market for battery energy storage systems (BESS), followed by China, Japan and Germany. With regard to the main companies, NGK Insulators, AES Energy Storage, Sumitomo Electric, LG Chem, Samsung SDI, NEC Energy, BYD, Toshiba, GE and Saft stand out.[4]

Energy storage has the ability to bring flexibility to the grid through a variety of end-use applications, "Its greatest advantages are that it improves the amount of energy that can be distributed, avoids fluctuations in renewable energy and time displacement of electricity, as well as a rapid short-term balance of electricity for secondary markets. "

Despite the bright future that they predict for energy storage systems, they point to a series of obstacles that this emerging market will need to overcome, among which the following stand out:

- High costs
- The low maturity of the technology
- The lack of a clear business model and value proposition
- Limited practical applications that support laboratory efficiency and safety standards
- Inadequate incentives and lack of supportive goals and policies
- Few market consolidations for turnkey solutions
- Low oil and gas prices

"In general, the attractive price, combined with an increase in manufacturing and support policies for the development of renewable energy, will increase the financing capacity of energy storage projects associated with renewable energy."

2.1. Characteristics of storage technologies

There are many techniques for storage, based on practically all forms of energy: mechanical, chemical and thermal.

The storage technologies used present technical and economic criteria, which vary considerably, depending on the specific needs and applications. These technologies can be divided into four categories, depending on the applications:

1. Low power applications in isolated areas, essentially to power transducers and emergency terminals.
2. Medium power applications in isolated areas, individual electrical systems and electrical supply to cities.
3. Network connection application with Peak leveling.
4. Power-quality control applications.

The first two categories are suitable for small-scale systems, where energy could be stored in the form of kinetic energy, chemical energy, compressed air, hydrogen, supercapacitors, and superconductors.

The third and fourth categories are suitable for large-scale systems, where energy could be stored as gravitational energy in hydraulic systems, thermal energy in the form of latent and sensible heat, chemical energy in accumulators and batteries, or compressed air.

2.1.1. Energy storage systems

The different modes of energy storage are set out below, considering the most relevant properties of the different technologies.[5]

A. **Flywheel** (Flywheels store energy in its kinetic form.)
B. **Superconducting Magnetic Energy Storage (SMES),** This consists of a large superconducting coil, which is kept at cryogenic temperature by a cooler or cryostat containing helium or liquid nitrogen.
C. **Supercapacitors** (They are used as storage units in power systems, vehicles, etc.)
D. **Compressed Air Energy Storage (CAES).** In this technology, the air is compressed and stored in underground tanks, aquifers or cavities. Stored energy is released during peak demand intervals, expanding the air through a turbine.
E. **Pumped Hydroelectric Energy Storage (PHES).** This technology enables the use of electrical energy in off-peak hours, to pump water from a lower reservoir to another located at a higher altitude.
F. **Battery Electric Storage System (BESS),** in this case we will develop the topic in greater depth below.

2.2. Battery Energy Storage Systems (BESS)

2.2.1. General characteristics

Reversible electrochemical accumulators, or batteries, are devices that store energy in chemical form inside. Basically, they are made up of a positive pole or cathode, a negative pole or anode, and an electrolyte, which is the substance in which the first two are submerged. The movement and

IOP Conf. Series: Materials Science and Engineering 1091 (2021) 012001 doi:10.1088/1757-899X/1091/1/012001

recombination of the ions and / or molecules that are part of these three elements allows a circulation of direct current to occur when an external circuit connected to the electrodes is closed, which flows through it from the cathode to the anode when the battery discharges, and vice versa when charged.[6]

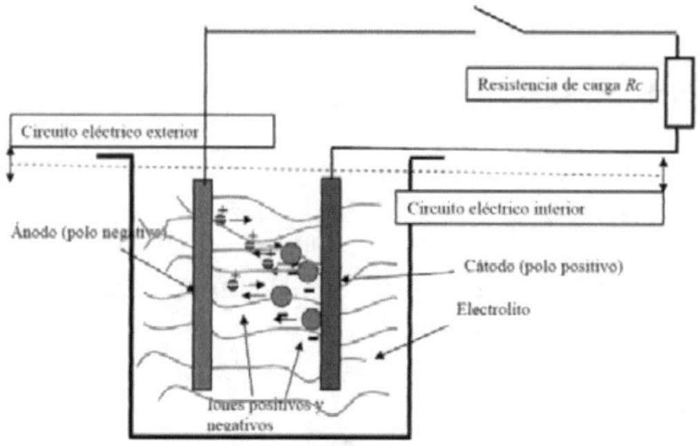

Fig. 2. Basic Battery Operation

There are many types of electric batteries, made of different materials, and in each one the chemical reactions that take place inside are different. In some, such as lead-acid, the electrolyte reacts with the active substances at the poles, thereby altering its composition. In others, such as lithium-ion batteries, the composition of the electrolyte remains unchanged and it only participates by transporting ions from one side to the other. Finally, there are batteries in which only chemical changes take place in the electrolyte, without the plates undergoing any change.

In any case, all batteries share certain common characteristics, which will be detailed below.

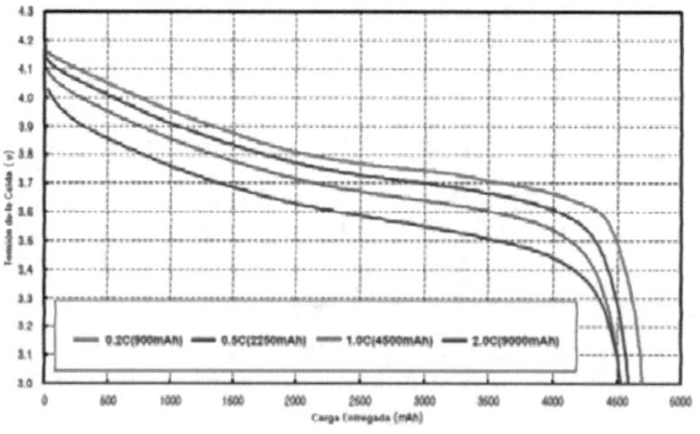

Fig. 3. Characteristic curve for a lithium-ion battery model

2.2.2. Main Types of Batteries

2.2.2.1. Lead-Acid

Lead-acid accumulators are the oldest kind of battery that exists, having been invented in 1859. They are characterized by being capable of providing a high instantaneous current, and have a high power density, due to which they have been used massively in the automotive industry, as a source of energy for the starter motors of gasoline vehicles, among other things

2.2.2.2. *Nickel*

Nickel-based batteries are another alternative for large-scale energy storage. There are several types, of which the most common is nickel-cadmium (NiCd) batteries, which are a mature technology. In them, a nickel oxide-hydroxide cathode, a metallic cadmium anode, and potassium hydroxide are used as the electrolyte.

2.2.2.3. *Ion-Lithium*

Lithium-ion batteries are well known for the great success they have had in the portable electronic device market, mainly due to their very high energy density. They are also becoming widely used in electric vehicles, and recently they have also been used as BESS in electrical networks. In particular, in Chile there are already projects in operation that use this type of battery to perform support functions for the Primary Frequency Control (CPF) in the SING.

2.2.2.4. *Sodium-Sulfur*

Sodium-sulfur (NaS) batteries, despite not being as well known as previous battery types, can be a very good alternative to BESS, and have been commercially available on the market for over ten years.

2.2.2.5. *Flow Batteries*

This is a newer technology than the previous ones. So-called "flow batteries" differ from conventional batteries in that the chemical elements or compounds that react in them are stored in external tanks, and are only introduced into the device during operation. This implies that the relationship between power and energy in these accumulators is not fixed. On the contrary, while the power is limited by the number of cells used, the energy and capacity of the battery depend on the amount of electrolyte in the tanks. Also, it means that to recharge the battery, you can simply replace the electrolyte.

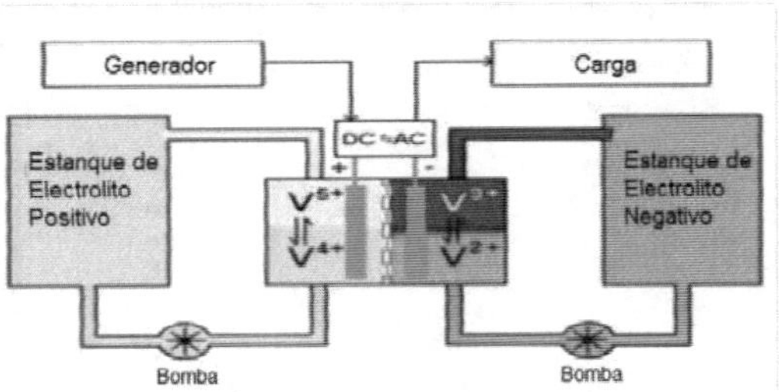

Fig. 4 Schematic of a Vanadium Redox Battery

2.2.2.6. *Other Technologies*

In addition to those described, there are other types of electrochemical accumulators being developed today that have the potential to become competitive in this market. These have mostly not been used commercially yet, but several appear to be promising. Here are some of the lines of research that are being followed today:

Zinc-bromine: A type of accumulator that has a lot in common with the vanadium redox battery, since it uses tanks to store the electrolyte just like this one. However, part of the active material is also incorporated into the cell itself, which is why it is known as a "hybrid flow battery". Its main advantage is its low cost. In addition, they have a useful life of more than 2,000 cycles with a discharge of 100% of their capacity.

Metal-air: They have a very good energy density and have the potential to be very low cost relative to other electrochemical accumulators. However, their life cycles are short, they are not very efficient, and, as their degree of development is low, it is very complex to recharge them.

Sodium ion: Similar to sodium-sulfur, with the difference that they can be manufactured on a small scale since they work at room temperature. They are also predicted to become relatively cheap in the future.

Magnesium-antimony (Liquid Metal Battery): A type of battery that could have a very low cost, due to the abundance of the elements that compose it. In it, both electrodes and the electrolyte are in a liquid state, and they stratify at different levels due to their difference in density. The negative electrode is made up of magnesium, the positive one of antimony, and the electrolyte is a molten salt. Experimental results in relation to this type of battery indicate that it would have an efficiency of 69%, a high power density and a low self-discharge rate.[7]. However, as in the case of NaS accumulators, maintaining the liquid state of metals requires high operating temperatures.[4]

2.2.2.7. Summary table with the characteristics of the main battery technologies

The following table presents a summary with the main characteristics of the most important types of batteries that have been mentioned in this section, and that are currently being marketed worldwide in the form of BESS. The technologies considered are: lithium-ion, lead-acid type VLA (Vented Lead-Acid), lead-acid type VRLA (Valve Regulated Lead-Acid), sodium-sulfur (NaS), nickelcadmium (NiCd) and Vanadium Redox (VRB).

Table 1: Characteristics of different types of batteries

TECNOLOGÍA	TENSIÓN CARACTERÍSTICA	EFICIENCIA (1)	VIDA UTIL (2) (3)	TASA DE AUTODESCARGA	DENSIDAD DE ENERGÍA (por volumen)	DENSIDAD DE ENERGÍA (por peso)
Ion-Litio	3,7 [V] para $LiCoO_2$; 3,3 [V] para $LiFePO_4$	90	3.000 ciclos para 100% DOD; 4.000 para 80% DOD; 20.000 para 30% DOD	2% mensual	200-400 [kWh/m^3]	80-160 [Wh/kg]
Plomo-Ácido, Tipo VLA	2,0 [V]	80	1.500 ciclos para 100% DOD; 2.000 para 80% DOD; 4.000 para 50% DOD; 6.000 para 30% DOD	5% mensual	25 [kWh/m^3]	25 [Wh/kg]
Plomo-Ácido, Tipo VRLA	2,0 [V]	80	750 ciclos para 100% DOD; 1.000 para 80% DOD; 2.000 para 50% DOD; 3.000 para 30% DOD	4% mensual	60 [kWh/m^3]	40 [Wh/kg]
NaS	1,7 [V]	83	2.500 ciclos para 100% DOD; 4.500 para 80% DOD; 6.500 para 65% DOD	-	170 [kWh/m^3]	100 [Wh/kg]
NiCd	1,2 [V]	70	2.000 ciclos para 100% DOD; 3.000 para 80% DOD	10% mensual	65 [kWh/m^3]	45 [Wh/kg]
VRB	1,4 [V]	72	5.000 ciclos para 100% DOD; 10.000 para 75% DOD	- (4)	25 [kWh/m^3]	25 [Wh/kg]

Notes:

(1). Including PCS losses.

(2). In many cases, particularly with lead-acid batteries, it is not technically feasible for the batteries to discharge to 100% of their capacity, as this could damage the cells.

(3). In addition to the number of charge / discharge cycles that the batteries endure, the life of the BESS may be limited by the useful life of other components, such as the PCS, or the pumps. and ponds for the case of Vanadium Redox batteries.

(4). Self-discharge is nil for applications that do not require a quick response in this type of BESS. However, if this is not the case, it would be of considerable value, as the electrolyte pumps would have to be kept running permanently.

2.2.3. Components of a BESS

Commercial Battery Storage Systems, as already mentioned, are almost always modular, consisting of several elements in addition to the batteries themselves. Some suppliers offer integrated systems in which all these components come inside containers designed to deliver certain power and energy, depending on the customer's requirements. In other cases, companies that manufacture the different elements work together to develop the system, and the different components can be physically separated into different blocks. The topology to be used varies from case to case, depending, among other things, on the battery technology used, the function that the BESS is intended to fulfill, and the physical and electrical characteristics of the place where the battery is carried out. installation.

The essential components found in any BESS are batteries, DC-AC converters, BMS [7], a master controller, and a system of protections and relays. In addition, a power transformer is almost always needed to make the interconnection with the main power bus of the place where the system is being installed. Additionally, the BESS could require a Thermal Management System, harmonic filters, DC-DC converters type "chopper", power boards for auxiliary services, communication systems with distant external elements (SCADA type), and other elements according to the particular requirements of each project.

Below is an example of two possible connection topologies between these elements that can be used for this type of system:

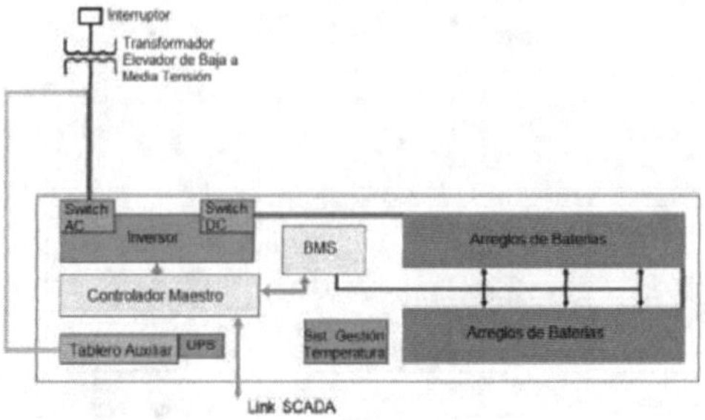

Fig. 6. BESS with all its elements in the same container

In this figure, all the elements, except the power transformer, are physically inside the same container. The common bus of the electrochemical cells is connected directly to the input of the inverter, which in turn is connected to the transformer. The main controller and BMS are also housed in the container, and since the inverter and batteries need a certain temperature to operate properly, a Thermal Management System is used to keep it at proper levels.[8]

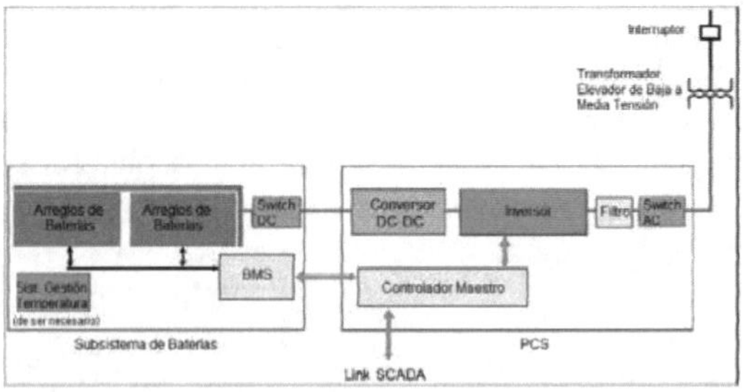

Fig. 7. BESS with Batteries and PCS physically separated

In the second case, instead, the inverter is in a separate container from the electrochemical accumulators, and, additionally, a DC-DC converter is used to change the direct voltage level before the input to the inverter. The set of converter, controller and filters in this case is usually called PCS ("Power Control System").

The basic functions of the mentioned BESS components are briefly explained below.

- **Batteries**

They are the most important element of the BESS, and are often the most expensive component of it. They consist of a series of cells that are connected in series and parallel arrangements. By connecting them in series, the continuous voltage with which the BESS operates is achieved, which varies according to the type of technology and characteristics of the project. Among the factors that determine the final voltage level chosen, it is worth mentioning that, on the one hand, it seeks to reduce the ohmic losses in the battery (which is achieved, for a given power level, by increasing the voltage and decreasing the current) , and, on the other hand, the maximum admissible levels must be respected for safety reasons. Typically, DC voltages are used in the common bus of the cells between 400 [V] and 1,200 [V].

- **Converter**

The DC-AC converter is the element that allows the interconnection of the batteries, which operate in direct current, with the electrical network, which operates in alternating current, allowing there to be power flows from one to the other, in both directions. They consist of power electronic circuits, which work from valves made up of semiconductor devices, such as IGBTs or IGCTs.[9]

- **BMS**

BMS ("Battery Management System") are an essential component of BESS. They are electronic equipment that performs monitoring, equalization and protection functions of electrochemical cells, as well as communication with other systems. They are responsible for measuring certain parameters of the cells, such as their voltage and temperature, estimating their state of charge and health, and sending certain orders that allow them to be loaded or unloaded properly.[7]

- **Controller**

This essential element of the system is in charge of ensuring the correct functioning of the BESS. It has multiple functionalities. For example, it sends modulating signals to the inverter valves that allow a control of active and reactive power to and from the batteries.[7]

2.2.4. Protection System

It includes the safety switches, measuring transformers and relays that the system incorporates to be able to act appropriately in case of failure. You generally have switches on both the DC and AC side of the inverter. The relays are in charge of comparing magnitudes such as currents, voltages and angles in the main bars, measured by current and potential transformers, and based on that they decide if the system is in a fault state, in which case a Open signal to switches or fuses.

2.2.5. Transformer

In most cases, it is necessary to incorporate a transformer to increase the voltage at the inverter output and to interconnect with the electrical grid. Generally, a transformer is used for this to bring the voltage from low to medium voltage (typical values in Chile are 12 [kV], 13.2 [kV] and 23 [kV]), similar to the transformers used in Distribution.[8]

There are few applications where it is necessary to raise to higher voltages. Only in some very large projects in which the BESS causes the power delivered from a plant or substation to increase considerably at certain times it becomes necessary to additionally incorporate a power transformer for high voltage.

The price of transformers, as in the case of inverters, is linked to their nominal power. Furthermore, in some cases the use of multiple inverters in parallel implies that multiple transformers in parallel, or three-winding transformers, must also be used for technical reasons.

2.2.6. Temperature Management System

As already mentioned, it is extremely important to keep the temperature of the batteries at optimal levels, because the useful life and the total charge that they are capable of delivering are negatively affected if the temperature goes too far from these values. Inverters also need to operate within a certain temperature range, although these are located far from batteries at times.

Most battery technologies operate well at temperatures close to room temperature, but high currents tend to cause them to heat up. For this reason, it is necessary to cool them using air conditioning systems, and, in some cases, more complex systems based on the circulation of some cooling liquid. Sodium-sulfur batteries, however, unlike most, only operate at high temperatures, so in this case the Temperature Management System is in charge of heating them.

2.2.7. Additional Civil Works

Finally, a battery energy storage project often requires a significant investment in infrastructure in many cases. While many commercially available BESSs are built into containers that can be installed outdoors, in other cases, and particularly with batteries that tend to take up more space, such as lead-acid and nickel-cadmium batteries, storage may be required. construction of special buildings to house the equipment. Within this item the wiring, conduits, and grounding of the structures are also considered. [11-12]

3. CONCLUSIONS

This document presents the most relevant properties of energy storage technologies currently being developed in the design of power systems. In this sense, the parameters that characterize the behavior of the different technologies are exposed, in order to properly use them.

Considering permanent low-power applications, where the key phenomenon is possible self-discharge, lithium-ion batteries are presented as the most successful technology.

In small electrical systems, of a few kWh, located in isolated areas and based on renewable resources, the most relevant characteristic for opting for one technology or another is autonomy, so the lead battery is the most appropriate.

REFERENCES

[1] «The era of high-capacity storage batteries is coming | Electricity Sector | Professionals in Electrical Engineering ».2016.
[2] «Energy Storage Technologies for Electric Applications ». [Online]. Available at: http://www.sc.ehu.es/sbweb/energias-renovables/temas/almacenamiento_1/almacenamiento_1.html. [Accessed: Dec 20, 2016].
[3] "Smart Energy Consulting: Batteries connected to the grid", *Smart Energy Consulting*, Sep 18, 2015. .
[4] EI Pedro Estévez, Maider Varela, «Energy storage systems for renewable energies», 2016.
[5] "Energy Storage Technologies for Electric Applications". 2016.
[6] ABM Shawkat Ali, *Smart Grids: Apportunities, Developments, and Trends*. 2013.
[7] MT Lawder *et al.*, "Battery energy storage system (BESS) and battery management system (BMS) for grid-scale applications", Proc. IEEE, vol. 102, no 6, pp. 1014-1030, 2014.

[8] S. Chouhan, D. Tiwari, H. Inan, S. Khushalani-solanki, and A. Feliachi, "DER Optimization to Determine Optimum BESS Charge / Discharge Schedule using Linear Programming", 2016.
[9] D. Rodríguez, R. Gómez, and A. Campos, «Battery management systems (BMS) and their importance for battery storage systems (BESS)», *Tecnura*, pp. 51-56, 2015.
[10] "Keys to understanding energy storage". [Online]. Available at: http://www.revistaei.cl/reportajes/claves-para-entender-el-almacenamiento-de-energia/.[Accessed: Dec 21, 2016].
[11] M. Arunkumar, M. Kannan, G. Murali,Experimental studies on engine performance and emission characteristics using castor biodiesel as fuel in CI engine,Renewable Energy,Volume 131,2019, Pages 737-744, ISSN 0960-1481
[12] Loganathan N., Arvin Tony A., Malini T., Gobhinath S. (2020) Comparison of Renewable Energy Generation in an Electrical Network with Energy Storage System. In: Lecture Notes in Electrical Engineering, vol 626. Springer, https://doi.org/10.1007/978-981-15-2256-7_3

Development of stitching jigs for sleeve button placket and sleeve buttonhole placket in men's formal shirt

Dr. D. Vijay Kirubakar Raj[1], Dr. M. Renuka Devi[2]
[1,2]Assistant Professor, Department of Textile Technology, Anna University, Chennai-25, India.

Abstract - The operations in assembling garment from their components, requires lot of energy and skill for the sewing operators. One of such operation is sleeve placket attachment in the construction of men's formal shirt. Currently, the apparel technologists are working in inventing new jigs for the operations involved such construction both to lower the level of operator skill required and to increase the level of automation possible. However the development of jigs for operations like sleeve placket attachment are difficult to achieve. This work is an attempt to develop two stitching jigs for use in stitching together the components of the sleeve.

I. INTRODUCTION

Automation is the process or technique of doing certain tasks by the use of automatic equipment in the place of human operators during product manufacture. Automation seeks to reduce human intervention to a minimum resulting in saving of labour and energy, improves precision, accuracy, quality of products, and productivity. It has the huge scope in industries where the wages for labour keeps increasing. Thus sewing units in developed countries are shifting to increasing levels of automation of labour-intensive clothing production.

Stitching jigs are work aids which are used where accurate stitching lines are required. It holds fabric layers during sewing which enables easier seam formation and allows semi-automated stitching for the specific garment constructing operations. The fabric is loaded first into the jig and then the jig loaded into the sewing machine. This reduces the complexity of the stitching operations, (Carr and Latham 1999).

However, stitching jigs are not available for many of the garment sub assembly operations. One among these is sleeve placket attaching. Developing a stitching jig(s) for attaching sleeve button plackets and sleeve buttonhole plackets will improve quality and reduces the difficulties of the sewing operation thus will help to increase the level of automation for the sewing operation, (RajkishoreNayak and Rajiv Padhye 2017).

II. OBJECTIVES

- To develop stitching jigs for the sleeve button and buttonhole placket creation operations.
- To make possible a retro fit of such jigs into existing sewing machines.
- To eliminate the need for ironing of the sleeve button and buttonhole placket components.
- To discuss the advantages of replacing analog jigs with digital control in the future.

Work aids are devices that are built into machines, added to them afterwards, attached alongside or made use to improve productivity, improve or maintain quality standards, reduce training time and minimize fatigue to the operator, (Carr and Latham 1999).

The collars where stitching jigs are successful have two identical points. The front and reverse edges of formal jackets and raincoats much match. Likewise patch pockets, cuffs and other small items such as epaulettes and tabs must be in perfect pairs. The only way to ensure this where die cutting is not possible is to control the sewing line in relation to the pattern shape of the garment rather than the edge of the cut material and this is what stitching jigs are meant to achieve (Carr and Latham 1999).

A stitching jig (Figure 1) is a type of work aid to be used where accurate stitching lines are required. It uses an edge guide to achieve accurate sewing assuming accurate cutting. Hence run-stitch or profile-stitch operations can be done using a jig or template, (Carr and Latham 1999).

Jigs consisting of two layers of a rigid material such as aluminum or plastic which are joined together by a hinge. A slot is cut out of both layers of material which conforms to the exact sewing line in the garment part and the bottom layer carries a shaped guide close to the hinge to which the edges of the garment parts are aligned, (Carr and Latham 1999).

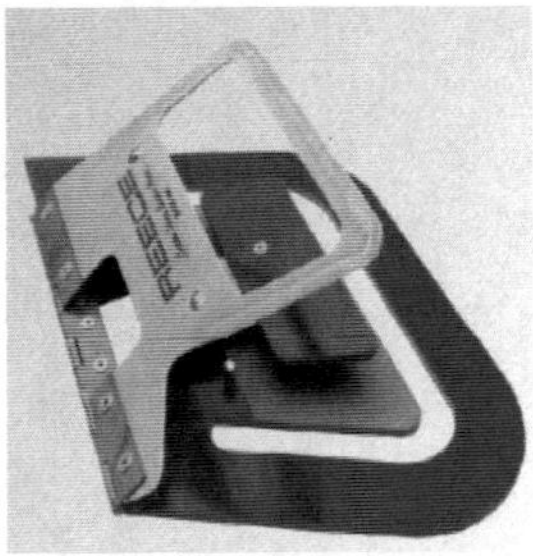

Fig. 1 Stitching jig used for cuff

The inside surface of the jig is covered with a strip of non-slip material like emery sheet to have a firm grip on the fabric. The fabric plies are sandwiched in the jig and the jig is sewn in a specially adapted sewing machine where the throat plate has a protrusion to glide through the slot; the double roller presser feet and rubber-covered feed dog guides the jig along the slot. Sewing stackers and bundle clamps are additional devices attached to or outside the sewing table to organize the garment components before sewing, during sewing or after sewing. (Solinger 1988).

The two pieces of the garment part are laid into the jig, along with the interlining if a sew-in interlining is being used, and the jig is closed to hold them firm. If fullness is needed on the top ply to give a piped effect on a part such as a pocket flap, a fulling bar would be included in the jig between the two plies (Figure 2). This ensures the inclusion of extra length in the top ply so that the lining cannot show on the right side of the finished part. The operator then sews round the slot on a machine with a specially adapted presser foot and feed dog. (RajkishoreNayak and Rajiv Padhye 2015).

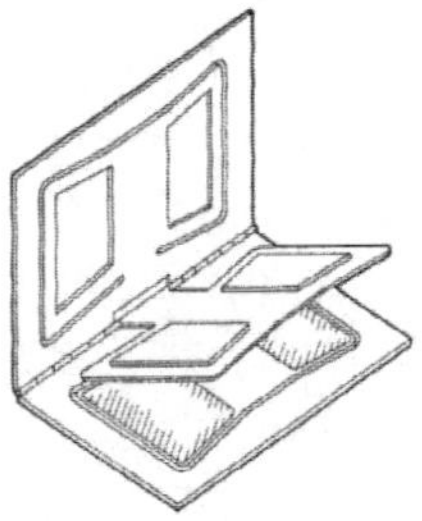

Fig. 2 Stitching jig for pair pocket flaps

List of Components that can be Sewn Using Jig (Figure 3)

1. Attach collar band
2. Collar
3. Cuff
4. Epaulette
5. Jacket forepart (SB)

6. Jacket forepart (DB)
7. Jacket collar
8. Pocket flap
9. Pointed collar
10. Raincoat forepart
11. Van Dyke flap
12. Vest Front
13. Yoke or gun patch

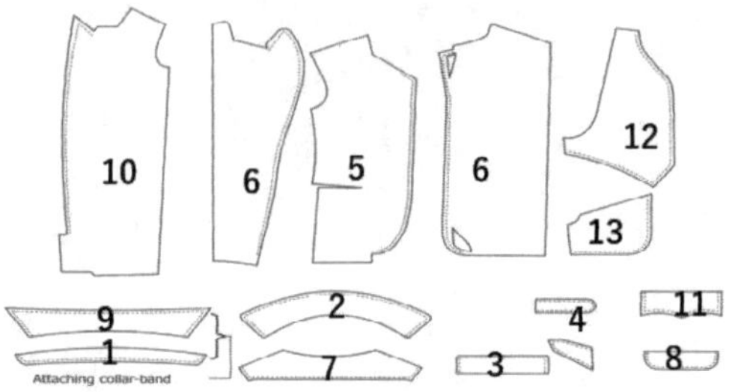

Fig. 3 List of components that can be sewn using jigs

Advantages of Stitching Jigs

- Ensure correct shape and size of the components.
- Outer edge of the component needs not be accurately cut.
- Quality is improved.
- If automated, productivity is more.
- Higher possibility of employing unskilled operators.
- Reduces labour cost.
- Reduces rejection rate.

Disadvantages of Stitching Jig

- A larger seam allowance around the outside of the garment part is needed so that the fabric edge does not become pushed down into the stitching slot.
- The extra fabric for seam allowance is costly and it must be trimmed off, often by hand, afterwards.
- Separate jigs are needed for every style and size and consequently they are only used for long production runs. A minor change of pattern shape can require a new set of jigs.
- Careful training and control are required if stitches are longer than 5 mm.
- Suitable only of sewing identical pieces. Example: Sewing two layers of collar.
- Conventional feeding system has to be replaced by a specially designed presser foot and feed dog.
- Maintaining constant Stiches per Inch (SPI) is difficult. (Harold Carr et al 2007).

Construction of Sleeve Placket

The construction of the sleeve placket involves following major operation.

a. Ironing of sleeve button placket and sleeve button hole placket fabric components.

b. Sewing of sleeve button placket.

c. Sewing of sleeve buttonhole placket.

The sleeve button placket attaching is always performed first and then followed by sleeve buttonhole placket.

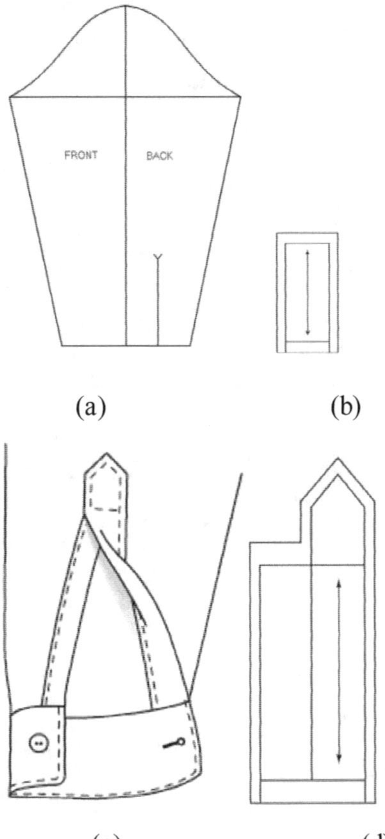

(a) (b)

(c) (d)

Fig. 4 Components of men's sleeve placket (a) men's sleeve pattern (b) sleeve button placket (c) men's sleeve (d) sleeve buttonhole placket

III. MATERIALS AND METHODS

Materials

The stitching jigs for sleeve buttonhole and sleeve button placket is made from acrylic. To create the groove along the edge of the slot, thin aluminum sheet of thickness 0.01mm is used along the edges of the slot.

A sliding plate is used for inserting and folding the seam allowance of sleeve buttonhole placket and sleeve button placket. The sliding plate is also made up of aluminum sheet of thickness 0.01mm.

To move to the acrylic jig during sewing, a rubber coated feed dog is used.

Jig

The jig is cut out in the acrylic sheet of thickness 1.5cm. The acrylic sheet is first cut to the exact dimensions of each components of the jig. For precision in cutting, laser cutting machine is employed.

Sleeve button placket jig

The sleeve button placket jig (Figure 5) is used for sewing sleeve button placket with sleeve.

Components of sleeve button placket jig

1) Bottom plate (Figure 6)
2) Top plate (Figure 6)
3) Middle fixed plate (Figure 7)
4) Middle movable plate (Figure 7)

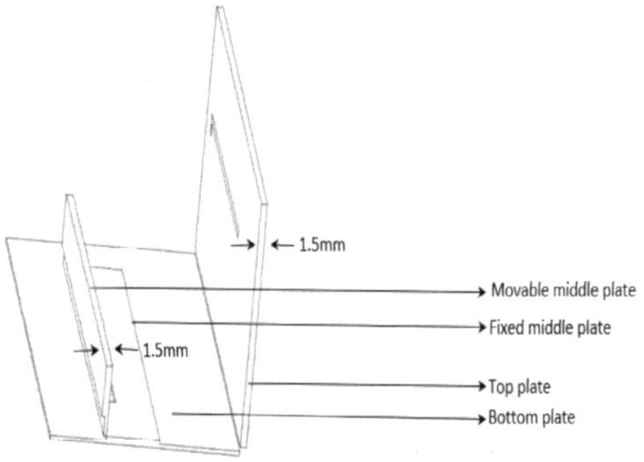

Fig. 5 Isometric view of sleeve button placket jig

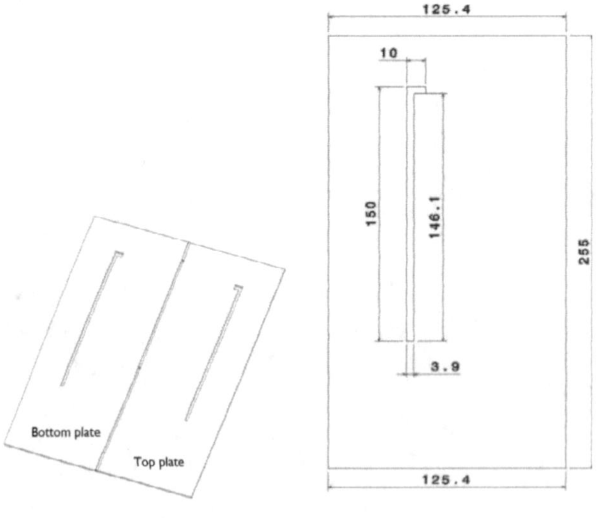

All dimensions are in mm

(a) (b)

Fig. 6 Bottom and top plates of sleeve button placket jig (a) Isometric view (b) Sketch of bottom / top plate.

The bottom plate (Figure 6) is the base plate of the jig that supports all the component such as middle plates and top plate. The middle-bottom plate is fixed to the bottom plate of the jig. The middle-bottom plate and middle-top plate are fixed through a hinge. The middle plate consists of grooves which are created by thin metal sheet (Figure 7).

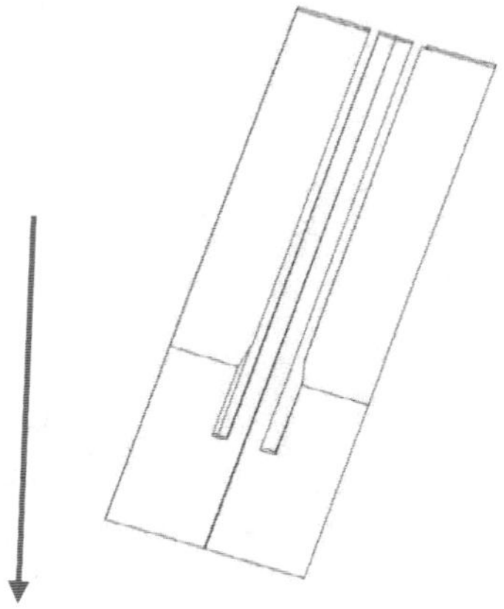

Thin metal sheet

(a)

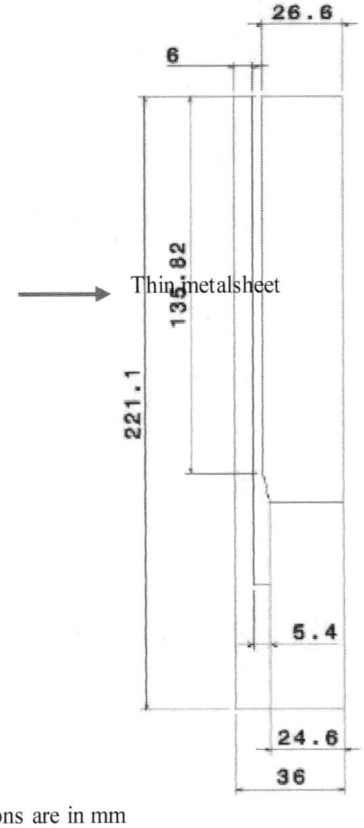

All dimensions are in mm

(b)

Fig. 7 (a) Middle plates of sleeve button placket jig (b) Sleeve buttonhole placket jig

The sleeve buttonhole placket jig (Figure 8) is used to attach sleeve buttonhole placket with sleeve.

Components of sleeve buttonhole placket jig

1) Bottom plate (Figure 9)
2) Top plate (Figure 9)
3) Middle fixed plate (Figure 10)
4) Middle movable plate (Figure 11)

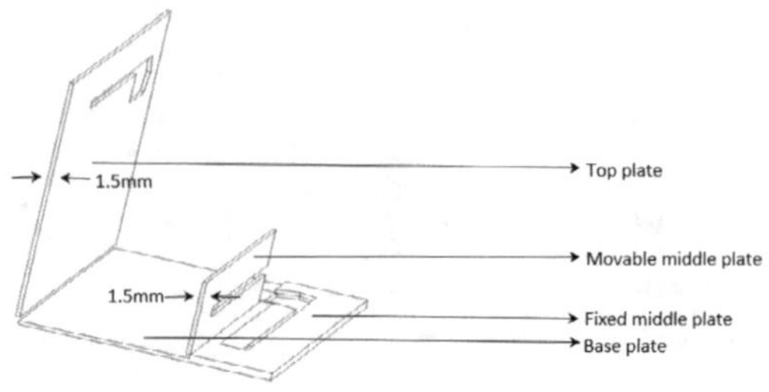

Fig. 8 Isometric view of sleeve buttonhole placket jig

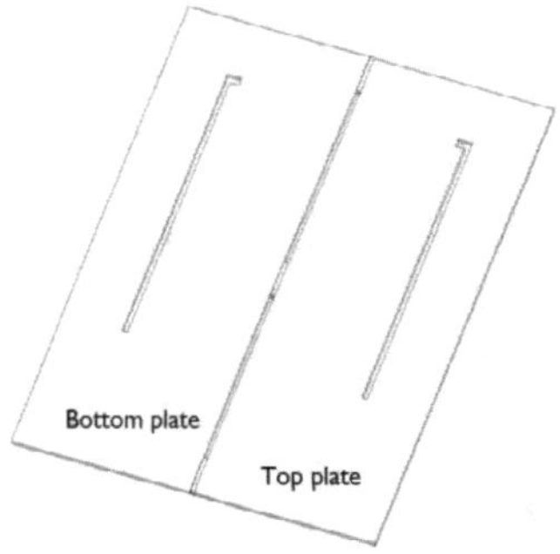

(a)

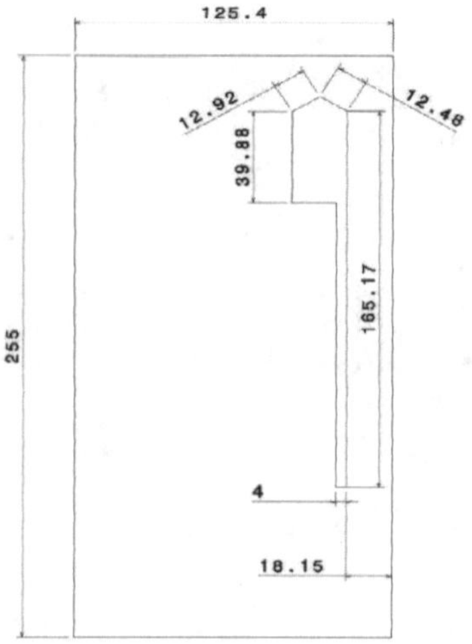

All dimensions are in mm

(b)

Fig. 9 Bottom and top plate of sleeve buttonhole placket jig
(a) isometric view (b) sketch of top / bottom plate

The bottom plate (Figure 9) is the base plate of the jig that supports all the component such as middle plates and top plate. The middle-bottom plate is fixed to the bottom plate of the jig. The middle-bottom plate and middle-top plate are fixed through a hinge. The middle plate consists of grooves which is created by thin metal sheet (Figure10).

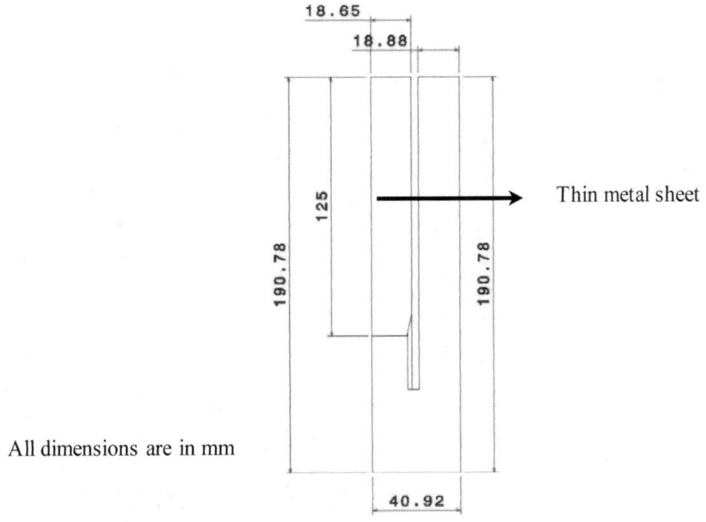

(a)

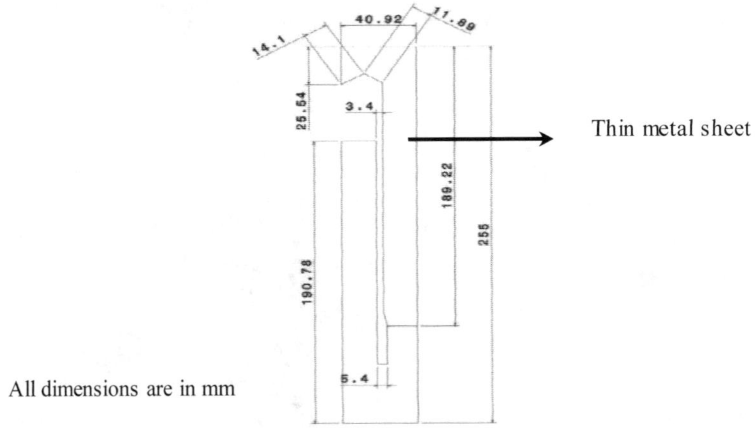

(b)

Fig. 10 Sketch of middle plates of sleeve buttonhole placket jig

(a) movable middle plate (b) fixed middle plate

Profiled flexible slider

A profiled aluminum sheet called slider of 0.01mm thickness is used to fold the seam allowances. The slider of sleeve button placket (Figure 11) and sleeve buttonhole placket (Figure 12) is guided by the groove present in the jig.

IOP Conf. Series: Materials Science and Engineering 1091 (2021) 012002 doi:10.1088/1757-899X/1091/1/012002

All dimensions are in mm

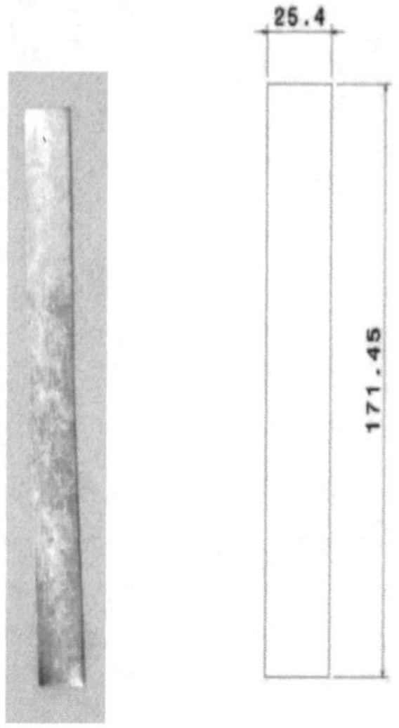

Fig. 11 Slider of sleeve button placket

All dimensions are in mm

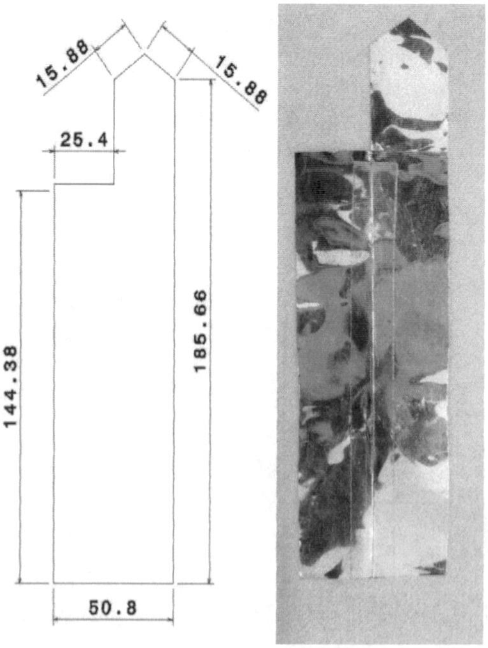

Fig. 12 Slider of sleeve buttonhole placket jig

Adhesive tape (Figure 13) is used to allow hinging action for middle movable plate.

IOP Conf. Series: Materials Science and Engineering 1091 (2021) 012002 doi:10.1088/1757-899X/1091/1/012002

Fig. 13 Polyester film acrylic tapes

Specifications

- Material – polyester adhesive film.
- Width – 1 mm.
- Thickness – 0.2 mm.

Hinge

The hinge (Figure 14) is attached between the top and bottom plates of the jigs which allows hinging action causes opening and closing of the jig.

Fig. 14 Hinge for jigs

Specifications of hinge

- Material – stainless steel.
- Thickness 3 mm.
- Length 2 cm.

Modified Feed Dog

- The modified feed dog (Figure 15 -a) for a moving stitching jigs comprises a zinc die cast body portion having a toothed surface region covered by a layer of elastomeric material.
- The elastomeric material is urethane coated over the surface of the feed dog (Figure 15 -b).
- The teeth are covered and hence results in flat toothless surface.

(a) Top view

(b) Side view

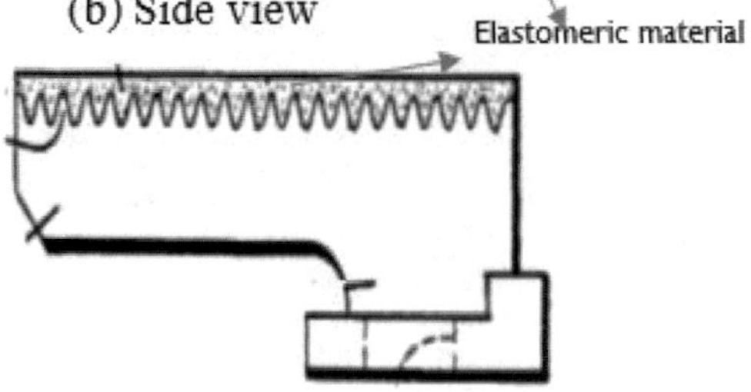

Fig. 15 Rubber composite coated toothless feed dog

Modified Pressure Foot

The presser foot is attached with a projection called stopper. The projection is cylindrical in shape, less than or equal to the thickness of the top plate of the jigs. In the sleeve button placket jig and sleeve buttonhole placket jig, the top plate is 1.5mm and hence the projection length is 1.5 mm. The outer diameter of the hollow cylindrical projection is 3.9 mm.

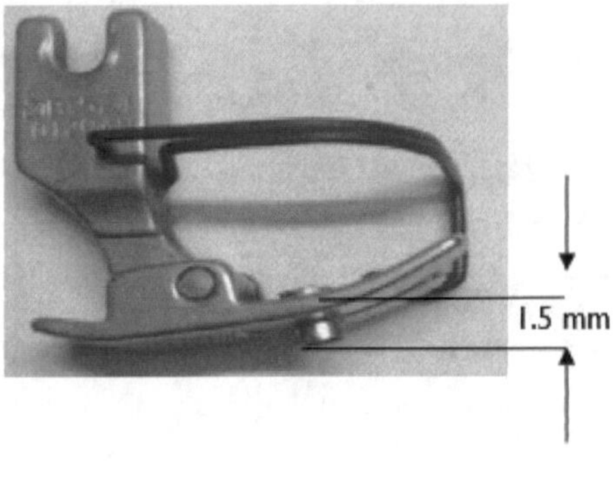

Fig. 16 Side view of modified presser foot

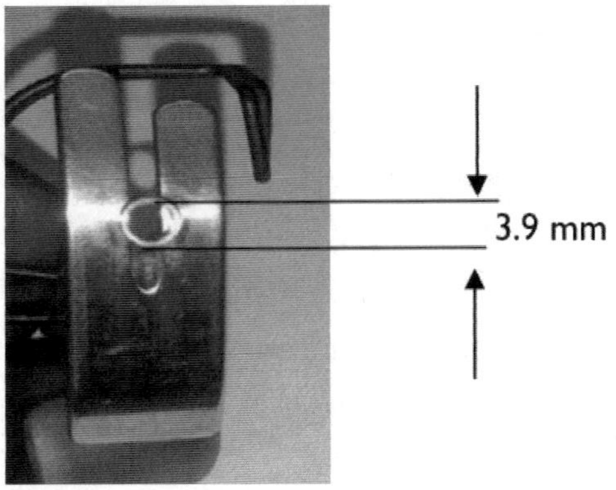

Fig. 17 Bottom view of modified presser foot

Role of the modified presser foot:

- The projection in the presser foot gets engaged with the slot of the top plate and guides the jig during sewing.
- The presser foot exerts pressure that is necessary to keep the jig in the closed position during sewing.

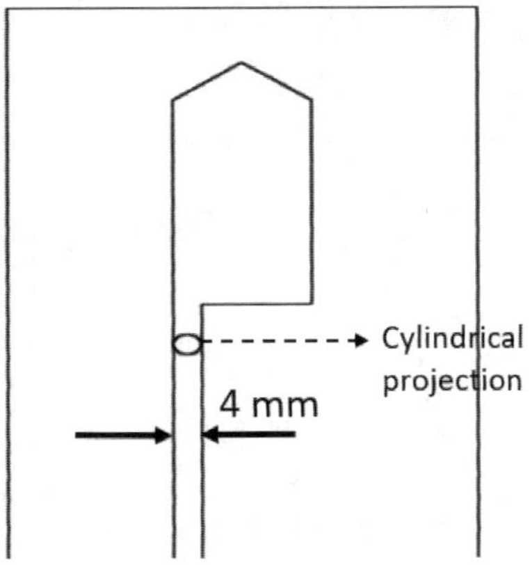

Fig. 18 Presser foot projection engaged with the slot of the jig

The cylindrical shaped projection can only have a maximum diameter of 4mm since the space between the two rows of feed dogs on either side of the needle hole is 4mm.

Methods

The sewing using jig is similar to the manual method yet more comfortable and also involves the stitching of sleeve button placket at first using sleeve button placket jig and followed by the stitching of sleeve buttonhole placket using sleeve button hole placket jig.

Jig Driving Mechanism

The jig is moved over the rubber coated feed dog and guided through the track of the slot by the projected stopper attached to the presser foot (Figure19).

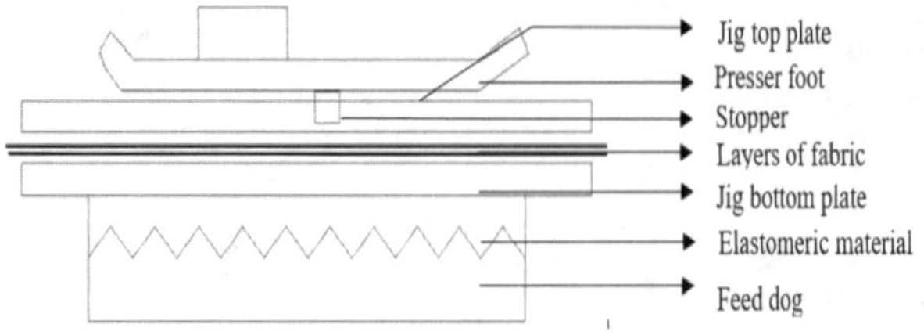

Fig. 19 Side view of the jig after loaded to the machine

Procedure for Sewing Sleeve Button Placket Using Sleeve Button Placket Jig

Loading of material into the jig

- Open the top plate of the jig and then open the middle top plate of the jig. Position the sleeve button placket fabric component to the location marks in the jig (Figure 20).
- Place the profiled flexible metal sheet for sleeve button placket over the fabric component and position it to the location marks (Figure 21).
- The top edge of the slider is pushed along with the flexible sheet to lock into the grooves (Figure 22).
- The sleeve component is placed in such a way that the slit is positioned properly and ensuring that the stitching is to be done on back of the sleeve.
- The middle plate is closed over the edge of the slit (Figure 23).
- Then the top plate is closed (Figure 24).

Fig. 20 Positioning of the sleeve placket component.

Fig. 21 Placement of slider

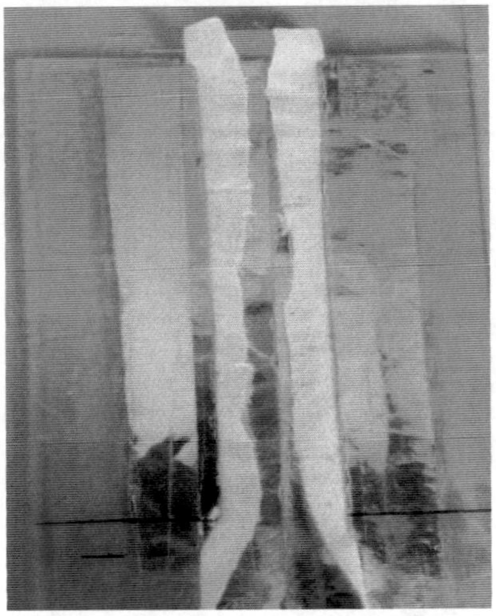

Fig. 22 loading of the fabric

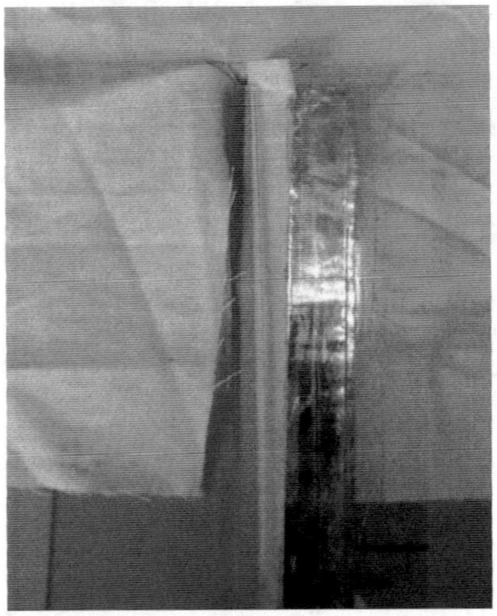

Fig. 23 Positioning of the sleeve into the jig and closing of the middle plate

Fig. 24 Closing the top plate of the top plate of the buttonhole placket jig

Loading jig to the machine

- Slide the loaded jig with the right hand on the top plate, towards the needle, lining up the start position.
- Once the jig is positioned properly, the presser foot is lowered and the projection in the presser foot will be locked into the slot of the jig.
- Start the sewing and stop when it reaches the right-angled corner of the slot.
- Rotate the jig to 90 degrees towards the side and sew until it reaches the end of the slot.

Unloading of material from the machine

- Lift the presser foot.
- Open the top plate and then open the middle top plate.
- Slide the sewn components up and pull the sleeve out of the jig.

Sewing Sleeve Buttonhole Placket Using Sleeve Buttonhole Placket Jig

Loading of material into the jig

- Open the top plate of the jig and then open the middle top plate of the jig. Position the sleeve buttonhole placket fabric component to the location marks in the jig (Figure 25).
- Place the profiled flexible metal sheet for sleeve button placket over the fabric component and position it to the location marks (Figure 26).
- The allowance at the top edge is folded as shown in the figure and pushed along with the flexible sheet to lock into the grooves (Figure 27).
- The sleeve component is placed in such a way that the slit is positioned properly and ensuring that the stitching is to be done on back of the sleeve.
- Then the sleeve is folded at the slit line (Figure 28 (a)).
- The sleeve is then folded at the top (Figure 28 (b)).
- The middle plate is closed over the edge of the slit (Figure 28 (c)).
- The sleeve is unfolded (Figure 28 (d)).

- Then the top plate is closed (Figure 29)

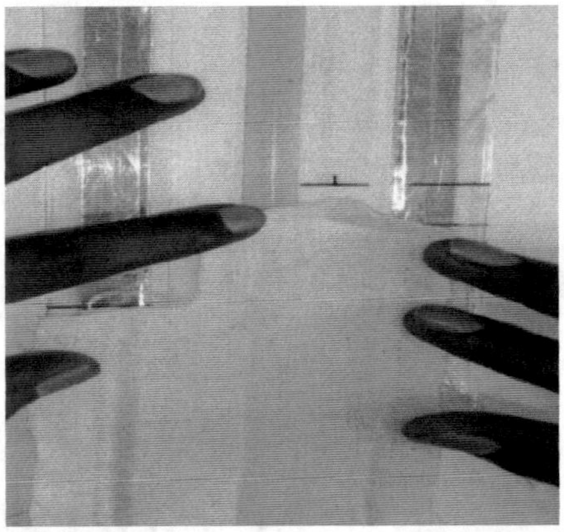

Fig. 25 Step 1 Positioning of the sleeve buttonhole placket component over the jig.

Fig. 26 Step 2 Pushing the slider to lock the folded edges into the jig

Fig. 27 Step 3 Removal of slider

(a)

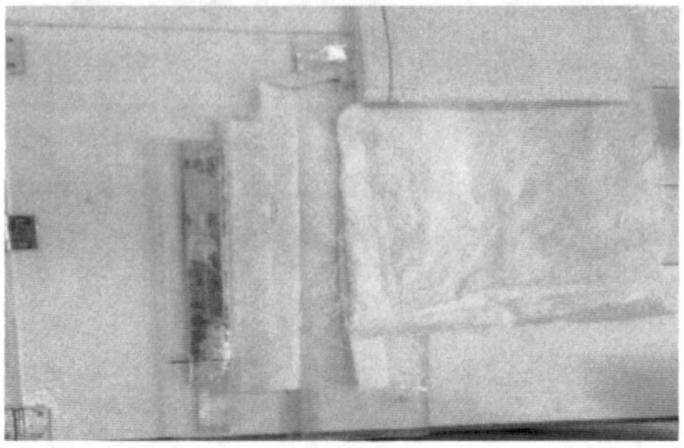

(b)

(c)

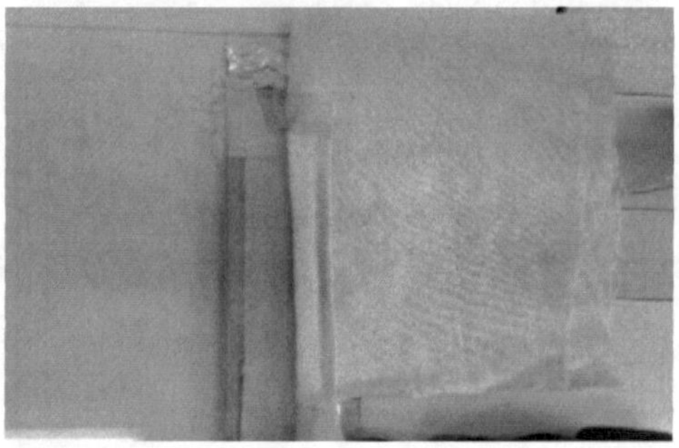

(d)

Fig. 28 Step 4 Sleeve positioning (a) Folded at the slit line
(b) Folded at the top (c) Closing of the middle plate (d) Unfolding the sleeve

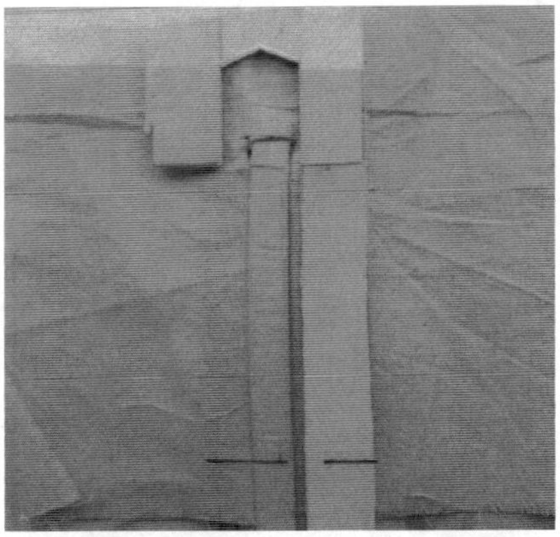

Fig. 29 Step 5 Closing the top plate

Loading jig to the machine

- Slide the loaded jig with the right hand on the top plate, towards the needle, lining up the start position.
- Once the jig is positioned properly, the presser foot is lowered and the projection in the presser foot will be locked into the slot of the jig.
- Start the sewing and stop when it reaches the right-angled corner of the slot.
- Rotate the jig to 90 degrees toward your side and sew until it reaches the end of the slot.

Unloading of material from the machine

- Lift the presser foot.
- Open the top plate and then open the middle top plate.
- Slide the sewn components downwards and pull the sleeve out of the jig

III. RESULTS AND DISCUSSIONS

Samples Comparison

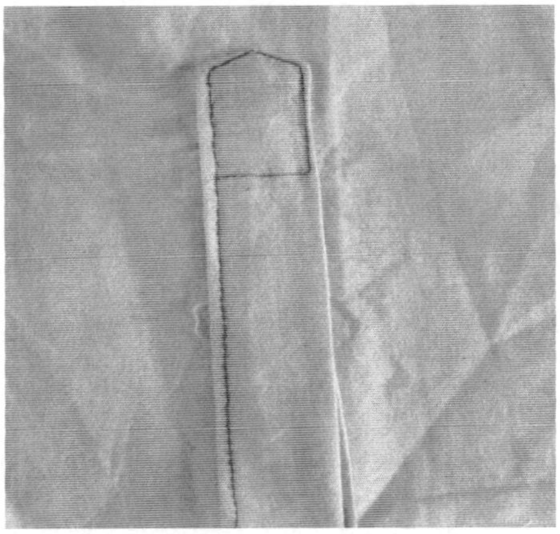

Fig. 30 Sample produced using jig

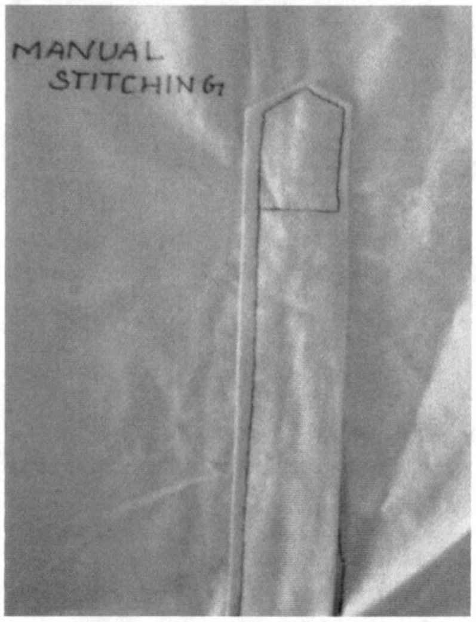

Fig. 31 Sample produced without using jigs

- The samples prove that the sleeve placket can be constructed using the stitching jig.
- From the comparison of samples, it is observed that the accuracy of the stitch line is good and the defect in the sample is negligible.

Time Study

A time study is conducted for the sewing of sleeve placket using jig and without jig. The results are given below.

Table 4.1 Comparison of time study for sleeve placket attaching

Operation cycles	Time taken for sewing (s)			
	Sewing without using jig		Sewing using developed jig	
	Unskilled operator	Semi-skilled operator	Unskilled operator	Semi-skilled operator
1.	278	241	217	213
2.	283	244	200	216
3.	287	231	221	213
4.	289	236	215	216
5.	290	236	206	218
Average time taken	285.4	238.67	212.33	215.2

The time study data was statistically analysed and it is observed that the time taken by the operator for sewing the sleeve placket using the jig is independent of the skill level of the sewing operator. Clearly the time taken by the operator for sewing the sleeve placket using jig is lower than that of operations carried out without using jig.

IV. CONCLUSION

In this work, the stitch jigs are developed for one of the critical sewing operations i.e. sleeve placket attaching. The developed stitching jigs include (i) stitching jig for sleeve button placket attaching and (ii) stitching jig for sleeve buttonhole placket attaching. The samples from the stitching jigs are taken out and compared. The time study for sleeve placket attaching operation using jig and without jig is done and compared. It is found that the average cycle time for the completion of sleeve placket is lower and independent of the skill level of the sewing operator using the developed jig. The stitching jig contributes greatly to perineal levels of quality,

REFERENCES

1. Anon (2012), 'Sewing automats for multi-product multi process manufacturing', Stitch World, Issue December.
2. Darren C. Peddle, Lower Sackvile Canada (1985), 'Adjustable jig assembly', US Patent Application Number US 4522380.
3. Carr H., Latham B. (1999), 'The Technology of Clothing Manufacture', Blackwell Science, Oxford, pp 147-152.
4. Guerreschi, Candidavid (1999), 'Clamp device or jig-guide and device for prepositioning and shifting cloth to be coupled with a programmable sewing machine or the like', European Patent Application Number EP 05658504B19.
5. Helen J. Armstrong (2006) 'Pattern making for fashion design', Pearson Education, pp 522 - 530.
6. Hubert Jinemann, Bielefeld (1987), 'Workpiece holder for sewing the tip area of Neckties', US Patent Application Number US 4643115.
7. Jana P. (2003), 'Automation in sewing: how does it work? And why it didn't?', Stitch world, Issue December.
8. Jason Prout, Malcolm Craven (2000), 'Drive mechanism for jig-fed sewing machine', US Patent Application Number US 4527492.
9. John A. Rose, Knaresborough; Keith Dyer (1980), 'Apparatus for Manufacture of Jigs', US Patent Application Number US 4779666.
10. John A. Rose, Knaresborough, Keith (1982), 'Method of Manufacturing of Jigs', US Patent Application Number US4507029.
11. John A.Rose, Knaresborough; Keith Dyer (1983), 'Apparatus for manufacture of jigs', US Patent Application Number US4379666.
12. Orland B. Reid Kenilworth, Joseph W. A. (Jan, 1971), 'Device for turning Fabric layers', US Patent Application Number US 3554354.
13. Peter R. Gill (1985), 'Jig for stitching fabric layers in a sewing machine', US Patent Application Number US 4527492.
14. RajkishoreNayak and Rajiv Padhye (2017), 'Garment Manufacturing Technology', Woodhead publishing,
pp 280 - 300.
15. RajkishoreNayak and Rajiv Padhye (2017), 'Automation in Garment Manufacturing', Woodhead publishing, pp 199-237.
16. Solinger J. (1988), 'Apparel Manufacturing Handbook', Bobbin Media Corp, Columbia, pp 34-50.
17. Yamashita, Fumio (2012), 'Sewing guide Jig', Japanese Patent Application Number JP 2011000084988.

Transforming Brackish Water into Ingestion Liquid Using Solar Still, Physico-Chemical Analysis

ShanmugasundaramK[a], SivasenthilE[b], ManikandanV[c*], GaneshRaju[c]

[a]Assistant Professor, Department of Physics,Government Arts and Science College,Thiruvennainallur, Tamilnadu, India - 607203.
[b]Assistant Professor, Department of Physics, Karpagam Academy of Higher Education, Coimbatore, Tamilnadu, India - 641021.
[c]Assistant Professor, Department of Physics, Sri Krishna College of Technology, Coimbatore, Tamilnadu, India – 641042.
email: [a]kssphd2010r@gmail.com, [b]esivasenthil@yahoo.co.in, [c*]manikadan.v@skct.edu.in, [c]ganeshraju.a.c@gmail.com,

Corresponding author: SivasenthilE, Shanmugasundaram K

Abstract: In this work, an endeavor has been made to purify waste water by using solar still and the purity of water has been examined using physico-chemical properties such as taste, odour, pH, electrical conductivity, TDS, total alkalinity, total hardness, magnesium, magnesium hardness, calcium, calcium hardness, sulphate, nitrate, iron, chloride, silica were shows the purity of water. Standard methods were used for determination of physical and chemical characteristics of the waste water and distilled water. The outcomes were equaled with standards prescribed by IS:(10500-1991). As a result, the physico-chemical parameters exceeded the water quality standards and the waste water (Brackish water) is unfit for drinking purpose as it was contaminated and the quality of water is very bad. It needs some degree of treatment before consumption. Hence the waste water is cleansed by means of solar still and found that the physicochemical parameters are underneath permissible limit of IS:(10500-1991). From these results it is clear that, purification of waste water by using solar still is an optimistic approach for the production of drinking water and also can be used as a purifier in brackish area.

Keywords: Water Purification, Water Quality, Solar Still, Physico-chemical parameters

Introduction

Fresh/Clean water is essential for human beings and life is impossible without water. The accessibility of drinking/clean wateris unattainable currentlyowing to rapid growth of population and development of industries. Only 1% of available water in Earth is fresh and in liquid state, and among thisavailability of water, it is polluted by both diseases and toxic chemicals. For this reason purification of water is necessary and in many countries the solar still working by utilizing solar energy which is renewable and sustainable has been used to produce drinking water [1-2]. Solar radiation is the source of energy for the solar still and in the process; water is evaporated and water vapor is separated from dissolved matter and condensed as pure water [3-4]. Drinking water quality management is a key pillar in the prevention and control of water born diseases [5]. Many researchers have analyzed the quality of drinking water using physico-chemical and microbial parameters. The drinking, borewell and sewage water from three different places of Sivakasi has been analyzed by Radha Krishnan *et al.,* [6].Pavan Kumar*et al.,*[7] have determined the quality of the Raman Padu balancing reservoir and Koilsagar reservoir by using bacteriological and physicochemical parameters. The outcome displayed the physico-compound parameters are inside as far as possible.The quality of 24 bore well water samples from Madyal and Vadgaon Villages of KagalTahsil MS, India has been investigated by Ashvin*et al.,* [8]. It has been recommended for drinking and domestic purposes, only after pretreatments like filtering, boiling, reverse osmosis and elctrodiolysis.Chemical characteristics of different sources of drinking water have been studied by Yousaf*et al.,* [9] in Nowshera district, Khyber Pakhtunkhwa. It has been inferred that the quality of water in Nowshera district is badly deteriorated and contaminated as a result of the flood of 2010. Sanjay *et al.,* [10] have studied the quality of Bansagar dam and nearby area, Madhya Pradesh, India using physico-chemical parameters and concluded that the quality of Bansagardam and nearby area is better. The physico-chemical parameters of ground water of 15 stations in Bhandara district has been studied by Asia *et al.,* [11] and found that the treatment of ground water is needed to minimize the contamination.The quality of different water samples from different sources in Coimbatore

district has been analyzed by various researchers [12-15] using physico-chemical and microbiological parameters and found that it is fit for drinking and domestic purposes.

Aim of this work is to purify the waste water by using solar still and quality of the water before and after distillation has been analyzed using physico-chemical parameters such as pH, electrical conductance, total hardness, total alkalinity, magnesium hardness, total dissolved solids, chloride etc.

Materialsand Method

Design of the solar still

A schematic diagram of the proposed passive distillation system is shown in Fig 1. It consists of a single basin type solar still. The still has an area of 1m^2. The bottom and sides are insulated against heat loss to surroundings by means of glass wool insulation. The surface of the basin facing the sun is painted black for maximumabsorption of solar radiation. The still cover is made of 0.004m thick ordinary glass and has an inclination of 11°.The yield is collected by an Aluminium channel attached to the lower end of the glass cover and measured directly by a measured jar.

Experiment have been carried out from 9am to 5pm with the dual purpose double slope single basin solar still at Department of Physics, Karpagam University, Coimbatore-641021 (latitude 11° N, long 77°52' E). Fig.2. shows the photograph of a single basin double slope solar still.

Results and discussion

The raw water sample was collected from Coimbatore District, Tamilnadu, India and purified by using solar still for drinking and domestic purposes. The quality of before and after purification of raw water have tested by using physic-chemical parameter analysis to know the quality of water. The results of the physicochemical parameters are given in Table.1.

The odor and taste has been tested for raw water and purified water and found that the before purification of raw water is objectionable but after the purification of it the results have shown that the odour and taste are agreeable.The pH values of before and after purification of raw water is shown graphically in Fig.3.It is seen that the pH value of raw water is 8.4 and after distillation or purification of this water is 7.4. From these results it is confirmed that the pH values of samples are within the permissible limit of IS:(10500-1991).The value electrical conductivity for before and after distillation of raw water is depicted graphically in Fig 4 and observed that the value of electrical conductivity of brackish water is 900μmhos/cm and after distillation it is 59. The electrical conductivity brackish water is higher than the distilled water due to high salt content is presented in brackish water. Fig.5. shows the turbidity for before and after distillation of raw water. The result shows that the turbidity of raw water is higher than the purified water and it is not exceed the prescribed limit of ISI(10500-1991). Fig.6. illustrates the values of total dissolved solid for brackish and distilled water and Compared these water samples the TDS value of raw water is higher than the distilled water. Because, high total dissolved solids might be due to the presence of large number organic salts, such as chloride, magnesium, calcium and carbonates.

Alkalinity is the name given to the quantitative capacity of water to neutralize an acid. The value of total alkalinity of brackish water and after distillation of it is depicted graphically in Fig.7. From these results it is confirmed that the water samples are within the permissible limit of ISI(10500-1991). Total hardness is represented by the value of $CaCO_3$ is present in the water. Fig.8. shows the total hardness of water samples and it is found that the value of total hardness is 356mg/l.

The magnesium hardness of the brackish water range is 36 mg/l and after purification of it is below detective level is shown in Fig.9. It is found that the water samples are below the permissible limit. The values of calcium hardness for water samples is shown graphically in Fig.10 and inferred that the value of calcium hardness is 68mg/l and below detective level for before and after distillation of brackish water and water samples are below the prescribed limit of Indian drinking water standard IS:(10500-1991).

Calcium is naturally present in water. Calcium is largely responsible for water hardness. The calcium ranges in the present water samples high before distillation of brackish water and it is shown in Fig.11. The value of calcium content in raw water is 25.6mg/l.

Magnesium and other alkali metals are responsible for water hardness. Fig.12. shows the magnesium content of water samples and inferred that magnesium content is not able to detect after distillation of brackish water. The value of magnesium content is found to be 11mg/l. The tolerance of magnesium is 30 mg/l.

The iron content of water samples is shown graphically in Fig.13. The result shows that the values of iron content are below the permissible limit of Indian drinking water standard. The tolerance range of the iron is 0.30 mg/l.

Sulphate is a substance that occurs naturally in drinking water. The content of sulphate for water sample is illustrated in Fig.14. It is found that the water samples are below the prescribed sulphate. The tolerance range for sulphate is 200 to 400 mg/l.

The value of phosphate is found to be 1.2 for before distillation of raw water and it is shown in Fig.15. The results revealed that the after distillation of brackish water the phosphate value is not detective level. Most chlorides in water come from salt (NaCl). The value of chloride for brackish water is 165mg/l and after distillation by solar still it is 10.2mg/l.

Fig.16. shows the chloride content of water samples and confirmed that the water samples are within the permissible limit of IS:(10500-1991). The tolerance range of the chloride is 200 to 1000 mg/l.

In Figs. 17 and 18expresses the presence of nitrate and silica for the water samples. Fig.17. illustrates the values of nitrate content for before and after distillation of raw water and found that the value of nitrate for brackish water is 1.2mg/l and in distilled water is below detective level. From these results it is confirmed that the water samples are below the permissible limit. Fig.16. shows the value of silica for raw and distilled water. The result indicates that the value of silica is 2mg/l which is higher than the distilled water 0.4 mg/l.

Conclusion

In light of these outcomes following conclusion has been drawn. The waste water (Brackish water) is polluted and the physico-chemical parameters surpass the water quality guidelines and the nature of water is exceptionally terrible and it is unfit for drinking reason. In this way the waste water is filtered by utilizing solar still and found that the physico-chemical parameters are inside the reasonable furthest reaches of ISI(10500-1991). From these outcomes plainly, sanitization of waste water quality by utilizing sunlight based still is great and fit for drinking reason and prescribed that the sun based still can be utilized as purifier in brackish area.

Table.1. Results of the Physico-Chemical parameters and comparison between Indian drinking water standards of IS: (10500-1991)

Parameters	**Before distillation (Raw Water)**	**After distillation**	**Indian Standard IS: (10500-1991)**
Odour	Objectionable	Unobjectionable	Agreeable
Taste	Not agreed	Agreeable	Agreeable
pH	8.4	7.4	6.5-8.5
Electrical conductivity(μmhos/cm)	900	59	-
Turbidity(NTU)	0.9	0.1	5
Total dissolved solids(mg/l)	501	45	500
Total alkalinity(mg/l)	350	10	300
Total hardness(mg/l)	356	BDL	300
Magnesium hardness(mg/l)	36	BDL	50-100
Calcium hardness(mg/l)	68	BDL	75-200
Calcium as Ca (mg/l)	25.6	BDL	75
Magnesium Mg (mg/l)	11	BDL	30
Iron Fe (mg/l)	0.13	BDL	0.30
Sulphate SO_4(mg/l)	26	BDL	200
Phosphate(mg/l)	1.2	BDL	
Chloride Cl_2(mg/l)	165	10.2	200
Nitrate(mg/l)	1.2	BDL	200
Silica SiO_2(mg/l)	2	0.4	100

Note: BDL-Blow Detective Level

IOP Conf. Series: Materials Science and Engineering 1091 (2021) 012003 doi:10.1088/1757-899X/1091/1/012003

Figures

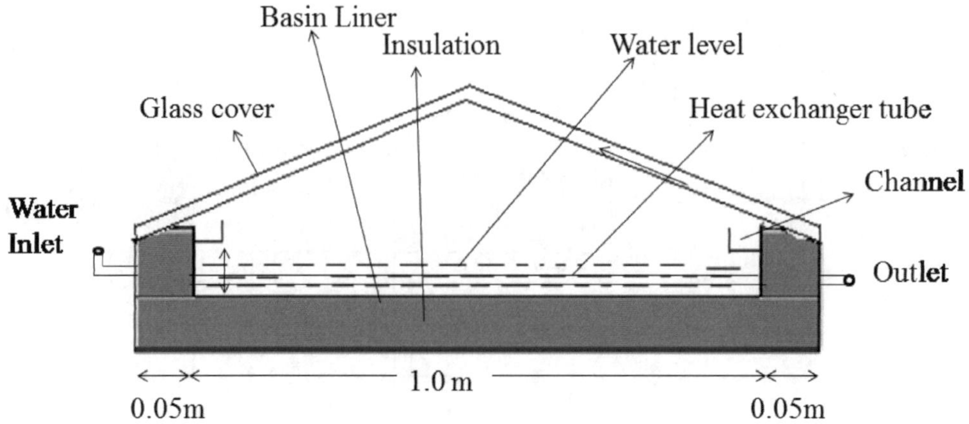

Fig 1. A schematic diagram of the single basin double slope solar still

Fig.2. Photograph of the single basin double slope solar still

Graphs

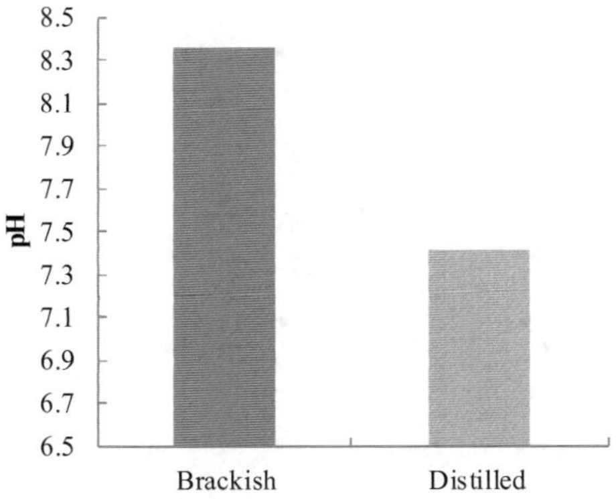

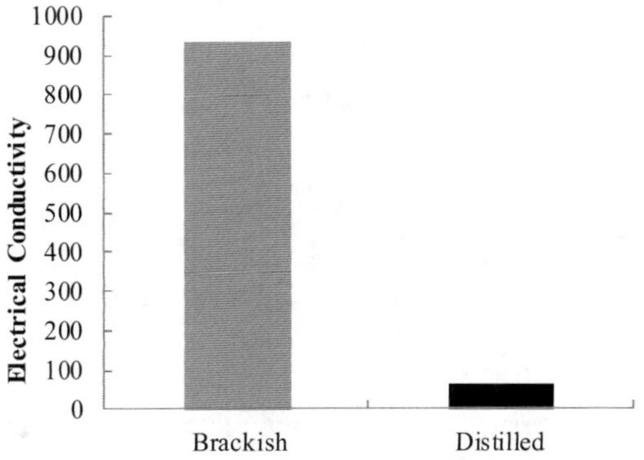

Fig.3. the pH value of water samples. Fig.4. Electrical conductivity for waters samples

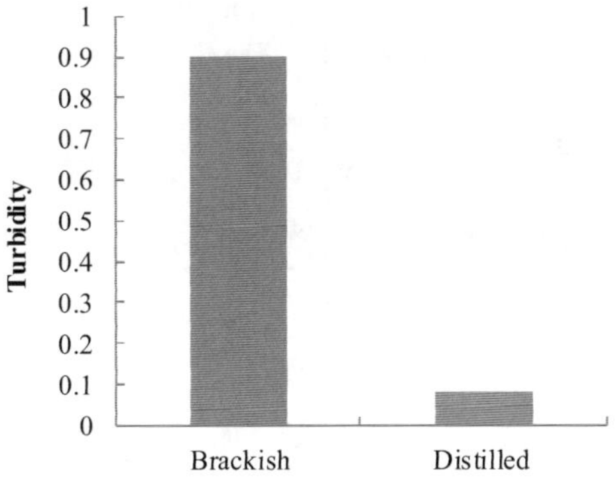

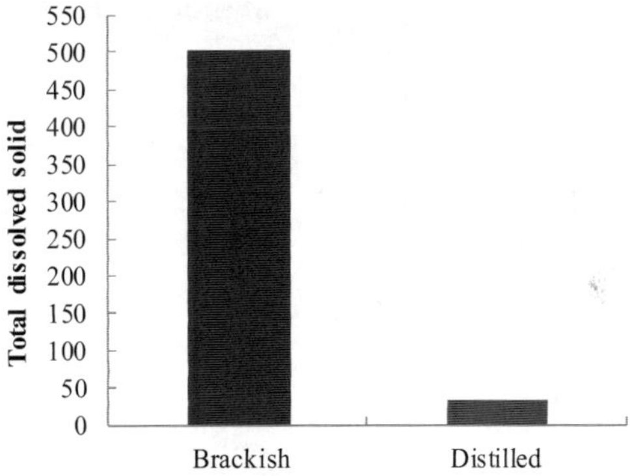

Fig.5. the turbidity of water samples Fig.6. Total dissolved solid of water samples

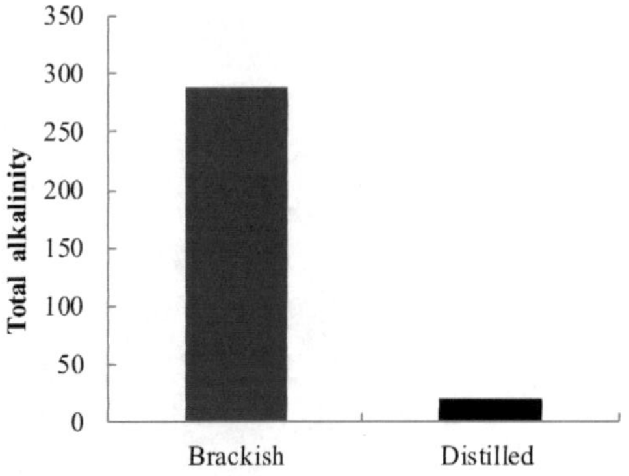

Fig.7. Total alkalinity of water samples

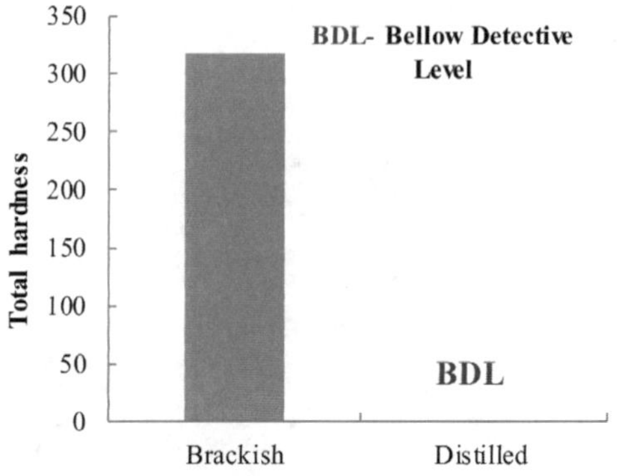

Fig.8. The total hardness for water samples

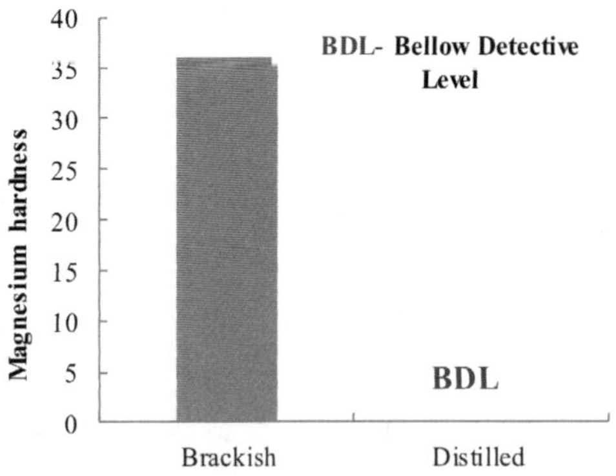

Fig.9. Magnesium hardness for the water samples

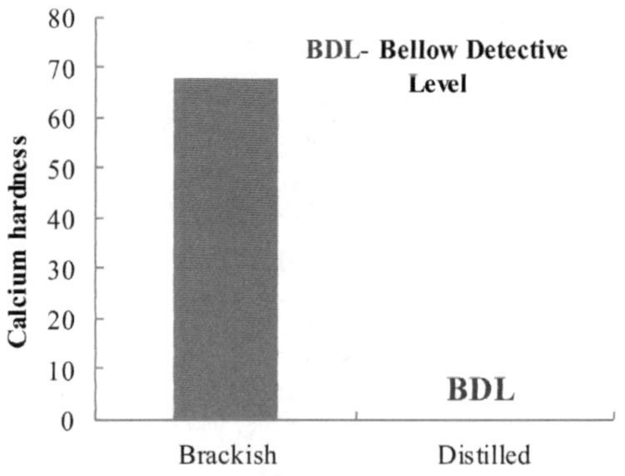

Fig.10. Calcium hardness of water samples

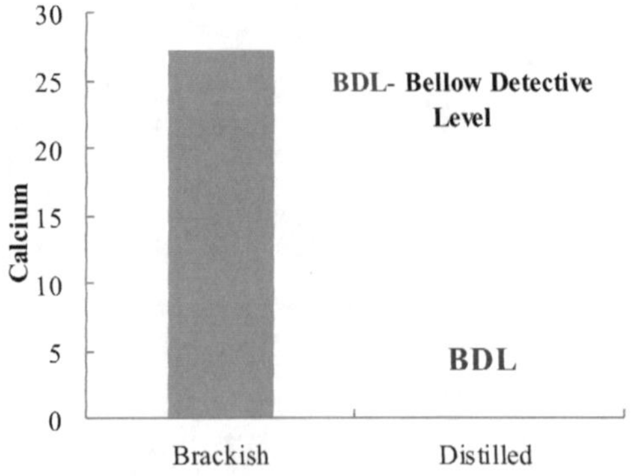

Fig.11. Calciumcontent of water samples

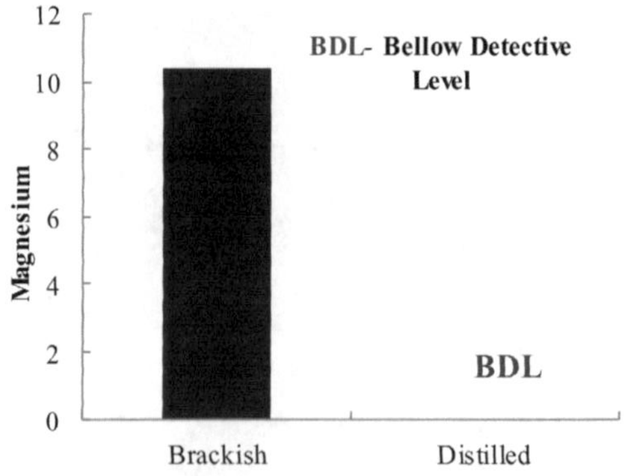

Fig.12. Magnesium content of water samples

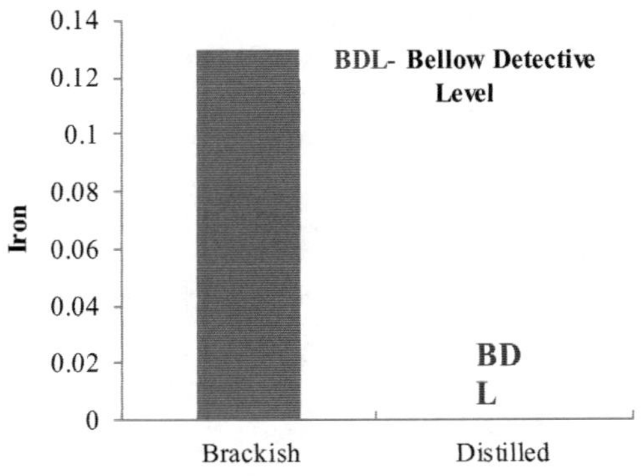

Fig.13. Iron content of water samples

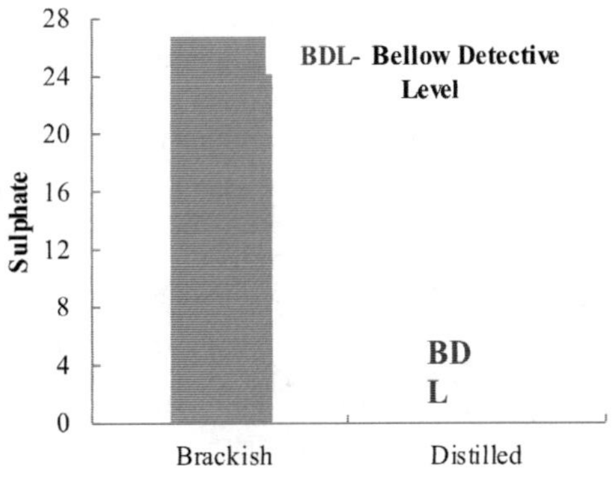

Fig.14. The values of sulphatefor the water samples

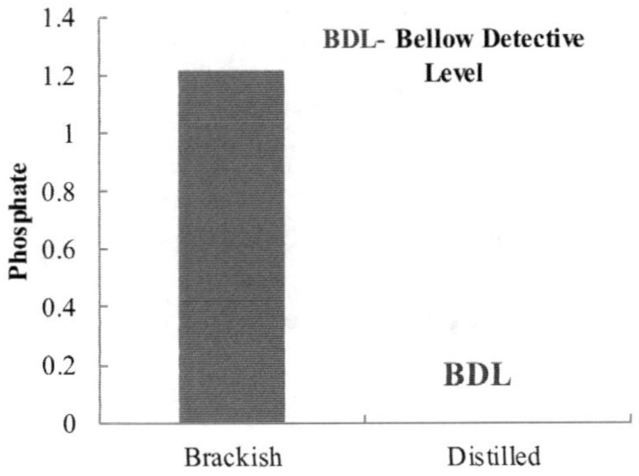

Fig.15. the value of Phosphate for water samples

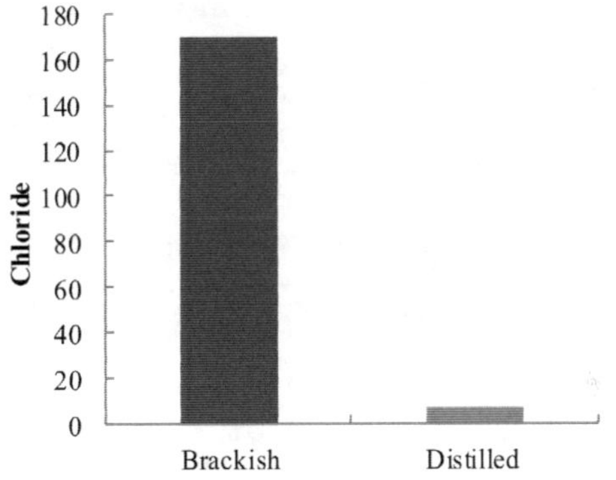

Fig.16. the chloride content of water samples

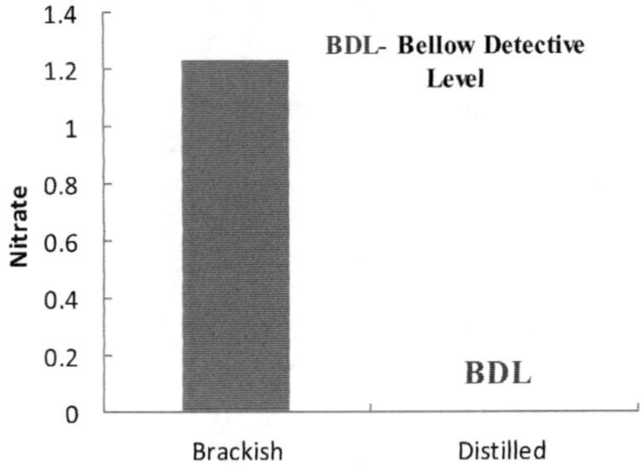

Fig.17. The value of Nitrate for water samples

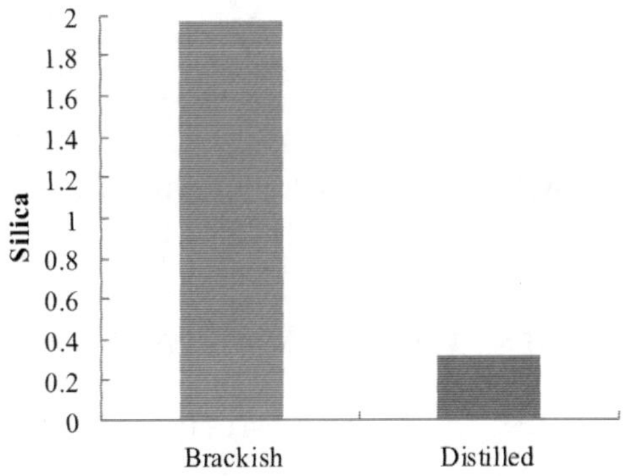

Fig.18. The content of silica for water samples

References

[1] MurhekarGopalkrushna, H.,Determination of physic-chemical parameters of surface water samples in and around Akot city, *International journal of research in chemistry and environment,* 2011, vol. 1, pp. 183-187.

[2] Chaibi, M.T., An overview of solar desalination foe domestic and agriculture water needs in remote arid areas, *Desalination,* 2000, vol. 127, no.1, pp. 119-133.

[3] McCluney,W.R., Solar distillation of water, Energy note, *Florida solar energy centre, Florida*, 1984.

[4] Al-Kharabsheh,S.S.,Goswami,D.Y., *Solar distillation and drying* - University of florida, NRGY: 00319, USA, 2004.

[5] Sidnei,M.T.,Fakio, A.L.T., Maria, C.R.,Francises,A.E.,Adaunto,F., Seasonal variation of some limnological factors of Lagoa does Guarana, a Varzea lake of the Rio paranana state of MatoGrosodosul Brazil,*Rev. Hydrobiol. Trop.,* 1992, vol. 25, no. 4, pp. 269-276.

[6] Radha Krishnan, R., Dharmaraj, K.,RanjithaKumari, B.D.,A comparative study on the physicochemical and bacterial analysis of drinking, borewell and sewage water in the three different places of Sivakasi, *Journal of environmental biology,* 2007, vol. 28, no. 1, pp.105-108.

[7] Pavankumar, P., RaghuveerYadav, P., AshwinKodaparthi, Bacteriological and physicochemical quality of main drinking water sources, *Polish journal of environmental studies,*2013, vol. 22, no. 3, pp. 825-830.

[8] Ashvin, G.G.,Rajaram S.S., ShobhaD.J., An evaluation of physic-chemical parameters to assess borewell water quality from Madyal and Vadgaon villages of KagalJahsil, MS, India,*International research journal of environmental sciences*, 2013, vol. 2, no. 5, pp. 95-97.

[9] Yousaf, S., Zada, A., Owais, M., Physic-chemical characteristics of potable water of different sources in district Now Shera: A case study after flood, *Journal of Himalayan earth sciences*, 2013, vol. 46, no. 1, pp. 83-87.

[10] Sanjay, G., Bhatnagar, M.K., Raviraj, S., Prachi, B., Studies on physico-chemical parameters of

Ban Sagar Dam and Nearby area Madhya Pradesh, India. *International journal of latest research in science and technology*, 2013, vol. 2,pp.20-22.

[11] Asia, S.J.,Inam, F.,Deo, S.,Malviya.N., Physico-chemical analysis of bore well water samples of Bhandara district, Maharashtra, India, *International journal of chemical science and technology*, 2011, vol. 2, pp. 202-204.

[12] Rajiv,P., Abdul Salam, H., Kamaraj, M., Rajeshwari, S., Balaji, R., Comparative physic-chemical and microbiological analysis of various ponds water in Coimbatore district, Tamilnadu, India. *Annals of biological research*, 2012, vol. 3, pp. 3533-3540.

[13] Remia,K.M., Logaswamy.S., Physic-chemical characteristics of ground water quality in Koundampalayampanchayat,Coimbatore district, Tamilnadu, India, *Recent research in science and technology*, 2010, vol. 2, pp. 14-18.

[14] Lenin Sundar, M., Saseetharan, M.K., Ground water quality in Coimbatore, Tamilnadu, along Noyyal River. *Journal of environmental science and engineering*, 2008, vol. 30, pp. 187-190.

[15] Usharani,K., Umarani, K.,Ayyasamy, P.M, Shanthi, K., Perumalsamy, P.L., Physico-chemical and bacteriological characteristics of Noyyal River and ground water quality of Perur, India, *Journal of applied science, management,* 2010, vol. 4, pp. 29-35.

IOP Conf. Series: Materials Science and Engineering 1091 (2021) 012004 doi:10.1088/1757-899X/1091/1/012004

Temperature influence on a quality on printing in a printing machine

Sedrak Martirosyan

National Polytechnic University of Armenia, e-mail: sedrakmartirosyanmm903@polytechnic.am

Abstract

In this paper the temperature influence on a quality printing in a printing machine is discussed. The case of contact printing shafts and paper is investigated, the cases, when temperature is neglected and it's effect on a quality of printing investigated as well. Beside that, three different type of papers investigated depending temperature influence. The stress-strain state in the printing layers is discussed, it is possible to define the printing quality in the contact zones.

Key words: printing machine, temperature influence, printing shaft, stress-strain state, ANSYS software

Introduction

In modern polygraph different printing machines are used depend on print run, colors of printings, paper quality. The type of machine is due to printed products. In [1] the presents research regarding influence of heat treatment on changes of prints on textile material which are obtained by inkjet printing technique. Total of 60 samples were analyzed, 20 for each material used in this experiment. Resistance of color to heat treatment and transfer to cotton cloth was determined for characteristic temperatures by using the gray scale. The surface changes of textile material before and after heat treatment ware monitored by SEM analysis. Heat can be transferred the textile material in three ways: conduction, convection and by electromagnetic radiation [2]. heat transfer can lead to changes in structure of materials (Michalak et al., 2009). In [3] the mechanical surface strength properties of offset prints were studied from six different papers and three different heat-set drying temperature profiles. The most common types of uncoated and coated papers used in the heat-set offset printing process were included in this research. The gloss of the dried prints compared to the gloss of the corresponding paper increased predictably. In [4] the Hybrid heat-set and cold-set printing is discussed and show positive and negative sides for both of them. But not only in technology. In [5] the Rapid Prototyping technology has been developed since its original introduction, leading to a significant improvement in the manufacture of high-quality finished products is discussed for analysis on temperature for 3D printing and found that liquefier temperature must be at 190 °C because higher temperatures cause the material to burn and lower temperatures may be detrimental for the extrusion process.

Not only printing technology is influence temperature, but mostly also in nodes of the printing machine. From this point of view it is possible to mention [5], where the stress-strain state occurring in the surface layers of hard and elastic printing cylinders working surfaces, as well as the statics of deformations and the relative displacements in the contact zones are considered, taking into account the stiffness, by whose generalized results, it is possible to ensure the printing quality in the contact zones.

Contact problems with temperature impact are investigated in [6]. So, here analytical solution of temperature flow between several solid body are presented. As particular cases taken into consideration cases, when the temperature flux acting on a body, constant temperature acting on body and both, flux and constant temperature acting on a body. Temperature distribution defined in different boundary conditions.

In [7] shaft is taken from the head stock of the lathe machine. In this analysis the shaft is connected with bearing and gear. This is the major important component to be taken into account while designing. In this case dynamic vibrations and comparing with static are presented.

So, well known, that during long printing process enough high temperatures are created, if printing machines without cooling system. Often, in small printing machines after some time adjustment the distance between printing shafts and paper is required. So, in current work the case of contact printing shafts and paper is view, the cases, when temperature is neglected and it's effect on a quality of printing is discussed. Beside that, three different type of papers investigated for temperature influence: thin, average and thick.

Problem statement

The printing machine Risograph TR1510 is discussed (Fig. 1). It is consists from First serve zone, Second serve zone, Pressure zone (printing), Paper delivery area, Drum area, Master clamping area, Master reset zone, Scan area, Master training area, Master download area [8].

Dimension for simulation are follow [9]. Hollow large cylinder (Fig. 3, part 1) external diameter D=155 mm, internal diameter is d=125 mm, small cylinder (Fig. 3, part 2) d1=25 mm, rubber thickness is around 2 mm. The rubber layer is consist of four layers (Fig. 2): chloroprene rubber, 2- fabric rubber , 3-porous rubber, 4 - fabric rubber

Two type of the problem discussed: one is without temperature impact, second is with temperature effect. The type of papers taken into consideration. Both of the problems solved in Finite Element method software. The ANSYS software used for solution of the described problem. The ANSYS software is based on Finite Element method (FEM) [10-12].

From the equilibrium condition of the node, as well as from the methods used in the models of all finite elements, the system of equilibrium equations of all finite elements of the model of the deformed body under study is formed. For quasi-static problems, it has follow form:

$$[K]\{U\} = \{P\} + \{P\}^{q} + \{P\}^{g} + \{P\}^{\varepsilon_0} + \{P\}^{\sigma_0}, \quad (1)$$

$$\{U\} = \begin{Bmatrix} U_1 \\ \dots \\ U_i \\ \dots \\ U_n \end{Bmatrix}, \{P\} = \begin{Bmatrix} P_1 \\ \dots \\ P_i \\ \dots \\ P_n \end{Bmatrix}, \quad (2)$$

where [K] is matrix of the stiffness, $\{U\}$ is vector of element node displacement, $\{P\}$ is general vector of external forces, $\{P\}^{q}, \{P\}^{g}, \{P\}^{\varepsilon_0}, \{P\}^{\sigma_0}$ are total vectors of nodal forces.

Discussed node of printing machine (Fig. 2) is from steel shaft (1), steel shaft (2), on that the rubber (3), which is in a contact with printing paper (4) [8].

Three types of paper take into consideration: light paper, with Specific Volume (SV) is 45 g/m^2), middle – SV is 60 g/m^2, heavy with SV 60 g/m^2.

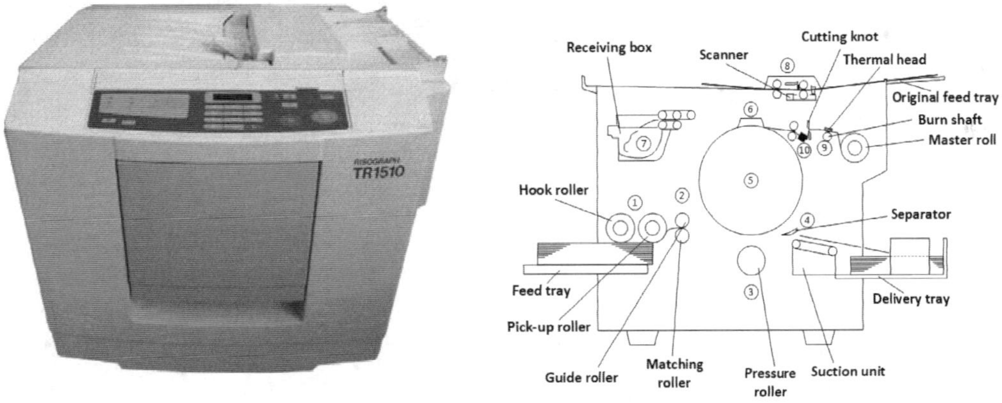

Fig. 1. Risograph TR1510 printing machine

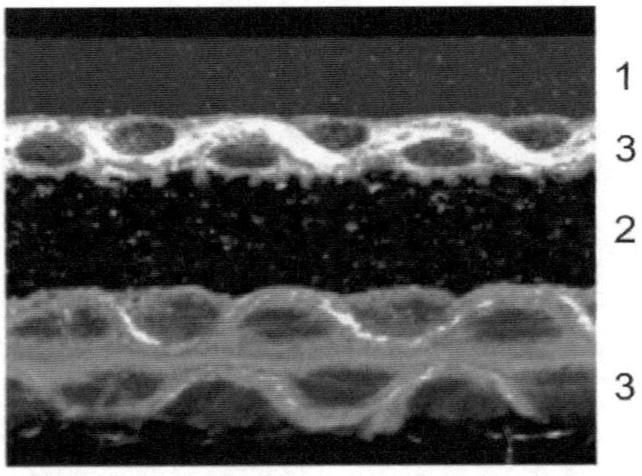

Fig.2. Rubber structure: 1- chloroprene rubber ($_1$ = 0.36 mm), 2- fabric rubber ($_2$ = 0.30 mm), 3-porous rubber ($_3$ = 0,49 mm) 4 - fabric rubber ($_4$ = 0.80 mm)

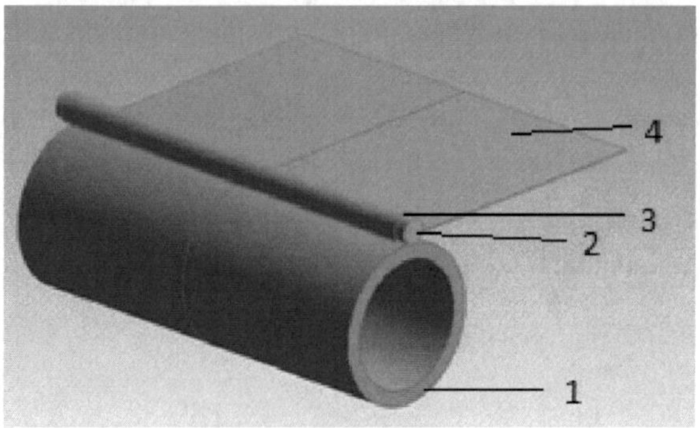

Fig. 3. Printing model

Mechanical properties for each of materials created in ANSYS according Table 1:

Table 1. Materials mechanical properties in a printing contact zone

1	Specific Volume, q, g/m^2	Density, $\rho, g/sm^3$	Thickness, ,mm	Elasticity module, E, Pa	Poisson's ratio, μ
1.	Cylinder (Steel 45)				
1.1.	-	7.85	-	$2 \cdot 10^{11}$	0.3
2.	Paper				
2.1.	45	0,642	0,07	9174,33	0,26
2.2.	60	0,750	0,08	9440,10	0,27
2.3.	120	0,800	0,15	11349,08	0,26
3	Rubber				
3.1.	-	1.2	0,36	$5,0 \cdot 10^6$	0.40
3.2.		1.1	0,30	$3,2 \cdot 10^6$	0.40
3.3.		1.0	0,49	$2,5 \cdot 10^6$	0.45
3.4.		1.1	0,80	$3,2 \cdot 10^6$	0.40

Thermal properties, used in thermal problem solution, are follow (Table 2):

Table 2. Thermal properties of materials

	Thermal conductivity, W/(m*K)	Specific heat, J/(kg*K)
Steel	60.5	502
Rubber	0.13	2092
Paper	0.05	1400

Problems are solved in quasi-static problem statement. As for the first problems used follow boundary conditions (BC). The shaft are fixed and the paper has displacement on 450 mm during 180 s. As for the second problem again The shaft are fixed and the paper has displacement on 450 mm during 180 s and temperature distribution acted as well. Total we have 38408 elements and 75950 nodes. As elements the rectangular elements have been chosen. In contact zone the sizes of the elements are decreased.

Results and Discussion

After running BC described above we have follow. On a Fig.4 are results for temperature distribution in 180 s starts from room temperature, we can see the temperature starts from the beginning of the paper and then continue along the axis. On Fig.5 the results with temperature are presented accordingly.

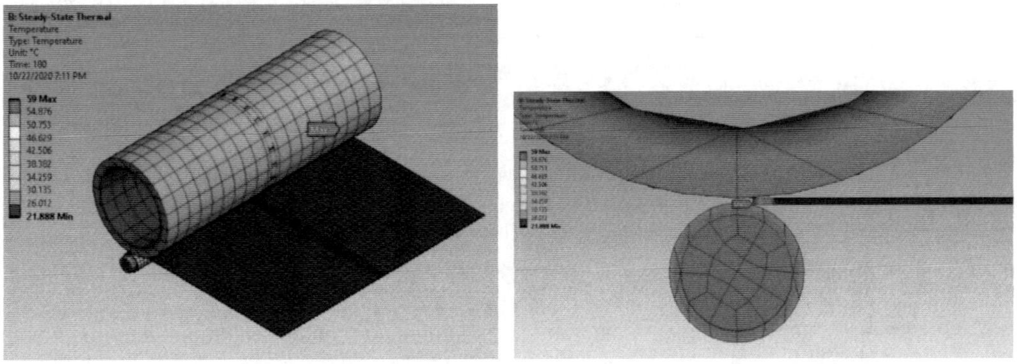

a) b)

Fig.4. Temperature distribution

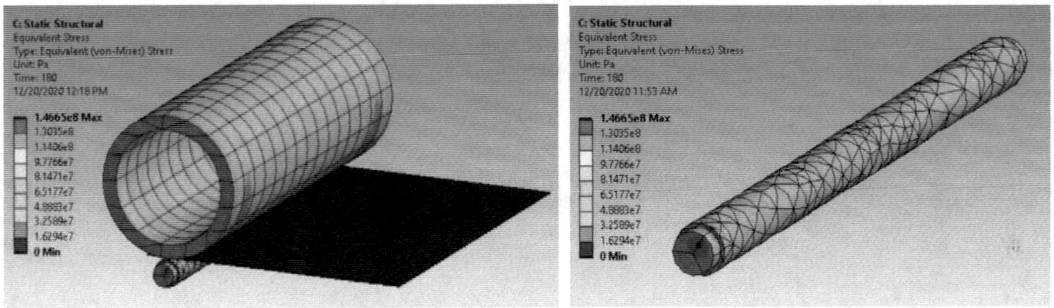

Fig.5. Stresses with temperature influence after 180 s.

Despite of the fact that paper is "blue" color, it is also under stresses, but that stresses is low. This is explained that the rubber on the bottom shaft and it is play damping role.

Table 3. Max stresses without and with temperature influence

Type of paper	Max Stress without temperature, MPa	Max Stress with temperature, MPa
1	141.24	146.65
2	142.15	147.27
3	143.83	147.95

As we can see from the table 3, the stresses are more in case with temperature effect. This effect come from friction during 180 s, so diameter of shafts a little bit expanded and influence on a pressure on the paper.

In the 180 s from temperature effect the gap between the shafts is smaller, than at the beginning of process.

Table 4.Gap between printing shaft

	Gap for paper in case without	Gap for paper in case with

	temperature, mm	temperature, mm
1	0,07	0,066
2	0,08	0,0775
3	0,15	0,14

Gap size decreases is result of the shaft expended and influence on a pressure on the paper (Table 4).

Conclusion

In this paper the temperature influence on a quality printing in a printing machine is discussed. The case of contact between printing shafts and paper is discussed, the cases, when temperature is neglected and it's effect on a quality of printing investigated as well. Beside that, three different type of papers investigated for temperature influence.

From results we can see, that after 180 s the shaft without cooling system stay more heated as 59^0 C from room temperature and hence, after thousands of cycles printing machine should be stopped, over wise the gap between two printing shafts decrease and quality of printing will be worse.

Acknowledgment

Author is thank to supervisor Prof. Levonyan H.L and Prof. Verlinski S.V. for consultancy in preparing this paper.

Reference

1. Nemanja Kašiković , Dragoljub Novaković , Gojko Vladić , Maja Klančnik. Influence Of Heat Treathment On Caracteristics Of Inkjet Prints On Textile Material. Journal of Graphic Engineering and Design, Volume 2 (1), 2011, 24-30.
2. Mao N, Russell S.J. (2007) The Thermal Insulation Properties of Spacer Fabrics with a Mechanically Integrated Wool Fiber Surface, Textile Research Journal, 77, 914-922.
3. Timo Hartus. Effect of drying temperature profile and paper on mechanical print quality in heat set offset printing. TAGA Journal, Volume 4, Issue number 4, 2008, 192-205.
4. Wells N., Hybrid heatset and coldset printing, Newspapers & Technology, 1-5, 2001.
5. A.A. Margaryan, S.V. Verlinski. Study of the stress-deformed condition of printed cylinders' contact zone by the method of finite elements. National Polytechnic University of Armenia, Bulletin Collection of scientific papers, part II, 2018, 15, V1.- 407-415 (in Armenian).
6. S. M. Mkhitaryan, L. A. Shekyan, S. V. Verlinski, D. Aidun & P. Marzocca (2012) Stationary Theory of Heat-Conductivity for an Axi-Symmetrical Piece Homogeneous Space with Circular Inclusion, Journal of Thermal Stresses, 35:5, 424-447.
7. K. Sathishkumar and N. Ugesh. Finite element analysis of a shaft subjected to a load. ARPN Journal of Engineering and Applied Sciences, VOL. 11, NO. 9, MAY 2016, pp.5996 - 6000
8. Risographs TR series, Technical guidance, version 1.0 (1), RISO Kagaku Corporation, October 1995.
9.**http://apg.in.ua/dir/offset rubber blade/contitech/ofcetnoe-polotno-conti-air-fsr.html**
10. O. C. Zienkiewicz, R. L. Yaylor, The Finite Element Method, Forth ed., McGraw-Hill (1994).
11. Segerlind, Larry J. Applied Finite Element Analysis, Applied Finite Element Analysis, 2nd Edition, October 1984, 448 Pages.
12. Bathe, K.J. Finite Element Procedures. Cambridge, MA: Klaus-Jürgen Bathe, 2006.

Thermal aspect of material selection for substrate board equipped with high heat flux IC chips

Shankar Durgam*, **Ajinkya Bhosale, Vivek Bhosale, Revati Deshpande, and Pankaj Sutar**

Department of Mechanical Engineering, College of Engineering Pune - 411005, India

E-mail: *sod.mech@coep.ac.in

Abstract. This paper investigates the material selection for substrate board equipped with nine silicon heaters that mimic IC chips. This includes computational research on laminar forced convection air cooling of silicon IC chips in electronic components mounted on substrate boards subjected to high uniform heat fluxes in a horizontal channel. Copper clad boards with single and multilayer having thermal conductivities of 8.8, 40.5, and 61.5 W/m K are the substrate materials used in this study. Three-dimensional steady-state conjugate heat transfer with laminar non-isothermal fluid flow module is used from COMSOL Multiphysics 5.4 to study fluid flow and heat transfer. Simulations performed with flowing air velocities of 1.5, 2.5, and 3.5 m/s with very high heat fluxes of 10000, 12500, and 15000 W/m^2. It showed that the temperature on subsrate board strongly depends upon placement of heaters, Reynolds number, substrate thermal conductivity. It also found that higher thermal conductivity substrate materials results in heat transfer enhancement and decrease chances of failures of electronic devices. A non-dimensional temperature based correlation is also devised in therms of non-dimensional heat flux, thermal conductivity, Nusselt number, and Reynolds number.

Keywords: Material Selection, Forced convection, Computational heat transfer, Heat sources, High heat flux, Conductive substrates, Thermal control.

1. Introduction

Due to continual miniaturization of electronic equipments proper heat dissipation has become challenging due to high heat fluxes. Forced convection is receiving increasing attention in recent times for dissipating high heat density. Along with forced convection material selection for high heat dissipation electronic devices is required. The current work is applicable for electronic devices like gaming laptops, desktop personal computers, and digital cameras which uses air cooling for guiding material selection. They uses customised PCB designs of about 3"× 3" subjected to high heat fluxes. Huang et al. [1] numerically investigated that change in incoming parabolic velocity and vortices zones significantly enhances heat transfer. Nemati [2] performed the fin heat sink's optimization using the effect of dimensionless parameters on entropy generation. Ismail et al. [3] found that 30% ethylene glycol is suitable cooling fluid for microchannel heat sink. Baris and Nagulapally [4] determined that power dissipation of PCB on trace correlates with the square of electric current. Heindel et al. [5] found that fully developed nucleate boiling is a preferred condition for electronic cooling. Choi et al. [6] studied loop heat pipe for high power electronics. Maghrabie et al. [7] numerically determined that third-row obstacles is best position for cooling of heated obstacles. Das et al. [8] studied conjugate

mixed convection from vertical non-isothermal heat sink and found that overall Nusselt number is a function of conductance, fin spacing, clearance. Anuwarullah et al. [9] experimentally investigated enhanced heat transfer from electronic equipment cooled by air jet impingement and presented the correlation of Nusselt number in terms of flow parameters and aspect ratio. Talukdar et al. [10] found 7-10 % heat transfer enhancement by gemetric modeifications in natural convection cooling of heat sources. Tsay et al. [11] numerically studied heat sources with slots for enhancement of heat transfer and found an 81% increment in Nusselt number for front blocks. Wang et al. [12] numerically conducted a parametric study for phase change material (PCM) based heat sink and found an increase in thermal performance under natural convection. Hotta et al. [13] introduced a geometric parameter for heat source configurations and found an optimal placement using an experimentally trained genetic algorithm. Bhowmik and Tou [14] experimentally studied transient forced convection from simulated electronic chips and developed correlation for Nusselt number in terms of Peclet number and Fourier number. Chen and Liu [15] carried out air cooling of nine identical heat sources with a wide range of velocities, there was a drop in the substrate board's temperature when the spacing between heat sources follows geometric ratio. Deng and Liu [16] performed liquid metal cooling for high-heater dissipation of CPUs and found energy saving compared with heat pipe and water cooling. Wang et al. [17] numerically studied the cooling of a chip with a flow-disturbing obstruction block and found enhanced heat transfer. Rosas et al. [18] performed experiments to investigate convective heat transfer in an electronic module using a curved deflector and found heat transfer enhancement. Hung and Fu [19] numerically studied the passive enhancement of electronic cooling through geometric modifications. Mathew and Hotta [20] numerically and experimentally studied forced convection cooling of IC chips in optimal configuration with different orientations of substrates and found 3 - 5 $^{\circ}C$ reduction in temperature. Durgam et al. [21] studied SVR technique for heat source temperature prediction. Moumni et al. [22] used nanofluid for enhancing heat transfer for heated blocks.
The literature shows that much of research based on laminar conjugate heat transfer under natural, forced, or mixed convection cooling is available. The studies mainly focusing on material selection of substrate board equipped with ICs is very limited in previous studies. Furthermore, forced convection air cooling of IC chips dissipating very high heat flux density is scarce. This work aims to select the material for substrate board by thermal aspect with study of nine silicon heaters subjected to very high fluxe densities. The present study is a unique work useful for a variety of heat transfer applications in the fields like military, biomedical, aircraft, space, and all kinds of consumer electronic devices subjected to high heat fluxes.

2. Numerical modeling and data reduction

The numerical model consists of a substrate board embedded with nine silicon heaters fixed on bottom walls of horizontal channel. The heat source material considered is silicon that mimics the IC chips on PCB. Substrate boards of different materials viz. single layer CCB (copper clad board), multilayer CCB (CCBML - 1), and multilayer CCB (CCBML 2) having thermal conductivities of 8.8, 40.5 and 61.5 W/m K, respectively have been considered. The different flowing air velocities of 1.5, 2.5, 3.5 m/s with very high uniform heat fluxes 10000 W/m^2, 12500 W/m^2, and 15000 W/m^2 are reviewed in the simulation study. This model, including the geometry and heat dissipation values, replicate an actual design. Thermal management techniques mainly heat spreaders and heat pipes are used for high heat fluxes, but for given heat flux values, air cooling systems are suitable. For methodology and necessary assumptions we have followed steps given by Durgam et al. [21]. Three dimensional steady state conjugate transfer with non-isothermal fluid flow module were selected from COMSOL Multiphysics 5.4 for finding numerical solution of present problem [23, 24]. The computational domain consists of a horizontal channel with heaters fixed on substrate board having specific arrangement of placement, coordinate axes, and the airflow direction have shown in Fig. 1(a).

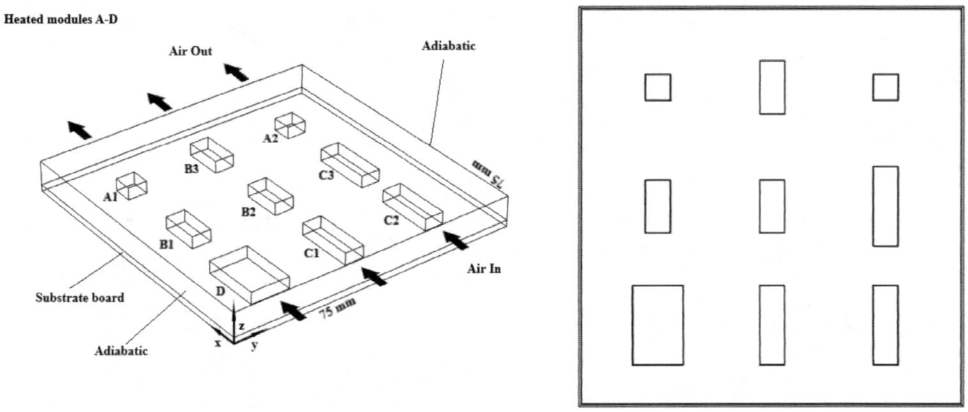

(a) Computational domain (b) Configuration

Figure 1. A computational domain and configuration.

Optimal placement of heaters used in the study for all simulations is shown in Fig. 1(b). This configuration was obtained from the simulation study of 100 random configurations for 1.5 m/s air velocity and 10000 W/m^2 heat flux. The configuration presented in Fig. 1(b) results in lowest maximum temperature of heat sources for the same input parameters; hence it is considered as optimal configuration but is limited to the 100 cases discussed in this study. This configuration is used to select material of substrate board according to thermal aspects. Length, width, and thickness of substrates board are 75×75×1.6 mm. The thickness of air above the board is assumed to be 6 mm. The heat source details and sizes are described in Table 1. The heat source sizes selected are used in practical application. The thickness of all heat sources is 3 mm.

Table 1. Heat source specifications

Heat source	Size, mm
A1, A2	5×5×3
B1, B2, B3	5×10×3
C1, C2, C3	5×15×3
D	10×15×3

2.1. Governing equations and boundary conditions

The governing equations used for finding numerical solution are Eqs. 1 - 5. Out of which Eq. 1 is continuity, Eq. 2, 3, and 4 are momentum equations used for fluids in x, y, z directions. Eq. 5 is energy equation used for solid.

Continuity equation for fluid:

$$\frac{\partial u}{\partial x} + \frac{\partial v}{\partial y} + \frac{\partial w}{\partial z} = 0 \tag{1}$$

x - momentum equation

$$u\frac{\partial u}{\partial x} + v\frac{\partial u}{\partial y} + w\frac{\partial u}{\partial z} = -\frac{1}{\rho}\frac{\partial p}{\partial x} + \frac{\mu}{\rho}\left[\frac{\partial^2 u}{\partial x^2} + \frac{\partial^2 u}{\partial y^2} + \frac{\partial^2 u}{\partial z^2}\right] \tag{2}$$

y - momentum equation

$$u\frac{\partial v}{\partial x}+v\frac{\partial v}{\partial y}+w\frac{\partial v}{\partial z}=-\frac{1}{\rho}\frac{\partial p}{\partial y}+\frac{\mu}{\rho}\left[\frac{\partial^2 v}{\partial x^2}+\frac{\partial^2 v}{\partial y^2}+\frac{\partial^2 v}{\partial z^2}\right] \tag{3}$$

z - momentum equation

$$u\frac{\partial w}{\partial x}+v\frac{\partial w}{\partial y}+w\frac{\partial w}{\partial z}=-\frac{1}{\rho}\frac{\partial p}{\partial z}+\frac{\mu}{\rho}\left[\frac{\partial^2 w}{\partial x^2}+\frac{\partial^2 w}{\partial y^2}+\frac{\partial^2 w}{\partial z^2}\right] \tag{4}$$

Solid region: Energy equation

$$u\frac{\partial T}{\partial x}+v\frac{\partial T}{\partial y}+w\frac{\partial T}{\partial z}=\alpha\left[\frac{\partial^2 T}{\partial x^2}+\frac{\partial^2 T}{\partial y^2}+\frac{\partial^2 T}{\partial z^2}\right] \tag{5}$$

The coefficient of heat transfer and flow parameters are calculated as shown in Eq. 6 - 8.

$$h=\frac{q}{T-T_\infty} \tag{6}$$

$$Nu=\frac{hL_h}{k_f} \tag{7}$$

$$Re=\frac{uL_h}{\nu} \tag{8}$$

The non-dimensional parameters viz. maximum temperature, thermal conductivity and heat flux are defined as given in Eq. 9 - 11.

$$\theta=\frac{\triangle T}{T_\infty} \tag{9}$$

$$k^*=\frac{k_b}{k_f} \tag{10}$$

$$q^*=\frac{qL_h}{T_\infty k_f} \tag{11}$$

The given cooling problem is solved for material thermophysical properties at $T_\infty = 27\ °C$ and 1 atm are shown in Table 2.

Table 2. Material thermophysical properties

Material	k W/m K	ρ kg/m^3	C_p J/kg K
CCBSL	8.8	2050	1370
CCBML1	40.5	2582	1276
CCBML2	61.5	2796	1219
Silicon	148	2329	700
Dry air	0.02675	1.165	1005

BC's used for given cooling problem are at inlet: x = 0, T = T_∞ and u = u_0 = 1.5, 2.5 and 3.5 m/s. at outlet: x = L, p = p_∞ (Atmospheric pressure) [24].

2.2. Thermal design for selecting substrate material

Printed circuit boards are the heart of electronic devices. Design of PCBs must be adequate form smooth operations, realibility, safety and life of equipments. By adequate thermal design failures due to improper heat dissipation in the electronic devices can be minimised. To achieve that material selection for substrate board is important. The present study is carried out for high thermal conductivity substrate materials equipped with high heat flux IC chips. To achieve high thermal conductivity substrate board are constructed using copper layers and with FR4 layer. As copper clad board is composite structure it has two type conductivities i.e. in plane thermal conductivity and through plane thermal conductivity. Out it these in-plane thermal conductivity is higher for copper clad boards, while it is very less along through plane direction. Hence two-dimensional heat dissipation is assumed in the present cooling problem. We have selected material namely, single layer copper clad board, multilayer copper clad board 1, multilayer copper clad board 2. Material properties are mentioned in Table 2. The material is selected are with increase in thermal conductivity value. Study is carried out for all the material to check the effect of substrate thermal conductivity for heat transfer enhancement which is required to dissipate heat properly. For calculation of effective in-plane thermal conductivity following relation were used.

$$k_{eff} = \frac{k_{FR4}t_{FR4} + k_{cu}t_{cu}}{t_{FR4} + t_{cu}} \quad (12)$$

As FR4 is homogeneous, effective thermal conductivity is substrate board thermal conductivity.

2.3. Grid independence test

The pre-processor generated free tetrahedral coarse mesh with 333579 grids has been employed for all simulations. Study mesh has 317696 number of domain elements, 32330 number of boundary elements, and 1728 number of edge elements. Study has free mesh time of 8.02 s and quality of 0.01872 for minimum elements. Comparative mesh study is carried out for grid numbers 309758, 333579, 360763, 385924, and 408750 that is from fine to extra coarse. It is noticed from the grid test that for all five grids employed with substantial difference in grid elements results in temperature $\leq 1\ ^{\circ}C$. Hence the grid with 333579 was used for all simulations, which saves the cost and time of the simulations. Table 3 shows the grid sensitivity test and the % change in temperature for different grids. A mesh used for heaters and substrate boards is shown in Fig. 2.

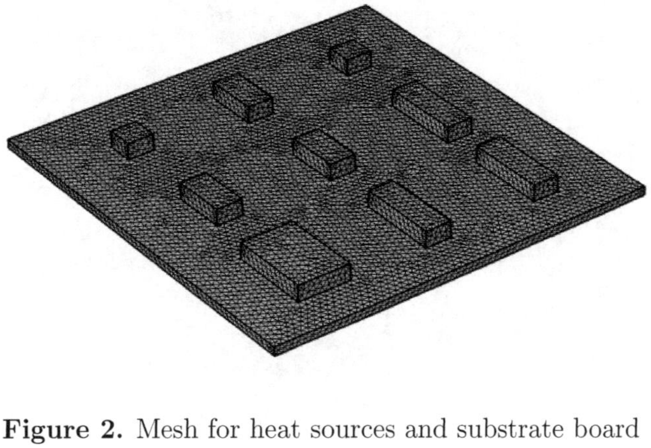

Figure 2. Mesh for heat sources and substrate board

Table 3. Maximum temperatures obtained for various grids.

Element growth	Cells	T °C	% change in T
1.4	30978	79.75	–
1.35	333579	80.20	0.56
1.3	360763	80.94	0.92
1.25	385924	82.10	1.43
1.2	408750	82.80	0.85

3. Results and discussion

The simulation results of optimal configuration found from 100 configurations using different Reynolds numbers, heat flux values, and substrate board materials are presented. A total of 27 number of cases (1 optimal configuration × 3 thermal conductivities × 3 heat fluxes × 3 velocities) were found. The flowing air velocities of 1.5, 2.5, 3.5 m/s corresponding to Reynolds numbers 938, 1563, and 2188, respectively. The different substrate board materials used are CCBSL, CCBML-1, CCBML-2, with thermal conductivity values of 8.8, 40.5, 61.5 W/m K, respectively subjected to very high uniform heat flux values of 10000, 12500, 15000 W/m^2 are used for the study. The laminar fluid flow forced convection numerical results of nine heat sources fixed on different substrates arranged in optimal placement (as shown in Fig. 1(b)). They fixed to the bottom walls of the horizontal channel are presented. Computer software generated temperature, and velocity contour plots are available after convergence of each simulation study. Few such representative plots to visualize clear insight into the physical phenomenon are given in these sections. Fig. 3 show the temperature, temperature contour, velocity and conductive heat flux plots for CCBSL at Re = 938, 1563, and 2188 using q = 10000, 12500, and 15000 W/m^2 respectively. CCBSL gives maximum temperature 86 °C at Re = 938 using heat flux of 10000 W/m^2. The maximum temperature is in a suitable limit hence safe. The heat flux value 12500 W/m^2 gives maximum temperature of 83.8 °C at Re = 1563 for 15000 W/m^2 heat flux and Re = 2188 required to keep the maximum temperature 85.1 °C. It indicates that for higher heat flux, the requirements of Reynolds number should be higher to maintain the temperature within a safe operating limit. In velocity plot the black arrowheads show the air flow direction and the red lines indicate the streamlines [Fig. 3(d)]. Conductive heat flux plot show the heat flux available on the IC chips and substrate [Fig. 3(e)]. The temperature trend of all heat sources in first, second, and third rows is similar to that in Ref. [15, 31]. The maximum temperature is under predicted in the computation. The reason for the deviation in the computational results from experimental results is the different Reynolds numbers and heat inputs. Fig. 4 the temperature, temperature contour, velocity and conductive heat flux plots for CCBML-1 at Re = 935 and 1563 for q = 10000, 12500, and 15000 W/m^2 respectively. CCBML-1 gives maximum temperature 75.8 °C for Re = 938 with heat flux of 10000 W/m^2. For heat flux of 12500 W/m^2, the maximum temperature of 87.6 °C at Re = 938 is observed. For 15000 W/m^2 heat flux and Re = 1563 is required to maintain the maximum temperature 82.2 °C. It indicates that for higher heat flux, the requirements of Reynolds number was increased. All the temperature for particular Reynolds number is suitable operating limits for all kinds of electronic equipment. The results for q = 12500, and 15000 W/m^2 for CCBSL, CCBML1 indicates that the requirement of Reynolds number is decreased for high thermal conductivity substrate board materials.

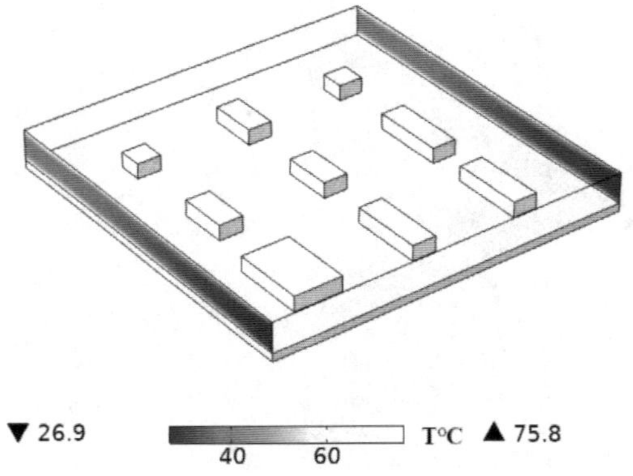

(a) Temperature plot, 10000 W/m^2, Re = 938

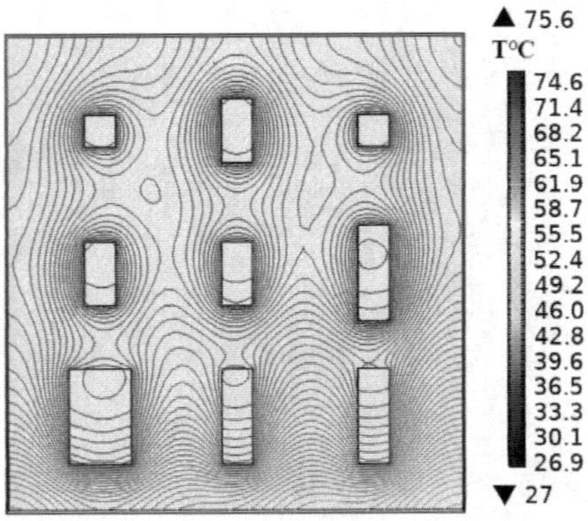

(b) Temperature contour, 10000 W/m^2, Re = 938

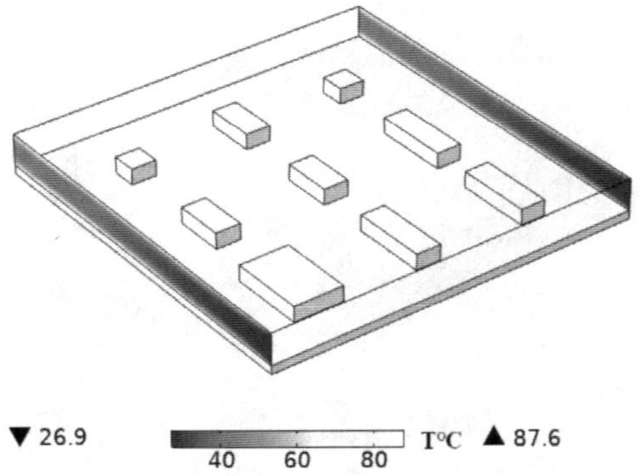

(c) Temperature plot, 12500 W/m^2, Re = 938

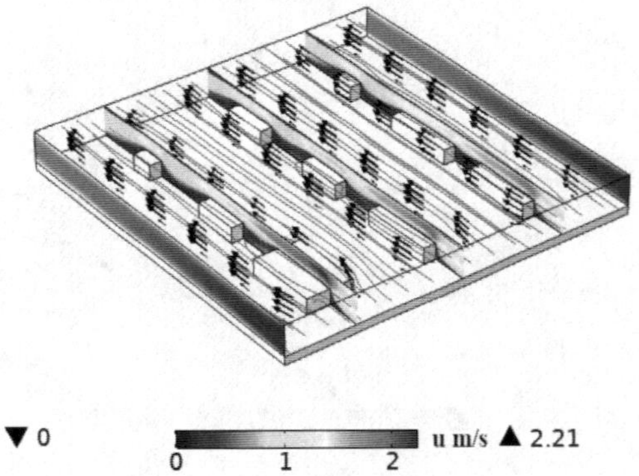

(d) Velocity plot, 12500 W/m^2, Re = 938

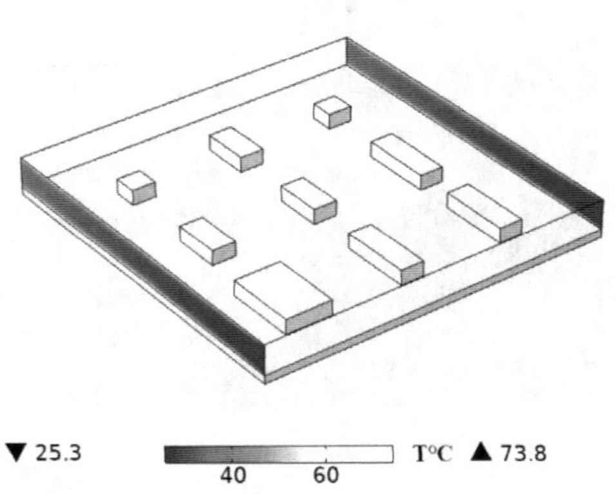

(e) Temperature plot, 15000 W/m^2, Re = 1563

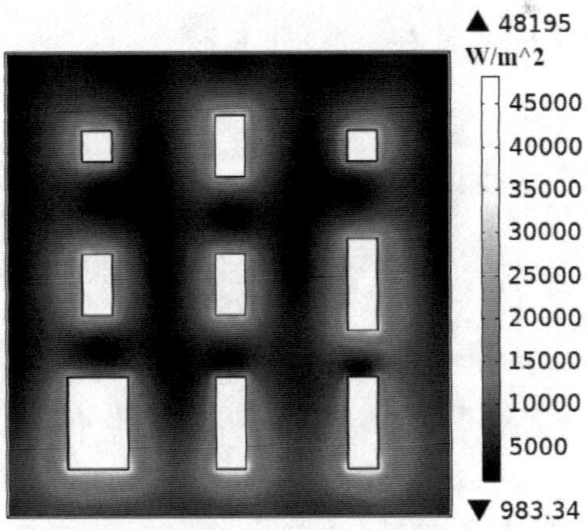

(f) Conductive heat flux, 15000 W/m^2, Re = 1563

Figure 3. Temperature, temperature contour, velocity and conductive heat flux plots for CCBML for Re = 938, 938, 1563, and q = 10000, 12500, 15000 W/m^2.

3.1. Comparative study of CCBML1, CCBML2

Thermal performance of higher thermal conductivity materials viz. CCBML-1 and CCBML-2 compared at Re = 938, 1563 and 2188 for q = 15000 W/m^2 has been shown in Fig. 5. It is seen that CCBML-1 gives maximum temperature 99.4 $°C$ while CCBML-2 gives 97.5 $°C$ for Re = 938. Whereas at Re = 1563 the maximum temperature obtained are 82.2 $°C$ and 80.6 $°C$ for

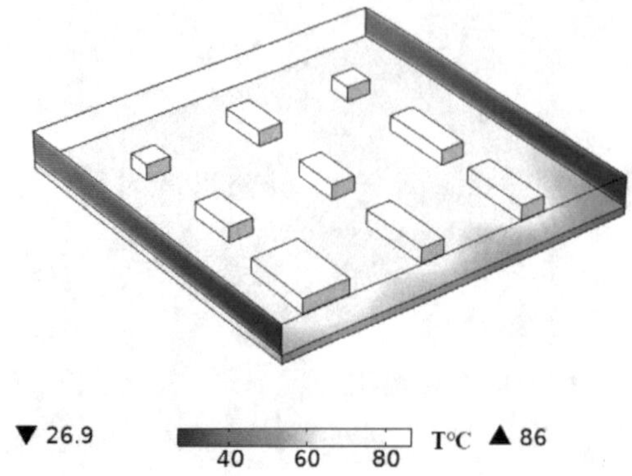

(a) Temperature plot, 10000 W/m^2, Re = 938

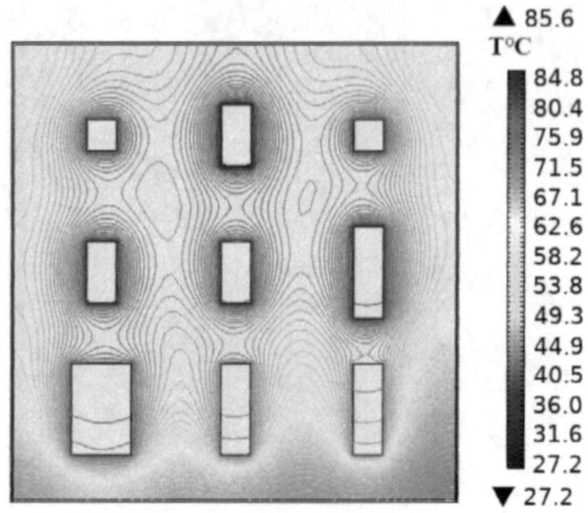

(b) Temperature contour, 10000 W/m^2, Re = 938

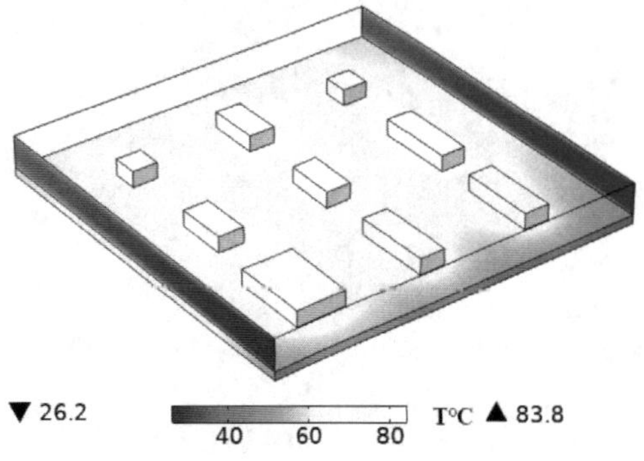

(c) Temperature plot, 12500 W/m^2, Re = 1563

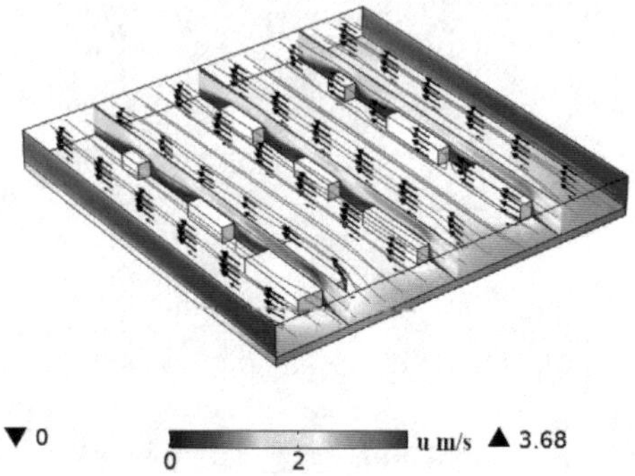

(d) Velocity plot, 12500 W/m^2, Re = 1563

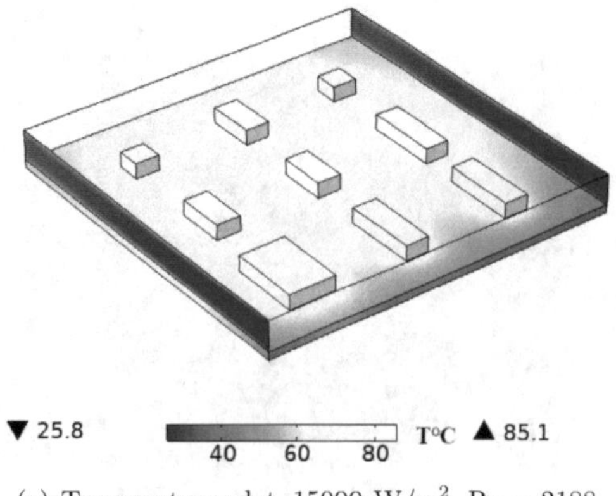

(e) Temperature plot, 15000 W/m^2, Re = 2188

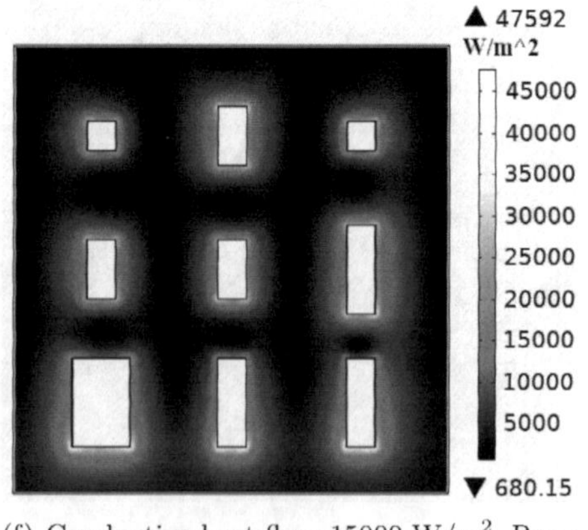

(f) Conductive heat flux, 15000 W/m^2, Re = 2188

Figure 4. Temperature, temperature contour, velocity and conductive heat flux plots for CCBSL for Re = 938, 1563, 2188 and q = 10000, 12500, 15000 W/m^2.

CCBML-1 and CCBML-2. At the maximum value of Re = 2188, CCBML1, and CCBML-2 results in 73.8 and 72.4 $^{\circ}C$. It indicates that temperature drop in CCBML-2 compared with CCBML-1 is 1.83,1.63, 3.47 $^{\circ}C$, respectively, for the given heat flux of 15000 W/m^2 and Reynolds numbers 938, 1563 and 2188, respectively. It shows that using higher thermal conductivity material, the decrease in maximum temperature is not very significant. Therefore CCBML-1 is ideally preferred over CCBML-2 based on cost, reliability, and performance for very high heat fluxes, purely based on heat transfer point of view. In general, from this study, it is clearly revealed that the minimum temperature occurs for heaters placed in the initial row.

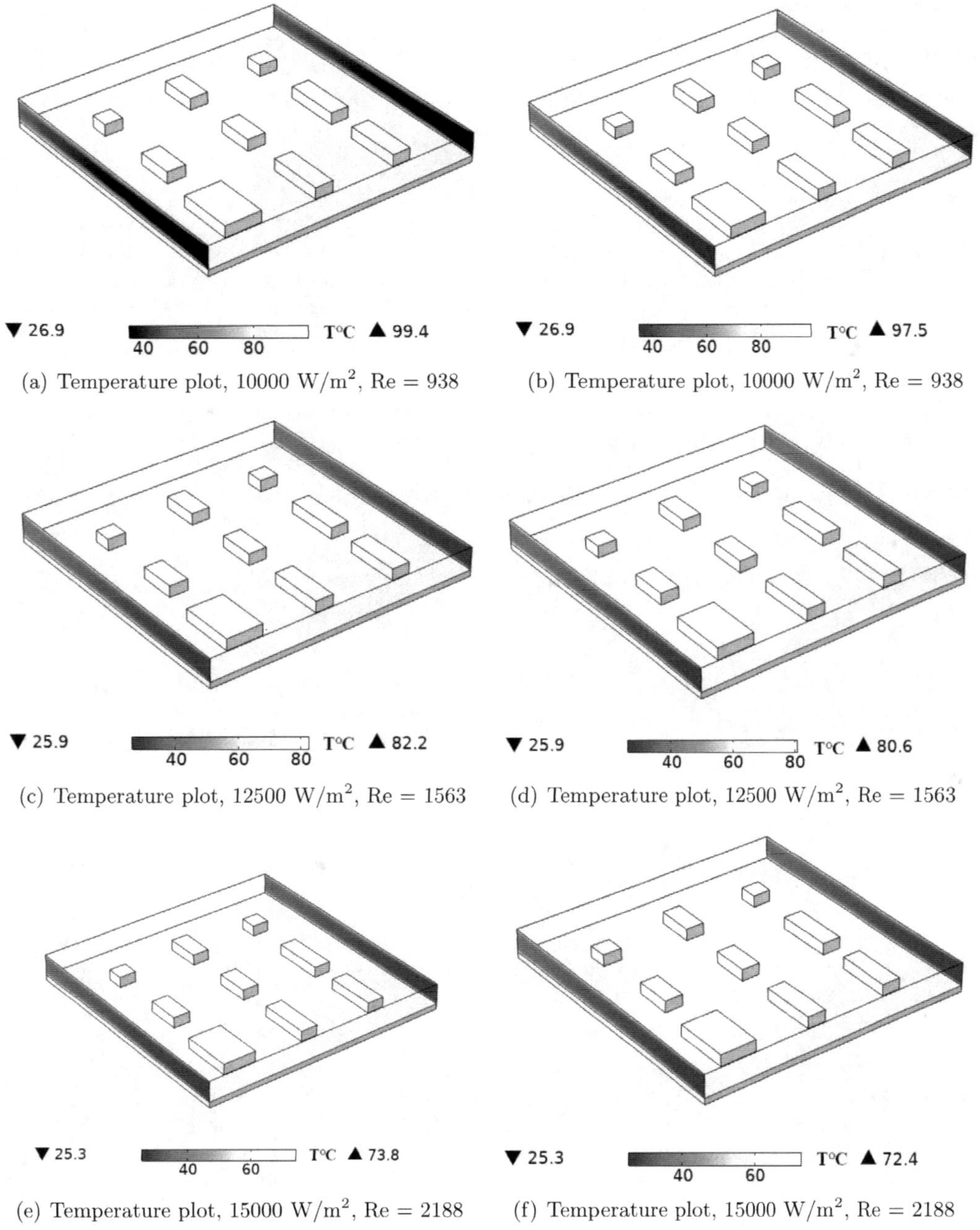

(a) Temperature plot, 10000 W/m^2, Re = 938

(b) Temperature plot, 10000 W/m^2, Re = 938

(c) Temperature plot, 12500 W/m^2, Re = 1563

(d) Temperature plot, 12500 W/m^2, Re = 1563

(e) Temperature plot, 15000 W/m^2, Re = 2188

(f) Temperature plot, 15000 W/m^2, Re = 2188

Figure 5. Temperature plots for CCBML-1, CCBML-2 at Re = 938, 1563, 2188 and q = 15000 W/m^2.

The temperature for heat source in the secodn row is greater than the initial row. The temperature in the third row heat sources was maximum for all Reynolds numbers and heat sources. It showed that temperatures obtained in the case of CCBSL are higher compared with CCBML-1 and CCBML-2. The temperature distribution on the substrate board is uniform for higher thermal conductivity substrates. Furthermore, it is common to observe that for all cases using all Reynolds numbers, heat flux values, and substrate board materials, the maximum temperature occurs in the middle of the top row i.e., heat source B3. The common hotspot on the substrate board in the vicinity of B3. From Fig. 3 - 5 it is observed that higher Reynolds numbers and higher substrate thermal conductivities result in lower temperatures. Whereas higher heat flux values exhibit higher temperatures for the same Reynolds number and substrate thermal conductivity. The results obtained are significant in practical applications.

3.2. Variation of excess temperature with substrate material

Forced convection simulation results of nine heat sources mounted on CCBSL, CCBML-1, CCBML-2 arranged in the optimal configuration is studied. The variation of maximum excess temperature for all substrate materials, 10000 W/m^2 heat flux and Reynolds number have been presented. Fig. 6 shows the variation of excess temperature for all the material for Reynolds numbers 938, 1563, and 2188, respectively, and 10000 W/m^2 heat flux. At Re = 938 the maximum excess temperatures are 59, 48.7, and 47.5 °C CCBSL, CCBML-1, CCBML-2, respectively. A similar trend shown for Re = 1563; the temperatures are 45.5, 37, and 36 °C respectively, and that for Re = 2188 are 38.9, 31.5, 30 °C respectively. It is an indication that the drop in the maximum temperature is significant when CCBML-1 used instead of CCBSL, as compared with the drop to CCBML-2 from CCBML-1. Therefore, it is found that the use of CCBML-1 is ideally preferred based on thermal performance and cost competency.

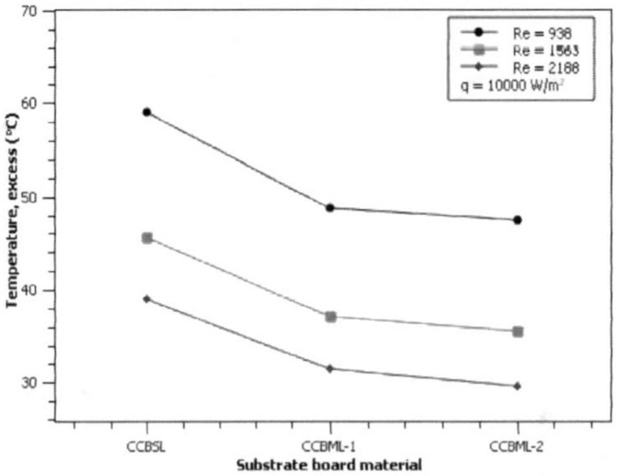

Figure 6. Excess temperature variation with substrate materials for different Reynolds number at q = 10000 W/m^2

3.3. Variation of heat transfer coefficient with substrate materials

Variation of heat transfer is studied with the substrate board materials at various uniform heat flux values with different Reynolds numbers of 938, 1563, and 2188, respectively. The heat transfer coefficient variation with substrate materials is as shown in Fig. 7 at Re = 938, 1563, 2188 and q = 15000 W/m^2. At Re = 938 heat transfer coefficient is 191, 217, 220 for CCBSL, CCBML-1 and CCBML-2 respectively. Similarly, at Re = 1563 and 2188 heat transfer coefficient are 244, 284, 289, and 287, 336, 343, respectively. There is increament in heat transfer coefficient by using higher thermal conductivity substrate and this increase is steep from CCBSL to CCBML-1, whereas from CCBML-1 to CCBML-2, it is flat. The heat transfer coefficient curves are asymptotic to each other for different Reynolds numbers.

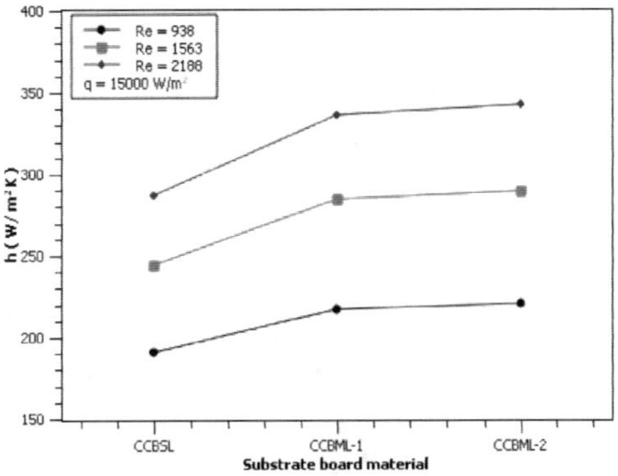

Figure 7. Heat transfer coefficient variation with substrate materials for different Reynolds number at q = 15000 W/m^2

3.4. Nusselt number variation with Reynolds number
The study is carried out for Nusselt number and Reynolds number associated with different substrate board materials viz. CCBSL, CCBML-1, CCBML-2. Fig. 8 shows variation of Nu with Re for different substrate materials. Nusselt number is increased with increasing Reynolds number as well as increase of substrate board thermal conductivity. The variation obtained is typical for forced convection cooling problem. CCBSL gives Nusselt number values of 80,103,120 while CCBML-1 gives 91, 118,140 and CCBML-2 gives 92, 120, 143 for Re = 938, 1563, 2188 respectively. The increment in Nusselt number is very high from CCBSL to CCBML-1, while it is less from CCBML-1 to CCBML-2. It is clearly visible that increment in heat transfer by use of higher thermal conductivity substrate.

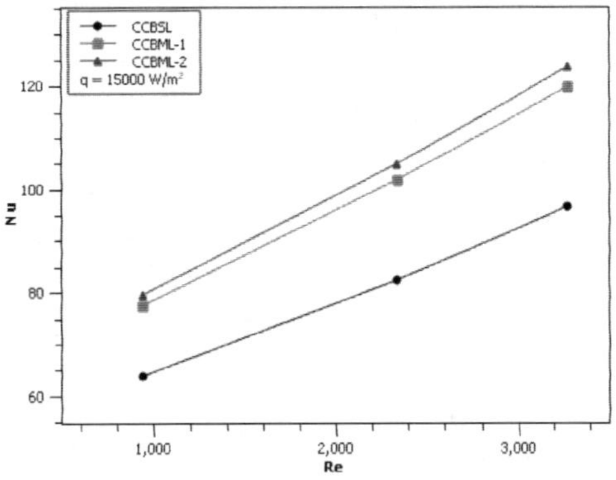

Figure 8. Substrate materials and Reynolds numbers effect on Nusselt number at q = 15000 W/m^2

3.5. Substrate thermal conductivity effect on temperature of heat sources
The temperature distribution of heat source is studied for various substrate board material, heat fluxes, and Reynolds numbers. Fig. 9 shows variation of heat source temperatures of optimal configuration for q = 12500 W/m^2 and Re = 938. It gives an understanding that temperatures of higher thermal conductivity substrate are less compared with lower thermal conductivity substrate. Furthermore, higher thermal conductivity substrate maintains uniform temperature distribution over the substrate board. The uniform temperature distribution is almost similar in the case of CCBML-1 and CCBML-2. Due to the uniform temperature distribution amount of thermal stresses are decreased, as a result, performance and life of PCB can be increased.

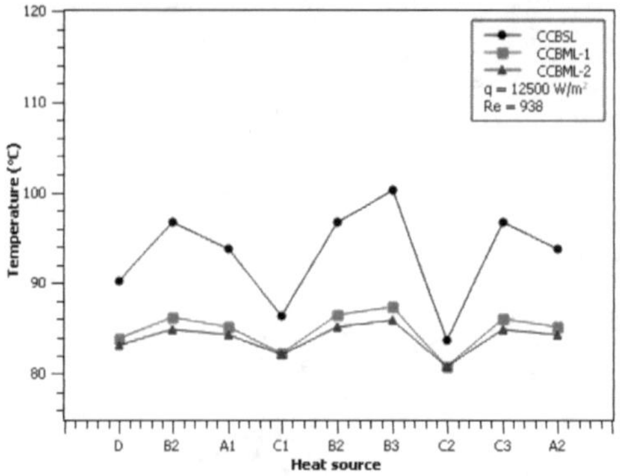

Figure 9. Temperatures of all heat sources for Re = 938 at q = 12500 W/m^2 using CCBSL and CCBML-1, CCBML-2

3.6. Validation of numerical results with experimental study

Fig. 10 show simulation results of temperature plots using bakelite, CCBSL, CCBML-1 for Re = 625 and q = 1500 W/m^2. The maximum temperatures are 53.1, 38.1, and 36.4 $°C$ in the case of bakelite, CCBSL, CCBML-1, respectively. The temperatures are reducing with use of higher thermal conductivity substrate material. However, the temperatures in the case of CCBSL and CCBML-1 are uniform compared with bakelite. The highest temperature is found for a heat source placed in the middle of the third row.

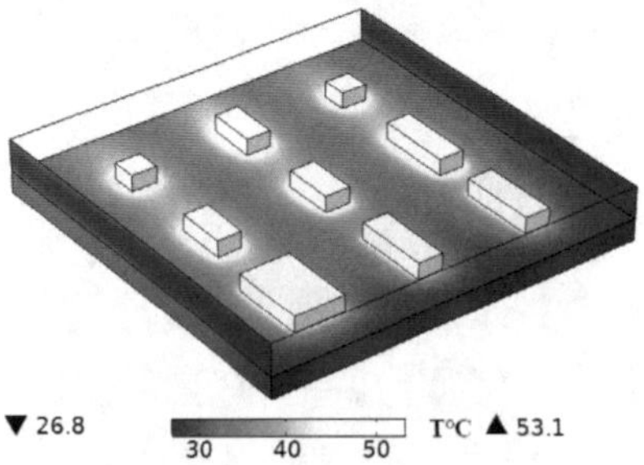

(a) Temperature, bakelite, 1500 W/m^2, Re = 625

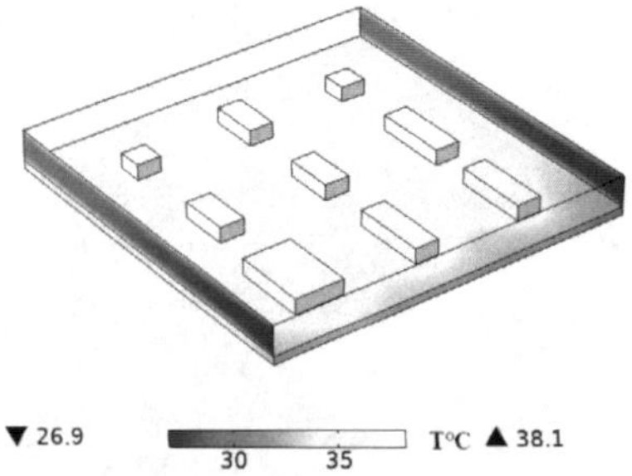

(b) Temperature oC, CCBSL, 1500 W/m^2, Re = 625

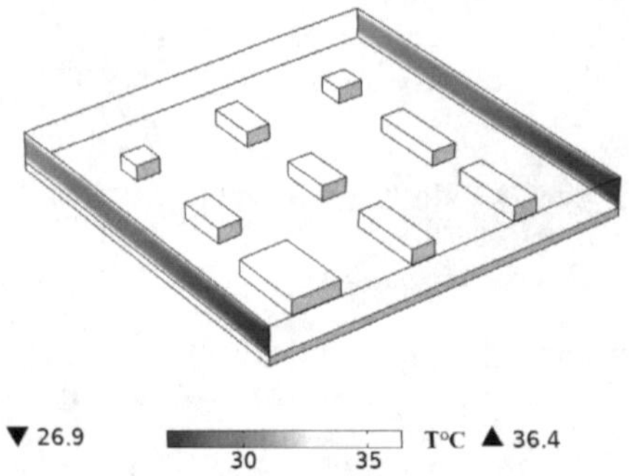

(c) Temperature plot, 12500 W/m^2, Re = 1563

Figure 10. Temperature $°C$, plots for bakelite, CCBSL and CCBML-1 for Re = 625 and q = 15000 W/m^2.

Chen and Liu [15] experimentally studied the configuration of nine identical heat sources with a 1500 W/m^2 heat flux. The study covered wide range of Reynolds numbers 465 - 1560, in the laminar regime. The experimental data of [Ref. [15]] are validated with the numerical results for nine heaters fixed on substrate board. For comparison with their results, 1500 W/m^2 heat flux and Re = 625 with bakelite substrate board of thickness 4 mm and CCBSL, CCBML-1 with a thickness of 1.6 mm are considered. The configuration is also analyzed for different substrate boards to check thermal conductivity effect on temperature distribution and increament in heat transfer. The current study results are compared with results obtained by [Ref. [15]] for the case S2/S1 = 1 (TYPE 1). It showed that the trend is exactly followed, and the excess temperature and heat transfer coefficient values are encouraging in the current study for the same heat flux, Reynolds number, and substrate board material. Fig. 11 show the results plotted for higher thermal conductivity materials CCBSL, and CCBML1. It indicates that for higher thermal conductivity substrate materials, the temperatures are significantly low. Substrate board size in the previous study [Ref. [15]] is about four times the size used in the present study. It shows that better results have been obtained with the consumption of less space. Miniaturization obtained is substantial. To remove same amount of heat as in [Ref. [15]] for maximizing heat transfer, a combination of figure four substrate should be placed one over the other. Due to this maximum temperature reduces and by which performance of electronic gadgets can be increased.

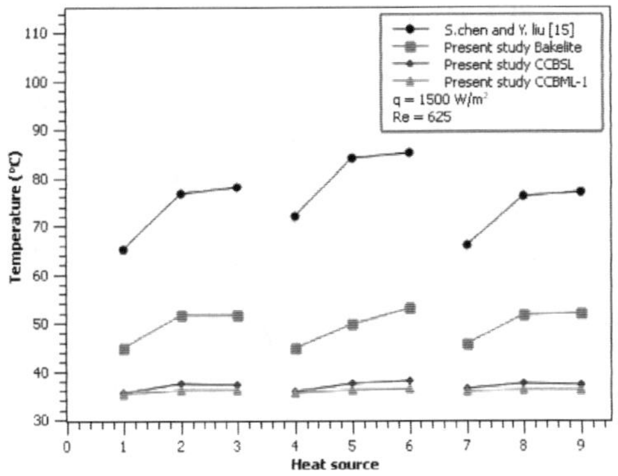

Figure 11. Temperatures vs heat sources for Re = 625 at q = 1500 W/m^2 using CCBSL and CCBML-1

3.7. Material selection for substrate board - Thermal aspect
Table 4 shows the selected substrate materials for different Reynolds numbers and heat flux densities for reliability, performance, and cost requirements of electronic devices. It is found from results that CCBSL is suitable for lower values of heat flux and Reynolds number. For higher values of heat flux requirement, high values of Re are required for low thermal conductivity substrate. For high heat flux values, CCBSL is suitable for higher Reynolds number, CCBML-1 and CCBML-2 are preferable for lower Reynolds number.

Table 4. Substrate materials selection for different heat flux and Re

Heat Flux W/m^2	Re	Selected material
10000	938	CCBSL
12500	938	CCBML-1
15000	1563	CCBML-1

3.8. Proposed correlation
A non-dimensional maximum temperature based correlation, in terms of non-dimensional heat flux, thermal conductivity, Nusselt number, and Reynolds number, has been developed as an

empirical relationship and is given in Eq. 13.

$$\theta_{max} = \exp[0.06q^* - 0.01Nu - 8.4 \times 10^{-6}k^* - 1.2 \times 10^{-5}Re] \quad (13)$$

Eq. 13 is based on 27 data points obtained from simulation results and is valid for variables in following range:

$12.5 \leq q^* \leq 19$ $29 \leq k^* \leq 2300$

$63 \leq Nu \leq 124$ $940 \leq Re \leq 2188$

Eq. 13 has coefficient of correlation 0.99 and a standard deviation 0.47. Fig. 12 shows a parity plot for non-dimensional temperature θ drawn between correlated and simulated values. It is found that the deviation in the data is inside $\pm 10\%$. The given correlation is useful for obtaining temperatures in electronic equipment having configurations similar to current study.

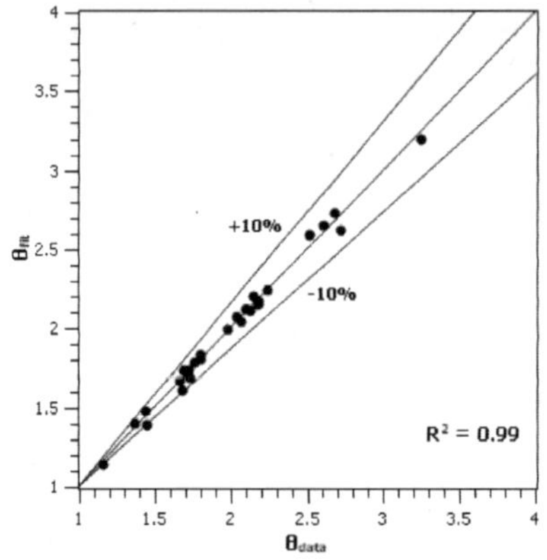

Figure 12. A parity plot

4. Conclusions

High thermal conductivity substrates mounted with nine heaters subjected to very high uniform heat fluxes and given Reynolds number in a horizontal channel has been carried out numerically. The conclusions are as follows: The highest temperature on the board is lowest in the case of CCBML-2, and it is increasing in order of CCBML-1, CCBSL for a given set of uniform heat flux, and Reynolds numbers. The temperature difference is very less in CCBML-2 and CCBML-1. Hence it is concluded that CCBML-1 is best suited material as substrate board for higher heat fluxes according to thermal considerations. The maximum temperature on the substrate board is determined by placement of heat sources, heat flux, Reynolds number, and substrate thermal conductivity. The temperature obtained in configuration is minimum when more substantial sized heat sources are at initial rows and sequentially decreasing sizes of heat sources. It is proposed that CCBML1 is optimum material as substrate board material for high heat fluxes as its performance is excellent compared with its cost i.e., CCBML-1 is more cost-effective compared with CCBML-2. The present work is beneficial for the electronic industry and gives useful guidelines for improving the performance and life of electronic gadgets subjected to very high heat fluxes. This study is suitable for electronic gadgets like laptops, digital cameras, and personal computers. With the use of higher Reynolds number, this material can also be used for higher heat fluxes. This configuration is suitable for electronic gadgets having heat dissipation in the range of 5 - 25 W. For high rates of heat dissipation; multiple substrates may be employed. Due to multiple substrate board, highest temperature on the board can be minimized. A non-dimensional maximum temperature correlation for thermal conductivity, non-dimensional heat flux, Nusselt number and Reynolds number shows a statistical goodness of fit.

References

[1] P. Huang, C. Yang, J. Hwang, M. Chiu, Enhancement of forced-convection cooling of multiple heated blocks in a channel using porous covers, *International journal of heat and mass transfer* 48 (3-4) (2005) 647–664.

[2] H. Nemati, A general equation based on entropy generation minimization to optimize plate fin heat sink, *Engineering Journal* 22 (1) (2018) 159–174.

[3] F. Ismail, A. Rashid, M. Mahbub, Cfd analysis for optimum thermal design of carbon nanotube based micro-channel heatsink, *Engineering Journal* 15 (4) (2011) 11–22.

[4] M. B. Dogruoz, M. K. Nagulapally, Effects of trace layers and joule heating on the temperature distribution of printed circuit boards: a computational study, *Journal of Thermal Science and Engineering Applications* 1 (2) (2009).

[5] T. J. Heindel, F. P. Incropera, S. Ramadhyani, Laminar Natural Convection in a Discretely Heated Cavity: II—Comparisons of Experimental and Theoretical Results, *Journal of Heat Transfer* 117 (4) (1995) 910–917.

[6] J. Choi, B. Sung, J. Yoo, C. Kim, D.-A. Borca-Tasciuc, Enhanced miniature loop heat pipe cooling system for high power density electronics, *Journal of Thermal Science and Engineering Applications* 4 (2) (2012).

[7] H. M. Maghrabie, M. Attalla, H. E. Fawaz, M. Khalil, Effect of Jet Position on Cooling an Array of Heated Obstacles, *Journal of Thermal Science and Engineering Applications* 10 (1) (07 2017). doi:10.1115/1.4036788.

[8] B. Das, A. Giri, S. Debnath, Conjugate mixed convection heat transfer from a shrouded vertical nonisothermal heat sink, *Journal of Thermal Science and Engineering Applications* 9 (4) (2017).

[9] M. Anwarullah, V. Vasudeva Rao, K. V. Sharma, Experimental investigation for enhancement of heat transfer from cooling of electronic components by circular air jet impingement, *Heat and Mass Transfer* 48 (9) (2012) 1627–1635. doi:10.1007/s00231-012-1005-y.

[10] D. Talukdar, C.-G. Li, M. Tsubokura, Investigation of compressible laminar natural-convection for a staggered and symmetric arrangement of discrete heat sources in an open-ended vertical channel, *Numerical Heat Transfer, Part A: Applications* 76 (3) (2019) 115–138.

[11] Y. Tsay, J. Cheng, Z. Chiu, Characteristics and enhancement of heat transportation from a block heat source module in three-dimensional cabinets to an ambient natural convective air stream, *Numerical Heat Transfer, Part A: Applications* 61 (1) (2012) 18–37.

[12] X.-Q. Wang, C. Yap, A. S. Mujumdar, A parametric study of phase change material (pcm)-based heat sinks, *International Journal of Thermal Sciences* 47 (8) (2008) 1055 – 1068.

[13] T. Kumar Hotta, C. Balaji, S. P. Venkateshan, Optimal Distribution of Discrete Heat Sources Under Mixed Convection—A Heuristic Approach, *Journal of Heat Transfer* 136 (10) (07 2014). doi:10.1115/1.4027350.

[14] H. Bhowmik, K. Tou, Experimental study of transient forced convection heat transfer from simulated electronic chips, *Heat and mass transfer* 41 (7) (2005) 599–605.

[15] S. Chen, Y. Liu, An optimum spacing problem for three-by-three heated elements mounted on a substrate, *Heat and Mass Transfer* 39 (1) (2002) 3–9. doi:10.1007/s00231-001-0286-3.

[16] Y. Deng, J. Liu, Design of Practical Liquid Metal Cooling Device for Heat Dissipation of High Performance CPUs, *Journal of Electronic Packaging* 132 (3) (09 2010). doi:10.1115/1.4002012.

[17] S. K. Wang, T.-C. Hung, B.-S. Pei, A.-F. Chen, J.-L. Du, A Numerical Study of the Enhancement of Chip Cooling via a Flow-Disturbing Obstruction Block, *Journal of Electronic Packaging* 127 (4) (2005) 523–529.

[18] A. Rosas, R. Ali, A. Abdel-Aziz, K. Elshazly, An experimental investigation of convective heat transfer enhancement in electronic module using curved deflector, *Heat and Mass Transfer* 53 (3) (2017) 985–994.

[19] T. Hung, C. Fu, Conjugate heat transfer analysis for the passive enhancement of electronic cooling through geometric modification in a mixed convection domain, *Numerical Heat Transfer: Part A: Applications* 35 (5) (1999) 519–535.

[20] V. Mathew, T. K. Hotta, Experiment and numerical investigation on optimal distribution of discrete ics for different orientation of substrate board, *International Journal of Ambient Energy* (2020) 1–8.

[21] S. Durgam, A. Bhosale, V. Bhosale, R. Deshpande, P. Sutar, Support vector regression method for predicting temperatures of heat sources cooled by forced convection in a horizontal channel, *Thermal Science and Engineering Progress* 20 (2020) 100725.

[22] H. Moumni, H. Welhezi, E. Sediki, Numerical investigation of heat transfer enhancement in a square ventilated cavity with discrete heat sources using nanofluid, *High Temperature* 55 (3) (2017) 426–433.

[23] S. Patankar, Numerical heat transfer and fluid flow, CRC press, 1980.

[24] Comsol, COMSOL Multiphysics: Version 5.4, Comsol, 2018.

Experimental investigation on multiferroic properties of Ti-doped $BiFeO_3$ bulk and nanoparticles

M Y Arafat[1,*], N Anjum[2,], S N E Lamia[3]**
[1]Department of Mechanical Engineering, Bangladesh University of Engineering and Technology, Dhaka 1000, Bangladesh
[2]Department of Mechanical Engineering, Shahjalal University of Science and Technology, Sylhet 3114, Bangladesh
[3]Department of Mechanical Engineering, University of Delaware, DE 19716, USA

[*] E-mail: yeashir.arafat45@gmail.com
[**] E-mail: nafizapro-mee@sust.edu

Abstract. Bismuth ferrite ($BiFeO_3$), an inorganic chemical compound, has added a new dimension to the multiferroic materials research. We previously reported the synthesis routes and improved photocatalytic properties of 10% Ti-doped $BiFeO_3$ ($BiFe_{0.9}Ti_{0.1}O_3$). In this study, we investigated the ferroelectric and ferromagnetism properties of $BiFe_{0.9}Ti_{0.1}O_3$ bulk and nanoparticles and compared them with those of undoped $BiFeO_3$ bulk ceramics. Elemental analysis was conducted to explore the chemical composition in the doped samples. The magnetic characterization and electrical measurements were performed in Vibrating Sample Magnetometer and Ferroelectric Loop Tracer respectively. Magnetization versus applied magnetic field, leakage current density, and polarization versus electric field measurements reveal improved multiferroic properties in Ti-doped bulk and nano samples compared to undoped $BiFeO_3$ bulk.

1. Introduction

The co-existence of (anti)ferroelectricity and (anti)ferromagnetism in the same phase of multiferroic materials has been of significant interest to researchers in the past few years [1,2]. 'Ferro'-orders in multiferroics allow them to be electrically polarized by applying an external magnetic field and vice versa [3]. Thus, multiferroic materials may provide a wide range of opportunities for their potential applications in data storage, spintronics, and sensors [4]. Bismuth ferrite having the rhombohedrally distorted perovskite ABO_3 (A = Bi, B = Fe) structure is one of the most investigated and promising multiferroic materials. It possesses a ferroelectric Curie point (T_C) of 1103K and Néel temperature (T_N) of 653K approximately and thus exhibits multifunctional behavior at room temperature (RT) [5]. However, the application of $BiFeO_3$ (BFO) in the multifunctional context is being limited ascribed to the presence of impurity phases, free moving charges, and oxygen vacancy defects in the bulk ceramics [6,7].

Remarkable attempts, including synthesis method modification, foreign element doping, and solid solution formation with other perovskites have been undertaken to date aimed at eliminating the existing defects in BFO, thereby stabilizing the perovskite structure [8-10]. The selection of Ti as a dopant to enhance the multiferroic features of the BFO ceramics has received special consideration owing to a closely similar ionic radius as that of Fe ion, partially filled 3d orbitals, and some other intriguing properties [11]. The behavior of Ti-doped $BiFeO_3$ (BFTO) materials largely depends on the synthesis parameters, which may lead to inconsistent outcomes among different researches conducted at variable process conditions [12]. In general, Ti doping results in improved magnetization along with reduced leakage current in BFO while literature exists revealing the opposite phenomena attributed to the structural defects originating from Ti^{4+} substitution, thereby leaves a scope to contribute further in this arena [13,14]. Though BFO microstructures and thin films have received extensive attention, very few studies have been conducted to investigate the multiferroic properties of BFTO nanoparticles.

The synthesis process of Ti-doped BFO bulk and nano counterparts involving solid-state reaction technique and sol-gel method, respectively, were reported in our previous investigations [15,16]. Ti exhibits the characteristics of a potential dopant by reducing impurity phases, oxygen vacancies, and particle size and enhancing photocatalytic properties of BFO [15,16]. In this paper, efforts are aimed to investigate the magnetic and electrical behavior of BFO bulk, BFTO bulk, and BFTO nanoparticles at RT. Elemental analysis was performed to confirm Ti doping in BFTO bulk and nano counterparts. Magnetization versus applied magnetic field measurements were conducted to explore the magnetic properties while leakage current density and polarization versus electric field measurements were carried out for electrical characterization of the synthesized materials.

2. Experimental Details

The synthesis of $BiFeO_3$ (BFO) as well as Ti-doped $BiFe_{0.9}Ti_{0.1}O_3$ (BFTO) bulk polycrystalline powder was accomplished by conventional solid-state reaction technique [17]. The bottom-up synthesis process involved calcination of compact oxide mixture at 750°C succeeded by pressing the calcined powders into pellets and sintering at 850°C for five hours [15]. To prepare the BFTO nanoparticles, sol-gel method, a wet chemical technique was followed [18]. The individual solutions of $Bi(NO_3)_3.5H2O$, $Fe(NO_3)_3.9H_2O$, and Ti Butoxide dissolved in 2-methoxy ethanol were mixed and stirred at 100-110°C for eight hours accompanied by drying and calcination at 650°C for one hour to get the desired nanoparticles [16]. The chemical compositions of BFTO bulk and nanoparticles were investigated in Energy Dispersive X-Ray Spectroscopy (EDS). The magnetization versus applied magnetic field (*M-H*) hysteresis loops of the synthesized materials were measured in Vibrating Sample Magnetometer (VSM). The ferroelectric measurements of the considered samples were conducted in Ferroelectric Loop Tracer (PE Loop Tracer).

3. Results and Discussion

3.1. Elemental analysis

EDS (Energy-dispersive X-ray spectroscopy) analysis was performed at three different positions of each BFTO bulk and nano samples to confirm successful Ti doping. The EDS spectra reveal that O, Ti, Fe, and Bi elements are homogeneously present throughout the prepared BFTO bulk and nano-materials, as shown in Figure 1. The absence of any foreign elements is also evident from total mass% (100%), as presented in Table 1. Calcination in the air at high temperature removed almost all other foreign components and eventually resulted in pure BFTO samples.

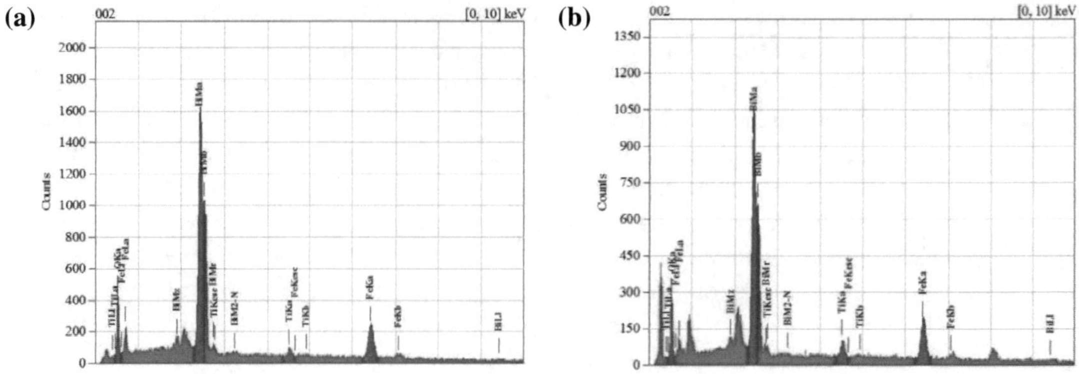

Figure 1. EDS images of the $BiFe_{0.9}Ti_{0.1}O_3$, (a) bulk material synthesized by solid-state reaction technique, (b) nanoparticles synthesized by sol-gel method

Table 1. Elemental mass percentage of $BiFe_{0.9}Ti_{0.1}O_3$ bulk and nanoparticle samples

Samples	Elements	Mass%
BFTO bulk prepared by solid-state reaction techniquc	O	2.44
	Ti	1.97
	Fe	16.83
	Bi	78.75
	Total	**100**
BFTO nanoparticles prepared by sol-gel method	O	2.58
	Ti	3.94
	Fe	19.66
	Bi	73.82
	Total	**100**

3.2. Magnetic characterization

The field directed magnetic measurements were conducted to explore the magnetic behavior of undoped $BiFeO_3$, $BiFe_{0.9}Ti_{0.1}O_3$ ceramics, and $BiFe_{0.9}Ti_{0.1}O_3$ nanoparticles. Figure 2 presents the *M-H* hysteresis loops of the synthesized samples investigated at room temperature (RT) with an applied magnetic field of up to ±10 kOe. The antiferromagnetic nature of undoped BFO is well demonstrated by the linear hysteresis loop having coercive field (H_c) and remanent magnetization (M_r) close to zero [19]. In contrast, 10% Ti-doped bulk and nanoparticle samples exhibited improved ferromagnetism with increased H_c and M_r values. The coercive field and remanent magnetization obtained in BFTO bulk were 58.97 Oe and 0.3041 emu/g, respectively, while those recorded in BFTO nanoparticles are 93.219 Oe and 0.0254 emu/g, respectively. Ti doping resulted in the demolition of spin cycloid structure within BFO associated with the release of latent magnetization inherently present inside the cycloid structure and eventually contributed to enhanced ferromagnetism in BFTO [20]. The H_c and M_r are calculated from the hysteresis loops adopting the following equations [21]

$$H_c = \frac{H_{c1} - H_{c2}}{2} \quad (1)$$

where H_{c1} and H_{c2} represent the left and right coercive fields respectively

$$M_r = \frac{|M_{r1} - M_{r2}|}{2} \quad (2)$$

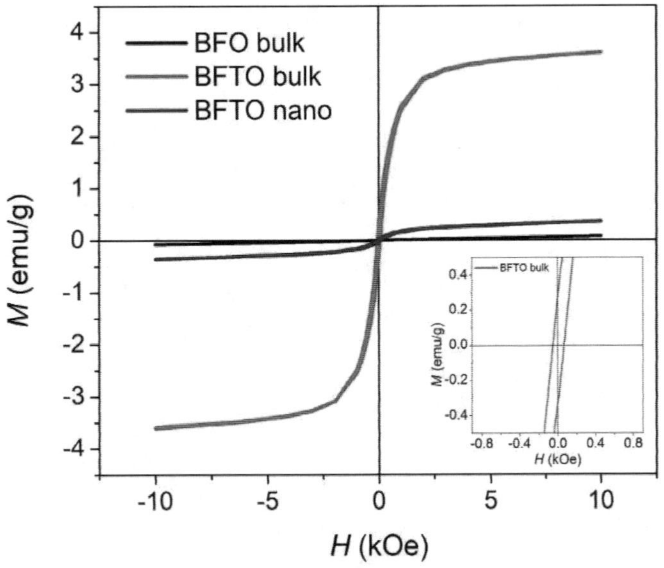

Figure 2. M-H hysteresis loops of BFO bulk, BFTO bulk, and BFTO nanoparticles. The inset shows a magnified view of BFTO bulk

where M_{r1} and M_{r2} denote the magnetization with positive and negative points of intersection respectively at $H = 0$

The sol-gel method yielded nanoparticles in the nanometer range, while the sizes of the bulk ceramics prepared by the solid-state technique fall in the micrometer range as reported in our previous investigations [15,16]. The considerable size reduction may impart higher coercivity in BFTO nanoparticles than its bulk sample due to effective spin contribution to the magnetic moment of the particle at an increased surface-to-volume ratio [22,23]. On the other hand, higher remanent magnetization is evident in BFTO bulk compared to the nano counterpart. The low microstrain and leakage current (discussed in the later section) in BFTO bulk may play the preeminent role in augmented remanent magnetization [24]. Substantially higher maximum magnetization (M_s) was observed as well in BFTO bulk (3.599 emu/g) compared to its nano sample (0.3564 emu/g).

3.3. Electrical Measurements

To investigate leaky behavior in bulk polycrystalline and nanoparticle samples, leakage current density (I_d) versus electric field (E) measurements were performed, as shown in Figure 3. It is perceived from Figure 3 that Ti doping suppressed leakage current density in both bulk and nano counterparts. The higher leakage current in undoped BFO bulk was dominated by impurity phases and oxygen vacancies [25]. X-ray diffraction (XRD) results reported earlier revealed that those impurity peaks were prominent in undoped BFO samples [15]. In contrast, the elimination of impurity phases, to a considerable extent, was noticed in the Ti-doped bulk and nano samples [15,16]. Ti being a less volatile element, may hinder the volatile nature of Bi along with the formation of secondary phases. Though BFTO bulk and nano counterparts exhibit close I_d values, slightly better results are observed in bulk ceramics, presumably due to the presence of lower oxygen vacancies.

The polarization versus electric field (P-E) hysteresis loops of the considered samples generated at ±5 kV electric field and 50 Hz driving frequency are illustrated in Figure 4. The undoped BFO bulk ceramics produced a round-shaped hysteresis loop ascribed to freely movable charge and oxygen

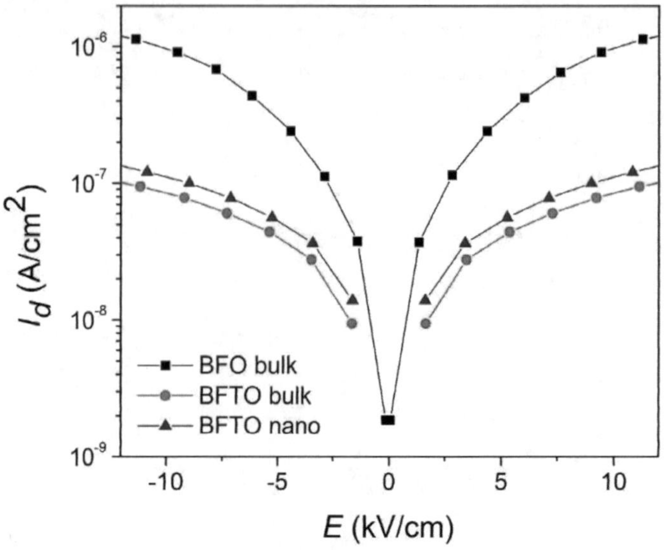

Figure 3. Leakage current density of undoped BFO bulk, BFTO bulk, and BFTO nanoparticles

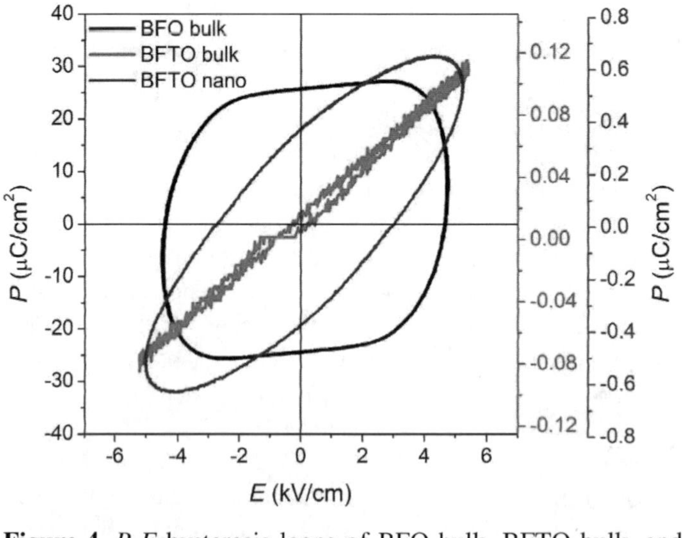

Figure 4. *P-E* hysteresis loops of BFO bulk, BFTO bulk, and BFTO nanoparticles. The black, red, and blue Y-axes correspond to BFO bulk, BFTO bulk, and BFTO nanoparticles respectively

vacancy defects [18]. The minor difference in leakage current densities of BFTO bulk and nano counterpart gave rise to a significant deviation in their hysteresis loops. The remanent polarization observed in BFTO bulk and nano counterpart were 0.005 μC/cm^2 and 0.373 μC/cm^2 respectively.

To explore further the ferroelectric polarization behavior in BFTO bulk and nano-materials *P-E* hysteresis loops measurements were conducted at variable electric fields and a constant driving frequency of 50 Hz. The obtained results depicted a gradual increase in remanent polarization with the electric field because the electric field of high magnitude imparts increased driving power to reverse the ferroelectric domains [26]. Ideal hysteresis loops are achieved in BFTO bulk owing to suppressed leakage current density, whereas the loops of BFTO nanoparticles are slightly distorted on account of

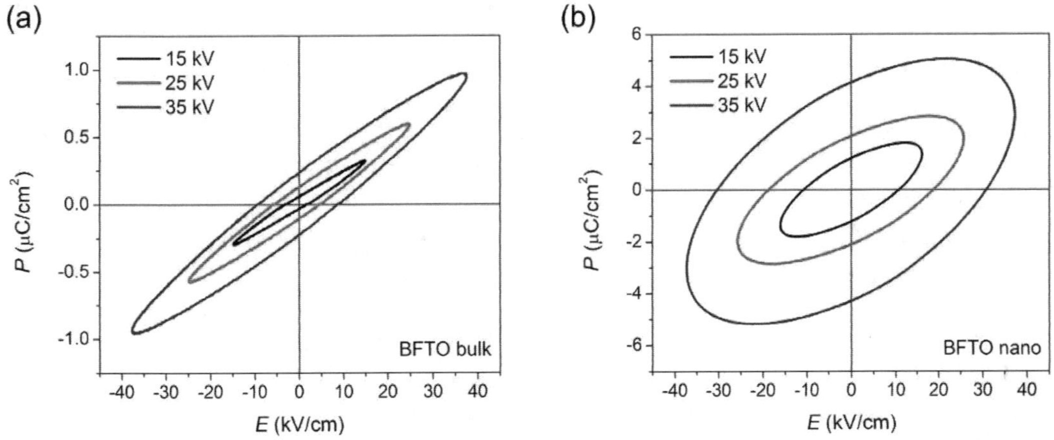

Figure 5. *P-E* hysteresis loops of BFTO samples at the variable electric field, (a) bulk ceramics, (b) nanoparticles

leaky behavior and space charge defect. Moreover, the agglomeration of particles during the Sol-gel synthesis process may cause deterioration of ferroelectric properties in nanoparticles [27].

4. Conclusion

Multiferroic properties of 10% Ti-doped $BiFeO_3$ bulk and nanoparticles synthesized by solid-state reaction technique and sol-gel method respectively have been reported. EDS analysis has confirmed the successful inclusion of the Ti dopant and the absence of any foreign elements in $BiFe_{0.9}Ti_{0.1}O_3$ samples. The *M-H* measurements reveal enhanced ferromagnetism in BFTO ceramics having increased coercive field, remanent magnetization, and maximum magnetization compared to undoped BFO bulk. Leaky behavior and rounder shape *P-E* hysteresis loops of BFO bulk have been ameliorated to a considerable extent in BFTO bulk and nano counterparts. Ti doping has resulted in the release of latent magnetization, reduced impurity phases as well as oxygen vacancy defects and thus yielded superior multiferroic materials. However, BFTO bulk ceramics exhibit improved magnetic and electrical properties compared to nano counterparts because of lower crystal microstrain and space charge defects. Hence, further research in the synthesis process of BFTO nanoparticles is being recommended to eliminate the existing defects, thereby enhancing the multiferroic aspects.

Acknowledgment

Part of this investigation was done under the patronage of Infrastructure Development Company Limited (IDCOL). EDS analysis was performed in the Department of Glass and Ceramic Engineering, BUET. Magnetic characterization was accomplished in the Materials Science Division, Atomic Energy Center, Dhaka. Electrical measurements were investigated in Nanotechnology Research Laboratory, Department of Physics, BUET.

References

[1] Heron J T, Schlom D G and Ramesh R 2014 *Appl. Phys. Rev.* **1** 021303.
[2] Das S, Sultana I, Bhuyan M D I and Basith M A 2019 *IEEE Magn. Lett.* **10** 2503704.
[3] Das S, Ahmmad B and Basith M A 2020 *AIP Adv.* **10** 095135.
[4] Lazenka V V, Zhang G, Vanacken J, Makoed I I, Ravinski A F and Moshchalkov V V 2012 *J. Phys. D: Appl. Phys.* **45** 125002.
[5] Catalan G, Scott J F 2009 *Adv. Mater.* **21** 2463.
[6] Munoz T, Rivera J P, Monnier A and Schmid H 1985 *Jpn. J. Appl. Phys.* **24** 1051.
[7] Agarwal R A, Ashima S S and Ahlawat N 2012 *J. Phys. D: Appl. Phys.* **45** 165001.

[8] Chen L, Ren W, Zhu W, Ye Z-G, Shi P, Chen X, Wu X and Yao X 2010 *Thin Solid Films* **518** 1637.
[9] Ke H, Wang W, Wang Y, Xu J, Jia D, Lu Z and Zhou Y 2011 *J. Alloy. Compd.* **509** 2192.
[10] Anwar A, Basith M A and Choudhury S 2019 *Mater. Res. Bull.* **111** 93.
[11] Bhalla A S, Guo R and Roy R 2000 *Mat. Res. Innov.* **4** 3.
[12] Bernardo M S, Jardiel T, Peiteado M, Mompean F J, Garcia-Hernandez M, Garcia M A, Villegas M and Caballero A C 2013 *Chem. Mater.* **25** 1533.
[13] Kim S J, Han S H, Kim H G, Kim A Y, Kim J S and Cheon C I 2010 *J. Korean Phys. Soc.* **56** 43.
[14] Murari N M, Thomas R, Melgarejo R E, Pavunny S P and Katiyar R S 2009 *J. Appl. Phys.* **106** 014103
[15] Anjum N, Lamia S N E, Arafat M Y, Mahboob M and M A Basith 2018 *AIP Conf. Proc.* **1980** 060004.
[16] Anjum N, Arafat M Y, Noman A A, Lamia S N E 2020 *IJERT* **09** 270.
[17] Basith M A, Kurni O, Alam M S, Sinha B L and Ahmmad B 2014 *J. Appl. Phys.* **115** 024102.
[18] Sushmita G, Subrata D, Amarnath S and Himadri S M 1349 *J. Am. Ceram. Soc.* **88** 1349.
[19] Han J T, Huang Y H, Wu X J, Wu C L, We W, Peng B, Huang W and Goodenough J B 2006 *Adv. Mater.* **18** 2145.
[20] Wang K F, Liu J M and Ren Z F 2009 *Adv. Phys.* **58** 321.
[21] Basith M A, Billah A, Jalil M A, Yesmin N, Sakib M A, Ashik E K, Yousuf S M E H, Chowdhury S S, Hossain M S, Firoz S H and Ahmmad B 2017 *J. Alloy Compd.* **694** 792.
[22] Huang F, Wang Z, Lu X, Zhang J, Min K, Lin W, Ti R, Xu T, He J, Yue C and Zhu J 2013 *Sci. Rep.* **3** 2907.
[23] Basith M A, Yesmin N and Hossain R 2018 *RSC Adv.* **8** 29613.
[24] Chowdhury S S, Kamal A H M, Hossain R, Hasan M, Islam M F, Ahmmad B and Basith M A 2017 *Ceram. Int.* **43** 9191.
[25] Mukherjee A, Basu S, Manna P K, Yusuf S M and Pal M 2014 *J. Mater. Chem. C.* **2** 5885.
[26] Wu M S, Huang Z B, Han C X, Yuan S L, Lu C L and Xia S C 2012 *Solid State Commun.* **152** 2142.
[27] Dan L, Zhongyang L, Yu C and Kefa C 2007 *J. Rare Earths* 2007 **25** 163.

Fatigue life prediction of cfrp composites in plane bending by modal analysis

Pravin Kulkarni[1,3,4], Akshay Potdar[1,3], V V Kulkarni[1,3] and N K Patil[2,3]

[1]Department of Mechanical Engineering
[2]Department of Civil Engineering
[3]Sanjay Ghodawat Group of Institutions, Kolhapur, Maharashtra, India, 416118
[4]sagitaria0460@gmail.com

Abstract. This is investigation of the fatigue behaviour of carbon fiber reinforced polymer laminates under various angles of deflections of cantilever beam in bidirectional (R=-1) plane bending. The Unidirectional 45° carbon fiber oriented CFRP composite were tested at different cantilever bending angles at frequency of 8 Hz. The non- destructive modal testing is conducted to evaluate reduction in structural stiffness in fully reversed plane bending in terms of natural frequency. The modal test at specific intervals is conducted by interrupting the fatigue test. The curve fitting technique for prediction of fatigue life over experimental analysis is used to formulate behaviour of CFRP in plane bending fatigue. It is shown that A Non-destructive Testing method (Modal Test) can be applied for the fatigue life estimation of CFRP laminates. The natural frequency reduction model is able to represent the overall reduction in stiffness due to fully reversed plane bending fatigue.

1. Introduction

Carbon Fiber Reinforced Polymer (CFRP) has various applications in aerospace, aeronautical, automobiles and sports goods due to their comparatively high strength, low specific weight and varieties in the types as per the applications than polymers, metals, alloys and foams etc. During its service life, it is subjected to different types of fatigue loading, especially plane bending fatigue failure occurs often. The fatigue performance of CFRP composites is still a big concern. A. Mohamed Ansar et al. [1] presented work on fatigue analysis of both aluminium and glass fiber reinforced composite material under vibrating or oscillating bending force. B. M. Faisal et al. [2] estimated the mechanical static and dynamic properties like S-N curve and fatigue limit of aluminium alloy and fiber glass-polyester composite material under static and fully reverse bending dynamic test (R= -1) and he found that the life cycles are significantly less for glass polyester composite than that for aluminium alloy at the same applied stress level. Bryan Harris [3] claimed that the fatigue phenomenon in composites is quite a complex one and need to be investigated for different parameters such as fibre orientation, thickness, type of fibre, matric etc. M. Abo-Elkhier et al. [4] investigated the capability of experimental modal analysis to characterize and quantify the fatigue behaviour and to determine modal parameters of glass fiber reinforced polyester laminated composite cantilevered beam by interrupting Plane bending fatigue tests at different fatigue life ratios (n/N*f*). Also curve fitting technique was used to correlate modal parameters to fatigue life. The value of damping ratio is more noticeable than the value of frequency. The extent of fatigue damage determines the damping ratio which indicates the fatigue life ratio at lower mode. T.ae-Chul Moon et al. [10] proposed the natural frequency reduction model for the composite laminates with matrix dominated fatigue damage. A flexural stiffness reduction model was also suggested for estimation of residual flexural properties. Farhad Adel et al. [11] applied modal testing for analysing the dynamic behaviour of hybrid composite/aluminium joints. A. N. Damir et al. [12] used the modal analysis for prediction of fatigue life of gray and ductile cast iron in rotating bending fatigue. Rui-Jie Wang et al. [13] estimated the fatigue damage for spot welded joints in terms of changes in natural frequency. It is seen that the natural frequency reduction model is a better technique for fatigue life estimation to present the overall stiffness reduction model. M M Shokrieh et al. [5] studied the effect of adding graphene nanoplatelets on mechanical properties of epoxy resin under flexural bending stress at displacement-controlled bending loading and different displacement amplitudes at room temperature. He observed the improvement in fatigue life of epoxy resin by addition of graphene nanoplatelets. P. K. Dash et al. [6] investigated that the residual strength of the cyclic stressed bidirectional CFRP produced significant stress reduction with respect to increased number of cycles and increased value of cyclic stress. Pritam. V. Kulkarni et al. [7] presented work on the development of fatigue testing set up for the fatigue life estimation of GFRP in plane bending with different orientation (0-degree, 45 degree and 90 degree) and suggested modal testing (NDT) for estimation of fatigue life. Raif Sakin et al. [8] tested aluminium alloys with different rolling direction to obtain tensile and fatigue characteristics at high and low cycle fatigue region by static and deflection-controlled plane-bending fatigue tests under fully reversed loading. H. Dong et al. [9] studied non-linear degradation in both the strength and stiffness through residual strength and stiffness models by fatigue theory to predict the fatigue life, residual strength and residual failure envelope of fibre reinforced composite laminates under multidirectional loadings.

In this paper the application of Non-destructive modal testing is analysed to predict fatigue life of CFRP composites under fully reversed plane bending fatigue (R= -1). The plane bending fatigue test interrupted by modal tests (Bump test) is conducted to analyse the behaviour of CFRP composites under bidirectional fatigue loading. An experimental Modal analysis is conducted to study modal parameters of cantilever structure under plane bending fatigue cycles. The reduction in stiffness of cantilever structure in terms of natural frequency is studied and a fatigue life prediction model is presented.

2. Experimental Analysis

2.1. Material Properties

The unidirectional 45^0 carbon fiber oriented T700/ HY3544 carbon/epoxy resin composite laminates is used. The material properties in GPa of selected CFRP are shown in table 1. The specimens for static flexure and tensile testing were prepared according to ASTM D 7264 and D 3039 respectively.

Table 1. Material Properties of CFRP

E_{11}	E_{22}	E_{33}	G_{12}	G_{13}	G_{23}	υ_{12}	υ_{13}	υ_{23}
123	8.3	8.3	4.8	4.8	2.84	0.3	0.3	0.46

2.2. Static tensile testing

Static tensile and three points bending flexural test is conducted on UTM. There are five specimens were tested to obtain peak load, ultimate strength, maximum displacement and deflection of CFRP under tensile and flexure loading as shown in table 2.

Table 2. Static Test Results of 45^0 CFRP

Test	Thickness (mm)	Width (mm)	CS Area (mm^2)	Peak Load (N)	Ultimate Strength (N/mm^2)	Maximum Displacement (mm)	Maximum Deflection (mm)

IOP Conf. Series: Materials Science and Engineering 1091 (2021) 012007 doi:10.1088/1757-899X/1091/1/012007

Tensile	2	25	50	12340	247	23.42	-
Flexure	2	20	40	300	243.75	-	6.72

2.3. Plane bending fatigue test

The dog bone shaped test specimens were prepared as per the guidance given by Japanese standard, JIS Z2275. The dimensions of the specimen in mm are shown in figure 1.

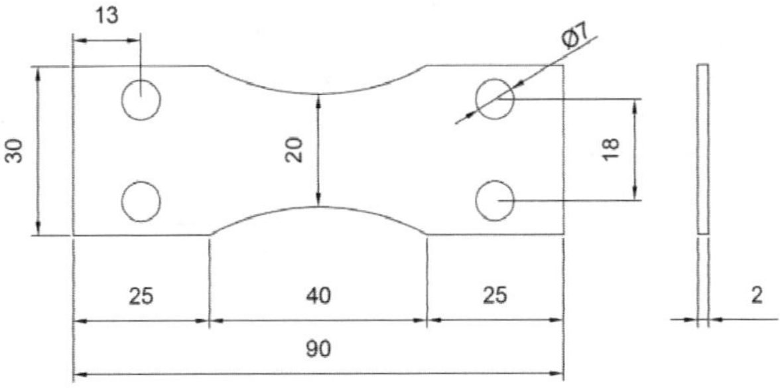

Figure 1. Test specimen

The fully reversible plane bending fatigue test was carried out on plane bending fatigue test machine at stress ratio R = -1 and frequency of 8Hz on 45°fibre orientated CFRP laminates as shown in figure 2. There are five test specimen were tested at $\pm 8^0$, $\pm 10^0$ and $\pm 12^0$ of angle of bending to obtain reduction in strength of cantilever structure.

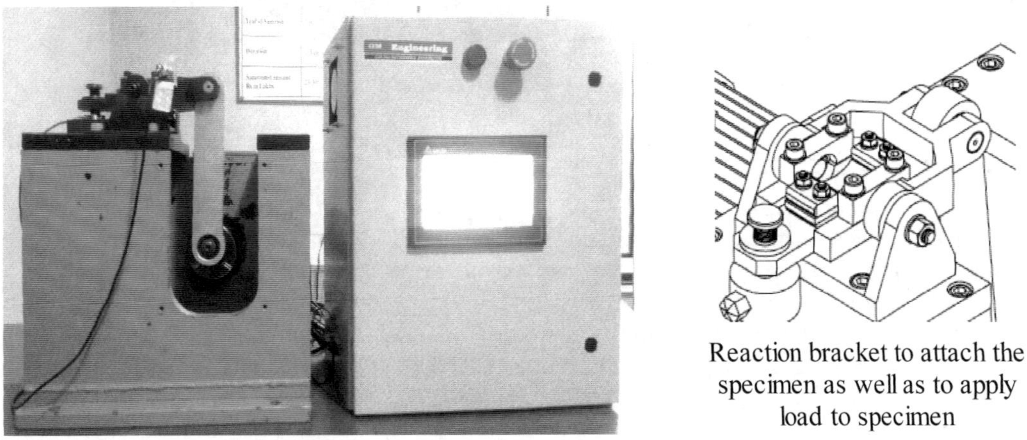

Figure 2. Plane bending Fatigue testing machine

2.4. Modal test

The modal test was carried out by interrupting fatigue test at specific number of cycles (n) to obtain frequency response i. e. Natural Frequency (f_n). The figure 3 shows experimental setup for this test. The FFT analyser and instrumented impact hammer (Piezoelectric sensor) was used to obtain natural frequency of test specimen at each stage of fatigue test. Five test specimens were tested for each level as shown in figure 4.

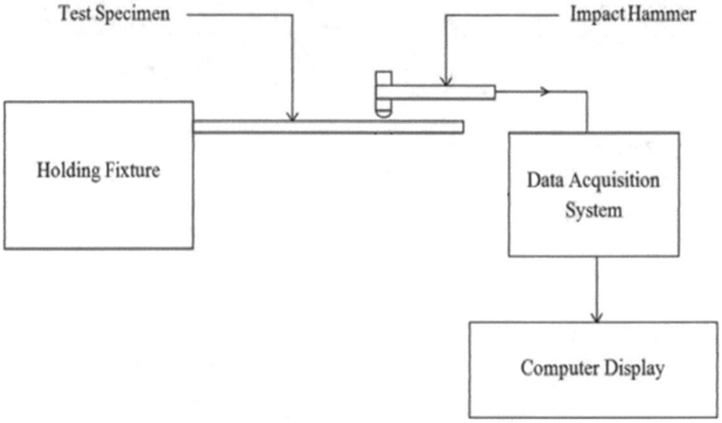

Figure 3. Experimental setup of modal testing

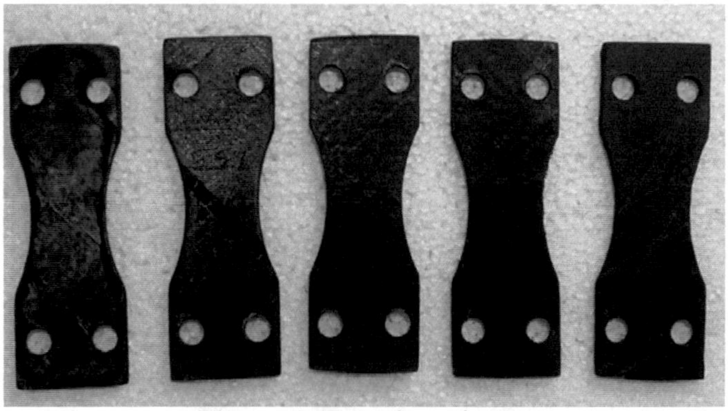

Figure 4. Tested specimens

3. Results and Discussions

3.1. Stiffness reduction model

Fatigue tests were conducted up to 10^6 numbers of cycles (Nf) at constant angle of bending and reduction in stiffness at each specified number of cycles was obtained. Five specimens were tested at each bending angle. The results are shown in table 3. The percent reduction in stiffness shows failure of specimen. Failure of Work piece at 12^0 of bending angle at 400000 numbers of cycles is shown in figure 5. The plot of Stiffness (K) verses Normalized Number of Cycles (n/Nf) at different angle of bending is shown in figure 6.

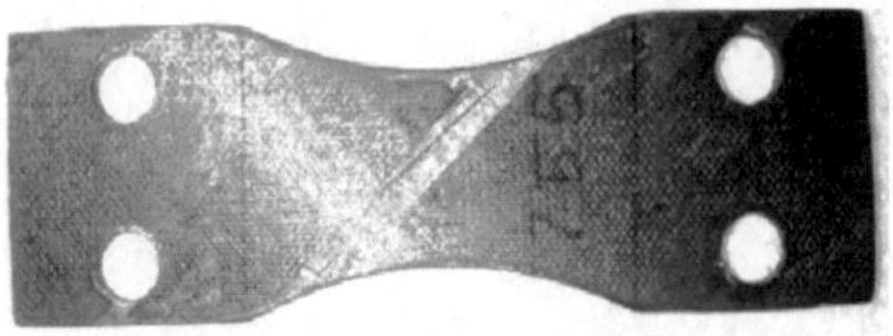

Figure 5. Failure of test specimen

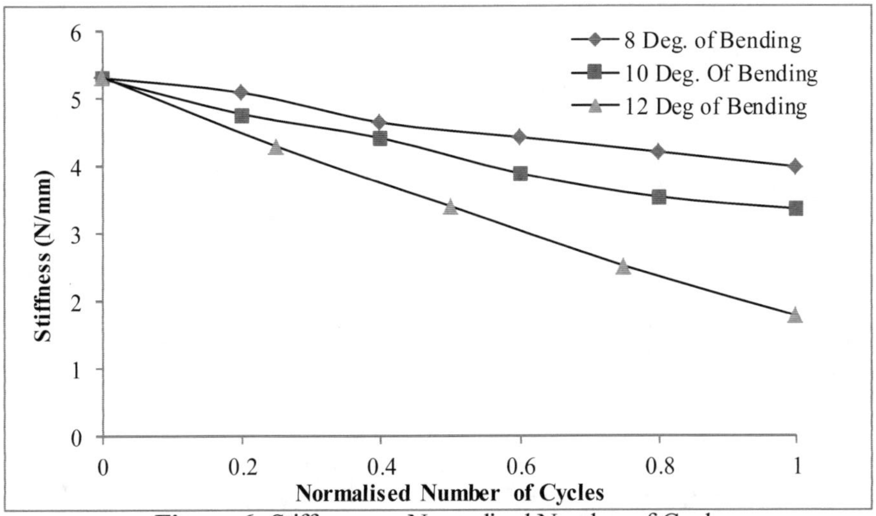

Figure 6. Stiffness vs. Normalized Number of Cycles

It was observed that quadratic equation is nicely fitted to experimental data. The quadratic equation at each angle of bending is given below

For 12^0

$$y = 0.7x^2 - 4.2x + 5.3 \quad (1)$$

For 10^0

$$y = 0.8x^2 - 2.8x + 5.3 \quad (2)$$

For 8^0

$$y = 0.4x^2 - 1.8x + 5.3 \quad (3)$$

Where, y = Stiffness N/mm (K)

x = (n/N*f*)

3.2. Frequency reduction model

The formation of frequency reduction model obtained by modal analysis indicates overall reduction in stiffness due to fatigue loading. The graph of Natural Frequency verses Normalised number of cycles at different angle of bending is mentioned in figure 7. The quadratic equations obtained by curve fitting technics are given below.

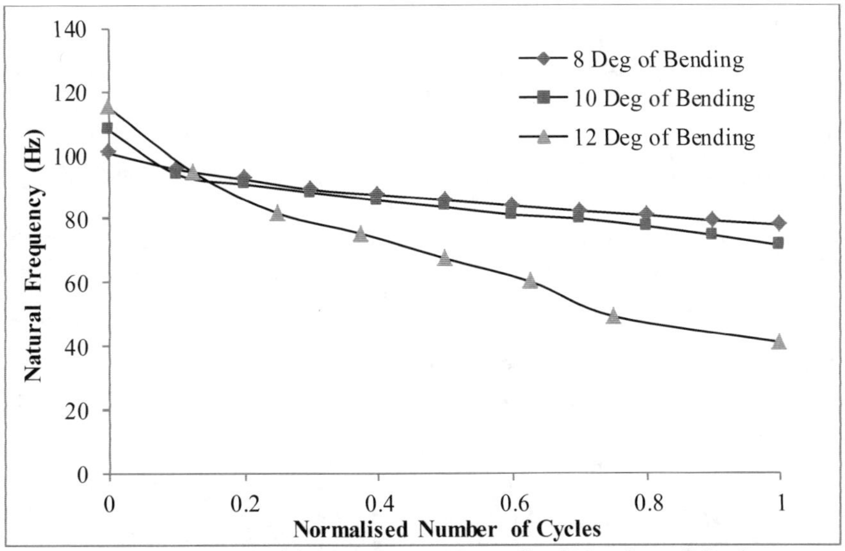

Figure 7. Natural frequency vs. Normalized Number of Cycles

For 12^0

$$f_n = 46x^2 - 116x + 112 \quad (4)$$

For 10^0

$$f_n = 23x^2 - 52x + 103 \quad (5)$$

For 8^0

$$f_n = 15x^2 - 36x + 100 \quad (6)$$

Where, f_n= Natural Frequency (Hz)
$x= n/Nf$ =Normalised Number of cycles

The Comparative results of reduction in stiffness as well as natural frequency of CFRP are shown in table 3.

Table 3. Test Results for Stiffness and Natural Frequency Reduction Model

Angle of Bending	Number of Cycles (n)	Load at peak point (N)	Stiffness (N/mm)	Natural Frequency (Hz)	% Reduction in Stiffness	% Reduction in Natural frequency
$\pm 8^0$	0	24	5.31	100.6	25%	22.76%
	200000	23	5.09	92.3		
	400000	21	4.65	87.3		
	600000	20	4.42	83.9		
	800000	19	4.20	80.8		
	1000000	18	3.98	77.7		
$\pm 10^0$	0	30	5.30	108.1	36.66%	34.04%
	200000	27	4.77	90.7		
	400000	25	4.42	85.7		
	600000	22	3.89	81.0		
	800000	20	3.53	77.3		
	1000000	19	3.36	71.3		
$\pm 12^0$	0	36	5.33	115.2	66.67%	64.49%
	100000	29	4.29	81.5		
	200000	23	3.40	67.2		
	300000	17	2.51	49.0		
	400000	12	1.78	40.9		

4. Conclusions

Based on the experimental load and frequency reduction models by modal test, following conclusions can be derived:

1. The testing for modal parameters is best Non-destructive fatigue life prediction method which can be useful to estimate fatigue behavior of CFRP laminates in plane bending fatigue.
2. The Reduction in Natural frequency and strength is depends on angle of bending of CFRP laminated.
3. As there is no complete fracture of 45^0 fiber oriented CFRP laminates so a natural frequency reduction model will be the appropriated approach to prediction fatigue life in plane bending.

5. Acknowledgements

The authors are grateful to financial support for this study granted by Defence Research & Development Organization DRDO, New Delhi, India (ERIP/ER/DG-NS&M/990916702/M/01/1720)

6. References

[1] A. Mohamed Ansar, Dalbir Singh and Balaji D Fatigue analysis of glass fiber reinforced composites. *International journal of engineering research and applications.*

[2] B M Faisal, A T Abass and A F Hammadi 2016 Fatigue Life Estimation of Aluminum Alloy 2024-T4 and Fiber Glass-Polyester Composite Material *International Research Journal of Engineering and Technology* **03** 1760.

[3] Bryan Harris 2013 Fatigue in composites, science and technology of the fatigue response of fibre-reinforced plastics *Woodhead Publishing Limited, England.*

[4] M Abo-Elkhier, A A Hamada A Bahei El-Deen 2014 Prediction of fatigue life of glass fiber reinforced polyester composites using modal testing *International Journal of Fatigue* **69** 28–35.

[5] M M Shokrieh, M Esmkhani, F Taheri-Behrooz and A R Haghighatkhah 2013 Displacement-controlled flexural bending fatigue behavior of graphene/epoxy Nano-composites *Journal of Composite Materials.*

[6] P K Dash and Ankita Dash 2018 An Experimental Study on Impact of Cyclic Damage on Residual Flexure Strength of Plain Weave Carbon/Epoxy Composite *International Journal of Composite Materials* **8(1)** 1-9.

[7] Pritam V kulkarni, P J Sawant and V V Kulkarni 2017 Design and development of plane bending fatigue testing machine for composite material *Proceedings of International Conference on Materials Manufacturing and Modeling*.

[8] Raif Sakin 2018 Investigation of bending fatigue-life of aluminium sheets based on rolling direction *Alexandria Engineering Journal* **57** 35–47.

[9] Dong H, Li Z, Wang, J and Karihaloo B L A 2016 New fatigue failure theory for multidirectional fiber-reinforced composite laminates with arbitrary stacking sequence *International Journal of Fatigue.*

[10] Tae-Chul Moon, Hyeung-Yun Kim and Woonbong Hwang 2003 Natural-frequency reduction model for matrix dominated fatigue damage of composite laminates *Composite Structures* **62** 19-26

[11] Farhad Adel, Saeed Shokrollahi, Majid Jamal-Omidi and Hamid Ahmadian 2017 A model updating for hybrid composite/aluminum bolted joint using modal test data *Journal of Sound and Vibrations* **396** 172-185

[12] A N Damir, A Elkhatib and G Nassef 2007 Prediction of fatigue life using modal analysis for gray and ductile cast iron *International Journal of fatigue* **29** 499-507

[13] Rui-Jie Wang, De-Guang Shang, Li-Sen Li and Cheng-Shan Li 2008 Fatigue damage model based on the natural frequency changes for spot welded joints *International Journal of fatigue* **30** 1047-1055

On modelling and adaptive control of a linear smart beam model interacting with fluid

Hayder F N Al-Shuka[1,*], **Kareem Jawad Kadhim**[1], **AbdnoorJameel Shaheed Al-hamadani**[1],

[1]Department of Aeronautical Engineering, University of Baghdad, Baghdad, Iraq

*dr.hayder.f.n@coeng.uobaghdad.edu.iq

Abstract. This paper deals with modelling and control of Euler-Bernoulli smart beam interacting with a fluid medium. Several distributed piezo-patches (actuators and/or sensors) are bonded on the surface of the target beam. To model the vibrating beam properly, the effect of the piezo-patches and the hydrodynamic loads should be taken into account carefully. The partial differential equation PDE for the target oscillating beam is derived considering the piezo-actuators as input controls. Fluid forces are decomposed into two components: 1) hydrodynamic forces due to the beam oscillations, and 2) external (disturbance) hydrodynamic loads independent of beam motion. Then the PDE is discretized using the Galerkin approach to obtain standard multi-modal equations. An adaptive approximation control structure is proposed to suppress the beam vibration. The controller consists of a proportional-derivative PD control plus an adaptive approximation compensator AAC with guaranteed stability. A simply supported beam with 2 piezo-patches interacting with fluid is simulated. The disturbance hydrodynamic force that excites the beam vibration is assumed as a harmonic force with 50 Hz frequency and 1 N amplitude. The results prove the efficacy of the proposed control architecture.

1. Introduction

Modelling, design and motion regulation of fluid-flexible structures witness much attention since they are used in many life applications such as underwater robotic systems, space vehicles or resonant beams for measurement purposes [1-3]. These structures would be destabilized if undesired disturbance loads are applied. To this end, active vibration attenuation is a powerful tool to resolve this dilemma. By bonding piezo-patches (actuators and sensors) on the surface of the vibrating beam, the oscillations of the dynamic system can be suppressed. However, a control structure is required to stabilize the fluid-beam oscillations [4]. Consequently, this paper is focused on modelling and adaptive control of a flexible beam immersed in a fluid under modelling uncertainty. Designing an appropriate control system requires proper modelling for the target dynamic system. Two key points should be considered in modelling of a smart beam-fluid system: 1) fluid hydrodynamics and 2) piezo-electric patches. Fortunately, the dynamics of piezo-patches integrated with the beam can be neglected due to the slight effect on vibrations characteristics of the system [5]. On the other hand, if the beam is imposed to oscillate with fluid vibrations then the hydrodynamic forces can be decomposed into two main forces: 1) hydrodynamic forces dependent of beam motion that can be represented by inertia, damping and stiffness forces and 2) external hydrodynamic forces independent of beam oscillations and these forces are of random signals and are determined experimentally [6]. For more details on fluid-structure interaction, see [1-3, 6-14].

IOP Conf. Series: Materials Science and Engineering 1091 (2021) 012008 doi:10.1088/1757-899X/1091/1/012008

For simplicity, linear modelling for the coupled beam-fluid system is considered and this eases the task of the controller; however, the parameters of the target system are assumed unknown. Therefore, conventional control structures such as conventional PID or optimal control etc. (see e.g. [15] and the references therein) maybe not useful for this type of system. Two control techniques are strong to deal with the uncertainty: adaptive control and robust bounded control, see e.g. [16-18] for more details. In general, two strategies of adaptive control are available: regressor and approximation-based control. The former is a physics-based control that is difficult to deal with disturbances if exist, while approximation-based control is a powerful tool to treat complex systems, the reader is referred to [15, 18-23] for more information.

In view of the above, this work is concerned with modelling and adaptive approximation control of a linear smart beam model interacting with fluid. The control architecture consists of three main terms: PD term, adaptive approximation compensator term and a robust sliding term for reducing the modelling error if exist. A simply supported beam provided with two piezo-patches is simulated. The smart beam system is immersed in a fluid media. The vibration source is assumed as a harmonic force that excites the coupled fluid-beam system. In effect, this disturbance results from the external fluid hydrodynamic forces that are independent of beam vibrations. The results show the strength of the proposed controller under uncertain modelling and impulse disturbances.

The remainder of the paper is structured as follows. Section 2 describes the dynamic modelling of the smart beam-fluid system while section 3 introduces the proposed control architecture. Simulation results and discussions are evaluated in Section 4. Section 5 concludes.

2. Dynamic modelling of a coupled beam-fluid system

Below modelling of smart beam immersed in a fluid is presented, see Figure 1 for a depiction of beam-fluid model. For modelling purposes, the following points are assumed [4, 6]:

(i) A linear fluid-structure model is considered, i.e. small deflections are imposed.

(ii) The hydrodynamic forces are decomposed into two terms: forces dependent on beam oscillations and external disturbance forces free from beam vibrations.

(iii) A sufficient number of piezo-sensors are available such that modal amplitudes are measurable.

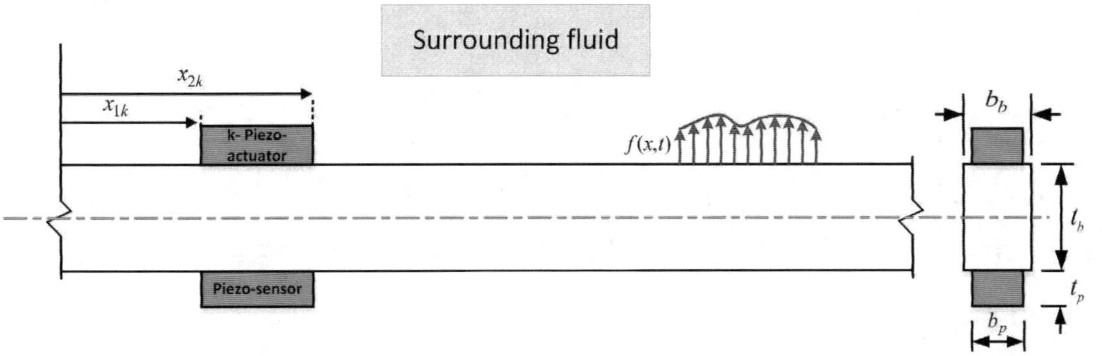

Figure 1. A flexible beam with piezo-patches interacting with fluid.

The equation of motion for the transverse vibration $w(x,t)$ of a smart beam-fluid system can be represented as[4,24]

$$E_b I_b \frac{\partial^4 w}{\partial x^4} + \rho_b A_b \frac{\partial^2 w}{\partial t^2} = f(x,t) - \frac{\partial^2 T_p}{\partial x^2} \quad (1)$$

$$T_p = Dv_a(t)\frac{\partial^2}{\partial x^2}\left[H(x-x_1)-H(x-x_2)\right] \quad (2)$$

$$f(x,t) = f_m(x,t)+f_e(x,t) \quad (3)$$

where E_b, I_b, A_b, and ρ_b are Young's modulus, moment of inertia, cross-sectional area, and density of the beam respectively. T_p is the moment per unit length exerted by the piezo-actuator, D is a constant depending on properties of the regular beam and smart materials [23]. $H(.)$ is a Heaviside step function of the beam displacement. On the other hand, $f(x,t)$ denotes to the hydrodynamic forces that composed of two terms: $f_m(x,t)$ denoting to the hydrodynamic force per unit length due to the beam motion and $f_e(x,t)$ referring to an external hydrodynamic force independent of beam motion.

Remark 1. As aforementioned, the overall hydrodynamic fluid forces are partitioned into two terms; $f_m + f_e$, where f_m is a function of acceleration and velocity of the beam oscillations (the concept of added mass and damping), while f_e is experimentally determined. In effect, this work assumes the vibration source results from f_e.
In order transfer the PDF of Equation (1) into multi-modal ODEs, the Galerkin approach is used, hence

$$w(x,t)=\sum_{i=1}^{N}\phi_i(x)q_i(t) \quad (4)$$

where $\phi_i(x)$ is the mode shape, $q_i(t)$ is the modal amplitude and N is the number of the mode shapes.Substituting Equation (4) into Equation (1), multiplying by an arbitrary $\phi_j(x)$, integrating along the beam length and using the orthogonal conditions for a simply supported beam with some manipulations, we obtain

$$m_j\ddot{q}_j + b_j\dot{q}_j + k_jq_j - g_{mj}(x,t) - g_{ej}(x,t) = u_j,\ j=1,2,3,...,N \quad (5)$$

where

$$m_j=\left(\frac{\rho_bA_bl_b}{2}\right), b_j=(2\zeta_jw_j)m_j, -1\le\zeta_j\le 1, k_j=\frac{E_bI_b\lambda_j^4}{2}, w_j^2=k_j/m_j, u_j=-\sum_{i=1}^{N_a}\mu_{ji}v_{ai}(t),$$

$$\mu_{ji}=D(\varphi_j'(x_{2i})-\varphi_j'(x_{1i}), g_{mj}(x,t)=-\int_0^{l_b}f_m(x,t)\varphi_j(x)dx, g_{ej}(x,t)=-\int_0^{l_b}f_e(x,t)\varphi_j(x)dx, \lambda_j=\frac{j\pi}{l_b}$$

wherel_b referring to length of the beam and N_a is the number of piezo-actuators. In a matrix form, Equation(5) is expressed as

$$\mathbf{M\ddot{q}}+\mathbf{B\dot{q}}+\mathbf{Kq}+\mathbf{g}_m+\mathbf{g}_e=\mathbf{u} \quad (6)$$

$$\mathbf{q}=\begin{bmatrix}q_1\\ \vdots\\ q_N\end{bmatrix}\in R^N, \mathbf{M}=\begin{bmatrix}m_1 & \cdots & 0\\ \vdots & \ddots & \vdots\\ 0 & \cdots & m_N\end{bmatrix}\in R^{N\times N}, \mathbf{B}=\begin{bmatrix}b_1 & \cdots & 0\\ \vdots & \ddots & \vdots\\ 0 & \cdots & b_N\end{bmatrix}, \mathbf{K}=\begin{bmatrix}k_1 & \cdots & 0\\ \vdots & \ddots & \vdots\\ 0 & \cdots & k_N\end{bmatrix}\in R^{N\times N}$$

$$, \mathbf{g}_m=\begin{bmatrix}-g_{m1}\\ \vdots\\ -g_{mN}\end{bmatrix}\in R^N, \mathbf{g}_e=\begin{bmatrix}-g_{e1}\\ \vdots\\ -g_{eN}\end{bmatrix}\in R^N, \mathbf{u}=\begin{bmatrix}u_1\\ \vdots\\ u_N\end{bmatrix}=-\begin{bmatrix}\mu_{11} & \cdots & \mu_{1N_a}\\ \vdots & \ddots & \vdots\\ \mu_{N1} & \cdots & \mu_{NN_a}\end{bmatrix}\begin{bmatrix}v_{a1}\\ \vdots\\ v_{aNa}\end{bmatrix}\in R^{N_a}$$

3. Control design

The objective of the controller is to regulate the beam oscillations due to external disturbances. The proposed control law consists of three terms: PD term, an adaptive approximation compensator term, and a robust sliding term for compensation of modelling errors if exist. Thus, the controller form can be represented as [18-23]

$$\mathbf{u} = \hat{\boldsymbol{\alpha}} - \mathbf{K}_p\mathbf{e} - \mathbf{K}_d\dot{\mathbf{e}} - \boldsymbol{\gamma}\,\mathrm{sgn}(\dot{\mathbf{e}}) \tag{7a}$$

$$\mathbf{e} = \mathbf{q} - \mathbf{q}_d, \hat{\boldsymbol{\alpha}} = \hat{\mathbf{M}}\ddot{\mathbf{q}}_d + \hat{\mathbf{B}}\dot{\mathbf{q}}_d + \hat{\mathbf{K}}\mathbf{q}_d + \hat{\mathbf{g}}_m + \hat{\mathbf{g}}_e,\ \hat{\boldsymbol{\alpha}} = \hat{\boldsymbol{\psi}}^T\boldsymbol{\theta} \tag{7b}$$

where $(\hat{\cdot})$ refers to estimation, $\mathbf{K}_{(.)} \in R^{N\times N}$ and $\boldsymbol{\gamma} \in R^{N\times N}$ are diagonal positive definite feedback gain matrices, $\mathbf{q}_d \in R^N$ is the desired reference vector, $\boldsymbol{\psi} \in R^{N\beta\times N}$ and $\boldsymbol{\theta} \in R^{N\beta}$ are the weighting coefficient and the basis function matrix/vector, and β is the number of basis function. Equating Equation (7a) to Equation (6) to get

$$\mathbf{M}\ddot{\mathbf{q}} + \mathbf{B}\dot{\mathbf{q}} + \underbrace{\mathbf{K}\mathbf{q} + \mathbf{g}_m + \mathbf{g}_e}_{\boldsymbol{\sigma}} = \hat{\mathbf{M}}\ddot{\mathbf{q}}_d + \hat{\mathbf{B}}\dot{\mathbf{q}}_d + \underbrace{\hat{\mathbf{K}}\mathbf{q}_d + \hat{\mathbf{g}}_m + \hat{\mathbf{g}}_e}_{\hat{\boldsymbol{\sigma}}} - \mathbf{K}_p\mathbf{e} - \mathbf{K}_d\dot{\mathbf{e}} - \boldsymbol{\gamma}\,\mathrm{sgn}(\dot{\mathbf{e}}) \tag{8}$$

Adding $(-\mathbf{M}\ddot{\mathbf{q}}_d - \mathbf{B}\dot{\mathbf{q}}_d - \boldsymbol{\mu})$ to the above equation to obtain

$$\mathbf{M}\ddot{\mathbf{e}} + (\mathbf{B} + \mathbf{K}_d)\dot{\mathbf{e}} + \mathbf{K}_p\mathbf{e} + \boldsymbol{\gamma}\,\mathrm{sgn}(\dot{\mathbf{e}}) = -\tilde{\boldsymbol{\alpha}} + \boldsymbol{\varepsilon} \tag{9}$$

where $\boldsymbol{\varepsilon} \in R^N$ is the modelling error. Equation (9) represents a closed-loop dynamics for the beam-fluid system. As we see from Equation (7), the unknown matrix $\hat{\boldsymbol{\alpha}}$ is required to be updated with guaranteed stability. To this end, consider the following non-negative function along Equation (9)

$$V = \frac{1}{2}\dot{\mathbf{e}}^T\mathbf{M}\dot{\mathbf{e}} + \frac{1}{2}\mathbf{e}^T\mathbf{K}_p\mathbf{e} + \frac{1}{2}tr(\tilde{\boldsymbol{\psi}}^T\mathbf{F}^{-1}\tilde{\boldsymbol{\psi}}) \tag{10}$$

Taking time derivative of Equation (10)

$$\dot{V} = \dot{\mathbf{e}}^T(-(\mathbf{B} + \mathbf{K}_d)\dot{\mathbf{e}} - \mathbf{K}_p\mathbf{e} - \boldsymbol{\gamma}\,\mathrm{sgn}(\dot{\mathbf{e}}) - \tilde{\boldsymbol{\alpha}} + \boldsymbol{\varepsilon}) + \dot{\mathbf{e}}^T\mathbf{K}_p\mathbf{e} - tr(\tilde{\boldsymbol{\psi}}^T\mathbf{F}^{-1}\dot{\hat{\boldsymbol{\psi}}}), \dot{\mathbf{M}} = \mathbf{0} \tag{11}$$

By manipulating Equation (11), we can obtain

$$\dot{V} = -\dot{\mathbf{e}}^T(\mathbf{B} + \mathbf{K}_d)\dot{\mathbf{e}} + \dot{\mathbf{e}}^T\boldsymbol{\varepsilon} - \dot{\mathbf{e}}^T\boldsymbol{\gamma}\,\mathrm{sgn}(\dot{\mathbf{e}}) - tr(\tilde{\boldsymbol{\psi}}^T(\boldsymbol{\theta}\dot{\mathbf{e}}^T + \mathbf{F}^{-1}\dot{\hat{\boldsymbol{\psi}}}) \tag{12}$$

To stabilize Equation (12), the appropriate update law is chosen as

$$\dot{\hat{\boldsymbol{\psi}}} = -\mathbf{F}^{-1}\boldsymbol{\theta}\dot{\mathbf{e}}^T \tag{13}$$

where $\mathbf{F} \in R^{N\beta\times N}$ is an adaptation gain matrix.
Equation (12) becomes

$$\dot{V} = -\dot{\mathbf{e}}^T(\mathbf{B} + \mathbf{K}_d)\dot{\mathbf{e}} + \dot{\mathbf{e}}^T\boldsymbol{\varepsilon} - \sum_i \gamma_i\left|\dot{e}_i\right| \tag{14}$$

Choosing $\gamma_i \geq \dot{e}_i + \delta_i$, where δ_i is a positive constant results in

$$\dot{V} = -\dot{\mathbf{e}}^T(\mathbf{B} + \mathbf{K}_d)\dot{\mathbf{e}} - \sum_i \gamma_i\left|\dot{e}_i\right| \tag{15}$$

Equation (15) is stable according to Lyapunov theory.

4. Results and discussions

To prove the efficacy of the proposed controller, a pinned-pinned beam with a couple of piezo-patches installed on its surface. The dimension and the properties of piezo-materials, beam and fluid are listed in Table 1. The oscillations source that irritates the smart beam motion is considered as a sinusoidal disturbance force with a frequency of 50 Hz with 1 N amplitude. This disturbance is developed due to the hydrodynamic fluid loading that is free of beam motion. Below, we will not investigate the configurations of the beam mode shapes and the frequency response for the coupled beam-fluid system, for more details see [4]. What is important here is the effectiveness of the proposed controller while the coupled beam-fluid system under harmonic oscillations.

Table 1. Properties of beam, piezo-materials, and fluid

Beam	$\rho_b = 8000kg/m^3, l_b = 0.4m, E_b = 190\times10^3 MPa, A_b = 0.04\times0.001m^2, b_1 = 007N.$ $b_2 = 0.03N.s/m$
Piezo-material	$l_p = 0.08m, A_p = 0.035\times0.0004m^2, E_p = 70\times10^3 MPa.$
Fluid (FC-72)	$c_m = 2.1, c_v/\mu = 2280, \mu = 0.4\times10^{-6}m^2s, \rho_f = 1.68\times10^3 kg/m^3.$

The suggested control structure is applied to the target beam system with the following control gains: $\mathbf{K_p} = 300\mathbf{I}_2, \mathbf{K}_d = 100\mathbf{I}_2, \boldsymbol{\psi} = 20I_{22}, \beta = 11, N = 2, \mathbf{F} = I_{22}$.

For proper tuning of control gains, a private strategy suggested in [25] was adopted. The key idea is to set values for the target gains gradually from zero to a value at which the system instability occurs and then the gain value is halved for safety. Besides, two important points should be noted:

1. The modelling error is neglected and hence the robust sliding term is nullified.
2. The Chebyshev polynomials are used as approximators for the adaptive approximation compensator.

The initial conditions for the beam oscillations and the weighting coefficient matrix are imposed zero. Investigating Figures 2 and 3 shows that the proposed controller strongly damps out the oscillations. Two experiments are tested: 1) beam oscillations without a controller, and 2) beam oscillations with the application of the proposed control law. It should be mentioned that the piezo-actuators would have limit value for the output voltage and hence a bounded control is preferred to be implemented that is lost in this work. Besides, we have simulated 2 piezo-actuators/sensors for the first two-mode shapes and this can ease the calculations. In practice, the number of actuators could be not equal to the investigated mode of shapes and hence pseudo-inverse matrix should be used, see [26] for more details.

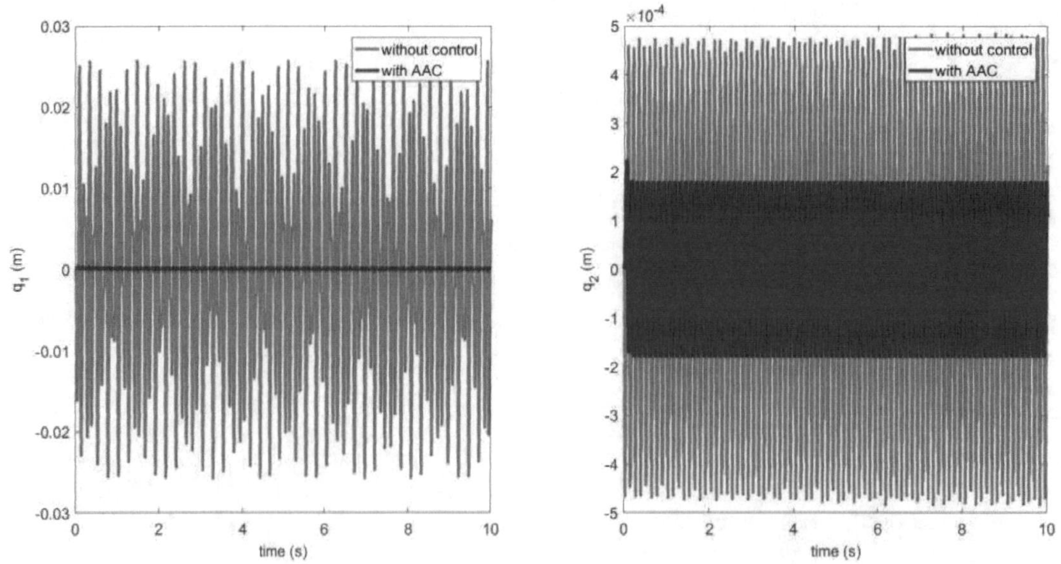

Figure 2. Modal coordinates for the coupled smart beam-fluid system considering the first two modes.

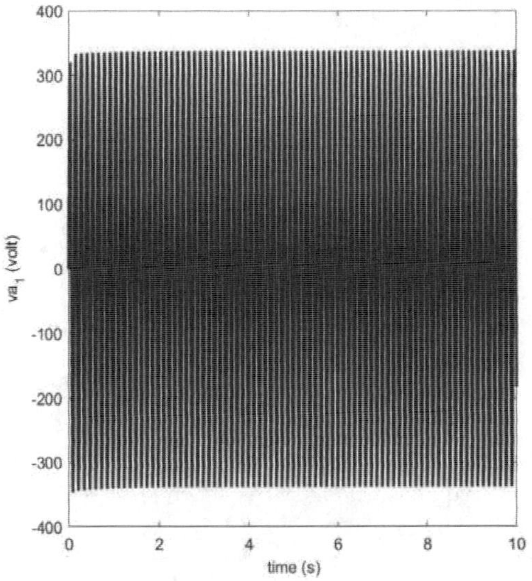

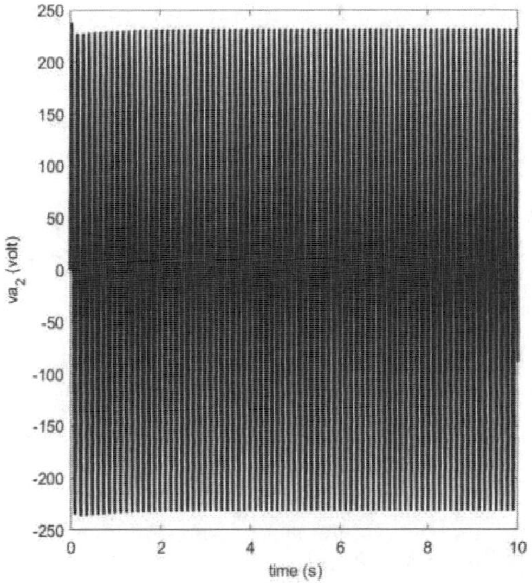

Figure 3. The input control for the piezo-actuators.

5. Conclusions

This work supposes adaptive approximation control with integrated PD for motion regulation of the beam-fluid system. The analysis is confined to the linear behaviour of the beam-fluid system. Despite the results show that the proposed controller is a powerful strategy for oscillation damping of flexible structures, the following points should be noted:

1. The nonlinear model for flexible structures should be considered.
2. The fluid hydrodynamic forces are evaluated based on the concept of added mass and damping, however, more complex theory should be investigated considering different fluid mediums.
3. The current controller is applied to a simple beam. The behaviour of the suggested controller should be tested on plate and shells.

References

[1] Naik T·, Longmire *E K and* Mantell *S C2003* Dynamic response of a cantile*ver in liquid near a solid wall* Sensors and Actuators A: Physical, **103**, *p 240-254*
[2] LinW and Qiao N 2008 Vibration and stability of an axiall*y moving beam immersed in fluid* International Journal of Solids and Structures, **45**,*p* 1445-*1457*
[3] FadaeeM *and* Talebitooti *M 2020* Active vibration control of carbon nanotube-reinforced composite beam submerged in fluid using magnetostrictive layers, Mechanics Based Design of Structures and Machines, *published online*
[4] Wagg D and Neild S 2010 Nonlinear vibration with control *(*SMIA, Springer Verlag*)*
[5] Zhang W, Meng G and Li H 2006 Adaptive vibration control of micro-cantilever beam with piezoelectric actuator in MEMS *Int J Adv Manuf Technol* **28**, p321–327
[6] Faltinsen O M 1990 *Sea loads on ships and offshore structures* [Cambridge University Press)
[7] Xing J T, Price W G, Pomfret M J, and Yam L H 1997 Natural vibration of a beam-water interaction system *Journal of Sound and Vibration* **199**, p 491–512
[8] Sader J E 1998 Frequency response of cantilever beams immersed in viscous fluids with applications to the atomic force microscope *Journal of Applied Physics* **84**, p 64–76
[9] Zhao S, Xing J T and Price W G 2002 Natural vibration of a flexible beam–water coupled system with a concentrated mass attached at the free end of the beam *Proceedings of the Institution of*

Mechanical Engineers, Part M: Journal of Engineering for the Maritime Environment **216**, 145–54

[10] Lin W and Qiao N 2008 Vibration and stability of an axially moving beam immersed in fluid. *International Journal of Solids and Structures* **45**, p 1445–57

[11] Shabani R, Hatami H, Golzar F. G, Tariverdilo S and Rezazadeh G 2013 Coupled vibration of a cantilever micro-beam submerged in a bounded incompressible fluid domain *Acta Mechanica* **224**, p 841–50

[12] Ni Q, Li M, Tang M, and Wang L 2014 Free vibration and stability of a cantilever beam attached to an axially moving base immersed in fluid *Journal of Sound and Vibration* **333**, 2543–55

[13] Eftekhari S and Jafari A A 2014 A mixed modal-differential quadrature method for free and forced vibration of beams in contact with fluid *Meccanica* **49**, p 535–64

[14] Rezaiee-Pajand M, Kazemiyan M S and Aftabi S A 2018 Solving coupled beam-fluid interaction by DTM *Ocean Engineering* **167**, p 380–96

[15] Al-Shuka H F N, Song R 2018 *Hybrid regressor and approximation-based adaptive control of piezoelectric flexible beams*. 2018 2nd IEEE Advanced Information Management, Communicates, Electronic and Automation Control Conference (IMCEC*)*, Xi'an, China, p 330-334

[16] Farrell J A and Polycarpou M M 2006 *Adaptive Approximation Based Control: Unifying Neural, Fuzzy and Traditional Adaptive Approximation Approaches* (John Wiley, USA)

[17] Ioannou P and Fidan B 2006 *Adaptive control tutorial* (SIAM, USA)

[18] Huang A-C, Chien M-C 2010 *Adaptive control of robot manipulators: a unified regressor-free approach* (World Scientific Publishing Co. Pte. Ltd. Singapore)

[19] Al-Shuka H F N, Corves B, Zhu W-H 2013 Function approximation technique-based adaptive virtual decomposition control for a serial-chain manipulator *Robotica* **32**, p 375-399

[20] Al-Shuka H F N 2018 On local approximation-based adaptive control with applications to robotic manipulators and biped robots*Int. J. Dynam. Control* **6**, p 339–353

[21] Al-Shuka H F N, Song R 2018 *Hybrid regressor and approximation-based adaptive control of robotic manipulators with contact-free motion* 2nd IEEE Advanced Information Management, Communicates, Electronic and Automation Control Conference (IMCEC), Xi'an, China, p 325-329

[22] Al-Shuka H F N and Song R 2019 Decentralized adaptive partitioned approximation control of high degrees-of-freedom robotic manipulators considering three actuator control modes. *International Journal of Dynamics and Control***7**, p 744-757

[23] Al-Shuka H F N 2021 FAT-Based Adaptive Backstepping Control of an Electromechanical System with an Unknown Input Coefficient *FME Transactions* **49**, p 113-120

[24] Dowell E H 1975 *Aeroelastic plates and shells* (Noordhoff International Publishing)

[25] Zhu W-H. 2010 *Virtual decomposition control: Toward Hyper Degrees of Freedom Robots* (Springer Verlag Berlin Heidelberg)

[26] Preumont A 2018 *Vibration control of active structures* (4th Edition, Springer International Publishing AG)

Sulphate and chloride resistance of vermiculite blended concrete

N Kiran Kumar[1]*, B Damodhara Reddy[2], and C.Sasidhar[3]

[1]Research Scholar (PhD), Jawaharlal Technological University, Anantapur, AP, India
[2]Professor, SVCET (Autonomous), Chittoor, Andhra Pradesh, India
[3]Professor, Jawaharlal Technological University, Anantapur, AP, India

kirsir.kumar@gmail.com

Abstract: In the construction industry lot of natural sand is being used in the preparation of concrete. Levels of natural sand are depleting day by day. The cost of natural sand is also increasing every day. Therefore, there is a need for alternative materials to natural sand. In this study, exfoliated vermiculite is replacing natural sand partially. Vermiculite at 5%, 10%, and 15% is used for replacement. The concrete cubes are cast for M 30 grade to determine the optimal percentage of replacement of natural sand by vermiculite. In this study, coarse aggregate used includes both normal aggregate and recycled aggregate. The water-cement ratio of 0.45 is adopted. The durability tests conducted include the sulfate resistance Test and the chloride resistance test. The cost analysis was also done and results are presented. The test results are promising when vermiculite was used to replace natural sand at 10%.

Key Words: concrete, natural sand, vermiculite, durability, normal aggregate and recycled aggregate, sulfate and chloride.

1. Introduction

The concrete, a versatile material consists of cement, natural sand, normal coarse aggregate, and water. Presently, natural sand is becoming costlier and also scarce. So to continue constructional activities, there is a need for an alternative material for Natural sand. So, M-sand, granite fines are being used to partially replace natural sand in constructions. Vermiculite is evaluated for replacing natural sand because of its superior properties of thermal insulation and sound insulation. The vermiculite acts as filler material and also make concrete lighter. Vermiculite when exfoliated can resist higher temperatures. When used in concrete, this exfoliated vermiculite makes concrete resist higher temperatures. In the present study, an attempt is made to assess the impact of sulfate and chloride on Vermiculite blended concrete.

2. Related Work

Sulfate resistance of concrete is studied over the years, but not properly understood. Out of two sources external and internal, External source is a natural occurring sulfates coming from fertilizers whereas internal source is a one resulting when introduced in the cement. Noor N.M.et.al [1] studied the effect of crumb rubber on concrete strength and chloride ion penetration resistance and proposed that resistance to chlorine is achieved by increasing the content of rubber in concrete at the water cement ratio of 0.35 compared to that of 0.5. Chiara F. Ferrari's et.al [2] during their study of concrete subjected to sulfate attack insisted that to quickly assess the concrete life, there is a need for simulation models. Sumaiya Binte Huda [3] when studying strength and durability of recycled coarse aggregates emphasized that strength improved after 7 weeks when concrete is subjected to sulfate attack the reason being that continuous hydration happens and voids get it covered by hydrated products. Yingwu Zhou et.al [4] when studying Strength deterioration of concrete in sulfate environment: An experimental study and theoretical modeling, revealed that when the concrete is subjected to dry-wet cycling action of sulfate corrosion that the strength of concrete showed a maximum at 60th day of soaking in sulfate solution.

Jackson Muthengia Wachira et.al [5] while studying Effect of Sulphate and Chloride Ingress on Selected Cements Mortar Prisms Immersed in Seawater and Leather Industry Effluent, revealed that concrete when exposed to sulphate and chloride, there was reduction in compressive strength when compared with conventional sample but compressive strength improved with age for all water cement ratios. Seong-kyum Kim et.al [6] studied about the resistance of reinforced concrete to sulfate attack. This study used flumes containing liquid crystal display waste glass powder. A Sulfate resistance test was carried by immersing specimens in sodium sulfate and magnesium sulfate solutions. They expressed that there is less reduction in weight and volume of the specimens when immersed in sodium sulfate as to that of magnesium sulfate. Peng Liu et.al [7] studied sulfate attack mechanism of cement concrete based on chemical thermodynamics expressed that effect of sulfates on cement concrete was magnificent. Bingol A.F et.al [8] emphasized that sulfate causes certain deteriorations like cracking, expansion and spalling through particular reactions inside concrete. By replacing 10% of cement and 5% of cement of fly ash and silica fume respectively, a better compressive and flexural strength results was achieved when compared to conventional concrete. Aiestein Rozario et.al [9] indicated that use of water soluble polymers in concrete has improved performance of concrete. Otieno, M,et.al [10] revealed that, due to possible incomplete filling of the original device caused the huge ingress of chloride in concrete. Park. J et al. [11] examined that chloride ion diffusion coefficient is less for water cured specimen when compared with air cured specimens. They also felt that with increase in GGBFS replacement ratio, chloride ion diffusion coefficient value decreased. The reason was attributed to higher fineness and hydraulic reactions. Niu, Ditao et.al [12] expressed that in view of attaining sufficient compressive strength and chloride ion resistance of concrete, hybrid fibers in the concrete should not exceed 0.15% by volume.

From the literature Review following objectives are drafted

1. To carry out preliminary tests on cement, sand, vermiculite, Normal coarse aggregate and recycled aggregate.
2. To carry out Slump Cone Test and Compaction Factor Test on Fresh concrete.
3. To carry out weight absorption Test and Split Tensile test on Vermiculite blended concrete with normal aggregate and recycled aggregate separately and compare it with conventional concrete when subjected to sulfate attack.
4. To carry out Carbonation chloride test on Vermiculite blended concrete with normal aggregate and recycled aggregate separately and compare it with conventional concrete when subjected to chloride attack.
5. To carry out cost analysis on various mixes and arrive at optimal mix that can be adopted.

3. Proposed Work

In this study, Vermiculite is used as a filler material. Vermiculite is a hydro-phyllo-silicate mineral which expands on Heating. Vermiculite is light in weight and possesses improved fire resistance. Vermiculite also possesses specific gravity of around 2.1. It exhibits chemical inertness and also resists Cracking and shrinkage on exfoliation. Here coarse aggregate used is plain normal gravel and recycled gravel, natural sand, vermiculite and Cement used is of 53 grade and its properties are shown in Table 1 to table 4.

Table 1 Physical properties of cement

Details	Value	**IS12269-1987 (Reaffirmed 2013)**
Normal consistency	31	-
Initial setting time	65 minutes	Not less than 30 minutes
Final setting time	310 minutes	Not greater than 600 minutes
Soundness	1mm	Does not exceed 10mm
Fineness	6.1%	Does not exceed 10%
Specific gravity	3.145	-

The physical properties of cement like Initial setting time, final setting time, soundness, fineness are confirming to IS 12269. Physical properties like normal consistency and specific gravity are also as per accepted norms.

Table 2 Sand Properties

S.no	**Details**	**Unit**	**Results**
1.	Specific gravity	-	2.61
2.	Bulking of sand	%	3.8
3.	Particle size variation	mm	0.16 to 4.55
4	Water absorption for sand	%	0.9
5	Bulk Density of Sand	kg/m^3	1450
6	Fineness modulus	%	3.1

The properties of sand like specific gravity of sand, Bulking of sand, particle size variation, water absorption, Bulk density of sand and Fineness modulus are as per accepted norms.

Table 3 Characteristics of Vermiculite

S.No	**Properties**	**Values**
1	Color	yellow,
2	Specific Gravity	2.1
3	Fineness modulus	3.2

The properties of Vermiculite, like specific gravity is 2.1 which is less than 2.61 of Natural sand, which, when replacing Natural sand makes concrete lighter in weight. The color of vermiculite considered for this study is yellow. The fineness modulus for vermiculite it was 3.2.

Table 4 Properties of Normal Gravel and Recycled Gravel

S.No	Properties	Values of Normal Gravel	Values of Recycled Gravel
1	Water Absorption	1 %	5.1
2	Specific Gravity	2.72	2.6
3	Crushing Strength	1620 kg/m2	1650 kg/m2
4	Color	Green	Brownish Green
5	Fineness Modulus	2.83	2.96

When comparing the Properties of Normal gravel and Recycled gravel, Specific gravity and crushing strength for recycled gravel is better than normal gravel. There is not much change in Fineness modulus and color for both Normal gravel and recycled gravel. However, water absorption of recycled gravel is more than normal gravel. The reason may be attributed to presence of dried cement paste attached to recycled gravel already. But when used in concrete, recycled gravel provides better interlocking and exhibits lesser water absorption after hardening of concrete.

The slump cone test apparatus was made according to IS: 7320-1974 and used for calculating normal consistency of concrete. Fresh concrete was filled in the slump cone by ramming every layer 25 times. As soon as the slump cone is lifted vertically, the settlement of concrete happens and slump of the concrete is measured by the scale. The compaction factor test setup consists of upper hopper, bottom hopper and the cylinder. First concrete is filled in upper hopper, and then later the trap door is opened. The concrete falls into the bottom hopper. Once the trap door of bottom hopper is opened, concrete deposits into the cylinder. Level the top of cylinder with a trowel and clean outside of cylinder. Weigh the cylinder (w1). Pour the concrete outside and fill it back in to cylinder in different layers. Compact each layer and trim the cylinder top using trowel. Weigh the cylinder (w2). The weight of empty cylinder is taken as w. Compaction Factor is the ratio between (w1-w2) to (w2-w).

For carrying out sulfate resistance test the procedure to be adopted is to choose Sodium sulfate or magnesium solutions using around 2.500 mg to 4.000 mg SO_4^2/l at room temperature. In present study Sodium sulfate solution is used. The impact of the sulfate solution is studied by measuring the tensile strength. Split Tensile Test procedure as per IS specifications IS 5816:1999 Split tensile test is to be conducted. The cylinder size is 150 mm diameter and 300 mm length. Cylinders are loaded horizontally in compression testing machine. Application of load is done till cylinder fails. The maximum load on the cylinder was to be observed. Split Tensile strength is given by the ratio of 2N and (π/d) where the d=diameter of cylinder in mm, l= length of cylinder in mm and the N= maximum load applied to the cylinder. The concrete cubes and cylinders are cast using concrete mix design, whose details are provided in table 5 to table 8.

Table 5 Concrete Mix design (As per IS: 456-2000 and IS: 10262-2009)

S.No	Criteria	Description
1	Grade of Concrete	M30
2	Cement	OPC 53
3	Maximum size of aggregate	20 mm
4	Exposure	Moderate

The grade of cement chosen for the study is an OPC 53. The maximum size of aggregate is 20 mm. The exposure conditions of concrete are moderate in preparation of M30 grade of concrete. The mix design for concrete is done as per IS 456 and IS 10262. The Target means strength required is 38.25 N/mm^2.

Table 6 Test data for Materials

S.No	**Test Data for materials**	
1	cement used	Zuari 53-Grade
2	specific Gravity of water	1
3	Chemical Admixture	plasticizer
4	Free(surface) Moisture of coarse aggregate	Nil
5	Free(surface) Moisture of fine aggregate	Nil

The brand of cement used is Zuari cement. The specific gravity of water chosen for the study is 1 and plasticizer is used as a chemical admixture to improve workability. The coarse and fine aggregate is free from moisture.

Table 7 Mix ratio of M30 Grade Concrete

Type of Coarse aggregate	**Water**	**Cement**	**Fine aggregate**	**Coarse Aggregate**
Normal Gravel (CC, M2, M4, M6)	197.16 (W/C Ratio 0.45)	438.13	619.44	1143.59
		1	1.41	2.61
Recycled gravel (M1, M3, M5, M7)	197.16 (W/C Ratio 0.45)	438.13	628.5	1151.37
		1	1.44	2.63

Here in this study, two sets of samples are prepared. The samples CC, M2, M4 and M6 are prepared using Normal gravel with a mix ratio of 1: 1.41:2.61. The samples M1, M3, M5 and M7 are prepared using Normal gravel with a mix ratio of 1: 1.44:2.63. The water cement ratio for both the set of samples is 0.45

Table 8 Specimen specifications

S.No	Sample No	Specifications
1	CC	Conventional concrete with 0% vermiculite and Normal gravel as coarse aggregate
2	M1	Concrete With 0% vermiculite and recycled gravel as coarse aggregate
3	M2	Concrete With 05% vermiculite and Normal Gravel as coarse aggregate
4	M3	Concrete With 05% vermiculite and recycled Gravel as coarse aggregate
5	M4	Concrete With 10% vermiculite and Normal Gravel as coarse aggregate
6	M5	Concrete With 10% vermiculite and recycled Gravel as coarse aggregate
7	M6	Concrete With 15% vermiculite and Normal Gravel as coarse aggregate
8	M7	Concrete With 15% vermiculite and recycled Gravel as coarse aggregate

Here Conventional concrete (CC) is prepared with normal gravel and without vermiculite. The M1 sample was prepared with recycled gravel and without vermiculite. Here for both CC and M1, natural sand was not replaced. For samples M2 and M3, natural sand was replaced by vermiculite by 5%. For samples M4 and M5, natural sand was replaced by vermiculite by 10%. For samples M6 and M7, natural sand was replaced by vermiculite by 15%.

4. Results and Analysis

The results on Fresh concrete and hardened concrete are presented here. Tests included on fresh concrete include workability tests in the form of slump test and compaction factor test. The tests on hardened concrete include split tensile strength test, and durability tests like water absorption test, Sulfate resistance test and carbonation Chloride test. The results are depicted in Table 9-13. The same is represented in Figure 1-4. The cost analysis is done and is shown in Table 14.

4.1 Tests on Fresh Concrete

Table 9 Results on fresh concrete

S.No	Test	Result (cm)	
		Normal Gravel	Recycled gravel
1	Slump Test	12	10
2	Compaction Factor Test	0.89	0.7

The slump test values showed a lesser slump for Recycled gravel when compared to normal gravel. Similar values were also noticed in the compaction factor test.

4.2 Split Tensile strength Test:

Table 10 Spilt Tensile Strength of Vermiculite concrete (0.45) for M30 Grade of Concrete

S.No	Replacement (Sample Number)	Spilt Tensile Strength (N/mm2) Before sulfate resistance test			Spilt Tensile Strength (N/mm2) After sufhate resistance test		
		7 Days (N/mm2)	14 Days (N/mm2)	28 Days (N/mm2)	7 Days (N/mm2)	14 Days (N/mm2)	28 Days (N/mm2)
1	CC	2.43	2.95	3.08	2.05	2.63	2.51
2	M1	1.85	2.12	2.52	1.41	1.83	2.17
3	M2	2.32	2.86	2.90	1.98	2.35	2.28
4	M3	2.46	2.91	3.21	2.08	2.45	2.93
5	M4	2.53	3.05	3.85	2.31	2.82	3.69
6	M5	2.71	3.21	3.97	2.45	2.91	3.84
7	M6	2.39	2.87	3.61	1.89	2.39	3.24
8	M7	2.45	2.98	3.72	2.05	2.56	3.47

Split Tensile strength of M5 Mix has the highest value after exposed to sulfate environment when compared with other mixes. The reason may be attributed to the usage of recycled gravel as coarse aggregate, which is effective in resisting sulfate ingress by possessing good interlocking between fine and coarse aggregate. The exfoliated vermiculite acted as filler material filling the voids thus making concrete specimen free from air voids. So 10% of vermiculite replacing natural sand are effective when compared with other mixes.

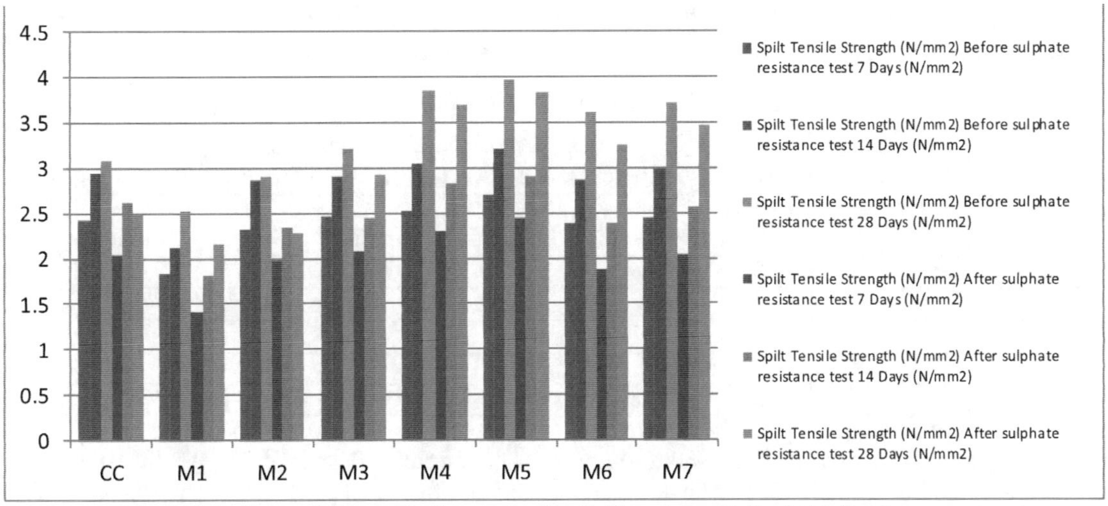

Fig 1: Split Strength Test (Before and after Sulfate Attack) for M30 Grade concrete

The figure 1 clearly depicts that sample M5 could effectively resist ingress of sulfate into concrete when compared with other mixes.

Table 11 Spilt Tensile Strength of Vermiculite concrete (0.45) for M30 Grade with Normal Gravel

S.No	Replacement (Sample Number)	Spilt Tensile Strength (N/mm2) Before sulfate resistance test			Spilt Tensile Strength (N/mm2) After sulfate resistance test		
		7 Days (N/mm2)	14 Days (N/mm2)	28 Days (N/mm2)	7 Days (N/mm2)	14 Days (N/mm2)	28 Days (N/mm2)
1	CC	2.43	2.95	3.08	2.05	2.63	2.51
3	M2	2.32	2.86	2.90	1.98	2.35	2.28
5	M4	2.53	3.05	3.85	2.31	2.82	3.69
7	M6	2.39	2.87	3.61	1.89	2.39	3.24

When plain normal gravel is used in concrete, after exposing to sulfate solution, M4 mix is showing maximum split tensile strength when compared with other mixes after 28 days. All the specimens exhibited a decrease in split tensile strength after soaking in sulfate solution. However, when vermiculite at 10% (M4) replaced Natural sand, better split strength values are obtained. The reason may be attributed to vermiculite which acted as filler material in removing air voids and improving split tensile strength.

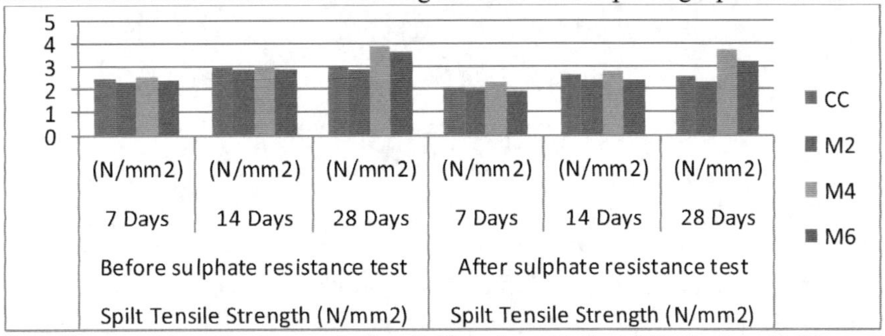

Fig 2: Split Strength Test (Before and after Sulfate Attack) for M30 Grade with Normal Gravel

The figure 2 clearly depicts that sample M4 could effectively resist ingress of sulfate into concrete when compared with other mixes using normal gravel as a coarse aggregate.

Table 12 Spilt Tensile Strength of Vermiculite concrete (0.45) for M30 Grade with recycled gravel

S.No	Replacement (Sample Number)	Spilt Tensile Strength (N/mm2) Before sulfate resistance test			Spilt Tensile Strength (N/mm2) After sulfate resistance test		
		7 Days (N/mm2)	14 Days (N/mm2)	28 Days (N/mm2)	7 Days (N/mm2)	14 Days (N/mm2)	28 Days (N/mm2)
1	CC	2.43	2.95	3.08	2.05	2.63	2.51
2	M1	1.85	2.12	2.52	1.41	1.83	2.17
4	M3	2.46	2.91	3.21	2.08	2.45	2.93
6	M5	2.71	3.21	3.97	2.45	2.91	3.84
8	M7	2.45	2.98	3.72	2.05	2.56	3.47

When recycled gravel is used in concrete, after exposing to sulfate solution, M5 mix is showing maximum split tensile strength when compared with other mixes after 28 days. All the specimens exhibited a decrease in split tensile strength after soaking in sulfate solution. However, when vermiculite at 10% (M5 Mix) replaced Natural sand, better split strength values is obtained. The reason may be attributed to vermiculite which acted as filler material in removing air voids. This in turn improved the split strength values better than control mix. The early strength increase could be attributed to continuous hydration and filling up of pores by compounds formed as a result of the reactions between sulfate and cement hydration products.

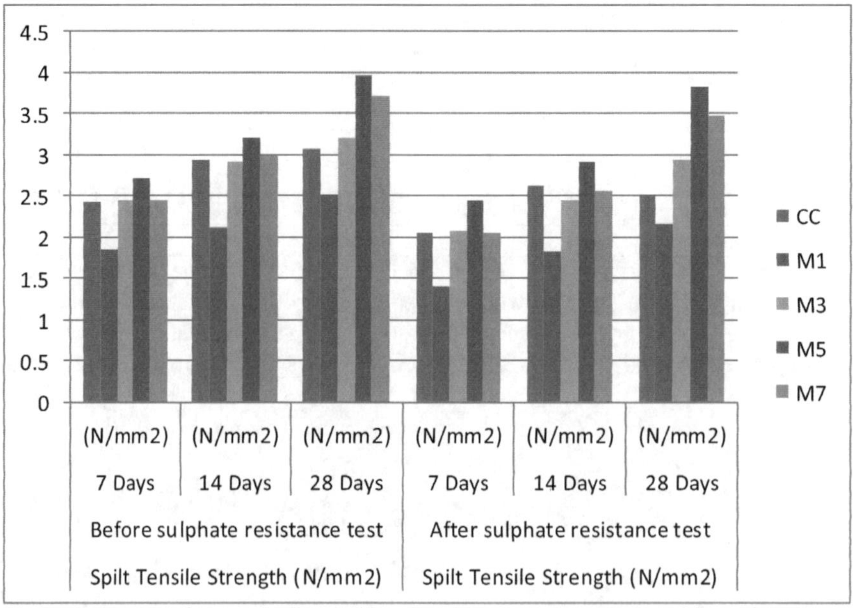

Fig 3: Split Strength Test (Before and after Sulfate Attack) for M30 Grade with recycled aggregate

The figure 2 clearly depicts that sample M5 could effectively resist ingress of sulfate into concrete when compared with other mixes using recycled gravel as a coarse aggregate.

IOP Conf. Series: Materials Science and Engineering 1091 (2021) 012009 doi:10.1088/1757-899X/1091/1/012009

Table 13 Chlorination Depth for different mixes of concrete specimens

S.NO	DESIGN MIX	CHLORINATION DEPTH (mm)
1	CC	**7.10**
2	M1	**7.02**
3	M2	**9.4**
4	M3	**9.15**
5	M4	**8.25**
6	M5	**7.15**
7	M6	**8.35**
8	M7	**7.95**

For M5 Mix Chlorination depth values are almost same as that of conventional sample. The reason may be attributed that usage of recycled aggregate has created good interlocking between fine and coarse aggregate. This in turn decreased the voids. Those available voids are also filled with exfoliated vermiculite which was used as filler material. As voids are filled effectively, the concrete specimen could resist penetration of chlorine when compared with other mixes, which is clearly shown in Table 13 above and fig 4 below.

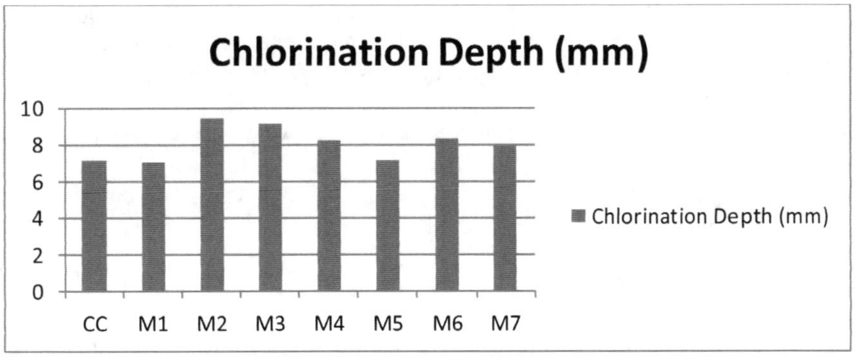

Fig 4: Chlorination Depth (mm)

Figure 4 clearly shows that chlorination depth is more for mix M2 and least for M1. However mix M5 values are almost near to Mix M1 and can be said to effectively resist the chlorine entry into concrete.

4.4 Cost Analysis

Table 14 Cost Analysis of concrete per cubic meter of concrete

Mix	Sand Content	Natural Sand cost	Vermiculite Cost	Recycled Gravel Cost	Normal Gravel Cost	Cement Cost	Total Final Cost (RS)
CC	100%	89	0	0	813	3111	4013
M1	100%	90	0	600	0	3111	3801
M2	95%	85	806	0	813	3111	4815
M3	95%	86	832	600	0	3111	4629
M4	90%	80	1612	0	813	3111	5616
M5	90%	81	1664	600	0	3111	5456
M6	85%	76	2418	0	813	3111	6418
M7	85%	77	2496	600	0	3111	6284

Here
Cost of Sand is Rs.0.143 /Kg
Cost of Crushed Recycled Gravel is Rs.0.52/Kg
Cost of Vermiculite is Rs.26/Kg
Cost of Crushed Normal Gravel is Rs.0.7107/Kg

Here when the cost of M1 mix is compared with that of conventional concrete, it is 6% less, because of lesser cost or recycled gravel when compared with Normal Gravel. When M4 and M5 are compared with conventional concrete the cost is around 40% and 38% more, but superior split tensile strength when compared to conventional concrete and also because of its superior properties like Thermal insulation and better acoustic properties. M4 or M5 mix may be adopted in keeping durability point of View.

5. Conclusion

In this study, because of scarcity of sand, vermiculite in exfoliated form is used to partially replace natural sand at 5%, 10% and 15%. The water cement ratio of 0.45 was adopted. The coarse aggregate used in this study is plain normal aggregate and recycled aggregate. The test conducted on concrete specimens cast includes, Split tensile strength test, Sulfate resistance test, carbonation chloride test and cost analysis.
On comparing the test results, the following conclusions are arrived at:

1. The slump test and compaction factor test indicate that the workability of concrete with recycled gravel is slightly weaker when compared with concrete made with plain normal gravel. However, the difference is very less, so if durability properties are satisfactory, then recycled gravel may be adopted in the place of plain normal gravel.
2. The Split tensile strength values for concrete specimens with recycled gravel are slightly better when compared with concrete specimens with plain normal gravel when tested after 28 days. It is clear from the results of the study that M5 mix was slightly better when compared with M4 mix. Both M4 and M5 mixes are made with vermiculite replacing natural sand at 10%. The test results of the M4 and M5 mixes are better than all other mixes.
3. When the concrete was exposed to sulfate attack and tested for Split tensile strength, the results show that concrete specimens with recycled gravel are slightly better when compared with concrete specimens with plain normal gravel when tested after 28 days. It is clear from the results of the study that M5 mix was slightly better when compared with M4 mix. Both M4 and M5 mixes are made with vermiculite replacing natural sand at 10%. The test results of the M4 and M5 mixes are better than all other mixes because of superior interlocking between aggregates and reduction of air voids, because of usage of exfoliated vermiculite.
4. When the carbonation chlorination test was conducted and depth of chlorine penetration was measured, the M5 mix gave lesser penetration depth when compared with all other mixes. The reason is that when recycled gravel was used in the place of normal plain gravel, because of superior interlocking between aggregates and reduction of air voids, as exfoliated vermiculite fills the air voids, thus not allowing penetration of chlorine in the concrete.
5. Even though the cost of concrete is increasing with the increase in vermiculite content, the special properties like Thermal insulation and acoustic properties. M4 or M5 mix may be adopted in keeping durability point of View.

Considering above results we can conclude that recycled gravel as a coarse aggregate can be used to replace plain normal gravel, when vermiculite is used to partially replace natural sand at 10%.

References

[1] Noor, N. M., Hamada, H., Sagawa, Y., & Yamamoto, D. (2015). Effect of crumb rubber on concrete strength and chloride ion penetration resistance *Jurnal Teknologi*, *77*(32), 171-178. https://doi.org/10.11113/jt.v77.7004

[2] Chiara F. Ferraris, Paul E Stutzman and Kenneth A.Snyder: Sulfate resistance of Concrete: A New Approach, *PCA*, (2006)

[3] Sumaiya Binte Huda: Mechanical and durability properties of Recycled and repeated Recycled coarse aggregate concrete, *The University of British Columbia,* pp 1-122, (2014)

[4] Yingwu Zhou, Hao Tian, **Lili Sui**, Feng Xing, Ningxu Han, "Strength Deterioration of Concrete in Sulfate Environment: An Experimental Study and Theoretical Modeling",*Advances in Materials Science and Engineering*, Vol. 2015, Article ID 951209, 13 pages, 2015, https://doi.org/10.1155/2015/951209.

[5] Jackson Muthengia Wachira, reginah Wangui Ngari, Joseph Karanja Thiong'o, Joseph Mwiti Marangu, "Effect of Sulfate and Chloride Ingress on Selected Cements Mortar Prisms Immersed in Seawater and Leather Industry Effluent", *Advances in Civil Engineering*, vol 2019, Article ID 8191689, 16 pages, 2019. https://doi.org/10.1155/2019/819168

[6] Kim, S.-K.; Hong, W.-K. High Sulfate Attack Resistance of Reinforced Concrete Flumes Containing Liquid Crystal Display (LCD) Waste Glass Powder. *Materials* **2019**, *12*, 2031.

[7] Peng Liu, Ying Chen, Zhiwu Yu, Lingkun Chen, Yongfeng Zheng, "Research on Sulfate Attack Mechanism of Cement Concrete Based on Chemical Thermodynamics", *Advances in Materials Science and Engineering*, vol. 2020, Article ID 6916039, 16 pages, 2020.https:// doi.org/10.1155/2020/6916039.

[8] Bingöl, A.F., Balaneji, H.H. determination of Sulfate Resistance of Concretes Containing Silica Fume and Fly Ash *Iran J Sci Technol Trans Civ Eng* **43,** 219–230 (2019).https://doi.org/ 10.1007/s40996-018-0160-x

[9] A. Aielstein Rozario, Dr.C.Freeda Christy, M.Hannah Angelin, 2013, Experimental Studies on Effects of Sulphate Resistance on Self-Curing Concrete, *International Journal of engineering & Technology(IJERT)* Volume 02, Issue 04 (April 2013)

[10] Otieno, M, & Alexander, M. (2015) chloride conductivity testing of concrete-past and recent developments. *Journal of the South African Institution of Civil Engineering*, *57*(4), 55-64. https://dx.doi.org/10.17159/2309-8775/2015/v57n4a7

[11] Park J, Park C, Joh S, Lee H. Effect of Curing Condition on Resistance to Chloride Ingress in Concrete Using Ground Granulated Blast Furnace Slag. *Materials (Basel)*. 2019;12(19):3233. Published 2019 Oct 2. doi:10.3390/ma12193233

[12] Niu, Ditao, Huang, Daguan, Fu, Qiang Experimental investigation on compressive strength and chloride permeability of fiber-reinforced concrete with basalt-polypropolene fibers *Advances in Structural Engineering.* Jul2019, Vol. 22 Issue 10, p2278-2288. 11p

Physical Device for Measuring Tension in a Wire

[1]Dr. Yusuf R, [2]Dr. Shashikiran N D, [3]Dr. Chanamallappa Ganiger, [4]Dr. Sadashiv Shetty
1Professor, Department of Orthodontics and Dentofacial Orthopaedics, School of Dental Sciences, Krishna Institute of Medical Sciences "Deemed To Be University", Karad
2Dean, School of Dental Sciences, Krishna Institute of Medical Sciences "Deemed To Be University", Karad
3Professor and Head, Department of Orthodontics and Dentofacial Orthopaedics, School of Dental Sciences, Krishna Institute of Medical Sciences "Deemed To Be University", Karad
4Director, Bapuji Dental College, Davangiri, Karanataka
Corresponding author's e-mail address: ryusuf54@outlook.com

Abstract. This paper provide a physical device for measuring tension in a wire of an orthodontic brace. The wire is having a closed circumference and capable of being wind in a winding position around brackets of two teeth. The device includes an anchoring member, a holding portion and a display. The anchoring member is having a first end attachable with the wire in the winding position and a second end attachable with the holding portion of the device. The device upon pulling the wire by anchoring the anchoring member along with its length, measures the tension exerted by the wire on the teeth and displays the measured data on the display of the device.

1. Introduction

Orthodontics in dentistry is associated with improvement of the general appearance of a patient's teeth and deals with the diagnosis, prevention and correction of misaligned or mal-positioned teeth and jaws. Orthodontics braces are the device which is used for correction of misaligned or mal-positioned teeth and jaws [1]. Orthodontics braces are provided with wires and brackets. The brackets are fixed with each of the teeth and wires are winded around the teeth and are fixed with the brackets [2]. The medical practitioner (orthodontist) adjusts the required tension on the wires depending on the complexity of misaligned teeth. Further, periodic adjustment of the wires is required for the proper alignment of the misaligned or mal-positioned teeth and jaws. The wires may be replaced with a rubber band or rubber chain or any other elastic material [3-5].
If the tension required on the wire arranged between the mal-aligned teeth is known to the practitioner at each stage of the treatment, the alignment period and the diagnosis term can be substantially reduced. Improper adjustment of the wire in varied tension may unnecessarily extend the term of diagnosis and thereby delaying the alignment period [4,6]. Wires or elastic materials are available with predefined tension which can be applied across the brackets [7]. However, these wires cannot be customized according to the need of the patient. Hence there is a requirement of a dental device which

can measure the tension of the wire / elastic wound around a bracket of two teeth in an orthodontic brace treatment which may overcome few or all drawbacks of the existing dental devices.

1.1. Research Objectives

The present study aims to:

a) To provide a device for measuring tension in a wire configured around a bracket of two teeth in an orthodontic brace treatment.

b) To provide a device for measuring tension in a wire of an orthodontic brace, which substantially shortens the time duration of treatment.

c) To provide a device for measuring tension in a wire of an orthodontic brace, which reduces the alignment period of the treatment.

2. Literature Review

Ravera et al. [8] designed a pendulum force measuring device, which simulated the oral environment and measured the effect of orthodontic force on the crown and root. In order to prevent undermining resorption light forces should be used during orthodontic treatment. The optimum force used in orthodontic treatment should be enough to produce tooth movement without tissue dam-age and with maximum comfort for the patient. Excessive forces can lead to severe pain, damage of the periodontal ligament and root resorption [9]. The improper teeth positioning is the most common clinical features [10]. It usually causes the uneven tooth alignment and dental arch deformity. And it not only affects the function of mastication, but also has a certain impact on the pronunciation. The tooth decay, dentin hypersensitivity, and more mouth diseases will be accompanied by it. Kumar et al. [11] planed an integrated bracket, which integrates the pressure pickup into the bracket and can directly measure the pressure between the archwire and the bracket, but it cannot truly reflect the actual relationship between the archwire and the bracket in the clinical state. Upadhyay et al. [12] have established a measuring device with a three-dimensional orthodontic force, which replaces the actual teeth with a simplified cylinder. It can preliminarily measure the orthodontic force of all the teeth on the dental arch and distinguishes the difference between passive ligation and other ligation methods. Jhiang et al. [13] realized the calculation of bending stress by using the hyper-elasticity of the archwire in the process of studying the bending properties of hollow nickel-titanium alloy wires.

3. Materials and Method

In the orthodontic braces treatment, wires, elastic members or the like are used to adjust the tension between the teeth. In the orthodontic braces, brackets are attached with each tooth, and the wires are configured around the brackets of two teeth. The wire is having a closed circumference and is capable of being wind in a winding position around the brackets of two teeth. The wires can be attached with the adjacent brackets. The wires are tightened according to the required tension. Once the tension is set across the brackets, the device can be used to verify the tension across the wire.

The device is a force measuring instrument having an anchoring member, a holding portion and a display. The anchoring member is adapted to attach with the wire. Specifically, a first end of the anchoring member which is the distal end of the device is attachable with the wire in the winding position. A second end of the anchoring member is attached to the holding portion of the device.

In an embodiment, the anchoring member can be detachable from the holding portion and can be replaced with anchoring members of different length and sizes. Upon pulling the device after anchoring the anchoring member with the wire along the length of the anchoring member and away from the wire facilitates the device to measure the tension across the wire. The device measures the tension across the wire and displays the measured data on the attached display. To calculate the forces on the lateral incisor we used the stresses which act on the bracket of the lateral incisors.

A device used for an orthodontic brace treatment in accordance with the orthodontics applications is presented in the paper. The device is specifically for measuring tension in a wire of an orthodontic

brace. In the orthodontic brace treatment, wires, elastic members or the like are used to adjust the tension between the teeth. If the tension is adjusted between the teeth according to the need, the teeth are likely to be aligned in a proper orientation which may help the patient an easy recovery.
In the orthodontic braces, brackets are attached with each tooth, and the wires are configured around the brackets of two teeth. The wire is having a closed circumference and are capable of being wind in a winding position around the brackets of two teeth. The wire can be attached with the adjacent brackets. The wire can be replaced with elastic members such as rubber bands, and the like. The wires are tightened according to the required tension. Once the tension is set across the brackets, the device can be used to verify the tension across the wire.
The device is a force measuring instrument having an anchoring member, a holding portion and a display. The anchoring member is adapted to attach with the wire. The anchoring member is an elongated member having a hookable portion configured on a first end of the anchoring member for anchoring the device with the wire. Specifically, the first end of the anchoring member which is the distal end of the device is attachable with the wire in the winding position. The tension across the wires are at the maximum in the winding position.
A second end of the anchoring member is attached to the holding portion of the device. In the present embodiment, the anchoring member is fixed with the holding portion of the device. In an embodiment, the second end of the anchoring member is connected pivotally and detachably with the holding portion to provide enough degree of freedom for the anchoring member. The anchoring member may be detachable from the holding portion and can be replaced with anchoring members of different length and sizes. The holding portion is for holding and providing sufficient gripping to the device.
Upon pulling the device after anchoring the anchoring member with the wire along the length of the anchoring member and away from the wire facilitates the device to measure the tension across the wire. Specifically, the device measures the tension across the wire and displays the measured data on the display attached therewith. The display can be either analogue or digital. The holding portion and the display are integral to the device.
Advantages
The advantages and features of the present research will be understood better with reference to the following detailed description and claims taken in conjunction with the accompanying drawings, wherein like elements are identified with like symbols, and in which:
1. The present research has an advantage of providing a device for measuring tension in a wire configured around a bracket of two teeth in an orthodontic brace treatment.
2. The device substantially shortens the time duration of treatment. It also provides accurate and precise historical data of the wire tension allowing the medical practitioner to ease the examination procedure.

4. Conclusion
In this paper a device used for an orthodontic (brace) treatment is provided. The device is specifically for measuring tension in a wire of an orthodontic braces. Moreover, this device provides measuring of tension in a wire of an orthodontic brace, which provides accurate and precise historical data of the wire tension allowing the medical practitioner to ease the examination procedure. This device is able to measure extra oral force accurately and precisely so that we can modify growing facial structures.

References
1. Alobeid A, El-Bialy T, Khawatmi S, Dirk C, Jäger A, Bourauel C. Comparison of the force levels among labial and lingual self-ligating and conventional brackets in simulated misaligned teeth. European journal of orthodontics. 2017;39(4):419-25.
2. Naziris K, Piro NE, Jäger R, Schmidt F, Elkholy F, Lapatki BG. Experimental friction and deflection forces of orthodontic leveling archwires in three-bracket model experiments. Journal of Orofacial Orthopedics/Fortschritte der Kieferorthopädie. 2019;80(5):223-35.

3. Jiang JG, Chen YH, Wang L, Zhang YD, Liu Y, Qian W. Modeling and Experimentation of the Unidirectional Orthodontic Force of Second Sequential Loop Orthodontic Archwire. Applied Bionics and Biomechanics. 2020;2020.
4. Jiang J, Huang Z, Ma X, Zhang Y, He T, Liu Y. Establishment and Experiment of Utility Archwire Dynamic Orthodontic Moment Prediction Model. IEEE Transactions on Biomedical Engineering. 2019.
5. Florian G, Gabor AR, Nicolae CA, Iacobescu G, Stănică N, Mărăşescu P, Petrişor I, Leulescu M, Degeratu S, Gîngu O, Rotaru P. Physical properties (thermal, thermomechanical, magnetic, and adhesive) of some smart orthodontic wires. Journal of Thermal Analysis and Calorimetry. 2018;134(1):189-208.
6. Alobeid A, El-Bialy T, Reimann S, Keilig L, Cornelius D, Jäger A, Bourauel C. Comparison of the efficacy of tooth alignment among lingual and labial brackets: an in vitro study. European Journal of Orthodontics. 2018;40(6):660-5.
7. Lai W, Midorikawa Y, Kanno Z, Takemura H, Suga K, Soga K, Ono T, Uo M. Development and modification of a device for three-dimensional measurement of orthodontic force system: The V-bend system re-visited. Dental materials journal. 2016 Nov 28;35(6):908-17.
8. Ravera S, Castroflorio T, Garino F, Daher S, Cugliari G, Deregibus A. Maxillary molar distalization with aligners in adult patients: a multicenter retrospective study. Progress in Orthodontics. 2016;17(1):12.
9. El-Bialy T, Alobeid A, Dirk C, Jäger A, Keilig L, Bourauel C. Comparison of force loss due to friction of different wire sizes and materials in conventional and new self-ligating orthodontic brackets during simulated canine retraction. Journal of Orofacial Orthopedics/Fortschritte der Kieferorthopädie. 2019;80(2):68-78.
10. Jiang J, Huang Z, Qian W, Zhang Y, Liu Y. Registration technology of augmented reality in oral medicine: A review. IEEE Access. 2019;7:53566-84.
11. Kumar K, Gurubaran N, Shoba T. Smart Thinker Brackets-Next Generation. Indian Journal of Public Health Research & Development. 2019;10(12):2452-4.
12. Upadhyay M, Shah R, Peterson D, Asaki T, Yadav S, Agarwal S. Force system generated by elastic archwires with vertical V bends: a three-dimensional analysis. European journal of orthodontics. 2017;39(2):202-8.
13. Jiang JG, Ma XF, Zhang YD, Han YS, Liu Y. Prediction model and examination of open vertical loop orthodontic force. Arabian Journal for Science and Engineering. 2019;44(2):1489-99.

Construction of a Physical Device for Abduction Movement Enhancement

[1] **Dr. Devayani Moghe,** [2] **Dr. Kashinath Sahoo,** [3] **Dr. G. Varadharajulu,** [4] **Dr. Suraj Kanase**
[1]Private Practitioner, Saish Physiotherapy Stroke Prevention and Rehabilitation Clinic, Satara.
[2]Associate Professor, Krishna College of Physiotherapy, Krishna Institute of Medical Sciences "Deemed To Be University", Karad.
[3]Dean, Krishna College of Physiotherapy, Krishna Institute of Medical Sciences "Deemed To Be University", Karad.
[4]Associate Professor, Krishna College of Physiotherapy, Krishna Institute of Medical Sciences "Deemed To Be University", Karad.
corresponding author's e-mail address: kashipo@gmail.com

Abstract. The present research has made an attempt to theorize a physical device using the scientific method which will provide support to shoulder joint of a user. This physical orthotic device includes a rigid support, at least one pair of electrodes, at least one strap and the at least one-second strap. The rigid support is having an outer surface and an inner surface and is resting against the shoulder joint. The at least one strap is arranged on the rigid support to wrap around an upper arm of the shoulder joint for securing the orthotic device on the shoulder joint of the user. The at least one-second strap is extending from the rigid support and adapted to wrap around an armpit of a second shoulder joint for efficiently treating a subluxation condition.

1. Introduction

Generally, orthosis is used to support body parts of patients under the circumstances such as injuries, dislocations, subluxation and the like. The orthosis is general term used for any orthotic device. These orthosis are also used to treat shoulder joint mis-alignments caused due to strokes (paralysis), shoulder subluxation and the like [1]. A stroke is an acute onset of neurological dysfunction caused due to the abnormality in a cerebral blood circulation with resultant sign and symptom that correspond to the involvement of focal areas of the brain. It can give the symptoms like paralysis (hemiplegia) or weakness (hemiparesis). Shoulder subluxation is a common problem in the stroke. The subluxation causes shoulder pain and hinders activity [1,2].

Presently existing orthosis or any such devices are not effective in reducing effects caused due to forces developed during movements of an arm of the user (abduction movement) and also in efficiently treating the subluxation condition during abduction movements [3].

A normal stirring action of the force couple of supra-spinatus and posterior fibers of the deltoid is affected due to a flaccid stage of the muscles. So, while abduction and flexion movement due to gravitational pull to the head of the humerus subluxates caudally [4,5]. Presently, Orthotic devices (shoulder orthosis) are used to support the shoulder joint to decrease the glenohumeral subluxation [6].

Further, when a patient (user) suffering from the Shoulder subluxation condition is wearing the existing orthosis at the shoulder joint, an upper arm of the patient can be moved away from the torso up to a maximum of 30 degrees from a vertical reference [7]. Whenever the patient moves the upper arm away from the torso (In medical terms, this movement is called an "abduction movement") there will be a set of forces acting on the shoulder joint and the upper arm.

Angle of abduction movement can be referred as a movement angle θm. These set of forces are caused due to movements along a direction away from the torso. These forces cause an enormous amount of pain to the patient even when the patient is wearing the orthosis [8]. Furthermore, these movements also effect the subluxation condition of the patient, thereby reduces efficiency of the orthosis in treating the subluxation condition.

Presently existing orthosis or any such devices are not effective in reducing effects caused due to forces developed during movements of an arm of the user (abduction movement) and also in efficiently treating the subluxation condition during abduction movements.

1.1 Objectives

i. To provide an orthotic device for supporting a shoulder joint of a user, which is simple in construction and easy to use.

ii. To provide an orthotic device for supporting a shoulder joint of a user, which nullifies forces caused on the user due to movements of an arm of the user while wearing the orthotic device.

iii. To provide an orthotic device for supporting a shoulder joint of a user for reducing the pain of the user caused due to movements of an arm of the user while wearing the orthotic device.

2. Literature Review

According to Pan et al. [9] the modified wheelchair arm-support is an ideal example of subluxation supportive device, designed to reduce hemiplegic pain through assisting in the maintenance of normal posture. Sharma et al. [10] suggested that pain associated with subluxation likely presents later after stroke as "fibrous changes or injury can occur in connective tissue of the ligaments and joint capsule due to incorrect alignment between the humerus and the scapula". As well, the lack of consistency among findings may be related to the heterogeneity of patient characteristics and method/timing of assessment. In Arm orthosis arm sling is used. Arm sling is used for scapular orhumeral fractures, rotator cuff injury, bicipital tendinitis, and hemiparesis with subluxation. It also includes cuff sling and glenohumeral support. Patient with proximal arm weakness involving the shoulder and arm, the functional arm orthosis is used. As tone returns to the shoulder muscle, the risk of shoulder subluxation decreases, and slings can then be withdrawn [11]. Slings tend to hold the limb in a poor position, which may accentuate the abduction and internal rotation posture and may contribute to shortening of tonically active muscles [12]. While some observational studies have reported a significant correlation between subluxation and pain in the hemiplegic shoulder [13,14], several others failed to find such a relationship.

3. Materials and Method

For this 20 studies (9 controlled tests and 11 pre and post-individual group studies) were finally selected for the review. The rehabilitation intervention: Functional electrical stimulation (FES) / electrical stimulation, orthosis / support, snapping / strapping, etc. were reviewed.

3.1. Orthotic Device Details

The orthotic device includes a rigid support, at least one pair of electrodes, one strap and at least one-second strap. The rigid support includes a shoulder section and an arm section. More specifically, the shoulder section covers a shoulder section of the user and the arm section covers an upper arm section of the user. The rigid support is having an outer surface and an inner surface b. The rigid support is resting against the shoulder joint.

In a preferred embodiment, the rigid support is configured according to a shape of a shoulder the user and the rigid support is arranged on the shoulder joint of the user with a surface contact of the inner surface of the orthotic device with the skin of the user. The rigid support is arranged on the shoulder joint of the user with a surface contact of the inner surface of the orthotic device with the skin of the user. The rigid support includes a cushioning layer arranged on the inner surface of the rigid support for providing comfort to the user while wearing the orthotic device. The cushioning layer can be made from materials, such as medicinal rubber or ethaflex and the like. The rigid support is made from materials, such as polypropylene and the like.

At least one pair of electrodes arranged on the inner surface of the rigid support for pain relief modality. The at least one pair of electrodes are connected to a power source and a control unit for supplying and controlling current flow thereto. The strap is wrapped around an upper arm of the shoulder for securing the orthotic device on the shoulder of the user. The strap is arranged with a pad. The pad is configured to provide support and pressure to the upper arm when the at least one strap is wrapped around the upper arm of the user.

The orthotic device includes two straps a & b. The straps a and b are having a securing arrangements, such as a snap lock, hook and loop arrangement or any such obvious securing engagements which are capable to secure the orthotic device on the upper arm of the user. The straps a and b are having a pad. The pad is arranged with the straps a and b. The pad is configured to provide support and pressure to the upper arm when the straps a and b are wrapped around the upper arm of the user. More specifically, the pad is arranged in such a way that, when the straps a & b are wrapped around the upper arm, an interior surface of the pad is in contact with the skin of the upper arm as shown in figure. In an alternative embodiment, the orthotic device can be configured without the pad.

The at least one-second strap is extending from the rigid support and adapted to wrap around an armpit of the second shoulder joint. In the present, the orthotic device includes a second strap. In an embodiment, the second strap is extending from the outer surface a of the rigid support. More specifically, the one-second strap is extending from a corner *c* of the shoulder profile of the rigid support. The corner here refers to a geometric area where the shoulder section and the hand section of the rigid support meets. The second strap includes securing arrangements such as a hook and loop arrangements, for securing the -second strap around the torso of the user at the armpit of the second shoulder joint.

3.2. Working Mechanism

When the strap is at a wrapped position, it does notallow the user to move his/ her arm beyond degrees (a movement angle θ_m) from a vertical reference. This restriction of movement helps in retaining a correcting position of the rigid support on the shoulder joint. Hence, the orthotic device corrects an affected shoulder efficiently. Hence, the second strap also enables the orthotic device in treating a subluxation condition of the user efficiently even if the user moves his/her arm while wearing the orthosis.

When the orthosis is worn by the user around the shoulder joint, by securing the strap a around the upper arm and securing the second strap around the torso around the armpit of the second shoulder joint, the orthotic device applies pressure on the shoulder joint according to a point pressure system principle. In medical industry, this point pressure system is known as a Jordon's principle. In the point pressure system, the applied force and two counteracting forces are in the opposite direction to each other.

Further, supplying therapeutic current through the pair of electrodes a & b results in pain relief of the user. This supplying therapeutic current is generally known as TENS (Trans-cutaneous Electrical Nerve Stimulation). More specifically, this TENS gives pain relief to the user in the shoulder subluxation condition. Therefore, the orthotic device is beneficial in reducing the shoulder subluxation condition along with the pain relief effect. Furthermore, when the user is wearing the orthotic device, if the user moves the arm away from the torso, the second strap distributes the force exerted on a shoulder and the upper arm of the user along the second strap towards the armpit of the second shoulder joint, thereby nullifying the forces occurred during movements of an arm of the user. When the forces occurred during movements of arm a of the user are nullified, the resultant forces acting on the user will be equal to zero. Therefore, the forces caused due to movement of the arm a of the user do not result in causing pain to the user or any such discomforts.

4. Results

Measurements have shown increased shoulder and arm performance when the subjects were wearing the prototype. Upward and forward movements were easier to perform. The arm support is easy to put on and remove. Moreover, the device felt comfortable for the subjects. Shoulder support or orthosis while *in situ* may reduce the subluxation condition. X-ray was the most commonly used assessment tool for the subluxation. Implication of the rehabilitation technique on motor recovery has not been investigated.

5. Conclusion

The present research has the advantage of providing the orthotic device for supporting the shoulder joint of a user. It is observed that the presented orthotic device nullifies forces caused on the user due to movements of an arm of the user. Morevoer, the orthotic device is simple in construction as well as the orthotic device is easy in use. Furthermore, the orthotic device is economical in construction and operations. The presented orthotic device reduces the pain of the user caused due to movements of an arm of the user while wearing the orthotic device. This orthotic device efficiently treats the subluxation condition of the patients even if any movements of an arm of a user is occurred while wearing the orthotic device.

References

[1] Moghe DM, Kanase SB. Effect of Therapeutic Shoulder Sling And Proximal Control Exercises on Shoulder Subluxation in Stroke Survivors. Indian Journal of Forensic Medicine & Toxicology. 2020;14(3):222-7.

[2] Nadler M, Pauls MM. Shoulder orthoses for the prevention and reduction of hemiplegic shoulder pain and subluxation: systematic review. Clinical rehabilitation. 2017;31(4):444-53.

[3] Ates S, Haarman CJ, Stienen AH. SCRIPT passive orthosis: design of interactive hand and wrist exoskeleton for rehabilitation at home after stroke. Autonomous Robots. 2017;41(3):711-23.

[4] Aydın T, Şen Eİ, Yardımcı MY, Kesiktaş FN, Öneş K, Paker N. Efficacy of ultrasound-guided suprascapular nerve block treatment in patients with painful hemiplegic shoulder. Neurological Sciences. 2019;40(5):985-91.

[5] Janssen MMHP, Bergsma A, Geurts ACH, de Groot IJM. Patterns of decline in upper limb function of boys and men with DMD: an international survey. J Neurol. 2014;261(7):1269–88.

[6] Pillastrini P, Rocchi G, Deserri D, Foschi P, Mardegan M, Naldi MT, Villafañe JH, Bertozzi L. Effectiveness of neuromuscular taping on painful hemiplegic shoulder: a randomised clinical trial. Disability and rehabilitation. 2016;38(16):1603-9.

[7] Lee JH, Baker LL, Johnson RE, Tilson JK. Effectiveness of neuromuscular electrical stimulation for management of shoulder subluxation post-stroke: a systematic review with meta-analysis. Clinical rehabilitation. 2017;31(11):1431-44.

[8] Ravichandran H, Janakiraman B, Sundaram S, Fisseha B, Gebreyesus T, Gelaw AY. Systematic Review on Effectiveness of shoulder taping in Hemiplegia. Journal of Stroke and Cerebrovascular Diseases. 2019;28(6):1463-73.

[9] Pan R, Zhou M, Cai H, Guo Y, Zhan L, Li M, Yang Z, Zhu L, Zhan J, Chen H. A randomized controlled trial of a modified wheelchair arm-support to reduce shoulder pain in stroke patients. Clinical rehabilitation. 2018;32(1):37-47.

[10] Sharma GS, Singh YN, Joy A, Singh B, Touthang AT, Devi T. Study of Effectiveness of

Shoulder Elbow Wrist Hand Orthosis in the Management of Glenohumeral Subluxation in Post-stroke Hemiplegic Patients.
[11] Sohn MK, Jee SJ, Hwang P, Jeon Y, Lee H. The effects of shoulder slings on balance in patients with hemiplegic stroke. Annals of rehabilitation medicine. 2015;39(6):986.
[12] Moghe DM, Kanase SB. Effect of Therapeutic Shoulder Sling And Proximal Control Exercises on Shoulder Subluxation in Stroke Survivors. Indian Journal of Forensic Medicine & Toxicology. 2020;14(3):222-7.
[13] Jeong YG, Jeong YJ, Kim HS, Lee KH. Predictors of the effect of an arm sling on gait efficiency in stroke patients with shoulder subluxation: a pre-post design clinical trial. Physiotherapy Theory and Practice. 2020:1-8.
[14] Arya KN, Pandian S, Vikas, Puri V. Rehabilitation methods for reducing shoulder subluxation in post-stroke hemiparesis: a systematic review. Topics in stroke rehabilitation. 2018;25(1):68-81.

Device customization with novel adhesive electrode

[1]Dr. Sameer Karpe, [2]Dr. Kashinath Sahoo, [3] Dr. G. Varadharajulu, [4] Dr. Suraj Kanase
[1] PG student, Krishna College of Physiotherapy, Krishna Institute of Medical Sciences "Deemed To Be University", Karad.
[2]Professor, Krishna College of Physiotherapy, Krishna Institute of Medical Sciences "Deemed To Be University", Karad.
[3]Dean, Krishna College of Physiotherapy, Krishna Institute of Medical Sciences "Deemed To Be University", Karad.
[4]Professor, Krishna College of Physiotherapy, Krishna Institute of Medical Sciences "Deemed To Be University", Karad.
corresponding author's e-mail address: kashipo@gmail.com

Abstract. The present research deal with the physical examination and analysis of an instrument that is ankle foot orthotic device. The physics of this device consist of calf piece; calf strap; a muscle stimulator; stimulator suspension including a press button with nylon strap; two adhesive electrodes; electrical wires; hinge joint; JBR outsole; foot piece; ankle strap; forefoot strap; rings; adjustable strap; press button; cold and hot pack pouch; a means to provide upward projection; & shank. The customized AFO device is characteristic in the sense that the adhesive strap being mounted on the rings so as to keep the plantar section of the foot piece in straight position and wherein the adhesive strap being made up by a combination of polyvinyl chloride, polypropylene and polyethylene wherein polyvinyl chloride, polypropylene and polyethylene is 1:1:2 by weight; wherein the cold and hot pack pouch being made up of 40.5 wt% water; 40.5 wt% ammonium nitrate, 4 wt% hydropropylmethyl cellulose and 15 wt% propylene glycol; and wherein the said electrode being made up of a hydrogel comprises of acrylic acid and N-vinylpyrrolidone. The overall result feedback is a feeling of increased muscle control, flexibility, and range of motion on using the device.

1. Introduction
Foot drop is common problem in stroke, multiple sclerosis, cerebral palsy patients, and in common peroneal nerve injury patient. Electrical stimulation and ankle foot orthosis (AFO) have been routinely used in individuals with foot drop to re-educate muscles which are weak and to keep ankle in neutral position. It is known that electrical stimulation is useful in treating individuals with foot drop. Sharif et al., [1] showed that the functional electrical stimulation (FES) is better in foot drop than conventional electrical stimulation (EMS) in stroke patients.

An ankle-foot orthosis, or AFO, is an orthotic device which is a support intended to control the position and motion of the ankle, compensate for weakness, or correct deformities of foot and ankle. AFOs can be used to support weak limbs, or to position a limb with contracted muscles into a more normal position [2,3]. In addition, AFOs are used to control foot drop caused by a variety of neurologic and musculo-skeletal disorders [4]. Due to the common use for addressing foot drop, AFO has become synonymous with the term "foot-drop brace" [5]. AFO are easy to wear, and can be easily available at orthotics.

1.1. Research Objectives

The present study aims to provide a customized ankle foot orthotic device with an effective hot and cold pouch unit for improving foot drop of a patient selected from strokes, multiple sclerosis, cerebral palsy patients and common peroneal nerve injury.

i. To provide a customized ankle foot orthotic device with novel adhesive electrode for the treatment of foot drop.

ii. To provide a customized ankle foot orthotic device with novel adjustable strap for the treatment of foot drop.

iii. To provide a novel customized foot orthotic device for improving gait and rehabilitation.

iv. To provide a device that could reduce the pain as compared to conventional AFO while treating foot drop.

2. Literature Review

Prenton et al., [6] discloses that the people with multiple sclerosis (MS) have difficulty walking. According to the authors, gait impairment, including the reduced ankle dorsiflexion of foot drop, is one of the most common indicators of disability early in the course of this progressive auto-immune disease of the central nervous system, affecting approximately 75% of people with MS. Assistive technology, such as ankle–foot orthosis (AFO) and functional electrical stimulation (FES), increases the safety of walking and the speed of ambulation (even then, only about one half of patients remain ambulatory 15 years after disease onset). Assistive technology also reduces the risk of injury to the knee and ankle and reduces the effort of ambulation. Walbran et al., [7] discloses that "Cerebral Palsy (CP) is a non-progressive neurological disorder which develops in-utero or after birth. Current treatment for CP includes physical therapy and braces used to increase ambulation. Ankle-Foot Orthoses (AFOs) are lightweight plastic braces that secure the lower leg, ankle, and foot in a predetermined position, commonly used to aide dorsiflexion in CP patients. In another study [8] the common treatment, Functional Electrode Stimulation (FES), is administered by physical therapists in order to build muscle tone and improve dorsiflexion. According to the study, FES uses low energy electrical stimulation to excite either the common peroneal nerve or the tibialis anterior muscle, causing the patient to actively dorsiflex, increasing foot-ground clearance. Carolus et al., [9] discloses neuro-muscular disorders and injuries such as cerebral palsy and stroke often result in foot-drop which can result in a person having great difficulty walking. Ankle foot orthoses (AFOs) or splints have been prescribed for many years now to limit the range of motion of the ankle, provide the patients with support and assist with rehabilitation. However the majority of AFOs require a long, labour-intensive manufacturing process which results in unacceptable waiting times for children that are rapidly growing and patients with varying conditions. Deckers et al., [10] discloses AFO designs vary in size, shape, and functional characteristics depending on the desired clinical application. Passive Dynamic (PD) Response ankle–foot orthoses (PD-AFOs) constitute a design that seeks to improve walking ability for persons with various neuromuscular disorders by passively (like a spring) providing variable levels of support during the stance phase of gait. This study compared the mechanical damping of the CF-AFO to PD-AFOs manufactured by SLS using three different materials. The study showed that a SLS-based framework is ideally suited for this application.

3. Materials and Method

3.1. Materials

The present paper provides a customised ankle foot orthotic device for improving foot drop of a patient selected from strokes, multiple sclerosis, cerebral palsy patients and common peroneal nerve injury. The device of the present invention consist of:

1. Calf piece; 2. Calf strap; 3. Portable Muscle stimulator; 4. Stimulator suspension consisting of press button with nylon strap; 5. Adhesive electrode; 6. Electrical wires; 7. 2D hinge joint consisting of 4mm MS nut and bolt; 8. JBR outsole; 9. Foot piece; 10.Ankle strap; 11. Forefoot strap; 12. Ring; 13.Adjustable strap; 14.Press button; 15.Cold and Hot pack pouch; 16.Upward projection; & 17.Shank;

A customized AFO device that is applied to the ankle for modifying functional characteristics of neuro-muscular conditions. Characterized in the context that the adhesive strap being mounted on the rings so as to keep the plantar section of the foot piece in straight position and wherein the adhesive strap being made up by a combination of polyvinyl chloride, polypropylene and polyethylene wherein polyvinyl chloride, polypropylene and polyethylene is 1:1:2 by weight; wherein the cold and hot pack pouch being made up of 40.5%wt. water; 40.5% wt. ammonium nitrate, 4%wt. hydropropylmethyl cellulose and 15%wt. propylene glycol; andwherein the said electrode being made up of a hydrogel comprises of acrylic acid and N-vinylpyrrolidone.

3.2. Details of the Device

The presented device mainly consists of two sections: a Calf piece and a Foot piece which are articulately joined on each side of the ankle by two hinge joint of MS nuts and Bolts. The calf piece comprises a calf strap with an upper portion which may be wrapped around the patient's calf and secured by a velcro strap. The strap is attached to one side of the greave while the other end is free and is designed to loop around the calf. Below the strap, there is a stimulator suspension which consists of press buttons in which portable muscle stimulator is mounted. The greave extends downward from the calf area to forward narrow shank below which the greave broadens at ankle area to match the contour of the ankle. The plantar section has a JBR (Johnson bros rubber) outsole and an upward project on which intimately wraps around heel & ankle areas of the patient. The hinge joints are mounted loosely so that a plantar section can rotate upward around the axis delineated by the two hinge. This movement of plantar section provides for dorsiflexion of foot during the swing phase of gait cycle. The downward movement of plantar section is stopped when the upper edge of the projection comes in contact with the on the inner side of greave thus, preventing the foot drop. The two sections and of the present device are made from thin-sheeted polypropylene material which are designed so as to counter the shape of the objects leg and foot. Ankle strap is looped around the ring ankle so as to fasten the strap tightly around the ankle. Forefoot strap which is looped around the forefoot so as to fasten the strap tightly around the forefoot. Adjustable straps is mounted on respective side by a ring for keeping plantar section in a stretched position.

Electrical muscle stimulator consist of two adhesive electrodes attached by electrical wire. Pouch which is the inner aspect of calf piece of AFO includes cold and hot pack. All the straps herein are embedded by press buttonsAll straps are embedded by press buttons.The self-adhesive hydrogel electrodes according to present invention is prepared by the method as given by Malesevic et al., [11]. The impulses are generated by the device and are delivered through electrodes on the skin near to the muscles being stimulated. The electrodes are generally pads that adhere to the skin. The impulses mimic the action potential that comes from the central nervous system, causing the muscles to contract.

The self adjusted strap herein is used for stretching purposes in which specific muscle or tendon (or muscle group) is deliberately flexed or stretched in order to improve the muscle's felt elasticity and achieve comfortable muscle tone. The result is a feeling of increased muscle control, flexibility, and range of motion.

3.*3 Experimental Study*

The experimental trial as to evaluate the efficacy of the present device for foot drop,was conducted in Krishna Institute of Physiotherapy, near Dhebewadi Road, Malkapur, Karad, Pin code-415110, Maharashtra, India. 10 patients whose average weight of 40-70 either gender, was selected for the following each groups as given:

Goup-I: strokes;

Goup-II: multiple sclerosis;
Goup-III: cerebral palsy patients; &
Goup-III: common peroneal nerve
EMS modes

There were two EMS modes of option, S (synchronous) or A (alternate) .Select a mode by pressing the mode control when a EMS mode is selected, the LCD shows EMS on the top. After a mode is selected, press SET control to enter next setting. The patient may adjust the setting only when it is flashing and then press the increment or decrement control to change the settings.

Ramp Time

The ramp time controls the time of output current that increase from 0 to the setting level, and from the setting value to 0. When the ramp time was set, each contraction was ramped up and down in order that the signals come on and come off gradually and smoothly. The ramp time was adjustable from 1 to 8 seconds. The on-off range is adjustable from 2 seconds to 90 seconds.

10-meter walk test:

The present device was evaluated by 10-meter walk test and the results were noted in metres/second. The individual was walked without assistance for 10 metres, with the time measured for the intermediate 6 metres to allow for acceleration and deceleration. Assistive devices may be used, but must be kept consistent and documented for each test. Count the start time when the toes pass the 2 metre mark and stoping time when the toes pass the 8 metre mark. It can be tested at either preferred walking speed or maximum walking speed (ensure to document which was tested). This test was performed for each group of disease three times and calculated te average of the same.

4. Results

The test results (Table 1-4), showed the superior effect of customized ankle foot orthotic device as compared to conventional AFO in view of both 10-meters walk test and the visual analogue scale for different diseases conditions.

Table 1: Group I (Stroke patients)

SN	Patients age	10 meter walk test SPEED (metres/sec)		Visual analogue scale	
		Customized AFO device (Proposed)	Conventional AFO#	Conventional AFO#	Customized AFO device (Proposed)
1	40	0.97	0.71	8	2
2	45	0.99	0.68	7	2
3	55	0.97	0.73	9	1
4	60	0.99	0.65	6	2
5	58	0.98	0.73	6	1
6	59	0.99	0.69	7	4
7	53	0.94	0.72	8	4
8	54	0.98	0.76	9	1
9	55	0.98	0.70	8	3
10	45	0.92	0.68	9	4

conventional AFO was prepared with the same components as in present device but without the adhesive electrode , the hot & cold pack unit and adjustable strap. Silicone electrode was used in the conventional AFO.

Table 2: Group II (multiple sclerosis)

SN	Patients age	10 meter walk test SPEED (metres/sec)		Visual analogue scale	
		Customized AFO device (Proposed)	Conventional AFO#	Conventional AFO#	Customized AFO device (Proposed)
1	55	0.98	0.69	9	1
2	45	0.96	0.70	6	2
3	56	0.89	0.73	7	4
4	95	0.99	0.75	8	1
5	58	0.96	0.68	9	3
6	65	0.98	0.78	6	2
7	40	0.97	0.77	5	1
8	42	0.94	0.71	8	3
9	49	0.96	0.77	7	4
10	54	0.94	0.74	8	2

Table 3: Group III (cerebral palsy patients)

SN	Patients age	10 meter walk test SPEED (metres/sec)		Visual analogue scale	
		Customized AFO device (Proposed)	Conventional AFO#	Conventional AFO#	Customized AFO device (Proposed)
1	14	0.85	0.65	5	2
2	15	0.89	0.68	7	3
3	25	0.90	0.70	6	1
4	21	0.87	0.68	9	2
5	11	0.90	0.69	4	1
6	25	0.91	0.65	7	1
7	18	0.92	0.66	5	3
8	14	0.85	0.68	6	1

9	19	0.86	0.70	3	2
10	12	0.87	0.75	2	1

conventional AFO was prepared with the same components as in present device but Without the adhesive electrode, the hot & cold pack unit and adjustable strap. Silicone electrode was used in the conventional AFO.

Table 4: Group IV (common peroneal nerve)

SN	**Patients age**	**10 meter walk test SPEED (metres/sec)**		**Visual analogue scale**	
		Customized AFO device (Proposed)	Conventional AFO#	Conventional AFO#	Customized AFO device (Proposed)
1	25	1.02	0.85	7	3
2	18	1.05	0.89	8	2
3	42	1.50	0.90	9	2
4	35	0.99	0.78	6	3
5	28	1.23	0.91	8	1
6	45	1.7	0.86	5	2
7	54	1.56	0.87	7	3
8	52	1.22	0.83	8	2
9	70	1.23	0.92	7	1
10	22	1.11	0.89	9	2

conventional AFO was prepared with the same components as in present device but without the adhesive electrode, the hot & cold pack unit and adjustable strap. Silicone electrode was used in the conventional AFO.

5. Conclusion

The orthotic device of the present invention improves the gait and rehabilitation ,in previously used orthosis there were not active dorsiflexion which is very important for gait training and rehabilitative purposes we just have to wear and do gait training ,thus our device is doing dorsiflexion of ankle with the help of functional electrical stimulation through adhesive pads which is fitted on tibialis anterior muscle, that also gives positive feedback. The proposed customized ankle foot orthotic device increases the speed of a foot drop patient in treadmill as compared to conventional AFO.

References

1. Sharif F, Ghulam S, Malik AN, Saeed Q. Effectiveness of functional electrical stimulation (FES) versus conventional electrical stimulation in gait rehabilitation of patients with stroke. J Coll. Physicians Surg. Pak. 2017;27(11):703-6.
2. Prenton S, Hollands KL, Kenney LP, Onmanee P. Functional electrical stimulation and ankle foot orthoses provide equivalent therapeutic effects on foot drop: a meta-analysis providing direction for future research. Journal of rehabilitation medicine. 2018;50(2):129-39.

3. Kim SJ, Na Y, Lee DY, Chang H, Kim J. Pneumatic AFO Powered by a Miniature Custom Compressor for Drop Foot Correction. IEEE Transactions on Neural Systems and Rehabilitation Engineering. 2020;28(8):1781-9.
4. Zamanian H. Application of Superplastic Materials in Ankle Foot Orthosis. JSM Foot & Ankle. 2018;3(1):1045.
5. Bulley C, Mercer TH, Hooper JE, Cowan P, Scott S, van der Linden ML. Experiences of functional electrical stimulation (FES) and ankle foot orthoses (AFOs) for foot-drop in people with multiple sclerosis. Disability and Rehabilitation: Assistive Technology. 2015;10(6):458-67.
6. Prenton S, Hollands KL, Kenney LP. Functional electrical stimulation versus ankle foot orthoses for foot-drop: a meta-analysis of orthotic effects. Journal of rehabilitation medicine. 2016 Sep 5;48(8):646-56.
7. Walbran M, Turner K, McDaid AJ. Customized 3D printed ankle-foot orthosis with adaptable carbon fibre composite spring joint. Cogent Engineering. 2016;3(1):1227022.
8. Bethoux F, Rogers HL, Nolan KJ, Abrams GM, Annaswamy TM, Brandstater M, Browne B, Burnfield JM, Feng W, Freed MJ, Geis C. The effects of peroneal nerve functional electrical stimulation versus ankle-foot orthosis in patients with chronic stroke: a randomized controlled trial. Neuro-rehabilitation and neural repair. 2014;28(7):688-97.
9. Carolus AE, Becker M, Cuny J, Smektala R, Schmieder K, Brenke C. The interdisciplinary management of foot drop. Deutsches Ärzteblatt International. 2019;116(20):347.
10. Deckers JP, Vermandel M, Geldhof J, Vasiliauskaite E, Forward M, Plasschaert F. Development and clinical evaluation of laser-sintered ankle foot orthoses. Plastics, Rubber and Composites. 2018;47(1):42-6.
11. Malesevic J, Dujovic SD, Savic AM, Konstantinovic L, Vidaković A, Bijelic G, Malesevic N, Keller T. A decision support system for electrode shaping in multi-pad FES foot drop correction. Journal of neuroengineering and rehabilitation. 2017;14(1):66.

Application of rubber pad electrodes and vacuum suction cups neuromuscular electrical stimulation

[1] **Dr. Prasannajeet P. Nikam,** [2] **Dr. G. Varadharajulu,** [3]**Dr. Kashinath Sahoo**
[1]Assistant Professor, Krishna College of Physiotherapy, Krishna Institute of Medical Sciences "Deemed To Be University", Karad.
[2]Dean, Krishna College of Physiotherapy, Krishna Institute of Medical Sciences "Deemed To Be University", Karad.
[3]Professor, Krishna College of Physiotherapy, Krishna Institute of Medical Sciences "Deemed To Be University", Karad.
corresponding author's e-mail address: deanphysiotherapy@kimskarad.in

Abstract. The present paper describes a wearable orthopaedic treatment unit. The wearable product requires a service member that is tailored to wear across the user's body. The support person shall have a bracelet and a lumbar belt to ensure fitting over the user's body. In addition, the support member is fitted with rubber pad electrodes and vacuum suction cups Neuromuscular electrical stimulation is provided by rubber pad electrodes which promote passive contraction and muscle relaxation. The system is particularly suited for the treatment of lower back pain. It also enables the absorption of waste materials such as lactic acid and thereby helps to enhance the functional properties of muscles such as flexibility, contraction, etc.

1. Introduction

Back pain affects persons despite all ages for a variety of reasons. The vast majority of back pain encountered by the general public is found in the lower part of the back, commonly referred to as the lumbar region. Lower back pain affects the mobility, flexibility and strength of the spine, resulting in pain, irritation and stiffness during everyday life tasks such as walking, driving or sleeping [1,2].

In most cases, some form of a support garment is a common expedient for reducing the back pain. One of the well-known garments for this purpose is the wrap-around corset of a non-flexible material. Physical therapy treatments are also widely used as a useful measure of pain relief [3,4]. Typical stretching and other conventional physical therapy treatments may not alleviate pain caused by radiculopathy and acute lower back pain. Till now; as a part of conventional physiotherapy, the first choice of treatment modality used for treating low back pain was "Transcutaneous Electrical Nerve Stimulation" (TENS) which is also called as "Neuro-muscular Electrical Stimulation" (NMES).

If LBP (low back pain) was accompanied with radiculopathy, then the modality used is "Interferential Therapy" (IFT). Both TENS/NMES and IFT work on the principle of "pain control" or "pain gate theory". Further, to correct posture, ergonomic advice is given to the patients. Advances like biofeedback are rarely used as it is not available on a global basis. For LBP cases secondary to PIVD (prolapsed intervertebral disc), spondylolisthesis, spondylitis, etc., exercise protocol like Mckenzie's Extension Regime is advised [5]. While administering modalities like TENS/NMES or IFT, it is not always taken into consideration that the patient is in anatomically and bio-mechanically correct posture. The management strategies are available for LBP focus only on the pain-relieving component [6].

Therefore there is a need for a concrete physiotherapy treatment option for low back pain based on medically explained mechanism of action which can partially or completely overcome few or all the drawbacks of the existing devices.

1.1 *Research Objectives*

i. To provide a wearable device for orthopaedics treatment, especially for treating lower back pain.

ii. To provide a wearable device which will not only focus on the pain relief component but also target posture-correction as well as soft-tissue off-loading simultaneously.

iii. To provide a wearable device, which provides soft-tissue decompression by lifting and off-loading the muscles of the back.

iv. To provide a wearable interface that enables the elimination of waste materials such as lactic acid and thereby helps to enhance the functional properties of muscles such as endurance, contractibility, etc.

2. Literature Review

Recent developments in health-related assessment technology might open up new opportunities beyond the hospital context for the implementation of therapeutic services, offering continuous control and guidance on critical interventions for both healthcare providers including patients[7]. While poor balancing is the major source of dropping, most studies have investigated state-of-the-art equipment to track equilibrium efficiency as well as provide immediate results in an attempts to enhance equilibrium. Pressure and movement detection mechanisms installed on the floor [8].

A number of research studies focused on evaluating the potential technological adoption of medical instruments to track neuro-chirurgical mobility and the mobility of orthopaedic patients. post - operative surgery, therapeutic effectiveness assessment as well as patient movement could be supported by weighted medical equipment which detect movement of the patients after spinal surgeries. The medical system evaluation by doctors is a dynamic procedure related to multiple requirements[9].

Argent et al. [10] introduced post-surgery with a sensor-based monitoring device to monitor motion range (ROM) normally consisting of an gyroscope, barometer, accelerometer, as well as a pressure gauge. Only the acceleration sensor and navigation systems involved were installed on the thigh for the study. After exercising and moving, knee ROM was measured utilizing data from the sensors. The feasibility analysis looked at the association among knee ROM as well as the BMI of patients, using (or not) of anaesthetic regulation and kind of hemostatic intervention in use at different periods both before and up to 6 weeks from the operation[11].

Jeldi et al., [12] assessed the transfer cycle UT (upright time) as well as (STS) sit-to-stand following thigh operation. The patients were examined for post-operative treatment in the clinic via a sensing accelerometer linked to proximal parts of the un-operated thigh. During the first 24 hours, data stream processing revealed a large variance in the STS performance. Findings show that a woman stays 20hrs considerably longer than their male counterparts, moreover, between the 1st as well as the last 24 hours the female patients had undergone low STS and UT [12].

However, no negative impacts from use of any of the equipment in the research conducted have been documented.

3. Materials and Method

Wearable device for orthopedic treatment is provided according to the present research. The wearable device has to be wore by the user. The wearable interface requires a support member that is tailored to wear over the user's torso. The help is fitted with an elastic bracelet and an adjustable lumbar belt to ensure fitting over the user's body. The support member is a jacket like wear arrangement around the user's body. The support member varies in size, shape and aesthetics according to the requirement of the user. In an embodiment, the device is configured in such a way that an inner circumferential area of the support member contacts with the back region of the user for providing body treatment. With rubber pad electrodes and vacuum suction cups, the support member is configured.

Specifically, the vacuum suction cups are operably mounted in between the rubber pad electrodes. The vacuum suction cups are removably configured from the support member such that the user can wear the support member for correcting the posture. The device provides soft-tissue decompression by lifting and off-loading the muscles of the body through the vacuum suction cups. In the presented orthopedic device, the vacuum suction cups are operated manually.

The wearable interface incorporates a support member that is adapted to be placed across the user's torso. To ensure fitting around the user's body, the support member has an adjustable arm-strap and an adjustable lumbar belt. The arm strap and the lumbar belt enables the user to properly fit the wearable device around the body despite the size and shape of the user. Two such arm straps are provided for holding the support member, each arm straps being wear around both the shoulders of the user. The arm strap has a locking member to be fastened with a corresponding locking element configured on the support member to hold the support member around the body, particularly around the shoulder of the user. Similarly, the lumbar belt is fastened around the waist or lumbar region of the body to hold the support member . It may be obvious to a person skilled in the art to fasten the arm strap and the lumbar belt using hook and loop arrangement and the like.

The support member is a jacket like an arrangement to wear around the body of the user. The support member varies in size, shape and aesthetics according to the requirement of the user. This helps the device to be easy to fit for a variety of individuals with diverse anatomical and physical variations. In the present embodiment, the device is configured in such a way that an inner circumferential area of the support member contacts with the back region of the user for providing body treatment. The support member ensures that the user maintains good posture throughout the treatment. The device also provides soft-tissue decompression by lifting and off-loading the muscles of the body through the vacuum suction cups. In the present embodiment, the vacuum suction cups are operated manually.

Furthermore, a neuro-muscular electrical stimulation (NMES) machine is arranged on the support member for providing neuro-muscular electrical stimulation through the rubber pad electrodes. To promote passive contraction and relaxation of the muscles, neuro-muscular electrical stimulation is delivered via the rubber pad electrodes. The rubber pad electrodes are controlled by a battery-operated controlling unit arranged inside a pocket over the lumbar belt.

The rubber pad electrodes are operated through a cable of electrodes connected between the neuro-muscular electrical stimulation machine and the rubber pad electrodes. The device provides neuro-muscular electrical stimulation along with soft-tissue decompression therapy in anatomically and bio-mechanically to correct the posture of the user. This would help to minimise pain and promote the elimination of waste materials such as lactic acid, while enhancing the muscle's physiological properties, such as flexibility, contractibility, etc.

4. Conclusion

The present research has the advantage of providing a wearable device for orthopedic treatment, especially for treating lower back pain. The device will not only focus on the pain relief component but also target posture-correction as well as soft-tissue off-loading simultaneously. Further, the device provides soft-tissue decompression by lifting and off-loading the muscles of the back. This device facilitates the removal of waste-products such as lactic-acid and thus helps improve the physiological properties of muscles such as flexibility, contractility, etc. Also, the device is simple and economical in construction.

References

1. Hogaboam L, Daim T. Technology adoption potential of medical devices: The case of wearable sensor products for pervasive care in neurosurgery and orthopedics. Health Policy and Technology. 2018 Dec 1;7(4):409-19.

2. Kim H, Park H, Lee W, Kim J, Park YL. Design of wearable orthopedic devices for treating forward head postures using pneumatic artificial muscles and flex sensors. In 2017 14th International Conference on Ubiquitous Robots and Ambient Intelligence (URAI), 2017:809-814.

3. Chuan Yen T, Mohler J, Dohm M, Laksari K, Najafi B, Toosizadeh N. The effect of pain relief on daily physical activity: In-home objective physical activity assessment in chronic low back pain patients after paravertebral spinal block. Sensors. 2018 Sep; 18(9):3048.

4. Yin P, Yang L, Du S, Qu S, Jia B, Zhao N. The Effect of Mobile Wearable Waist Assist Robot on Lower Back Pain during Lifting and Handling Tasks. Mobile Networks and Applications. 2020:1-9.

5. Van Dieën JH, Reeves NP, Kawchuk G, Van Dillen LR, Hodges PW. Analysis of motor control in patients with low back pain: A key to personalized care? Journal of orthopaedic & sports physical therapy. 2019 Jun; 49(6):380-8.

6. Yin P, Yang L, Wang C, Qu S. Effects of wearable power assist device on low back fatigue during repetitive lifting tasks. Clinical Biomechanics. 2019 Dec 1; 70:59-65.

7. Argent R, Slevin P, Bevilacqua A, Neligan M, Daly A, Caulfield B. Wearable sensor-based exercise biofeedback for orthopaedic rehabilitation: a mixed methods user evaluation of a prototype system. Sensors. 2019 Jan; 19(2):432.

8. Ianculescu M, Andrei B, Alexandru A. A Smart Assistance Solution for Remotely Monitoring the Orthopaedic Rehabilitation Process Using Wearable Technology: Re. flex System. Studies in Informatics and Control. 2019 Sep 1;28(3):317-26.

9. Lee TJ, Galetta MS, Nicholson KJ, Cifuentes E, Goyal DK, Mangan JJ, Fang T, Schroeder GD, Kepler CK, Vaccaro AR. Wearable Technology in Spine Surgery. Clinical spine surgery. 2020 Jul 1;33(6):218-21.Other paragraphs are indented (BodytextIndented style).

IOP Conf. Series: Materials Science and Engineering 1091 (2021) 012014 doi:10.1088/1757-899X/1091/1/012014

Parameteric estimation in doppler's effect forecasts

[1]Dr Asif Tamboli, [1]Dr Prakash Patil, [1]Dr Amol Gautam, [1]Dr Amol Mohite, [1]Dr R R Kumbhar
[1]Department of Radio diagnosis, KIMS, Karad
corresponding author's e-mail address drkumbhar@outlook.com

Abstract. In present study, we have attempted the Umbilical artery evaluation which will further help us to recognize the high risk IUGR cases by the AEDV or the REDV. Objective of this paper is to estimate the utility of the umbilical artery including mid cerebral artery (MCA) parameters in Doppler's effect forecasts of perinatal negative effects in medically diagnosed IUGR pregnancies. PI of MCA was studied in all our 50 cases. It is found that PI of MCA had the sensitivity of 87.5% & the low specificity of 46% in forecasting negative perinatal outcomes. In this study it is observed that in contrast to 88% in UA-PI and 66% in MCA-PI, the diagnostic precision of the MCA / UATPI ratio (90 per cent) was higher when taken separately.

1. Introduction

An elevated risk of perinatal death, morbidity, and abnormal neuro development is correlated with Fetal Intra Uterine Growth Restriction (IUGR) [1-3]. The Ultra-sonographic (US) bio-metre helps to classify the heterogeneous community of small for gestational age foetuses that involve IUGR foetuses, small-constitution foetuses, as well as (misdiagnosed as small) foetuses with acceptable development. A primary goal of antenatal treatment is the proper diagnosis of the unhealthy IUGR foetus to allow for prompt intervention.

A statistical-analysis of controlled, randomized UA Doppler velocimetry trials in risky pregnancies (principally linked pregnancies as well as presumed IUGR) showed that its implementation is related to a prenatal and postnatal morning decrease pattern of Doppler velocimetry in highly risky (principally connected risky pregnancy and presumed IUGR) pregnancies. Doppler US human foetal circulation tests have demonstrated that there is a substantial drop in the Mid Cerebral Arterial (MCA) and umbilical arterial pulsatility index (UA-PI) in foetuses with IUGR relative to those in normal foetuses [4].

Findings from a set of interventions show that the Doppler Function of MCAPI / UAPI is much more reliable than that of the US Doppler alone [5-7] when estimating poor pregnancy consequence. However, our study aims at identifying the function of UA as well as MCA Doppler US in predicting the negative impacts of prenatal and postnatal pregnancies and the role of Doppler velocimetry in clinical progress.

1.1. Aims and Objectives

To measure the utility of the umbilical artery including mid cerebral artery (MCA) parameters Doppler's effect forecasts of perinatal negative effects in medically diagnosed IUGR pregnancies.

Prognosticate the outcome. To follow-up & monitor moderate & mild IUGR cases under obstetrician's care.

2. Review of Literature

The cerebral & umbilical blood flow movements feature in healthy and prolonged pregnancies in development have been studied by Wladimiroff et al. [8]. They find that the cerebral-placental ratio only varies considerably from 26 and 38 weeks of GA. They also found that before the 25th week, there was no substantial variation between the cycles perceived in the comparative analysis of the blood speed types of the MCA and umbilical artery.

The study of FMC flow in 81 SMS foetuses (SSFs) using coloured-flow analysis including Pulsed Doppler tests by Vyas et al. [9]. The flow-pulsatility (PI) impedances were observed to be slightly lower as well as the blood supply rate averaging the corresponding benchmarks of the IUGR.

The ultrasound Doppler offered a tool that was not invasive to study foetal haemodyname, as observed by Campbell, Vyas, Nicolaides [10]. In the samples collected from small cordocentesis for gestational age (SGA) foetuses, they analysed changes in foetal haemodynamics with respect to the foetal blood stress.

The findings of UA Doppler velocity were analysed in three classes of pregnancy cases by Karsdorp et al. [11] reported 214 cases with positive end-diastolic velocity(PEDV), 178 with absent fin-diastolic velocity(AEDV) and in 67 cases with a reverse-end-diastolic velocity (REDV). In their view, absence and REDV waveforms were more likely to evolve in the pregnancies that were complicated by IUGR. The gross perinatal mortality in their study was 28 percent. There have been considerably more neonates of the NICU of the AEDV flow community.

Soregaroli, Bonera & Danti [12] retrospectively studied 578 singleton pregnancies with the diagnosis of IUGR. From this population, they formed: 1) the normal NUAPI fetuses -334; 2) the increased PI of UA but with normal diastolic flow - AUAPI fetuses-137; 3) the absent end-diastolic flow AEDF fetuses -70; 4) the reversed end diastolic flow; REDV fetuses -37. Fetal biometry, amniotic fluid and fetal-maternal Doppler velocimetry were done in all the above patients.

3. Materials And Methods

This study was conducted for a period of two years from April 2007 to March 2009. The patients for this study were those who attended the department of Obstetrics of Shadan Hospital. The gestational Age of pregnancy was assessed by an accurate documentation of normal menstrual cycle & the last menstrual period (LMP). The antenatal patients were referred for Ultrasound Scanning (USG) with a clinical suspicion of IUGR by the Obstetricians initially. In most patients the USG biometry was performed before 20th gestational week for purposes of accurately assess the gestational age (GA) & Estimated Fetal Birth Weight (EFB.W).

4. Observation and Results

Appropriate wave modes of all the 50 patients have been collected out of the 50 pregnancies examined. Seven Doppler studies have been followed up. Among these 24 fetuses (48%) had abnormal outcomes, of those; 8 had more than one abnormal outcome. But 26 (52%) were born with no complications.

Table 1: Spectral Characteristics of Absent Vs Reversed End Diastolic Flow (EDF)

Spectral Characteristics	No. of Cases	IUD	Post Natal Mortality
Absent EDF	09	03	33.33%
Reversed EDF	04	4	100%

As shown in table no. 1, the EDF had direct relationship to the post natal mortality rate of IUGR fetuses. It Can be seen that all the 4 foeti with reversed EDF died (100%) of the 9 foeti with absent EDF only 3 died (33.33%). The Umbilical artery evaluation helped us to recognize the high risk IUGR cases by the absence of end diastolic flow (AEDV) or the reversed end diastolic flow volume (REDV).

Table 2: Adverse Outcomes

Adverse Outcomes	No. of Cases
Emergency CS	12
Intrauterine deaths	7
Low APGAR score	7
Admission to NICU	8

As seen in table no. 2, of the IUGR fetuses imaged, Seven intrauterine deaths have occurred and 43 live births were reported. Of these 43 live births, seven newborns had score less than 7 in the 5 min APGAR ranking. NICU was entitled to 8 newborns. 12 babies were delivered by emergency caesarian sections. There were 7 IUDs among the 50 IUGR cases in this study. Of these 7 cases, 4 had the reversal of diastolic flow and another 3 had absent diastolic flow ante mortem.

Table 3: Diagnostic accuracy of Doppler Indices

Doppler indices	Diagnostics reliability
MCA PI	66%
UA PI	88%
MCA/UA PI ratio (CPR)	90%

As shown in table no. 3, CPR value has more diagnostic accuracy (90%) than PI of Umbilical Artery or PI of MCA in Predicting the adverse outcome.

5. Discussion

If there has been intrauterine delay in development (IUGR) during the third quarter in pregnancy, the obstetrician must decide if the fetus is either short or "constitutional" as a result of defective placental infusion. When resolving this problem, Doppler flow velocity analysis could be reliable. Hemodynamic changes are implicated in fetal developmental retardation, particularly in the fetuses and the umbilical cord. A Doppler index which represents each of these areasvascular beds can be useful for detecting foetuses with enhanced placental and/or decreased cerebral resistance.

Table 4: Comparison of present study with different studies for mortality rate in case of absent/reversal of EPF

Authors	No	Mortality
Battaglia etal [13](1993)	26	58%
Karsdorp et al [11](1994)	245	50%
Present study	13	53%

This table shows that the our research findings are close to those of previously published reports on perinatal mortality in cases of absent/reversal EDV of umbilical artery.

6. Conclusions

An ominous symptom is the presence of absent/reversal EDV in the umbilical artery as it holds a severe prognosis and high mortality. This gives a sign for serious consideration of early delivery is the finding of AEDV or REDV in pregnancy with established IUGR. In managing pregnancies at increased risk, such as those with proven IUGR, the Doppler study has the greatest utilization. Umbilical artery tests have helped to identify at-risk foetuses for hypoxia and acidosis. UA-PI had a sensitivity of 91.6% and a specificity of 84.6%. The specificity of UA-PI was equal to specificity of CPR. PI of MCA had high sensitivity of 87.5%) and low specificity of 46% in predicting adverse perinatal outcome. In contrast to UA-PI and CPR, PI of MCA had low specificity (46%) in prediction of adverse perinatal effect. In contrast with 88% for UA-PI and 66% for MCA PI, the MCA/UA PI proportion was 90% predictive precision. Moreover, 95.8% for CPR whereas 84.6% more for UA-PI & MCA-PI were diagnosis variable.

References:

1. Dobson PC, Abell DA, Beischer NA. Mortality and morbidity of fetal growth retardation. Aust N Z J Obstet Gynaecol 1981; 21:69-72.
2. Gilbert WM, Danielsen B. Pregnancy outcomes associated with intrauterine growth restriction. Am J Obstet Gynecol 2003; 188:1596-1599.
3. Wennergren M, Wennergren G, Vilbergsson G. Obstetric characteristics and neonatal performance in a four-year small for gestational age population. Obstet Gynecol 1988; 72:615-620.
4. Kok JH, den Ouden AL, Verloove-Vanhorick SP, Brand R. Outcome of very preterm small for gestational age infants: the first nine years of life. Br J Obstet Gynaecol 1998; 105:162-168.
5. Arduini D, Rizzo G. Prediction of fetal outcome in small for gestational age fetusr comparison of Doppler measurements obtained from different fetal vessels. J Perinat Med 1992; 20:29-38.
6. Gramellini D, Folli MC, Raboni S, Vadora E, Merialdi A. Cerebral-umbilical Doppler ratio as a predictor of adverse perinatal outcome. Obstet Gynecol 1992; 79:416-420.
7. Arias F. Accuracy of the middle-cerebral-to-umbilical-artery resistance index ratio in the prediction of neonatal outcome in patients at high risk for fetal and neonatal complications. Am J Obstet Gynecol 1994; 171:1541-1545.
8. Wladimiroff JW,Wijngaard JAGW,Degani S,Noordam MJ, Eyck J, Tonge HM. Cerebral umbillical artery blood flow velocity wave forms in normal and growth retarded pregnancies.Obstet Gynecol 1987;69:705-709.
9. Vyas S, Nicolaides KH, Bower S, et al. Middle cerebral artery flow velocity waveforms in fetal hypoxemia. Br J Obstet Gynaecol 1990; 97:797-803.

10. Campbell S, Vyas S, Nicolaides KH. Doppler investigation of the fetal circulation. J Perinat Med. 1991; 19(l-2):21-6.
11. Karsdorp VH, van Vugt JM, van Geijn HP, Kostense PJ, Arduini D, Montenegro N, Clinical significance of absent or reversed end diastolic velocity waveforms in umbilical artery. Lancet. 1994 Dec 17; 344(8938): 1664-8.
12. Soregaroli M, Bonera R, Danti L, Dinolfo D, Taddei F, Valcamonico A et al. Prognostic role of umbilical artery Doppler velocimetry in growth-restricted fetuses. J Matern Fetal Neonatal Med. 2002 Mar;l 1(3): 199-203.
13. Battaglia C, Artini PG. Galli PA, D'Ambrogio G, Droghini F, Genazzani AR. Absent or reversed end-diastolic flow in umbilical artery and severe intrauterine growth retardation: an ominous association. Acta Obstet Gyneco/ Scand 1993;72:167-171.

IOP Conf. Series: Materials Science and Engineering 1091 (2021) 012015 doi:10.1088/1757-899X/1091/1/012015

Study of green advertisements in indian automobile sector for environmental sustainability

Madhuranjan Vatsa[a]*, Satendar Singh[b], Sarita Chaudhary [c]
[a]Sharda University, Greater Noida, India
[b]Associate Professor, School of Business Studies, Sharda University, India
[c]School of Business, Galgotias University, Greater Noida, India
Corresponding author's e-mail address :madhujeh@yahoo.co.in

Abstract. Worsening state of environment is a major concern for the world today. One major sector which is supposed to be the most notorious in degrading environment is the transportation sector. It is therefore important for the transportation sector to think of ways to reduce its carbon footprint. One of the popular way is the adoption of green marketing which incorporates all the elements namely product, price, place and promotion. It means that that the objective of the transportation sector should not just to develop the environmental friendly products but they should give the environment friendly identity too. The present paper has attempted to understand the green initiatives taken by the car manufacturing firms. For this purpose we have studied the advertisements shown on the television. The advertisements shown in the Television during 2012-2016 have been collected. Cues have been developed to interpret the advertisements with the help of previously conducted studies. Some of the cases have been discussed in the present paper too. The results indicate that the car making firms are still hesitating to establish the identity of cars as environment friendly.

1. Introduction

One important aspect to understand the green marketing is the green advertisement which also constitutes one element of marketing mix of green marketing. Here it is important to understand that the essence of advertising is brand building approach [1]. It can therefore be said that automobile manufacturer when advertise they wish to convey a particular image of the product to the final consumers. Similar meaning got portrayed when Barabba and Zaltman [2] attempted to clear the meaning of advertising. Advertising for them too is a means to state that what the advertiser can provide to the consumers. Considering the same meaning of advertising, one might expect that environmental concern would be a primary theme in vehicle advertising, as our world these days are facing the pollution problems [3]. Our study has attempted to analyse the advertising messages and themes that could have found whether the firms are using the environmental themes in the advertisements or not. This study examines car-advertising themes with an aim of finding the prevalence of green advertising in India from June 2012 to December 2016, focusing on the national television audience.

1.1. Green Advertising

Green advertising usually involves the practice of claims that either underlines the association between the product/service with the environment, or the commitment of brand towards the ecosystems [4]. It may also understand as the conception of promotional communication that appeals to the requirement of environment friendly consumers [5]. Green advertisements usually emphasize the environmental attributes or ecological implications of the product [6].

During 1990's in western countries, development of taxonomies for classifying green advertisements using content analysis was the main focus. In these researches, Specificity and validity of the claims are primarily investigated. Most of these studies indicated that environmental claims made in green advertisements were vague, unsubstantiated, omitted factual information and served as an image building exercise for most firms [7-9]. Most green advertisements depicted the green image of an organization without any substantiation [3-9]. Green advertisements from developing countries have been analysed only recently. Green automobile advertisements in China [10], green advertisements in Malaysia [11] are studies that use samples from developing countries.

1.2. Green Marketing and Advertising in India
Indian companies recently have started taking steady steps to be ecologically conscious [12]. But some of the studies found that irrespective whatever has been the feature of the automobile firms they are still hesitant of showing themselves to be environment friendly. One probable reason may the lower level of consumer awareness. Business houses like the Tata, Unilever, Mahindra, Wipro and ITC are active participants in green and sustainability initiatives [12]. Additionally, it is also known that only 5% of environmental claims made in green campaigns are entirely true in India [13] and they are not always based on authentic scientific evidence [14]. This is an important consequence of a lack of regulation of environmental claims in India.

1.3. Regulation of green advertising claims in India
Efficient regulation of green advertising claims has been the objective of a number of public policy studies [7], [15]. Most developed countries have governing bodies that have strict regulations regarding green washing. The FTC (Federal Trade Commission) in the U.S. DEFRA (Department for Environment, Food and Rural Affairs) in the U.K. provides standards for the usage of green advertising claims. However, green advertising is considered to be in the nascent stage in India with no accepted guidelines to standardize environmental claims [16].

1.4. Environmental knowledge and green advertising
Most researchers have investigated the role of environmental concern in influencing consumer responses to green advertisements [6] [17-21] and have largely ignored environmental knowledge. Environmental knowledge can be defined as the ability to distinguish between the symbols, concepts and behaviour related to environmental protection [22]. Some researchers say that the link between environmental knowledge and pro-environmental behaviour is not well-established [22-23]. Yet, some say that the value of environmental education cannot be undermined in motivating environmentally responsible behaviour and purchase intentions [24] or encouraging environmentally concerned decision making in the market place [25]. Consumer knowledge can be measured as "subjective" or "objective" knowledge. Subjective knowledge is measured as a self-assessed report of one's knowledge and the consumers rate their own knowledge level. Objective knowledge reflects what is stored in memory and is an actual measure of what an individual knows [26]. People who report high subjective environmental knowledge may not have the requisite knowledge to make correct ecological decisions [27]. Manrai et al. [28] use "subjective environmental knowledge" measures by asking respondents to rate how much they knew about environmental issues and found a positive relationship between purchase intention and subjective knowledge. They were not able to relate subjective knowledge to brand attitudes. When the consumer is knowledgeable on specific issues like pollution, he develops a favorable attitude towards purchasing green products [29]. Rashid [30] found that consumer awareness of eco-labels is low in Malaysia and educating the consumer is necessary to increase their purchase intentions towards green products. Low consumer awareness is one of the reasons for failure of green marketing tools like eco-labels in developing countries Gallastegui [31-

32]. Consumer education on environmental issues or eco-literacy is therefore essential for encouraging green purchases [33]. Tilikidou [34] found moderate correlation between objective environmental knowledge and pro-environmental purchasing behaviour. Recent research has determined that higher levels of objective environmental knowledge determined purchase intentions among students. Researchers either use measures of subjective environmental knowledge [28] [35] or objective environmental knowledge [36-37] in their studies. Marketing studies have also found that consumer's expertise is based on objective knowledge as they tend to overestimate their subjective knowledge [38]. Some researchers also feel that subjective knowledge might be an incorrect measure of environmental issues [35].Consumers with low levels of knowledge or awareness have a low ability to process green advertisements. Again, if they are not concerned about the environment, then too, they have low motivation to process a green advertisement. Advertisement processing is characteristically a low involvement scenario and attitude persistence can be achieved only when the message is elaborated systematically [39]. Only when the message is processed in a more elaborate manner, memory is also enhanced for the message information [40]. Therefore green advertising studies should focus on increasing involvement in low motivation and awareness situations.

2. Methodology

To analyse the advertisements, it was collected from online source which use to keep the record of advertisements as per the date of its publication. The research method is shown in figure 1. All the advertisements that were available for the period of 2012-2016 has been downloaded for further analysis. While selecting the advertisements only car advertisements were chosen. The total number of such TV ads which were unique and were published in the time frame of 2012-16 were 156. A new set of 22 themes were defined (Table.1) as per the need of the present research. The work of Ferguson et al. 2003 who have worked on the study of themes of advertisements was very instrumental in deciding the themes too. All the themes were further broken down into many cues which was based on previous studies and suggestion from the people of advertisement people. Based on the theme and their cues, 156 advertisements were analysed.

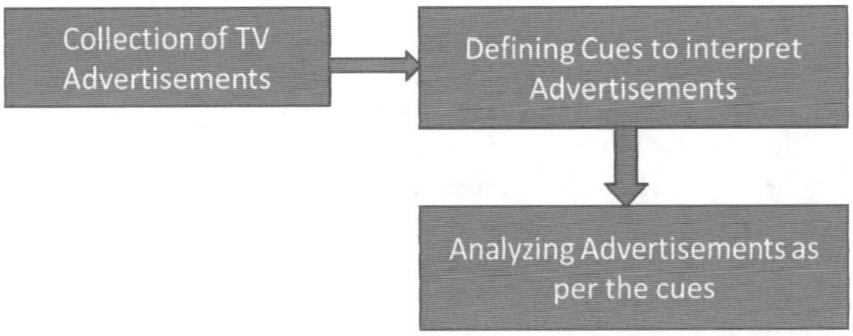

Fig. 1. Flowchart of method

3. Results and Discussions

The analysis showed the distribution of primary themes of advertisements (2 of the 22 themes are not included in the table because they were not mentioned). Performance was the most frequent primary theme depicted in the advertisements (44%), followed by Luxury/prestige themes (21%), economy/good value (16%), and themes that stressed the Attractive styling of the vehicle (14%). Exciting/fun to drive was the theme of 13% of the advertisements, and for younger people was featured in about 13%.Comfort/convenience featured in 12 % of the selected advertisements whereas for families were shown in 9% of the advertisements. Among other themes, new technology (4%), safety (3%), new/different (3%), incentives/sales (2%), environment friendly (2%), an escape (1%), award winner (1%), heritage (1%), quality, reliability, durability (1%), well-engineered (1%), others (6%) were being found.

Table 1. Appearance of themes in selective TV advertisements in numbers and percentage

	Themes	Number of appearance in selective TV Advertisements	% Appearance in selective TV advertisements
1	Performance	68	43.59
2	Luxury/prestige	32	20.51
3	Economy/good value	25	16.03
4	Attractive styling	22	14.10
5	Exciting/fun to drive	21	13.46
6	For younger people	19	12.18
7	Comfort/convenience	18	11.54
8	For families	14	8.97
9	New technology	7	4.49
10	Safety	5	3.21
11	New/different	4	2.56
13	Incentive/sales	3	1.92
14	Environment friendly	3	1.92
15	An escape	2	1.28
16	Award winner	2	1.28
17	Heritage	2	1.28
18	Quality, reliability, durability	2	1.28
19	Well-engineered	1	0.64
20	Other	9	5.77

The analysis (table 1), shows that most of the car's advertisement highlights the performance aspect of the car followed by luxury, exciting and fun to drive. But the most important aspect is that very few of the car's advertisements are talking about the environment features, clearly reflecting that car marketers are least concerned about advertising the green features. The probable reason is the demand which is reflected in the next chapter, the result here also are not in line of green manufacturing as being claimed by the car manufacturing firms. Two cases of the advertisements analysis based on the cues and theme have also been stated below.

3.1. Case study of Maruti Suzuki Ciaz

In an attempt to exemplify the analysis of TV advertisement we have taken the case studies. First case is that of Maruti Suzuki Ciaz which has been featured by the Bollywood actor Ranveer Kapoor. He has been shown as a well-dressed gentle man (luxury/prestige) and he has been dressing for a new car which states itself as the India's first diesel hybrid car (environment friendly). In the advertisement one can easily find the cues as for example a well-dressed gentle man, the statement of hybrid diesel car, statement of new technology, vehicle moving at speed. These cues indicate that marketers are focusing on themes like luxury/prestige, environment friendly, new technology, and performance.

3.2. Case study of Hyundai Creta

The perfect SUV (Hyundai) advertisement starts with a demonstration of its fit and finish (quality/reliability/durability) of the interior of the car. Young couple (for younger people) has been shown running on high speed and cornering at speed in this advertisement. These three cues indicate that marketers are focusing on three themes that is performance and quality/reliability/durability and for younger people.

3.3. Case study of other cars

Advertisement of Mahindra Verito Vibe wherein it has been shown that some male and female friends are enjoying the ride. The advertisement ends with special mention of its price. In this advertisement, the marketer is intended to show the enjoyable ride with the dear ones at a low price. In another example of Maruti Suzuki Alto, it has been shown as complete set of performance. In this advertisement, when two musician friends one young male and a female, are not able to get the perfect music composition. They take a ride of Alto K10 which is having an engine capacity of 1000 cc, 68 PS power with sleek look with a mileage of 20.92 km/l. In yet another example of cost benefit is Beat of Chevrolet which is having features like electronic power steering fitted with air bags for safety, foldable seat in the ratio of 60:40, with a mileage of 25.44 km/litre, and 3 year free servicing. In the advertisement of Fiat an extra discount of up to 78 thousand in their cars has been shown. The cues of well-dressed gentle man in the advertisement of Etios with special mention of style or in the advertisement of Renault Pulse where the Bollywood actor Anil Kapoor is a well-dressed gentleman taking care of day to day life with full energy.

In case of Eco (Maruti Suzuki) full family of bride and groom both are moving in a car which can adjust the full family. This car is also fitted with AC and a strong engine at the same it is also available in CNG variant which provides cheaper mobility. In another advertisement of Honda Amaze, the full family is getting ready for a ride. Here the cues of full family in the car and quoting the car as the 'best in class cabin space' shows the marketer's intention of showing the comfort. In the advertisement of Ford Figo Aspire the car marketers have emphasized the availability of six airbags in the car.

4. Conclusion

With the increasing pressure on our environment, companies are supposed to be follow green marketing however the present content analysis of the Indian TV advertisements show a contrary result. The content analysis shows that the performance is the most important theme during all the years taken under consideration for the present study followed by luxury, economy and attractive styling. At the same time eco-friendly themes are least shown in the Indian TV advertisements. This indicate the present state of the automobile sector with reference to seriousness toward degrading environment. This is an alarming finding which leads to a number of new questions toward all the stakeholders that whether it is the automobile players, the government and the users are really concerned about the environment.

References

[1] Gardial, S. F., Schumann, D. W., Petkus Jr, E., & Smith, R. Processing and retrieval of inferences and descriptive advertising information: The effects of message elaboration. Journal of Advertising, 22(1), 25-34 (1993)

[2] Barabba, Vincent, and Zaltman, Gerald: Hearing the Voice of the Market, Harvard Business School Press, Boston, MA. 1991.

[3] Vatsa, M., Chakrabarti, M., & Kumar, R.. Analysis of Green Marketing Trend in Passenger Car Segment of Indian Automobile Industry. International Journal of Management (IJM), 7(6), 181-188 (2016)

[4] Banerjee, Subhabrata, Charles S. Gulas, and Easwar Iyer. Shades of green: A multidimensional analysis of environmental advertising. Journal of advertising 24(2) 21-31(1995)

[5] Zinkhan, G. M., & Carlson, L. Green advertising and the reluctant consumer. Journal of Advertising, 24(2), 1-6(1995).

[6] Schuhwerk, M. E., & Lefkoff-Hagius, R. Green or non-green? Does type of appeal matter when

advertising a green product?. Journal of advertising, 24(2), 45-54. (1995)
[7] Kangun, N., Carlson, L., & Grove, S. J. Environmental advertising claims: A preliminary investigation. Journal of public policy & marketing, 10(2), 47-58. (1991)
[8] Mohr, K., Backman, K. F., Gahan, L. W., & Backman, S. J. An investigation of festival motivations and event satisfaction by visitor type. Festival Management and Event Tourism, 1(3), 89-97. (1993)
[9] Carlson, L., Grove, S. J., Laczniak, R. N., & Kangun, N. Does environmental advertising reflect integrated marketing communications?: An empirical investigation. Journal of Business Research, 37(3), 225-232. (1996).
[10] Dai, X., Goh, T. T., & Cheng, S. A Content analysis of Green advertising in China. Journal of Marketing Research and Case Studies, 1-19. (2014)
[11] Saabar, S. S., Fiah, A. F. M., Halim, A. A., & Nor, A. M. (2011, June). A content analysis of creative execution on green advertisements in Malaysia. In 2011 3rd International Symposium & Exhibition in Sustainable Energy & Environment (ISESEE) (pp. 152-156). IEEE.
[12] Madhuranjan, V., Milindo, C., & Rajesh, K. Singh, JV, GMS is Powerful Force in Passenger Car Segment of Indian Automobile Industry to Enhance Innovation and Widen Profits. International Journal of Engineering Research & Technology (IJERT), 4(9), 278-295 2015.
[13] Mishra, P., & Sharma, P. Green marketing in India: Emerging opportunities and challenges. Journal of Engineering, Science and Management Education, 3(1), 9-14 (2010).
[14] Nair, S. R., & Menon, C. G. An environmental marketing system–a proposed model based on Indian experience. Business Strategy and the Environment, 17(8), 467-479. (2008).
[15] Abd Rahim, M. H., Zukni, R. Z. J. A., Ahmad, F., & Lyndon, N. Green advertising and environmentally responsible consumer behavior: The level of awareness and perception of Malaysian youth. Asian Social Science, 8(5), 46 (2012)
[16] de Freitas Netto, S. V., Sobral, M. F. F., Ribeiro, A. R. B., & da Luz Soares, G. R.. Concepts and forms of greenwashing: a systematic review. Environmental Sciences Europe, 32(1), 1-12. (2020)
[17] Kim, Youn-Kyung, Judith Forney, and Elizabeth Arnold. "Environmental messages in fashion advertisements: Impact on consumer responses." Clothing and Textiles Research Journal 15.3 (1997): 147-154.
[18] Newell, Stephen J., Ronald E. Goldsmith, and Edgar J. Banzhaf. "The effect of misleading environmental claims on consumer perceptions of advertisements." Journal of Marketing Theory and Practice 6.2 (1998): 48-60.
[19] Chang, Hua, Lingling Zhang, and Guang-Xin Xie. "Message framing in green advertising: The effect of construal level and consumer environmental concern." International Journal of Advertising 34.1 (2015): 158-176.
[20] Bickart, Barbara A., and Julie A. Ruth. "Green eco-seals and advertising persuasion." Journal of advertising 41.4 (2012): 51-67.
[21] Do Paço, Arminda Maria Finisterra, and Rosa Reis. "Factors affecting skepticism toward green advertising." Journal of advertising 41.4 (2012): 147-155.
[22] Laroche, Michel, Jasmin Bergeron, and Guido Barbaro- Forleo. "Targeting consumers who are willing to pay more for environmentally friendly products." Journal of consumer marketing (2001).
[23] Duerden, Mat D., and Peter A. Witt. "The impact of direct and indirect experiences on the development of environmental knowledge, attitudes, and behavior." Journal of environmental psychology 30.4 (2010): 379-392.
[24] Manrai, Lalita A., and Ajay K. Manrai. "A field study of customers' switching behavior for bank services." Journal of retailing and consumer services 14.3 (2007): 208-215.
[25] Roberts, James A., and Donald R. Bacon. "Exploring the subtle relationships between environmental concern and ecologically conscious consumer behavior." Journal of business

research 40.1 (1997): 79-89.
[26] Brucks, Merrie. "The effects of product class knowledge on information search behavior." Journal of consumer research 12.1 (1985): 1-16.
[27] Ellen, Pam Scholder. "Do we know what we need to know? Objective and subjective knowledge effects on pro-ecological behaviors." Journal of Business Research 30.1 (1994): 43-52.
[28] Manrai, Lalita A., et al. "How green- claim strength and country disposition affect product evaluation and company image." Psychology & Marketing 14.5 (1997): 511-537.
[29] D'Souza, Clare, Mehdi Taghian, and Peter Lamb. "An empirical study on the influence of environmental labels on consumers." Corporate communications: an international journal (2006).
[30] Rashid, N. R. N. A. "Awareness of eco-label in Malaysia’s green marketing initiative." International Journal of Business and Management 4.8 (2009): 132-141.
[31] Galarraga Gallastegui, Ibon. "The use of eco- labels: a review of the literature." European Environment 12.6 (2002): 316-331.
[32] Mehta, Hitesh. "Towards an internationally recognized ecolodge certification." Quality assurance and certification in ecotourism (2007): 415-434.
[33] Cheah, Isaac, and Ian Phau. "Attitudes towards environmentally friendly products." Marketing Intelligence & Planning (2011).
[34] Tilikidou, Irene. "The effects of knowledge and attitudes upon Greeks' pro- environmental purchasing behaviour." Corporate Social Responsibility and Environmental Management 14.3 (2007): 121-134.
[35] Schlegelmilch, Bodo B., Greg M. Bohlen, and Adamantios Diamantopoulos. "The link between green purchasing decisions and measures of environmental consciousness." European journal of marketing (1996).
[36] Maloney, Michael P., Michael P. Ward, and G. Nicholas Braucht. "A revised scale for the measurement of ecological attitudes and knowledge." American psychologist 30.7 (1975): 787.
[37] Bartiaux, Françoise. "Does environmental information overcome practice compartmentalisation and change consumers' behaviours?." Journal of Cleaner Production 16.11 (2008): 1170-1180.
[38] Alba, Joseph W., and J. Wesley Hutchinson. "Knowledge calibration: What consumers know and what they think they know." Journal of consumer research 27.2 (2000): 123-156.
[39] Haugtvedt, Curtis P., Richard E. Petty, and John T. Cacioppo. "Need for cognition and advertising: Understanding the role of personality variables in consumer behavior." Journal of Consumer Psychology 1.3 (1992): 239-260.
[40] Hawkins, Scott A., and Stephen J. Hoch. "Low-involvement learning: Memory without evaluation." Journal of consumer research 19.2 (1992): 212-225.

Fabrication of PET/PU/ZnO NPs based Polymer Nanocomposite for Uniform Dispersion of ZnO Nanoparticles by Bar Spreading and Chemical Bath Deposition Method

Jayant Pawar [1], Rushikesh Deshmukh [2], Shilpa Ruikar[3], Sharmilee Pisal [4], Snehal Masurkar[3], Mihir Bhayani [4], S. C. Kale[3*]
[1]Directorate of Research, Krishna Institute of Medical Sciences (Deemed to be University), Malkapur, Karad-415539, Maharashtra, India
[2]Department of Mechanical Engineering, Jawaharlal Nehru Institute of Technology, Pune-411043, Maharashtra, India
[3]Dipartment of Allied Sciences, Krishna Institute of Medical Sciences "Deemed to be University, Karad, Maharashtra, India
[4]TERA Research Institute India Pvt. Ltd., Pune- 411057, Maharashtra, India.
Corresponding author's e-mail address: deankibb@kimsuniversity.in

Abstract. Currently, nanomaterials are one of the materials being used commercially almost in all domains to improve the quality of products due to its physical, chemical, optical, antimicrobial and catalytic properties. In present paper, one of its roles in polymer chemistry has been explored, in this regards, the fabrication method developed for uniform coating of PET films with ZnO nanoparticles. In this direction, ZnO nanomaterials are very important as they are biocompatible, non- cytotoxic, cheap and easy to produce. The ZnO nanoparticles were obtained by facile hydro-thermal route using zinc acetate and urea as precursors in water. The surface coating of nanomaterials on PET/PU films were achieved by bar spreader and chemical bath deposition method. Further, the structural and morphological analysis reveals uniform coating of ZnO nanostructures on to the PET/PU films.

1. Introduction

The barrier properties of existing polymers like polypropylene (PP), low-density polyethylene (LDPE), Polystyrene (PS), polyethylene (PE), polyethylene terephthalate (PET), polyvinyl chloride (PVC) and ethylene vinyl alcohols (EVOH) can improve by incorporation of nanomaterials into existing polymers. These polymer-nanofiller interactions tend to change with size of nanomaterials, which subsequently affect performance of resulting nanocomposites. Initially corrugated materials, paperboard cartons, glass bottles and paraffin wax paper container were developed for packaging of various industrial products [1]. The wax and petroleum based material were used for making packets of dry cereals and biscuits during World War II [2]. This eventually replaced metal, glass and paperboard packaging materials, which helped industries and consumers to protect their goods from extreme conditions [3], [4]. Introduction of polymers as packaging materials revolutionized the packaging industry, as they are light weight, transparent and economical. Although all these packaging

materials raised the standard of packaging, they suffered from major drawbacks like innate permeability to gases and other small molecules. In this regard migration and permeability are the important issues found in existing polymers because no known polymeric material is totally impermeable to atmospheric gases and moisture [5], [6], [4].
Despite the numerous advantages offered by polymers, the present scenario of industries demands the path breaking approach where polymers not only perform the passive role of food packaging, but also they must actively participate in controlling microbial growth, migration of gases and moisture into packets. Therefore, the development of active polymers by incorporation of nanomaterials is need of the day. In this context, nanofillers may improve thermal, mechanical and barrier properties [7, 8]. Additionally, nanomaterials like metal and metal oxide nanoparticles are found to have antimicrobial property which may give extra functionalities to the polymer films to be used in food and medicine packaging [9, 10, 11, 12, 13]. However, the dispersion of nanomaterials onto the matrix is not uniform, which may create problems of agglomeration and leads to affect the overall performance of polymer. In this regards, the PET/PU films and ZnO NPs in different proportion were used to develop polymer nanocomposite (PNC) by bar spreading and chemical bath deposition method to achieve uniform distribution of nanofillers on to the matrix.

2. Materials and Methods

A. Chemicals

All the bulk chemicals, solvents used in the study were of analytical reagent (AR) grade and procured from suppliers including SD fine-chemicals, Loba Chemie, Sisco Research Laboratories and Qualigenes, India.

B. Synthesis and characterization of ZnO nanomaterials

Synthesis of ZnO nanomaterials were carried out by hydro- thermal route. All the analytical grade chemicals were used without any further purification. In typical experimental procedure, 0.1M zinc acetate Zn $(O_2\ CCH_3\ )_2\ (H_2\ O)_2$ was dissolved in 50 mL DI water. 0.6M urea $[CO(NH_2\ )_2\ ]$ solution was prepared in 50 mL DI water. Both solutions were mixed under constant stirring for 120 minutes at 60 °C then allowed to cool down at room temperature. The resultant settled precipitate was washed by repeated and alternative centrifugation of resultant product at 5000 rpm for 15 minutes each with DI water. The final product was dried in hot air oven at 60 °C for 4 hours to obtain white colored powder of ZnO nanomaterials.
The structural and morphological features of resultant ZnO were characterized by UV-Vis spectroscopy, XRD and Stereoscope for identifying various crystalline phases present in the powder as well as their morphological and optical information.

C. Development PET/PU/ZnONPs active polymer nanocomposite films

The PET/PU/ZnONPs PNC films were obtained by coating MONs (125 µg/mL) dispersed in absolute ethanol, which was then mixed with PU. A one side polyurethane coating on PET films was achieved by dispersing MONs in mixture of PU and isophorone diisocyanate (IPDI) as a hardener component in 4:1 proportion and then resultant mixture were applied on to PET films using bar applicator. The coated films were allowed to cure completely for two hours at room temperature. The same film made out of cured PET/PU films and dispersed MONs (1000 µg/mL) in DI water by chemical bath deposition (CBD) under constant stirring at 60 °C for 2 hours.

3. Results and discussion

A. Characterization of ZnO nanomaterials

The X-ray diffractogram of the hydro- thermally synthesized powder is shown in figure 1 (a). The peak broadening indicates the formation of nanoscale structures and noise indicates the impurities in powder which demands further purification before fabrication of nanocomposites. The powder sample shows the peaks corresponding to (100), (002), (101), (102), (110), (103), (112) and (207) planes

which indicates formation of hexagonal crystal structure of ZnO. The similar structure observed for ZnO nanomaterial synthesized by semi-solvo thermal method [9].
The absorption spectrum of ZnO powder disperses in ethanol is shown in figure 1 (b). The absorption edge is found to be shifted to the lower wavelength which can be due to size quantization effect. The changed peak intensities have reported with increased reaction time due to the size quantization effect [14].

B. Fabrication of PET/PU/ZnONPs films

The developed PET/PU/ZnONPs films were studied under stereoscope at 400X magnification for understanding surface morphology. *Georgopoulos et al.,* reported the use of stereoscopic optical microscope for study the thermoplastic polymers reinforced with fibrous agricultural residues [15]. Figure 2 reveals the surface of polymer coated with ZnO NPs by bar spreader and CBD method. The more uniform coating of nanomaterials was achieved by CBD method but found to be weakly attached to the matrix of polymer, whereas bar spreading method found to be more effective with respect to surface adherence. The average composition of PET, PU and ZnO NPs in PNC is 40, 30 and 30 % respectively for film coated by bar spreading method and 45, 25 and 30 %, respectively for film coated by CBD method.
Further, it discloses that the coating on films by CBD was not 100 % successful as it was removed by friction while, also did not form uniform coating. Therefore, PET/PU films coated with ZnO NPs by bar spreading was found to have complete adhesion and uniform distribution of ZnO NPs throughout the matrix.

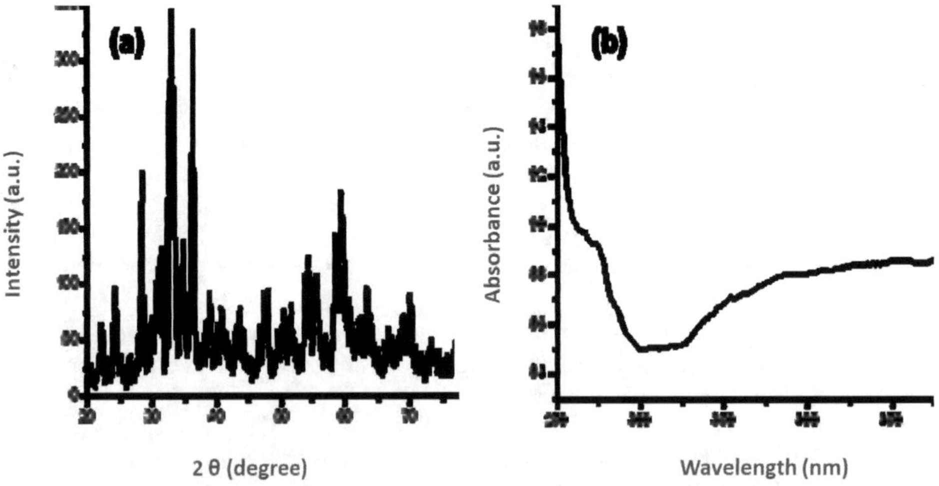

Figure 1: (a) XRD graph, (b) UV-Vis Spectrophotometry of ZnO nanomaterials

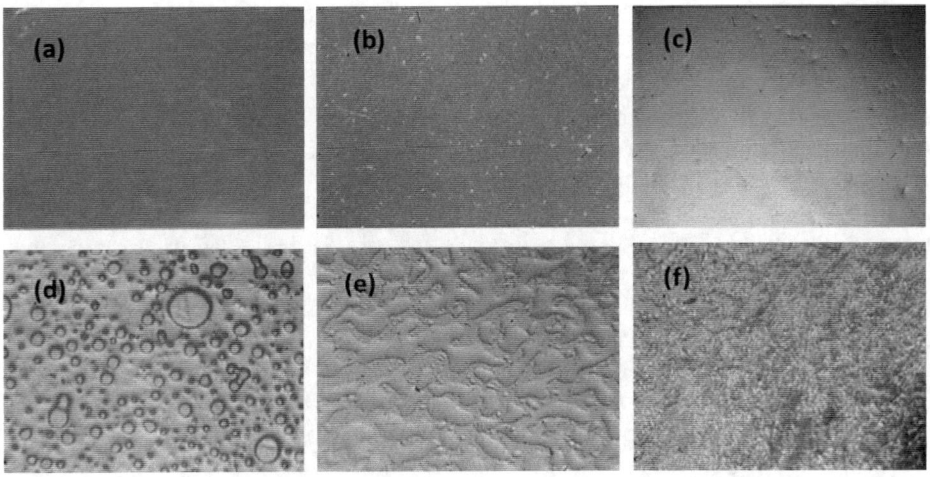

Figure 2: Surface morphology of PET/PU/ZnONPs under stereoscope at 400X (a) Blank PET; (b) PET/ZnONPs (CB); (c) Blank PET/PU (BS); (d) PET/PU/MONs in water (125 μg/mL) (BS); (e) PET/PU/MONs in ethanol (125 μg/mL) (BS); (f) PET/MONs (CBD) (1mg/mL)
Note: BS- bar spreading method; CBD- chemical bath deposition

4. Conclusion

The X-ray diffractogram of the hydro- thermally synthesized powder was studied. The peak broadening indicates the formation of nanoscale structures and noise indicates the impurities in powder which demands further purification before fabrication of nanocomposites. The absorption edge is found to be shifted to the lower wavelength which can be due to size quantization effect. The synthesized ZnO nanomaterials found to be adhere more in proportion on the matrix of PET/PU films by bar spreading than the CBD method. Therefore, in future such films can be used for packaging of various goods in industries, which demands active properties like antimicrobial activity, improved barrier potential for gases and moisture, improved sensing of factors like humidity and moistures, improved mechanical properties etc.

References

1. Coles, R. and Kirwan, M.J. eds., 2011. Food and beverage packaging technology (p. 303). Hoboken, NJ: Wiley-Blackwell.
2. Twede, D., Selke, S.E., Kamdem, D.P. and Shires, D., 2014. Cartons, crates and corrugated board: handbook of paper and wood packaging technology. DEStech Publications, Inc.
3. Brody, A. L., Bugusu, B., Han, J. H., Sand, C. K., and McHugh, T. H., Innovative food packaging solutions. Journal of Food Science, 2008.
4. Duncan, T. v., Applications of nanotechnology in food packaging and food safety: Barrier materials, antimicrobials and sensors. Journal of Colloid and Interface Science, 2011, 363, 1–24.
5. Robertson, G., Food packaging: principles and practice 2005.
6. Finnigan, B., Barrier polymers. The Wiley Encyclopedia of Packaging Technology, John Wiley and Sons, Inc., New York, 2009, 103-109.
7. Alexandre, M., Reports, P. D.-M. S. and E. R., and 2000, undefined, Polymer-layered silicate nanocomposites: preparation, properties and uses of a new class of materials. Elsevier,.
8. Azeredo, H. M. C. de, Nanocomposites for food packaging applications. Food Research International, 2009, November, pp. 1240–1253.

9. Pawar J, Shinde M, Chaudhari R, Singh EA. Semi-solvo thermal synthesis and characterization of zinc oxide nanostructures against food borne pathogens. Int J Pharm Bio Sci. 2017;8(2):311-5.
10. Pawar, J., Henry, R., Viswanathan, P., Patwardhan, A., & Singh, E. A. (2019). Testing of antibacterial efficacy of CuO nanoparticles by methylene blue reduction test against Bacillus cereus responsible for food spoilage and poisoning. Indian Chemical Engineer, 61(3), 248-253.
11. Jain, N., Bhosale, P., Tale, V., Henry, R. and Pawar, J., 2019. Hydrothermal assisted biological synthesis of silver nanoparticles by using honey and gomutra (Cow Urine) for qualitative determination of its antibacterial efficacy against Pseudomonas sp. isolated from contact lenses. EurAsian Journal of BioSciences, 13(1), pp.27-33.
12. Shirsat, S., Pawar, D., Jain, N., Pawar, J., Tale, V.S. And Henry, R., 2019. Synthesis Of Copper Oxide Nanoparticles By Chemical Precipitation Method For The Determination Of Antibacterial Efficacy Against Streptococcus Sp. And Staphylococcus Sp. Synthesis, 12(5).
13. Prakash, V., Pawar, J., Vinchurkar, P., Henry, R. and Patwardhan, A., 2017, March. Chemi-thermal method of synthesis of ZnO nanomaterial and its characterization. In Int. Conf. Innov. Sust. Dev. Sci. Manag. Technol (Vol. 1, p. 36).
14. Sendi, R.K. and Mahmud, S., 2012. Quantum size effect on ZnO nanoparticle-based discs synthesized by mechanical milling. Applied surface science, 258(20), pp.8026-8031.
15. Georgopoulos, S.T., Tarantili, P.A., Avgerinos, E., Andreopoulos, A.G. and Koukios, E.G., 2005. Thermoplastic polymers reinforced with fibrous agricultural residues. Polymer Degradation and Stability, 90(2), pp.303-312..

Investigation of a highly sensitive vortex beam-shined optical sensor for the detection of glucose concentration in urine

Arijit Datta*, **Mukta Chaturvedi**

Department of Electrical & Electronics Engineering, CMR Institute of Technology, Bengaluru, Karnataka-560037, India.

*arijit.datta048@gmail.com

Abstract. The present work investigates the notion of a highly sensitive optical sensor for the detection of urinary glucose concentration by shining an optical Vortex Beam. The proposed sensing scheme relies on the theory of multimode interference where the transmitted output power varies due to the change in the refractive index of cladding as formed by various concentrations of glucose in urine. With a significant amount of energy at the beam's periphery, Vortex beam can energize a massive number of higher order modes inside the sensing system. In support of theoretical analysis, an Eigenmode expansion solver is employed in Lumerical Mode solution software to visualize the full transmission spectrum of the sensor. In contrast to the conventional sensor, the proposed sensor proffers a 3.65 times superior sensitivity of around 1.116 dB/gm/dL over the glucose concentration range from 0-10 gm/dL. The presence of minimum 0.008 gm/dL glucose level in urine could be detected using the proposed technique; whereas the traditional Gaussian-beam based technology could detect the presence of around 0.032 gm/dL of the same. Due to the high sensing performance, the present paradigm can be a useful tool for monitoring the elevated glucose levels which is responsible for triggering Diabetes Mellitus.

1. Introduction

The physical characteristics of the urine including the refractive index, viscosity and specific gravity are affected by the glucose concentration contained in it. In general, glucose is not found in the urine under normal human body condition. Though, glucose can often be found in a limited range from 0-15 mg/dL and this level is named as 'Renal Threshold of Glucose (RTG)' [1]. However, due to kidney and liver dysfunction; blood glucose level increases and the presence of glucose is seen in the urine. This disease is called as 'Glycosuria' [1]. According to the World Health Organization (WHO), there had been roughly 422 million people living with diabetes globally in 2014 [2]. WHO also predicts that diabetes will be the world's seventh leading disease. The accumulation of glucose in the urine is also responsible for gestational diabetes which mostly occurs because of changes in the filtration mechanism of the kidney.

Till date, diverse approaches have been attempted towards the detection of glucose levels in human body. Among them, the traditional invasive method examines of the glucose content in blood as collected by the process of pinpricking [3, 4]. Another technique uses "Test strips" where the strip colour will be altered by the amount of urinary glucose, but the results may be erroneous due to the contamination from air/finger and light source interference [4]. The method of fluorescent signal analysis is not label-free and also requires costly equipment for the photon detector [5, 6]. Besides this, the performance of surface plasmon resonance-based sensor is strongly dependent on the deposited metal's stability [7].

In recent times, beam shaping has been an alluring theme which renders non-Gaussian beam profiles like Bessel-Gauss, Mathieu-Gauss, Airy, Laguerre–Gaussian, and Vortex. With this enticing arrival of non-diffracting beams, the Vortex beam has captivated an immense interdisciplinary attention and has made prodigious advances in optical manipulation, atom optics, and non-linear optics [8, 9]. In our prior works, the benefits of a 0th order Bessel-Gauss beam and Vortex beam have already been explored in the usage of fiber-optic sensing [10-19].

In this report, the authors introduce a new fiber-optic probe for the urinary glucose detection by injecting a Vortex beam inside a No-core fiber. The analysis indicates that such glucose level in urine contributes to changing the sample's refractive index and therefore leads to a change in the optical power loss. By following the classical Wave optics, proper excitation of different linearly polarized modes in the sensing system was evaluated. Moreover, a Beam propagation method is employed in Mode Solutions simulation tool (Lumerical Inc) to envisage the transmission property of the sensing device. The corresponding simulation outcomes have proved the bettered sensitivity of the proposed sensing device in comparison to the conventional one utilizing the Gaussian beam.

2. Proposed Sensing Scheme

In this work, the realization of an ultra-sensitive fiber-optic sensor for detecting the urinary glucose concentration has been achieved by exploiting a special class of Laguerre-Gaussian beam called 'Vortex beam'. As illustrated in Figure 1, the suggested method of sensing includes a No-core fiber (NCF) and two specially engineered air-core fibers (called as 'Vortex fiber'). The unique structural characteristic of this NCF is used to mold the evanescent field at the core-cladding boundary. Here, the Vortex fiber is used for guiding the Vortex beam up-to the sensing region so that effective higher order modal excitation can be achieved inside the NCF [11, 20, 21]. Initially, a Gaussian beam from a single-mode pigtailed laser diode (1550 nm) impinges on the spiral phase plate. By controlling the phase pattern of the spiral phase plate, the input Gaussian beam is allowed to modulate with some phase delays which transforms to a Vortex beam having a dark central spot with an adjacent luminous area. Afterward, this Vortex beam is focused by using a convex lens and the focused beam is coupled inside the waveguide structure with the help of an input fiber coupler. When the NCF is submerged into the sensing liquid, the transmitted optical power will vary because of variations in the cladding refractive index as produced by different urine samples. Finally, the transmitted output power is collected by the use of an output fiber coupler and is read-out from an optical power meter.

In this study, we have conducted the theoretical investigation with the following finite-power input Vortex beam [11]:

$$E_s\left(r,\theta,0\right)=\left(\frac{r\sqrt{2}}{w}\right)^{L} L_0^L\left(\frac{2r^2}{w^2}\right)e^{-\left(r^2/w^2\right)}e^{-iL\theta} \tag{1}$$

Where r is the radial and θ is the azimuthal coordinate. The parameter $L_0^L\left(x\right)$ and w is the Laguerre function and beam waist, respectively. The factor L signifies the order of Vortex beam that determines the helicity of the beam throughout the fiber transverse plane.

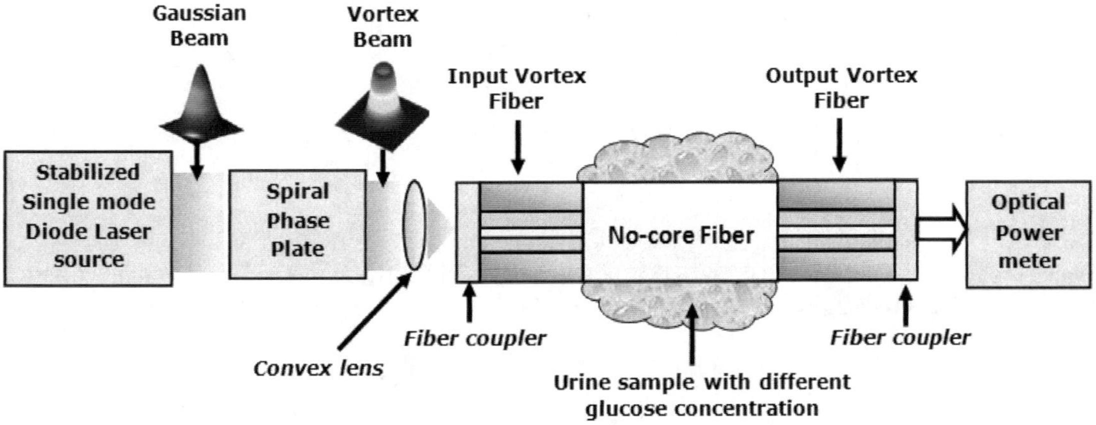

Figure 1. Design of the proposed glucose sensor by shining a Vortex beam.

The power coupling coefficient signifies the coupling of energy from the input field to different order modes as supported by the multimode fiber. To estimate the modal excitation inside the NCF, the power coupling co-efficient for various linearly polarized ($LP\mu,v$) mode can be estimated by doing the overlap integration between the applied input field ($E_s(r,\theta)$) and the electric field of $LP\mu,v$ mode ($E_{v,\mu}(r,\theta)$) as [22]:

$$\eta_{v,\mu} = \frac{\left| \int_0^{2\pi} \int_0^{\infty} E_s(r,\theta) E_{v,\mu}(r,\theta) r dr d\theta \right|^2}{\int_0^{2\pi} \int_0^{\infty} |E_s(r,\theta)|^2 r dr d\theta \int_0^{2\pi} \int_0^{\infty} |E_{v,\mu}(r,\theta)|^2 r dr d\theta} \tag{2}$$

The suffixes v and μ correspond to the indexes for the radial and azimuthal components, respectively. It is noteworthy that as the injected Vortex beam possesses topological charge (L) of 6; thus the modal fields of NCF are also influenced by the azimuthal variation of $\mu = L = 6$. This implies that only those modes which are having the azimuthal component of $\mu = \pm 6$ will be energized inside the NCF. Also, there will be excitation of corresponding $LP\text{-}\mu,v$ mode in combination with $LP\mu,v$ mode which is analogous but having the reverse polarity. Under such condition, the modal fields of NCF can be described as [23]:

$$E_{v,6}(r,\theta) = \frac{J_6\left(\frac{u_{v,6} r}{a}\right)}{J_6\left(u_{v,6}\right)} \cos(6\theta); \quad r \le a$$

$$= \frac{K_6\left(\frac{w_{v,6} r}{a}\right)}{K_6\left(w_{v,6}\right)} \cos(6\theta); \quad r > a \tag{3}$$

Where $u_{v,6}$ and $w_{v,6}$ is the normalized transverse wave number inside the core and cladding, respectively. Further, Eq. 2 can be explicitly solved as:

$$\eta_{v,6} = \frac{\left| \int_0^{2\pi} \int_0^{\infty} \left(\frac{r\sqrt{2}}{w} \right)^6 L_0^6 \left(\frac{2r^2}{w^2} \right) e^{-\left(r^2/w^2\right)} J_6 \left(\frac{u_{v,6} r}{a} \right) e^{-i6\theta} \cos(6\theta) r dr d\theta \right|^2}{\int_0^{2\pi} \int_0^{\infty} \left| \left(\frac{r\sqrt{2}}{w} \right)^6 L_p^6 \left(\frac{2r^2}{w^2} \right) e^{-\left(r^2/w^2\right)} \right|^2 r dr d\theta \left[\int_0^{2\pi} \int_0^{a} J_6 \left(\frac{u_{v,6} r}{a} \right)^2 \cos^2(6\theta) r dr d\theta + \left(\frac{J_6(u_{v,6})}{K_6(w_{v,6})} \right)^2 \int_0^{2\pi} \int_a^{\infty} K_6 \left(\frac{w_{v,6} r}{a} \right)^2 \cos^2(6\theta) r dr d\theta \right]} \quad (4)$$

Due to the multimodal property of NCF, the electrical field distribution can be formulated as the summation of all propagating modes as:

$$E(r,\theta,z) = \sum_{v=1}^{N} c_{v,6} E_{v,6}(r,\theta) \exp(i\beta_{v,6} z) \quad (5)$$

Where $c_{v,6}$ can be directly calculated through $c_{v,6} = \sqrt{\eta_{v,6}}$ and $\beta_{v,6}$ implies the propagation constant. The total transmission loss of the sensor can be represented as [24]:

$$L_s(z) = 10\log_{10}\left(\left| \sum_{\mu=-L}^{L} \sum_{v=1}^{N} c_{v,\mu}{}^2 \exp\left(-i\beta_{v,\mu} z\right) \right|^2 \right) \quad (6)$$

Based on the concept of multimode interference, the self imaging length (L_z) can be formulated by combining the effect of center wavelength (λ), refractive index of NCF (n_{co}), diameter of NCF (*D*), and penetration depth of evanescent field (*P*) as [24]:

$$L_z = \frac{4n_{co}(D+2P)^2}{\lambda} \quad (7)$$

Here, the value of penetration depth is decided by the refractive index of cladding. In our sensing technique, various urine samples with different glucose concentration behave like a cladding layer of the NCF. As mentioned in Eq. 7, the self image length tends to shift horizontally with varying cladding material as caused by the various urinary glucose levels. Owing to the higher energy density at the edge of Vortex beam, energy coupling in the higher order modes takes place. The quantity of energy flows in the sensing medium also increases for these higher order modes as they are having extended decaying energy tail. As a result, an efficient evanescent coupling and major modulation are detected in response to the urinary glucose concentration. Therefore, the designed Vortex beam-based sensor has an exceptionally high sensitivity as evaluated against the standard Gaussian beam-shined sensor.

3. Results & Discussions

By exploiting the computer-aided simulation, we have plotted the power coupling coefficients for different linearly polarized modes as supported by a 125 micron NCF having a refractive index of 1.4446. The wavelength of single-mode diode laser source is presumed as 1550 nm and the value of axicon apex angle is taken as 130°. The beam waist for both Gaussian and Vortex beam is considered as 7 μm. In this research, five different urine samples with glucose concentration of 0 gm/dL, 0.625 gm/dL, 1.25 gm/dL, 2.5 gm/dL, 5 gm/dL, and 10 mg/dL have been considered which corresponds to a refractive index of 1.335, 1.336, 1.337, 1.338, 1.341, and 1.347, respectively [1]. As revealed in Ref. [11], a Vortex beam can be guided by appropriately selecting the refractive index profile of the air-core Vortex fiber. It is inferred from Fig. 2(a) that the four most dominating modes for Gaussian beam excitation are the $LP_{0,3}, LP_{0,4}, LP_{0,5}, LP_{0,6}$ with relative power coupling coefficient of

0.131, 0.147, 0.144, and 0.127, respectively. Notably, the four most dominating modes for Vortex beam excitation are the $LP_{6,6}$, $LP_{6,7}$, $LP_{6,8}$, $LP_{6,9}$ (as shown in Fig. 2(b)) having estimated power coupling coefficient of 0.077, 0.099, 0.095, and 0.078, respectively. Moreover, for each $LP\mu,v$ mode, there will be excitation for corresponding $LP\text{-}\mu,v$ mode with same power coupling coefficient value but with opposite polarity.

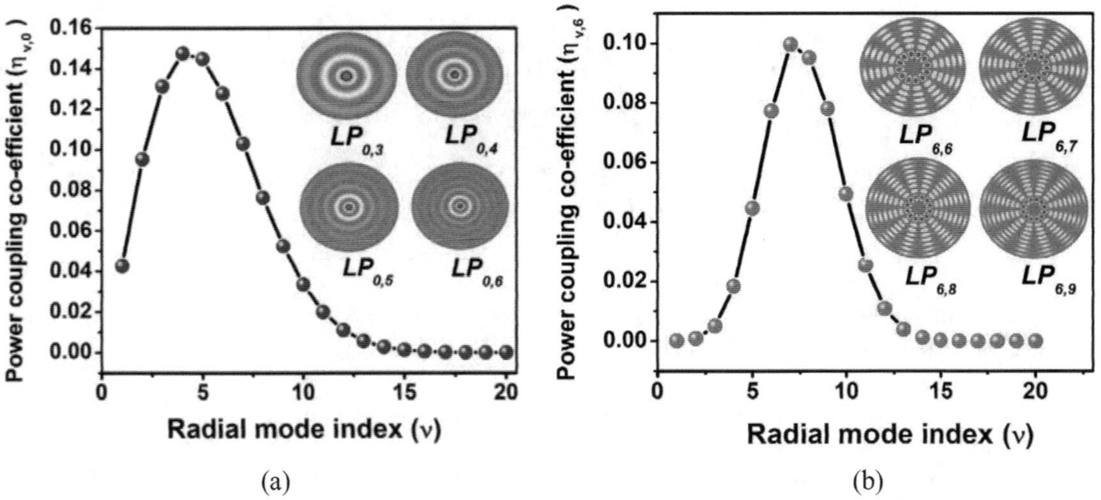

Figure 2. Variation of power coupling coefficient with mode order when the input is (a) Gaussian beam and (b) Vortex beam of order 6. Inset shows the four most dominating modes.

By following Eq. 5, the lateral field profiles are simulated subjected to various propagating distances along the NCF and are plotted in Figure 3. It is apparent that the lateral field distribution at self-imaging length (z =59.139 mm) is identical to the input Vortex mode, when the glucose content in urine sample is 10 gm/dL.

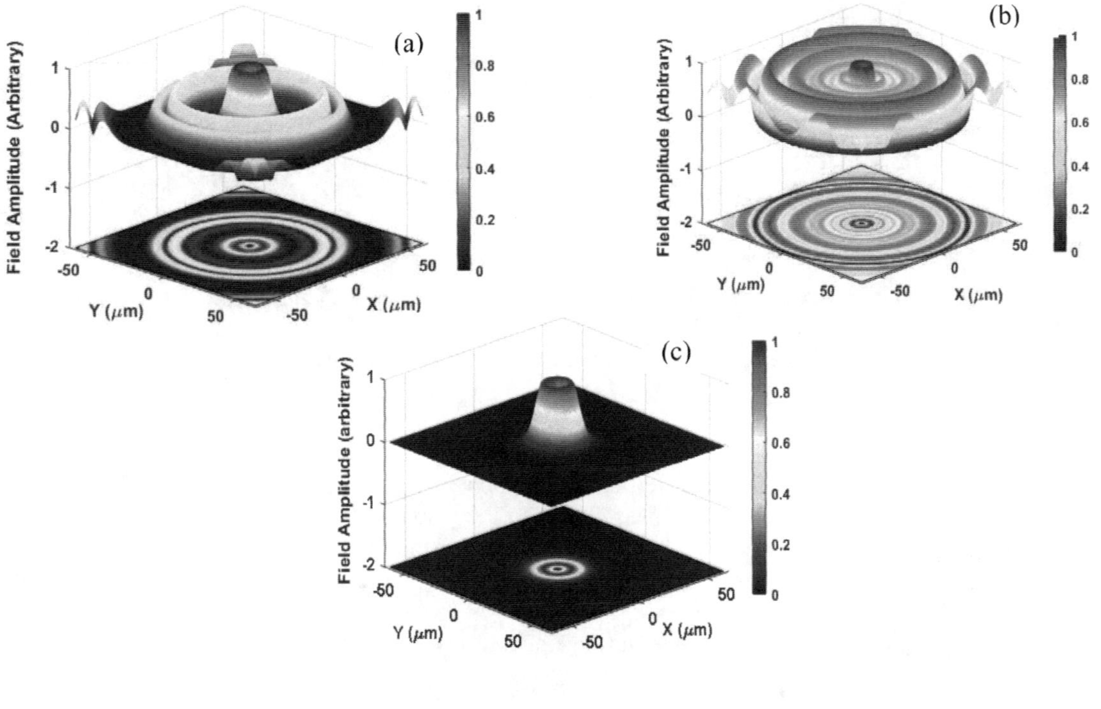

Figure 3. Simulated Elecric field profiles at (a) z=5 mm, (b) z= 40 mm, and (c) z= 59.139 mm (self-imaging length)

To envisage the transmitting property of the proposed device, we have simulated the proposed framework in Mode solution software (Lumerical Inc) by using an Eigenmode expansion solver. A perfectly matched layer (PML) boundary condition is incorporated in our study with a simulation mesh accuracy of 0.1 μm. The resultant field evolution and the interference carpet inside the NCF have been rendered in Figure 4 when the glucose content in urine sample is 10 gm/dL.

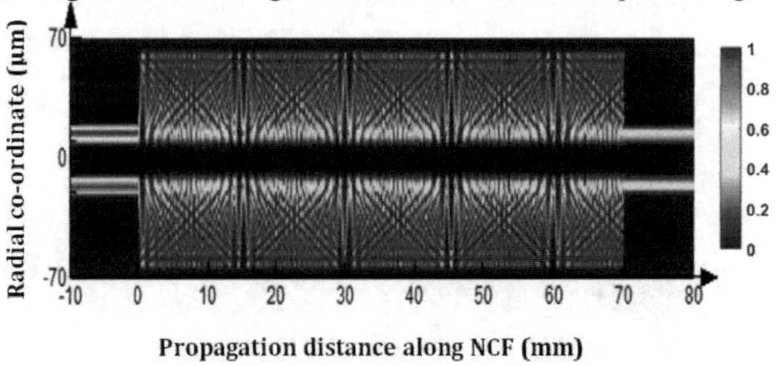

Figure 4. Transmission and propagation of Vortex beam through the waveguide structure when the glucose content in urine sample is 10 gm/dL.

Figure 5 illustrates the simulated transmission loss with respect to propagation distance along the NCF when the glucose concentration is 10 gm/dl. It is discerned that the suggested fiber-optic glucose sensor manifests a greater transmission loss of about -127.8 dB since there are several higher order modes within the NCF. But, it is worth mentioning that commercially available Femtowatt level optical power meter can measure low power accurately up to -100 dB. To comply with practical feasibility, the length of NCF will be chosen in such a way so that the transmission loss or sensor response lies within the range of minimum delectability of optical power meter.

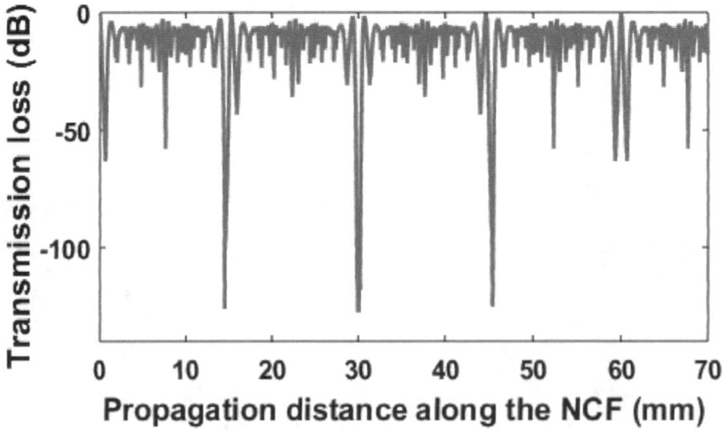

Figure 5. Transmission loss vs. propagation distance

In vision of Figure 6, it is anticipated that the transmission loss for the proposed glucose sensor fluctuates from -80.13 dB to -68.77 dB while the glucose level in urine sample varies from 0-10 gm/dL. On the contrary, a narrower discrimination range from -38.72 dB to -35.59 dB is observed for the Gaussian beam-based sensor. Through the study of the curve-fitting, a sensitivity of ~1.116 dB/gm/dL is obtained for the proposed glucose sensor. Whereas, it is monitored that the conventional glucose sensor (using the Gaussian beam) has a much lower sensitivity of ~0.305 dB/gm/dL. Thus, our sensing system envisages a 3.65-fold higher sensitivity. The sensing resolution can also be calculated by using a high-resolution optical power meter with 0.01 dB accuracy and is presented in Table 1. Because of this excellent sensitivity, the proposed sensor provides a new degree of freedom that finds intense implication in the field of biomedical engineering.

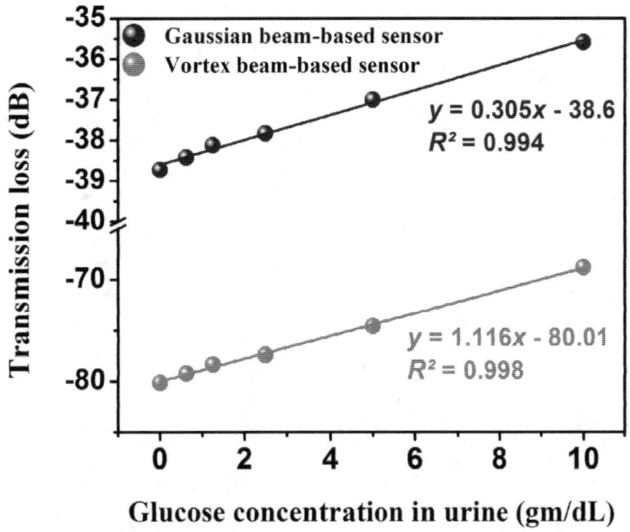

Figure 6. Sensor response when the NCF length is 29.30 mm.

Table 1. Comparative analysis of sensor's performance

Length of NCF (mm)	Sensitivity (dB/gm/dL)		Sensing Resolution (gm/dL)	
	Conventional Sensor	Proposed sensor	Conventional Sensor	Proposed sensor
29.30	0.305	1.116	0.032	0.008

4. Conclusion

The research mentioned in this paper offers the possibility of using Vortex beam in fiber-optic sensing to detect the glucose levels in urine. Such type of non-Gaussian beam will maximize the sensor's performance by exciting many higher order modes within the framework of fiber. The enticing aspect of such system is that it challenges the traditional wisdom and significantly improves the sensitivity without the need of any intricate fabrication techniques like in tapering, bending etc. With the intervention of Mode Solution software, the modal interference and the transmission of the overall sensor structure was visualized by using a Beam propagation method. While the sensing length is 29.30 mm, the proposed sensor proffers a maximum sensitivity of 1.116 dB/gm/dL which is 3.65 times greater than the standard Gaussian beam-based sensor.The sensor being proposed is extremely sensitive as it can detect even minute variation in the glucose concentration present in urine. Such novel urinary glucose sensor is very much effective in the diagnosis of diabetes or diseases associated with liver and kidney failure.

Acknowledgment

The authors would like to thank the Department of Electrical & Electronics Engineering, CMR Institute of Technology, Bengaluru, India, for the valuable motivation and support.

References

[1] Sharma P and Sharan P 2015 *IEEE Sensors J.* **15** 1035.
[2] Robinson S, Dhanlaksmi N 2017 *Photonic Sensors* **7** 11–19

[3] Bratlie K M, York R L, Invernale M A, Langer R, and Anderson D G 2012 *Advanced Healthcare Materials* **1** 267.
[4] Bruen D, Delaney C, Florea L, and Diamond D 2017 *Sensors(Basel)* **17** 1866.
[5] Ganesh A B and Radhakrishnan T K 2005 *Sensors & Transducers* **60** 439.
[6] Rosenzweig Z and Kopelman R 1996 *Sensors and Actuators B: Chemical* **36** 475.
[7] Chiu M H, Wand S F, and Chang R S 2005 *Opt. Lett.* **30** 233.
[8] Yao A M and Padgett M J 2011 *Advances in Optics and Photonics* **3** 161.
[9] Banerji A, Singh R P, Banerjee D, and Bandhopadhyay A 2016 *Optics Communications* **380** 492.
[10] Datta A and Saha A 2017 *J. Opt. Soc. Am. B* **34** 1327.
[11] Datta A, Saha A, and Shukla A 2017 *J. Opt. Soc. Am. A* **34** 2034.
[12] Datta A, Babu A M, and Saha A 2019 *Optical Engineering* **58** 056112.
[13] Saha A, Datta A, and Kaman S 2018 *Optical Engineering* **57** 036118.
[14] Datta A and Saha A 2020 *Optik* **218** 165006.
[15] Saha A, Datta A, and Kaman S 2016 *Proc.of International Conference on Fiber Optics and Photonics (Optical Society of America*) paperTh4F.
[16] Saha A and Datta A 2017 *Proc of Conference on Lasers and Electro-Optics /Pacific Rim (Optical Society of America)* paper s1796.
[17] Datta A and Saha A 2020 *Proc of 2nd International Conference on Innovative Mechanisms for Industry Applications (IEEE Xplore, 2020)* pp. 515-519.
[18] Datta A and Saha A 2020 *AIP Conference Proceedings* **2281** 020023.
[19] Datta A, Karmakar S, and Saha A 2020 *Proc. SPIE* **11525** 115250F.
[20] Ramachandran S and Kristensen P 2013 *Nanophotonics* **2** 455.
[21] Gregg P, Kristensen P, and Ramachandran S 2015 *Optica* **2**, 267.
[22] Mohammed W S, Mehta A, and Johnson E G 2004 *J. Light. Technol.* **22** 469.
[23] Ghatak A and Thyagarajan K 1998 *An Introduction to Fiber Optics (* Cambridge University Press).
[24] Wang Q, Farrell G, and Yan W 2008 *J. Light. Technol.* **26** 512.

Modelling the dynamic behaviour of car hydraulic dampers

A Dobre[1]

[1] Faculty of Transports, University Politehnica of Bucharest, Bucharest, Romania

alexandru.dobre@upb.ro

Abstract. When the vehicle travels on a road, that usually isn't perfectly smooth, this having a series of irregularities, occur shocks and vibrations due to the interaction of the wheels with the irregularities of the road, as well oscillations of sprung and unsprung masses that characterize comfort. This paper presents the theoretical and experimental researches undertaken by the author on the dynamic behaviour of hydraulic shock absorbers, in order to improve the comfort and dynamics of vehicles. The main objective of the paper is to analyse the damping characteristics of hydraulic dampers using simulation with AMESim specialized program. To validate the model, a set of experimental determinations was used with the help of a hydraulic dampers test stand. The experimental researches undertaken allowed to determine the real dynamic characteristics of the damper and to identify the influence of the parameters on these characteristics.

1. Introduction

The existence of the vibrations and oscillations at cars influences negatively the comfort of passengers, causing them great fatigue and unpleasant physiological sensations. Due to the vibrations and oscillations, the wear of some components of the car's transmission increases, but also the possibility of failure by breaking some of them due to the dynamic loads that appeared in the transmission. They also affect the integrity of the commodities transported. They negatively affect the comfort, stability, manoeuvrability, but also the reliability of the roads.

The appearance of vibrations and oscillations leads to an increase in fuel consumption, because the additional resistances in the elastic elements of the suspension must be overcome, but also to the increase of the additional energy losses in the tires [1], [2].

The dynamic behaviour of the hydraulic shock absorber and its damping characteristics greatly influence the overall response of the suspension system to input signals. To achieve a satisfactory level of comfort and manoeuvrability, the shock absorbers must be designed and adjusted according to various criteria, such as force, maximum displacement, speed etc. [3], [4].

The vibrations and oscillations in cars depend on the type of road they travel on and the type of vehicle. Thus they fall into two broad categories:

- Vibrations and oscillations produced by external sources – occur especially due to the road irregularities, but also to the wind gusts, braking etc.
- Vibrations and oscillations whose origin is located inside the car, being caused by the powertrain and steering system.

According to the causes that produce them, the oscillations are divided into:

- Free (or own) oscillations – occurs when a body or oscillating system is removed from its equilibrium position and then left free, oscillating around the equilibrium position.

- Forced (or maintained) oscillations – manifests itself when on a body or an oscillating system acts in addition to the tensile or compressive force of the suspension spring an external periodic force, which is called disturbing force (or excitation force).

Free oscillations can be:

- Unamortized – when theoretically it continues to infinity.
- Amortized – when after a certain time it stops by itself.

In conclusion, in the suspension system, the damper has the following roles:

- Dissipates quickly the energy of the vertical oscillations of the sprung mass (body car, chassis etc.), resulted from the deformation of the suspension;
- Determines the fast decrease of the oscillations of the unsprung masses (wheels, axles etc.) in order to ensure the continuous contact of the wheels with the road, by limiting the amplitude of the forced oscillations at the resonant frequency.

The shock absorbers are of two categories, friction (solid elements) and hydraulic (fluid elements). The friction shock absorber are distinct due to associated problems, such as sensitivity to water or oil contamination. Other problems of the friction shock absorber are due to the Coulomb friction characteristics of the shock absorbers [5]. There have been numerous studies on different aspects of hydraulic shock absorbers characteristics and performance [6], [7], [8], [9], [10], [11].

The influence of the damper on the vibrations of the sprung masses (m_1) and unsprung (m_2) is shown schematically in Figure 1, where A represents the amplitude of the vibrations, the blue line represents the non-damped vibration, and the magenta line the damped one.

The mass-spring and the spring-damper systems without the source term were studied in [12], [13]. Other applications of fractional calculus to mechanical oscillators are given in [14], [15], [16].

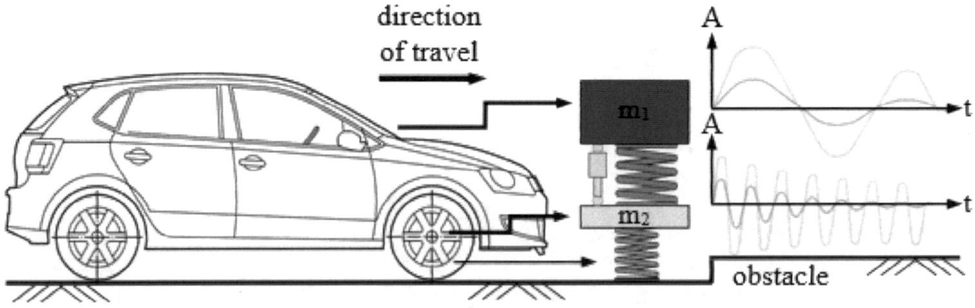

Figure 1. The influence of the shock absorber on the car.

Excitations are directly responsible for the oscillations of the cars. The main excitations that have major implications on the comfort are given by the irregularities of the road, to which are added the excitations due to the non-uniformity and unbalance of the wheels. The excitations can also be given by the change of the movement regime (acceleration, braking, turning), but these have low implications on the comfort.

2. Material and Methods

In this chapter I will present some the experimental results of the tests of hydraulic dampers, in order to analyse their dynamic behaviour. An example of a typical and practical simulation of a suspension equipped with a hydraulic damper is also presented in this chapter.

The test bench consists of the main components: electrohydraulic servomechanism, position transducer, force transducer, velocity transducer, accelerometer, hydro pneumatics accumulator, hydraulic damper, servo valve, servo cylinder etc. For the control of the electrohydraulic servomechanism and for the acquisition of the measured data, a PXI type industrial electronic computer is used, provided with a data acquisition board produced by the National Instruments Corporation, assisted by the LabVIEW program produced by the same corporation (Figure 2).

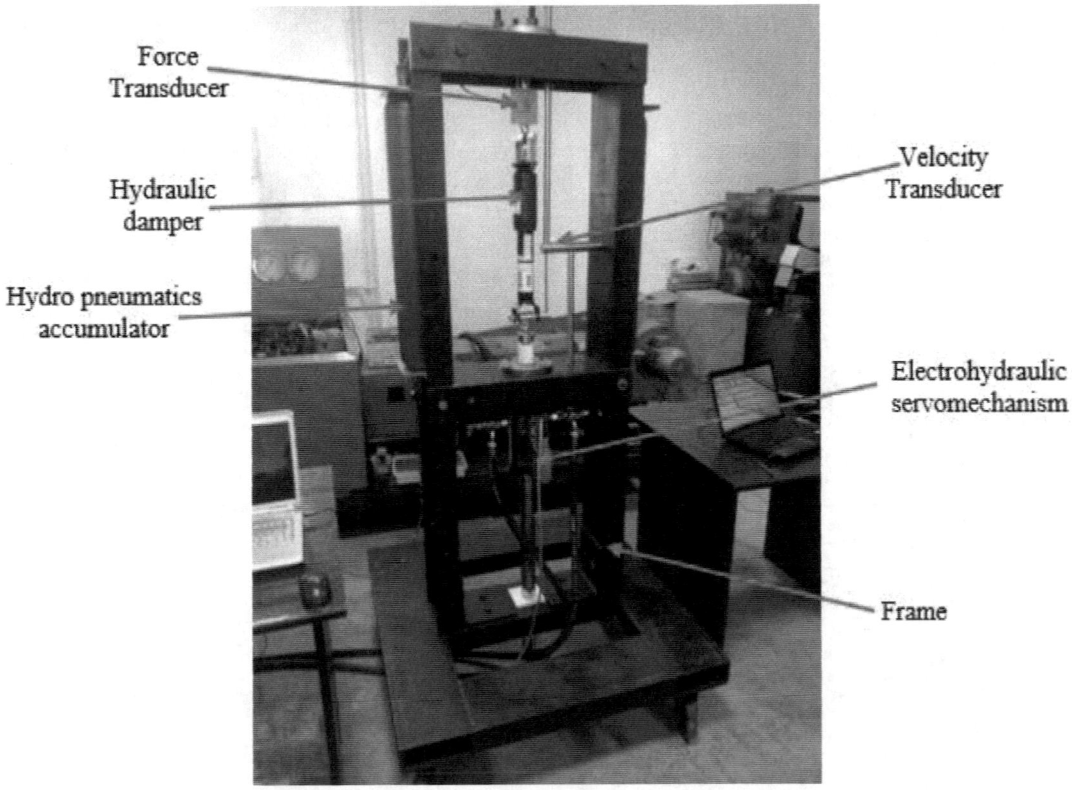

Figure 2. A part of the installation for testing hydraulic dampers.

To determine the regulation characteristic, the reference signal used was a triangular signal with an amplitude of 9 V and a frequency of 0.02 Hz, which provided the servomechanism with a quasi-stationary operating mode (piston speed was 3.6 mm/s). The sampling frequency was in this case 10 Hz. The registered characteristic is presented in Figure 3.

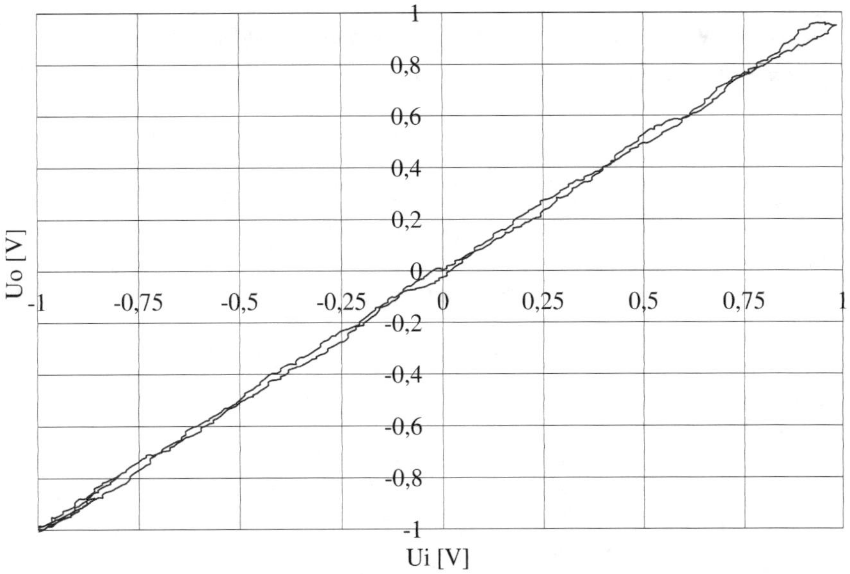

Figure 3. Schematic regulation characteristic of the electrohydraulic servomechanism.

The principle scheme of the tested electrohydraulic servomechanism is presented in Figure 4. The servomechanism electrohydraulic is compound from: 1 - assembled body; 2 - piston; 3 - composite sealing; 4 - anti-friction ring; 5 - anti-friction ring; 6, 7, 10 - "O" rings; 8 - compound seal; 9 - lid; 11 - felt ring; 12 - lid; 13 - grower washer; 14 - screw; 15 - position transducer; 16 - nut; 17 - lid; 18 - transducer support; 19 - screw; 20 - grower washer; 21 - limiter; 22 - screw; 23 - grower washer; 24 - support ring; 25 - lid, 26 - gasket; 27 - screw; 28 - servo valve.

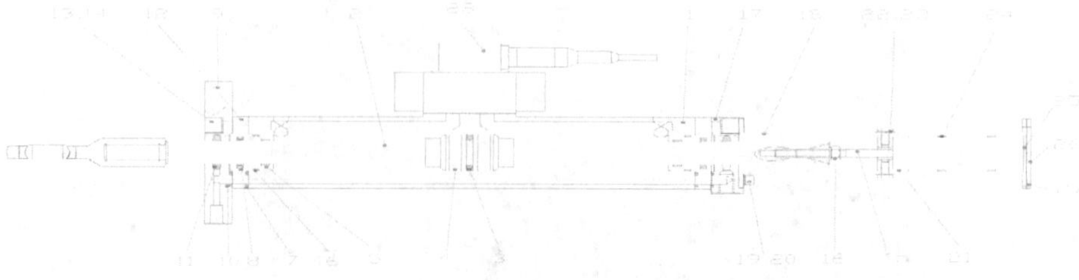

Figure 4. The principle scheme of the tested electrohydraulic servomechanism.

Figure 5 shows the interface of the LabVIEW work program, with the aid of which the main data acquisition program was developed.

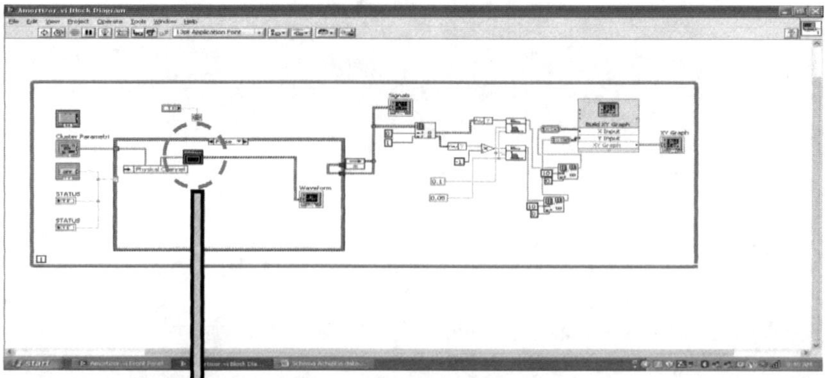

a

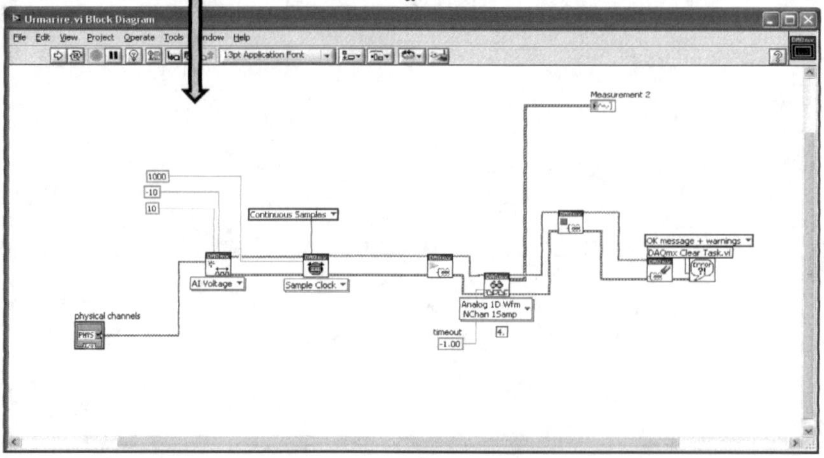

b

Figure 5. The data acquisition model in LabVIEW: a - The main data acquisition program; b - The data acquisition subprogram.

3. Result analysis

In recent years, a new language for modelling and numerical simulation of technical systems has emerged, called AMESim, developed by professors Lebrun and Richards in 1997. The language has been used successfully in both industrial and academic settings. The language allows the assembly of the mathematical models of the studied processes from correct models of technical components stored in the libraries written in the C programming language [17]. The simulation was performed using the AMESim program. The shock absorber was modelled according to the closed loop control law. Using a closed loop control law, we can consider a nonlinear compensation law. This compensation law imposes that we need to measure the pressure in the actuator cylinder (or the force) and the actuator velocity. Indeed, the aim is to adapt the gain of the system to the time constant due to hydraulic volumes of the cylinder and the restriction. The AMESim model of the actuator with the variable gain and the non-linear compensation law is presented in Figure 6 and response damper in Figure 7 and Figure 8.

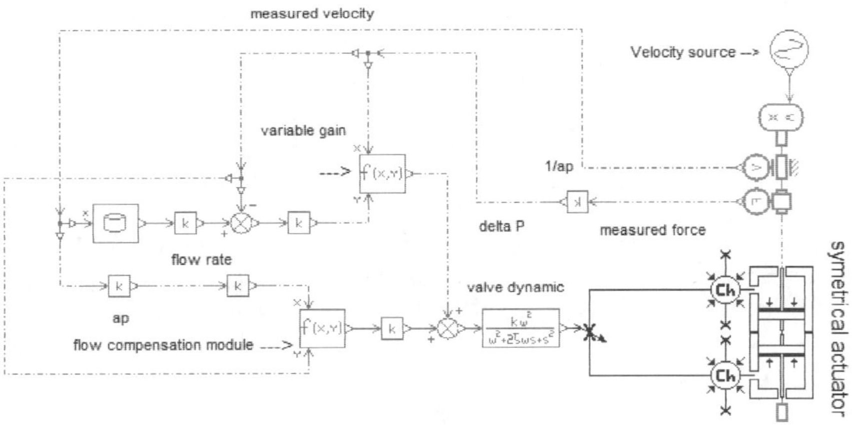

Figure 6. Model of a force control actuator based on a non-linear close loop control law.

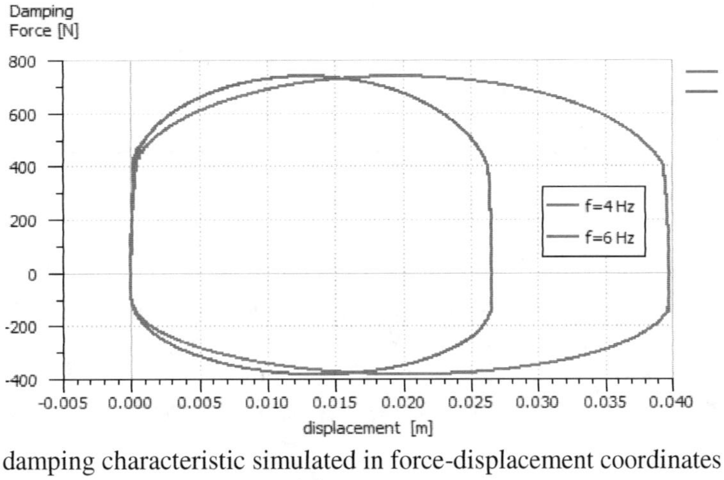

Figure 7. The damping characteristic simulated in force-displacement coordinates for a hydraulic damper.

The road excitation signal (input signal) was approximated with a sinusoidal signal with a frequency of 4 Hz and 6 Hz. At low frequencies (up to about 4 Hz) the reference is very well. At high frequencies (starting at about 6 Hz) the response may be better, but instead oscillations occur, especially when changing the slope. The force/velocity characteristic generated by the closed loop actuator has been improved, compared to the open loop one. The modelling of the vehicle suspension is of great interest for road vehicle engineers, but also for vibration engineers.

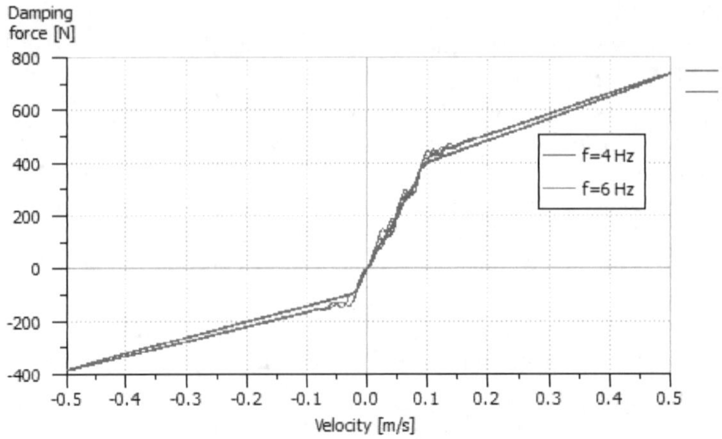

Figure 8. The damping characteristic simulated in force-velocity coordinates for a hydraulic shock absorber.

During the shock absorber testing, the two main characteristics were determined: damping characteristic in force-velocity coordinates and damping characteristic in force-displacement coordinates; which define the behaviour of the shock absorber. The characteristics were determined using the three transducers: speed, force and position. The damper simulation model was validated by experimental test. Some experimental results obtained for a hydraulic damper are presented in the Figure 9 and Figure 10. The maximum force sites in the range (-1100 N ... +1000 N) for a velocity of the piston sited in the range (-200 mm/s…+200 mm/s). The hydraulic damper tested is new and it is equipping a Chevrolet Corvette car.

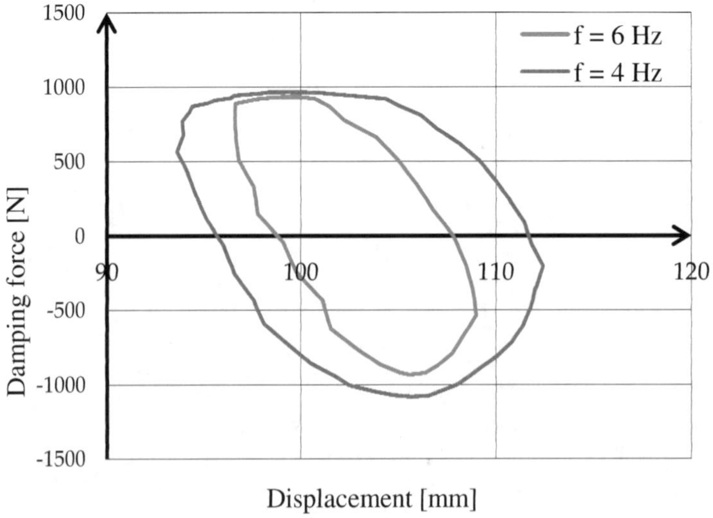

Figure 9. The damping characteristic determined experimentally in force-displacement coordinates for a classic shock absorber.

The energy dissipated by the hydraulic damper in a compression-rebound cycle is equal to the surface area delimited by the force-displacement characteristic (Figure 9). The characteristics presented above define the shock absorber itself and according to them the shock absorber produced by a manufacturer specialized in this field is chosen. The chosen shock absorber must provide a suspension damping characteristic, respective a damping characteristic required on the wheel. To determine the damping characteristic, the required damping range for the car body and wheels must first be determined.

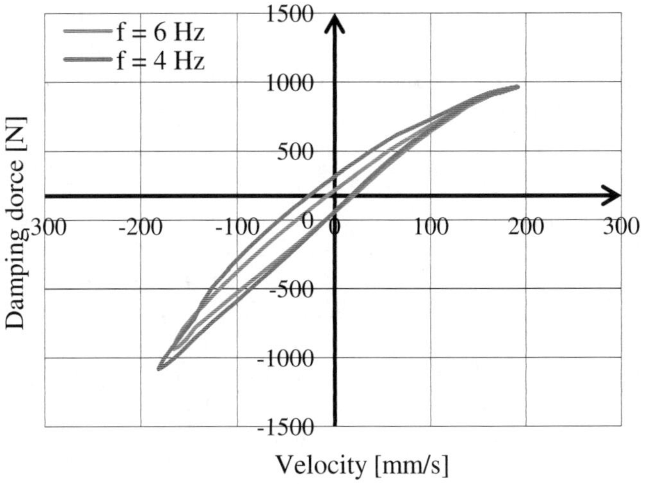

Figure 10. The damping characteristic determined experimentally in force-velocity coordinates for a classic shock absorber.

A sinusoidal signal were chosen as excitation signal with the excitation frequency is 4 Hz, respectively 6 Hz (Figure 11 and Figure 12). The variation in time of the damping force, respectively of the velocity of movement of the piston for a period of 0.25 s, respectively 0.16 s is observed.

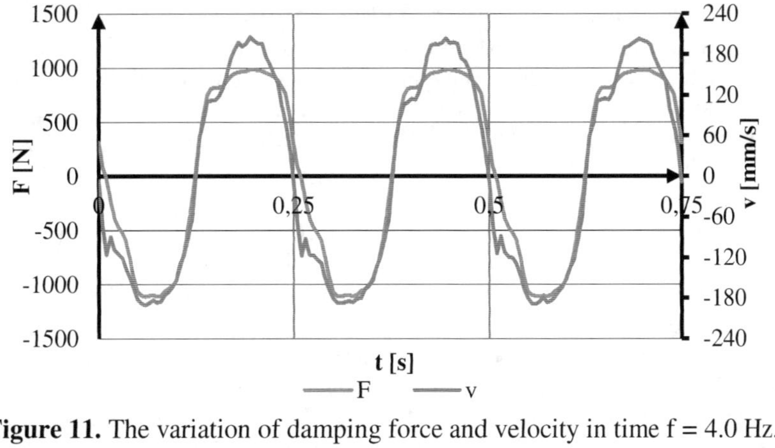

Figure 11. The variation of damping force and velocity in time f = 4.0 Hz.

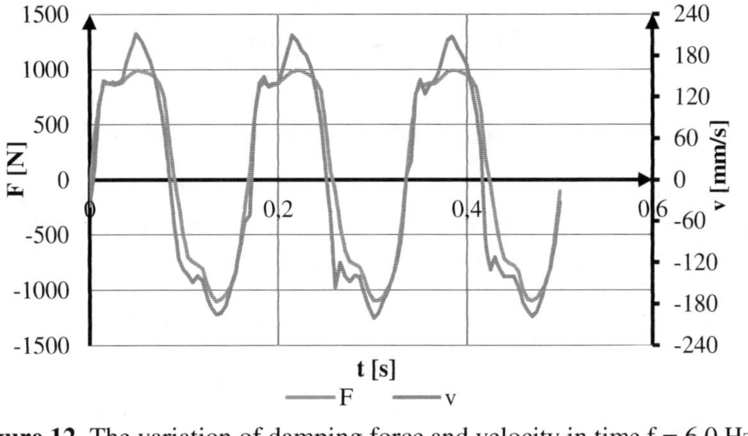

Figure 12. The variation of damping force and velocity in time f = 6.0 Hz.

The excitation of the hydraulic damper is done with a position signal, similar to the real operating situation, a signal which depending on the type of test performed can be sinusoidal, triangular or compound. The output parameter is the damping force developed by the hydraulic damper corresponding to the different load speeds.

4. Conclusions

Modern hydraulic mono-tubular dampers have two-disc valves with different diameters, which are insensitive to temperature variation and thus to viscosity variation. At the moving the vehicle on roads with long length bumps, when the damping force decreases, the comfort also decreases. The change in stiffness of the suspension associated with the change in damping (so that the relative damping remains unchanged) leads to small variations in the car's behaviour in terms of comfort and safety of wheel contact with the road. When moving the car, the dynamic forces taken over by the shock absorbers are quite high. As the load decreases, there is an increase in relative damping, which is essential for low frequency oscillations. When the sprung mass decreases, the natural frequency increases, favouring the appearance of the resonance phenomenon at the usual travel speeds. If the damping factor is maintained when the load decreases, the comfort of oscillation at resonance regimes decreases.

References

[1] Savaresi M S, Poussot-Vassal C, Spelta C, Sename O and Dugard L 2010 *Semi-active Suspension Control Design for Vehicles* (Elsevier Ltd.)

[2] Guglielmino E, Sireteanu T, Stammers C W, Ghita G and Giuclea M 2008 *Semi-active Suspension Control – Improved Vehicle Ride and Road Friendliness* (London: Springer-Verlag)

[3] Cao D, Song X and Ahmadian M 2011 Editors' Perspectives: Road Vehicle Suspension Design, Dynamics, and Control, *Veh. Sys. Dyn.* **49** 1-2 pp 3–28

[4] Farjoud A, Ahmadian M, Craft M and Burke W 2012 Nonlinear Modeling and Experimental Characterization of Hydraulic Dampers: Effects of Shim Stack and Orifice Parameters on Damper Performance, *Non. Dyn.* **67** 2 pp 1437–1456

[5] Dixon J C 2007 *The shock absorber handbook* 2nd edition (London: John Wiley & Sons Ltd.)

[6] Duym S W R 2000 Simulation Tools, Modelling and Identification, for an Automotive Shock Absorber in the Context of Vehicle Dynamics *Veh. Sys. Dyn.* **33** 4 pp 261–285

[7] Eyres R D, Champneys A R and Lieven N A J 2005 Modelling and Dynamic Response of a Damper with Relief Valve *Non. Dyn.* **40** pp 119–147

[8] Simms A and Crolla D 2002 The Influence of Damper Properties on Vehicle Dynamic Behaviour *SAE Technical Paper* 2002-01-0319

[9] Talbott M S and Starkey J 2002 An Experimentally Validated Physical Model of a High-Performance Mono-Tube Damper *SAE Technical Paper* 2002-01-3337

[10] Van Kasteel R, Cheng-Guo W, Lixin Q, Jin-Zhao L and Guo-Hong Y 2005 A New Shock Absorber Model for Use in Vehicle Dynamics Studies *Veh. Sys. Dyn.* **43** 9 pp 613–631

[11] Lu Y, Li S and Chen N 2013 Research on Damping Characteristics of Shock Absorber for Heavy Vehicle *Res. J. Appl. Sci. Eng. Technol.* **5** 3 pp 842-847

[12] Gomez-Aguilar J F, Yepez-Martinez H, Calderon-Ramon C, Cruz-Orduna I, Escobar-Jimenez R F and Olivares-Peregrino V H 2015 Modeling of a Mass-Spring-Damper System by Fractional Derivatives with and Without a Singular Kernel *Entropy* **17** 9 pp 6289-6303

[13] Gomez-Aguilar J F, Rosales-Garcia J J, Bernal-Alvarado J J, Cordova-Fraga T and Guzman-Cabrera R 2012 Fractional Mechanical Oscillators *Rev. Mex. Fis.* **58** pp 348-352

[14] Li Z H and Au F T K 2014 Damage Detection of a Continuous Bridge from Response of a Moving Vehicle *Sh. and Vib.* vol. 2014 Article ID 146802

[15] Naranjani Y, Sardahi Y, Chen Y Q and Sun J Q 2015 Multi-Objective Optimization of Distributed-Order Fractional Damping *Commun. Non. Sci. Numer. Simul.* **24** 1-3 pp 159–168

[16] Tavazoei M S 2015 Reduction of Oscillations via Fractional Order Pre-filtering *Sign. Proc.* **107** pp 407–414

[17] Vasiliu N, Vasiliu D, Calinoiu C and Puhalschi R 2018 *Simulation of Fluid Power Sistem with Simcenter Amesim* (CRC Press Taylor & Francis Group)

Predictions of fugacity coefficients of pure substances from equations of state

S Kaur[1*] and B P Akhouri[2]
[1]Department of Physics, Nirmala college, Ranchi-834002, Jharkhand (India)
[2]Department of Physics, Birsa College, Khunti-835210, Jharkhand (India)
[*]Corresponding author. Email: sumitbharaj@gmail.com

Abstract. This aim of this paper is to predict fugacity coefficients from newly developed equations of state. The fugacity coefficients have been calculated for Methane and Caron di-oxide from equations of state of Soave Redlic Kwong (SRK), Peng Robinson (PR) and Twu Sim Tassone (TST) taking reduced temperatures, T_r of interest from each of the three ranges $0<T_r<1$, $1<T_r<3$ and $1<T_r<5.5$ together with the reduced pressures, P_r in steps of 0.2 in the range $0<P_r<10$. The calculated results of fugacity coefficients were compared with the available fugacity coefficient data for Beattie-Bridgeman (BB) equation. The calculations suggest that the TST equation of state is most suitable in predicting the fugacity coefficient of pure substances.

Keywords: Cubic Equations of State, Compressibilty Factor, Fugacity Coefficients, Beattie-Bridgeman (BB) equation

1. Introduction

In the petroleum engineering applications, the cubic equations of state are widely used because of its simplicity, solvability, accuracy and generalization [1-10]. Some cubic equations and their modifications have been proposed by many investigators. A simple equation of state such as van der Waals, cannot accurately model the dense or complex fluids mixtures. After formulation of vdW EOS (1973) numerous modifications have been made to improve its ability by modifying the repulsive and attractive terms. In all van der Waals type equation of state, we recognize a repulsion term that takes into account the volume of the molecules using parameter b, or covolumes, and secondly an attraction term, or internal pressure that is dependent on the parameter a. The unlikely hypothesis of van der Waals was that the repulsive term, expressed by a constant b is more effective in predicting the fluid properties than the attractive term. But, modified attractive term a for all the van der Waals type equation of state is found to be more promising in the theory of equations of state. The values for these parameters are determined by applying critical constraints. The equations of state is satisfied at the critical point as: $\left(\frac{\partial P}{\partial V}\right)_T = \left(\frac{\partial^2 P}{\partial V^2}\right)_T = 0$, for $T=T_c$ and $P=P_c$.

1.1 Equations of state:

A unified form for most commonly used two parameter cubic equations of state with modified representation of attractive term is:

$$P=\frac{RT}{V-b}-\frac{a=a_c\alpha(T_r)}{V^2+\Omega_1 bV+\Omega_2 b^2} \tag{1}$$

where Ω_1 and Ω_2 are two integer values specific for each cubic equation and are given in Table.1. The numerical values for the dimensional parameters Ω_1 and Ω_2 , and the critical compressibility factor are given in Table 1. The actual values of the critical compressibility factor are, however, between 0.26 and 0.28 for popular hydrocarbons. It is noted that the values of Z_c from the Soave-Redlich-Kwong [8, 9] and Peng-Robinson [10] models are both larger than 0.3 (0.333333 and 0.307401, respectively), but that for TST is slightly below 0.3 (0.297), which is closer to the typical value of critical compressibility factor for most compounds [1, 2]. The EOSs which are tabulated in Table 1 are more useful because they provide an analytical solution for the density.

Table 1. Constants in equation (1) for cubic equations of state [6]:

	Ω_1	Ω_2	a_c	α	b	z_c
VdW [7]	**0**	**0**	$\frac{27}{64}\frac{R^2T_c^2}{P_c}$	**1**	$\frac{RT_c}{8P_c}$	**0.375**
RK [8]	**1**	**0**	$\frac{0.42748R^2T_c^2}{P_c}$	$T_r^{-1/2}$	$\frac{0.08664RT_c}{P_c}$	**0.333**
SRK [9]	**1**	**0**	$\frac{0.42748R^2T_c^2}{P_c}$	$\left[1+f_w\left(1-T_r^{1/2}\right)\right]^2$ $f_\omega=0.48+1.57\omega-0.176\omega^2$	$\frac{0.08664RT_c}{P_c}$	**0.333**
PR [10]	**2**	**-1**	$\frac{0.45724R^2T_c^2}{P_c}$	$\left[1+f_w\left(1-T_r^{1/2}\right)\right]^2$ $f_\omega=0.37464+1.54226\omega-0.2699\omega^2$	$\frac{0.07780RT_c}{P_c}$	**0.307**
TST [1,2]	**0.5**	**3**	$\frac{0.470507R^2T_c^2}{P_c}$	$T_r^{N(M-1)}e^{L\left(1-T_r^{NM}\right)}$	$\frac{0.0740740RT_c}{P_c}$	0.296
LM [11]	**2** -1	1	$\frac{0.4572355R^2T_c^2}{P_c}$	For $T_r<1.0$ $1+p(T^*)^q$; $T^*=\frac{T_c/T-1}{T_c/T_b-1}$ For $T_r>1.0$; $1-p(T^*)^q$ $T^*=\frac{T/T_c-1}{T_c/T_b-1}$	$\frac{0.077960RT_c}{P_c}$	0.307

p, q, L, M and N are substance dependent constants[1,6].

The parameter a is a function of temperature. The value of a at any temperature $\alpha(T_r)$ can be calculated from $a(T)=\alpha(T_r)a_c$, where the alpha, $\alpha(T_r)$, is a function only of reduced temperature, $T_r=T/T_c$. We use the Twu alpha function[1, 2] which is given in Table 1 has three parameters, L, M and N. These parameters are unique to each component and can be determined from the regression of pure component vapor pressure data[1]. Table 2 lists the L, M and N parameters for

CH_4, CO_2, N_2, H_2S and H_2O. In this work we have only interest on the pure substance CH_4 and CO_2.

1.2 The Beattie-Bridgeman (BB) equation:

The graphical and the analytical methods are the mainly tools for the calculation of fugacities. In both the methods the knowledge of P-V-T data is necessary. The condition of employing analytical method is that the P-V-T data must be fitted to an analytical expression for the equation of state so that it can be integrated. It is found that the Beattie-Bridgeman equation [3, 4, 5] is of such a form and with the proper constants closely represents the P-V-T data for many gases over a wide range of temperatures and pressures.

It is written as:

$$PV^2 = RT\left(1-\frac{c}{VT^3}\right)\left(V+B_0-\frac{bB_0}{V}\right)-A_0\left(1-\frac{a}{V}\right) \tag{2}$$

where A_0, B_0, a, b, and c are empirical constants to be determined experimentally for a particular gas and P, V, R, and T have their usual meanings. Values of the constants of the Beattie-Bridgeman equation of state for CH_4: $A_0 = 2.2769$ $a = 0.01885$ $B_0 = 0.05587$ $b = -0.01587$ $c = 12.83\times10^4$ and for CO_2: $A_0 = 5.0065$ $a = 0.07132$ $B_0 = 0.10476$ $b = 0.07235$ $c = 66\times10^4$, respectively [3, 4, 5]. Equation (1) can be rearranged in a quadratic form in V which can be further solved for one real positive root must found close to the experimental value. In this EOS, the fugacity, f, can be determined by the expression: $\ln\left(\frac{f}{P}\right) = -\frac{1}{RT}\int_0^P \alpha dP$, where, $\alpha = \frac{RT}{P} - V$. The equations used for calculating the fugacity coefficients, $\phi = f/P$ are given in Table 2. The paper is organized in such a way that the second section represents the theory for calculating the fugacity coefficients for pure substances and the third section explains the results and discussion and finally the conclusions.

2. Theory

In statistical mechanical theory, the fugacity defines the grand canonical ensemble (a system that may exchange particles with the environment). G.N. Lewis (1923), the American chemist, introduced Gibbs energy using an auxiliary property fugacity. Its logarithm, multiplied by k_B, is the chemical potential, $\mu = k_B T \log f$ where, k_B is the Boltzmann constant, and T is the temperature. In other words, fugacity, $f = \exp\left(\frac{\mu}{k_B T}\right)$. Actually, fugacity coefficient is relatively abstract value and is used in place of fugacity for pure substance. For Gibbs energy, fugacity plays the same role for a real fluid as pressure plays for ideal gas. This means that when a system approaches the ideal gaseous state

$$\frac{f}{P} \to 1 \quad \text{if } P \to 0 \tag{3}$$

Thus fugacity is equal to effective pressure which is useful in order to preserve the expressions for ideal gas when it is substituted for pressure.

2.1 Fugacity coefficients using available experimental data:

When sufficient data on the compressibility factor, Z, and pressure, P, are available, the fugacity coefficient can be obtained from the numerical integration:

$$\ln\phi = \int_0^P \frac{Z-1}{P}dP \tag{4}$$

Table 2. Physical property of the component and parameters of Twu alpha function [6]:

Component	T_c(K)	P_c(atm)	T_b(K)	L	M	N	ω
CO_2	304.21	72.80	194.70	0.945951	0.888652	0.65000	0.225
CH_4	190.564	45.99	111.70	0.081382	0.905296	2.13000	0.008
N_2	126.20	34.00	77.40	0.064994	0.892385	2.34000	0.040
H_2S	373.53	89.62	212.875	0.231877	0.784346	1.12000	0.100
H_2O	647.13	22.05	373.15	0.430058	0.870932	1.67211	0.344

Table 3. Expression of fugacity coefficients for different EoS :

EOS	$\log\phi$
VW[7]	$Z-1-\ln\left\{Z\left(1-\frac{b}{V}\right)\right\}-\frac{a}{RTV}$
RK[8]	$Z-1-\ln\left\{Z\left(1-\frac{b}{V}\right)\right\}-\frac{a}{aRT}\ln\left(1+\frac{b}{V}\right)$
SRK[9]	$Z-1-\log(Z-B)+\frac{a}{bRT}\ln\left(\frac{Z}{Z+B}\right)$
PR[10]	$Z-1-\log(Z-B)+\frac{A}{2\sqrt{2}B}\ln\left(\frac{Z-0.414B}{Z+2.414B}\right)$
ML[11]	$Z-1-\log(Z-B)+\frac{A}{2\sqrt{2}B}\ln\left(\frac{Z-0.414B}{Z+2.414B}\right)$
TST[1,2]	$Z-1-\log(Z-B)+\frac{A}{3.5B}\ln\left(\frac{Z-0.5B}{Z+3B}\right)$
BB[4]	$\ln\left(\frac{RT}{V}\right)+\left(B_0-\frac{c}{T^3}-\frac{A_0}{RT}\right)\frac{2}{V}+\left(bB_0-\frac{cB_0}{T^3}-\frac{aA_0}{RT}\right)\frac{3}{2V^2}+\frac{cbB_0}{T^3}\frac{4}{3V^3}$

Here, $A = \frac{aP}{R^2T^2}$ and $B = \frac{bP}{RT}$ can be determined from the three parameters of the gas: critical pressure, P_c; critical temperature, T_c and the eccentric factor ω.

Table 4. Represents the coefficients of the equation, $Z^3 + a_2Z^2 + a_1Z + a_0 = 0$, used for the calculation of largest root of compressibility factor Z for different EOS:

EOS		$Z^3 + a_2Z^2 + a_1Z + a_0 = 0$	
VW[7]	$a_2 = B - 1$	$a_1 = -A$	$a_0 = -AB$
SRK[9]	$a_2 = -1$	$a_1 = A - B - B^2$	$a_2 = -1$
PR[10]	$a_2 = B - 1$	$a_1 = A - 2B - 3B^2$	$a_0 = B^3 + B^2 - AB$
ML[11]	$a_2 = B - 1$	$a_1 = A - 2B - 3B^2$	$a_0 = B^3 + B^2 - AB$
TST[1,2]	$a_2 = 1.5B - 1$	$a_1 = A - 2.5B - 4B^2$	$a_0 = 1.5B^3 + 1.5B^2 - AB$

3. Results and discussion

We know that the curves are the guides for the eye. Plotting has been performed for every inspection for the nature of the drawn curve. For plot of these curves, the fugacity coefficients have been calculated for Methane and Caron di-oxide from equations of state of Soave Redlic Kwong (SRK), Peng Robinson (PR) and Twu Sim Tassone (TST) taking reduced temperatures, T_r of interest from each of the three ranges $0 < T_r < 1$, $1 < T_r < 3$ and $3 < T_r < 5.5$ together with the reduced pressures, P_r in steps of 0.2 in the range $0 < P_r < 10$.Let us first consider the plot of calculated fugacity coefficients of methane from considered EOSs for $T_r = 3.149$ and $T_r = 5.248$, belonging to the range $1 < T_r < 5.5$, and varying P_r from 0 to 8. By a close look into the Fig. 1 we find that the continuous curve represents the TST data, which agrees to some extent to the data of BW for $T_r = 3.149$ and $T_r = 5.248$ both. For the same T_r values, the dashed curve of SRK and the dotted curve of PR EOS show a complete disagreement behavior with BB data [4, 5]. The SRK EOS is predicting well for $T_r = 5.248$, but for $T_r = 3.149$ it completely fails to predict the data of BW. The reverse result is obtained in case of the PR EOS, which is predicting to some extent for $T_r = 3.149$ while for $T_r = 5.248$ it fails completely to predict the data of BB [4,5]. In Fig. 2 the plot has been performed for $T_r = 1.574$ and $T_r = 2.099$ which belonging to $1 < T_r < 3$. For both of these T_r values the TST EOS predicting well with the BB data. The dashed curve of SRK EOS gives the coefficient values more than the BB data for each T_r. The curve PR has exact coincidence with BB data these reduced temperature and for $0 < P_r < 1$. In Fig. 3, for $T_r = 0.84$ and $T_r = 0.945$, the SRK EOS again reproducing the fugacity coefficients value more than the BB data for the considered pressure range i.e., $0 < P_r < 0.6$. But both the TST EOS and SRK EOS have an exact match with the BB data for the considered pressures range.

Similarly, for plot of CO_2 fugacity coefficient curves, we find in Fig. 4 that the continuous curve represents the TST data, which agrees to some extent to the data of BW for $T_r = 2.17$ and

$T_r = 3.287$ both and the range of P_r. For the same T_r values, the dashed curve of SRK and the dotted curve of PR EOS show a complete disagreement behavior with BB data. The SRK EOS is predicting well for $T_r = 2.17$, but for $T_r = 3.287$ it completely fails to predict the data of BW. The reverse result is obtained in case of the PR EOS, which is predicting to some extent for $T_r = 2.17$ while for $T_r = 3.287$ it fails completely to predict the data of BB. In Fig. 5 the plot has been performed for $T_r = 1.315$ and $T_r = 2.972$ which belonging to $1 < T_r < 3$. For both of these T_r value the TST EOS predicting well with the BB data. The dashed curve of SRK EOS gives the fugacity coefficient values more than the BB data for each T_r. The curve PR has exact coincidence with BB data these T_r values and for $0 < P_r < 1$. In Fig. 6, for $T_r = 0.986$ and $T_r = 0.945$, the SRK EOS again reproducing the fugacity coefficients value more than the BB data for the considered pressure range i.e., $0 < T_r < 0.6$. But both the TST EOS and SRK EOS have an exact match with the BB data for the considered range of P_r. In Fig. 7, for $T_r = 3.149$ and $T_r = 5.248$, the comparative plots has been performed for methane from each equation of state to observe it as a eye wink.

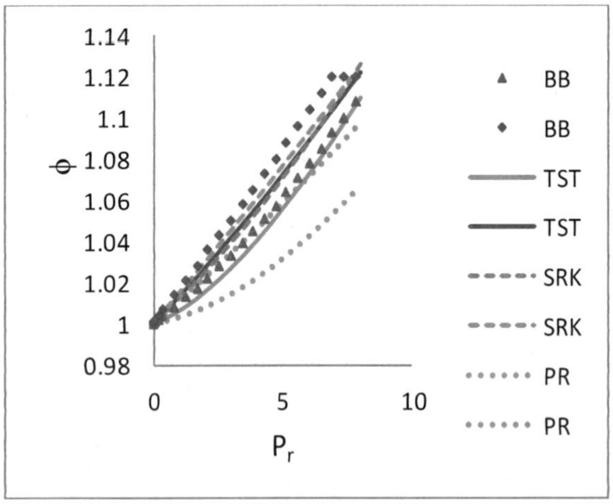

Figure 1. Plot for the pure substance methane: for $T_r = 3.149$ curve [5], its plots against P_r represent the TST EOS (first continuous curve), the SRK EOS (first dashed curve), and the PR EOS (first dotted curve), respectively and for $T_r = 5.248$ curve [5], its plots against P_r represent the TST EOS (second continuous curve), the SRK EOS (second dashed curve), and the PR EOS (second dotted curve), respectively. For higher values of T_r ($3 < T_r < 5.5$) the values of fugacity coefficient increase with P_r.

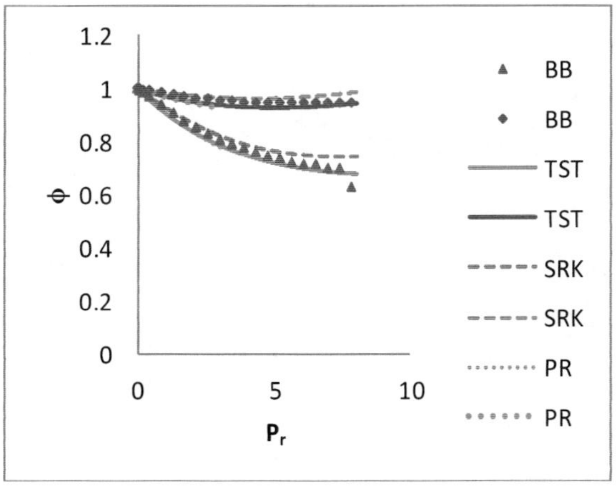

Figure 2. Plot for the pure substance methane: for $T_r = 1.574$ curve [5], its plots against P_r represent the TST EOS (first continuous curve), the SRK EOS (first dashed curve), and the PR EOS (first dotted curve), respectively and for $T_r = 2.099$ curve [5], its plots against P_r represent the TST EOS (second continuous curve), the SRK EOS (second dashed curve), and the PR EOS (second dotted curve), respectively. For middle values of T_r $(1 < T_r < 3)$ the values of fugacity coefficient decrease in small steps with P_r.

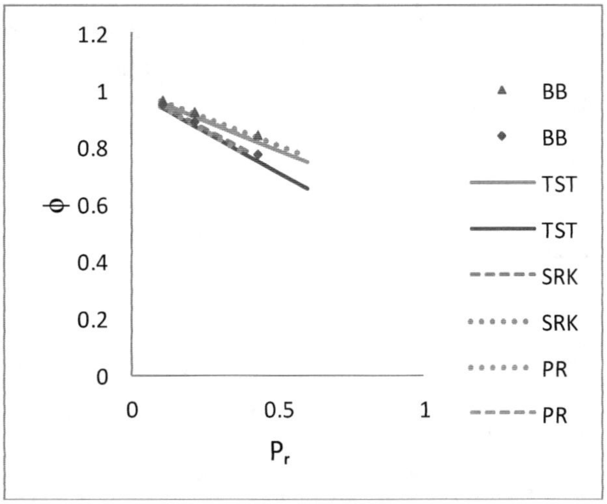

Figure 3. Plot for the pure substance methane: for $T_r = 0.84$ curve [5], its plots against P_r represent the TST EOS (first continuous curve), the SRK EOS (first dashed curve), and the PR EOS (first dotted curve, respectively and for $T_r = 0.945$ curve [5], its plots against P_r represent the TST EOS (second continuous curve), the SRK EOS (second dotted curve), and the PR EOS (second dashed curve), respectively. For lower values of T_r $(0 < T_r < 1)$ the values of fugacity coefficient decrease in big steps with P_r.

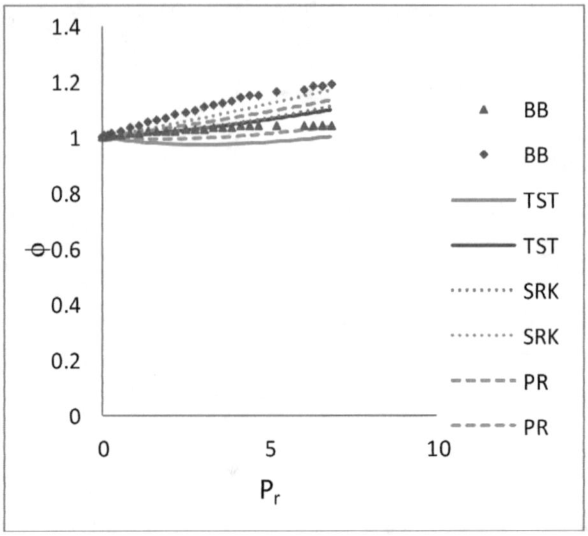

Figure 4. Plot for the pure substance CO_2: for $T_r = 2.17$ curve [5], its plots against P_r represent the TST EOS (first continuous curve), the SRK EOS (first dotted curve), and the PR EOS (first dashed curve), respectively and for $T_r = 3.287$ curve [5], its plots against P_r represent the TST EOS (second continuous curve), the SRK EOS (second dotted curve), and the PR EOS (second dashed curve), respectively. For higher values of T_r $(2 < T_r < 5.5)$ the values of fugacity coefficient increase with P_r.

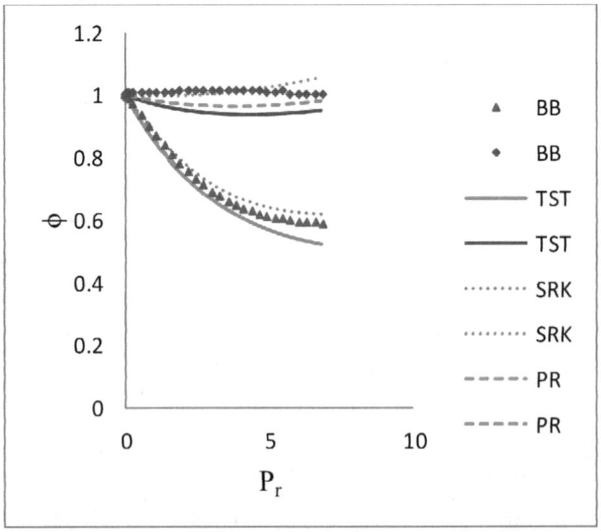

Figure 5. Plot for the pure substance CO_2 : for $T_r = 1.315$ curve [5], its plots against P_r represent the TST EOS (first continuous curve), the SRK EOS (first dotted curve), and the PR EOS (first dashed curve), respectively and for $T_r = 1.972$ curve [5], its plots against P_r represent the TST EOS (second continuous curve), the SRK EOS (second dotted curve), and the PR EOS (second dashed curve), respectively. For middle values of $T_r = 1.315$ curve [5] $(1 < T_r < 3)$ the values of fugacity coefficient decrease in small steps with P_r and for $T_r = 1.972$ curve [5] $(1 < T_r < 3)$ the values of fugacity coefficient decrease in somewhat larger steps with P_r.

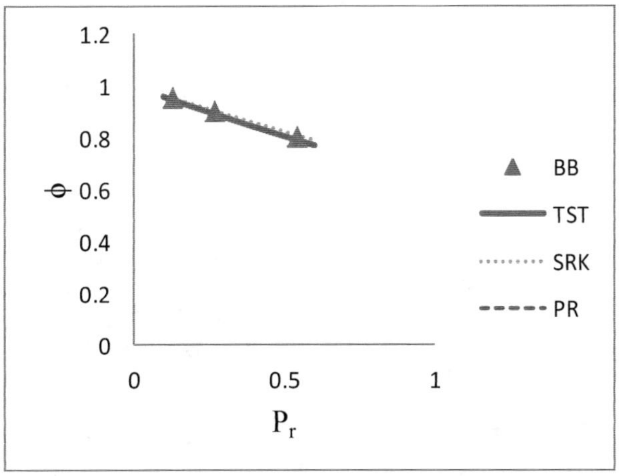

Figure 6. Plot for the pure substance CO_2: for $T_r = 0.986$ curve[5], its plots against P_r represent the TST EOS (continuous curve, the SRK EOS (dashed curve), and the PR EOS (dotted curve), respectively. For lower values of T_r ($0 < T_r < 1$) the values of fugacity coefficient decrease in big steps with P_r.

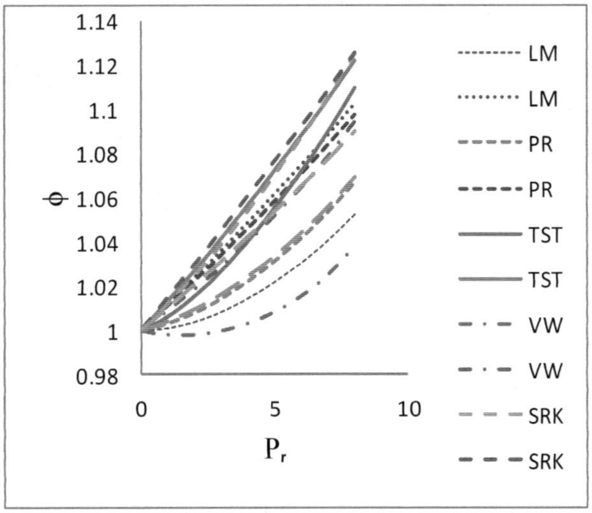

Figure 7. Plot for the pure substance methane: for $T_r = 3.149$ curve [5], its plots against P_r represent the LM EOS (first dotted curve), the PR EOS (first dashed curve), the TST EOS (first continuous curve), the VW EOS(first dashed dotted curve), and for SRK EOS (first bigger dashed curve) , respectively. For $T_r = 5.248$ curve [5] its plots against P_r represent the LM EOS (first dotted curve), the PR EOS (first dashed curve), the TST EOS (second continuous curve), the VW EOS(second dashed dotted curve), and for SRK EOS (second bigger dashed curve) , respectively. For higher values of T_r ($3 < T_r < 5.5$) the values of fugacity coefficient increase with P_r.

4. **Conclusions**

The fugacity coefficients from different equations of state and with different T_r values have been compared with the calculated values of BB [3, 4, 5]. The outcome of this study is that the calculated fugacity coefficients for selected different T_r values in the region of interest were found either of increasing or decreasing nature with P_r. It is interesting to note that the TST EOS among all others shows its ability to reproduce the data as predicted by BB. Finally, the comparison of the fugacity coefficient values obtained for CH_4 and CO_2, it can be concluded that among all the equations of state the newly developed EOS of Twu Sim Tassone the is most authentic.

5. **References**

[1] Twu C H, Tassone V and Sim W D 2005 *Fluid Phase Equilibria* **228** 213
[2] Twu C H, Tassone V and Sim W D 2001 *Fluid Phase Equilibria* **183** 65
[3] Beattie J A and Bridgeman O 1928 *American Academy of Arts & Science* **63** 229
[4] Beattie J A, Hadlock C and Offenber P 1935 *J. Chem. Phys* **3** 93
[5] Holley C E, Worlton W J Jr. and Zeigler R K 1958 "Chemistry General, compressibility factors and fugacity coefficients calculated from the Beattie and Bridgeman equation of state for hydrogen, nitrogen, oxygen and carbon dioxide, ammonia, methane and helium", **14th** Ed. California
[6] Polling B E, Prausnitz J M and O' Connell J P 1958 *The Properties of Gases and liquids,* **5th** Edition, McGRAW-Hill
[7] van der Waals J D 1873 *doctoral dissertation*, Leiden, Holland
[8] Redlic O and Kwong J N S 1949 *Chem. Rev.* **44** 223
[9] Soave G 1972 *Chem. Eng. Sci* **27** 1197
[10] Peng D Y and Robinson D B 1976 *Ind. Eng. Chem. Fundam.* **15** 59
[11] Lielmezs J and Merriman L H 1972 *Thermochimica Acta* **197** 131

Modified Fluorescent PEG-Based Carrier System for *In-vitro* Imaging

Aswathi Thomas[1], and Dr. Aravind Kumar Rengan[1]*

[1]Department of Biomedical Engineering, Indian Institute of Technology Hyderabad, Kandi, Sangareddy- 502 28, Telengana, India.

Corresponding auther *e-mail* id: aravind@bme.iith.ac.in

Abstract. A facile, one-pot synthesis of polymer based fluorescent coating agent was developed. Polyethylene glycol (PEG), known to be biocompatible and biodegradable, has been modified with ascorbic acid (AA) to yield modified fluorescent PEG based carrier system (PEGAA). The fluorescence obtained was tuned using varying concentrations of ascorbic acid. With increasing concentration of ascorbic acid, a red shift in fluorescence was observed which could be utilized for effective *in vitro* imaging. The cellular uptake studies reveal the potential use of this novel material for bioimaging applications.

Keywords: Polyethylene glycol, PEGAA, Bioimaging

1. Introduction

Owing to the *in vivo* compatibility and degradability, biodegradable polymers are widely opted for various biomedical applications such as tissue engineering, drug delivery, gene delivery etc. They have been used as coating material for implants to eliminate the direct interaction between implants and biomolecules in order to reduce toxicity. [1-2] Biocompatible carrier systems, especially polymeric carrier systems are often used in drug delivery to reduce drug associated toxicity. The type and nature of the polymeric coating up on the drug molecule, influences the particle size, drug release and the physiological circulation time. [3-4] Among polymeric carrier systems, low molecular weight Polyethylene glycol (PEG) molecules exhibiting enhanced biocompatibility and optical properties are well preferred for diagnostic applications. [5] PEG, well recognized for its high biocompatibility and resistance to opsonization due to inherent immune response, has paved the way for its use in the medical research field as efficient coating agents. [6] Additionally, PEG coating provides an enhanced circulation time under physiological conditions when compared to other polymeric coatings. The enhanced circulation time increases the enrichment of drug entities within the tumour site through enhanced permeability and retention (EPR) effect. Literature reports indicated that PEG of varying molecular weights, or their conjugates such as lipo-polymer carrier systems, are well considered for encapsulation of drugs. [7] PEG-based carrier systems in conjugation with fluorescent dyes are used for cancer/tumour diagnosis to eliminate the toxic side-effects of these dye molecules. Though the complex conjugation results in reduced dye toxicity, it fails to control photo-bleaching and particle size. Moreover, the complex coating over dye along with photo-bleaching deduces the fluorescence signal strength leading it to poor choice for bioimaging. [8] Further, fluorescing metallic nanoclusters such as gold nanoclusters are investigated for their better bio-imaging properties compared to chemical dyes. The surface interaction of these particles with biomolecules changes the hydrophobicity-hydrophilicity balance leading to aggregation followed by fluorescence quenching. [9,10] In our previous work, we have reported the synthesis of fluorescently modified PEG to overcome the shortcomings of current fluorescent carrier systems. [11] The biocompatible PEG exhibited excellent fluorescence upon heating at optimum conditions. This modified fluorescent PEG, termed as FL PEG, possessed good biocompatibility, enhanced circulation time and tunable fluorescence. Though *in vitro* bioimaging showed a tunable fluorescence, the fluorescent emission intensity of FL PEG was found to be maximum when it was excited with a wavelength of 450 nm (blue light). The corresponding emission wavelength was obtained at 515 nm (Green emission).

This work is aimed to improve the fluorescence intensity throughout the visible spectra. PEG of molecular weight 300 (PEG M.w 300), used in the synthesis of PEGAA, owing to its better biocompatibility. In order to tune the emission intensity, Ascorbic acid was used to modify PEG resulting in formation of PEGAA. The absorbance spectra and fluorescence emission spectra of the particle recorded, showed the tunability of inherent fluorescence of the novel carrier system. Further, *in vitro* bioimaging studies were performed to authenticate the bioimaging application of the developed complex.

2. Materials and Methods

2.1. Materials

Polyethylene glycol (PEG, Mw:300) and Ascorbic acid were purchased from SRL (India). Phosphate buffer pH 7.0, Trypsin-EDTA, DMEM, Bovine Serum (US origin. All chemicals were reagent grade and used as received. Ultra-pure de-ionized water (Milli Q) was used for all solution preparations and experiments.

2.2. Synthesis of PEGAA sample:

PEG (Mw :300) was heated at 95 ºC for about 30 minutes, followed by the addition of predetermined concentrations of ascorbic acid (5 mg, 10 mg, 20 mg and 30 mg). Upon further heating for 5-7 minutes, the bright yellow coloured solution obtained was monitored under a UV lamp of 365 nm. The green fluorescence observed indicated the formation of PEGAA complex.

2.3. Characterization:

The PEGAA sample was evaluated for its fluorescence under UV lamp (365 nm, Analytik Jena, CA, U.S.A) as initial characterization. Excitation and emission properties of PEGAA samples were analyzed using fluorescence spectroscopy (RF 6000, Shimadzu, Japan). The UV-Visible absorbance was analyzed by UV-visible spectroscopy (UV 1800, Shimadzu, Japan),

2.4. In-vitro studies

2.4.1 Cell culture & maintenance: Human breast cancer cell lines (MCF-7) were obtained from CSIR-Centre for Cellular and Molecular Biology, Hyderabad, India. The cell line was cultured in DMEM medium supplemented with 10% (v/v) fetal bovine serum (FBS), 1% L-Glutamine, and 100 U/ml penicillin/streptomycin and were maintained at 37 ºC in a humidified atmosphere containing 5% CO_2 under sterile conditions.

2.4.3 In-vitro imaging: Human breast cancer cells (MCF-7) were used for bio imaging of PEGAA samples. Briefly, cells were seeded in a 24 well plate, 24 hours prior to the treatment, at a density of 5×10^4 cells per well. On the following day, the cells were treated with PEGAA prepared with varying ascorbic acid concentration (5 mg, 10 mg, 20 mg and 30 mg). The cells were incubated with PEGAA samples for 24 hours. After the incubation period, the culture media was removed, cells were washed twice with PBS and the fluorescence was recorded with fluorescent microscopy (Olympus CKX-53, U.S.A). [12,13]

3. Results and Discussion

3.1. Synthesis and characterization of PEGAA samples

Upon heating PEG in presence of ascorbic acid, the solution turned bright yellow in color and was initially characterized for its fluorescence. The samples showed a green emission when showered with UV-lamp of wavelength 365 nm. Figure 1a shows the bright light imaging of PEGAA prepared with various concentrations of AA (5 mg/ml, 10 mg/ml, 20 mg/ml and 30 mg/ml). It was observed that with increase in the concentration of ascorbic acid from 5 mg/ml to 30 mg/ml, colour of the solutions turned darker. Figure 1b shows the corresponding fluorescence of PEGAA samples. The fluorescence observed under the UV lamp was validated with fluorescence spectroscopy. The normalized fluorescence spectra (Figure 1c) shows the shift in emission peak as the concentration of AA increases.

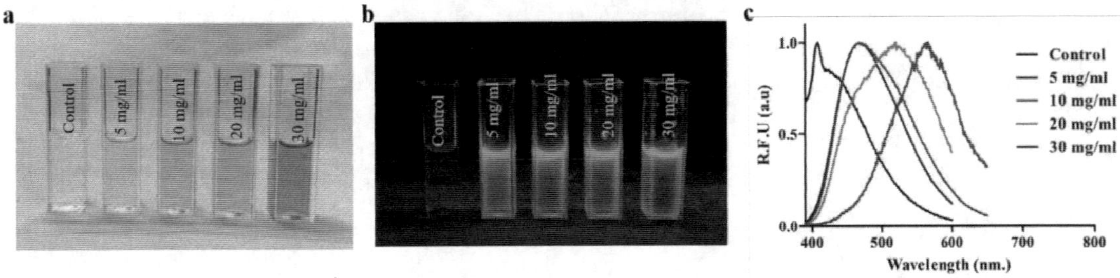

Figure 1: Characterization of PEGAA sample: Digital photographic image under a) Bright light, b) UV lamp of 365 nm and c) Normalized emission spectra for the excitation 365 nm.

UV Visible absorption spectroscopy was further used to evaluate the absorbance (Figure 2a). The absorbance spectra obtained were similar to that of our previous work. [11] Black solid line representing the UV absorbance of PEG diluted against water was considered as experimental control. The absorbance spectra revealed formation of a broad spectra, in place of the characteristic absorbance peak, at high frequency region (190-400 nm). The UV Vis absorbance spectra indicates a red shift in absorbance with increase in the concentration of AA. This could be attributed to the formation of molecular aggregates caused by elevated temperature. Due to the absence of characteristic UV Vis absorbance peak, fluorescence of the sample was analyzed with excitation and emission scan. Both Emission (Figure 2b) and Excitation (Figure 2c) spectra showed a red shift in peaks as the concentration of AA increases.

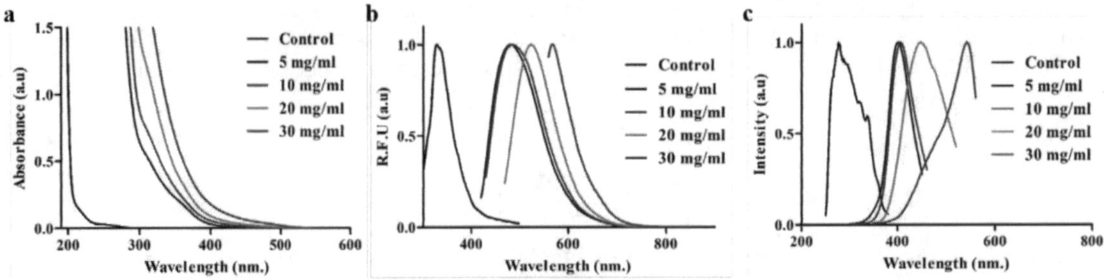

Figure 2: Spectral analysis: a) UV Visible Absorbance spectra b) Normalized Emission spectra and c) Normalized excitation spectra of PEGAA samples upon various concentrations of Ascorbic acid (5 mg/ml, 10 mg/ml, 20 mg/ml and 30 mg/ml).

An experiment was performed to analyze the fluorescence behaviour of the PEGAA sample. PEGAA containing 10 mg/ml of AA, having a better emission intensity, was taken for this experiment. 1 ml of the sample was added with equal volume of nHexane. nHexane was used accounting for its immiscibility and lower density when compared with the PEGAA sample. The emission spectra showed (red solid line) a slight increase in intensity when compared to the control sample (green solid line). Further, addition of nHexane (2 ml) showed a significant decrease in fluorescence (Figure 3a). Similar effect was also observed in the corresponding excitation spectra (Figure 3b). The experimental results indicate that though the PEGAA samples can obtain an increased intensity through aggregation induced emission (AIE), further molecular aggregation can result in quenching of the fluorescence (AIQ).

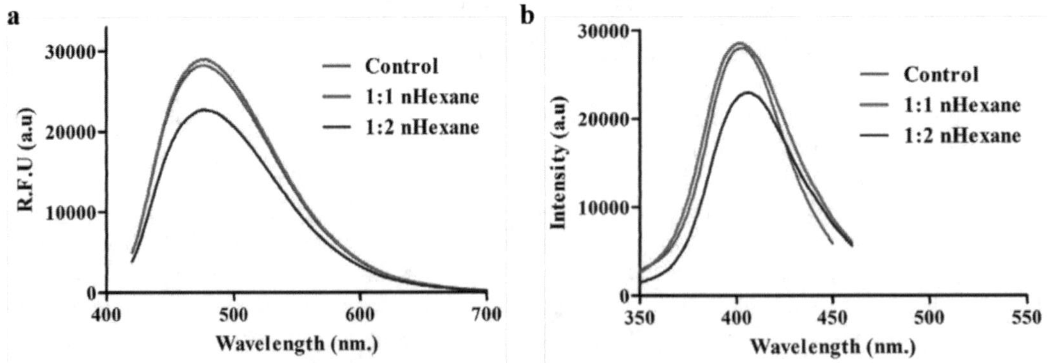

Figure 3: Fluorescence analysis: a) Emission and b) Excitation spectra.

3.2. *In vitro studies*

A qualitative in vitro bioimaging of PEGAA samples was evaluated in Human breast cancer cells (MCF-7) cells. Upon 24 hours of incubation, the cells treated with predetermined concentration of PEGAA were analyzed for cellular uptake using fluorescence microscopy.

Figure 4 shows the fluorescence of PEGAA samples of various concentrations of ascorbic acid. The biocompatible, self-fluorescing complex exhibited fluorescence under green light as shown. The cells were able to exhibit fluorescence at an ascorbic acid concentration of 10 mg/ml. A faint red fluorescence could be observed when the sample was excited using green light. A better fluorescence due to cellular uptake was visible for the ascorbic acid concentration of 30 mg/ml. In all the cases of varying ascorbic acid concentrations, the cell morphology remains unaffected, indicating the biocompatible nature of PEGAA samples. This further points out that the material has the potential to replace the existing non fluorescent carrier systems and potentially toxic chemical dyes.

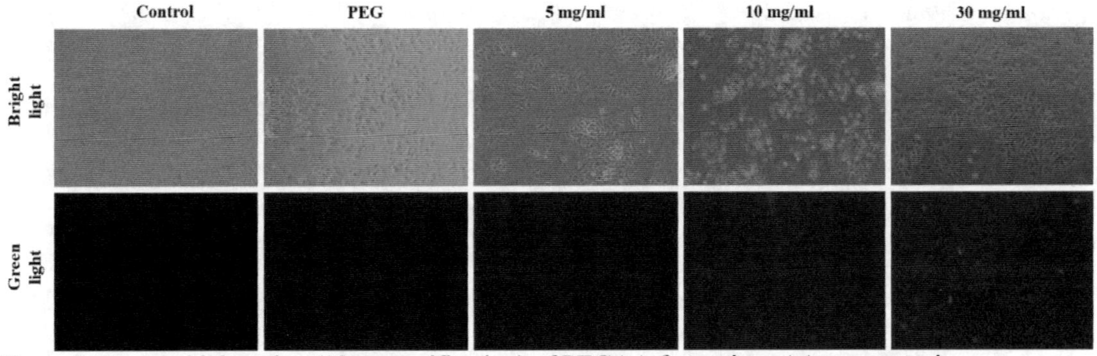

Figure 4: *In vitro* bioimaging (10x magnification) of PEGAA for various AA concentrations.

4. Conclusion

In summary, we have developed a facile one-pot approach for the synthesis of PEGAA using PEG and ascorbic acid. The spectral analysis confirms the shift in self fluorescence of PEGAA from green to yellow upon varying concentrations of ascorbic acid. *In vitro* bioimaging studies reveal the possibility of its application as fluorescent encapsulating agent for medical devices and toxic drug molecules.

Acknowledgments

The authors thank the Department of Science & Technology (DST), the Department of Biotechnology (DBT), the Ministry of Human Resources & Development (MHRD), Government of India, for funding the following projects: DST-Inspire (DST/INSPIRE/04/2015/000377), DBT-NNT (BT/NNT/28/1386/2017), DST-AMT (DST/TDT/AMT/2017/227(G)), and MHRD IMPRINT (4291). The authors also thank Vasudha Foundations, Hyderabad, for their generous financial support. Author AT would like to acknowledge MHRD for providing her fellowship. Author AT would like to thank Sri Amruthaa for her valuable help.

References

[1] H. Zhou, Z. Fan, P.Y. Li, J. Deng, D.C. Arhontoulis, C.Y. Li, W.B. Bowne, H. Cheng, Dense and dynamic polyethylene glycol shells cloak nanoparticles from uptake by liver endothelial cells for long blood circulation, (2018) *ACS nano* **12(10)** 10130-10141.

[2] W.J. Kelley, C.A. Fromen, G. Lopez-Cazares, O. Eniola-Adefeso, PEGylation of model drug carriers enhances phagocytosis by primary human neutrophils, (2018) *Acta biomaterialia* **79** 283-293.

[3] Y. Zhang, H.F. Chan, K.W. Leong, Advanced materials and processing for drug delivery: the past and the future, (2013) *Advanced drug delivery reviews* **65(1)** 104-120.

[4] Y. Wang, L. Feng, S. Wang, Conjugated Polymer Nanoparticles for Imaging, Cell Activity Regulation, and Therapy, (2019) *Advanced Functional Materials* **29(5)** 1806818.

[5] A. Dreiss, L.A. Dailey, Low molecular weight PEG – PLGA polymers provide a superior matrix for conjugated polymer nanoparticles in terms of physicochemical, (2019*) J. Mater. Chem. B*, **7**, 5115-5124

[6] A.A. D'souza, R. Shegokar, Polyethylene glycol (PEG): a versatile polymer for pharmaceutical applications, (2016). *Expert Opin. Drug Deliv.* **13** 1257–1275.

[7] M. Zhang, X.H. Li, Y.D. Gong, N.M. Zhao, X.F. Zhang, Properties and biocompatibility of chitosan films modified by blending with PEG, (2002) *Biomaterials.* **23** 2641–2648.

[8] A. Reisch, A.S. Klymchenko, Fluorescent Polymer Nanoparticles Based on Dyes: Seeking Brighter Tools for Bioimaging, (2016) *Small.* **12** 1968–1992.

[9] E. Porret, L. Sancey, A. Martín-Serrano, M.I. Montañez, R. Seeman, A. Yahia-Ammar, H.

Okuno, F. Gomez, A. Ariza, N. Hildebrandt, J.B. Fleury, J.L. Coll, X. Le Guével, Hydrophobicity of Gold Nanoclusters Influences Their Interactions with Biological Barriers, (2017) *Chem. Mater.* **29** 7497–7506.

[10] S. Zhu, X. Wang, Y. Cong, L. Li, Regulating the optical properties of gold nanoclusters for biological applications, (2020) *ACS Omega.* **5** 22702–22707.

[11] A. Thomas, T. Appidi, A. Jogdand, S. Ghar, K. Subramaniyam, G. Prabusankar, J.R. Mohanty, A.K. Rengan, Facile Synthesis of Fluorescent Polymer Encapsulated Metal (PoeM) Nanoparticles for Imaging and Therapeutic Applications, (2020) *ACS Appl. Polym. Mater.* **2** (*Preprint.*) 1388–1397.

[12] X. Qiu, L. Xu, Y. Zhang, A. Yuan, K. Wang, X. Zhao, J. Wu, H. Guo, Y. Hu, Photothermal ablation of in situ renal tumor by PEG-IR780-C13 micelles and near-infrared irradiation, (2016) *Molecular pharmaceutics* **13(3)** 829-838.

[13] D. Pemmaraju, T. Appidi, G. Minhas, S.P. Singh, N. Khan, M. Pal, R. Srivastava, A.K. Rengan, Chlorophyll rich biomolecular fraction of A. cadamba loaded into polymeric nanosystem coupled with Photothermal Therapy: A synergistic approach for cancer theranostics, (2018) *International journal of biological macromolecules* **110** 383-391.

Investigation of Performance Parameters Affecting the Efficiency of Solar Water Heater: A Review

Swapnil R. Mane [a,*], Dr. Rajesh V. Kale [b]
[a] Assistant Professor, Department of Mechanical Engineering, Vidyavardhini's College of Engineering & Technology, Vasai 401202, Maharashtra, India.
[b] Head, Department of Mechanical Engineering, MCT's Rajiv Gandhi Institute of Technology, Mumbai 400053, Maharashtra, India.

ABSTRACT

The need for installation of Solar Water heater is increasing in society because of several factors such as rapid urbanization, government interventions, low cost of installations, and environment-friendly application acting as a direct replacement to fossil fuels. During the past few years, the research and development associated with the technological enhancement of the utilization of solar energy have increased exponentially. However, there are various challenges involved in the selection of proper solar technology to provide a high-performance energy harvesting application for domestic water heating requirements. There is a wide literature available on various performance parameters required to develop an efficient Solar Water Heating System. This paper investigates a state-of-the-art review of the performance parameters affecting the efficiency of Solar Water Heaters by component-wise analysis of parameters divided broadly into the design, operational and external parameters. The technological advancements in solar water heaters are classified based on performance parameters and the paper summarizes the possibility of combining different performance parameters to achieve more efficient and cost-effective solar water heaters for the society as future scope of the review for researchers working in the similar domain.

Keywords:
Solar Water Heating System, Performance parameters, Solar Efficiency, Flat Plate Collector, Evacuated Tube Collector, Phase Change Materials, Nanoparticles

1. Introduction

Global climate change has resulted to hunt for non-conventional energy sources in place of conventional fossil fuel-based energy sources. As per the Indian context and with the government favoring the use of renewable energy, the utilization of solar energy has increased by leaps and bounds. Solar technologies can be grouped into two broad domains, Solar Photovoltaic (SPV) which generates electricity, Solar Thermal (ST) which produces heat directly from the energy of the Sun and new technology which involves both generations of electricity and process heat which is Solar Photovoltaic and Thermal (SPV/T). All these choices have their advantages and disadvantages touted in the literature. To achieve higher solar contribution by selecting the best solar technology, one has to study the dominant factor responsible for altering the efficiency of the system [1]. Even though Solar energy is available in abundance, the technology required to harness the power of the Sun is costly.

The residential sector has an 80% share of Solar Water Heating (SWH) systems, the commercial sector which includes hotels, hospitals, others have 6%, 3%, 5% respectively and the industrial sector contains a 6% share of SWH systems. The percentage of share data shows that the residential sector is one of the largest sectors among commercial and industrial sectors for India. The residential sector is also the largest sector for the whole of the world, with China dominating at 10% of the total global share of SWH systems [2]. About 31.3 gigawatts-thermal (GWth) of glazed (including flat plate and vacuum tube technology) and unglazed solar collectors are being added globally in 2019 and by year's end about 479 GWth in operation [3]. With the announcement of National Solar Mission by the Government of India, there is increase in the number of SWHS deployment across India. The phase wise target is set to achieve 20 million sq. m area of solar water heater installations by the financial year 2021-22 [4].

This data illustrates that the global trend is towards installing a greater number of SWH systems for residential heating, however, the efficiency of the system proves a major setback among manufacturers. In recent times, there have been increased technological advancements by researchers to improve the efficiency of the system.

This review paper discusses different performance parameters affecting the efficiency of the SWH system. Section 2 presents an overview of components associated with a domestic SWH system. There are two types of SWH systems typically installed in residential applications, namely Flat Plate Collector Solar Water Heater (FPCSWH) and Evacuated Tube Collector Solar Water Heater (ETCSWH). Section 3 provides the recent advancements made to increase the efficiency of the SWH system by bifurcating the system into components being segregated into different parameters. Section 4 is dedicated to the discussion and future scope of this review, which will be helpful for the researchers actively involved in improving the efficiency of the SWH system.

2. Types of Solar Water Heating System for Residential Application

2.1 Flat Plate Collector Solar Water Heating System

Flat Plate Collector comprises an insulated metallic box covered on the top with a glass sheet. Inside there are blackened metallic absorber (selectively coated) sheets with built-in channels or riser tubes to carry water. The absorber absorbs the solar radiation and transfers the heat to the flowing water. A typical layout of a flat plate collector solar water heating system is shown in Fig. 1

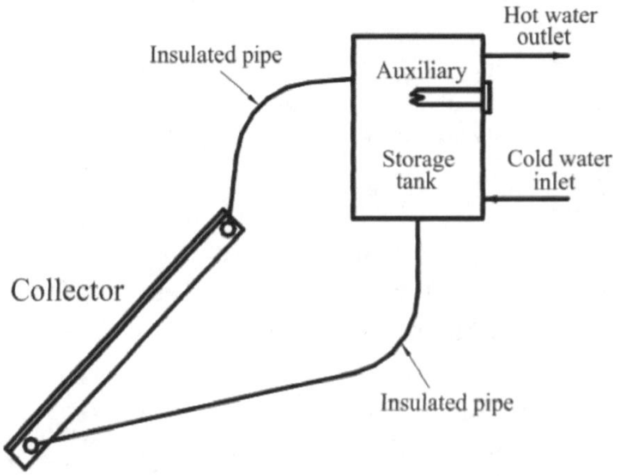

Fig. 1. A typical layout of flat plate collector solar water heating system [5]

2.2 Evacuated Tube Collector Solar Water Heating System

Evacuated Tube Collector (ETC) as the name suggests comprises concentric glass tubes separated by a spacer at the end of the inner tube which has U shape at the end as shown in Fig. 2. The mouth of both tubes is closed and the space inside the two tubes is evacuated to minimize the effect of radiation heat loss. ETC is highly efficient with excellent absorption over 93% and minimum emittance less than 6% as the tubes are round and sun rays strike the tubes at right angles thus minimizing reflection. In some cases, the evacuated tubes are provided with parabolic reflectors at the bottom to enhance the utilization of solar energy. The entire system is controlled and monitored by an automatic control panel. There is no scaling in the glass tubes thus suitable for areas with hard water, which is one drawback of the FPCSWH system.

With the government promoting the use of Solar thermal technologies, countries like China have increased the production capacity to supply this global surge in demand of the ETC's.

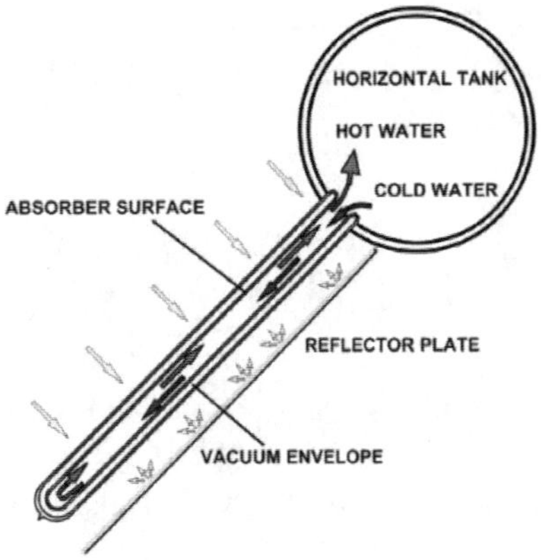

Fig. 2. A Typical Layout of evacuated tube solar water heater [6]

3. Performance Parameters Affecting the Efficiency of Solar Water Heating System

The performance of the solar water heating system can be classified into three broad parameters, first design parameters which comprise the basic components required to serve the purpose of heating and the enhancements in the efficiency caused by alterations in the geometry, design, and external features. The second parameter which affects the performance of the SWH system is operational parameters comprising a working fluid, type of flow inside the system, use of internal modifications in the system which include heat pipe and inserts. The third parameter is the external parameter is related to the climatic conditions, inclination, solar irradiation, and wind speed which also affects the performance of the SWH System.

3.1 Design Parameters

3.1.1 Tank

As far as the solar energy system is concerned, from the lower temperature section of the water storage tank, solar radiation heats the cold water circulating through the collectors, where it becomes the hot water and returns to the storage tank. If the hot water is allowed to mix with the cold water in the tank, the supplied temperature to the load is lowered and it degrades the useful quality of energy. Also, because of mixing, the whole water temperature in the tank tends to the even, the amount of energy collected may be decreased if the collector inlet fluid temperature is higher than the unmixed storage temperature. Therefore, to get the maximum efficiency of stored energy, thermal stratification technology is introduced [7]. The efficiency of the SWH system is directly proportional to the volume of the storage tank and has a negligible effect on the configuration either horizontal or vertical [8]

A study conducted by Muhammad Nadjib concluded that the SWH system that involves Phase Change Material PCM as thermal energy storage has formed thermal stratification inside the tank. This condition is represented by the Richardson number parameter, where the average value of Richardson numbers during the experiment was greater than 0.615 [9]. A design space approach is applied by Indrajeet Jadhav, to optimize the solar water-heating system which mainly focuses on two major decision variables, the solar collector area, and the storage tank volume, to reduce capital investment. A thermocline tank offers benefits like the uniformity of the output temperature and reduction in thermal losses from the solar collector, through the establishment of thermal stratification It is observed from the study that the solar collector area requirement and the overall capital investment can be reduced by 57% and 28%, by optimally integrating a thermocline tank [10]. Dileep Karuthedath reported by carrying out a 3D transient CFD analysis using Ansys software, that with an increase of fluid-flow during discharge, the stratified layers disorient and lead to rapid mixing, which eventually results in an earlier drop in the outlet water temperature [11].

3.1.2 Manifold

The manifold is that part of the SWH system that is placed in the storage tank and it is a link between the evacuated tubes and tank to maintain stratification in the storage tank. The principal function of a manifold is to channelize the flow and reduce the additional pumping power required to circulate the working fluid. There are two types of design models followed by researchers while designing a manifold namely, bifurcation type and conservative type manifolds. A comprehensive review of basic geometries on manifold designs is discussed by O. K. Siddiqui et al. and recommend the bifurcation type model for achieving better performance. They have also provided future work directions related to analyzing temperature-dependent parameters like viscosity, density, etc. and to develop complex geometries of the manifold and test it experimentally. [12] While designing manifold, the Reynolds number of the inlet flow, Richardson Number, Biot Number (to calculate heat loss from the tank), Darcy Number (permeability of the porous manifold), location of the baffle, and the thickness of the porous tube is to be considered [13]. G. L. Harding and Yin Zhiqiang in their study showed that with no extra pumping power proper manifold design is capable to transfer the heat from the absorber pipes to the main header pipe from the manifold [14].

3.1.3 Collector

A collector is that component of the SWH system which receives solar radiation in the form of heat. The function of a collector is to collect maximum heat from solar radiation and pass it on to the working fluid with minimum loss. An experimental study concluded by assessing the performance of double glazed solar flat plate water heated with different geometries of absorber plates like flat plate, v-grooved and square pulse, that the thermal efficiency of double glazed flat absorber plate is higher [15]. Reinforced cement concrete (RCC) based collector slabs with embedded aluminum pipes can provide hot water between 38°C to 58°C during winter season [16].

A flat plate collector with a modified pattern of tubes carrying the working fluid is compared in two different modes namely serpentine and zig-zag arrangements in an experimental study conducted by Maram et al. and concluded that the zig-zag arrangement of the tubes gives efficiency more than compared to conventional and serpentine flow [17]. There are two parameters by which the performance of a collector can be tested, first the collector efficiency factor and second the overall heat transfer coefficients [18]. A spherical solar collector reported by Mahmoud Ahmed Elhefnawy, et al. claims to have the collector efficiency factor of about 0.924 and overall heat transfer coefficient as 7.8 W/m2°C [19]. A Concentrated Solar Power (CSP) setup incorporating a linear Fresnel Lens concentrating the solar radiation to an absorber can provide a temp of about 255°C [20]. The performance of an absorber is affected by dust accumulation and wind velocity, which is inversely proportional to the absorber efficiency factor [21].

3.1.4 Integrated Design

As per the review conducted by Ramkishore el. al., Integrated collector storage solar water heater (ICSSWH) systems are compact type, cost-effective, aesthetically attractive, and environmentally friendly. It has all the components like collector, absorber, and storage integrated with one compact assembly as shown in Fig. 3. There are several designs of ICSSWH with the inclusion of PCM, reflective coating on absorbers, double cylindrical tank designs but these modifications in the design proves to be costly. The review suggests with a double cylindrical storage tank using thermal diode might improve the overall thermal performance significantly [22]. Based on the insulation coverage strategy, proposed by [23], parabolic trough collector with insulated tanks performs better in term of retention of useful heat as compared to ICSSWH.

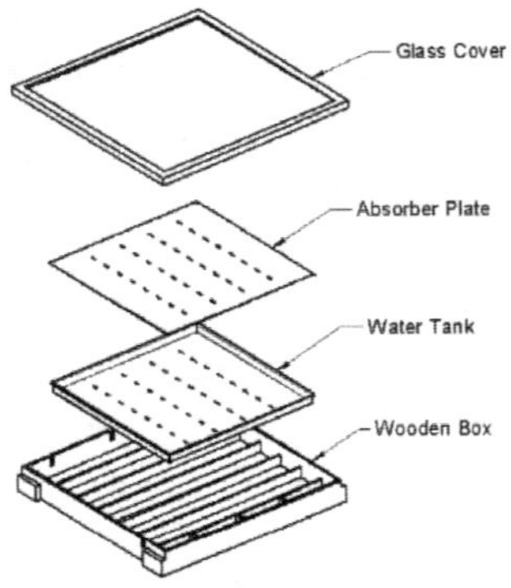

Fig. 3. Exploded view of Heater Assembly of a typical ICSSWH [24]

Anis Messaouda proposed a novel Integrated Collector Storage using a vacuum between double-glazing covers and studied the effects of reflectivity, vacuum, and night insulation on the system performance and found out that the daily efficiency reaches 58.62% and the loss coefficient decreases to 1.68W/Km2 [25]. In a recent development at Ulster University by Mervyn Smyth the design of a horizontal cylindrical ICSSWH prototype that has a double vessel, thermal diode features (to enhance heat retention during non-collection periods) achieved efficiency over 55% by incorporating a liquid-vapor phase change material (PCM) with a very low-pressure annular cavity[26]. It is seen from the literature that ICSSWH has a limitation of heat loss during night time in the form of radiation loss, which can be addressed by implementing new technologically advanced design incorporating insulated storage tanks, double cylindrical tanks using diodes and absorber with evacuated space between the glass cover.

As the technological advancements in ICSSWH will continue all across the globe, the cost associated with the manufacturing and insemination in the market will increase in near future. The only disadvantage of this system is the bulkiness, which makes it challenging to use it in domestic applications.

3.2 Operational Parameters

3.2.1 Working Fluid

From recent studies, it is found that the working fluid can influence the performance of a solar collector significantly. Water, oil, and air are the most common working fluids used in the solar energy system, but the thermal conductivity of these fluids is relatively low [5]. Nanofluids comprise base liquid and nanomaterials that have enhanced thermophysical properties such as higher thermal conductivity, thermal diffusivity, and convective heat transfer coefficients [27] [28]. Besides improving the effectiveness of heat transfer, nanofluids also improve optical properties, transmittance, and extinction coefficient of solar collectors. By using nanomaterials, the efficiency of an FPC increase by at least 10 than a conventional FPC [29]. The most studied nanofluids as working fluids are Al2O3, Water-based CuO nanofluid, TiO2, Carbon nanotube nanofluid, Deionized water, and water-based CuO nanofluids, CO2 as working fluid having efficiency range from 35% to 61% [30].

It is seen from the literature that; nanoparticles increase the thermal performance of the SWH system as they impart high thermal conductivity to the working fluid without an increase in friction factor, so the pumping power required is also not considerably increase [31]. However, many authors have reported that the efficiency decreases as the weight fraction and material size increase and there is an optimal mass flow rate at which the efficiency of the system is maximum [32]. Another challenge associated with the nanofluid reported by several research articles is of agglomeration. Agglomeration of the nanoparticle is the assemblage of mass particles to grow from nanoscale dimensions to larger dimensions because of the presence of intermolecular forces present in the material [33]. The problem of agglomeration of nanoparticles is found less in glycol-based nanoparticles like ethylene glycol, propylene glycol and Therminol VP-1 where no agglomerations were reported [34].

3.2.2 Flow dynamics

In a thermally stratified tank, the working fluid coming from the absorber region must follow natural convection. This compromises the efficiency of the system as the heat received by solar radiation should be absorbed and transmitted quickly so that the water gets heated quickly. Nanomaterial with an appropriate base fluid is one such solution as discussed in the above section. Any alternative working fluid will have a combined effect of the SWH system. While some parameters give better advantage to other problems like agglomeration, the thermal conductivity of nanomaterial, etc. will be worth researching. The surface area and wavelength of the system plays an important role in improving the efficiency of the system [35].

There are several softwares like ANSYS CFD, Fluent, Gambit which are used to validate the results conducted. A dynamic simulation is validated with experimental analysis to find the efficiency of the system [36] The location of the inlet and outlet pipe of the system also plays an important part in improving the efficiency of system an arrangement is required for thermal stratification in the storage tank [37].

The experimental investigation of designed and developed system has been carried out for two modes i.e. mid-day charging mode and full-day charging mode. It has been observed that for considered mass flow rates, thermal efficiency

of the system was varied in the range of approximately 52–62% for full-day charging mode while for mid-day charging mode, it was varied between 55 and 72%. The maximum value of thermal efficiency was approximately 72.52% at mass flow rate of 24 LPH for mid-day charging mode [38].

3.2.3 Heat Pipe

A Heat pipe is a device used to transfer heat from hot source to cold source separated at a short distance. A heat pipe has working fluid contained in the pipe and the space is evacuated and sealed from both ends the heat pipe is divided into three regions namely evaporator section, adiabatic zone and condenser region as shown in Fig. 4.

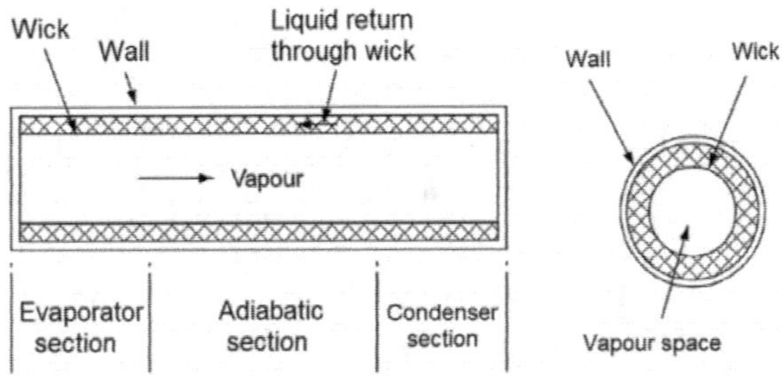

Fig. 4. Longitudinal section and cross-sectional view of a typical heat pipe [39]

The working fluid comes in contact with the hot zone in the evaporator region and vaporizes to reach the other end of the tube and transfers this heat to the condenser region as shown in Fig. 5. The thermal conductivity of heat pipe can be increased 100 times depending upon the pipe material, working fluid and distance between condenser and evaporator region [40].

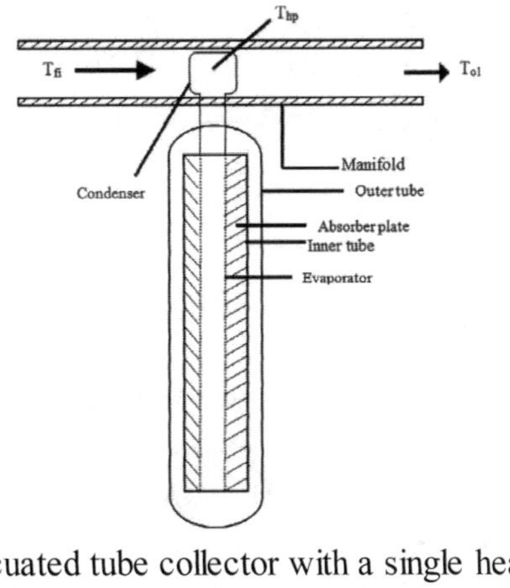

Fig. 5. Evacuated tube collector with a single heat pipe [41]

Using heat pipe for solar application dates back to the 1980s [42] and since then the advancements in the heat pipe technology have increased tremendously. Table 1 below summarizes the technological advancements in the materials used for heat pipes, working fluid and nanomaterials used in heat pipe for enhancement of performance parameters of SWH systems.

Table1 Summary of studies focused on heat pipe

Material of Heat Pipe	Copper [43], Stainless Steel, Aluminum
Working fluid	Chloroform, Hexane, Methanol, Petroleum Ether, Acetone, and Ethanol, Ammonia, Pentane, R22, R134a [44]
Nanomaterial as	Carbon nanotubes with water, Water based CuO, Propanol and (TiO_2 + DI De Ionized)

working fluid	water [45], MgO nanofluid [46], Al_2O_3 and CuO/acetone nanofluids [47], Sodium Dodecyl Benzene Sulfonate (SDBS) with Al_2O_3 nanofluid systems [48].

Due to several advantages of using Heat Pipe Solar Collectors like high overall efficiency, capability to absorb solar radiation higher than conventional and high heat transfer, quick start operation and minimal effect of inclination angle [49]. Also, the cost of manufacturing is coming down as there are more players entering to fabricate technologically advanced design, configuration and specific to operation products.

3.2.4 Inserts

In the previous section, the use of heat pipe increases the efficiency of the SWH system, but the cost of the system also increases. As the stratification in the storage tank is essential to get a higher temperature at the outlet, the absorbers are so designed that there must be natural convection in the absorber region of the water heater. However, to increase the heat transfer rate in the absorber section, evacuated tubes are incorporated with inserts. The role of inserts is to inculcate near turbulent flow inside the tubes so that there is an increase in the heat transfer rate. It is proved that Nusselt number, a measure of convective heat transfer increases with the augmentation of inserts a thermosyphon type solar water heater [50].

Research interest among researchers is gaining momentum as the installation of inserts is easy, cost-effective, and can be fabricated in various shapes and sizes. The technological advancements in this field can be seen from the increase in the number of research publications for the past decade. Few of the important include the inserts such as twisted tape [51] [52], nanofluid with twisted tape [53] are metallic strips twisted to give swirl effect to the flow which causes an enhancement in heat transfer rate, square and V-cut twisted tape [54] is another variation in twisted tapes where additional boundary layer separation is induced because of sudden change on the flow profile, helical screw tape inserts [53], wire coil inserts [55] [56] [57] which do not reduce the volume flow rate inside the tube keeping the capacity of the SWH system intact.

3.3 External Parameters

3.3.1 Solar Intensity

Nearly 58% of the geographical area in India has global solar radiation of more than 5kWhr/m2/day, and such zones are called "Solar Hotspots" of India [58]. In the places where the solar intensity is higher and the average ambient temperature is higher, are the potential places to install SWH systems. Higher solar intensity yields a greater rate of heat absorption by the collector and by a working fluid. This is because of the decrease in ambient temperature, the heat transfer rate from the outer surface of the glass tube to the working fluid decreases when the ambient temperature is less. This implies that the SWH system works on the principle of the difference of temperature which causes the neat heat to transfer from solar energy to the working fluid. Hence in many research articles, we find that the significance of ambient temperature is very important at high solar intensity regions and the study becomes insignificant at low ambient temperature zones [59].

The solar radiation is also blocked by scaling on the inside and dust accumulation [60] on the outside of the absorber glass cover for FPC and tubes of ETC's. The instantaneous efficiency drops by approximately 63% because of the formation of a 3.7mm scale inside the collector [61]. Scaling is one factor that makes ETC's more popular than FPC's for domestic Solar Water Heating.

There are two forms of solar absorbers, focusing and non-focusing. The focusing type of absorbers concentrates the solar energy at a single point (circular Fresnel lens) or along a line (linear Fresnel lens) to collect the incident and diffuse solar radiation falling onto the absorbers [62]. Fresnel lens is being used in Solar water heating, Concentrated Solar Power (CSP) for cooking, on PV panels, etc. and one such novel approach uses a Thermo Electric Generator (TEG) with the help of Fresnel lens to achieve higher efficiency in water heating [63].

3.3.2 Wind

In India, the way Solar intensity and ambient temperature vary, wind speed also varies over a wide range in a year. The average wind speeds recorded in India for the entire year lie between 7m/s to 10m/s [64]. So, it would be interesting to know how it affects the performance of the SWH system. Even though there is a lot of literature available for the parameters related to design and operations, there is very limited work done in analyzing the effect of wind speed on the performance of the SWH system. The difference between ambient temperature and collector temperature is the driving force for heat transfer to take place, and variation in the wind speed affects the ambient temperature and so is the rate of heat transfer from absorber surface to working fluid [65]. The major concern of wind velocity affecting the solar water heater is during cyclones, the model suggested by Chung et al. is to lift the structure to a suitable height depending upon wind velocity history and to provide a guide plate just behind the storage tank inclined at an angle 'α' with the horizontal [66]. They further investigated by experimental results and found the relationship between mean longitudinal and spanwise pressures associated with tilt angle, guide plate, and freestream turbulent intensities [67].

3.3.3 Inclination

The angle made by the absorber with the horizontal is known as the collector tilt angle and is usually denoted by 'β' in most of the literature. As discussed in the previous section about the importance of thermal stratification in the storage tank, collector tilt plays an important role in the stratification in the storage tank of a thermosyphon type water-in-glass SWH [68]. Bracamonte et al. suggested that the inclination angle should be equivalent to the latitude of the location so that the solar radiations are almost normal to the absorber mostly in a year.

The absorber tilt can also affect the flow regime inside the evacuated tube to develop a stagnation region if the inclination is over 45° [69]. Stagnation region is developed when there is no effect of heat transfer to cause convection currents in the fluid [70] because of which there will be no improvement in the efficiency. In the afternoon time, the incoming cold water is usually hot which try to rush into the evacuated tube because of the inclination and mixing with the already present hot water and because of this sudden turbulence the thermally stratified region breaks which are reverted in the evening [71].

4. Discussion and Future Scope

The following points are worth discussing from the literature review about the parameters that affect the thermal performance of a SWH system. Emphasis is being made to provide insights about future scope by suggesting to combine two parameters and study its effect of the overall system of domestic water heating.

- To improve the efficiency of the system, the tank should be made capable to achieve a thermal barrier between the incoming cold fluid and outgoing hot fluid known as thermal stratification. Further research related to developing prototypes with adiabatic thermal barrier materials capable of floating inside the tanks can be considered and tested prior to modeling using CFD and TRNSYS.

- In places where the frequency of discharging the hot water is less, an auxiliary storage tank with PCM material can be installed which can be monitored with sensor-based technology to sense the external parameters like whether forecast to optimize the collection rate in the tank.
- Complex geometries of manifold and novel collector design incorporating triangular, pentagonal collector tubes with integration of solar panels can be experimentally investigated and validated by using simulation.
- Increasing the length of the evacuated tube collectors generate stagnation region which makes it inefficient and the addition of inserts results in decreased volume flow rate so further research in inserts should be taken by increasing the diameter of the tube. The variation of tube diameter along with various inserts may be studied for performance improvements.
- Very less literature is available for external parameters affecting the performance parameters of SWH system. The wind velocity is inversely proportional to the efficiency so in places where the wind velocity if more than 7m/s the system can be incorporated with auxiliary air heaters and this hot air can be incorporated over the absorbers. This may open a new field of study in the design of microchannels over the absorbers for effective movement of hot air.

The salient feature of this review paper is to provide an insight to the components classified broadly into three basic performance parameters and their effect on the performance of SWH system

5. Conclusion

This paper presents an overview of recent studies on performance parameters affecting the efficiency of the Solar Water Heating System and revealed that there are three different parameters, namely design, operational and external parameters in which the components of the solar water heating systems can be classified. The potential of each component is studied in detail in this review and it is found that the efficiency of the system is influenced by varying the parameters. The paper provides some inputs related to future research in the discussion section. This review will be very useful for industries involved in installations of domestic solar water heaters, and for research organizations.

References

[1] O. M. Bany Mousa and R. A. Taylor, "A Broad Comparison of Solar Photovoltaic and Thermal Technologies for Industrial Heating Applications," *J. Sol. Energy Eng.*, vol. 141, no. 1, p. 011002, Feb. 2019, doi: 10.1115/1.4040840.

[2] S. Singh, A. Anand, A. Shukla, and A. Sharma, "Technical, financial, and environmental feasibility of solar water heater for residential, commercial, and industrial application: A theoretical approach," *Mater. Sci. Energy Technol.*, p. S2589299120300392, Jul. 2020, doi: 10.1016/j. mset.2020.07.001.

[3] "REN21," *Renew. 2020 Glob. Status Rep.*, 2020.

[4] ABPS Infrastructure Advisory Private Limited, "Scheme and Framework for Promotion of Solar Water Heating Systems by Utilities and Regulators," Ministry of New and Renewable Energy, 2010.

[5] O. Mahian, A. Kianifar, S. A. Kalogirou, I. Pop, and S. Wongwises, "A review of the applications of nanofluids in solar energy," *Int. J. Heat Mass Transf.*, vol. 57, no. 2, pp. 582–594, Feb. 2013, doi: 10.1016/j.ijheatmasstransfer.2012.10.037.

[6] I. Budihardjo and G. L. Morrison, "Performance of water-in-glass evacuated tube solar water heaters," *Sol. Energy*, vol. 83, no. 1, pp. 49–56, Jan. 2009, doi: 10.1016/j.solener.2008.06.010.

[7] Y. M. Han, R. Z. Wang, and Y. J. Dai, "Thermal stratification within the water tank," *Renew. Sustain. Energy Rev.*, vol. 13, no. 5, pp. 1014–1026, Jun. 2009, doi: 10.1016/j.rser.2008.03.001.

[8] A. Hasan, "Thermosyphon solar water heaters: effect of storage tank volume and configuration on efficiency," *Energy Convers. Manag.*, vol. 38, no. 9, pp. 847–854, Jun. 1997, doi: 10.1016/S0196-8904(96)00099-4.
[9] M. Nadjib, "Thermal Stratification Analysis on a SWH Tank Filled with PCM," p. 11.
[10] I. B. Jadhav, M. Bose, and S. Bandyopadhyay, "Optimization of solar thermal systems with a thermocline storage tank," *Clean Technol. Environ. Policy*, vol. 22, no. 5, pp. 1069–1084, Jul. 2020, doi: 10.1007/s10098-020-01849-4.
[11] K. Dileep, A. Raj, D. Dishnu, A. Saleel, M. Srinivas, and S. Jayaraj, "Computational fluid dynamics analysis on solar water heater: Role of thermal stratification and mixing on dynamic mode of operation," *Therm. Sci.*, vol. 24, no. 2 Part B, pp. 1461–1472, 2020.
[12] O. K. Siddiqui and S. M. Zubair, "Efficient energy utilization through proper design of microchannel heat exchanger manifolds: A comprehensive review," *Renew. Sustain. Energy Rev.*, vol. 74, pp. 969–1002, Jul. 2017, doi: 10.1016/j.rser.2017.01.074.
[13] C. K. Yee and F. C. Lai, "Effects of a porous manifold on thermal stratification in a liquid storage tank," *Sol. Energy*, vol. 71, no. 4, pp. 241–254, Jan. 2001, doi: 10.1016/S0038-092X(01)00043-3.
[14] G. L. Harding and Y. Zhiqiang, "Thermosiphon circulation in solar water heaters incorporating evacuated tubular collectors and a novel water-in-glass manifold," *Sol. Energy*, vol. 34, no. 1, pp. 13–18, 1985, doi: 10.1016/0038-092X(85)90088-X.
[15] J. Manikandan and B. Sivaraman, "Experimental Analysis of Double Glazed Flat Plate Solar Water Heater with Various Absorber Plate Geometries," *Int. Energy J.*, vol. 15, no. 4, Art. no. 4, May 2016, Accessed: Jul. 28, 2020. [Online]. Available: http://rericjournal.ait.ac.th/index.php/reric/article/view/1371.
[16] P. B. L. Chaurasia, "Solar water heaters based on concrete collectors," *Energy*, vol. 25, no. 8, pp. 703–716, Aug. 2000, doi: 10.1016/S0360-5442(99)00091-2.
[17] M. B. Sidky, K. Haroun, N. Shukri, and K. Yacoob, "Performance Study of Different Types of Solar Water-Heaters Collectors," *Appl. Model. Simul.*, vol. 3, no. 3, Art. no. 3, Nov. 2019.
[18] L. Ma, Z. Lu, J. Zhang, and R. Liang, "Thermal performance analysis of the glass evacuated tube solar collector with U-tube," *Build. Environ.*, vol. 45, no. 9, pp. 1959–1967, Sep. 2010, doi: 10.1016/j.buildenv.2010.01.015.
[19] M. A. Elhefnawy, M. M. A. Awad, and A. R. A. Ahmed, "Experimental investigation on the performance of spherical solar collector for water heating," *Bull. Fac. Eng. Mansoura Univ.*, vol. 41, no. 2, pp. 27–37, Jun. 2020, doi: 10.21608/bfemu.2020.98818.
[20] M. Oulhazzan, D. Saifaoui, S. Ettami, and A. Lilane, "Design and development of small-scale linear Fresnel solar concentrator for medium temperature applications," *Mater. Today Proc.*, p. S2214785320330029, May 2020, doi: 10.1016/j.matpr.2020.04.380.
[21] M. Mani and R. Pillai, "Impact of dust on solar photovoltaic (PV) performance: Research status, challenges and recommendations," *Renew. Sustain. Energy Rev.*, vol. 14, no. 9, pp. 3124–3131, Dec. 2010, doi: 10.1016/j.rser.2010.07.065.
[22] R. Singh, I. J. Lazarus, and M. Souliotis, "Recent developments in integrated collector storage (ICS) solar water heaters: A review," *Renew. Sustain. Energy Rev.*, vol. 54, pp. 270–298, Feb. 2016, doi: 10.1016/j.rser.2015.10.006.
[23] M. Chaabane, H. Mhiri, and P. Bournot, "Thermal performance evaluation and enhancement of a parabolic trough collector," *J. Renew. Sustain. Energy*, vol. 12, no. 4, p. 043702, Jul. 2020, doi: 10.1063/1.5145257.
[24] D. Henderson, H. Junaidi, T. Muneer, T. Grassie, and J. Currie, "Experimental and CFD investigation of an ICSSWH at various inclinations," *Renew. Sustain. Energy Rev.*, vol. 11, no. 6, pp. 1087–1116, Aug. 2007, doi: 10.1016/j.rser.2005.11.003.
[25] A. Messaouda, M. Hamdi, M. Hazami, and A. A. Guizani, "Analysis of an integrated collector storage system with vacuum glazing and compound parabolic concentrator," *Appl. Therm. Eng.*, vol. 169, p. 114958, Mar. 2020, doi: 10.1016/j.applthermaleng.2020.114958.

[26] M. Smyth *et al.*, "Experimental characterisation of different hermetically sealed horizontal, cylindrical double vessel Integrated Collector Storage Solar Water Heating (ICSSWH) prototypes," *Sol. Energy*, vol. 206, pp. 695–707, Aug. 2020, doi: 10.1016/j.solener.2020.06.056.
[27] F. Sarrafzadeh Javadi, S. Rahman, and M. Kamalisarvestani, "Investigating performance improvement of solar collectors by using nanofluids," *Renew. Sustain. Energy Rev.*, vol. 28, pp. 232–245, Dec. 2013, doi: 10.1016/j.rser.2013.06.053.
[28] W. Yu and H. Xie, "A review on nanofluids: preparation, stability mechanisms, and applications," *J. Nanomater.*, vol. 2012, 2012.
[29] Z. Abdin *et al.*, "Solar energy harvesting with the application of nanotechnology," *Renew. Sustain. Energy Rev.*, vol. 26, no. C, pp. 837–852, 2013.
[30] M. A. Sabiha, R. Saidur, S. Mekhilef, and O. Mahian, "Progress and latest developments of evacuated tube solar collectors," *Renew. Sustain. Energy Rev.*, vol. 51, pp. 1038–1054, 2015.
[31] K. M. Pandey and R. Chaurasiya, "A review on analysis and development of solar flat plate collector," *Renew. Sustain. Energy Rev.*, vol. 67, pp. 641–650, 2017.
[32] A. J. Moghadam, M. Farzane-Gord, M. Sajadi, and M. Hoseyn-Zadeh, "Effects of CuO/water nanofluid on the efficiency of a flat-plate solar collector," *Exp. Therm. Fluid Sci.*, vol. 58, pp. 9–14, Oct. 2014, doi: 10.1016/j.expthermflusci.2014.06.014.
[33] R. Ranjbarzadeh, A. Akhgar, S. Musivand, and M. Afrand, "Effects of graphene oxide- silicon oxide hybrid nanomaterials on rheological behavior of water at various time durations and temperatures: Synthesis, preparation and stability," 2018, doi: 10.1016/J.POWTEC.2018.05.036.
[34] T. Yousefi, E. Shojaeizadeh, F. Veysi, and S. Zinadini, "An experimental investigation on the effect of pH variation of MWCNT–H2O nanofluid on the efficiency of a flat-plate solar collector," *Sol. Energy*, vol. 86, no. 2, pp. 771–779, Feb. 2012, doi: 10.1016/j.solener.2011.12.003.
[35] R. Nasrin, M. A. Alim, and A. J. Chamkha, "Effects of physical parameters on natural convection in a solar collector filled with nanofluid," *Heat Transf.-Asian Res.*, vol. 42, no. 1, pp. 73–88, Jan. 2013, doi: 10.1002/htj.21026.
[36] H. Taherian, A. Rezania, S. Sadeghi, and D. D. Ganji, "Experimental validation of dynamic simulation of the flat plate collector in a closed thermosyphon solar water heater," *Energy Convers. Manag.*, vol. 52, no. 1, pp. 301–307, Jan. 2011, doi: 10.1016/j.enconman.2010.06.063.
[37] M. R. Assari, H. Basirat Tabrizi, and M. Savadkohy, "Numerical and experimental study of inlet-outlet locations effect in horizontal storage tank of solar water heater," *Sustain. Energy Technol. Assess.*, vol. 25, pp. 181–190, Feb. 2018, doi: 10.1016/j.seta.2017.12.009.
[38] K. Chopra, V. V. Tyagi, A. K. Pathak, A. K. Pandey, and A. Sari, "Experimental performance evaluation of a novel designed phase change material integrated manifold heat pipe evacuated tube solar collector system," *Energy Convers. Manag.*, vol. 198, p. 111896, Oct. 2019, doi: 10.1016/j.enconman.2019.111896.
[39] D. Reay, R. McGlen, and P. Kew, *Heat Pipes: Theory, Design and Applications*. Butterworth-Heinemann, 2013.
[40] S. W. Chi, *Heat pipe theory and practice*. Washington, D.C., Hemisphere Publishing Corp.; New York, McGraw-Hill Book Co, 1976.
[41] E. Azad, "Theoretical and experimental investigation of heat pipe solar collector," *Exp. Therm. Fluid Sci.*, vol. 32, no. 8, pp. 1666–1672, Sep. 2008, doi: 10.1016/j.expthermflusci.2008.05.011.
[42] S. Kitajima, K. Iwamura, and Y. Takeda, "Heat-pipe type solar water heater," US4438759A, Mar. 27, 1984.
[43] R. Daghigh and A. Shafieian, "Theoretical and experimental analysis of thermal performance of a solar water heating system with evacuated tube heat pipe collector," *Appl. Therm. Eng.*, vol. 103, pp. 1219–1227, Jun. 2016, doi: 10.1016/j.applthermaleng.2016.05.034.

[44] A. Shafieian, M. Khiadani, and A. Nosrati, "A review of latest developments, progress, and applications of heat pipe solar collectors," *Renew. Sustain. Energy Rev.*, vol. 95, pp. 273–304, 2018.
[45] A. K. Hussein, D. Li, L. Kolsi, S. Kata, and B. Sahoo, "A Review of Nano Fluid Role to Improve the Performance of the Heat Pipe Solar Collectors," *Energy Procedia*, vol. 109, pp. 417–424, Mar. 2017, doi: 10.1016/j.egypro.2017.03.044.
[46] M. S. Dehaj and M. Z. Mohiabadi, "Experimental investigation of heat pipe solar collector using MgO nanofluids," *Sol. Energy Mater. Sol. Cells*, vol. 191, pp. 91–99, Mar. 2019, doi: 10.1016/j.solmat.2018.10.025.
[47] A. A. Eidan, A. AlSahlani, A. Q. Ahmed, M. Al-fahham, and J. M. Jalil, "Improving the performance of heat pipe-evacuated tube solar collector experimentally by using Al2O3 and CuO/acetone nanofluids," *Sol. Energy*, vol. 173, pp. 780–788, Oct. 2018, doi: 10.1016/j.solener.2018.08.013.
[48] A. Shafieian, M. Khiadani, and A. Nosrati, "Strategies to improve the thermal performance of heat pipe solar collectors in solar systems: a review," *Energy Convers. Manag.*, vol. 183, pp. 307–331, 2019.
[49] Z. He, "Advanced evacuated tube solar collectors," p. 6.
[50] P. Sivashanmugam and P. K. Nagarajan, "Studies on heat transfer and friction factor characteristics of laminar flow through a circular tube fitted with right and left helical screw-tape inserts," *Exp. Therm. Fluid Sci.*, vol. 32, no. 1, pp. 192–197, Oct. 2007, doi: 10.1016/j.expthermflusci.2007.03.005.
[51] A. Kumar and B. N. Prasad, "Investigation of twisted tape inserted solar water heaters—heat transfer, friction factor and thermal performance results," *Renew. Energy*, vol. 19, no. 3, pp. 379–398, Mar. 2000, doi: 10.1016/S0960-1481(99)00061-0.
[52] K. Yao, T. Li, H. Tao, J. Wei, and K. Feng, "Performance Evaluation of All-glass Evacuated Tube Solar Water Heater with Twist Tape Inserts Using CFD," *Energy Procedia*, vol. 70, pp. 332–339, May 2015, doi: 10.1016/j.egypro.2015.02.131.
[53] L. S. Sundar, M. K. Singh, V. Punnaiah, and A. C. M. Sousa, "Experimental investigation of Al2O3/water nanofluids on the effectiveness of solar flat-plate collectors with and without twisted tape inserts," *Renew. Energy*, vol. 119, pp. 820–833, Apr. 2018, doi: 10.1016/j.renene.2017.10.056.
[54] A. Saravanan, J. S. Senthilkumaar, and S. Jaisankar, "Performance assessment in V-trough solar water heater fitted with square and V-cut twisted tape inserts," *Appl. Therm. Eng.*, vol. 102, pp. 476–486, Jun. 2016, doi: 10.1016/j.applthermaleng.2016.03.088.
[55] R. H. Martín, J. Pérez-García, A. García, F. J. García-Soto, and E. López-Galiana, "Simulation of an enhanced flat-plate solar liquid collector with wire-coil insert devices," *Sol. Energy*, vol. 85, no. 3, pp. 455–469, Mar. 2011, doi: 10.1016/j.solener.2010.12.013.
[56] A. García, R. H. Martin, and J. Pérez-García, "Experimental study of heat transfer enhancement in a flat-plate solar water collector with wire-coil inserts," *Appl. Therm. Eng.*, vol. 61, no. 2, pp. 461–468, Nov. 2013, doi: 10.1016/j.applthermaleng.2013.07.048.
[57] R. H. Martín, A. García, and J. Pérez-García, "Influence of wire-coil inserts on the thermo-hydraulic performance of a flat-plate solar collector," *J. Phys. Conf. Ser.*, vol. 395, p. 012057, Nov. 2012, doi: 10.1088/1742-6596/395/1/012057.
[58] T. V. Ramachandra, R. Jain, and G. Krishnadas, "Hotspots of solar potential in India," *Renew. Sustain. Energy Rev.*, vol. 15, no. 6, pp. 3178–3186, Aug. 2011, doi: 10.1016/j.rser.2011.04.007.
[59] B. K. Naik, A. Varshney, P. Muthukumar, and C. Somayaji, "Modelling and Performance Analysis of U Type Evacuated Tube Solar Collector Using Different Working Fluids," *Energy Procedia*, vol. 90, pp. 227–237, Dec. 2016, doi: 10.1016/j.egypro.2016.11.189.

[60] V. Gupta, M. Sharma, R. K. Pachauri, and K. N. Dinesh Babu, "Comprehensive review on effect of dust on solar photovoltaic system and mitigation techniques," *Sol. Energy*, vol. 191, pp. 596–622, Oct. 2019, doi: 10.1016/j.solener.2019.08.079.
[61] A. umesh Chandavar, "PERFORMANCE DETERIORATION OF THERMOSIPHON SOLAR FLAT PLATE WATER HEATER DUE TO SCALING," *IIUM Eng. J.*, vol. 12, no. 3, Dec. 2011, doi: 10.31436/iiumej.v12i3.157.
[62] V. K. Soni, R. L. Shrivastava, S. P. Untawale, and K. Shrivastava, "Performance Analysis and Optimization of Fresnel Lens Concentrated Solar Water Heater," *J. Sol. Energy Eng.*, vol. 141, no. 3, p. 031010, Jun. 2019, doi: 10.1115/1.4041846.
[63] H. Hazama, Y. Masuoka, A. Suzumura, M. Matsubara, S. Tajima, and R. Asahi, "Cylindrical thermoelectric generator with water heating system for high solar energy conversion efficiency," *Appl. Energy*, vol. 226, pp. 381–388, Sep. 2018, doi: 10.1016/j.apenergy.2018.06.015.
[64] C. Draxl, A. Purkayastha, and Z. Parker, "Wind Resource Assessment of Gujarat (India)," NREL/TP-5000-61741, 1149654, Jul. 2014. doi: 10.2172/1149654.
[65] X. Q. Kong, D. Zhang, Y. Li, and Q. M. Yang, "Thermal performance analysis of a direct-expansion solar-assisted heat pump water heater," *Energy*, p. S0360544211006724, Oct. 2011, doi: 10.1016/j.energy.2011.10.013.
[66] K. Chung, K. Chang, and Y. Liu, "Reduction of wind uplift of a solar collector model," *J. Wind Eng. Ind. Aerodyn.*, vol. 96, no. 8–9, pp. 1294–1306, Aug. 2008, doi: 10.1016/j.jweia.2008.01.012.
[67] K.-M. Chung, C.-C. Chou, K.-C. Chang, and Y.-J. Chen, "Wind loads on a residential solar water heater," *J. Chin. Inst. Eng.*, vol. 36, no. 7, pp. 870–877, Oct. 2013, doi: 10.1080/02533839.2012.747062.
[68] J. Bracamonte, J. Parada, J. Dimas, and M. Baritto, "Effect of the collector tilt angle on thermal efficiency and stratification of passive water in glass evacuated tube solar water heater," *Appl. Energy*, vol. 155, pp. 648–659, Oct. 2015, doi: 10.1016/j.apenergy.2015.06.008.
[69] V. Dabra, L. Yadav, and A. Yadav, "The effect of tilt angle on the performance of evacuated tube solar air collector: experimental analysis," *Int. J. Eng. Sci. Technol.*, vol. 5, no. 4, p. 100, Mar. 2018, doi: 10.4314/ijest.v5i4.9.
[70] G. L. Morrison, I. Budihardjo, and M. Behnia, "Water-in-glass evacuated tube solar water heaters," *Sol. Energy*, vol. 76, no. 1, pp. 135–140, Jan. 2004, doi: 10.1016/j.solener.2003.07.024.
[71] R. Tang, Y. Yang, and W. Gao, "Comparative studies on thermal performance of water-in-glass evacuated tube solar water heaters with different collector tilt-angles," *Sol. Energy*, vol. 85, no. 7, pp. 1381–1389, Jul. 2011, doi: 10.1016/j.solener.2011.03.019.

A study on 6063aa elliptical and semi-circular cross-section tpct with fe_3o_4 nano fluid

D. Arokiya Pushparaj[1], Dr. N. Karunakaran[2], Dr. N. Alagappan[3]
[1]Assistant Professor, Department of Mechanical Engineering, Annamalai University, Annamalai Nagar, India.
[2]Professor, Department of Mechanical Engineering, Annamalai University, Annamalai Nagar, India.
[3]Assistant Professor, Department of Mechanical Engineering, Annamalai University, Annamalai Nagar, India.
corresponding author e-mail: samphuspa2012@gmail.com

Abstract.: Thermosyphon is the most preferred device to transfer heat for it costs low, requires comparatively low maintenance, besides being adaptable to any size. The ship transfer characteristics in this cooling system get enhanced by the use of the working fluid this paper deals with the elliptical and semicircular Two-Phase Closed Thermosyphon (TPCT) range of efficiency when it is added with Fe_3O_4 nanofluid. A number of tests are conducted with elliptical and semicircular TPCT with Fe_3O_4 with the inevitable input parameters such as heat input, inclination and pure water flow rate to obtain the output result of efficiency. The results of the experiments with thermosyphons proved the fact that the thermal performance has improved with the incorporation of elliptical TPCT combined with Fe_3O_4 Nano.
Key words: Al 6063 elliptical cross-section, Al 6063 semi-circular cross-section, Fe_3O_4 Nano fluid, efficiency.

Introduction

The world in its move towards prosperity with the support of ever developing technologies needs energy and the quantum of energy consumption and their need increases every day. The BP Company estimated the energy consumption is increasing by 2.9% in 2018, a double fold increase of the average consumption recorded with last 10 years. The energy information administration (EIA) projected the energy consumption, for the year 2050, as 50% increase from today's Annul consumption level (international energy Outlook 2019). Energy consumption although indicates countries march towards progress, it does have its own negative aspects such as carbon emissions and other pollutants which naturally lead to environment related problems [1]. Unless attention is not paid on the control of pollutants the life on the earth would become only a dream. In order to count the ever-increasing energy resources man has been trying to tap energy from various renewable resources, but optimum use of the fossil fuels to generate energy remains the prioritized and viable one. However, the use of fossil fuels not only fulfils our energy requirement, but causes air pollution and a short lived one. So, to save our environment and to find newer energy resources we need to identify the energy efficient methods [2].

Energy consumption happens at a higher level in the industrial sector from the various energy sources available in the world both natural and manmade. Thermal energy is an important one as it converts the excessive heat energy into electric energy involving the heat recovery system equipped with heat exchangers, heat pipes and so on. The heat transfer applications become the most preferred as they employ heat pipes which cost less and the process of installation maintenance and operation are easy [3, 4]. The charged liquid undergoing a continuous process of evaporation and condensation two phase flow in the heat pipe is a process on which the heat pipe works [5].

Evaporation, the chief mechanism, allows the heat in the plain heat pipe in the process of heat transfer, but, sometimes nucleate boiling occurs when the input rate of heat is too high. Using the pressure in the sub atmospheric state the system is able to provide faster and potential heat transfer in low temperature too. Working fluids and their particles hamper the efficiency of the heat pipes. Nanofluids, containing particles such as base fluid, Nano particles and a surfactant, have long been used as working fluid although distilled water is quite popular as a working fluid. However, the Nano particles enhance the heat transfer characteristics and thus promote the thermal conductivity which improves the thermal efficiency. The properties of Nano fluid depend on the factors like concentration, size and morphology of the Nano particles [6]. Investigators have employed a few Nano particles such as copper, silver, etc. (metals), oxides of zinc, Silicon, etc. (Metal oxides), Graphene, carbon nanotube, diamond (carbon based), boron nitride, aluminum nitrite (metal nitrites) and silicon carbides (metal carbides), for preparing the nanofluid suspensions [7].

The investigators have found the efficiency of nanofluid in the process of heat transfer the traditional fluids with solid particles mixed as colloids enhance the thermal conductivity of the base liquid owing to the fact that the solid particle has greater thermal conductivity power than the base liquids. The metallic nanoparticles suspended in the traditional fluid improve the specific heat capacity of the liquid and broaden the surface area. while the effective heat capacity decreasing due to pressure of suspended particles, the contact surface between the flow passage and the fluid gets improvement in its quantity owing to the interactions and collisions occur between the nanoparticles [8]. A number of studies analyzed the utilization of fluids containing nanoparticles and the improved level of heat pipe thermal efficiency; a few of such studies are summarized. Peyghambarzadeh et al. [9] analysed the thermal performance caused by three working fluids in a 40cm heat pipe. Using low heat flux (2500Wm^{-2}) experiments were conducted with water, ethanol and methanol. Of the three liquids water showed the best results and suited well for such studies. At lower condenser temperatures, the authors found the heat transfer coefficient and increments as declining. Employing heating power, tilt angle and cooling water mass flow rate with the working fluid such as water, ethanol and ethylene glycol, Gedik attempted to study the performance of tpct. The distribution of temperature or at surface level temperature Measured both entry and exit points of cooling water in the thermosyphon were quantified and adopted in order to calculate the thermal resistance and efficiency the experiments underlined the print that the thermal efficiency of the TPCT was found varying between 30% and 95% of the three liquids utilized water yielded result of under 200W and 10 l/hr condition, Whereas ethanol gave under 600W and 10 litre/hour and ethylene glycol recorded under 200W and 30 litre per hour conditions[10].

Jafari, et.al. [11] used distilled water to investigate the heat transfer realized with evaporation-condensation in the heat pipe fabricator on the tpct model. While comparing the results with those of the experimental studies in the literature, the authors found the length of the pipe as 50 mm the evaporator as 150 mm the condenser as 150 mm and the adiabatic as 200mm along with the feeling ratio as 16%, 35%, 50% and 100% at the evaporation region. The obtained results were incongruent with the equation found in the literature by +_30% [11]. Xu et.al., did experiments on TPCT using both single and hybrid nanofluids the nanoparticle induced by Author include Al_2O_3, TiO_2 along with the three mixtures of two particles namely alumina 75% + titanium oxide 25%, Alumina 50% + titanium dioxide 50% and alumina 25% + titanium dioxide 75%. The power input for heating had a range of 200 to 400 watts with the increasing increment of 50W. Three different flow rates of cooling water and same number of filling ratios were the other conditions observed in the experiments. The results of the experiments, conducted with the mixture of alumina 25% and titanium dioxide 75% - water and titanium dioxide - water nanofluids was proved to be yielding the best performance results. The heating power of 400W loading rate of 50% and the flow of cooling water at 0.4 l/m, determine the thermal resistance which decreases by 26.8% for the mixture of water nanofluids alumina 25% + titanium dioxide 75% under similar conditions, there was 22.8 decrease in the thermal resistance of titanium dioxide - water nanofluid. The 25% alumina + 75% titanium dioxide - water nanofluid recorded 25.78% upgraded heat transfer coefficient with 10.6% increment ratio in the thermal efficiency.

Similarly, an upgraded heat transfer coefficient of 21.62% was noticed with the increment of 9.2% thermal efficiency of titanium - water nanofluid [12]. Despite the Iron (III) oxide and alumina the metal oxide nanoparticles - providing considerable chemical stability and diffusion, these metal oxides are not apparent choice in enhancing the thermal characteristics, particularly, thermal conductivity. The base fluid gets its thermal conductivity enhanced when the metal nanoparticles, such as iron, copper and others, are employed, but a stable nanofluid suspension is not guaranteed with these metal nanoparticles. Therefore, a combination of metal oxide nanoparticles and metal nanoparticles is the only method of making Nanofluid solution which could bear chemically stable as well as enhanced thermal properties [13].

Materials and Methods

In order to narrow down on the choice of material to be used in this study, the chemical composition test is performed and it is presented in the Table 1. From the chemical composition, it is identified that the selected container material is alloyed with magnesium and silicon in a high quantity by weight percentage (wt%). Therefore the grade of the material chosen for the study is Elliptical 6063AA TPCT and Semi-Circular 6063AA TPCT. Some of the important properties used for the study are listed in the Table 2.

Table. 1: Chemical composition (wt%) of Aluminium (6063 AA) as container metal

Mg	Si	Fe	Cu	Mn	Zn	Cr	Al
0.7	0.532	0.35	0.1	0.7	0.02	0.1	Remainder

Table.2: Key properties of 6063 AA

Density	2.7 g/cm^3
Thermal conductivity K	200 W/mK
Thermal expansion coefficient	23 µm/m°C

1. Experimental Set Up

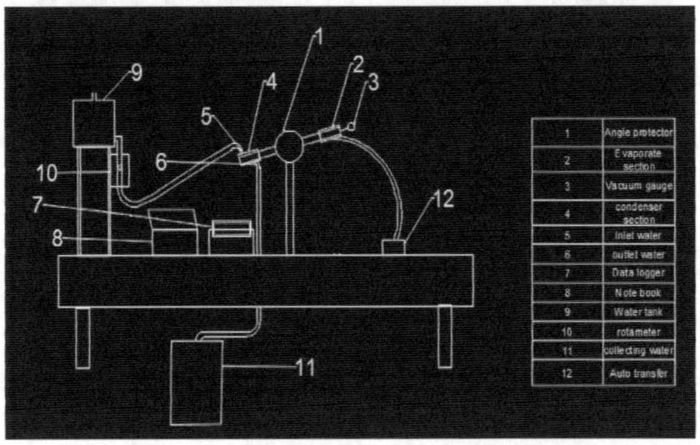

Figure 1. Layout of experimental setup

An experimental setup was built to carry out the thermal performance analysis of Two-Phase Closed Thermosyphon (TPCT) with Elliptical 6063AA TPCT and Semi-Circular 6063 AA TPCT. The schematic layout of the experimental setup is shown in Fig 1. The setup consists of the following equipment and facilities. The inclinometer (Bevel Protector) provision is made on the test rig to rotate the TPCT through 0° - 180° angle.

To supply heat as output to the evaporator an electrical heater with 200 watts capacity is wrapped around the evaporator based on the requirement a regulated power supply with three phase variac 1800 watts 230 volts and 20 amps is provided. The wattmeter attached measures the wattage of the power supply provided to the heater in the evaporator, using the various having three phases with 0 - 250 watts and 230-volt range. The load value is shown on the digital display of the wattmeter on 8 points that k-type Thermocouple are Soldered on the TPCT and connected to the data logger in order to yield better results while measuring the temperature Condenser jacket, through which water passes, is employed to measure the power output of TPCT. This experiment uses the elliptical and semicircular 6063 AA TPCT with a 6063 AA jacket having 30 mm and 250 mm diameter and length respectively along with diagonally positioned inlet and outlet. So that the swirl flow gets induced, besides a rotometer controls the flow rate of the coolant water.

The evaporator section of the TPCT is connected with heat source of plate type heater with a maximum power output of 200W at 220V and the condenser section is cooled by tap water. The mass flow rate of condenser section is controlled by rotometer. The adiabatic section is insulated by glass wool to avoid heat energy interaction with the ambience. The working fluid is charged after evacuating the tpct enclosure by maintaining 99kPa to 100 kPa vacuum.

The container material was tested with at filling ratio of 50 percent. The wall temperature on the TPCT container materials is measured by eight thermocouples of K-type. Two thermocouples were mounted on the evaporator section, two on the adiabatic section and four on the condenser section. The location of thermocouples is shown in Fig 4.2. All the thermocouples (K-Type) are connected and monitored using 8-channel data logger system. The flat plate type heater of the evaporator section is connected to the three phase variac of 20 amps. The heat input is varied by using a variac. The heating power input can be observed from watt mater.

Preparation of nanofluids

This study employs the two step method in order to prepare the nanofluids. The Fe3O4 nano particles (30mm) bought from the USA mixed with DI water at 0.09% of concentration. Then, the dispersed nano particle is allowed to undergo the sonication process for 12 hours with Ultrasonic homogenizer (Olee Pvt Ltd, $42K_{HZ}$). Similarly Fe_3O_4 nano-fluid prepared in the concentration of 0.09% by volume and sonicated for 15 hours it is shown in fig2. Prepared Fe_3O_4 nano-fluid shown in fig3.

Figure 2: Ultrasonic Homogenizer

Figure : 3. Fe_3O_4 nano fluid

Test Procedure

The Design of Experiments (DOE) is an effective process that helps us in conducting experiments as well as in analyzing the collected data easily. In order to optimize the input parameters for the desired performance, the RSM is employed for the Design of Experiments. Box – Behnken Design is used with three varying input parameters were used, namely heat input (A), angle of inclination (B) and flow rate of water in the condenser section (C) to obtain the output responses of thermal resistance (R_{th}), and overall heat transfer coefficient ($U_{overall}$).Table. 3 shows the process parameters and their levels. Table. 4 shows the design of matrix.

Table. 3 : Shows Process parameters and their levels

Parameters	Level		
	-1	0	1
Heat Input, W	60	80	100
Angle of inclination, °	30	60	90
Flow Rate, ml/min	60	90	120

Table.4: Shows the Design of Matrix

Std	Run	Factor 1 A: Heat Input	Factor 2 B: Angle of Inclination	Factor 3 C: Flow Rate
2	1	80	60	90
1	2	80	90	120
7	3	60	90	90
5	4	100	60	120
16	5	100	90	90
9	6	100	30	90
10	7	80	60	90
4	8	60	60	60
14	9	80	30	60
13	10	80	30	120
8	11	80	60	90
11	12	60	30	90
12	13	60	60	120
17	14	80	60	90
3	15	100	60	60
15	16	80	60	90
6	17	80	90	60

The BBD resulted in 17 runs of simulation arranged by three factors as listed in Table 4. The 17 runs were conducted as per the resulting design matrix by varying, the mass flow rate of pure water flowing through the condenser section using rotometer, the inclination angle of TCPT and heat input. Approximately it took 15 minutes for Fe_3O_4 nanofluid to attain the steady state. The temperature at each trial is recorded after the attainment of steady state condition using data logger system.

Data reduction

Data reduction is a process in which the information collected, in numerical and alphabetical forms, usually derived from empirical or experimental analysis, gets corrected, ordered and entered in a simplified manner. Reducing a huge amount of data into small units of meaningful information is concept behind data reduction.

The thermal efficiency of the thermosyphon (η%), is evaluated by,

$$\eta\% = \frac{Q_{out}}{Q_{in}}$$

$$Q_{out} = \dot{m}_c . c_p . (T_{out} - T_{in})$$

where,

$\dot{m}_c$ = Mass flow rate of pure water in condenser section, kg/sec.
C_p = Specific heat of water, J/kg°C
T_{out} = Temperature at outlet of the water condenser section, °C
T_{in} = Temperature at inlet of the water condenser section °C
Q_{in} = Heat input in watts

Uncertainty for $\eta\%$ is evaluated by,

$$e=\frac{\partial\eta}{\eta}=\left[\sum_{t-1}^{N}\left(\frac{1}{\eta}\times\frac{\partial R}{\partial xi}\partial xi\right)^2\right]^{1/2}$$

$$=\left[\left(\frac{\partial m}{m}\right)^2+\left(\frac{\partial Q_{out}}{Q_{out}}\right)^2+\left(\frac{\partial Q_{in}}{Q_{in}}\right)^2\right]^{1/2}$$

By calculating η_1 for the entire experimental range, the maximum uncertainty is associated with the resulting of $\eta_{\%}$ values that found to be around 1.97 percent.

Result and Discussions

Response 1 EFFICIENCY (elliptical) Fe_3O_4
ANOVA for Response Surface Quadratic Model
Analysis of variance table [Partial sum of squares - Type III]

Source	Sum of Squares	df	Mean Square	F Value	p-value Prob > F	
Model	0.25	9	0.028	243.13	< 0.0001	significant
A-HEAT INPUT	4.770E-003	1	4.770E-003	41.26	0.0004	
B-ANGLE	0.018	1	0.018	153.47	< 0.0001	
C-FLOW RATE	0.15	1	0.15	1330.58	< 0.0001	
AB	4.381E-004	1	4.381E-004	3.79	0.0926	
AC	7.009E-003	1	7.009E-003	60.63	0.0001	
BC	0.000	1	0.000	0.000	1.0000	
A^2	0.014	1	0.014	124.35	< 0.0001	
B^2	2.693E-003	1	2.693E-003	23.30	0.0019	
C^2	0.047	1	0.047	405.57	< 0.0001	
Residual	8.092E-004	7	1.156E-004			
Lack of Fit	8.092E-004	3	2.697E-004			
Pure Error	0.000	4	0.000			
Cor Total	0.25	16				

The model “F-value” being 243.13 it is considered as significant the model of value of this magnitude occurs, owing to the noise, since there is 0.01 % probability.
If the prob>F records 0.0500 values, it will indicate the models terms as significant.
Here, A, B, C, AC, A^2, B^2 and C^2 are found as significant model terms.
If the values are found greater than 0.1000 then the model terms are not considered as significant. incase the significant model terms (except those considered for maintaining hierarchy),there will be improvement in the production model.

Std. Dev.	0.011	R-Squared	0.9968
Mean0.70	Adj R-Squared	0.9927	
C.V. %	1.55	Pred R-Squared	0.9490
PRESS	0.013	Adeq Precision	45.475

The "Pred R-Squared" of 0.9490 is in reasonable agreement with the "Adj R-Squared" of 0.9927. "Adeq Precision" measures the signal to noise ratio. A ratio greater than 4 is desirable. Your ratio of 45.475 indicates an adequate signal. This model can be used to navigate the design space.

Factor	Coefficient Estimate	df	Standard Error	95% CI Low	95% CI High	VIF
Intercept	0.78	1	4.808E-003	0.77	0.80	
A-HEAT INPUT	0.024	1	3.801E-003	0.015	0.033	1.00
B-ANGLE	0.047	1	3.801E-003	0.038	0.056	1.00
C-FLOW RATE	0.14	1	3.801E-003	0.13	0.15	1.00
AB	-0.010	1	5.376E-003	-0.023	2.247E-003	1.00
AC	0.042	1	5.376E-003	0.029	0.055	1.00
BC	0.000	1	5.376E-003	-0.013	0.013	1.00
A^2	-0.058	1	5.240E-003	-0.071	-0.046	1.01
B^2	-0.025	1	5.240E-003	-0.038	-0.013	1.01
C^2	-0.11	1	5.240E-003	-0.12	-0.093	1.01

Final Equation in Terms of Coded Factors:

EFFICIENCY	=
+0.78	
+0.024	* A
+0.047	* B
+0.14	* C
-0.010	* A * B
+0.042	* A * C
+0.000	* B * C
-0.058	* A^2
-0.025	* B^2
-0.11	* C^2

Final Equation in Terms of Actual Factors:

EFFICIENCY	=
-1.39010	
+0.019360	* HEAT INPUT
+6.33714E-003	* ANGLE
+0.020145	* FLOW RATE
-1.74417E-005	* HEAT INPUT * ANGLE
+6.97667E-005	* HEAT INPUT * FLOW RATE
+0.000000	* ANGLE * FLOW RATE
-1.46074E-004	* HEAT INPUT^2
-2.81005E-005	* ANGLE^2
-1.17247E-004	* FLOW RATE^2

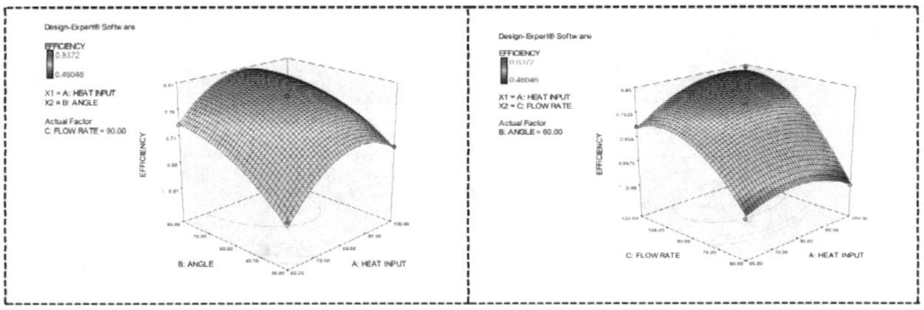

IOP Conf. Series: Materials Science and Engineering 1091 (2021) 012022 doi:10.1088/1757-899X/1091/1/012022

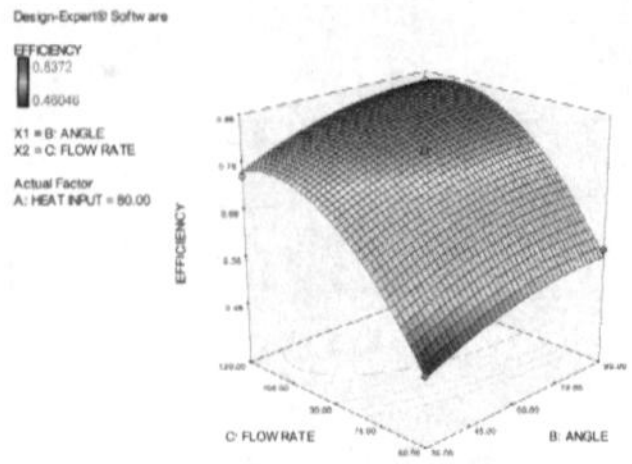

Fig 4: Wire – Mesh Plot for Elliptical 6063 TPCT along with Fe_3O_4 Nanofluid

The plots show the interactive effect of heat input and inclination angle of the elliptical with thermal efficiency.

From the figure it is seen that the interactive effect of heat input and inclination angle of the elliptical 6063 TPCT along with Fe_3O_4 nanofluid increase in heat input gradually increases the thermal efficiency upto 80 W and thereafter decrease. Whereas increase of inclination angle the thermal efficiency increases with from 30° to 90°. Similarly the efficiency is increases with the flow rate. Therefore the thermal efficiency of the elliptical TPCT 6063 with Fe_3O_4 nanofluid is affable after 80W of heat input. Hence, the influential parameter is heat input. Nanofluid utilization increases the efficiency 85% under 90W and 120 lit/min cooling water mass flow rate condition.

Response 1 EFFICIENCY Semi Circular (Fe_3O_4)
ANOVA for Response Surface Quadratic Model
Analysis of variance table [Partial sum of squares - Type III]

Source	Sum of Squares	df	Mean Square	F Value	p-value Prob > F	
Model	0.14	9	0.016	54.56	<0.0001	significant
A-HEAT INPUT	0.016	1	0.016	57.27	0.0001	
B-ANGLE	6.037E-003	1	6.037E-003	21.02	0.0025	
C-FLOW RATE	0.040	1	0.040	140.68	< 0.0001	
AB	4.381E-004	1	4.381E-004	1.52	0.2567	
AC	1.947E-004	1	1.947E-004	0.68	0.4375	
BC	6.845E-004	1	6.845E-004	2.38	0.1666	
A^2	0.014	1	0.014	47.83	0.0002	
B^2	4.266E-003	1	4.266E-003	14.85	0.0063	
C^2	0.053	1	0.053	184.07	< 0.0001	
Residual	2.011E-003	7	2.873E-004			
Lack of Fit	2.011E-003	3	6.703E-004			
Pure Error	0.000	4	0.000			
Cor Total	0.14	16				

The model "F-value" being 54.56 it is considered as significant the model of value of this magnitude occurs, owing to the noise, since there is 0.01 % probability.
If the prob>F records 0.0500 values, it will indicate the models terms as significant.
Here, A, B, C, AC, A^2, B^2 and C^2 are found as significant model terms.
If the values are found greater than 0.1000 then the model terms are not considered as significant. incase the significant model terms (except those considered for maintaining hierarchy),there will be improvement in the production model.

Std. Dev.	0.017	R-Squared	0.9859
Mean	0.61	Adj R-Squared	0.9679
C.V. %	2.77	Pred R-Squared	0.7751
PRESS	0.032	Adeq Precision	21.435

The "Pred R-Squared" of 0.7751 is in reasonable agreement with the "Adj R-Squared" of 0.9679.
"Adeq Precision" measures the signal to noise ratio. A ratio greater than 4 is desirable. Your ratio of 21.435 indicates an adequate signal. This model can be used to navigate the design space.

Factor	Coefficient Estimate	df	Standard Error	95% CI Low	95% CI High	VIF
Intercept	0.71	1	7.580E-003	0.69	0.72	
A-HEAT INPUT	0.045	1	5.992E-003	0.031	0.060	1.00
B-ANGLE	0.027	1	5.992E-003	0.013	0.042	1.00
C-FLOW RATE	0.071	1	5.992E-003	0.057	0.085	1.00
AB	-0.010	1	8.474E-003	-0.031	9.574E-003	1.00
AC	6.977E-003	1	8.474E-003	-0.013	0.027	1.00
BC	-0.013	1	8.474E-003	-0.033	6.958E-003	1.00
A^2	-0.057	1	8.260E-003	-0.077	-0.038	1.01
B^2	-0.032	1	8.260E-003	-0.051	-0.012	1.01
C^2	-0.11	1	8.260E-003	-0.13	-0.093	1.01

Final Equation in Terms of Coded Factors

EFFICIENCY	=
+0.71	
+0.045	* A
+0.027	* B
+0.071	* C
-0.010	* A * B
+6.977E-003	* A * C
-0.013	* B * C
-0.057	* A^2
-0.032	* B^2
-0.11	* C^2

Final Equation in Terms of Actual Factors:

EFFICIENCY	=
-1.87149	
+0.025116	* HEAT INPUT
+7.86328E-003	* ANGLE
+0.024724	* FLOW RATE
-1.74417E-005	* HEAT INPUT * ANGLE
+1.16278E-005	* HEAT INPUT * FLOW RATE
-1.45347E-005	* ANGLE * FLOW RATE
-1.42804E-004	* HEAT INPUT^2
-3.53678E-005	* ANGLE^2
-1.24514E-004	* FLOW RATE^2

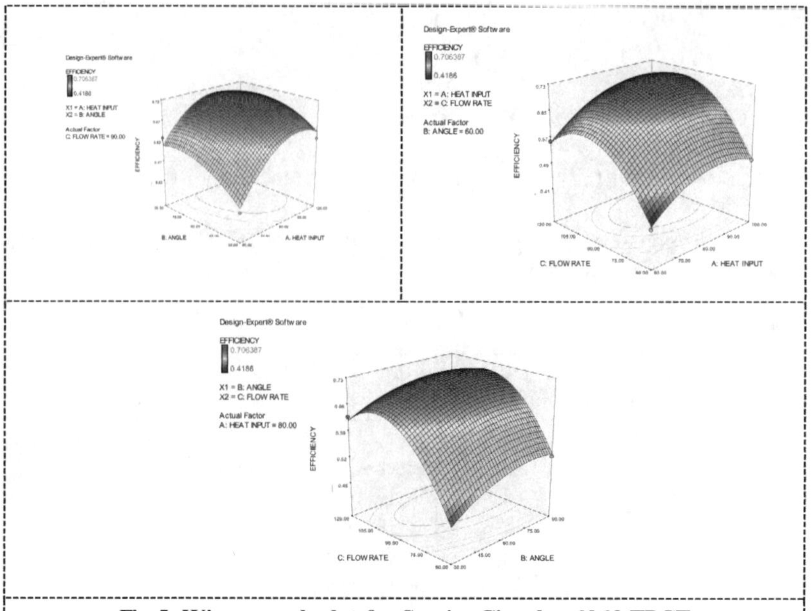

Fig 5: Wire – mesh plot for Semi – Circular 6063 TPCT along with Fe_3O_4 Nanofluid

The thermal efficiency of the TPCT is calculated as the ratio of heat rejected at the condenser section to heat input at evaporator section From Fig shows the effect of heat input and inclination angle on thermal efficiency. From the figure, it is observed that an increase in heat input results in a gradual increase of the thermal efficiency initially and then slightly decreases whereas an increase in the flow rate which causes a steep increase in the thermal efficiency. The interactive effect of flow rate and angle increase the thermal efficiency. The thermal efficiency of the heat pipe increases with the increase in heat flux, due to the fact that temperature gradient between the evaporator and condenser section increases. For higher values of heat input in evaporator section, the heat generated in the surface is more and working fluid which is in form of vapor moves vigorously in to the condenser section The circulating water in condenser section absorb this excessive heat and as a result, the efficiency of heat pipe increases. Hence, the influential input parameters are the heat input at evaporator inclination angle and flow rate. The thermal efficiency is maximum up to 90° inclination angle and then decreases.

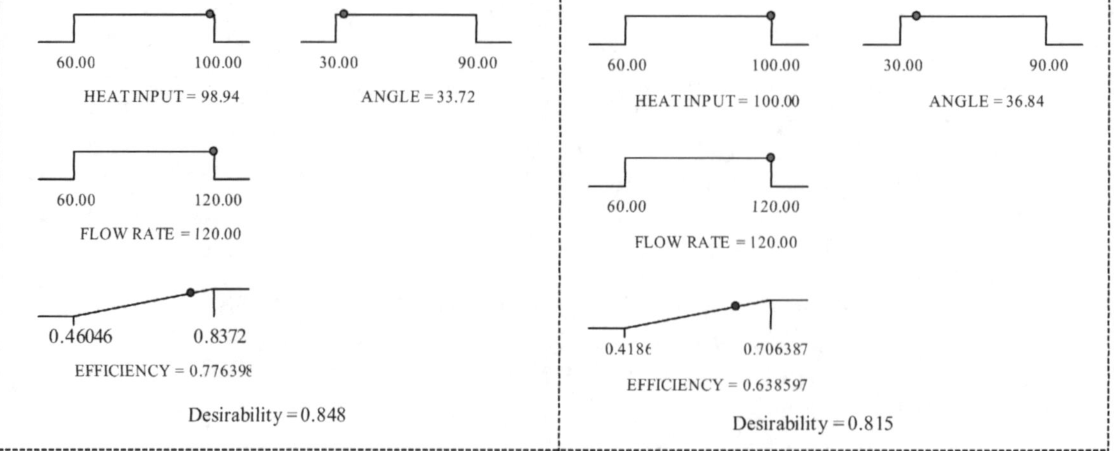

Elliptical Cross Section	**Semi Circular Cross Section**

Fig 6: Optimization Plot

Fig 6 shows the optimization plot,

For elliptical cross-section the optimized values

Heat Input : 99W
Flow rate : 120 l/m
Angle : 35°
Efficiency : 77.63%

For Semi – Circular cross-section the optimized values

Heat Input : 100W
Flow rate : 120 l/m
Angle : 40°
Efficiency : 65%

For the above optimized values, the experimental reading has also taken the Efficiency Elliptical 79% and for the Semi – Circular 67%.

CONCLUSION

With the aim of improving the heat transfer aptitude of the 6063 alloy TPCT with cross section as elliptical and semi circular with the working fluid as Fe_3O_4 has been investigated in a test rig. Experiments were accomplished under changing cross-section and then experimental results were compared to each other. The findings acquired from this study on the assessment of TPCT performance of a give rise to the consequent remarks:

1) Distribution of the wall temperature was uniform to a elliptical pipe with Fe_3O_4 nanofluid, when compare to semi circular cross – section. The difference between the wall temperature of the evaporator and the condenser sections was smaller when employing the elliptical cross section.
2) The maximum enhancement in efficiency 77.63% was obtained under 99W heating power and 120 l/m cooling water mass flow rate conditions.
3) Surface area of the pipe plays a major role with the addition of nanoparticles inside it.
4) From the experiment elliptical cross-section gives the best result.

References

1. BP. BP Statistical Review of World Energy Statistical Review of World. https://www.bp.com/content/dam/bp/business-sites/en/global/corporate/pdfs/energy-economics/statistical - review/bp-stats -review-2019-full-report.pdf. (2019)
2. U.S. Energy Information Administration (EIA). International Energy Outlook 2019 with projections to 2050. (2019)
3. X.K. Yan, Y.N. Duan, W. Wang, J. Li, J.H. Yang, Thermal characteristics of a sealed glass-water heat pipe from 0 °C to 60 °C. Int. J. Thermophys. 32, 224–236 (2011)
4. A. Merlone, S. Giunta, A. Tiziani, A new mercury gas-controlled heat pipe for temperature ampli- fier and as calibration facility. Int. J. Thermophys. 29, 1876–1886 (2008)
5. E. Çiftçi, A. Sözen, E. Karaman, TiO_2 İçeren Nanoakışkan Kullanımının Isı Borusu Performansına Etkisinin Deneysel Olarak İncelenmesi. J Polytech. 19, 367–376 (2016)
6. A. Khanlari, A. Sözen, H. İbrahim Variyenli, M. Gürü, Comparison between heat transfer charac- teristics of TiO_2/deionized water and kaolin/deionized water nanofluids in the plate heat exchanger. Heat Transf. Res. 50, 435–450 (2019)
7. H. Babar, H.M. Ali, Towards hybrid nanofluids: preparation, thermophysical properties, applica- tions, and challenges. J. Mol. Liq. 281, 598–633 (2019)
8. A. Sözen, H.I. Variyenli, M.B. Özdemir, M. Güru, Upgrading the thermal performance of parallel and cross-flow concentric tube heat exchangers using MgO nanofluid. Heat Transf. Res. 48, 419– 434 (2017)

9. S.M. Peyghambarzadeh, S. Shahpouri, N. Aslanzadeh, M. Rahimnejad, Thermal performance of different working fluids in a dual diameter circular heat pipe. Ain Shams Eng. J. 4, 855–861 (2013)
10. E. Gedik, Experimental investigation of the thermal performance of a two-phase closed thermosy- phon at different operating conditions. Energy Build. 127, 1096–1107 (2016)
11. D. Jafari, P. Di Marco, S. Filippeschi, A. Franco, An experimental investigation on the evapora-tion and condensation heat transfer of two-phase closed thermosyphons. Exp. Therm. Fluid Sci. 88, 111–123 (2017)
12. Q. Xu, L. Liu, J. Feng, L. Qiao, C. Yu, W. Shi et al., A comparative investigation on the effect of dif- ferent nanofluids on the thermal performance of two-phase closed thermosyphon. Int. J. Heat Mass Transf. 149, 119189 (2020)
13. H.M. Ali, *Hybrid Nanofluids for Convection Heat Transfer* (Elsevier, Amsterdam, 2020), pp. 18–20. https://doi.org/10.1016/C2018-0-04602-2.

IOP Conf. Series: Materials Science and Engineering 1091 (2021) 012023 doi:10.1088/1757-899X/1091/1/012023

Characterization of CRYO AND RT ECAPed Aluminium 6063 Followed by Aging Using CHARPY IMPACT TEST

Srinivasulu Arnuri[1], Swami Naidu Gurugubelli[2]
[1]Department of Metallurgy, University College of Engineering, JNTUK, Vizianagaram, Andhra Pradesh, India srinivasulu.arnuri@gmail.com
[2]Department of Metallurgy, University College of Engineering, JNTUK, Vizianagaram, Andhra Pradesh, India gsnaidujntuk@gmail.com

ABSTRACT

Equal Channel Angular Pressing (ECAP) is one of the severe plastic deformation techniques that have attracted much attention from the material community in present days. This technique has ability to produce ultra fine-grained microstructures, without a significant change in geometry. Present work describes the effect of CRYO treatment on impact strength of post ECAP aging for aluminium 6063 alloy studied using Charpy impact test. For conducting ECAP process, a channel having angle of 108^0 is used. For ECAP purpose, samples having dimensions of 9.5mm X 9.5mm X 100mm chosen and they were given, solution treatment at 520^0C for 120 minutes. These solution treated samples divided into two groups. One set of samples were given by CRYO treatment by dipping in liquid nitrogen for 20 minutes and ECAP processes carried out and obtained CRYO ECAP samples. Another set of solution treated samples subjected directly to ECAP operation and obtained RT ECAP samples. These two sets of samples are again, reduced to Micro Vickers and Charpy Impact test dimensions and aged at 180^0C for different timings. Micro Vickers hardness tests and Impact tests are conducted on these two sets of samples. The obtained results are correlated with Macro and Micro examinations.

1. INTRODUCTION

Severe Plastic deformation causes very fine gain size in metallic alloys. In those processes Equal Channel Angular Process (ECAP) is one process, to get nano-grain size material without changing dimensions [1-3]. The fine grain size of the metals improves various mechanical properties like hardness, toughness; fatigue life etc., Alloy 6063 is one of many commercially available aluminium alloys, which can be hardened by aging treatment. Emanuela Cerri. et al [4] studied the microstructure, hardness and electrical conductivity influence on by post ECAP aging. Ananda babu et al. [5-6] studied that after fourth pass microstructure and mechanical properties were highly improved, when the billets were pressed in route A through the dies having channel angle of 120^0. This is done for aluminium 5083 with and without copper casing subjected to ECAP at room temperature.

Sushanta Kumar Panigrahi et al. [7] compared the mechanical properties for cryogenic rolled and room temperature rolled Al 6063 alloys. Precipitation hardening and grain coarsening mechanisms forming in the aging treatment of the severely deformed Al 6063 alloy causes improved strength and ductility. Suppression in grain dynamic recovery happening occurs in cryogenic temperature on ECAP. This phenomena cause strength, tensile ductility, as well as uniform strain [8-10]. Potential recrystallization sites formation because of suppression dynamic recovery is likely to preserve the high density of defects, causes production of fine grain structures. Hence, in the achievement of enhanced mechanical properties, ECAP at cryogenic temperature has a promising role. However, scarcity of literature on precipitation hardenable alloys, especially on 6000 series alloys is present. Cryogenic rolling followed by annealing and aging treatment on improvement in the mechanical behaviour of some aluminium alloys like AA2219, and AA6061can be in literature [10-13].

The Charpy impact test is one of the impact testing techniques, that is used to determine the resistance towards, a sudden load or impact. As a first steo, before gpoing to ECAP processing, the alloy is solution treated, quenched in water, placed in liquid nitrogen for some time and subjected to ECAP process. For this material aging treatment is given for different timings. Comparison purpose another set of samples are made, without placing the samples in liquid nitrogen before ECAP process. These prepared materials are characterised with Micro Vickers Hardness test and Charpy Impact test and th effect of aging phenomena on Cryogenic ECAP and RT ECAP studied. The samples subjected to macro and micro examination and results were analyzed.

2. MATERIALS AND METHOD

2.1 Material

The material chosen for the present study is commercially available Aluminium 6063 having 9.5 mm X 9.5 mm. Before going to experiments, the material was tested by spectrometer and the obtained composition is shown in the Table1.

Mg	Si	Fe	Cu	Zn	Mn	P	Al
0.63 %	0.43 %	0.15 %	0.049 %	0.03 %	0.054 %	0.005 %	98.6 %

Table1. Composition of Aluminium 6063

2.2 ECAP Process

The material processing method for making the samples, for required tests is shown in Fig.1. The Aluminium 6063 bars were cut into dimensions of 9.5mm X 9.5mm X 100 mm and they were shown in Fig.2, which are suitable for ECAP process. Equal channel Angular Pressing (ECAP) die which has dimensions of 130 mm channel length, cross sectional area 9.5mm X 9.5mm, channel intersection angle (Φ) 108^0 and outer curvature angle (Ψ) 36^0 (Fig.3). Initially, all the samples were subjected solution treatment at 520^0C for 120 minutes. To get super saturated solid solution (SSSS), the samples were immediately quenched in water. These samples were divided into two groups.

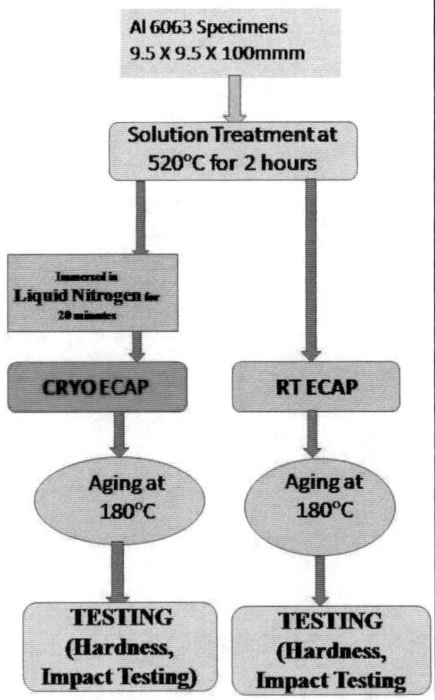

Fig.1 Process diagram

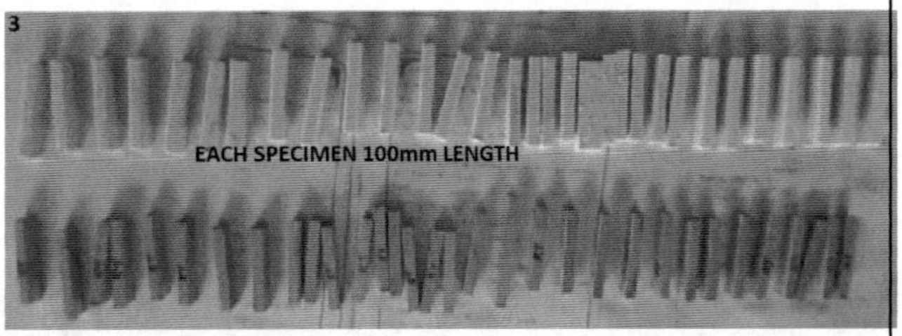

Fig.2. Specimens used for ECAP

Fig.3. ECAP Die

First group of samples dipped in liquid nitrogen (temperature –196°C) for 20 minutes, before each ECAP operation. This process continued up to four passes. The obtained samples were named as CRYO ECAPed (Cryogenic Equal Channel Angular Pressed) samples. Second group of samples after solution treatment, directly subjected to ECAP operation at room temperature and these samples were named as RT ECAPed (Room Temperature Equal Channel Angular Pressed) samples. The computer assisted universal testing machine (UTM) was used to perform the ECAP operation for both groups of samples. This ECAP process using UTM can be seen in Fig.4. Molybdenum disulfide (MoS_2) lubricant was used, for easy passage of billet through the channel. All the samples were processed in route-A (no rotation of billet after each pass) for four passes. Fig.5 shows the cryogenic treatment of solution treated samples before ECAP process. Fig.2 (above) and Fig.5 show the difference in surface appearance of samples, immediately after dipping in liquid nitrogen. The deformed billet in the die, after ECAP process can be seen in Fig.6. These two groups of samples were given aging treatment at 180^0C.

CRYO ECAP samples showed more pressing force in the UTM operation and shiny surfaces than the RT ECAP samples. From the ECAP process Samples of 1:1 aspect ratio were obtained. These two sets of samples were aged for 15 minutes increment timings, till 500 minutes at 180^0 C. These T8 heat treated (Solution treated + Cold worked (ECAP) + Aging) samples were again made suitable for Micro Vickers Hardness test and Charpy Impact Test.

2.3 Micro Vickers Hardness test and Charpy Impact Test

Micro Vickers Hardness tester used for Hardness measurements and the applied were: applied load - 300 gm and dwell time - 10 seconds. At each aging time point, hardness test conducted for both sets of samples and average value of ten readings were taken. By using these values, age hardening curves were drawn.

Fig.4. ECAP process using UTM

Fig.5. Cryogenic treatment of Solution treated samples before ECAP

Fig.6. Deformed specimen in die after ECAP

Certain points were chosen on these age hardening curves obtained by Micro Vickers Hardness test, to cover all the regions of under aging, peak aging and over aging. These aging time points were sleeted for both CRYO ECAPed and RT ECAPed aging curves. The samples obtained from ECAP process were subjected to aging for these selected timings to conduct Charpy Imapct test. They were cut carefully into dimensions of. 9.5mm*9.5mm*55mm and made notch at exact dimension according to ASTM specifications for Charpy Impact test. At each aging point five specimens were chosen, to take average value. By using conventional Charpy Impact testing machine, Charpy Impact tests were conducted and the impact testing machine used for testing can be seen Fig.7.

Fig.7. Impact test Machine

Fig.8. Specimen placed on parallel jaws

Fig.9. V-notched Charpy Impact test samples (ECAPed)

In Charpy impact test the specimen is in cantilever position (Horizontally). The sample is placed in such a way that, the notch face and hammer direction are in the same direction (Fig.8). Fig.9 shows some V-notched specimens of RT ECAPed followed by aging at different time intervals.. Before releasing the hammer from the Initial height towards the sample, for taking the readings of Impact test, the pointer is set up to its maximum value. The values of energy observed are tabulated. The same process was repeated for all the other samples in the same manner.

2.4 Macro and Micro Examinations:

After Charpy Impact test, the specimens were subjected to macro and micro examinations. After that, the samples were prepared for Scanning Electron Microscopic Examination. While cutting and metallographic preparation of the samples for SEM examination, proper care had taken, to reveal the features of Cryogenic ECAPed and RT ECAPed samples, followed by aging. All these examinations were discussed to correlate the results obtained by Micro Vickers and Charpy Impact test results.

3. RESULTS AND DISCUSSIONS:

ECAP process improves the strength of the material due to grain refinement and strain hardening. In addition, the dislocation density increase during severe plastic deformation forms dislocation tangles, which results in further strengthening of the alloy. The age hardening and Charpy Impact Test behaviour of the alloy after four passes of ECAP at Cryogenic (CRYO) and Room Temperature (RT) are evaluated. SEM structures were obtained and the Vickers Micro Hardness and Charpy Impact strength behaviour of the alloy is correlated with the microstructures.

3.1 AGE HARDENING

The hardness curves of T8 heat treated (solution treated + cold worked (ECAP + Aging) of un deformed, room temperature ECAPed and cryogenic ECAPed alloys are shown in Fig10. Without ECAP process, the samples, which were solution treated and immediately aged at 180^0 C shown the peak hardness of 75 HV after 480 minutes. The solution treated alloys ECAPed at cryogenic and room temperature have shown peak hardness at 90 minutes of aging.

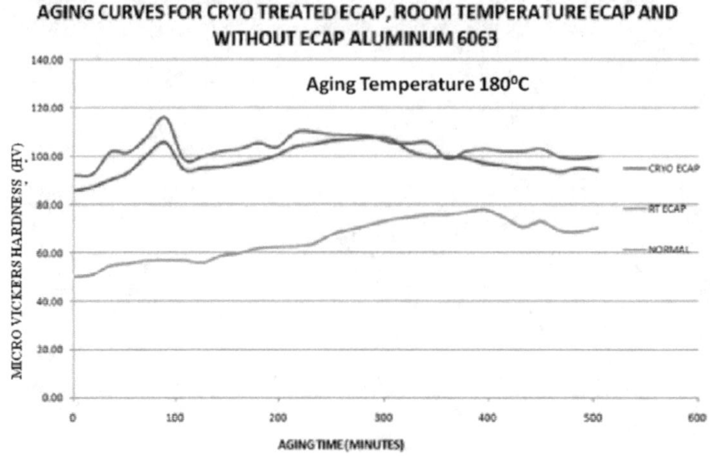

Fig10. Age hardening curves for CRYO ECAP, RT ECAP and Normal Al6063 samples at 180^0C

The peak hardness value for the CRYO ECAP age hardening curve can be observed as 116 HV and that for the RT ECAP age hardening curve is 105 HV. ECAP. These higher hardness values because of the formation of ultra-fine grain structures during ECAP and higher dislocation pinning by precipitates formed during aging. Higher hardness values observed for the cryogenic ECAPed alloy.

3.2 Charpy Impact test:

The amount of energy absorbed during the impact test determines the toughness of the material. It is interpreted that the higher amount of energy absorbed, the higher is the toughness and impact strength. In this present study, cryogenic treatment before ECAP process has been proposed to enhance the properties of Aluminium 6063 .The samples were given aging treatment with an incremental value of 30 minutes for both CRYO ECAPed and RT ECAPed specimens and then tested by Charpy Impact Testing equipment. The results obtained were tabulated in Table2. The graphs obtained from these values can be seen in Fig.11.

CHARPY IMAPACT TEST VALUES (JOULES)										
Aging Time (Minutes)	0	30	60	90	120	150	180	240	360	480
CRYO ECAP	128	142	148	174	**182**	150	152	158	120	108
RT ECAP	120	136	132	**168**	150	125	120	124	110	90

Table2. Charpy Impact test values

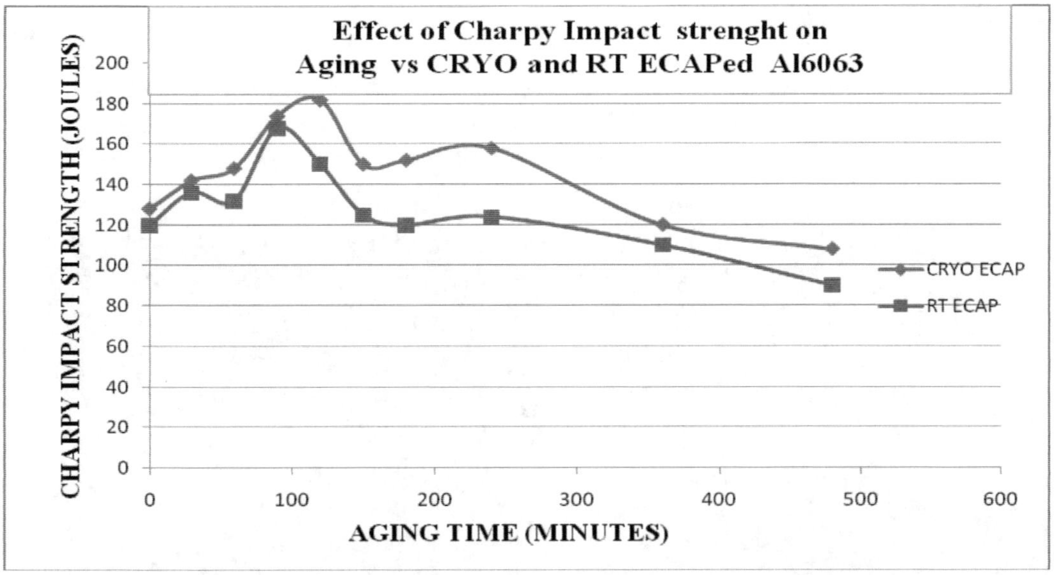

Fig11. Effect of Charpy test strength for CRYO and RT ECAPed Al6063 Vs Aging Time (180°C)

3.3 MACRO EXAMINATION:

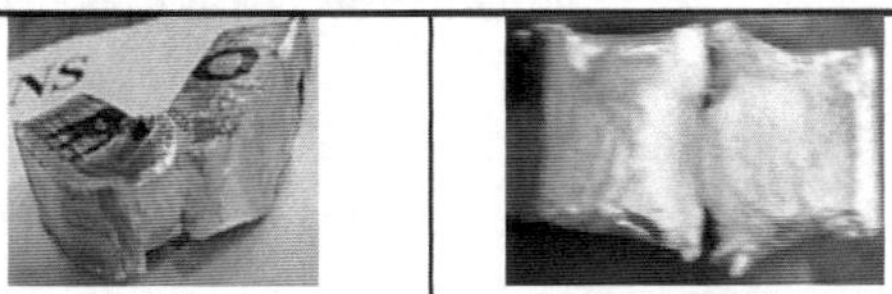

(a) RT (b) **CRYO**

Fig.12. 0 minutes aging

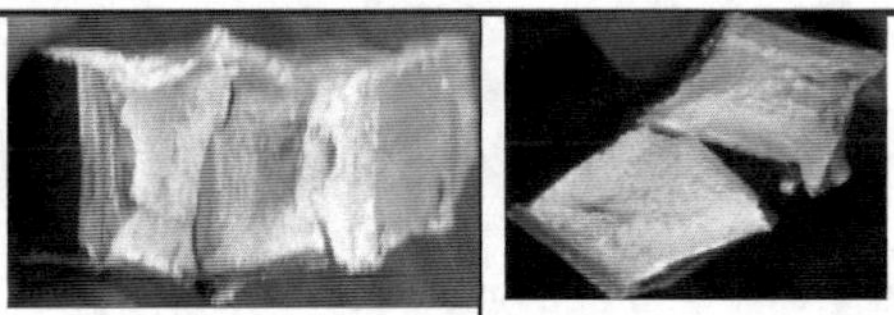

(a) RT **(b) CRYO**

Fig.13. 60 minutes aging

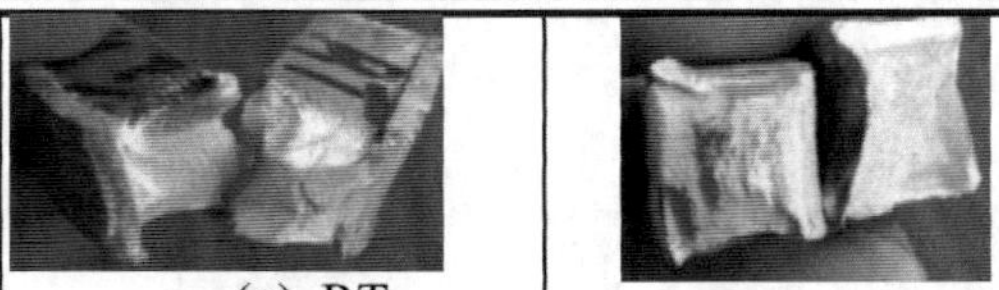

(a) RT (b) CRYO

Fig.14. 120 minutes aging

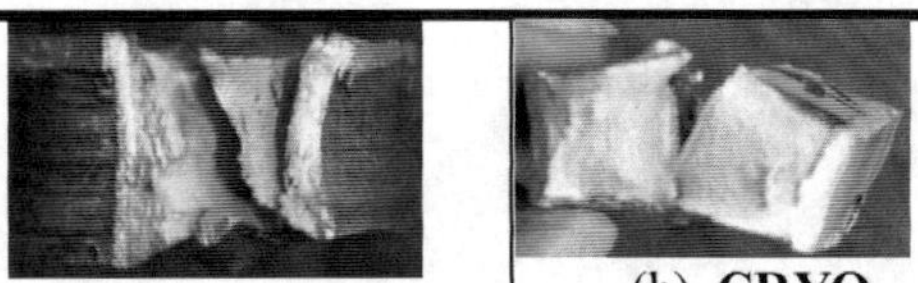

(a) RT (b) **CRYO**

Fig.15. 150 minutes aging

Macroscopic examination of Impact tested specimens showed more brittle nature of CRYO ECAPed samples than RT ECAPed samples at every aging time. Fig.12(a), Fig.13(a), Fig.14(a) and Fig.15(a) shows RT ECAP samples aged for 0 min, 60 min, 120 minutes and 150 minutes. At 120 minutes showed more brittleness, which is nearer to peak aging time compared to other aging times. Fig.12(b), Fig.13(b), Fig.14(b) and Fig.15(b) shows CRYO ECAP samples aged for 0 min, 60 min, 120 minutes and 150 minutes. At 120 minutes, Charpy Impact test value showed the peak value with more brittle surface and shiny surface than all other samples. While breaking the CRYO ECAP samples, more flat surfaces observed, where as in RT ECAP samples, cup like surfaces were observed.

3.4 SEM analysis of the alloy during age hardening

The Scanning Electron Microscope (SEM) is used to perform structural analysis on the precipitates formed in CRYO and RT-ECAPed Al6063 specimens which were aged at 180^0C. The SEM images were obtained at 0 min, 90 min, and 360 min of aging time for both types.

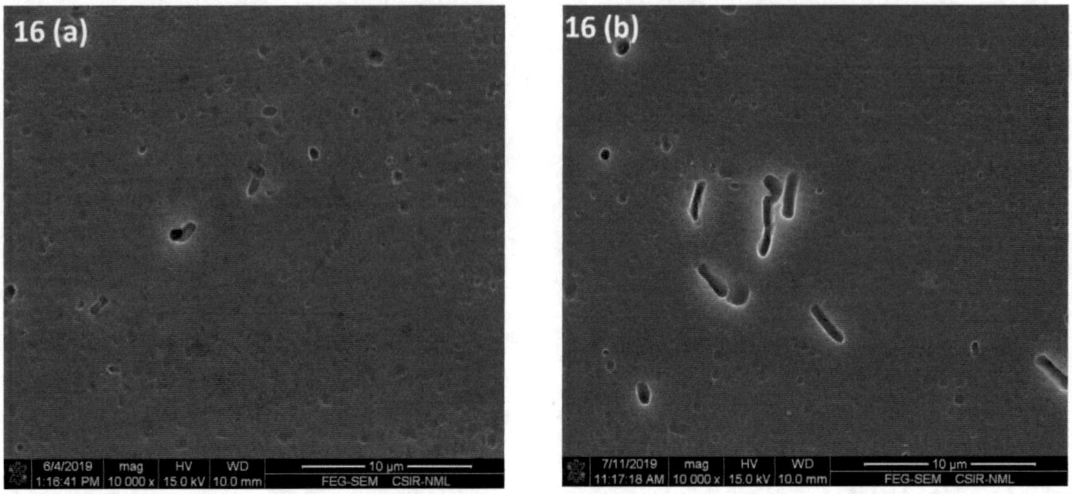

Fig.16 SEM images of Al6063 at 0 minutes of aging (a) RT ECAPed alloy (b) CRYO ECAPed alloy

Fig.16(a) and Fig.16(b) shows SEM images for 0 min of aging for RT and CRYO ECAPed specimens. It was observed more precipitation took place in CRYO ECAPed specimens. This is because of the dynamic precipitation taking place during ECAP process.

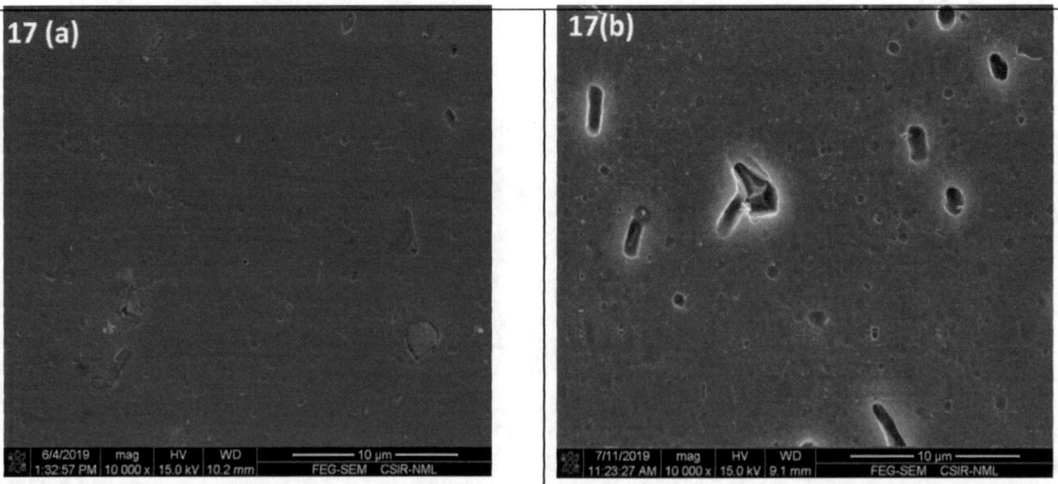

Fig.17 SEM images of Al6063 at 90 minutes of aging (a) RT ECAPed alloy (b) CRYO ECAPed Alloy

Fig.17 (a) and Fig.17 (b) shows SEM images for 90 minutes of aging for RT and CRYO ECAPed specimens. During the process of aging, precipitates are formed and the coherency of precipitates is more here. This is in agreement with the higher hardness values and Chapy Impact strength values obtained for the CRYO ECAPed alloy than RT ECAPed specimens..

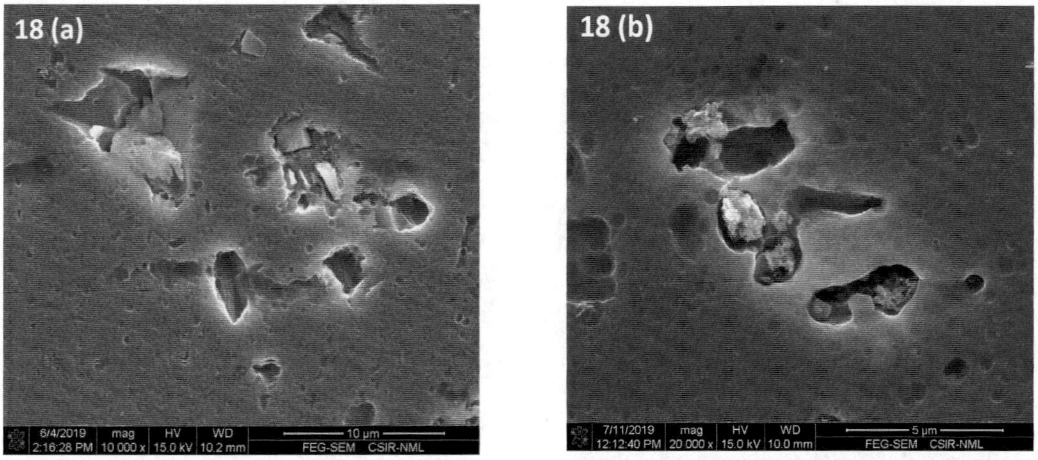

Fig.18 SEM images of Al6063 at 360 min of aging (a) RT ECAPed alloy (b) CRYO ECAPed alloy

Fig.18(a) and Fig.18(b) shows SEM images of 360 min of aging for RT and CRYO ECAPed specimens. It was observed more destruction in the regions, where precipitates are present. Due to over aging, the precipitates were grown and their coherency is also deviated in both the cases and the deviation is more for the RT ECAPed alloy. Further, after 360 min of aging it has been observed tremendous growth of the precipitates in RT-ECAPed alloy and hence the hardness is not maintained consistent

Post ECAP aging causes, two micro structural phenomena including precipitation and restoration. During the aging process, precipitates formed will pin dislocations. This obstruct the glide motion of dislocations. Recovery and recrystallizaion takes because of the restoration of dislocations introduced by ECAP processes. During recrystallization within the deformed structure, nucleation and growth of dislocation-free grains occurs. This reason causes, the T8 hardness of the ECAPed alloy is more compared to that of the undeformed alloy. The peak hardness occurring much earlier during aging of ECAPed alloy than that of the alloy aged without ECAP, because of the two processes. First one is during severe plastic deformation, increase in dislocation density occurs and second one is enhancement of the diffusion rate of Mg and Si through dislocation pipes. The dislocation sites act as nucleation sites and promote the formation of finely dispersed phases like precipitates of β'' and β' in the microstructure. This phenomena causing, at 90 minutes of peak aging values observed for both CRYO and RT ECAPed specimens. The hardness of CRYO ECAPed alloy observed as 120 HV and Charpy Impact strength observed as 174 Joules. The hardness of RT ECAPed alloy observed as 112 HV and Charpy Impact strength observed as 168 Joules.

4. CONCLUSIONS:

The effect of T8 heat treatment (Solution treatment + ECAP + Aging) on Vickers Micro Hardness and Charpy Impact test along with microstructure of 6063 aluminium alloy is studied and the Charpy Impact strength is correlated with microstructures. The given conclusions are drawn from the present experimental investigations.

1. The peak values of hardness and Charpy Impact were obtained much earlier timings nearer to 90 minutes for both CRYO ECAPed and RT EAPed specimens, whereas peak hardness for undeformed alloy is obtained at 480 minutes.. These phenomena happened because of the two reasons and they are (i) increase in dislocation density due to severe plastic deformation (ii) enhancement of the diffusion rate of Mg and Si through dislocation pipes.
2. paek hardness was 120 HV and for the RT ECAPed alloy followed by aging showed hardness was 112 HV. The higher T8 hardness was observed for the CRYO ECAPed alloy compared to the RT ECAPed alloy.
3. The CRYO ECAPed alloy followed by aging shown peak Charpy Impact test value of 182J at 120 minutes of aging time, whereas the solution treated alloy ECAPed at room temperature has shown peak Charpy Impact test value of 168J at 90 minutes of aging time.
4. Excellent agreement between the Micro Vickers Hardness, Charpy Impact test behaviour and microstructures observed from the present investigations.

5. References:

[1]. Valiev RZ, Langdon TG. Principles of equal-channel angular pressing as a processing tool for grain refinement. Progress in Materials Science [Internet]. 2006 Sep 1 [cited 2018 Apr 23];51(7):881–981.Available from: https://www.sciencedirect.com/science/article/pii/S0079642506000120
[2]. Ananda Babu Varadala, Swami Naidu Gurugubelli Sateesh Bandaru. Equal Channel Angular Extrusion of Al 5083 Alloy with Copper Shielding. Emerging Materials Research [Internet]. 2016;7(3). Available from: https://www.icevirtuallibrary.com/doi/pdf/10.1680/jemmr.18.00043
[3]. Hayoune A. Thermal Analysis of the Impact of RT Storage Time on the Strengthening of an Al-Mg-Si Alloy. Materials Sciences and Applications. 2012;
[4]. Cerri E, Leo P. Influence of severe plastic deformation on aging of Al-Mg-Si alloys. Materials Science and Engineering A. 2005;
[5]. Varadala AB, Gurugubelli N, Bandaru S. Equal channel angular extrusion of semicircular AA 5083 covered with copper casing. 2016;
[6]. Ananda Babu Varadala a, , Swami Naidu Gurugubelli b , Sateesh Bandaru c "Enhancement of structural and mechanical behavior of Al-Mg alloy processed by ECAE" Materials Today: Proceedings, https://www.sciencedirect.com/science/article/pii/S2214785319320279?via%3Dihub
[7]. Panigrahi SK, Jayaganthan R, Pancholi V. Effect of plastic deformation conditions on microstructural characteristics and mechanical properties of Al 6063 alloy. Materials and Design. 2009;
[8]. Baldissera P, Delprete C. Deep Cryogenic Treatment: A Bibliographic Review. The Open Mechanical Engineering Journal. 2008;
[9]. Bouzada F, Cabeza M, Merino P, Trillo S. Effect of Deep Cryogenic Treatment on the Microstructure of an Aerospace Aluminum Alloy. Advanced Materials Research. 2012;
[10]. Kalia S. Cryogenic processing: A study of materials at low temperatures. Journal of Low Temperature Physics. 2010;
[11]. Chen Y, Li Y, He L, Lu C, Ding H, Li Q. The influence of cryoECAP on microstructure and property of commercial pure aluminum. Materials Letters. 2008;
[12]. Shanmugasundaram T, Murty BS, Subramanya Sarma V. Development of ultrafine grained high strength Al-Cu alloy by cryorolling. Scripta Materialia. 2006;

[13]. Srinivasulu Arnuri, Swami Naidu Gurugubelli. "NDT evaluation of age hardening kinetics of RT and Cryo ECAPed Aluminum 6063 alloy from attenuation coefficient and longitudinal velocity measurements obtained by ultrasonic testing", Materials Today: Proceedings, 2020

Electrochemical characters of α-Fe_2O_3 hexagonal platelets synthesized by a facile hydrothermal route

Bui Thi Hang* **and Vu Manh Thuan**

International Training Institute for Materials Science,
Hanoi University of Science and Technology (HUST), Hanoi, Vietnam

hang@itims.edu.vn/hang.buithi@hust.edu.vn

Abstract. Fe_2O_3 materials with various morphologies were synthesized from iron (III) chloride ($FeCl_3.6H_2O$) by a facile hydrothermal route for Fe-air battery anodes. The crystal structure and morphology of the obtained Fe_2O_3 powder were studied by X-ray diffraction (XRD) and scanning electron microscopy (SEM). The morphologies, shapes and sizes of Fe_2O_3 particles synthesized via the hydrothermal method were controllable by the concentration of precursors. The composite electrodes were fabricated by mixing the synthesized Fe_2O_3 powders with nanocarbon and were then subjected to electrochemical measurements. The results indicated that the shape and size of iron oxide affected the electrochemical properties of Fe_2O_3/carbon composite electrodes. The concentration of electrode components also significantly influenced the cyclability of Fe_2O_3/carbon composite electrodes

1. Introduction

Rechargeable iron-based alkaline batteries are particularly attractive for large electrical energy storage because they have high theoretical capacity, long cycle life, and environment friendly [1–7].
In an alkaline solution, the iron electrode is oxidized from metallic iron to iron hydroxide (Eq.1) and then further oxidized to iron oxyhydroxide (Eq. 2) and/or magnetite (Eq.3):

$Fe + 2OH^- \rightleftarrows Fe(OH)_2 + 2e$ $E^0 = -0.975$ V vs. Hg/HgO [8] (1)

$Fe(OH)_2 + OH^- \rightleftarrows FeOOH + H_2O + e$ $E^0 = -0.658$ V vs. Hg/HgO [8] (2)

$3\,Fe(OH)_2 + 2OH^- \rightleftarrows Fe_3O_4.4H_2O + 2e$ $E^0 = -0.758$ V vs. Hg/HgO [9, 10] (3)

The standard hydrogen evolution reaction potential is a little positive to that of iron electrode, the hydrogen evolution reaction occurs together with the reduction of iron hydroxide to iron (Eq. 1):

$2H_2O + 2e \rightleftarrows 2OH^- + H_2$ $E^0 = -0.926$V vs. Hg/HgO [11–13] (4)

The hydrogen evolution occurring simultaneously with the reduction reaction of iron during charging process reduces the charging efficiency of the iron electrode. In addition, the passivation caused by $Fe(OH)_2$ electrical insulator formed during discharge (Eq.1) restrict their practical applications. Numerous investigations have focused on improving the performance of iron [7, 14–19].

Our previous works showed that the cyclability of iron electrodes is significantly improved by the modification of the iron electrode structure and using additives [20–24]. We also demonstrated that morphology, shape and size of iron particles strongly affected on the electrochemical properties of Fe/C electrodes [25]. In this study, by changing the preparation conditions, iron oxide particles with the desired morphology and size were synthesized via a hydrothermal route. Nanocarbon was applied as an additive to improve the conductivity, cyclability and cycle performance of iron electrodes.

2. Experimental

2.1. Synthesis of Fe_2O_3 powder

To synthesize Fe_2O_3 particles with various morphologies by hydrothermal method, we used analytical grade chemicals. Deionized water is used for dissolving the salts and rinsing the precipitates.

The hydrothermal process of synthesizing α-Fe_2O_3 was carried out as follows: 0.01 mol of $FeCl_3.6H_2O$ was dissolved in 30 ml of deionized water to form a uniform solution. Then 15 ml NaOH 2M was added slowly to the above solution to form a $Fe(OH)_3$ suspension. The resultant mixture was continuously stirred to obtain the brown suspension. The brown $Fe(OH)_3$ precipitates are filtered and washed with deionized water several times to remove Cl^-, Na^+ ions. To synthesize the various morphologies of Fe_2O_3 materials, the brown $Fe(OH)_3$ precipitates were added to 40 ml NaOH aqueous solutions with various concentrations of 2.5, 7.5 and 15 M under stirring for 30 min to obtain various mixtures. Subsequently, the mixture has been transferred and sealed into a 100 mL Teflonlined stainless steel autoclave, maintained at 160 °C for 20 h in a typical hydrothermal process. Then the autoclave was cooled down gradually to room temperature. The reddish brown precipitates were collected by centrifugation, washed with deionized water and ethanol for several times to remove the remaining NaOH, and subsequently dried at 60 °C for 12 h to obtain the Fe_2O_3 powder.

2.2. Characterizations of Fe_2O_3 particles

The crystal structure of the synthesized powder was examined by an X-ray diffractometer (XRD; Rigaku). The morphologies and particle size of the as-prepared materials were observed by a scanning electron microscopy (SEM) using JEOL JSM-6060LA/VI (JEOL, Japan).

Two types of electrode, with and without acetylene black (AB) additive, were prepared to investigate the electrochemical properties of the synthesized Fe_2O_3 materials. The Fe_2O_3 electrode sheet free AB additive was fabricated by mixing of 90 wt% Fe_2O_3 and 10 wt.% polytetrafluoroethylene (PTFE; Daikin Co.) followed by rolling. The Fe_2O_3/AB composite electrodes were prepared by the same procedure with the mixing ratios of Fe_2O_3: AB: PTFE = 45: 45: 10 wt.% and Fe_2O_3: AB: PTFE = 70: 20: 10 wt.%. The electrodes were punched into pellets with a diameter of 1 cm.

Cyclic voltammetry (CV) measurements were carried out using a three-electrode glass cell assembly with Fe_2O_3/AB composite as the working electrode, Pt mesh as the counter electrode, and Hg/HgO as the reference electrode. The electrolyte was 8 mol dm^{-3} aqueous KOH solution. The CV measurements were recorded at a scan rate of 2 mV s^{-1} and within a range of –1.3 V to –0.1 V.
All these measurements were described in our previous work [25].

3. Results and discussion

3.1. Morphological and structural properties of Fe_2O_3 particles

The X-ray patterns of the as-prepared materials synthesized at various concentrations of NaOH are presented in Figure 1. For convenience, the α-Fe_2O_3 samples synthesized at NaOH 2.5, 7.5 and 15 M are entitled as NaOH-2.5M, NaOH-7.5M and NaOH-15M. In all cases of various NaOH concentrations, the XRD patterns exhibit some typical diffraction peaks at 2 theta values of 24.15°, 33.16°, 35.63°, 40.86°, 49.46°, 54.07°, 62.44°, and 64.0°, which correspond to the (012), (104), (110), (113), (024), (116), (214), and (300) planes, respectively. These diffraction peaks are well matched with those of the α-Fe_2O_3 phase reported in ICSD No. 71194. No other peaks for impurities are observed in the XRD patterns, suggesting that the as-synthesized materials are pure α-Fe_2O_3.

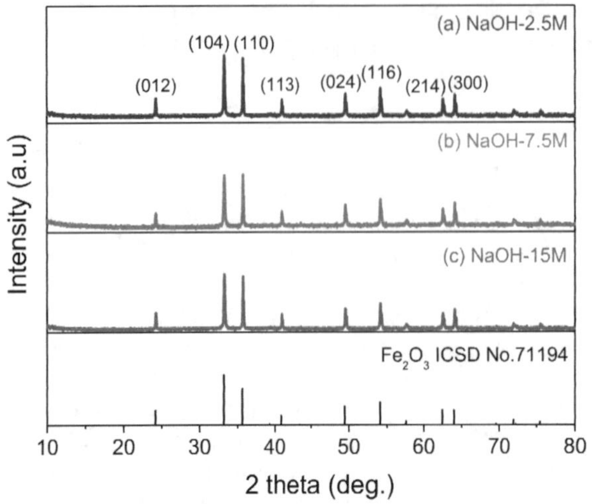

Figure 1. XRD patterns of the as-prepared α-Fe_2O_3 powders with various amounts of NaOH: (a) 2.5 M, (b) 7.5 M and (c) 15 M.

Figure 2. SEM images of Fe_2O_3 powder synthesized at various amounts of NaOH in a typical hydrothermal process.

The size and shape of the synthesized materials were examined by SEM as shown in Figure 2. The as-prepared samples have a hexagonal platelet shape, un-uniform morphology and the dimension, thickness increases with increasing the NaOH amount. The diameter of most hexagonal platelets is less than 1 µm at NaOH 2.5 M, more than 1 µm at NaOH 7.5 M, and more than 2 µm at NaOH 15 M. The thickness of hexagonal platelets also increase when NaOH increases from 2.5 M to 15 M. Therefore, the size of Fe_2O_3 hexagonal platelets at NaOH 15 M is largest but they are relative uniform. These results demonstrated that the NaOH amount influenced on the morphology and dimension of the synthesized products. In this work, only amount of NaOH was changed while the other conditions were kept constantly. Thus, the size and shape of Fe_2O_3 synthesized via hydrothermal route can be controlled by changing NaOH amount.

3.2. Electrochemical characterization of synthesized Fe_2O_3

To define the effect of morphology and dimension of Fe_2O_3 particles on their electrochemical properties, the CV measurements were carried out with Fe_2O_3 electrodes using synthesized Fe_2O_3, the results are shown in Figure 3. A couple redox reaction peak was observed at around −0.85 V (a_1) upon oxidation and at around − 1.15 V (c_1) upon reduction, together with a small couple peak at around −0.65 V (a_2) upon oxidation and around −1.0 V (c_2) upon reduction. The anodic peak (a_1) and cathodic peak (c_1) corresponds to the Fe/Fe(II) redox couple (Eq.1) and peak couple (a_2)/(c_2) corresponds to the Fe(II)/Fe(III) redox couple (Eq. 2 and/or 3). A small oxidation peak a_0 observed at around −1.0 V was ascribed to the oxidation of iron to form $[Fe(OH)]_{ads}$ before forming Fe(II) [10]. The hydrogen evolution appeared at around −1.2 V.

The Fe (II)/Fe(III) (a_2/c_2) redox couple was smaller compared to the Fe/Fe(II) (a_1/c_1) one. The iron deposition peak Fe(II)/Fe (c_1) was completely covered by the hydrogen evolution peak in the samples NaOH-2.5M (Figure 3a) and NaOH-7.5M (Figure 3b), whereas it was separated from the hydrogen evolution in the sample NaOH-15M (Figure 3c). The separation of hydrogen evolution and Fe(II)/Fe reduction peak will support the charge efficiency and discharge capacity of the iron electrode. However, the redox current of the NaOH-7.5M sample decreased with cycling, whereas they increased in the case of NaOH-2.5M and NaOH-15M samples. As shown in SEM images in Figure 2, Fe_2O_3 particles synthesized at NaOH-15M are hexagonal platelets, relative uniform and thicker than those at NaOH-2.5M and 7.5M. These results demonstrated that shape and size of Fe_2O_3 particles influenced to their electrochemical characterizations.

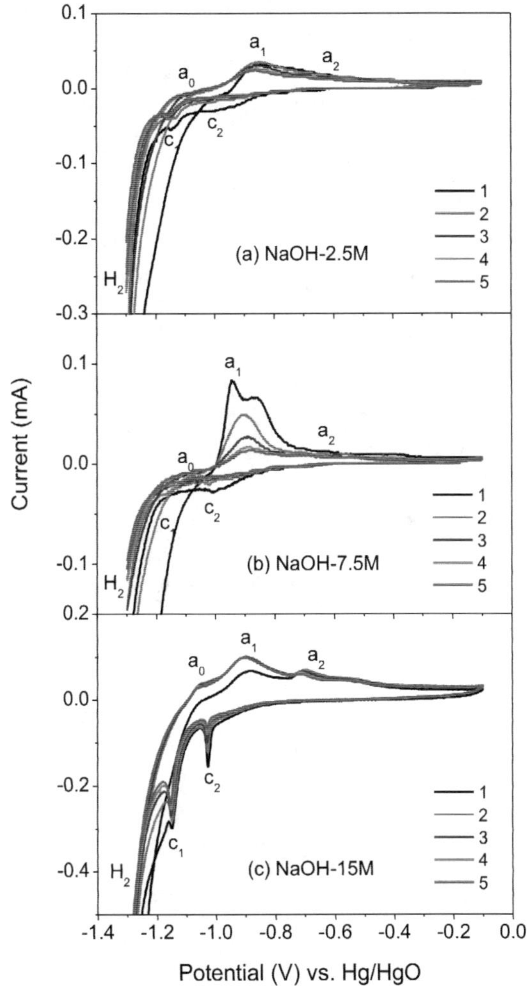

Figure 3. Cyclic voltammograms of Fe_2O_3 electrodes (Fe_2O_3: PTFE = 90: 10 wt.%) using Fe_2O_3 synthesized at various NaOH amounts in KOH solution.

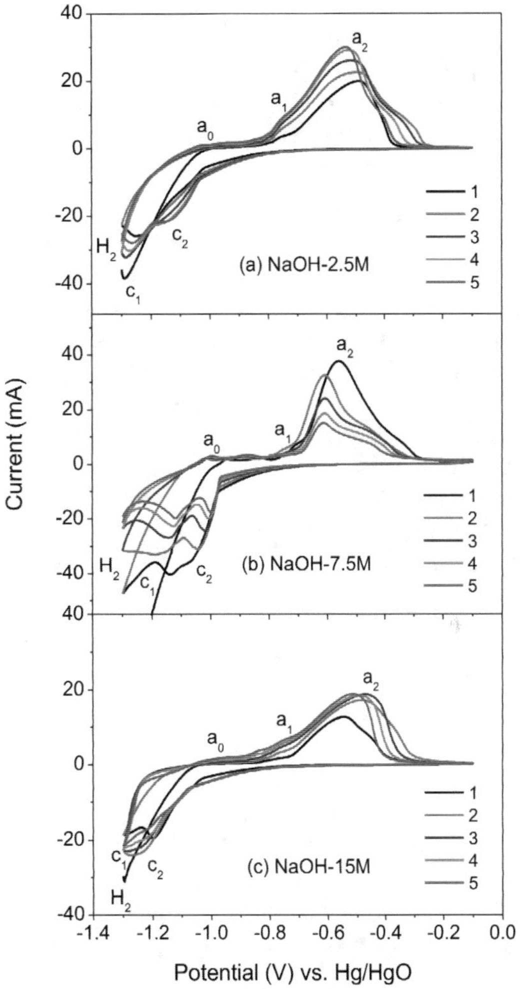

Figure 4. Cyclic voltammograms of Fe_2O_3/AB composite electrodes (Fe_2O_3:AB:PTFE=45:45:10 wt.%) using Fe_2O_3 synthesized at various NaOH amounts in KOH solution.

To evaluate the applicability of synthesized Fe_2O_3 materials, the Fe_2O_3/AB composite electrodes with Fe_2O_3: AB: PTFE = 45: 45: 10% have been fabricated and measured CV, the results are shown in Figure 4. For all samples, the CV profiles depicted two pairs of redox peaks of Fe/Fe(II) (a_1/c_1) and Fe(II)/Fe(III) (a_2/c_2). The redox peaks of Fe/Fe(II) (a_1/c_1) were observed at around −0.8 V (a_1)/−1.2 V (c_1) on the forward and backward scan respectively while the redox couple of Fe(II)/Fe(III) (a_2/c_2) were seen at around −0.5 V (a_2) and −1.1 V (c_2), respectively. Along with the appearance of these peaks, a small anodic peak a_0 was observed at around −1.0 V on the forward scan, and hydrogen evolution occurred at around −1.25 V. Opposite to the Fe_2O_3 electrode, the Fe/Fe(II) (a_1/c_1) redox couple was very smaller compared to the Fe(II)/Fe(III) (a_2/c_2) one. This suggests that a_2/c_2 couple contained both the redox reactions of Fe/Fe(II) (Eq.1) and Fe(II)/Fe(III) (Eq.2 and/or 3). However, the overpotential of two couples was increased and the reduction peak c_1 was still immerged in the hydrogen evolution when AB was added. This will cause the negative effect on the cycle performance of Fe_2O_3/AB electrode.

Comparison with the CV results of the Fe_2O_3 electrodes without AB (Figure 3), the redox peaks of the Fe_2O_3/AB electrodes (Figure 4) were higher and larger. This demonstrated that the reaction rate of Fe/Fe(II) and Fe(II)/Fe(III) couples increased in presence of AB and confirmed the positive effects of AB on the electrochemical behavior of Fe_2O_3. AB has good electrical conductivity, small particle size, and large actual surface area, thus it increased the conductivity of the Fe_2O_3/AB electrode and supported for redox reaction of Fe_2O_3.

During the discharge process, due to the dissolution-deposition mechanism of the redox reaction of iron (Eq.1), the soluble $HFeO_2^-$ species, which were formed in the intermediate step of Eq.1 would precipitate and distribute on the AB surface. This process reduced the passivity of the iron electrode caused by the $Fe(OH)_2$ layer, enhanced the reaction rate of the Fe/Fe(II) and Fe(II)/Fe(III) couples, and leads to improved cycleability of Fe_2O_3/AB electrode. Comparing CV results of these samples, we can see that there is a difference in CV profiles even though the sample composition and measuring conditions were the same. This confirmed that the shape and size of Fe_2O_3 particles affected the electrochemical properties of Fe_2O_3/AB electrodes.

To reduce the additive AB in electrode, the electrodes with 70 wt.% Fe_2O_3, 20 wt.% AB, and 10 wt.% PTFE were prepared and performed CV measurement, the results are showed in Figure 5. For the Fe_2O_3-NaOH 2.5M and NaOH 7.5M samples, the oxidation peaks decreased rapidly with repeated cycling while the reduction peaks were completely covered by hydrogen evolution. The CV profiles of the Fe_2O_3/AB electrode using Fe_2O_3 synthesized at NaOH 15M with Fe_2O_3 70 wt.% (Figure 5) are similar to those with Fe_2O_3 45 wt.% (Figure 4). This result suggested that increasing the active material of the Fe_2O_3/AB electrode to 70 wt.% seems to be not suitable for these Fe_2O_3 products. It may be due to the conductivity of the Fe_2O_3/AB composite electrode decreased when reduced AB.

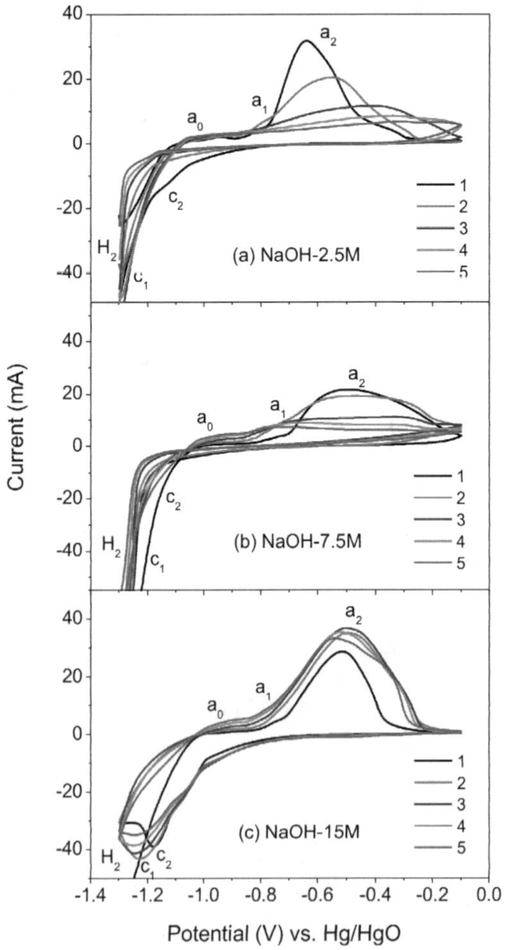

Figure 5. Cyclic voltammograms of Fe_2O_3/AB composite electrodes (Fe_2O_3:AB:PTFE=70:20:10 wt.%) using Fe_2O_3 synthesized at various NaOH amounts in KOH solution.

Thus, the electrochemical characteristics of Fe_2O_3/AB electrode were affected by not only the shape and size of Fe_2O_3 particles, but also by the concentration of electrode components. The adequate ratio of the electrode components in this case is Fe_2O_3:AB:PTFE=45:45:10 wt.%.

To find the most suitable Fe_2O_3 among the synthesized materials, the special discharge capacities of the Fe_2O_3/AB electrodes with Fe_2O_3 45 wt.% are calculated from the CV profiles in Figure 4, the results are shown in Figure 6. Among the Fe_2O_3/AB electrodes, the electrode with Fe_2O_3 synthesized at NaOH 2.5M gave the largest capacities. As seen in the SEM images (Figure 2), the Fe_2O_3 material synthesized at NaOH 2.5 M has smaller dimension than those at NaOH 7.5 M and NaOH 15 M. In the Fe_2O_3/AB electrodes, all the electrode components are the same, only size and shape of Fe_2O_3 particles are different. In the case of Fe_2O_3/AB electrode using Fe_2O_3 synthesized at NaOH 2.5M, the distance between iron oxide and carbon species is expected to be shorter than those using Fe_2O_3 prepared at NaOH 7.5 M and 15 M, thus diffusion path of mobile iron between an iron particle and AB carbon is shorter resulting in the higher redox reaction rate, larger capacity. These results confirmed the size and shape of Fe_2O_3 significantly affected the cycle performance of Fe_2O_3/AB electrodes. This finding reveals that Fe_2O_3 sample synthesized at NaOH 2.5 M is the most suitable material for Fe_2O_3/AB electrode. The charge-discharge measurement was conducted on the Fe_2O_3/AB electrode using this active material, the result is shown in Figure 7. The discharge capacity increased in few initial cycles, achieved high value and then gradually decreased with subsequent cycling. Although the large discharge capacity obtained after few cycles, but the degradation of discharge capacity rapidly occurred. Optimization of fabrication conditions to further improve the capacity retention, hexagonal platelet shape α-Fe_2O_3 particles synthesized via a facial hydrothermal method are potential active materials for iron-air battery anodes.

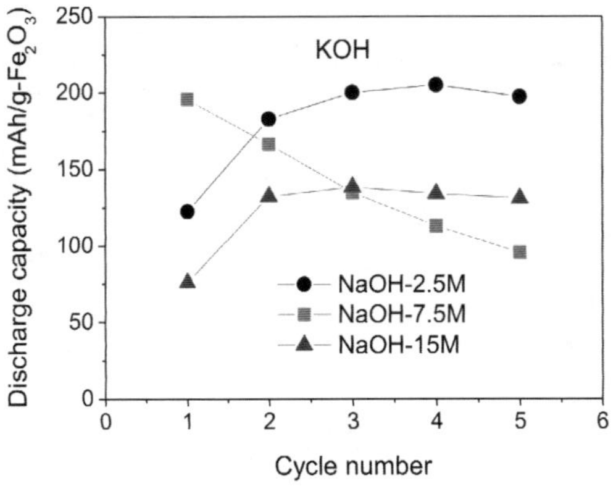

Figure 6. Discharge capacity of Fe_2O_3/AB electrodes (Fe_2O_3:AB:PTFE=45:45:10 wt.%) in KOH solution.

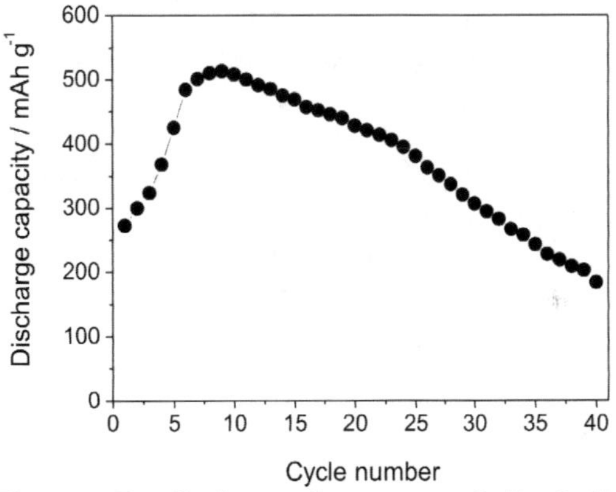

Figure 7. Cycle performance of Fe_2O_3/AB electrodes (Fe_2O_3:AB:PTFE=45:45:10 wt.%) using Fe_2O_3-NaOH 2.5M in KOH+K_2S solution.

4. Conclusions

In this study, hexagonal platelet shape α-Fe_2O_3 particles with different dimension were synthesized via a facile hydrothermal route by changing the concentration of the precursors. SEM measurement showed that the size and morphological characteristics of α-Fe_2O_3 particles were dependent on the NaOH amount during the hydrothermal treatment.

Investigation of the electrochemical properties of the Fe_2O_3/AB composite electrodes using synthesized Fe_2O_3 showed that the electrochemical properties of Fe_2O_3/AB electrodes were affected by the shape and size of the Fe_2O_3 materials. Among the various synthesized α-Fe_2O_3, the sample obtained at NaOH 2.5 M provided the highest discharge capacity. These iron oxides are potential materials for iron-based rechargeable batteries when further improved capacity and cycle performance.

Acknowledgment

This research is funded by the Hanoi University of Science and Technology (HUST) under project number T2020-SAHEP-018.

References

[1] Narayanan SR, Prakash GKS, Manohar A, Yang B, Malkhandi S, Kindler A. Materials challenges and technical approaches for realizing inexpensive and robust iron–air batteries for large-scale energy storage. *Solid State Ionics*. 2012;**216**:105-109.

[2] Ravikumar MK, Rajan AS, Sampath S, Priolkar KR, Shukla AK. In Situ Crystallographic Probing on Ameliorating Effect of Sulfide Additives and Carbon Grafting in Iron Electrodes. *J Electrochem Soc*. 2015;**162**(12), A2339-A2350.

[3] McKerracher RD, Ponce de Leon C, Wills RGA, Shah AA, Walsh FC. A Review of the Iron-Air Secondary Battery for Energy Storage. *Chempluschem*. 2015;**80**(2):323-335.

[4] Yang C, Manohar AK, Narayanan SR. A High-Performance Sintered Iron Electrode for Rechargeable Alkaline Batteries to Enable Large-Scale Energy Storage. *J Electrochem Soc*. 2017;**164**(2):A418-A429.

[5] Cui B, Xin H, Liu S, et al. Improved Cycle Iron Molten Air Battery Performance Using a Robust Fin Air Electrode. *J Electrochem Soc*. 2017;**164**(2):A88-A92.

[6] Paulraj AR, Kiros Y, Skårman B, Vidarsson H. Core/Shell Structure Nano-Iron/Iron Carbide Electrodes for Rechargeable Alkaline Iron Batteries. *J Electrochem Soc*. 2017;**164**(7): A1665-A1672.

[7] Manohar AK, Yang C, Narayanan SR. The Role of Sulfide Additives in Achieving Long Cycle Life Rechargeable Iron Electrodes in Alkaline Batteries. *J Electrochem Soc*. 2015;**162**(9): A1864-A1872.

[8] Chakkaravarthy C, Periasamy P, Jegannathan S, Vasu KI. The nickel/iron battery. *J Power Sources*. 1991;**35**(1):21-35.

[9] Micka K, Zábranský Z. Study of iron oxide electrodes in an alkaline electrolyte. *J Power Sources*. 1987;**19**(4):315-323.

[10] Černý J, Micka K. Voltammetric study of an iron electrode in alkaline electrolytes. *J Power Sources*. 1989;**25**(2):111-122.

[11] Shukla AK, Ravikumar MK, Balasubramanian TS. Nickel/iron batteries. *J Power Sources*. 1994;**51**(1-2):29-364.

[12] Periasamy P, Ramesh Babu B, Venkatakrishna Iyer S. Performance characterization of sintered iron electrodes in nickel/iron alkaline batteries. *J Power Sources*. 1996;**62**(1):9-14.

[13] Souza CAC, Carlos IA, Lopes M, Finazzi GA, De Almeida MRH. Self-discharge of Fe-Ni alkaline batteries. *J Power Sources*. 2004;**132**(1-2):288-290.

[14] Manohar AK, Yang C, Malkhandi S, Prakash GKS, Narayanan SR. Enhancing the Performance of the Rechargeable Iron Electrode in Alkaline Batteries with Bismuth Oxide and Iron Sulfide Additives. *J Electrochem Soc*. 2013;**160**(11):A2078-A2084.

[15] Shangguan E, Li F, Li J, et al. FeS/C composite as high-performance anode material for alkaline nickel–iron rechargeable batteries. *J Power Sources*. 2015;**291**:29-39.

[16] Posada JOG, Hall PJ. Towards the development of safe and commercially viable nickel–iron batteries: improvements to Coulombic efficiency at high iron sulphide electrode formulations. *J Appl Electrochem*. 2016;**46**(4):451-458.

[17] Posada JOG, Hall PJ. Controlling hydrogen evolution on iron electrodes. *Int J Hydrogen Energy*. 2016;**41**(45):1-11.

[18] Wang Q, Wang Y. Overcoming the Limiting Step of Fe 2 O 3 Reduction via in Situ Sulfide Modification. *ACS Appl Mater Interfaces*. 2016;**8**(16):10334-10342.

[19] Figueredo-Rodríguez HA, McKerracher RD, Insausti M, et al. A Rechargeable, Aqueous Iron Air Battery with Nanostructured Electrodes Capable of High Energy Density Operation. *J*

Electrochem Soc. 2017;**164**(6):A1148-A1157.

[20] Hang BT, Eashira M, Watanabe I, et al. The effect of carbon species on the properties of Fe/C composite for metal-air battery anode. *J Power Sources*. 2005;**143**(1-2):256-264.

[21] Hang BT, Watanabe T, Eashira M, et al. The electrochemical properties of Fe2O3-loaded carbon electrodes for iron-air battery anodes. *J Power Sources*. 2005;**150**(1-2):261-271.

[22] Hang BT, Watanabe T, Egashira M, Watanabe I, Okada S, Yamaki J. The effect of additives on the electrochemical properties of Fe/C composite for Fe/air battery anode. *J Power Sources*. 2006;**155**(2):461-469.

[23] Hang BT, Yoon S-H, Okada S, Yamaki J. Effect of metal-sulfide additives on electrochemical properties of nano-sized Fe2O3-loaded carbon for Fe/air battery anodes. *J Power Sources*. 2007;**168**(2):522-532.

[24] Hang BT, Hayashi H, Yoon SH, Okada S, Yamaki J ichi. Fe2O3-filled carbon nanotubes as a negative electrode for an Fe-air battery. *J Power Sources*. 2008;**178**(1):393-401.

[25] Trinh TA, Bui TH. α-Fe2O3 Urchins Synthesized by a Facile Hydrothermal Route as an Anode for an Fe-Air Battery. *J Mater Eng Perform*. 2020;**29**(2):1245-1252.

Comparative studies on strength properties of flyash based geopolymer concrete incorporating ggbs and alccofine subjected to elevated temperature

Chandrakanth V [1*] and Srikanth Koniki [2#]
[1] PG student, GRIET, Hyderabad, Telangana, India.
[2] Associate Professor, GRIET, Hyderabad, Telangana, India.
*Corresponding Author: vck1729@gmail.com
[#] corresponding author: sri.sri001@gmail.com

Abstract. Geopolymer concrete is a novel material trying to be implemented in the construction field. Efforts being made by the researchers to unfurl the advantages and application of Geopolymer concrete in engineering fields. Geopolymer concrete does not require cement in making and utilizes industrial waste products. Hence replacing the OPC concrete by GPC, we not only reduce the pollution caused by manufacturing cement, but also use calcium, silica and alumina rich industrial waste. It is important to study the performance of GPC subjected to elevated temperature before considering it for structures like industrial buildings, complexes which are vulnerable to fire accidents. Present study investigates the properties like compressive strength, weight loss/gain, and Thermal energy absorption for GPC subjected to elevated temperature from 100^{0}C, 200^{0}C, 400^{0}C, 600^{0}C and 800^{0}C for exposure time of 1hour and 2hours. We compare these properties of GPCs, one made using Flyash and GGBS as binders named FGGPC and the other using Flyash and Alccofine 1203 named FAGPC. From the results we found that strength of GPC increases with the rise in temperature up to 200^{0}C and then rapidly decreases with further rise in temperature. Use of FAGPC is advantageous than using FGGPC on field vulnerable to fire.

Keywords: GPC, Flyash, GGBS, Alccofine, Alkaline Activator, FGGPC, FAGPC, Elevated Temperature, Thermal energy.

1. Introduction

Concrete being mostly used construction material and its consumption increasing every day, its use cannot be restricted. Cement is the major component for making concrete but the production of cement contributes a major role in increasing greenhouse gases causing global warming. To limits the usage of concrete made from OPC, researchers investigated and identified a noble material called Geopolymer concrete [1,19]. Geopolymer concrete utilises calcium, alumina and silica rich industrial waste by products as binders which completely eliminates ordinary cement usage. To develop strength in geopolymer concrete an alkaline solution is required which chemically and physically reacts with these calcium, silica and alumina rich mineral admixtures [2,8]. The several mineral admixtures used for producing GPC are Flyash, GGBS, Metakaolin, Silicafume, Nanosilica, Alccofine, Rice husk ash etc.

Polymerization and Alkaline activation are two different processes but combine help in producing geopolymer concrete [9,20]. Polymerisation is a chain reaction process to form polymers from its monomers. Alkaline activation is the process of mixing alumina silicates minerals with alkaline activator solutions to enhance setting and hardening [3,10]. The alkaline activators used for GPC are KOH, NaOH solutions mixed with K_2SiO_3 or Na_2SiO_3 solutions. There is no exact mix design to develop GPC, but researchers developing several mix design methods which are in pending to be

approved by the concrete institutions [4,6]. Several applications of GPC which are already being implemented such as railway sleepers, tooling moulds, moulds for thermoplastics etc. [5,7].
The properties of the GPC depends on the several parameters like chemical and physical properties of mineral admixtures used as binders, molarity of alkaline activator solution, ratio of Na_2SiO_3 to NaOH solution, molarity of NaOH solution, alkaline binder ratio etc. [11,16]. At elevated temperature exposure regime the properties of GPC also depend on exposure temperature, time of exposure, rate of rise in temperature, rate of cooling and cooling methods [12,13].

When comes to structures, GPC is less implemented due to lack of proper research and demerits like early setting etc., but mostly used for precast members. But when comes to structures strength, durability properties of concrete need to be studied. It's also important for a proper investigation is need to be done for GPC at high temperature exposure regime to counteract while at fire accidents. As a part of it, in the present study we compare the strength characteristics of two different GPCs subjected to elevated temperature. The concrete specimens are casted, ambient cured at room temperature then subjected to elevated temperature at 100^0C, 200^0C, 400^0C, 600^0C and 800^0C for 1hr and 2hr exposure time. Air cooling and Water quenching are adopted for cooling the specimens after fire exposure and then tested. The properties like residual strength, weight loss/gain, thermal energy are observed. The surface temperature and core temperatures immediately after taking out from furnace are observed.

2. Significance of the study

To examine whether the GPC is suitable for high temperature exposures or not. The GPC made with only flyash takes much time to give strength than OPC concrete, hence by replacing it by GGBS and Alccofine separately to some extent, the change in the properties of GPC are observed. To check whether the betterment of performance of GPC with addition of GGBS and Flyash. Effort done to make the GPC concrete to be made to use in field applications in Civil and Structural Engineering.

3. Experimental procedure

The experimental procedure implemented for this project has been adopted from the available literature. In this work we compare the strength properties of two GPCs made from Flyash as major binder and GGBS and Alccofine as minor binders replacing flyash by 20% for economic considerations [2, 18]. Hence we name the GPC made from Flyash and GGBS as FGGPC and GPC made from Flyash and Alccofine as FAGPC.

3.1. Materials

Flyash, GGBS and Alccofine 1203 are used as binders, Alkaline Activator has been used to activate these calcium, silica and alumina rich materials to strengthen and to be concrete [14,15]. Locally available natural fine aggregate and 20mm coarse aggregate are used. Low calcium Flyash and GGBS are collected from the nearby RMC plant and their chemical compositions and physical properties are given in table 1 and table 2. Alccofine 1203 is a low calcium silica based mineral additive manufactured by the Ambuja cements ltd. The percentage replacement of GGBS and Alccofine for Flyash is about 50% from previous works, but we set it to only up to 20% for economic considerations [17].

Figure 1. Flyash, GGBS and Alccofine from left to right

Table 1: Chemical composition of Flyash, GGBS and Alccofine.

Chemical composition In Mass (%)	Flyash	GGBS	Alccofine
SiO_2	53-57	33.28	30-35
Al_2O_3	25-35	17.47	20-22
CaO	4-6	35.59	30-36
Fe_2O_3	3-7	0.65	1.2-2.5
MgO	1.0-2.5	8.10	4-8
Na_2O	0.10-0.25	0.35	0-0.5
K_2O	0.10-1.43	0.48	0.0-0.75
SO_3	--	--	0.1-0.3
TiO_2	1.08	0.75	--

Table 2: Physical properties of Flyash, GGBS and Alccofine.

Property	Flyash	GGBS	Alccofine 1203
Specific Gravity	2.4	2.8	2.86
Fineness (m^2/kg)	425	395	1200

The Alkaline activator is a mixture of sodium silicate (Na_2SiO_3) and sodium hydroxide solution (NaOH) at a proportion of 2.5:1 and the NaOH solution taken is of molarity 4. The Alkaline activator solution is rested for 24 hours after preparation to ensure all the heat exhausts which produced due to NaOH pellets reaction with water [7, 20]. As per the previous studies the alkaline activator solution should use as early as possible, later which may not give estimated results. It is a colourless sticky liquid.

3.2. Equipment.

Electric furnace with temperature rise up to 1200^{0}C is used for elevated temperature exposure. Compressive testing machine with a loading capacity of 2000kN is used. Digital thermometer with capturing capacity of 1050^{0}C with the help of thermo couple is used. Weighing machines and Concrete mixer are the other major equipment are used.

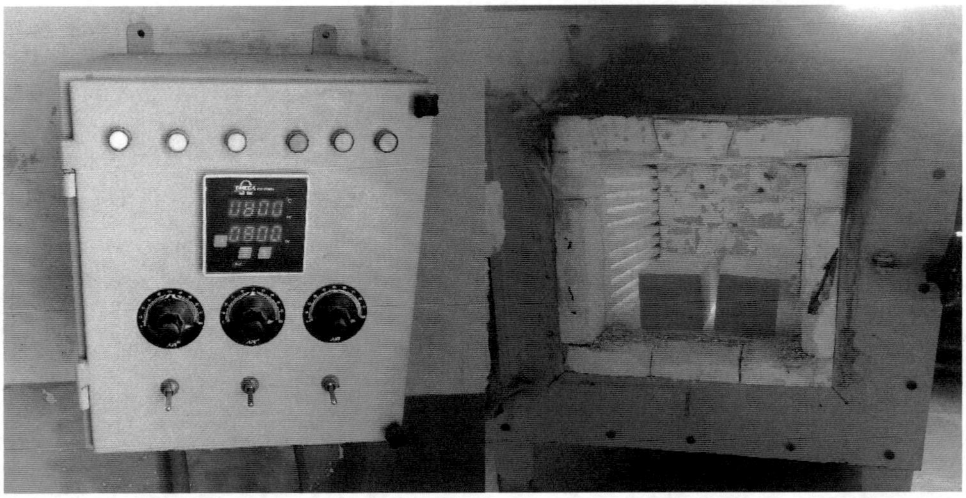

Figure 2. Electric muffle furnace and its digital meter

3.3. Trail mix

The mix design for preparing GPC of grade equal to M30 being adopted with reference to the work of Dr Mallikarjun Rao [10, 18]. Initially trail mix was done to ensure that the adopted mix design satisfies as per our requirement given in table 3. Test specimens are casted, ambient cured at room temperature (25^{0}C) for 28 days and then tested for compressive strength [2, 11]. The preliminary tests are done for two different alkaline binder ratios for both FGGPC and FAGPC to analyse the workability of the GPCs in the fresh stage. This is to verify the usage and role of superplasticizer in the GPC. The mix design adopted is given in table below. The percentage of Flyash to GGBS/Alccofine is 80% to 20%. The ratio of Na_2SiO_3 to NaOH is 2.5. Slump and Compressive strength results are given in table 4.

Table 3: Trail mix proportions for the FGGPC and FAGPC.

Mix Design	Alkalibinder ratio	Fly-ash (A)	GGBS (B)	Alccofine (C)	Binder (A+B+C)	Na_2SiO_3 (D)	NaOH (E)	Alkaline solution (D+E)	Coarse aggregate	Fine aggregate
FGGPC	0.4	304	76	-	380	152	43.4	108.6	1081	787
	0.5	284	71	-	355	177	50.6	126.4	1081	787
FAGPC	0.4	304	-	76	380	152	43.4	108.6	1081	787
	0.5	284	-	71	355	177	50.6	126.4	1081	787

Table 4: Slump and Compressive strength for Specimens casted for Trail mix.

Results of Trail mix design	Alkali binder ratio	Slump in mm	Compressive strength (28 days)
FGGPC	0.4	100	31.64
	0.5	107	33.26
FAGPC	0.4	103	33.51
	0.5	111	34.83

From the above results, to further proceed we take alkaline binder ratio to be 0.4 keeping in view of economy in producing concrete. As the increase in alkali binder ratio may increase slump and concrete has better workability for both FGGPC and FAGPC, but we can avoid it by using superplasticizer to some extent. Fosroc conplast SP430 sulphonated naphthalene based super plasticizer is used at 5% to the binder weight [12].

3.4. Elevated temperature exposure regime
Once ensuring and adopting a mix design then we proceed for mass production of specimens as per our requirement. After the 28 days curing, the specimens are subjected to elevated temperatures at 100^0C, 200^0C, 400^0C, 600^0C and 800^0C for duration of 60 min and 120 min at a rate of temperature rise being 9^0C/min. The heated specimens are cooled in two methods which are Air Cooling and Water quenching. Residual Strength properties like Uniaxial Compressive strength, Strength loss/gain and Thermal energy consumption are observed and compared for the FGGPC and FAGPC.

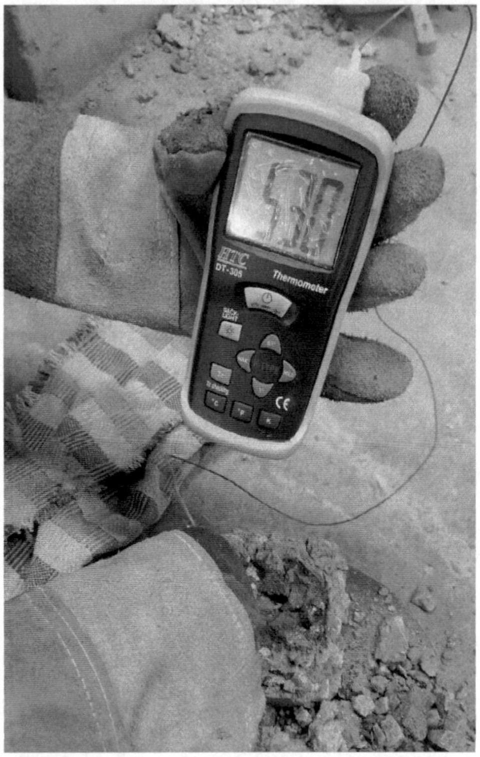

Figure 3: Core temperature reading of 530^0C for concrete exposed to 600^0C for 2 hours.

Figure 4. Colour change, unheated left specimen and heated right specimen 400^0C

Figure 5. Cracks on the specimen subjected to a temperature of 800^0C

3.4.1. Air cooling. The specimens are removed from the furnace and then they are kept at room temperature for 24 hours and then tested for weight loss and compressive strength test which gives us residual compressive strength

3.4.2. Water quenching. Immediately after removing specimens from the oven, we immerse them in 5 liters of water for each specimen and the temperature of water is noted at 1, 3, 5, 10, 15, 30 and 60 minutes. Initially the temperature of water rises rapidly and then it falls gradually. Left the specimen in the water until the water comes to room temperature and remove specimens and left it in room temperature for 24 hours and then tested.

4. Results, Observations and Discussions

Surface temperature and core temperature are found to know the change in core temperature to surface temperature for exposure time of 1hour and 2 hours. The values are listed in the table 5. From the observation it states that the temperature rise and increase in the time of exposure to fire will increase the difference between the surface temperature and core temperature both for FGGPC and FAGPC.

Table 5: Difference between Surface temperature and Core temperature

Time of exposure	Exposure Temperature	FGGPC			FAGPC		
		Surface temp	Core temp	Difference	Surface temp	Core temp	Difference
1 hour	100^0C	70	80	10	69	80	11
	200^0C	135	154	19	137	157	20
	400^0C	249	300	51	254	303	49
	600^0C	342	485	143	340	487	147
	800^0C	458	591	133	463	598	135
2 hours	100^0C	71	84	13	72	85	13
	200^0C	143	161	18	140	163	23
	400^0C	272	346	74	276	342	66
	600^0C	345	534	189	346	530	184
	800^0C	490	620	130	485	623	138

The surface temperatures and core temperatures for FGGPC and FAGPC are almost similar. A slight colour change is observed from 400^0C and above and a dark black colour is observed in the inner core at a temperature of 600^0C and above for both FGGPC and FAGPC.

4.1. Air Cooling and Weight Loss:

The specimens are removed from the furnace and then they are kept at room temperature for 24 hours and then tested for weight loss and compressive strength test which gives us residual compressive strength.

Table 6. Weight loss percentage for air cooling method

Weight loss %		FGGPC				FAGPC			
Exposure period	Exposure temp	Initial weight	Final weight	Change in weight	Change in %	Initial weight	Final weight	Change in weight	Change in %
1 hour	100^0C	2.379	2.367	0.012	0.51	2.393	2.345	0.047	1.97
	200^0C	2.413	2.390	0.023	0.93	2.415	2.350	0.065	2.69
	400^0C	2.394	2.310	0.084	3.50	2.350	2.265	0.085	361
	600^0C	2.283	2.283	0.104	4.34	2.348	2.232	0.116	5.94
	800^0C	2.411	2.291	0.120	4.97	2.379	2.244	0.135	5.67
2 hours	100^0C	2.422	2.404	0.018	0.73	2.414	2.357	0.057	2.33
	200^0C	2.384	2.345	0.039	1.62	2.363	2.294	0.069	2.92
	400^0C	2.399	2.299	0.100	4.15	2.356	2.226	0.130	5.51
	600^0C	2.407	2.295	0.112	4.64	2.346	2.205	0.141	5.98
	800^0C	2.389	2.259	0.130	5.42	2.387	2.232	0.155	6.49

Weight loss in FAGPC is dominant than that of FGGPC for all the temperature exposures. From this we can say that the FAGPC forms more pores by relieving moisture more than that of FGGPC. More pores may be lead to loss in strength but sometimes the pressure formed in the pores contributes to enhance the properties. For FGGPC a sudden gap in weight loss is observed from 200^0C to 400^0C but for FAGPC a gradual weight loss is observed with rise in temperature. Time of exposure also increase the weight loss in both GPCs.

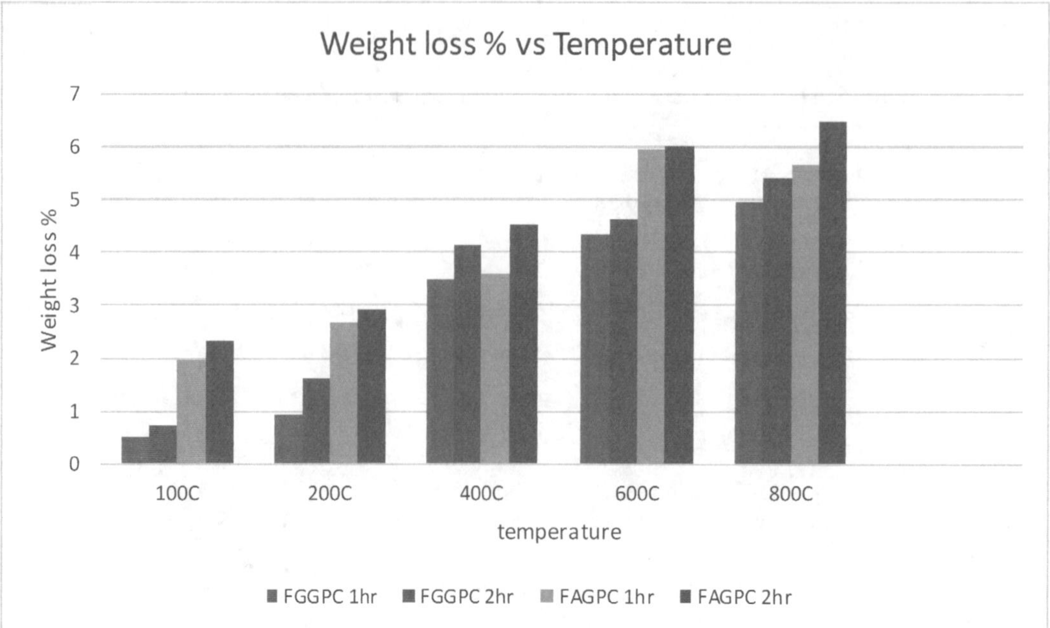

Figure 6. Bar chart showing Weight loss% vs Temperature

4.2. Water quenching and Weight Gain:

Immediately after removing specimens from the oven, we immerse them in 5 litres of water for each specimen and the temperature of water is noted at 1, 3, 5, 10, 15, 30 and 60 minutes. Initially the temperature of water rises rapidly and then it falls gradually. Left the specimen in the water until the water comes to room temperature and remove specimens and left it in room temperature for 24 hours and then tested.

Figure 7. Water temperature recording while water quenching method.

Table 7. Weight gain percentage for water quenching method.

Weight gain %		FGGPC				FAGPC			
Exposure period	Exposure temperature	Initial weight	Final weight	Change in weight	% change	Initial weight	Final weight	Change in weight	% change
1 hour	100^0C	2.393	2.448	0.055	2.31	2.376	2.432	0.047	1.97
	200^0C	2.421	2.496	0.075	3.13	2.331	2.406	0.075	3.20
	400^0C	2.385	2.522	0.137	5.75	2.356	2.498	0.142	6.03
	600^0C	2.389	2.553	0.164	6.89	2.343	2.510	0.167	7.09
	800^0C	2.420	2.593	0.173	7.16	2.394	2.584	0.190	7.92
2 hours	100^0C	2.412	2.478	0.066	2.77	2.350	2.408	0.058	2.46
	200^0C	2.395	2.504	0.109	4.55	2.363	2.467	0.104	4.41
	400^0C	2.405	2.567	0.162	6.75	2.307	2.468	0.161	6.99
	600^0C	2.392	2.562	0.170	7.10	2.400	2.583	0.183	7.63
	800^0C	2.398	2.586	0.188	7.83	2.413	2.610	0.197	8.20

From the results of water quenching the weight gain for FAGPC is more than that of FGGPC. This may be due to more pores formation in FAGPC for elevated temperature exposure regime. The weight gain percentage is gradually increasing with rise in temperature for both GPCs. As the time of exposure to fire increases the weight gain percentage is also increased.

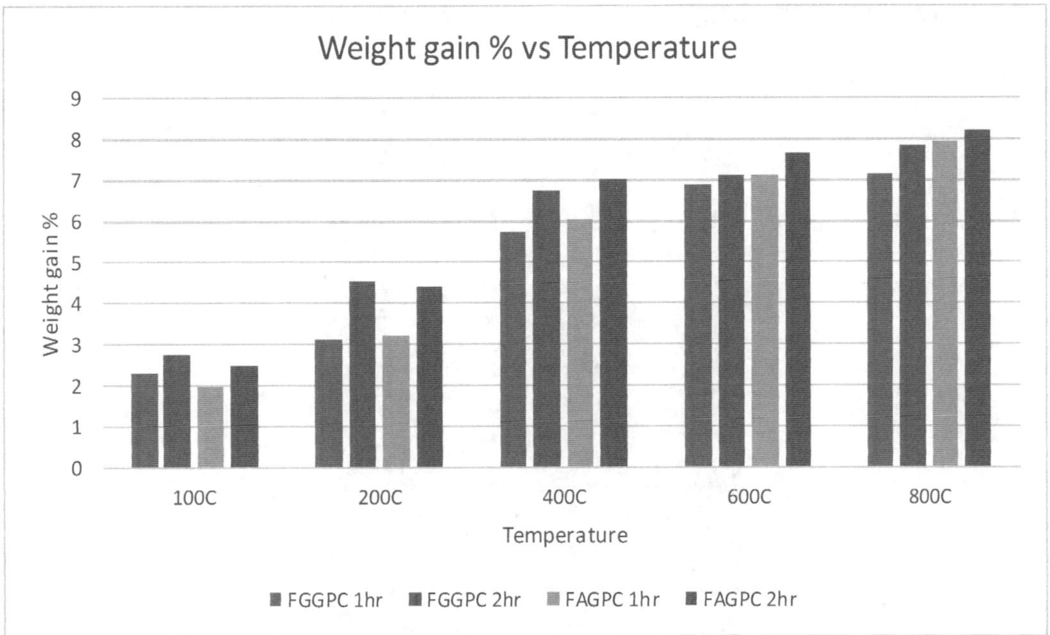

Figure 8. Bar chart showing Weight Gain % vs Temperature

4.3. Residual compressive strength:

The uniaxial Compression test is done for the specimens of Air cooling and water quenching after they are kept at room temperature for 24 hours. The test results and observations are given in table.

Table 8. Residual compressive strength for GPC post elevated temperature

Compressive strength (MPa)		FGGPC		FAGPC	
Exposure period	Exposure temperature	Air cooling	Water quenching	Air cooling	Water quenching
1 hour	100^0C	33.4	32.9	32.6	31.5
	200^0C	39.0	29.4	39.5	32.0
	400^0C	33.7	21.8	38.0	26.4
	600^0C	28.8	14.3	28.2	15.6
	800^0C	13.0	10.7	15.2	13.6
2 hours	100^0C	33.6	33.3	34.0	33.2
	200^0C	43.1	25.6	40.3	34.0
	400^0C	31.5	19.4	37.0	22.2
	600^0C	26.2	13.3	26.5	14.4
	800^0C	12.2	9.1	14.8	13.0

4.3.1. Residual compressive strength for Air Cooling:

From the observations residual strength of FGGPC increased up to a temperature rise of 200^0C and further rise in temperature, the strength decreased gradually up to 600^0C and a sudden drop is observed at 800^0C. For FAGPC the strength increased up to 200^0C and then decreased but a two-step drop in strength is observed at 600^0C and 800^0C. FAGPC has better performance for high temperatures and Air Cooling than that of FGGPC. The observation are similar for both exposure times of 60 minutes and 120 minutes, although the increase in exposure time decreases the strength.

Figure 9. Horizontal crack profile on specimen post elevated temperature testing

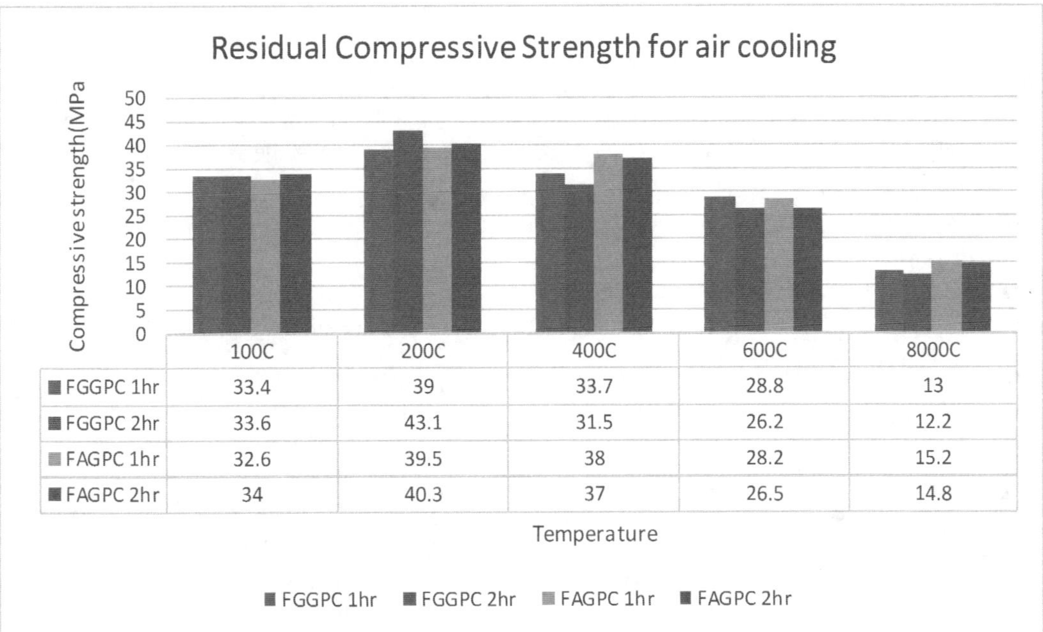

	100C	200C	400C	600C	8000C
FGGPC 1hr	33.4	39	33.7	28.8	13
FGGPC 2hr	33.6	43.1	31.5	26.2	12.2
FAGPC 1hr	32.6	39.5	38	28.2	15.2
FAGPC 2hr	34	40.3	37	26.5	14.8

Figure 10. Bar chart showing Residual Compressive strength for Air cooling

4.3.2. Residual compressive strength for Water Quenching:

When water quenching was done, strengths of FGGPC gradually decreases with rise in temperature and with increase in exposure time. Strength of FAGPC increases up to 200^{0}C and further gradually decreases. However only at 100^{0}C FGGPC strength is more than that of FAGPC. Further rise in temperature the strength of FAGPC is dominant. It shows that FAGPC is better at high temperatures than that of FGGPC for water quenching. Similar to the Air cooling observations, the increase in exposure time decreases the strength.

For both the GPCs Air cooling method gives better results than water quenching only except for FAGPC at 800^{0}C. Hence water quenching is not a preferable cooling method for GPCs exposed to elevated temperature. FAGPC is a better concrete for a temperature exposure of at 400^{0}C.

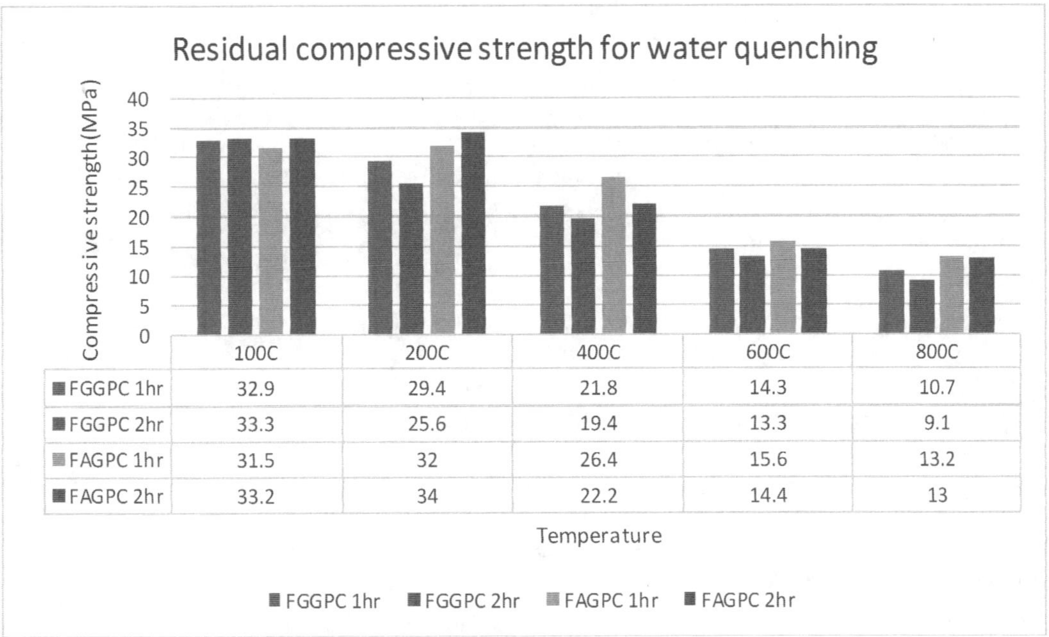

	100C	200C	400C	600C	800C
FGGPC 1hr	32.9	29.4	21.8	14.3	10.7
FGGPC 2hr	33.3	25.6	19.4	13.3	9.1
FAGPC 1hr	31.5	32	26.4	15.6	13.2
FAGPC 2hr	33.2	34	22.2	14.4	13

Figure 11. Bar chart showing Residual Compressive strength for Water quenching

4.4. Thermal energy:

The thermal energy concept is done to know the heat absorption of the concrete when exposed to fire. The time at maximum thermal energy for FGGPC is 15 minutes and for FAGPC is 10 minutes. This shows that the FAGPC absorbs energy at a faster rate. The maximum thermal energy values and their corresponding time is given in the table 9. The expression to calculate the thermal energy is given by

Thermal Energy (kcal) = Mass of water (litres) × Specific heat of water × Rise in Temperature

Mass of water taken for cooling = 5 litres

Specific Heat of Water = 1.0 cal/g/°C

Rise in Temperature = [Initial temperature of water – Temperature of water after immersing heated specimen in water]

Table 9. Maximum Thermal energy and corresponding time

Thermal energy		FGGPC		FAGPC	
Exposure period	Exposure temperature	Time at max thermal energy(min)	Max thermal energy(Kcal)	Time at max thermal energy(min)	Max thermal energy(Kcal)
	100^0C	15	50	10	45
	200^0C	15	70	10	55
1 hour	400^0C	15	130	10	110
	600^0C	15	225	10	225
	800^0C	15	265	10	270
	100^0C	15	60	10	50
	200^0C	15	85	10	70
2 hours	400^0C	15	165	10	150
	600^0C	15	240	10	240
	800^0C	15	280	10	275

The observations tells us that, although FAGPC has a faster rate of energy absorption but the maximum thermal energy for FGGPC is more than FAGPC. It shows that the use of Alccofine improves the pore structure of GPC than using GGBS and further studies on pore structure of the GPCs subjected to temperature rise is important.

4.5. Physical observations:

We can observe a surface color change in specimen subjected to elevated temperature of 200^0C and above for both FGGPC and FAGPC as shown in figure 4. The inner core color of the specimen changed to dark black color for FGGPC and FAGPC at and above 600^0C as shown in figure 3. We can observe serious cracks on the surface of the specimens at 800^0C as shown in figure 5. The specimen showed a horizontal crack while testing for uniaxial compression test using CTM as shown in figure 7.

5. Conclusions

- The Geo polymer concrete can be used for the structures vulnerable to fire exposure.
- The FAGPC gives better performance than FGGPC at and above the exposed temperature of 400^0C.
- Air cooling method gives better results than water quenching which did not give expected results.
- The rate of thermal energy consumption for FAGPC is at a faster rate but total thermal energy consumption for FGGPC is more than FAGPC.

6. Future scope of the study

There is much scope for the future studies regarding GPC subjected to elevated temperatures. Durability studies, microscopic analysis, porosity studies on GPC post fire need to be studied. The fly ash GPC properties are studied replacing different mineral admixtures with different proportions are studied. Deep studies on property recovery of concrete post elevated temperature is required.

7. References

[1] Davidovits J., Global warming impact on the cement and aggregates industries, *World resource review*, ***6*(2)**, 263-278 (1994).

[2] Rao G. M, Rao T. G, Reddy M. S. N & Seshu D. R., A Study on the Strength and Performance of Geopolymer Concrete Subjected to Elevated Temperatures, *Recent Advances in Structural Engineering,* **Volume 1** (pp. 869-889), Springer, Singapore (2019).

[3] Siddique R & Kaur D., Properties of concrete containing ground granulated blast furnace slag (GGBFS) at elevated temperatures, *Journal of Advanced Research*, ***3*(1)**, 45-51(2012).

[4] Davidovits J., Geopolymers and geopolymeric materials, *Journal of Thermal Analysis and Calorimetry*, ***35*(2)**, 429-441 (1989).

[5] Davidovits J., Geopolymers: inorganic polymeric new materials, *Journal of Thermal Analysis and calorimetry*, ***37*(8)**, 1633-1656 (1991).

[6] Shaikh F. U. A, & Vimonsatit V., Compressive strength of fly- ash- based geopolymer concrete at elevated temperatures, *Fire and materials*, ***39*(2)**, 174-188 (2015).

[7] Lăzărescu A. V, Szilagyi H, Baeră C, & Ioani A., The effect of alkaline activator ratio on the compressive strength of fly ash-based geopolymer paste, In *IOP Conference Series: Materials Science and Engineering* **(Vol. 209, No. 1,** p. 012064). IOP Publishing (June,2017).

[8] Duxson P, Provis J. L, Lukey G. C, & Van Deventer J. S., The role of inorganic polymer technology in the development of 'green concrete', *Cement and Concrete Research*, *37*(12), 1590-1597 (2007).

[9] Kodur V., Properties of concrete at elevated temperatures, *ISRN Civil engineering* (2014).

[10] A. Ehsani, M. Nili, K. Shaabani., Effect of nanosilica on the compressive strength development and water absorption properties of cement paste and concrete containing Fly Ash, KSCE J. Civ. Eng. **21 (5),** 1854–1865 (2017).

[11] H. Chao-Lung, B. Le Anh-Tuan, C. Chun-Tsun, Effect of rice husk ash on the strength and durability characteristics of concrete, Constr. Build. Mater. **25 (9)**, 3768–3772 (2011).

[12] A. Mehta, R. Siddique, Properties of low-calcium fly ash based geopolymer concrete incorporating OPC as partial replacement of fly ash, Constr. Build. Mater. **150,** 792–807 (2017).

[13] T. Yang, X. Yao, Z. Zhang, H. Wang, Mechanical property and structure of alkali-activated fly ash and slag blends, J. Sustain. Cement-Based Mater. **1 (4),** 167–178 (2012).

[14] P. Nuaklong, V. Sata, P. Chindaprasirt, Properties of metakaolin-high calcium fly ash geopolymer concrete containing recycled aggregate from crushed concrete specimens, Constr. Build. Mater. **161**, 365–373 (2018).

[15] Niveditha, M., & Koniki, S. (2020). Effect of Durability properties on Geopolymer concrete–A Review. In *E3S Web of Conferences* (Vol. 184, p. 01092). EDP Sciences.

[16] Chandrakanth, V., & Koniki, S. (2020). Effect of Review elevated temperature on geo-polymer concrete–A Review. In *E3S Web of Conferences* (Vol. 184, p. 01090). EDP Sciences.

[17] Ramesh, V., & Srikanth, K. (2020). Mechanical Properties and Mix Design of Geopolymer concrete–A review. In *E3S Web of Conferences* (Vol. 184, p. 01091). EDP Sciences.

[18] Mallikarjuna Rao G & Gunneswara Rao T. D, A quantitative method of approach in designing the mix proportions of fly ash and GGBS-based geopolymer concrete, *Australian Journal of Civil Engineering*, ***16*(1),** 53-63 (2018).

[19] Shetty M. S., Concrete technology. *S. chand & company LTD*, (2005).

[20] Neville A. M., *Properties of concrete* (Vol. 4). London: Longman (1995).

Acknowledgements

My Sincere thanks to the Management and Civil Engineering department of GRIET, Hyderabad for the support and Special regards to my guide Dr Srikanth Koiniki, Assosiate professor and my colleagues M Niveditha and V Ramesh for your part in completion of my project.

Effectiveness on mechanical properties of m60 grade scc with partial replacement of cement by different mineral admixture like ggbs, lime powder and metakaolin at various percentages

Polampally Vinod kumar[1*], Dr. G.V.V Satyanarayana[2#]
[1] PG student, GRIET, Hyderabad, Telangana, India.
[2] Professor, GRIET, Hyderabad, Telangana, India.
*Corresponding Author: polampallyvinodkumar1994@gmail.com
[#] Corresponding author: satyanarayanag42@gmail.com

Abstract Production of cement brings about the discharge of green house gases which at last prompts an unnatural weather change on one hand. On other hand consumption of crude materials like lime stone. So by utilizing elective materials we can lessen the creation of cement. Presently a day's the majority of the analysts are centered around replacement of cement with other mineral admixtures like Ground granulated blast furnanace slag(GGBS),Metakaolin and lime powder and so forth without bargaining its mechanical properties and durability. Self compacting concrete doesn't need vibration for setting and compaction. This paper presents the outcomes on various mechanical properties of self compacting concrete utilizing the standard Portland cement, GGBS, lime powder and metakaolin as binding material in making concrete. The hardened concrte properties like compressive strength, split tensile strength and flexural strength are found in trial work and comparison made with normal concrete.

Key words: M60 grade, SCC, GGBS, Lime powder and Metakaolin.

1. Introduction

Day by day development in human advancement keeps an eye on fast development in development movement with the end goal that elevated structures, honor shafts, express high ways and gigantic developments. Self-compacting concrete (SCC) become first brought in Japan at some point of 1980's, given that then it has been the trouble to a few examinations in order to achieve the favored homes of concrete structures. Simultaneously the makers of parts have created more noteworthy and more prominent advanced plasticizers and stabilizers custom-made for the precast and the equipped mix industry. Self compacting concrete (SCC) has created as a progressive time, equipped for achieving the notoriety of being a spectacular advancement inside the field of concrete time. No vibration is crucial for SCC which could stream around obstacles, exemplify the reinforcement and top off the system totally under its own one of a kind self weight. Making concrete system without vibration were executed inside the past. For example, placement of concrete submerged is performed by means of the utilization of termite without vibration.

Self compacting concrete requires high slump esteems which can be created by utilizing super plasticizer expansion to the solid combination. In this current examination an exploratory ingestion was led to decide the functionality, compressive strength, split rigidity and flexural strength of M60 grade SCC by utilizing various rates of GGBS, Lime powder and Metakaolin, the blend extents utilized are 75%C+25%GGBS (M1), 75%C+15%GGBS+10%LP(M2), 75%C+15%GGBS+10%MK (M3), 75%C+10%GGBS+15%LP, and 75%C+10%GGBS+15%MK.

2. Materials used for the study

For this experimental study initially collect the necessary materials to make self compacting concrete the details of materials collection and specification are discussed in the below

Cement

Ordinary Portland cement of 53 grade was used in this experimentation changing in accordance with I.S. – 12269-1987. For the current assessment I was assembled ultra tech concrete of 53 evaluation from neighborhood. The model concrete pack used in this examination is showed up in the underneath figure 1. The particular gravity 3.14 and consistency 28%.

Figure 1: OPC 53 grade Cement

Sand (FA)

Locally accessible zone II with specific gravity 2.53, water absorption 2.5% and fineness modulus 2.92, changing according to I.S. – 383-1970. It is the aggregates which passes through 4.75 mm IS sieve and contains fundamentally quite a phenomenal extent of coarser as is permitted by detail which are appeared in the figure 2.

Figure 2: Fine aggregates (Sand)

Gravel(CA)

Coarse aggregates are those stones which are retained on IS 4.75mm IS sieve number the coarse aggregates which are used in this experimental work is shown in below figure 3 the specific gravity is 2.6 and bulk density is 1523 Kg/m^{3}with water absorption is 2%.

Figure 3: 10mm size coarse aggregates

Mineral admixtures

GGBS

The Ground granulated blast furnace slag (GGBS) which is collected from JSW Allahabad is used in this experimental work.

Metakolin

Metakolin is the fine powder material which is utilized to improve the concrete strength and durability. The metakolin is appeared which is accessible at Adhipathi minerals kothapally.

Lime powder

Lime is fine powder which is collected from local areas used in this study.

Super plasticizer

Super plasticizer of master ease 3709 is used in this current study which is 1% of weight of the cement content.

3. Experimental investigation

For determine the mechanical properties of M60 grade self compacting mix cube, cylinders and beams are casted. The cubes of 150mmX150mmX150mm, cylinders of 150mm dia and 300mm length and beams of 150mmX150mmX500mm are utilized. All the specimens are permitted to curing for 7days, 28 days at ambient temperature. Different mechanical properties of M60 SCC is determined. The compressive strength and split tensile of concrete is find by using compression testing machine (CTM).

The flexural strength is controlled by utilizing two point load test.

The different mix proportions used for this study is shown in the below table. Different proportions of GGBS, metakolin and lime powder is used as replacement for cement.

Table 1: Mix proportions used for the study

Mix	Cement	GGBS	Metakolin (MK)	Lime powder (LP)
M1	75%	25%	0%	0%
M2	75%	15%	0%	10%
M3	75%	15%	10%	0%
M4	75%	10%	0%	15%
M5	75%	10%	15%	0%

4. Testing procedure

Fresh concrete tests

Workability of concrete is determined by using EFNARC guidance for M60 SCC for different mix proportions. The workability tests like slump cone, L box, V funnel and U box test is determined the test results are shown in the below table.

Table 2: Workability tests results on concrete

Property	M1	M2	M3	M4	M5
Slump (mm)	650	630	615	590	570
L-box (ratio)	0.9	0.86	0.81	0.75	0.7
V-funnel (sec)	10	11	12.5	14	16
U-box (ratio)	0.8	0.75	0.77	0.8	0.79

Harden concrete tests

The hardened tests like compressive, split ductile and flexural strength of M60 grade Self compacting concrete is determined for 7 days and 28 days restoring by utilizing different extents of GGBS, metakolin and lime powder as a substitution for concrete.

Compressive strength

The compressive strength is resolved for 7 days and 28 days examples age by utilizing compression testing machine (CTM). The outcomes are appeared in the below table and chart.

Table 3: Compressive strength test results

Mix	Compressive strength(N/mm2)	
	7days	28days
Mix1	48.2	71.33
Mix2	46.3	69.66
Mix3	39.23	59
Mix4	45.27	67.33
Mix5	20	29

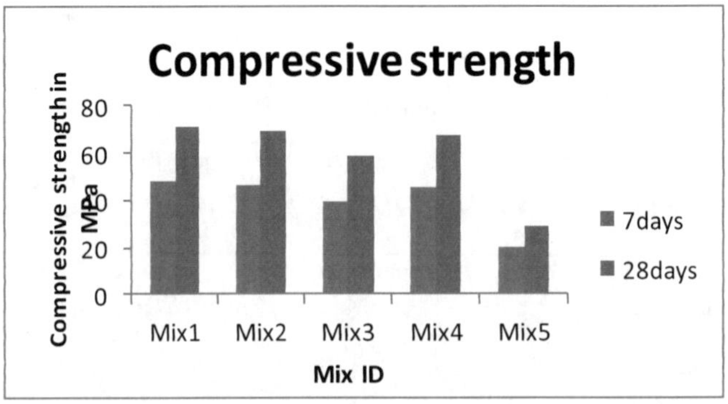

Graph 1: Comparison of compressive strength of concrete

Split tensile strength

The split tensile is resolved for 7days and 28 days restoring. The split tensile is controlled by utilizing compression testing machine (CTM) the test outcomes are appeared in below table and chart

Table 4: Split tensile strength test results

Mix	Split tensile strength(N/mm2)	
	7days	28days
Mix1	4.02	4.29
Mix2	3.98	4.01
Mix3	3.03	5.091
Mix4	3.18	3.34
Mix5	3.32	4.45

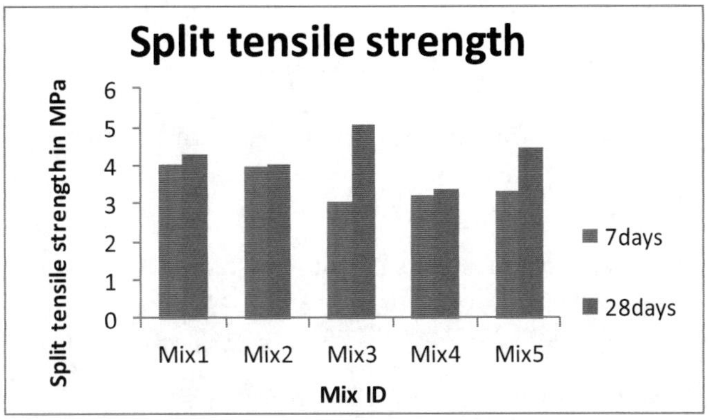

Graph 2: Comparison of compressive strength of concrete

Flexural strength

The flexural strength of SCC is resolved for 7 days and 28 days restoring days to radiates example. The flexural strength is dictated by two point load test machine. The test outcomes are appeared in the below table and chart.

Table 5: Flexural strength test results

Mix	Flexural strength(N/mm2)	
	7days	28days
Mix1	6.2	9.2
Mix2	5.6	6
Mix3	6	6.2
Mix4	5.7	6.5
Mix5	4.6	5.6

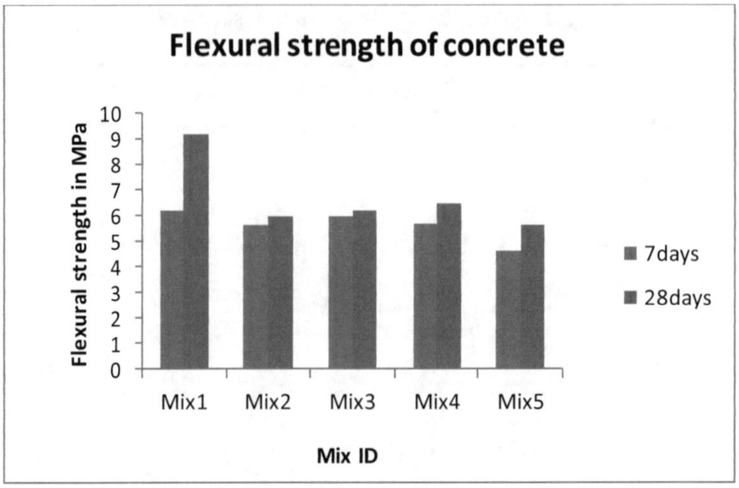

Graph 3: Comparison of flexural strength of concrete

5. Conclusions of the study

1. Mix 1 obtained maximum compressive strength when cement is partially replace with 25% of GGBS compressive strength of M60 grade SCC Concrete.
2. For Mix 2 also at par with conventional concrete when cement was partially replace with GGBS.
3. For Mix 3 Split tensile strength is highest when compared.
4. The maximum flexural strength occurred for mix1 i.e addition of ggbs to 70% of cement without adding lime powder and metakaolin.

References

[1]. N.Ravi teja and K.V .Ramana "Investigation of solidarity properties of self Compacting Concrete with Rise Husk Ash and Lime-stone Powder as Admixture". vol. 5, Issue 12, December 2016.
[2]. N.Bouzoubaa and M.Lachemi " Self Compacting Concrete Incorporating High-Volumes of Class F Fly Ash". Cement And Concrete Research, Vol. 31, No. 3, Mar 2001,pp. 413-420.
[3]. Gainster R. furthermore, Gibbs J., (2001) "Self-compacting solid section 1. The material an its property" Journal concrete, July/August 2001.
[4]. Kodama Y; "Current state of self-compacting concrete", cement shimbun, No.2304,Dec1997.
[5]. Koing G; Holschemacher K; Dehn F;Weibe D."Self-compacting solid Time of material properties and bond behavior".Proceedings of the second International Symposium on self-compacting concrete,Tokyo,(2001),pp507-516.
[6]. Nagataki S and Fujiwara H, "Self-compacting cement of high-flow able concrete", ACI Special Publications SP-154,June 1995, pp 301-314.
[7]. Okamura Hajime,"self-compacting superior concrete",(1997) solid global.
[8]. Ozawa K,"Development of high performance concrete dependent on the strength plan of solid structures".
[9]. Ozawa K, Sakara N, and Okamura H,"Evaluation of self-compatability of new solid utilizing the Funnel test".
[10]. Seto K, Okada K, Yanai S, and Nobuta Y, "Development and uses of self-compacting concrete".
[11]. Bilodeau, A. furthermore, Malhotra, V. M. (2000). "High-volume fly debris framework:

Concrete answer for practical turn of events." ACI Materials Journal, Vol. 97, No. 1, pp. 41-48.
[12]. Dinakar, P., Babu, K. G., and Santhanam, M. (2008). "Sturdiness properties of high volume fly debris self compacting cements." Cement Concrete Composites, Vol. 30, No. 10, pp. 880-886.
[13]. Gesoglu, M., Güneyisi, E., and Özbay, E. (2009). "Properties of self compacting cements made with parallel, ternary, and quaternary cementitious mixes of fly debris, impact heater slag, and silica rage." Construction and Building Materials, Vol. 23, No. 5, pp. 1847-1854.
[14]. Gesoglu, M. furthermore, Ozbay, E. (2007). "Impacts of mineral admixtures on new and solidified properties of self-compacting cements: Binary, ternary and quaternary frameworks." Materials and Structures, Vol. 40, No. 9, pp. 923-937.

Calcite mineral characterization and quantification of urease enzyme mediated sand columns

[1] V Srinivasa Reddy, [2] K Hema Latha, [3] Ravulaparthi Sudha Lahari, [4] V Mallikarjuna Reddy

[1]Professor of Civil Engineering, GRIET, Hyderabad, India.
[2]Assistant Professor of Civil Engineering, GRIET, Hyderabad, India.
[3] M.Tech (Structural Engineering), Department of Civil Engineering, GRIET, Hyderabad, India.
[4]Professor of Civil Engineering, GRIET, Hyderabad, India.

Corresponding Author: vempada@gmail.com

Abstract. In this paper urease enzyme facilitated calcium carbonate precipitation for soil strength improvement is characterized using X-ray diffraction and the amount of calcium carbonate precipitated is quantified based on acid leaching, thermal decomposition and CO_2 volume evaluation approaches. These techniques are adopted to evaluate the spreading of calcium carbonate crystals within the sand matrix after treatment with urease enzyme and nutrients. Sand columns of 60mm diameter and height 60 mm are prepared with the help of PVC pipes. Nutrient solution made up of urea, urease enzyme producing Sporosarcina pasteurii bacterium and calcium supplement in the form of calcium chloride is introduced into the sand columns ensuring the uniform distribution. Later treated samples of sand is examined for calcium carbonate precipitation by Sporosarcina pasteurii bacterium by XRD and quantified.

Keywords: Urease enzyme; soil-stabilization, Sporosarcina pasteurii, Thermal decomposition, CO_2 volume evaluation

1. Introduction

Urease enzyme facilitated calcium carbonate precipitation technique (UEFCCP) has found many applications such as detoxification of ground water from contaminants, crack remediation in concrete, restoration and preservation of limestone structures, controlling the swelling nature of clay soils, to stabilize soil from liquefaction during earthquakes etc. In UEFCCP technique, the calcium carbonate is precipitated due to reversible microbiological reactions in which the urea is hydrolyzed by urease enzyme into ammonia and carbonate ions. Ammonia will maintain the alkalinity of the green concrete and carbonate ions combine with calcium ions from supplied calcium nutrient to form calcite minerals. Due to its significant cementing capability this technique have been used for many civil engineering applications such as healing the cracks in concrete, improving the mechanical properties of the soil by binding its particles. The major challenge for this UEFCCP technique is to extend for in-situ application scale from the laboratory scale because UEFCCP technique needs sterilized approach for its bacterial activity. Few issue of this technique are (1) application to soils is that in case of fine grained soil the movement of bacteria is restricted (2) if the nutrients solution is used more than required then it may affect the bacterial activity (3) use of bacteria in soil with nutrients has problems with enzyme fixation and culturing.

So the precipitation of calcite crystals can be done using microbial method or by enzymatic method. In first method bacteria and nutrients are introduced into the soil where as in enzymatic method enzyme is injected along with the reagent. Calcite mineral diffusion in soil sample plays a major role so means to disperse urease enzyme produced calcite crystals are to be paid special attention. Unconfined compression tests will be conducted to assess the strength of the soil sample.

Nutrients broth consisting of urea and calcium supplement is used to culture different concentrations of Sporosarcina pasteurii which produces enzyme that decomposes urea. The one optimal concentration of Sporosarcina pasteurii will be added to PVC column soil samples. The soil getting stiff is observed.

2. Calcite quantification methods

2.1 Acid Leaching Method (ALM)

The precipitated calcite mineral can be quantified using this acid leaching method in which oven dried out bacteria treated soil samples are washed with HCL acid numerous times so that calcite mineral present in the soil sample will be washed out. Later the soil sample is washed with water and oven dried. So the loss of mass of acid washed and oven dried bacteria treated soil sample gives the amount of calcite mineral present in the soil sample. This method may overestimate the amount of calcite mineral precipitation because when the bacteria treated soil sample is washed with acid and water for at least 3 to 5 times during which not only calcite mineral is washed away but the soluble materials present in the soil sample may also get removed resulting in little more loss of mass.

Also this method is little time consuming because it has to be washed and oven dried many times. So to quantify the bacterially generated calcite crystals more accurately the thermal decomposition and CO_2 volume evaluation approaches are experimented to evaluate the reliability and speed with which these methods can quantify calcite crystals. Past researchers assessed the calcite precipitation even by measuring the S-wave velocity.

Three similar soil samples are chosen and treated with bacteria and are quantified for calcite mineral precipitation using three approaches such as acid leaching, thermal decomposition and CO_2 volume evaluation methods. Commercially available calcium carbonate of 98% purity is also obtained for calibration purpose.

Table 1. Results of Acid leaching method

Sample Type	Mass of Sand (gm)	Mass of $CaCO_3$ (gm)	Total Mass (gm)	Total Mass after acid and water wash (gm)	Loss of mass (gm)	Percentage Loss
1	50.0	0.0	50.0	49.936	0.064	0.13
2	49.5	0.5	50.0	49.748	0.252	0.50
3	49.0	1.0	50.0	49.648	0.352	0.70
4	48.5	1.5	50.0	49.464	0.536	1.07
5	48.0	2.0	50.0	49.292	0.708	1.42
6	47.5	2.5	50.0	49.124	0.876	1.75
7	47.0	3.0	50.0	48.852	1.148	2.30
8	46.5	3.5	50.0	48.756	1.244	2.49
9	46.0	4.0	50.0	48.444	1.556	3.11
10	45.5	4.5	50.0	48.356	1.644	3.29
11	45.0	5.0	50.0	48.264	1.736	3.47

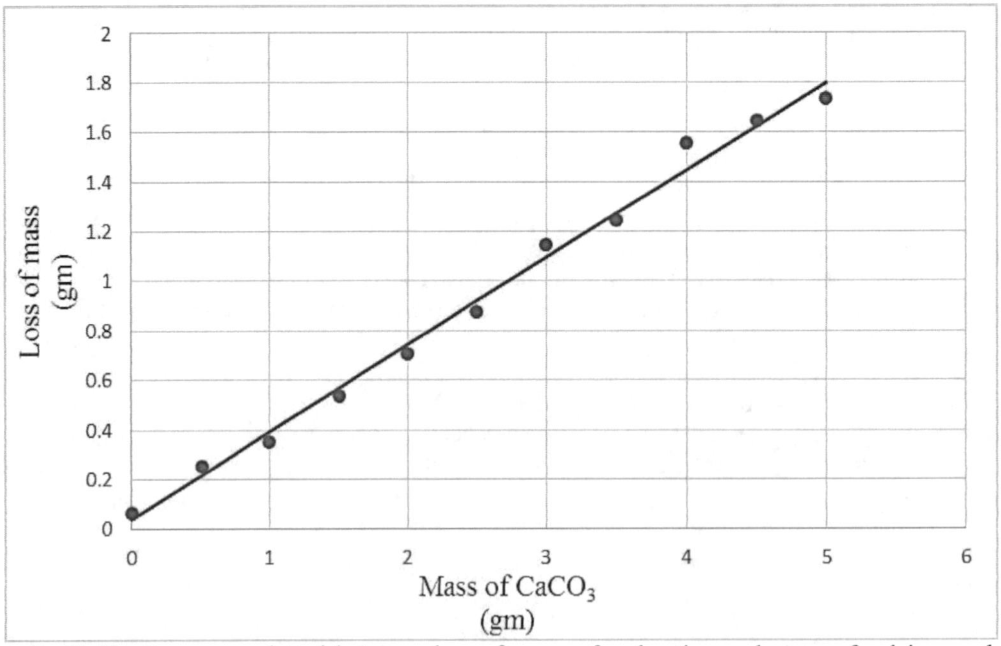

Figure 1. Calibration curve plotted between loss of mass after heating and mass of calcium carbonate mixed used for Acid leaching method to quantify calcite.

2.2 Thermal decomposition method (TDM)

This method is based on the principle that the calcium carbonate mineral will decompose at temperature of around 591°C into calcium oxide and carbon dioxide and may lose nearly 45% of its mass.

Table 2. Results of thermal decomposition method

Sample Type	Mass of Sand (gm)	Mass of $CaCO_3$ (gm)	Total Mass of mixture (gm)	Total Mass after thermal exposure (gm)	Loss of mass (gm)	Percentage Loss
1	50.0	0.0	50.0	49.920	0.080	0.16
2	49.5	0.5	50.0	49.685	0.315	0.63
3	49.0	1.0	50.0	49.560	0.440	0.88
4	48.5	1.5	50.0	49.330	0.670	1.34
5	48.0	2.0	50.0	49.115	0.885	1.77
6	47.5	2.5	50.0	48.905	1.095	2.19
7	47.0	3.0	50.0	48.565	1.435	2.87
8	46.5	3.5	50.0	48.445	1.555	3.11
9	46.0	4.0	50.0	48.055	1.945	3.89
10	45.5	4.5	50.0	47.945	2.055	4.11
11	45.0	5.0	50.0	47.830	2.170	4.34

The table 2 shows various mixture combinations of over-dried sandy soil and commercial calcium carbonate powder. A known composition of calcium carbonate and sand are mixed to make mixture of total mass 50 gm and placed in muffle furnace. After exposed to 600 °C temperature for 6 hrs. , the loss of mass is computed which will indirectly give the mass of carbon dioxide lost as gas. So for various compositions of calcium carbonate and oven-dried sand samples, loss of masses of mixture after heating for 6 hrs. at 600 °C are recorded and a calibration curve is plotted between loss of mass after heating and mass of calcium carbonate mixed. Due to high temperature there may be loss in mass of sand also due to decomposition of impurities. It was found that 0.080 gm of loss of mass was observed when the sand is heated to stipulated temperature. This error can be incorporated by making suitable corrections in the loss of masses of other mixtures.

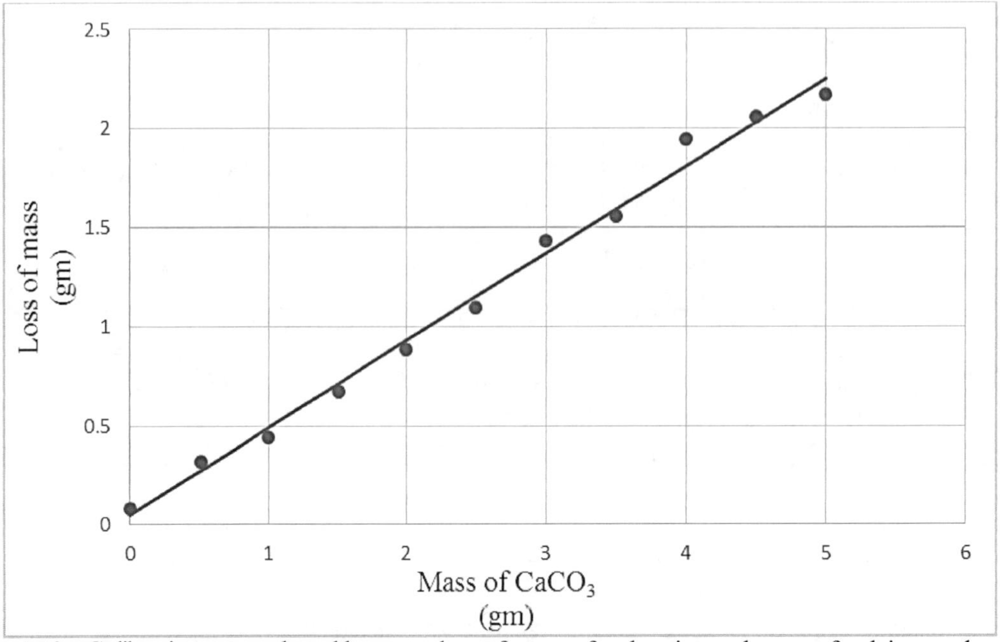

Figure 2. Calibration curve plotted between loss of mass after heating and mass of calcium carbonate mixed used for thermal decomposition method to quantify calcite.

The calibration curve plotted between loss of mass after heating and its corresponding mass of calcium carbonate mixed used can be used to quantify calcite using thermal decomposition method. It is apparent that the mass loss in the thermal decomposition method is more than the acid leaching method due to decomposition of other minerals or impurities present in the sand sample.

2.3 CO_2 volume evaluation method (CVEM)

The principle behind this method is that when HCL acid is added to the calcium carbonate, carbon dioxide gas will be evolved along with calcium salt. During the neutralization of calcium carbonate, carbon dioxide gas evolved is directly correlated to the quantity of calcium carbonate disintegrated. So if all the calcite minerals are expended then the gas collected and measured will quantify the amount of calcite precipitated.

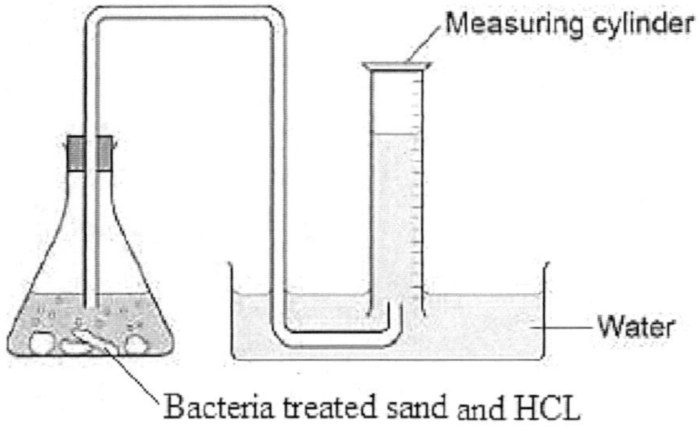

Figure 3. Carbon dioxide gas volume measurement test setup

This method was suggested to be most reliable as far as calcite quantification is concerned. A 50 grams of calcium carbonate mixed sand is taken into the flask and neutralized with the 50 mL HCL of 2.5 mol/L. It was estimated that the 50 mL HCl of 2.5 mol/L will decompose 12.5 gm of calcium carbonate. Shake the flask so that calcium carbonated sand and HCL to mix thoroughly so that gas bubbles of carbon dioxide will evolve. The evolved CO_2 gas passes through the tube and moves the water down in the measuring jar. So the amount of HCL acid added will produce gas which will displace equal amount of water in the measuring jar. Even in sand with no calcium carbonate also some amount of water is displaced means that some impurities also reacts and releases gas so this amount of correction is applied in subsequent measurements of gas from sand with calcium carbonate. The table XXX shows the amount of gas released for corresponding quantity of calcium carbonate added to sand.

Table 3. Amount of gas released after reacting with HCL

Sample Type	Mass of Sand (gm)	Mass of $CaCO_3$ (gm)	Total Mass of mixture (gm)	Volume of gas cm^3
1	50.0	0.0	50.0	50
2	49.5	0.5	50.0	115
3	49.0	1.0	50.0	225
4	48.5	1.5	50.0	360
5	48.0	2.0	50.0	460
6	47.5	2.5	50.0	570
7	47.0	3.0	50.0	670
8	46.5	3.5	50.0	755
9	46.0	4.0	50.0	850
10	45.5	4.5	50.0	945
11	45.0	5.0	50.0	1025

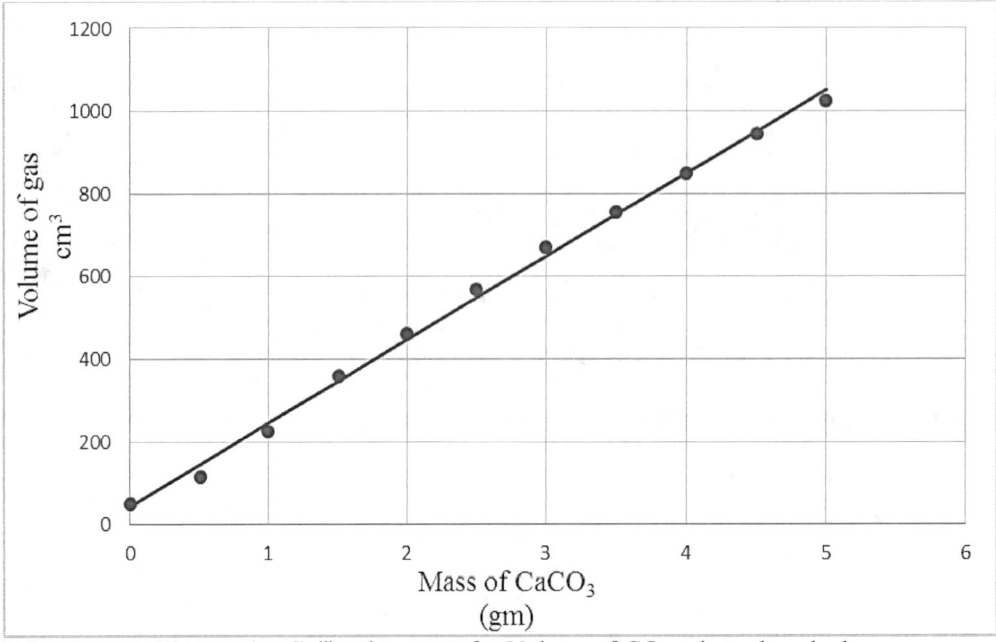

Figure 4. Calibration curve for Volume of CO_2 released method

When the bacteria treated sand samples were quantified for calcite using acid leaching method, thermal decomposition method and carbon dioxide volume evaluation method, it was observed that the calcite quantified using these methods are varying considerably. Amount quantified by acid leaching method is more due to presence of flushed soluble impurities other than calcite mineral. Similar observation is made in thermal decomposition method where ammonium chloride which is formed along with the calcium chloride evaporates at 600 °C temperature. Ammonium chloride formed with calcite mineral precipitation is inert in HCL in case of acid leaching method but it is soluble so it has no effect on the strength properties.

3. Conclusion

Based on three methods of quantifying the precipitated calcite mineral in sand samples. The following conclusions are established-

1. Urease enzyme facilitated calcium carbonate precipitation, has been appraised for its likely use in improving the soil properties.
2. Three calcite quantification methods that employ the elementary philosophies of chemistry have been assessed.
3. The tests conducted show that the Measuring volume of CO_2 method seems to be appropriate for measuring the precipitated calcite mineral.

4. References

[1] Al-Thawadi, S.M., 2011. Ureolytic bacteria and calcium carbonate formation as a mechanism of strength enhancement of sand. J. Adv. Sci. Eng. Res. 1 (1), 98–114.

[2] Burbank, M.B., Weaver, T.J., Green, T.L., Williams, B.C., Crawford, R.L., 2011. Precipitation of calcite by indigenous microorganisms to strengthen liquefiable soils. Geomicrobiol. J. 28 (4), 301–312.
[3] Fukue, M., Ono, S., Sato, Y., 2011. Cementation of sands due to micro- biologically-induced carbonate precipitation. Soils Found. 51 (1), 83–93.
[4] Ivanov, V., Chu, J., 2008. Applications of microorganisms to geo-technical engineering for bioclogging and biocementation of soil in situ. Rev. Environ. Sci. Biotechnol. 7, 139–153.
[5] Mortensen, B.M., DeJong, J.T., 2011. Strength and stiffness of MICP treated sand subjected to various stress paths. Geo-Frontiers 2011, 4012–4020 http//dx.doi.org/10.1061/41165(397)410.
[6] Nemati, M., Greene, E.A., Voordouw, G., 2005. Permeability profile modification using bacterially formed calcium carbonate: comparison with enzymic option. Process Biochem. 40 (2), 925–933.
[7] Neupane, D., Yasuhara, H., Kinoshita, N., Unno, T., 2013. Applicability of enzymatic calcium carbonate precipitation as a soil-strengthening technique. J. Geotech. Geoenviron. Eng. 139 (12), 2201–2211.
[8] van der Star, W.R.L., van Wijngaarden-van Rossum, W.K., van Paassen, L.A., van Baalen, L.R., van Zwieten, G., (2011). Stabilization of gravel deposits using microorganisms. Proceedings of the 15th European Conference on Soil Mechanics and Geotechnical Engineering, http://dx.doi.org/http//dx. doi.org/10.3233/978-1-60750-801-4-85.
[9] van Paassen, L.A., Ghose, R., van der Linden, T.J.M., van der Star, W.R.L., van Loosdrecht, M.C.M., 2010. Quantifying bio-mediated ground improvement by ureolysis: large-scale bio-grout experiment. J. Geotech. Geoenviron. Eng. 136 (12), 1721–1728.
[10] Whiffin, V.S., Van Paassen, L.A., Harkes, M.P., 2007. Microbial carbonate précipitation as a soil improvement technique. Geomicrobiol. J. 24 (5), 417–423.
[11] Yasuhara, H., Neupane, D., Hayashi, K., Okamura, M., 2012. Experiments and predictions of physical properties of sand cemented by enzymetically- induced carbonate precipitation. Soils Found. 52 (3), 539–549.

Behaviour of bacterial concrete subjected to elevated temperatures of various exposure times

V Srinivasa Reddy[1], Polina VVSSSR Krishna[2], G Sai Karthik[3], S Shrihari[4]

[1]Professor of Civil Engineering,GRIET, Hyderabad, Telangana, India. Email: vempada@gmail.com
[2]Assistant Professor of Civil Engineering, GRIET, Hyderabad, India.
[3]M.Tech (Structural Engineering), Department of Civil Engineering, GRIET, Hyderabad, India
[4]Professor of Civil Engineering, VJIT, Hyderabad, India

Abstract. *In this paper the behaviour of M25 grade bacterial concrete exposed to various elevated temperature and thermal cycles of different durations of exposure. Introducing the bacteria Sporosarcina pasteurii into the concrete will employ the bio-mineralization phenomena to precipitate calcite minerals in the cavity pores and dried gel pores of the concrete improving the internal void structure of the concrete. So the current study will try to understand the integrity of the bacterial concrete when subjected to a different elevated temperatures of 200 ℃, 400 ℃ and 600 ℃ exposed for 2 hrs., 4 hrs. and 6 hrs. duration. Compressive strength loss and weight loss is evaluated and internal structure quality is also assessed using non-destructive techniques as first part. Results suggest that till 500 -600 ℃ bacteria mixed concrete produces superior results beyond that the behaviour is almost similar to that of normal concrete*

Keywords- Bacterial concrete, self-healing concrete, elevated temperatures, Sporosarcina pasteurii, Bio-concrete

1. Introduction

Concrete, basically, is a good fire resistant material so it could be used as an excellent fire proof substance. But when exposed to elevated temperatures its chemically developed structure will get disintegrated leading to loss of bond in the concrete matrix. The rate of weakening of the concrete depends on the time of exposure and degree of heat. Due to deferential temperature changes, CSH gel may get transformed leading to loss of water leading to desiccation of concrete exhibiting the cracks and spalls in concrete.

IOP Conf. Series: Materials Science and Engineering 1091 (2021) 012028 doi:10.1088/1757-899X/1091/1/012028

2. Objectives

1. Compressive strength loss and weight loss is evaluated for the bacterial concrete when exposed to a different elevated temperatures of 200 ℃, 400 ℃ and 600 ℃ exposed for 2 hrs. , 4 hrs. and 6 hrs. duration.
2. Internal structure quality is also assessed using non-destructive techniques for the bacterial concrete when exposed to a different elevated temperatures of 200 ℃, 400 ℃ and 600 ℃ exposed for 2 hrs. , 4 hrs. and 6 hrs. duration.

3. Test Results and Discussion

Normal concrete and bacteria treated concrete cubes of 100 mm size are heated to temperatures of 200, 400 and 600 ℃ for 2, 4 and 6 hrs. in the muffle furnace. Compressive strengths and weights are recorded for every combinations temperature and exposure times are tabulated in table 1 and 2.

Figure 1. Electric muffle furnace

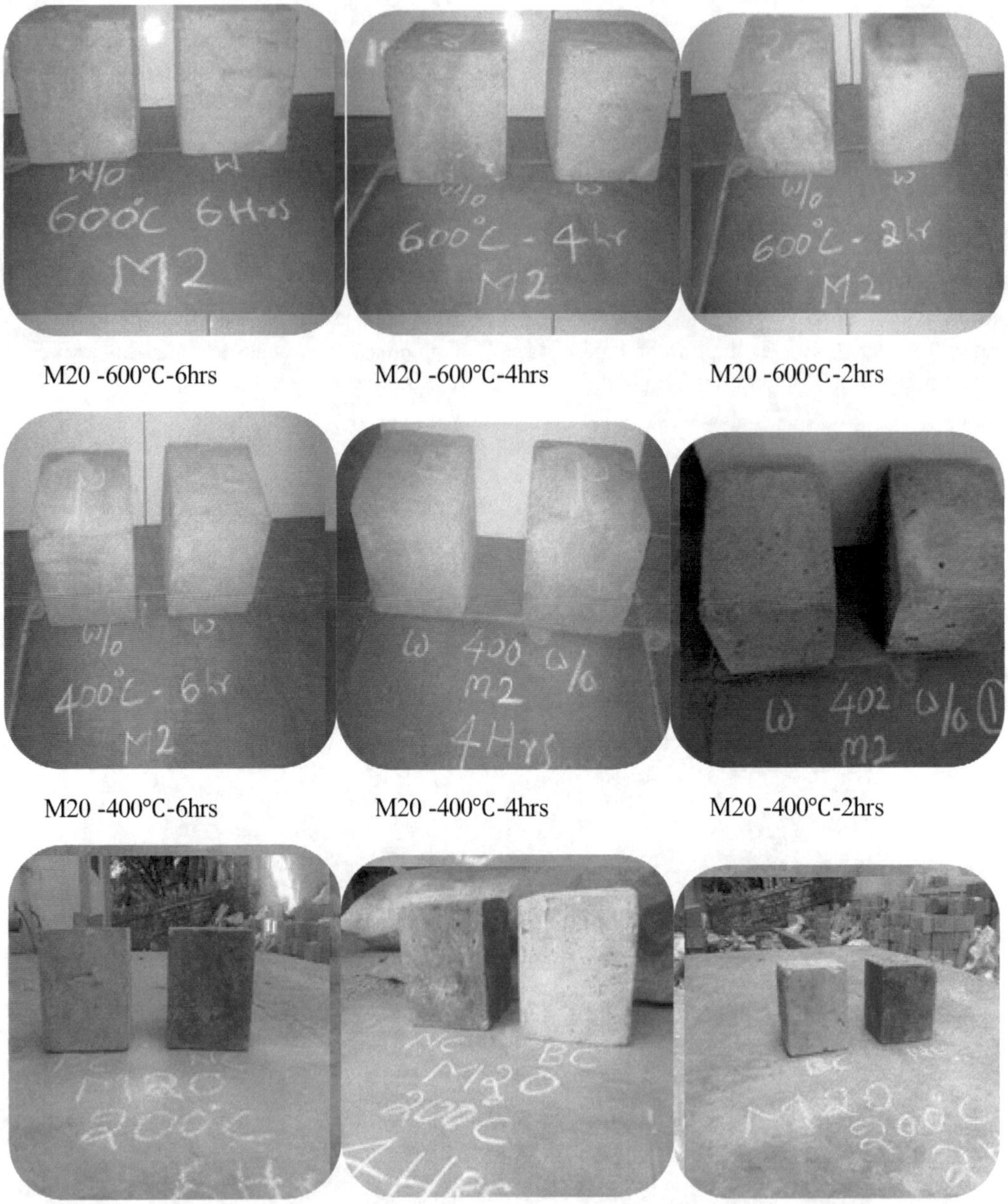

M20 -600°C-6hrs M20 -600°C-4hrs M20 -600°C-2hrs

M20 -400°C-6hrs M20 -400°C-4hrs M20 -400°C-2hrs

M20 -200°C-6hrs M20 -200°C-4hrs M20 -200°C-2hrs

Figure 2. Images of concrete cube specimens subjected to elevated temperatures for different periods of exposures

Table 1. Compressive strengths of bacteria mediated concrete at various temperatures for various times of exposure

		M25 Reference Concrete	M25 Bacteria mediated concrete
Compressive strength at Room Temperature	-	32.14 MPa	36.70 MPa
Compressive strength at 200 ℃	2 hrs. exposure	27.16 MPa	33.11 MPa
	4 hrs. exposure	26.89 MPa	32.89 MPa
	6 hrs. exposure	26.33 MPa	32.14 MPa
Compressive strength at 400 ℃	2 hrs. exposure	25.12 MPa	31.11 MPa
	4 hrs. exposure	25.01 MPa	31.02 MPa
	6 hrs. exposure	24.90 MPa	30.77 MPa
Compressive strength at 600 ℃	2 hrs. exposure	21.11 MPa	25.99 MPa
	4 hrs. exposure	20.19 MPa	25.34 MPa
	6 hrs. exposure	20.08 MPa	25.09 MPa

Table 2. Weight loss of bacteria mediated concrete at various temperatures for various times of exposure

			M25 Reference Concrete	M25 Bacteria mediated concrete
Weight Loss	Room Temperature	-	2.49	2.53
	200 ℃	2 hrs. exposure	2.46	2.51
		4 hrs. exposure	2.45	2.50
		6 hrs. exposure	2.44	2.48
	400 ℃	2 hrs. exposure	2.34	2.44
		4 hrs. exposure	2.33	2.43
		6 hrs. exposure	2.32	2.42
	600 ℃	2 hrs. exposure	2.13	2.22
		4 hrs. exposure	2.11	2.21
		6 hrs. exposure	2.10	2.20

Table 3. Rebound numbers of bacteria mediated concrete at various temperatures for various times of exposure

			M25 Reference Concrete	M25 Bacteria mediated concrete
Rebound numbers from Rebound Hammer Test	Room Temperature	-	31	38
	200 °C	2 hrs. exposure	28	33
		4 hrs. exposure	24	32
		6 hrs. exposure	22	30
	400 °C	2 hrs. exposure	19	27
		4 hrs. exposure	16	26
		6 hrs. exposure	15	24
	600 °C	2 hrs. exposure	11	17
		4 hrs. exposure	11	15
		6 hrs. exposure	10	14

Table 4. USPV measurements of bacteria mediated concrete at various temperatures for various times of exposure

			M25 Reference Concrete	M25 Bacteria mediated concrete
USPV measurements	Room Temperature	-	4.33	4.56
	200 °C	2 hrs. exposure	4.12	4.34
		4 hrs. exposure	4.11	4.31
		6 hrs. exposure	4.01	4.30
	400 °C	2 hrs. exposure	3.66	3.99
		4 hrs. exposure	3.54	3.89
		6 hrs. exposure	3.50	3.87
	600 °C	2 hrs. exposure	2.34	2.98
		4 hrs. exposure	2.13	2.90
		6 hrs. exposure	2.11	2.87

Due to high amount of pore pressure caused by high temperatures, the bind between aggregate and cement matrix is loosened. Thermal stresses induced will create strain energy leading to brittle failure. Concrete heated to high temperature loses water in capillary cavities to evaporation between 0 and 200 °C. Between 200 °C and 400 °C, concrete loses gel water leading to self-desiccation. When heated to temperature between 400 °C to 500 °C, calcium hydroxide can break down making more free lime availability giving the expansive behaviour for the concrete. So use of SCMs will control this mechanism by utilizing the calcium hydroxide positively. Calcium carbonate will be decomposed between 500 to 700 °C, increasing the porosity of the concrete. Due to disintegration of the calcium carbonate the compressive strength will get affected severely and get reduced by nearly 50%. Above 400 °C, negative thermal strain occurs manifesting cracks or scaling of the cover. Explosive spalling is very rarely observed in M25 grade because the pore interconnectivity is evidently established leading to gradual release of pore pressure due to high temperatures. In case of bacteria based concrete the loss of compressive strength is observed to be more than the normal concrete due to dense matrix structure in bacteria mixed concrete.

4. Conclusions

1. Bacteria cells count of 10^5 per ml of mixing water is adopted as optimum for enhanced performance due to formation of calcite minerals in the micro and macro pores of the concrete altering the microstructure for betterment.
2. Till 591°C bacteria mixed concrete produces superior results beyond that the behaviour is almost similar to that of normal concrete.
3. Compressive strength loss and weight loss is less in bacteria treated concrete compared to normal concrete
4. Beyond 591°C the calcite mineral gets decomposed into quick lime and carbon dioxide. Carbon dioxide expels out and quick lime may get dehydrated and hardens itself but have expansive nature.

5. References

[1] Wang, H. Y. (2008). The effects of elevated temperature on cement paste containing GGBFS. Cement and Concrete Composites, 30, 992–999
[2] Mohamedbhai, G. T. G. (1986). Effect of exposure time and rates of heating and cooling on residual strength of heated concrete. Magazine of Concrete Research, 38, 151–158.
[3] Arioz, O. (2007). Effects of elevated temperatures on properties of concrete. Fire Safety Journal, 42, 516–522.
[4] "EN 1991-1-2: actions on structures. Part 1-2: general actions—actions on structures exposed to fire," Eurocode 1, European Committee for Standardization, Brussels, Belgium, 2002.
[5] H. Buchanan, Structural Design For Fire Safety, John Wiley and Sons, Chichester, UK, 2002.
[6] V. R. Kodur and N. Raut, "Performance of concrete structures under fire hazard: emerging trends," The Indian Concrete Journal, vol. 84, no. 2, pp. 23–31, 2010.
[7] K.-Y. Shin, S.-B. Kim, J.-H. Kim, M. Chung, and P.-S. Jung, "Thermo-physical properties and transient heat transfer of concrete at elevated temperatures," Nuclear Engineering and Design, vol. 212, no. 1–3, pp. 233–241, 2002.
[8] Z. P. Bažant and M. F. Kaplan, Concrete at High Temperatures: Material Properties and Mathematical Models, Longman Group Limited, Essex, UK, 1996.
[9] L. T. Phan, "Fire performance of high-strength concrete: a report of the state-of-the-art," Tech. Rep., National Institute of Standards and Technology, Gaithersburg, Md, USA, 1996.
[10] V. R. Kodur and M. A. Sultan, "Thermal propeties of high strength concrete at elevated temperatures," American Concrete Institute, Special Publication, SP-179, pp. 467–480, 1998.

Comparative study on non-linear time history analysis of a building with and without base isolation using etabs

J.Selwyn Babu[1], J.Rex[2], V. Priya Reddy[3], M.S. Britto Jeyakumar[4]

[1]Professor, Civil Engineering Department , Malla Reddy Engg. College (A), Hyderabad. India
[2]Associate Professor, Department of Civil Engineering, Malla Reddy Engg. College (A), Hyderabad, India
[3]PG Student, Department of Civil Engineering, Malla Reddy Engineering College (A), Hyderabad, (Telangana), India
[4]Assistant Professor, Department of Civil Engineering, GRIET (Autonomous), Hyderabad, India

Corresponding Author: selwynbabu@gmail.com

Abstract. Base isolation is a device which is used to discrete the building from its foundation. Therefore, all through seismic activity, the building does not change due to the earthquake. Even if the earth moves violently, buildings tend to move like rigid bodies instead of collapsing In the current study, reinforced concrete framed structures of G + 8 and G + 16 (with and without base isolation) are considered with a 3 m high floor in the seismic zone V. The base isolator for both structures is designed to determine the stiffness and physical magnitudes of the lead rubber bearing core (LRB) specified at the base of the frame. The buildings were analyzed using nonlinear time history technique in ETABS. The effect of Base isolator is studied and building reactions like storey displacements, storey shear, storey drifts, and overturning moments can be seen in graphs. A general examination of G+8 and G+16 buildings with and without base isolator is watched.

Keywords-Base isolation,LRB, Nonlinear time history technique, Storey drift, Storey shear

1. Introduction

Earth tremors are unexpected phenomena when the building is in earthquake areas. A civil engineer has to stand up to save lives and at least damage buildings during an earthquake. The latest development in earthquake protection projects is basic insulation, which cannot reduce ground movement, but helps to minimize the effects of earthquake movement.

The basic isolation system is the most widely used seismic protection system. Reduces the effects of ground motion, thereby eliminating the effects of earthquakes on structures. Basic insulation has become popular in buildings and bridges over the last few decades. Foundation isolation has become a customary idea in the plan of building structures and bridges in many high-hazard territories. Many are now under construction. At the point when soil isolation happens all through the formation of seismic structures because of earthquakes, the ground moves sideways and harms the structure. The answer for this issue is to isolate the structure starting from the earliest stage presenting an adaptable protection framework between the establishment and the structure. This uses the adaptability of the framework to ingest tremor vibrations. In this manner the seismic energy moved to the structure is better decreased and the structure stays stable for a generally lengthy time span. Base protection expands the characteristic time of the structure and consequently lessens dislodging of the structure during seismic occasions. Elastic and lead direction are main considerations in presenting underlying adaptability. This increases the natural period of the structure and the displacement of the base beyond predetermined limits. However, base isolators do not always compensate for strong earthquakes, as it can cause more displacement at the bottom of the structure.

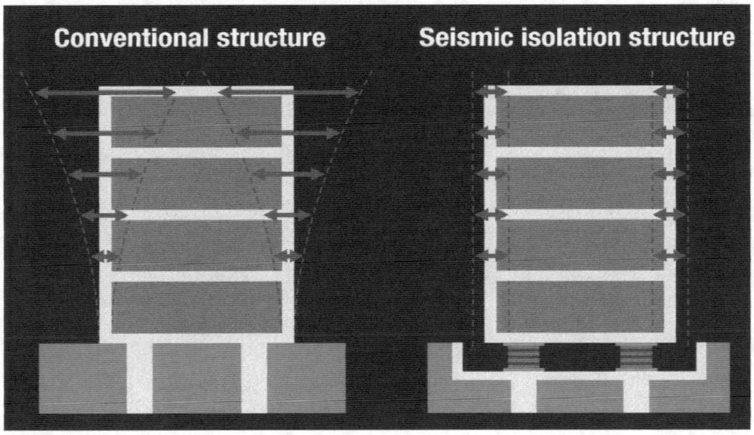

Figure 1. Performance of conventional building Vs seismic isolated building

2. Scope of the study

1. The scope of the present study is limited to analysis of reinforced concrete framed structures of G+8 and G+16 with and without base LRB base isolator using nonlinear time history technique in ETABS.
2. The seismic behavior of G+8 and G+16 buildings in region V with and without base isolators were studied.
3. The above buildings were compared by taking parameters like storey shears, floor displacements, storey drifts into consideration.

3. Methodology

Nonlinear time history technique

The only strategy to depict the genuine nature of a structure during tremor is through non-linear dynamic investigation or nonlinear time-history investigation.The strategy depends only on the direct numerical combination of differential equations of motion, taking into account the elasto-plastic deformation of a structural element. This is a significant strategy for structural seismic investigation, particularly when the assessed basic reaction is nonlinear. To conduct such an investigation, a agent seismic time history is needed for a structure being assessed. Time history examination is a bit by bit investigation of dynamic reaction of a structure to a predefined loading that may differ with time. Time history investigation is utilized to decide the seismic response of model to the dynamic loading of a typical earthquake. Time-history investigation is a bit by bit examination of the dynamical reaction of a structure to a particular loading that may differ with time. The investigation might be linear or nonlinear.

For nonlinear direct-combination time history investigation, the entirety of the material and geometric nonlinearities might be taken into account. For general nonlinear model time history examination, just only the nonlinear conduct of the link or support components is incorporated. On the off chance that the modes utilized for these examinations were calculated utilizing the firmness from the finish of a nonlinear investigation, all different kinds of nonlinearities are secured in the state that existed toward the finish of that nonlinear investigation.

4. Modelling

A.BuildingS data

S.No.	Description	Model A,B(G+8)	Model C,D (G+16)
1	**Building Details of the Structure:**		
i)	No. of Storeys	G+8	G+16
ii)	Structure Frame System	S.M.R.F	S.M.R.F.
iii)	Structure Type	Symmetrical and Regular	Symmetrical and Regular
iv)	Plan Area	6mx6m	6mx6m
v)	Storey Height- Bottom Storey	3.3m	3.3m
	Typical Storey	3m	3m
vi)	Height of the Building	24.3m	48.3m
vii)	Seismic Zone	V	V
vii)	Thickness- Outer Wall	230mm	230mm
	Inner Wall	115mm	115mm
2	**Material Properties**		
i)	Grade of Concrete	M40	M40
ii)	Grade of Steel	Fe415	Fe415
iii)	Density of Concrete	25kN/m^3	25kN/m^3
iv)	Young' Modulus (E_c)	31622776.6kN/m^2	31622776.6kN/m^2

IOP Conf. Series: Materials Science and Engineering 1091 (2021) 012029 doi:10.1088/1757-899X/1091/1/012029

v)	Young' Modulus (E_s)	2x10^8 kN/m^2	2x10^8 kN/m^2
3	**Loads Considered**		
i)	Floor Finish	1kN/m^2	1kN/m^2
ii)	Live Load	3kN/m^2	3kN/m^2
iii)	Parapet Wall Load	1kN/m^2	1kN/m^2
4	**Seismic Properties**		
i)	Zone Factor	0.3	0.3
ii)	Soil Type	Hard Rock	Hard Rock
iii)	Response Reduction Factor	8.5	8.5
iv)	Importance Factor	1.5	1.5
v)	Time History Function	Elcentro	Elcentro
vi)	Damping Ratio	5%	5%

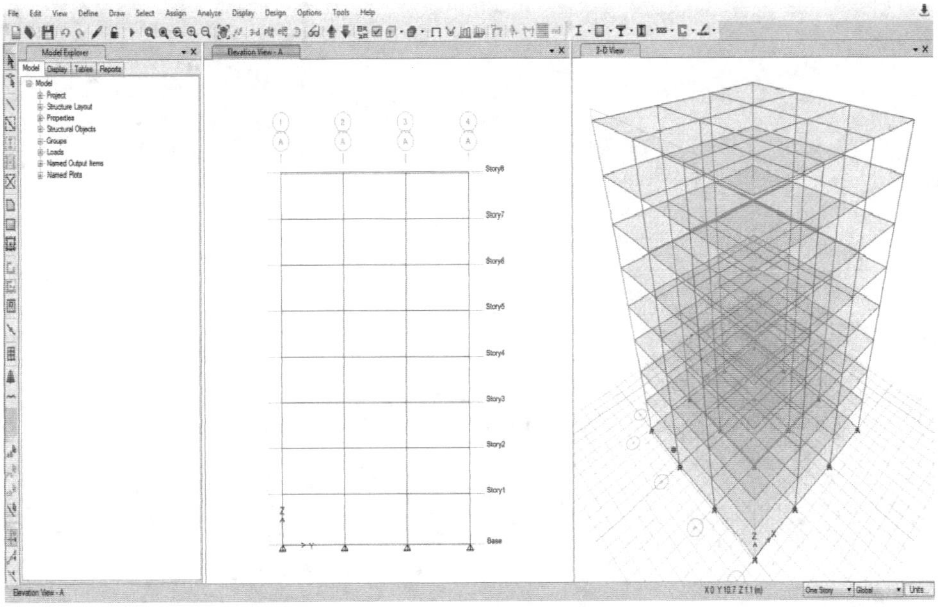

Figure 2. Model A: Elevation of G+8 building without LRB base isolator

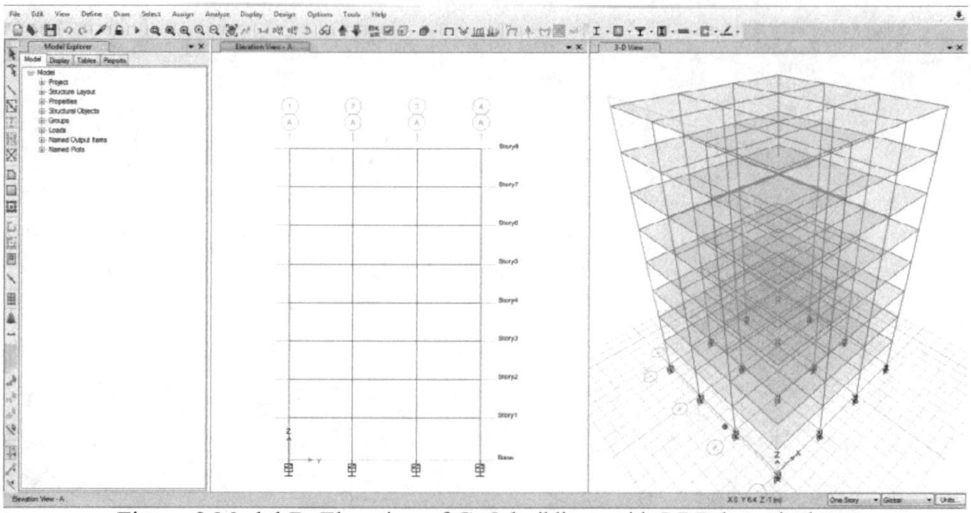

Figure 3.Model B: Elevation of G+8 building with LRB base isolator

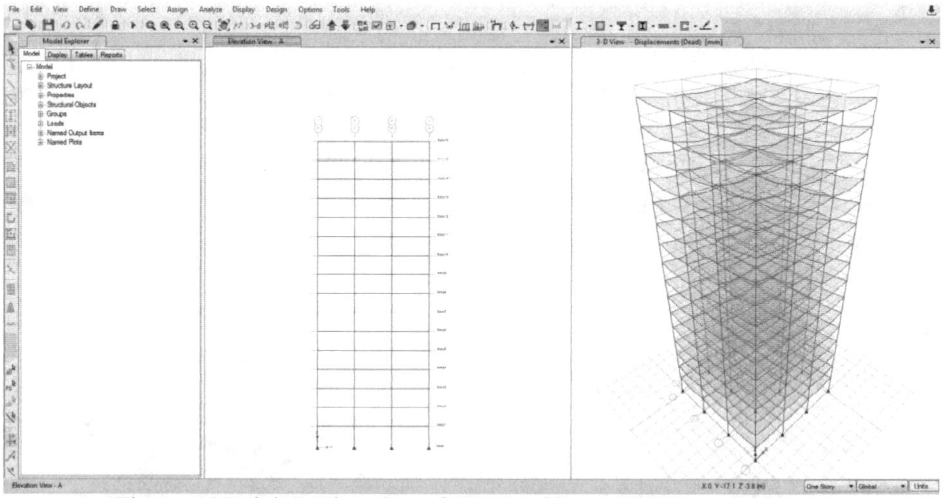

Figure 4.Model C: Elevation of G+16 without LRB Base Isolator

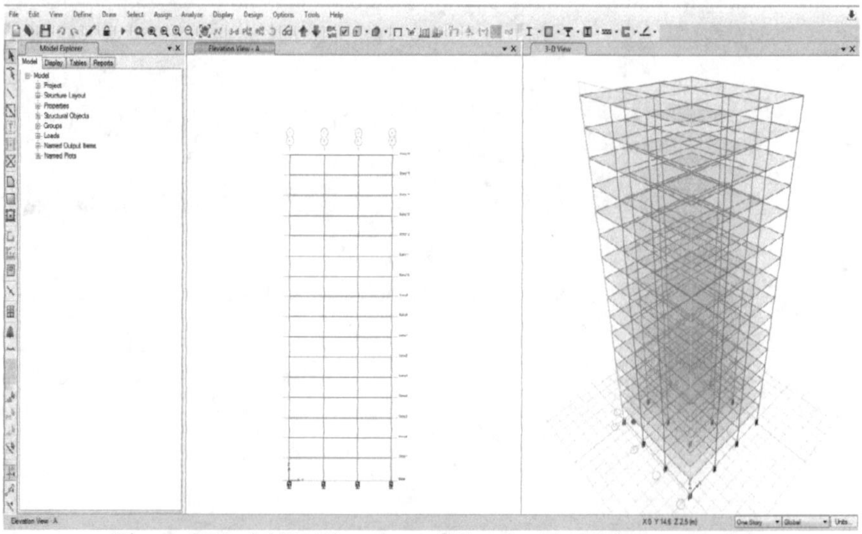

Figure 5.Model D: Elevation of G+16 with LRB Base Isolator

5. Results and Discussions

A. *Storey shear*

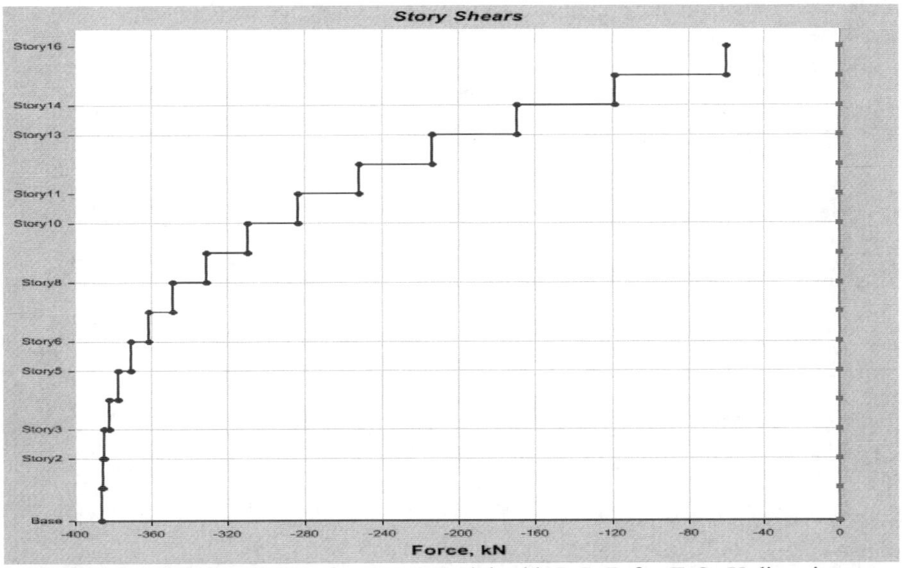

Figure 6. Storey Shears for G+16 model with L.R.B for E.Q. X direction

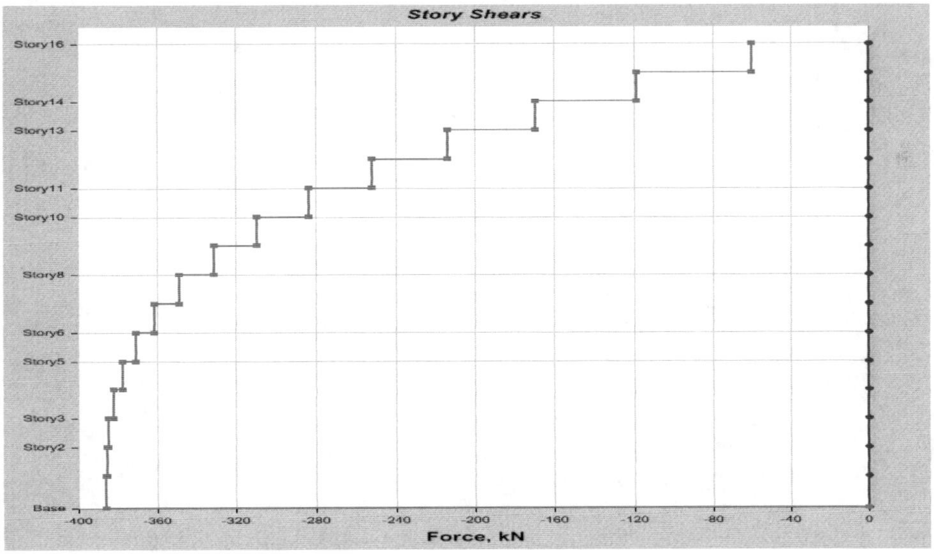

Figure 7. Storey Shears for G+16 model with L.R.B for E.Q. Y direction

Storey shears are diagrams showing how parallel (read: uniform) loads, wind or seismic, act on a plot. The lower you go, the more noticeable the shift is (see image below the shift plot below). The story float is again the next float chart on each floor. Storey shears with L.R.B for G+16 model is lesser than G+16 model without L.R.B. From the above Fig.6&7 it was observed that storey shear was same in both E.Q .X and Y direction.

B. *Storey drift*

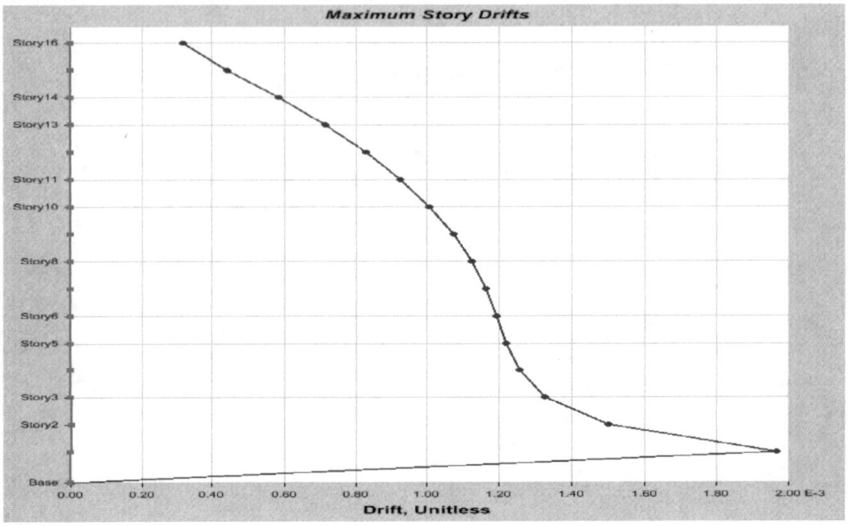

Figure 8. Storey Drifts for G+16 model without L.R.B for E.Q. X direction

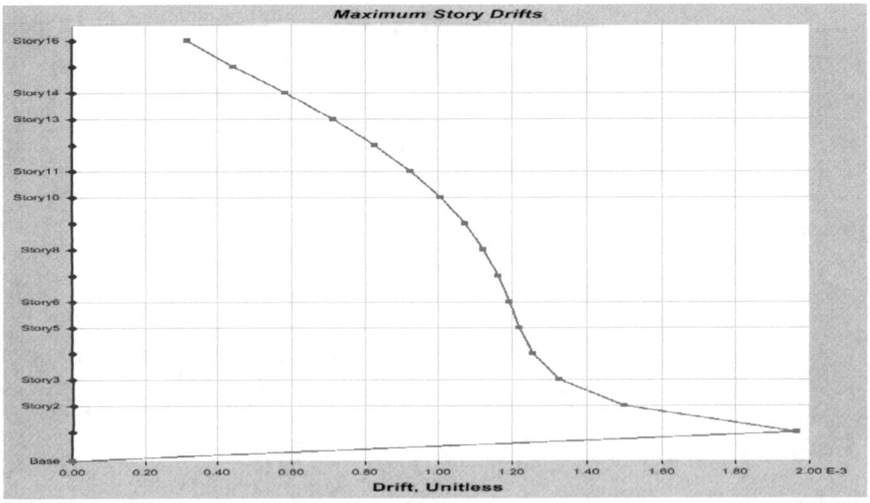

Figure 9. Storey Drifts for G+16 model without L.R.B for E.Q. Y direction

Story drift is the sidelong relocation of one level comparative with the level above or beneath. Story drift ratio is the story float isolated by the story tallness. Storey drifts without L.R.B for G+16 model is more than G+16 model with L.R.B. From the above Fig.8&9 it was observed that storey drifts were same in both E.Q .X and Y direction.

C. *Storey Displacements*

Storey displacement is all out dislodging of ith story regarding ground and there is greatest admissible cut-off recommended in IS codes for structures. Storey displacements with L.R.B for G+8 model is higher than G+8 model without L.R.B. From the above Fig.10& 11 it was observed that storey displacements were same in both E.Q .X and Y direction.

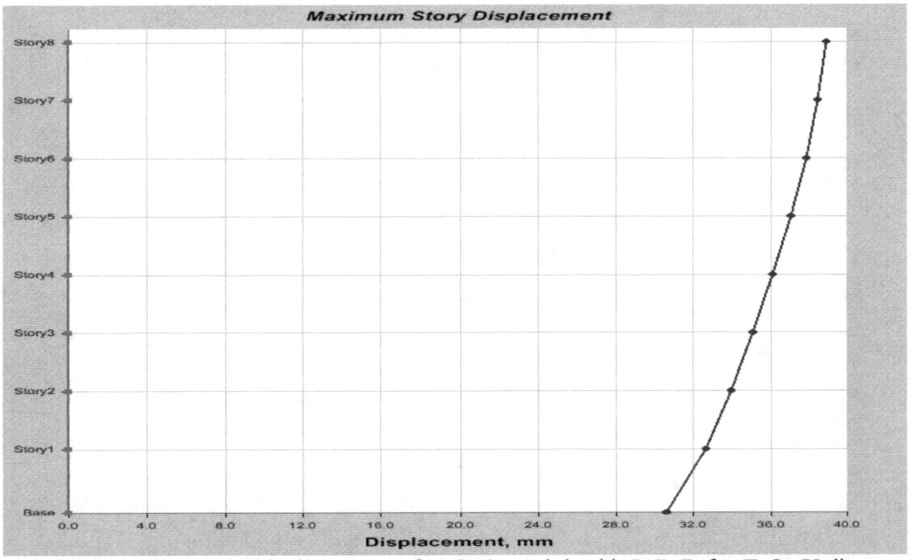

Figure 10. Storey Displacements for G+8 model with L.R.B for E.Q. X dir.

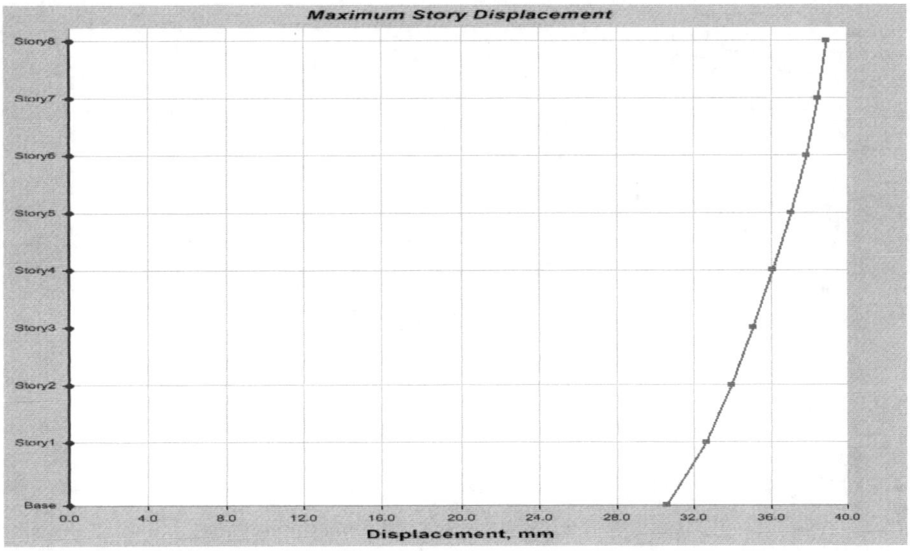

Figure 11. Storey Displacements for G+8 model with L.R.B for E.Q. Y dir.

D. *Overturning moments*

The Overturning Moment is taken as the measure of the minutes on the segment and any shear on the fragment expanded by the great ways from the base of the section to the base of the equilibrium. If there is lift on the part, by then the second associated with that center point load is furthermore considered in the Overturning Moment. Overturning Moments without L.R.B for G+8 model is higher than G+8 model with L.R.B. From the above Fig.12 & 13 it was observed that Overturning Moment was same in both E.Q .X and Y direction.

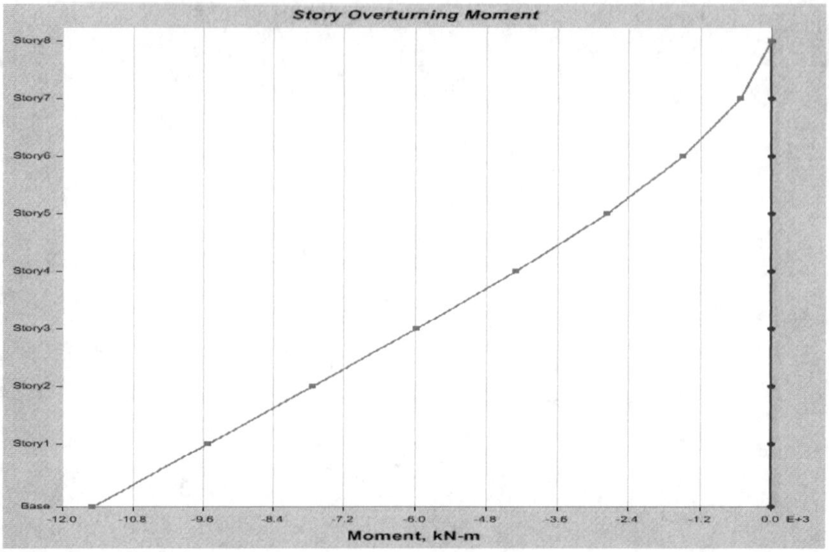

Figure 12. Overturning Moments for G+8 model without L.R.B for E.Q .X dir.

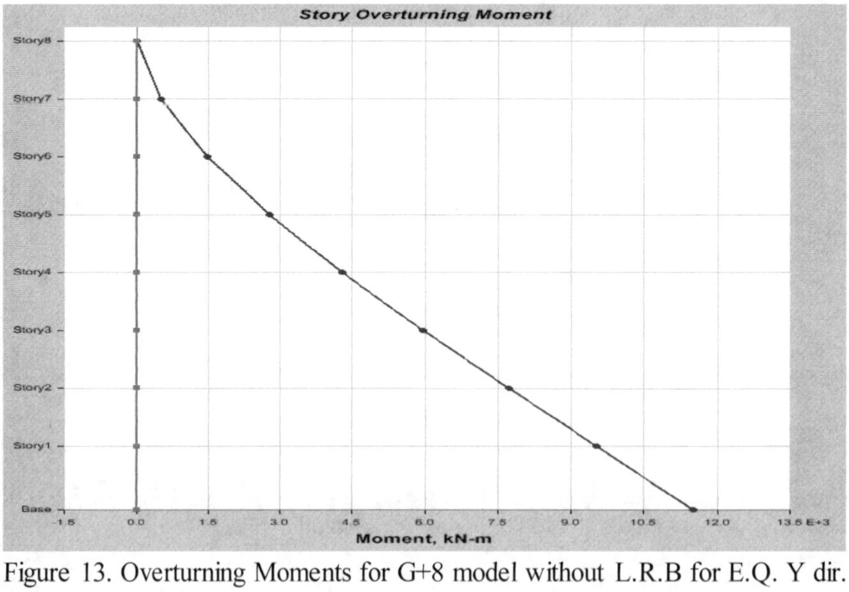

Figure 13. Overturning Moments for G+8 model without L.R.B for E.Q. Y dir.

6. Conclusions

The conclusions drawn from the analysis of G+8 and G+16 buildings with and without LRB are as follows

1. As the storeys increases storey displacements, drifts, overturning moments and base shears increases.
2. Storey displacements are increased in the wake ofgiving LRB which is important to make a structure flexible during earthquake.
3. Overturning moments of G+8 and G+16 buildings with LRB are 68% and 40% lesser than the G+8 and G+16 buildings without LRB respectively.
4. Storey displacements of G+8 and G+16 buildings with LRB are 49% and 66% higher than the G+8 and G+16 buildings without LRB respectively.

5. Storey drifts of G+8 and G+16 buildings with LRB are 25% and 30% lesser than the G+8 and G+16 buildings without LRB respectively.
6. Storey shears of G+8 and G+16 buildings with LRB are 69% and 40% lesser than the G+8 and G+16 buildings without LRB respectively.
7. The maximum base shears values of G+8 and G+16 buildings with LRB are 73% and 92.4% lesser than the G+8 and G+16 buildings without LRB respectively.
8. At long last,it is inferred that after LRB is given as base isolation system it increases the structures steadiness against earthquake and diminishes reinforcement subsequently make structure economical.

7. References

[1] **Mohammad Taghi, EshanKazeminezhad, Mirhosseini,**“ Altered technique of lead rubber isolator configuration utilized in the fortified concrete structure”, https://doi.org/10.1016/j.istruc.2020.07.056.
[2] **CSI study suggestion handbook for ETABS,** for modeling, investigation, and design of structures by means of ETABS, https://wiki.csiamerica.com/display/etabs/Home.
[3] **Farzad Naeim and Ronald L Mayers,** “Seismic isolated structures design”.
[4] **Wilkinson S and Hiley R (2006),** “A Nonlinear reaction history model for the seismic examination of tall outlined structures, PCs and structures”, volume 84.
[5] **Mario Paz, William Leigh,** “Structural dynamics hypothesis and calculation”, edition 5.
[6] **M. AmroyniFarissi and Ir. R. Bambang Budiono,** “Plan and Analysis of Base Isolated Structures”, DOI:10.2495/SAFE130761.
[7] **Shashikant K. Duggal,**“Tremor resistant design of structures”, edition 2.
[8] **Uniform Building Code 1997, Volume 2,**reference code book for structural design requirements, http://site.iugaza.edu.ps/sshihada/files/2012/02/UBC-1997.pdf
[9] **O.V. Mkrtychev, G.A. Dzhinchvelashvili , A.A. Bunov,** “Investigation of lead rubber bearing with changing stature structures at seismic tremor”, Volume 91, 2014, Pages 48-53, Procedia Engineering (ELSEVIER), DOI: https://doi.org/10.1016/j.proeng.2014.12.010.
[10] **Omkar Sonawane, Swapnil B. Walzade,** “Impact of base isolation in multistoried RC normal and irregular building utilizing time history technique”, ISSN: 2395-6992, volume 4, issue 5, IJOE.

Behaviour of Structural Elements made of Geopolymer Concrete with Recycled Aggregates

Srinivas.T[1], Abhignya.G[2]

[1] Professor, Civil Engineering Department, GRIET, Hyderabad, India.
[2]M.Tech Student, Civil Engineering Department, GRIET, Hyderabad, India.

Corresponding Author: srinu.tummala@gmail.com

Abstract. In the current scenario, concrete construction is rapidly becoming an uneconomical and unsustainable operation because of the lack of raw materials and environmental degradation caused by the production of cement. In this study, an attempt was made to propose recycled aggregates from demolition waste as coarse aggregates in geopolymer concrete (GPC). Experimental studies have been carried out to determine an optimum percentage of recycled aggregates (RA) in GPC by replacing 20 percent, 30 percent, 40 percent, 50 percent, and 60 percent and behaviour of structural elements also, such as beams and columns, when these are subjected to bending and axial compression . Beams and columns of 150 mm × 150 mm × 1200 mm were prepared and tested under two point loading on beams and axial loading on columns of OPC, GPC and RGPC elements. From this study, it is found that an optimum replacement percentage of recycled aggregate was 40 percent based on mechanical properties and flexural behaviour of RGPC, GPC are relatively similar, but better than OPC element.

Keywords: *Fly Ash, Geopolymer Concrete, GGBS, GPC, RGPC*

1. Introduction

Concrete is the most used tool for construction in the world. Ordinary Portland cement (OPC) has traditionally been used as the binding material for concrete. The production of OPC includes the burning of vast quantities of fossil fuels and the decomposition of limestone, resulting in high atmospheric CO2 emissions of carbon dioxide. The primary cause of heating is this CO2 pollution, which has become a serious concern. In order to scale back this, Geo polymer technology was developed.

"In 1978, Professor Davidovits used the term Geo polymer to explain the inorganic polymeric alumino-silicate gel resulting from the reaction of amorphous alumino-silicates with alkali hydroxide and silicate solutions." Unlike ordinary hydraulic cement, geo polymer does not shape calcium silicate hydrates for matrix formation and strength but uses silica and alumina poly condensation to achieve strength. Two main constituents of Geo polymers are the source and alkaline solvent. The source material should be alumino-silicate-based and rich in both silica and alumina. In Geo polymer concrete, additional cementing materials such as Fly ash, Silica fume, Rice Husk ash, Ground Granulated Blast Furnace Slag (GGBS) and metakaolin are used as alternative binders to hydraulic cement and Fly ash and Ground Granulated Furnace Slag (GGBS) are used during this project as alternative binders.

Geo polymer is a superb replacement material that transforms concrete binder into items of industrial waste such as GGBS and ash. Geo polymer binders are used alongside aggregates to provide Geo polymer concrete. They are suitable for building and repairing infrastructures and for pre-casting units, as they require very high early power. Their setting times can be monitored, and for a very long time they remain intact without the need for maintenance. Geopolymer, with characteristics such as adequate raw materials, low CO2 emissions, lower oil consumption, low cost, high early power, and fast setting.

2. Experimental Programme

In the experimental program, the fresh and hardened properties of geopolymer concrete with and without replacement of coarse aggregates with recycled aggregates were examined and the optimum percentage of recycled aggregates was determined for the materials. Beams and columns of various sizes have been cast and tested for analysis and comparison of the flexural behavior of beams and columns of geopolymer beams and columns with recycled aggregates.

2.1. Materials used

Materials used in this program are cement conforming to IS: 269-2015, coarse aggregates of size conforming to IS: 383-1970 and tested in accordance to IS: 2386-1963, fine aggregates confirming to IS: 383-1970, recycled aggregates were taken from construction and demolition waste, class-F fly ash, which is a by-product of coal is taken and replaced with cement up to 85% which confirms to IS: 3812-2003, ground granulated blast furnace slag (GGBS), which is a by-product from blast furnace used to make iron is heated at a temperature of 1500°C is taken and replaced with cement up to 15% by mass in geopolymer concrete, chemical admixtures like alkaline solution is used in this project which is a combination of sodium silicate and sodium hydroxide solutions and combination of these solutions give the molarity and concentration for activating the solution. Super plasticizers like high range water reducing agents are used to improve the workability of geopolymer concrete. A naphtha based super plasticizer known as Naphthalene sulphonated formaldehyde is taken for acquiring better strength for geopolymer concrete.

2.2. Mix design

Mix design is done as per the Indian Standards IS:10262-2009 for both conventional and Geopolymer concrete. Many methods were adopted and by trial-and-error method, calculation of conventional and geopolymer concrete is done. Water-cement ratio of 0.45 is adopted in this project. With all the calculations and results obtained from the experiment a target compressive strength of 30MPa is obtained. Ratio of the materials mix proportions of cement (fly ash + GGBS) : Fine aggregates : coarse aggregates : water is obtained as 1: 0.89: 3.27: 0.45. GPC strength characteristics have been compared to cement concrete with the same compressive strength. The GPC and OPC concrete proportioning mix are given in the table below.

Table-1 Mix proportions of GPC and OPC

S.no	Material	OPC (kg/m^3)	GPC (kg/m^3)
1	Cement	362	-
2	Fly ash	-	307.7
3	GGBS	-	54.3
4	Coarse aggregates	1182.79	-
5	Recycled aggregates	-	1182.79
6	Fine aggregates	694.6	694.6
7	Super plasticizer	9.05	9.05

3. Specimen Details and Test Setup

Structural elements with steel reinforcement like beams and columns are casted in this experiment with dimensions of 1200mm × 150mm × 150mm. As per Indian standards, reinforcement detailing for beams and columns are calculated according to IS: 456-2000 guidelines. The nominal diameter of the reinforcing bars was 12 mm and 8 mm respectively for the primary and lateral links. The appropriate quantity of Fly ash, GGBS, fine aggregate and coarse aggregate based on the proportion of the mix was mixed manually for one minute in dry conditions. Alkaline solution and super plasticizer were applied and well combined. Using reinforcement for beams and columns, 6 specimens were casted (2 RAGPC + 2 GPC+ 2 OPC). In the figure below, reinforcement of beams and columns before casting is seen. After 24 hours, all the specimens were demolded and shown in Fig. GPC specimens were cured under ambient curing conditions, OPC specimens were cured under water.

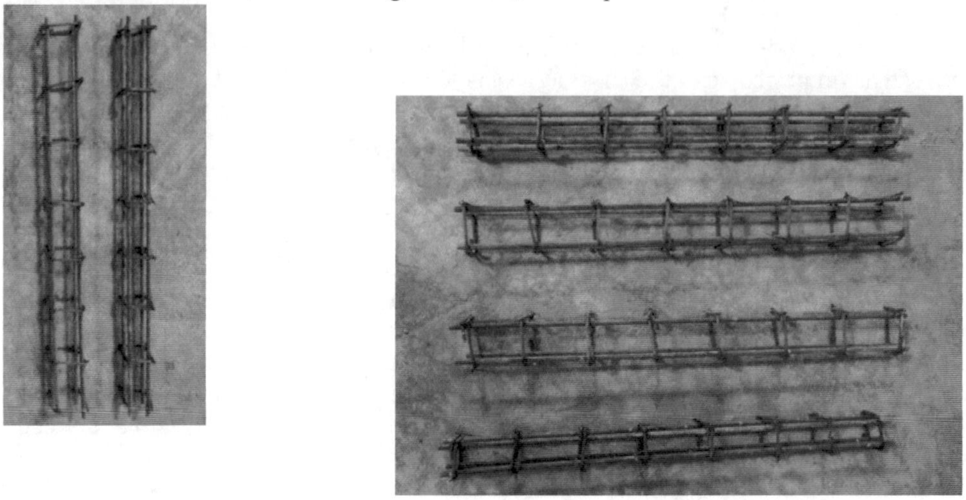

Fig.1 Reinforcement for beams and columns for OPC, GPC and RAGPC

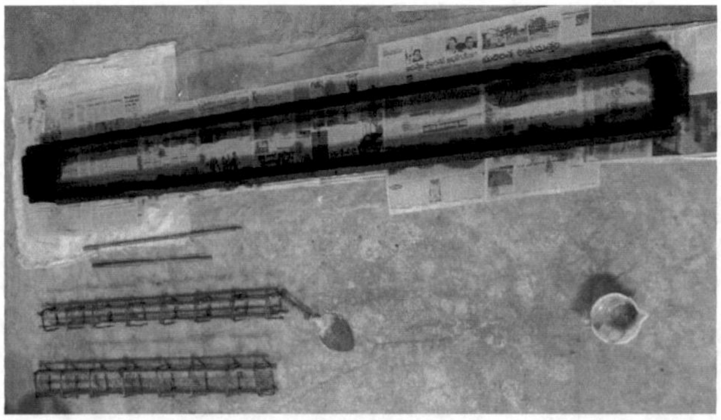

Fig.2 shows the mould and reinforcement to cast beams and columns

Testing of structural elements is done in loading frame machine. Loading frames are provided with all the safety features like over loading protection, over pressure protection etc. All the loading frames machines have the standard loading capacities. They are available for different loads like 50, 100, 200 & 300 tons, respectively.

Loading frame machine have standard frame size available as 10 feet width × 10 feet depth × 10 feet height. Loading frame machine of 200 tons capacity is used in this experimental program. They used reinforced beams and columns measuring 150*150*1200 mm. In accordance with IS 456-2000 guidelines, column and beam reinforcements are detailed. For the primary and lateral links, the nominal diameter of the reinforcing bars was 12 mm and 8mm, respectively. Ambient curing is done for all the beams and columns of OPC, GPC and RAGPC. No water is required in the process of curing for beams and columns. Fig.3 shows the equipment to test beams and columns i.e., loading frame machine.

Fig.3 showing casting of beam and Loading frame machine

4. Results and Discussion

4.1. Load carrying capacity of beams and columns

In loading frame method, beams are tested for flexure strength and columns are tested for compression strength.

Table 2 First Crack Load, Ultimate Load, Deflection, Strains of Beams

Beam designation	First crack (KN)	Ultimate load (KN)	Deflection (mm)	Strain (micro strains)
OPCB	28.43	75.66	0.6	29
GPCB	25.56	66.8	0.56	26
RAGPCB	24.89	72.34	0.45	24

Table 3 first crack, deflection, load of column

Column designation	First crack (KN)	Ultimate load (KN)	Deflection (mm)	Strain (micro strains)
OPCC	31	42.1	0.13	20.069

GPCC	30	40.02	0.26	20.08
RAGPCC	28	39.22	0.15	20.67

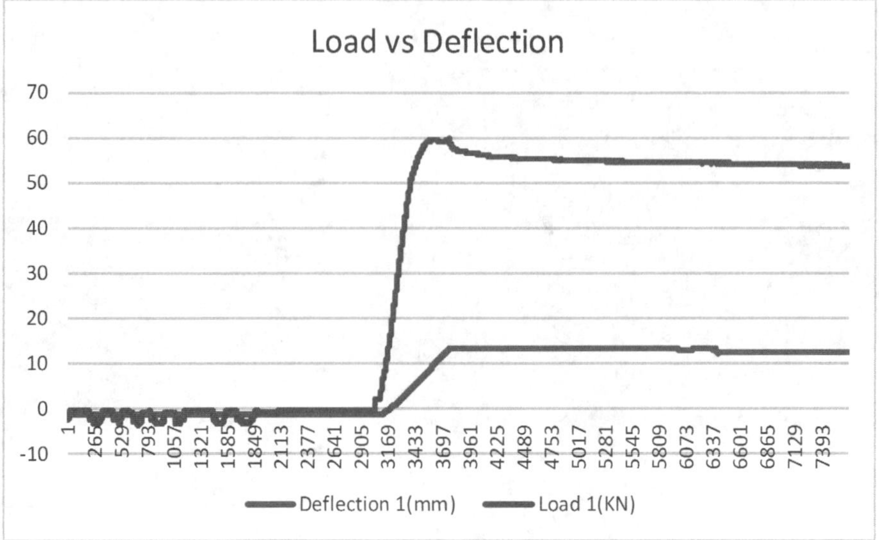

Figure showing load vs deflection behavior of beam

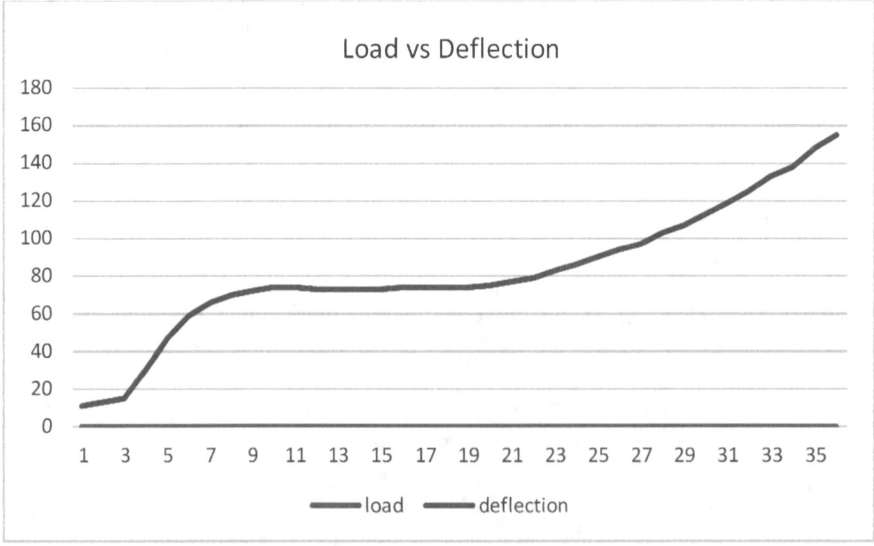

Figure showing load vs deflection behaviour of column

Figure showing crack patterns of beams and columns

5. Conclusions

From the test results, the following conclusions were drawn.

- The optimum percentage of recycled aggregate in geopolymer concrete is 40% based on mechanical properties and workability.
- The crack pattern and failure mode RGPC beams and GPC beams were almost similar but the width of crack in RGPC beams is slightly more than that of GPC beams.
- RGPC column with 40% RA behaves almost similar in axial compression when compared with GPC column.
- Since the strength reduction of RGPC is negligible compared to GPC, hence it can be recommended as an environment friendly construction material.

6. References

[1] Amar S, Byakodi and Srinivas N: “Effect of Curing Temperature on Compressive Strength of Geopolymer Concrete”, International Journal of Recent Scientific Research volume-7, Issue-7, July 2016, pp12377-12381.

[2]Anuradha R et al: "Relationship between compressive strength and split tensile strength of Geo polymer concrete", Indian Concrete Journal pp.18-24 Nov'2011.

[3] Baljot Kaur, Jagdish Chand; "Replacement of concrete by geo polymer concrete by using fly ash and GGBS", International Journal of Innovative Technology and Exploring Engineering (IJITEE), Volume 8, Issue 6, April 2019, ISSN:2278-3075.

[4]Chau-Khun M, Abdullah Zawawi Awang, Wahid Omar (2018), "Structural and material performance of geopolymer concrete", (2018) ELSEVIER 186 (2018) 90-102.

[5] Dattatreya J.K et al; "Flexural behaviour of reinforced geo polymer concrete beams", International Journal of Civil and Structural Engineering. Vol.2, No.1. pp. 138-159, Sep'2011.

[6] Deepa Raj S, JithinBhoopesh (2017), "Strength and behaviour of recycled aggregate geopolymer concrete beams" Advances in Concrete Construction: Vol. 5, No. 2 (2017) 145-154.

[7] M.K. Thangamanibindhu, Dr.D.S.Ramachandra Murthy; " Flexural Behavior of Reinforced Geopolymer Concrete Beams Partially Replaced with Recycled Coarse Aggregates" International Journal of Civil Engineering and Technology (IJCIET), Volume 6, Issue 7, July (2015), pp. 12-21.

[8] Malathi V et al: "Strength study on fly ash based Geo-polymer concrete under heat curing", International Journal of Emerging trends in Engineering and Development. Issue 2. Vol a, May'2012.

[9] N. Ganesan, R. Abraham, S. Deepa Raj, K. Namitha, "Effect of fibres on the strength and behaviour of GPC columns", Mag. Concr. Res. 68 (2), 99-106, (2015).

[10] Nella Shiva Kumar, Gone Punneshwar (2018), "Strength Properties of the Geopolymer Concrete as Partial Replacement of Recycled Aggregate", (2018) IJETER volume 6 issue-2 February 2018.

[11] Pradeep J et al: "Experimental study on flexural behaviour of fly ash and GGBS based geo polymer concrete beams in bending", International Journal of Engineering and Development, Issue 2, Vol.6, Sep'2012.

[12] S.Kumaravel, S.Thirugnanasambandam(2013), "Flexural Behavior of Geopolymer Concrete Beams", research paper E-ISSN2249–8974 (2013).

[13] Sanjay R, Dr. M. U. Aswath, Smita Singh (2012), "An Experimental Study on Flexural Behaviour of Reinforced Geopolymer Concrete Beams with Recycled Aggregate in Bending", Research Gate: Issue 2, Vol.6 (September 2012) ISSN 2249-6149. (2012).

[14] Shaswat Kumar Das and Jyotirmoy Mishra, "An Overview of Current Research Trends in Geopolymer concrete", International Research Journal of Engineering and Technology (IRJET), Volume 05, Issue 11, Nov 2018, e-ISSN: 2395-0056, p-ISSN:2395-0072.

[15] T. Srinivas and N V Ramana Rao; "Development and Optimization of Mix Design of Low Calcium Fly Ash and Slag Based Geopolymer for Standard Grade", IOSR Journal of Mechanical and Civil Engineering (IOSR-JMCE), volume 13, Issue 4, Jul-Aug 2016, 39-47.

[16] Vijai k et al: "Effect of types of curing on strength of Geo-polymer concrete", International Journal of Physical Sciences Vol.5(9),1419-1423, Aug 2010

Exploring Nano Biotechnology for detecting specific disease in Medical diagnosis and Therapeutic

M. Thangamani[1], S. Kavitha Bharathi[2] and N. Suresh Kumar[3]
[1] Associate Professor, Information Technology, Kongu Engineering College, Tamilnadu, India
[2]Associate Professor, Computer Technology, Kongu Engineering College, Tamilnadu
[3] Assistant Professor (SL.GR), Information Technology, Sri Ramakrishna Engineering College, Tamilnadu

E-mail: manithangamani2@gmail.com

Abstract. Nanobiotechnology connects the scientific openings between chemistry, physics and biology on the nanoscale. These guides to a lot of modern techniques help and create good result in medical-phtherapeutic appliances. Now a day nanoparticles good aspirants for drug discovery and new therapeutic applications. It consists of the nanotechnology and biotechnology. Nanobiotechnology will help the combination of therapeutics with diagnostics and make easy to improve the specific medicine suitability for an individual and prevent the immune system from side effect drugs delivery. This article explores the nano biotechnology y for detecting specific disease in medical field.

Keywords: Nanobiotechnology, Nanmaterial, Nanotechnology, Medical diagnosis and Therapeutic

1. Graphical abstract

Nanobiotechnology mainly contains materials such as nanoparticle, nanoemulsions, nanocomposites and nonostructured materials and products of delivery, formulation and packaging. Nanosensors and nanotracers are used for safety purposed. Manufacturing side after processing the materials and get products from industry. The physical chemistry and surface science are essential for fabrication of structure in carbon, silicon and inorganic materials. Nanobiotechnology combines medical diagnosis and

Medical therapeutic to improve the patient health and its shown by Figure 1 and figure 2.

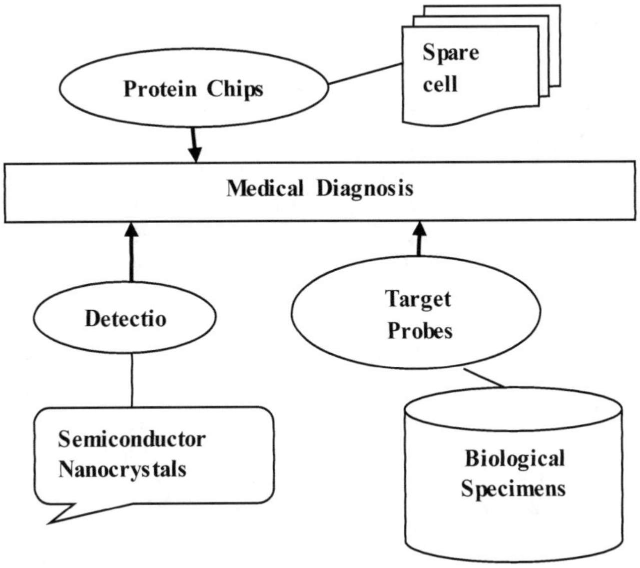

Figure 1 Medical diagnosis with contribution of Nano technology

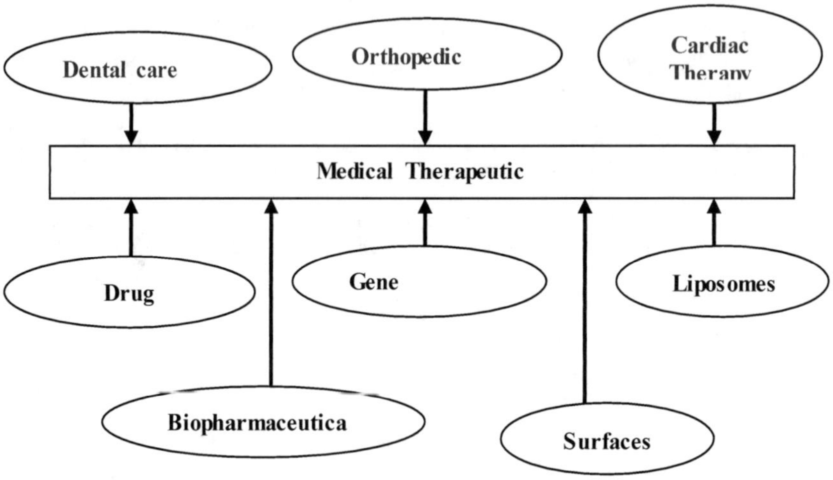

Figure 2 Medical Therapeutic with contribution of Nano Technology

2. Introduction

Nanotechnology is exploitation of substance on atomic, molecular and supramolecular range. It has wide collection of research in medicine and health care and drug delivery, molecular biology, chemical sensing, materials, transportation and clinically significant in surgical instrument, orthopedic surgery and drug therapy. Nanotechnology recovers the patients from diseases and prevent by early diagnosis along with treatment.

Normal polypeptides contain 20 amino acids and four similar nucleotides structures with the nucleic acids. Nanotechnology can divided into interphase, protein-based nanostructures, DNA (deoxyribonucleic acid) -based nanostructure and nanoanalytics. DNA-based nanostructure is main usage of manufacturing and constructing of nano devices and nanostructures. It reduces the requirement of biosensors, probes optical device and biological electronics [1, 2]. The protein-based nanostructures are functional adaptability and price valuable and sustainable production techniques offer powerful incentive for promote the fabrication in this way.

3. Related Works

Due to the increasing expenses of nanocharacterization and nanoprototyping, modeling and simulation is vital role in nanotechnology advancement [3]. Essential photons are biophotonics and neurophotonics in biotechnology, biology, medicine and engineering. For developing new technologies, need transformative engineering with fundamental research and science. The optical waveguides, optoelectronics and lasers fabricate medical therapies and diagnostics. The authors inspects Quantum outcome of microscopic and macroscopic devices [4]. For disease modeling and drug development, focused organ-on-a-chip approach [5]. The authors proposed mathematical technique [6-8] for capturing the dynamical characters of physics relations among biomolecules and nanopore due to electrostatic potential variation, ion screening effects, nanopore membranes surface variations and creation of DNA fluctuations.

The result of nanodevice usages are enormous significant. Various nanotheranostic techniques [9,10] unique characteristics and some intelligences task compare to sensitivity, biocompatibility, biodegradability, solubility in conventional medicines It reduce the partial physical dimensions and increase the performance of the biological properties and physic-chemical characteristics. Divagar Murugan et. Al [11] investigated fiber-optic absorbance biosensor to confine the spread of the COVID-19 disease at low cost and nanobiotechnologies applied in [12].

Diabetes disease [13] primarily grouped into Type 1 and Type 2. Along type 1 and type2, gestational diabetes take place when pregnancy. The protected cells of our body are accountable for harass the foreign particles. Nano biotechnology utilizes the improved surface properties, catalytic, catalytic and nanoparticle to identify monitors and takes care of a disease. The viral anoparticles are treated as bio-nonomaterials used in biomedical imaging field [14, 15]. Top down and bottom approach are suggested for nano particle fabrication. Genome sequence [16] improves the area of therapeutics and diagnostics. Nanotechnologies enhance the health care through hearing aids and remote device. Manufacturing the testing kits and Molecularly imprinted polymer technique is presented [17] to identify the taste found in tea. Machine learning algorithm applied [18,19] for predicting drugs and disease. Emily M.Miller et al.[20] nanoparticle-based therapies are used to rectify the ovarian cancer drug resistance for women. ElisabeteFernandes et al. [21] applied magnetive resistance of biosensor for stroke patients and diagnosis, therapeutric , immunization and vaccine production discussed [22].

Hong Wang et al. [23] proposed tumor immunosuppression technique to suppress the tumor development and provide a good nanostand with well-known immunosuppression-relieving ability for helpful cancer therapy. Raquel et al.[24] suggested new vascular system of nanobiotechnological advances for patient safety. The mechanical and biological way tissue engineering problems are solved by [25]. Magnetic nanoparticles [26] are broadly worn for diagnostic, therapeutic and drug delivery. Implantable vascular interface device [27,28] applied for vascular tissues. Iron-based nanozymes [29] are scientific inorganic nanoparticles for virus analysis and cure. Joloudari et al. [30] used nanobiomaterial device to handling coronary diseases.

4. Conclusion

This article exploited the nanobiotechnology usage and enhancement in this area. It illustrated the diagnosis and therapeutic in medical can be attain to greatest with least side effects by ways of the under tissue-specific clinical involvement. The selection of nano size, integration with device and molecules biocompatibility are very challenging one. In future can be need careful clinical traits for successful completion of medical diagnosis.

References

[1] Khalid M. Abu-Salah, Anees A. Ansari, and Salman A. Alrokayan· DNA-Based Applications in Nanobiotechnology, Journal of Biomedicine and Biotechnology, 2010, 1-15.

[2] Helena Gradišar & Roman Jeral, Self-assembled bionanostructures: proteins following the lead of DNA nanostructures, No.4, 2014.

[3] Bing Sheu;Xiaoning Jiang, Modeling and Simulation: New Roles in Nanotechnology, IEEE Nanotechnology Magazine, 2020, 2 – 2.

[4] Sergey Edward Lyshevski , ano-, NanoBio- and NanoBioMedical- Technologies: Enabling sensing, communication and processing paradigms, 2012 12th IEEE International Conference on Nanotechnology (IEEE-NANO), IEEE, 2012.

[5] Neda Azizipour, Rahi Avazpour, Derek H. Rosenzweig, Mohamad Sawan, Abdellah Ajji. Evolution of Biochip Technology: A Review from Lab-on-a-Chip to Organ-on-a-Chip. Micromachines 2020, *11* (6),599., IEEE Nanotechnology Magazine, 2020, 42 – 51.

[6] S. K. Tabatabaei et al., "DNA punch cards for storing data on native DNA sequences via enzymatic nicking," Nat. Commun. vol. 11, 2020.

[7] H. Qiu, A. Sarathy, K. Schulten, and J.-P. Leburton, “Detection and mapping of DNA methylation with 2D material nanopores,” npj 2D Mater. Appl., vol. 1, no. 1, p. 3, 2017.
[8] A. Sarathy, N. B. Athreya, L. R. Varshney, and J.-P. Leburton, “Classification of epigenetic biomarkers with atomically thin nanopores,” J. Phys. Chem. Lett., vol. 9, no. 19, pp. 5718–5725, Oct. 2018.
[9] Peredkov K.Ya,Nnanobiotechnology in modern medicine, Medico Research Phronicles, VOL. 5 NO. 6 (2018).
[10] Rieznichenko L.S., Rybachuk A.V., Bilous S. B., nanoparticles: synthesis, effectiveness in the treatment of purulent inflammatory diseases of the maxillofacial area, development of dosage forms, Journal of Chemical and Pharmaceutical Research, 2016. – Vol. 8, No.1, P. 332- 338.
[11] Divagar Murugan, Himanshu Bhatia, V. V. R. Sai, Jitendra Satija. P-FAB: A Fiber-Optic Biosensor Device for Rapid Detection of COVID-19. Transactions of the Indian National Academy of Engineering 2020, *5* (2), 211-215.
[12] Amna Batool, Farid Menaa, Barkat Ali Khan and Bushra Uzair, Progress and Prospects in Translating Nanobiotechnology in Medical Theranostics, Vol.15, No.5, 685-707.
[13] Antara Mukhopadhyay and Prosenjit Mondal, Application of Nano-biotechnology for Improvement in Therapeutic Approaches for the Treatment of Diabetes, 2018.
[14] Petrak K (2018) Essential properties of drug targeting delivery systems. Drug Discov Today 10: 1667-1673.
[15] Mariappan N. Recent trends in Nanotechnology applications in surgical specialties and orthopedic surgery. Biomed Pharmacol J 2019;12(3).
[16] James R. Heath, Nanotechnologies for biomedical science and translational medicine, PNAS, 2015 112 (47) 14436-14443 Trisita Nandy Chatterjee, Rajib Bandyopadhyay. A Molecularly Imprinted Polymer-Based Technology for Rapid Testing of COVID-19. Transactions of the Indian National Academy of Engineering 2020, *5* (2), 225-228.
[17] Jafar Ali Ibrahim, Dr. M. Thangamani, Prediction of novel drugs and diseases for hepatocellular carcinoma based on multi-source simulated annealing based randomwalk, Journal of Medical System, Springer, Vol.42, No.188, August 2018.
[18] S.Jafar Ali Ibrahim, Dr. M. Thangamani, Enhanced singular value decomposition for prediction of drugs and diseases with Hepatocellular carcinoma based on multi-source Bat Algorithm based Random walk, Journal of Measurement, Elsvier, Vol.141, July 2019.
[19] Emily M.Miller, Timothy M.Samec and Angela A. Alexander-Bryant, Nanoparticle delivery systems to combat drug resistance in ovarian cancer, Nanomedicine: Nanotechnology, Vol 31, 2021.
[20] ElisabeteFernandes, TomásSobrino, Verónica, Martins, Point-of-care quantification of serum cellular fibronectin levels for stratification of ischemic stroke patients, Nanomedicine: Nanotechnology Vol.30, 2020.
[21] Amr El-Sayed & Mohamed Kamel, Advances in nanomedical applications: diagnostic, therapeutic, immunization, and vaccine production, Environmental Science and Pollution Research, 27, pages19200–19213,(2020).
[22] Hong Wang, Jie Li, Zhiwan Wang, Yuqi Wang and Yaping Li, Tumor-permeated bioinspired theranostic nanovehicle remodels tumor immunosuppression for cancer therapy, Biomaterials, vol.269, 2021.
[23] Raquel C. Gonçalves, Andrea Banfi, Mariana B. Oliveira, João F. Mano, Strategies for re-vascularization and promotion of angiogenesis in trauma and disease, Biomaterials, Vol.269, 2021.
[24] Panel Krishna Mohan Agarwal, , UtkarshMohan, ShyamalMandal, DineshBhatia, Comprehensive study related to advancement in biomaterials for medical applications, Sensors International, Vol. 1, 2020.
[25] Saba Khoshbakht, Farzin Asghari-Sana, Anahita Fathi-Azarbayjani and Yaeghob Sharifi, Fabrication and characterization of tretinoin-loaded nanofiber for topical skin delivery, Biomaterials Research , Vol. 24, No. 8, 2020.
[26] Rachel Ghasemi Goorbandi, Mohammad Reza Mohammadi and Kianoosh Malekzadeh, Synthesizing efficacious genistein in conjugation with superparamagnetic Fe3O4 decorated with bio-compatible carboxymethylated chitosan against acute leukemia lymphoma, Vol. 24, No. 9, 2020.
[27] Dakota M. Binkley, Kathryn Grandfield. Advances in Multiscale Characterization Techniques of Bone and Biomaterials Interfaces. ACS Biomaterials Science & Engineering 2018, *4* (11), 3678-3690.
[28] Kaitlyn R. Ammann, Maxwell Li, Syed Hossainy, Marvin J. Slepian. The Influence of Polymer Processing Methods on Polymer Film Physical Properties and Vascular Cell Responsiveness. ACS Applied Bio Materials 2019, *2* (8), 3234-3244.

[29] Chu Shi , Dr. Yan Li , Iron-Based Nanozymes in Disease Diagnosis and Treatment, chemBiochem, combining chemistry and biology, 2019.
[30] Joloudari, Saadatfar, GhasemiGol and Razavi, coronary Artery disease diagnosis and ranking the significant features using a random tree model, International journal of environ.res. public health, 731,17, 2020

Dry and underwater friction stir welding of aa6061 pipes - a comparative study

I. Sabry[1] **and N. Zaafarani** [2]
[1]Manufacturing Engineering Department, Modern Academy for Engineering and Technology, P.O. Box, Cairo 11571, Egypt
[2]Department of Production Engineering and Mechanical Design, Tanta University Tanta 31527, Egypt

Abstract. This work compared the parameters offriction stir welding (FSW) and under water friction welding (UWFSW) on the weld joint, such as tool rotation speed, transverse speed, and wall thickness, and compare the ultimate tensile strength (UTS) weld joint using the experimental work for FSW and UWFSW was performed using a new modified fixture to eliminate the post-process problems. Experiments were performed at three levels of three parameters: Wall thickness, tool rotational speed and travel speed using milling machine center by UWFSW and traditional FSW. The full factorial approach was introduced for statistical research. UTS is measured as a reaction, meaning that the ultimate tensile strength achieved by welding the UWFS weld was greater than the conventional FSW weld. As a predictive research tool, regression analysis and variance analysis were used. From the study, it was observed that with minimal tool transverse speed, high tool rotation speed for UWFSW, maximum tensile strength is given, and that of traditional FSW.

1. Introduction

Four stages during friction stir welding (FSW) have been carried out: the plunging stage, the dwelling stage, the welding stage, and the escape or retracting stage of the tool [1]. The FSW procedure is essentially carried out by plunging a spinning FSW mechanism unit through the interface of two rigidly clamped sheets before the shoulder meets the surface of the material being welded [2] [3]. The travel of the unit progresses along the weld line allowing the material to displace from the advancing side to the retracting side. Meanwhile, the tool shoulder consolidates the material at the back of the pin, resulting in a stable state [4]. Compared to other conventional welding processes, FSW process uses a significant amount of energy. The absence of flux or cover gas makes the process environmentally safe [5].

The key parameters in the FSW process are welding speed, tool rotational speed, penetration depth, vertical and axial force, instrument tilt angle. They are critical in producing heat generation, forging strain, the flow of material, welding appearance, and performance. As an outcome of dislocation density increase, the nugget area has high stiffness at higher speeds [6]. Furthermore, the axial forces also shift at varying rotational speeds [7]. The softened area gets narrower as welding speed increases, and precipitate corrosion is weakened [8]. The ultimate tensile strength (UTS) of the welded joint is boosted by lower heat generation and high volume fraction of precipitate as a result of greater travel speeds, grain boundary reinforcement [9]. Maximum penetration demands maximum plunge depth, otherwise it will lead to a decrease in loadbearing potential. The depth of penetration is seen to influence the consistency of the welds. There is a linear relationship between the shoulder force and torque and the penetration depth [10]. FSW is recommended for maximum depth [11]. The mechanical and metallurgical characteristics are depending on the environments, the base material and the consumables in general. The working environment to which the joint is exposed is often critical. There is a chance of crack expanding as the joint is exposed to high-pressure and high-temperature fluids, which stops the device itself from operating [12].

Khourshid et al [13] conducted FSW trials on AA6063 pipes using drilling machine. The purpose was to determine the feasibility of welding two pieces of aluminum pipes by the method of friction stir welding and to investigate the effect of the welding process on the mechanical properties of welding joints. The findings of using various mechanical tests on the welding joints reveal that aluminum pipes (AA6063) can be welded by the FSW process with ultimate tensile efficiency (78.7 %) based on the stir welding experiments carried out in this report, using a rotational speed of 1400 rpm, 4 mm/min travel speed. As a general result, the mechanical performance of FSW welds increases with increasing tool rotating speed and lowering the tool linear feed [13].

A major concern in the FSW is the significantly low weld strength. This happens as the continuous increase of the temperature causes the work material to soften and plasticized which indirectly leads to a decline in the mechanical properties of the welded joints. By increasing the cooling rate and lowering pick temperature this drawback effect can be minimized [14]. It was found that the frequency of further plasticized weld joint zone and poor mechanical characteristic arising from the breakdown of the reinforcing sediment in FSW is by contrast, less in the case of using the underwater friction stir welding (UWFSW) technique leading to less faulty welds compared to traditional FSW [15]. Another investigation [16] shows that using UWFSW method to join aluminum alloys and magnesium alloys significantly suppresses the formation of intermetallic compounds because of the lower peak temperature.

In order to refine the parameters involved in the UWFSW process, researchers have used computational methods and approaches such as GRA, Taguchi, ANN and RSM [17] [18]. The detection of defects is important for correlating experimental results with the model findings of the optimum weld quality parameters [19] [20]. Taguchi technique and ANOVA technique are considered to be a robust effective tool for carrying out experimental work and further processing the experimental data for optimization [2]. As outcome of abrupt cooling, the peak temperature produced is lower in UWFSW. However, the primary cause for the presence of lack of fill, tunnel defect and lack of penetration is the cooling of the weld area before filling the joint.

Although UWFSW is recognized recently as a hot research topic, few literatures are available on underwater welding of flat base metals. Literature on the method of UWFSW designed for soldering hollow cross-sectional tubing is very unusual. A research on the influence of parameters in the method of UWFSW is therefore important. In the past, experiments carried out on flat plates made of various aluminum alloys using UWFSW instead of FSW discussed an improvement in the ultimate tensile strength [23]. When the AA7075-T351 sample was welded using UWFSW, Zhao et al [22] found that the tensile strength of 75 % of the parent material was substantially higher in the weld region than in the FSW phase. A substantial reduction in both the diameter of the softened region and the softening ratio in the case of the AA6061 weld joint as a result of decreased peak temperatures with a joint performance of 86 % shows the importance of UWFSW in the joining industry [24]. The mechanical properties are closely related to metallurgical transformation [25]. Further grain refining is provided by liquid to sample contact in UWFSW, which creates a high cooling rate [26]. A UWFSW process will create the mechanical properties that are important for the homogeneity of welding joint [27].

In industrial practice welded AA6061 pipes is widely used, where different welding techniques, such as resistance welding, welding of inert metal steel, welding of inert tungsten gas, etc. are used. FSW and UWFSW can be a good alternative welding technique for this product. The current research considers the comparison of the UTSs of the weldments using both UWFSW and FSW for AA6061 pipes. In this context the ANOVA analysis is utilized as a modelling tool to identify the better method to be used from the UTS of the weldment point of view.

2. Experimental work

2.1 Material and Experimental Setup

The AA6061 pipes were chosen as work material to conduct the welding. The pipe has a diameter of 30 mm and a wall thickness of 2, 3 and 4 mm. Table 1 shows the chemical composition of the AA6061 material of the pipes and Table 2 shows the mechanical properties of AA6061.

Table 1 Chemical composition (wt.%) of AA6061

Element	wt.%
Al	Bal.
Si	0.61
Fe	0.69
Cu	0.26
Mn	0.14
Mg	0.19
Ti	0.14
Cr	0.19
Zn	0.26

Table 2 Mechanical Properties of AA6061

Description	AA6061
UTS	240 MPa
El %	15 %

2.2. Machine

A vertical milling machine (VMM) prepared and equipped for the FSW and UWFSW for pipe process was used as in [28]. The preparation of the machine usually involves a fixture design that can accommodate a rotating motion for the pipe. The rotating motion of the pipes here acts as the travel speed (F) to allow the advancing motion of the tool over the weld line figure 1.

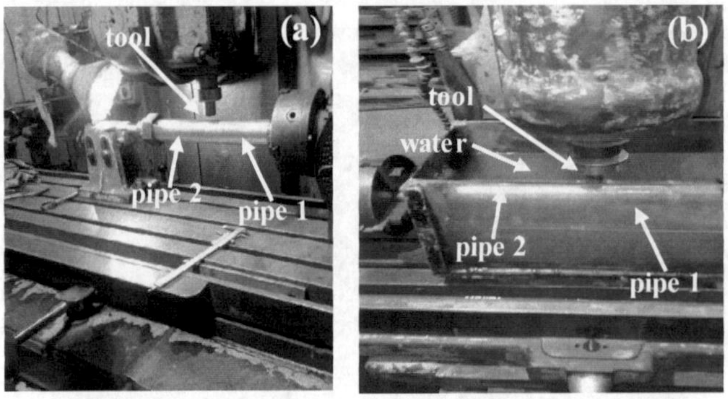

Figure 1 Experimental setup for (a) FSW (b) UWFSW

2.3 Fixture

The first fixture used in the FSW and UFSW for pipes process was composed of a solid rod back ended with a threaded part as shown in Figure 2(a) [29]. The two pipes are assembled on the rod and tightened with a washer and a nut. This fixture was efficient and provided several good quality samples. However, the issue was always to get off the pipes from this fixture after welding. Internal axial hammer support was used to perform this hard task. That stimulates to design a new fixture to eliminate this post welding hard task. In the new design, the solid rod is decomposed into two parts connected with alike gear teeth Figure 2 (a). After welding and loosing the nut, the rod two parts are disassembled, and every part is removed easily from inside the pipe welded.

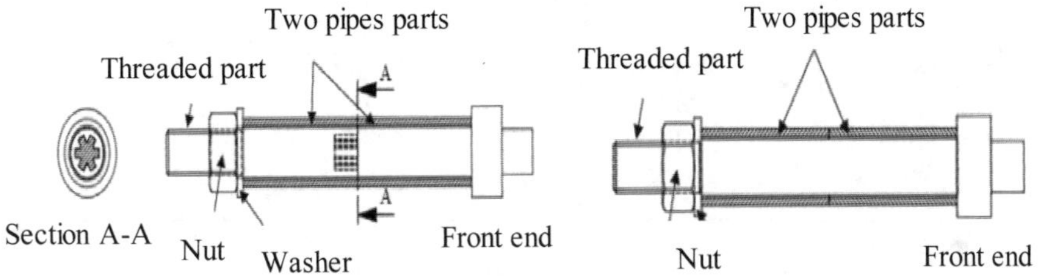

Figure 2 Fixing device (a) Old design (b) Modified design

2.4. Tool Design

Three special, non-consumable instruments constructed of 6 mm pin diameter Stainless steel 316 L [6] utilized to weld aluminium pipes. The pin profile was tapered with root diameter equal to 5 mm and the tip diameter was 1 mm and the pin length was 2, 3 and 4 mm. The pin was attached to the main tool body by a shoulder and a tip of conical shape with root diameter of 5 mm and head diameter of 1 mm. The tip height was 5 mm. The tool used is shown in Figure 3.

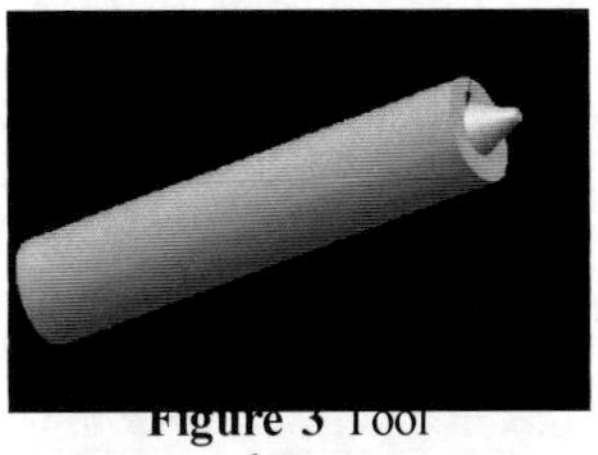

Figure 3 Tool

2.5 FSW and UWFSW process variables

The independent process parameters influencing the ultimate tensile strength (UTS) were defined as rotation speed (N) and travel speed (F), based on preliminary trials and previous studies. The friction-stir welding parameters are shown in Table 3.

Table 3 Parameters for the AA6061 FSW and UWFSW

Parameter	Value
Tool rotation [rpm]	485, 710, 910, 1120, 1400, 1800
Wall thickness [mm]	2, 3, 4
Transverse speed [mm/min]	4, 8, 10

Trial runs continued to evaluate the upper and lower limit of AA 6061 method parameters by modifying just one parameter at a time. A parameter set has been calibrated in such a way that no flaws were recognised by visual screening in the final welded joint. The upper limit was coded as 1 and the lower limit was coded as -1. The coded intermediate values were determined using equation (1).

$$X_i = 2X - \frac{X_{max}+X_{min}}{X_{max}-X_{min}} \quad (1)$$

where X_i, X, X_{max} and X_{min} are the coded value desired, the value of the variable, the lower limit of the variable, and the upper limit of the variable, respectively [20]. The process parameters considered, with their respectively boundaries, units and notations, are defined in Table 4.

Table 4 Method parameters and their levels in both FSW and UWFSW

Process Parameters	Unit	Levels		
		-1	0	1
Tool rotation (N)	[rpm]	485	1400	1800
Wall thickness (W)	[mm]	2	3	4
Tool feed rate (F)	[mm/min]	4	8	10

In order to carry out metallurgical experiments, specimens of the appropriate size were cut from the welded plate, as explained in figure 4(a). The FSW and UWFSW samples welded at 10 mm/min and 1800 rpm are illustrated in Figure 4(b) and Figure 4(c).

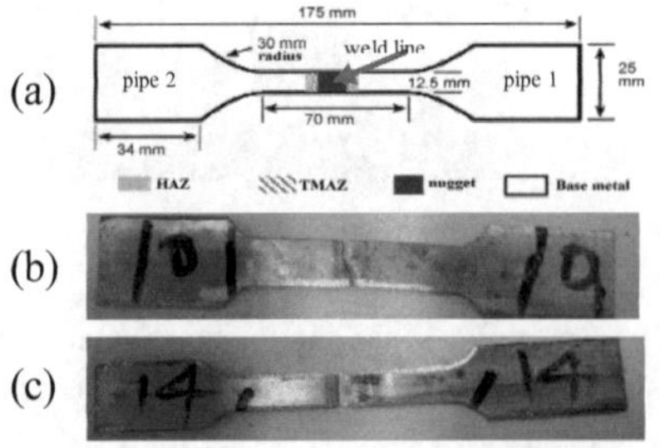

Figure 4 (a) Sample geometry of tensile test (b) A welded FSW sample (c) welded UWFSW sample welded

The experimental work was conducted on the milling machine. The welded sample's UTS were tested in compliance with ASTM D638-14[28]. In order to find relevant variables, ANOVA was used. Utilize a statistical analysis technique for and parameter at all stages, the main impact plots and their interaction plots were plotted to analyse the parametric effects on the features of the response. Utilize "Minitab 18" statistical program, all the findings were evaluated (Table 3).

IOP Conf. Series: Materials Science and Engineering 1091 (2021) 012032 doi:10.1088/1757-899X/1091/1/012032

3. Results and discussion

The goal of the analysis is to define the influence of FSW and UWFSW parameters using the full factorial system on the ultimate tensile strength. Primary impact plots and interaction plots were plotted utilizing ANOVA. The key impact plots have been used to analyse the parametric effects on the properties of the responses. All findings were analysed using statistical tools, "Minitab 18". The welded specimen UTSs were tested in accordance with ASTM D638-14. The outcome of the inspection welded component is as shown in Table 5.

Table 5 Experimental design matrix and results

Run No.	N (rpm)	W (mm)	F (mm)	UTS (MPa) FSW	UTS (MPa) UWFSW
1	1800	2	4	205.0	218
2	1800	2	8	199.2	210
3	1800	2	10	195.0	202
4	1800	3	4	198.0	204
5	1800	3	8	193.8	200
6	1800	3	10	189.0	192
7	1800	4	4	190.9	196
8	1800	4	8	186.4	192
9	1800	4	10	183.0	186
10	1400	2	4	193.0	201
11	1400	2	8	187.2	194
12	1400	2	10	182.4	190
13	1400	3	4	186.6	189
14	1400	3	8	181.0	188
15	1400	3	10	176.0	184
16	1400	4	4	179.9	186
17	1400	4	8	174.1	182
18	1400	4	10	169.3	179
19	1120	2	4	184.0	190
20	1120	2	8	178.6	186
21	1120	2	10	173.0	181
23	1120	3	4	177.3	184
24	1120	3	8	171.0	180
25	1120	3	10	166.0	179
26	1120	4	4	169.8	181
27	1120	4	8	164.0	176
28	1120	4	10	159.0	168
29	910	2	4	176.0	180
30	910	2	8	170.0	175
31	910	2	10	164.0	170
32	910	3	4	170.5	175
33	910	3	8	164.0	171
34	910	3	10	158.5	169
35	910	4	4	162.4	170
36	910	4	8	157.0	165
37	910	4	10	152.0	163
38	710	2	4	168.0	173
39	710	2	8	161.0	166
40	710	2	10	156.0	165
41	710	3	4	162.8	167
42	710	3	8	156.0	164

43	710	3	10	150.0	161
44	710	4	4	155.4	166
45	710	4	8	145.0	160
46	710	4	10	144.8	158
47	485	2	4	157.0	165
48	485	2	8	150.0	159
49	485	2	10	145.0	155
50	485	3	4	153.0	158
51	485	3	8	146.0	151
52	485	3	10	141.0	154
53	485	4	4	145.0	157
54	485	4	8	138.0	153

3.1 Mathematical Model

The ultimate tensile strengths of the FSW and UWFSW joints are functions of rotation speed, travel speed and wall thicknesses and can be expressed as in equation (2).

$$Y = f(N,F,W) \quad (2)$$

The UTS value may be considered as a response Y in equation 2. The selected polynomial equation could be expressed as an equation for the three variables as follows:

$$Y = \beta_0 + \beta_1 N + \beta_2 F + \beta_3 W + \beta_{11} N^2 + \beta_{22} F^2 + \beta_{33} W^2 + \beta_{12} NF + \beta_{13} NW + \beta_{32} WF \quad (3)$$

where β_0 is the free term of the regression equation; β_1, β_2 and β_3 are the linear terms; the coefficients β_{11}, β_{22} and β_{33} are quadratic terms; the coefficients β_{12} , β_{13} and β_{32} are the interaction terms. The values of the coefficients of the polynomial were calculated by regression analysis with the help of the following four equations [30]:

$$\beta_0 = 0.1663 \sum(Y) - 0.0568 \sum\sum(X_{ii} Y) \quad (4)$$
$$\beta_j = 0.0732 (X_i Y) \quad (5)$$
$$\beta_j = 0.0625 \sum(X_{ii} Y) + 0.00689 \sum\sum(X_{ii} Y) - 0.0568 \sum(Y) \quad (6)$$
$$\beta_{ij} = 0.1250 \sum(X_{ij} Y) \quad (7)$$

where i and j = 1, 2, 3 and i < j

The software package DESIGN Specialist 8.0.4 was used to approximate the values of these coefficients for various answers, and the results are shown in Table 6.

Table 6 Calculated statistical model regression coefficients for FSW and UWFSW

Regression factor	Tensile strength (MPa)	
	(FSW)	(UWFSW)
β_0	174.3	148.0
β_1	23.81	0.06387
β_2	-6.576	-6.341
β_3	-5.405	-0.02761
β_{11}	-2.989	-0.0000051
β_{22}	-0.7667	0.8056
β_{33}	-1.215	-0.09722

β_{12}	-0.1676	-0.00479
β_{13}	0.7726	-0.0005347
β_{32}	0.1723	0.1815

Minitab software was used to formulate the regression equations (8 and 9) that predict the average response of UTS of the UWFSW joints as a function of UWFSW processing parameters (wall thickness, rotational speed, and transverse speed). The levels of the UWFSW process parameters used in the regression model and their corresponding codes are shown in Table 6. By residual analysis, the adequacy of the suggested regression models is tested. The residual maps, histogram and standard probability for the UTS are seen in Figure 5. For both weld methods the standard probability map and the frequency histogram of the residuals show that errors are typically distributed. Moreover, residual versus fitted value or observation order plots do not suggest any significant inadequacies in the model.

$$UTS(FSW) = 146.2 + 0.05003N - 2.08W - 0.5320\,F - 6.914E006N^2 - 0.7667W^2 - 0.1350F^2 - 0.000254\,NW + 0.00039NF + 0.05744WF \quad (8)$$

$$UTS(UWFSW) = 148 + 0.06387N - 6.341W - 0.0276\,F - 5.181E006N^2 + 0.8058\,W^2 - 0.097F^2 - 0.000479\,NW - 0.000534NF + 0.1815WF \quad (9)$$

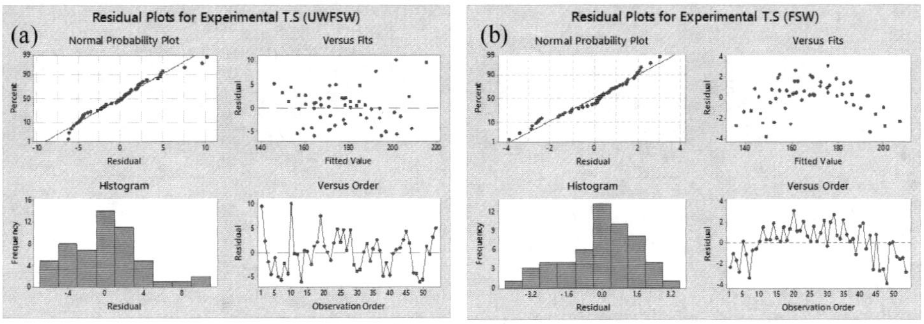

Figure 5. Residuals plot for UTS (a) UWFSW (b) FSW

UTS was maximum at higher tool rotation speed, and UTS was minimal at lower tool rotation speed. At lower tool transverse speed, the full UTS value was reached. For wall thickness, at higher wall thickness, the UTS was maximally reached Figure 6 (a). Tool rotation speed of 1800 rpm, tool transverse speed of 4 mm/min, and wall thickness of 2 mm were the optimal process parameters for the FSW without using water to gain the highest weld strength. It is seen in Table 7 that the parameters influencing UTS were investigated using ANOVA.

Figure 6(b) depicts the plots of the main effects of UTS. It shows that travel speed has a higher significant effect on UTS than the other parameters and the effect of this factor is directly proportional to UTS responses. It can be stated that the increasing of travel speed causes the UTS to increase significantly. The travel speed of 10 mm/min shows lower UTS values. The rotational speed of 1800 rpm shows higher UTS, wall thickness also shows an effect on UTS.

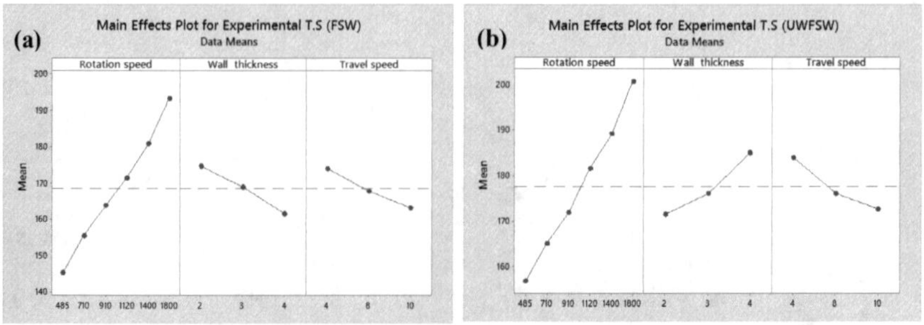

Figure 6. Main effect plot for **(a)** FSW **(b)**UWFSW

3.2 Checking the adequacy of the built models for FSW using ANOVA

The adequacy of the specialised models hires the variance methodology analysis (ANOVA). The percentage of contribution (p) signals the parameters for the linear square and interaction phase with the response functions. These (p) values are used to classify the significant response function parameters[31]. Table 7 displays the effects of ANOVA's tensile strength model. The 1929 F-value (FSW) model and the 312 F-value (UWFSW) for tensile strength indicate that the model is significant. Due to noise, there is only a 0.01 percent probability that a model F-value will occur. Values outside the interval [0 1] imply that the terms of the model are not important. The coefficient of determination R^2 values gives the adequacy of fitness of the model. Table 7 describes the determined values of the established model.

Table 7. Tensile strength coefficient of decision values for FSW and UWFSW

	F-value	R^2	*Adjusted* R^2	*Predicted* R^2
Tensile strength (UWFSW)	312	0.9494	0.9463	0.9391
Tensile strength (FSW)	1929	0.9914	0.9905	0.9898

3.3 Comparison of all Parameters with Respect to UTS

Comparison of all processes is recommended in order to determine the best properties obtained by the process with or without the use of water. A graphical approach was used for greater interpretation or better analysis of both methods. For both processes, the bar chart was created. The graph is as shown in Figure 7. It is evident from the graph that the UTS attained by the UWFSW is greater than the UTS attained by the FSW [6]. The overall UTS obtained using UWFSW and FSW were 225 MPa and 205.4 MPa, respectively.

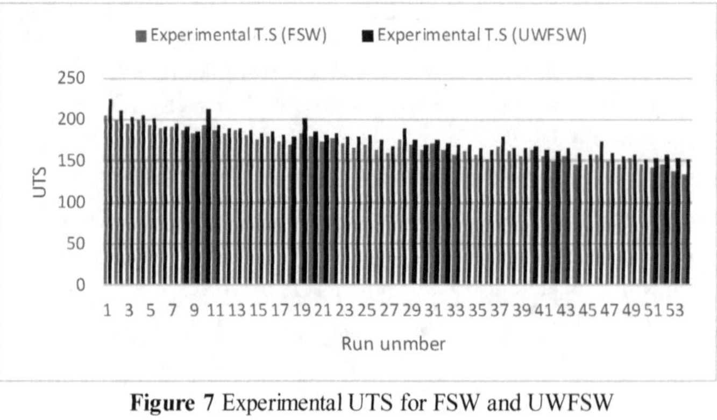

Figure 7 Experimental UTS for FSW and UWFSW

3.4 Influences of process parameters from FSW and UWFS

Using the experimental findings presented in figure 8, the tensile strength of FSW and UWFSW welded aluminium alloy 6061 was predicted by mathematical simulations, demonstrating the general patterns between the predicted and experimental performance. The ANOVA results are given in Table 7. By drawing scatter diagrams presented in Figure 8, the validity of regression models established is further checked (a-b). The observed values and the expected response values are scattered close to the 45° axis, suggesting that the established empirical models match almost perfectly [19]. At a greater rotation speed of 1800 rpm, the rise in UTS for UWFSW is important, although a slight decrease in UTS for FSW at a rotation speed of 1800 rpm is observed.

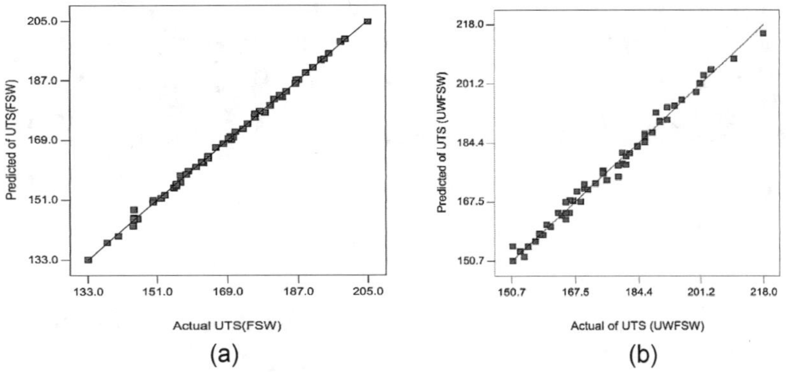

Figure 8 Scatter diagram of predicted and experimental UTS (a) FSW (b) UWFSW

3.5 Analysis of results

In Figure 9 (a and b), showing the general patterns between cause and effect, the effects of the various process parameters on the mechanical properties of FSW and UWFSW welded aluminum alloy AA6061 are estimated from the mathematical models using experimental observations. The contour graph of tensile strength, for FSW and UWFSW, is shown in Figure 9.

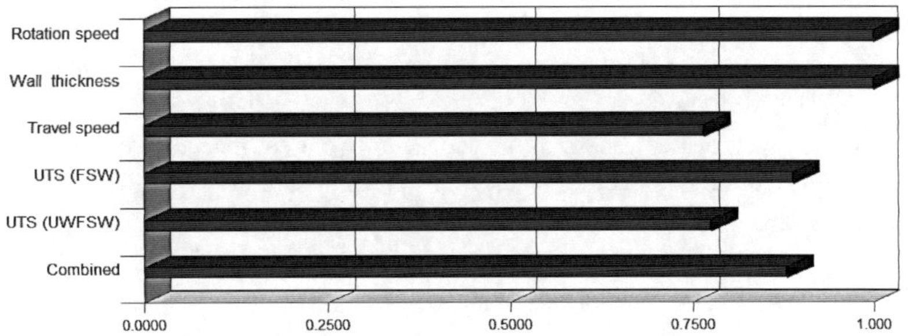

Figure 9 Bar chart displaying the overall desirability value of 0.915 for the combined desirability value

3.6 Microstructure Study

Because of the thermomechanical mechanism of FSW, and UWFSW, both the material flow and the heat dissipation rate will induce metallurgical transformations, altering the mechanical properties. Figure 10 (a) demonstrates the microstructure of welded zone under various welding stipulation. It is shown that from FSW to UWFSW, grain size is decreased. Thus, at the same time, the application of cooling results in high angle grain boundaries because higher dislocation density and dynamic recovery and finally fine grain size. Figure 10 (b) shows the optical microscopic image of the heat effected zone (HAZ) and nugget zone (NZ) . Two zones - the nugget zone and heat affected zone by heat were clearly distinguishable from the joint weld 's optical microscopic image. This implied the need to examine the mechanical properties of the joint because of the unusual distribution of the material in the weld zone.

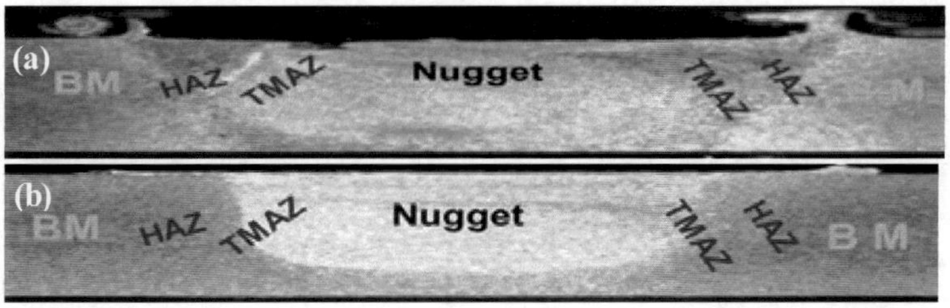

Figure 10 The microstructure of weld cross-section of AA6061 using (a) FSW (b) UWFSW UTS

4. Conclusions

An experimental distinction between FSW and UWFSW on Al6061 alloys is provided in this research paper. For the analysis of the work, ANOVA was carried out and regression was also performed; the following conclusions were reached:

1. UTS decreases as the speed of travel increases and UTS increases as the speed of travel decreases in UWFSW and FSW.
2. UTS has attained speed at a lower tool transverse speed of 4 mm/min, and tool transverse speed increases by 10 mm/min and UWFSW UTS decreases.
3. Wall thickness at 4 mm in UWFSW, UTS increases up to a certain level after that level and then decreases at wall thickness 2 mm and on the other hand in FSW at wall thickness 2mm leads to UTS increases up to a certain level after that level and then decreases at wall thickness 4 mm
4. In FSW and UWFSW, at higher tool rotation speed, UTS increases, and as tool rotation speed decreases, UTS decreases.
5. UWFSW add additional hardness in the sample in two areas-nugget zones and heat-affected zone. Because of the existence of small grains this is believed as in specimens welded utilizing standard FSW.
6. Water decreases the distribution of temperature in together UWFSW conditions but during the FSW phase, the effect of in-temperature is evident.

References

[1] I. S. A.M.El-Kassas 2019 A New Quality Monitoring System for Friction Stir Welded Joints of Aluminum Pipes *Int. J. of Eng. and Tech* **11** 78–87.
[2] I. S. A.M.El-Kassas 2018 Comparative Study on Different Tool Geometrics in Friction Stirred Aluminum Welds Using Response Surface Methodology *4th Int. Conf. on Weld. and Fail. A. of Eng. Mat.* , Aswan, Egypt.

[3] E. H. a. R. I. A. Ibrahim Sabry 2019 Improvement of Mathematical Model to Predict the Mechanical Properties and Corrosion rate of Friction Stir Welded 2024 Aluminum Alloy in *2nd Int. Conf. on Mater. Sci. and Eng.* Egypt.

[4] Z. A. a. N. Mohd. Atif Wahid 2018 Review on underwater friction stir welding: A variant of friction stir welding with great potential of improving joint properties *Trans. of Nonferr. Meta. Soc. of China* **28** 193-219.

[5] I. S. A.M.El-Kassas 2017 A Comparison between FSW, MIG and TIG Based on Total Cost Estimation for Aluminum Pipes *Europ. J. of Adv. in Eng. and Tech.* **4** 158–163.

[6] A.-H. I. M. D. T. T. Ibrahim Sabry 2020 Comparison of Mechanical Characteristics of Conventional and Underwater Friction Stir Welding of AA 6063 Pipe Joints *Inter. Rev. of Aerosp. Eng.* **14** 64-70.

[7] P. B. M. H. a. C. . H. Papahn 2015 Study on governing parameters of thermal history during underwater friction stir welding *Inte. J. of Adva. Manuf. Techn.* **78** 1101–1111, 2015.

[8] H. J. Z. a. L. Y. H. J. Liu 2011 Effect of welding speed on microstructures and mechanical properties of underwater friction stir welded 2219 aluminum alloy *Mat. and desig.* **32** 1548–1553.

[9] S. M. a. V. B. S. S. Sabari 2016 Influences of tool traverse speed on tensile properties of air cooled and water cooled friction stir welded AA2519-T87 aluminium alloy joints *J. of Mat. Proce. Techn.* **237** 286–300.

[10] A. S. I. M. A. T. D. El-Kassas 2019 Characteristics of Potential Sources - Vertical Force, Torque and Current on Penetration Depth for Quality Assessment in Friction Stir Welding of AA6061 Pipes *Inter. Review of Aerosp. Eng.* **4** 195-204.

[11] M. E.-S. M. K. A. Youssif 2016 Influence of Critical Process Parameters on the Quality of Friction Stir Welded Nylon 6 *Inter Review of Mech. Eng.* **7** 501-507.

[12] S. N. a. A.-H. I. M. S. K. Maiti 2008 A scheme for finite element analysis of mode I and mixed mode stable crack growth and a case study with AISI 4340 steel *Nucl. Eng. and Desig* **238** 787–800.

[13] I. S. F. S. w. S. o. A. P. A.M. Khourshid 2013 friction stir welding study on aluminum pipe *Int. J. of Mech. Eng. and Robot. Resea.* **2** 331-339.

[14] N. G. M. A. G. a. M. A.-M. Ibrahim Sabry 2020 Optimization of Process Parameters to Maximize Ultimate Tensile Strength and Hardness of Underwater Friction Stir Welded Aluminium Alloys using Fuzzy Logic *Mod. Concep. in Mater. Sci.* **3** 73–78.

[15] I. S. A.M. Khourshid 2013 Analysis and Design of Friction Stir *Int. J. of Mech. Eng. and Robot. Resea.* **2** 333-341.

[16] A.-Z. A. G. F. G. C. Mofid MA 2012 Submerged friction-stir welding (SFSW) underwater and under liquid nitrogen: an improved method to join Al alloys to Mg alloys *Miner. Met. Mater. Soc. ASM International* **43** 5106–5114.

[17] A. M. E.-K. a. A. K. Ibrahim Sabry 2015 Integration between Artificial Neural Network and Responses Surface Methodology for Modeling of Friction Stir Welding *Int. J. of Adv. Eng. Rese. and Sci.* **1** 67-73.

[19] I. S. Ahmed M. El-Kassas 2019 Optimization of the Underwater Friction Stir Welding of Pipes Using Hybrid RSM-Fuzzy Approach *Int. J. of App. Eng. Rese.* **14** 4562-4572.

[20] Ibrahim Sabry and M.Abdel Ghafaar Nabil Gadallah 2019 A Summarized Review on Fiction Stir Welding for Aluminum Alloys in *3rd Int. Conf. Arch., Eng. and tech. (AET)*, Cairo, Egypt.

[21] H. Z. Y. H. a. Y. U. L. . H.-J. Liu 2010 Mechanical properties of underwater friction stir welded 2219 aluminum alloy *Trans. Nonferrous Met. Soc. China* ***20*** 1387–1391.

[22] Q. W. H. C. a. K. Y. Y. Zhao 2014 Microstructure and mechanical properties of spray formed 7055 aluminum alloy by underwater friction stir welding," *Mat. and desig.,* **56** 725– 730.

[23] M. E.-S. M. K. A. Youssif 2016 Influence of Critical Process Parameters on the Quality of Friction Stir Welded Nylon 6 *Inter Review of Mech. Eng.* **7** 501-507.

[24] K. K. a. H. T. . D. Sakurada 2002 Underwater friction welding of 6061 aluminum alloy," *J. Japa. Instit. of Light Meta.* **52** 2-6.

[25] M. A. a. A. E. D. A-H. I. Mourad 2014 Study on the mechanical behavior of aluminum alloy 5083 friction stir welded joint *ASME Pressure Vessels and Piping Conf*, PVP2014-28556, V06AT06A014.

[26] D. C. H. a. K. S. Vecchio 2007 Thermal history analysis of friction stir processed and submerged friction stir processed aluminum *Mate. Sci. and Eng. A* **465** 165–175.

[27] H. J. Z. a. L. Y. H. J. Liu, 2011 Homogeneity of mechanical properties of underwater friction stir welded 2219-T6 aluminum alloy *J. of Mater. Eng. and Perf.* **20** 1419–1422.

[28] E.-K. A. Ibrahim Sabry 2019 An appraisal of characteristic mechanical properties and microstructure of friction stir welding for Aluminium 6061 alloy – Silicon Carbide (SiCp) metal matrix composite.," *J. of Mech. Eng. and Sci.* **13** 5804-5817.

[29] A. M. E.-K. A.-H. I. M. D. T. T. . J. A. Q. Ibrahim Sabry 2019 Friction Stir Welding of T-Joints: Experimental and Statistical Analysis *J. of Manuf. and Mater. Proce.* **3** 2-23.

[30] W. G. Cochran and G. M. Cox 1957 Experimental Designs", 2nd edition, New York, John Wiley & Sons Inc., 350.

[31] I. S. Ahmed M. El-Kassas 2019 Using Multi Criteria Decision Making in Optimizing the Friction Stir Welding Process of Pipes: A Tool Pin Diameter *Int. J. of App. Eng. Rese.* **14** 3668-3677.

Selection of optimal parting line for forging in various equipments using numerical analysis

Mathala Prithvi Raj[1], Manoj Kumar[2*], Ajit Kumar Pramanick[3]

[1,2,3]Department of Forge Technology, National Institute of Foundry and Forge Technology, Ranchi - 834003, Jharkhand, India.

*mknifft@gmail.com

Abstract. While designing of dies for forging, one of the most important criteria which has to be considered is the selection of parting line/plane. Selection and location of the parting line greatly affect the grain flow direction in the component, load which is coming on the dies and it also determines the location of the burr which is formed in the component after trimming. The main objective of this paper is to select a suitable parting line/plane, for hot forging a circular component in mechanical press and drop hammer. For studying the effect of parting line, four dies are designed with different parting lines (top parting line, bottom parting line, central parting and inverted dies) location for hot forging of an automotive component coupling flange. Two commonly used equipments (mechanical press and drop hammer) are considered for determining the effect of parting line location on load requirement which finally decides the die wear and die life. Numerical analysis is carried out with these dies using a finite element based software DEFORM-3D. Top parting line gave best results with minimum load, minimum effective-stress and favourable grain flow in mechanical press. Whereas, in case of hammer, inverted dies yielded in minimum load, minimum effective stress and favourable grain flow.

1. Introduction

In closed die forging the components are manufactured by deforming the raw material between a set of dies in which impression of the component is cut in. Designing of the dies for forging involves various steps namely selection of parting line, adding allowances which include machining allowances, finishing allowances and technological allowances, incorporating draft allowances, adding fillet and corner radii to avoid sharp corners, design of flash and gutter. After incorporating all these allowances, the impression is cut in the dies blocks of suitable dimensions and suitable material. The first and most significant step in the die design is the selection of suitable parting line. Parting line is the plane where the two die halves meet along the forging. The selection and location of parting line affect many parameters like load and energy requirement, grain flow, amount of draft, die sinking, trimming operation, cost of the die, defect formation, die wear and die life. The basic criteria for selection of parting line is to avoid deep impressions, to avoid side thrust, favourable grain flow, maximum periphery [1],[2],[3],[4]. Even after fulfilment of these basic criteria, still there is lot of scope for variation in the selection of parting line. The selection of correct parting line reflects the skill and knowledge of die designer. In closed die forging process mainly two basic equipments are used. First one is drop hammer which is least expensive and very versatile whereas second one is mechanical press which is more costly than hammer but production rate is high and more accurate as compared to

drop hammer. The die design for these two equipments may not be same even though the geometry(component) is same. Earlier practice for production was based on experience and shop floor trial and error which was very time consuming and expensive. Now the manufacturing concept has changed. Industry wants first time right that is they want to minimize the wastage, reduce product development time and ultimately minimize the cost of the forgings with all producing quality forgings from starting of the production. In this case numerical simulation is very useful tool which not only helps in optimum die design but also helps in controlling process parameters as a result of which load on dies are reduced, forging defects are eliminated and wastage of materials in the form of flash is greatly reduced. In this paper it has been shown how numerical simulation helps in selecting correct parting line for same component (same geometry) which are produced in two different forging equipments (drop hammer and mechanical press). This correct parting line helps in reducing forging load and hence die wear. The location of these correct parting lines are different for different equipment used for same forging component.
Many researchers have carried out the numerical analysis to predict the effects of variation in die design on output parameters. Zhenchao Qi et.al, [5] proposed a new forming method for manufacturing of bevel gears wherein they used finite element method of simulate the entire process. They also tried to optimize the design parameters using Taguchi method. Doriana M.D Addona et.al, [6] optimized the die shape of a circular forging, in conjugation with Neural Networks. The effect of flash allowances on the load has been studied by F. Fereshteh-Saniee et.al, [7]. Yahui Liu et.al, [8] optimized the forging process for production of two connecting rods using a RSA method and validated experimentally. Xinhai Zhao et.al, [9] optimized the die shape of the preform for reduction of material cost, using numerical analysis where the shape of the die is defined by a B-spline curve. The authors [12] tried to improve the yield in forging of a differential spider using numerical analysis and they have provided complete step by step of die design procedure in forging. Ibrahim Khoury et al., [13] in their work optimized the shape of the turbine engine disc. S.H.Chung et.al., [14] formulated the design problem on by integrating thermo-mechanical finite element process model to optimize the dies shape so that more design variables can be considered for optimization. The fillet radius and draft angle of the rib are varied and simulation is carried out in the finite element package which was developed by mechanical system and concurrent engineering laboratory (LASMIS). From the literature it is evident that most of the researches tried to optimize the shape of the dies for less load or for material saving but the effect of the parting line section in various equipments is least studied. Hence the current study focuses on the effect of parting line selection and its location in forging dies which are used in mechanical press and drop hammer.

In this paper forging of a coupling flange, an automotive component is considered for studying the effect of parting line selection on load, effective-stress and grain flow in two equipments – mechanical press and drop hammer. A three stage (Upsetter, Blocker and finisher) forging process is considered for carrying numerical analysis of coupling flange. For this purpose three different parting lines are selected in flange thickness namely - top parting line, bottom parting line and central parting line. Three die sets are designed with the selected parting lines and numerical analysis is carried out using a FEM based software DEFORM V12.0.2 [10]. The analysis is carried out in two different equipments – mechanical press and drop hammer. Also an inverted die is designed in such a way that the impression is cut completely in the top die.

2. Experimental Procedure

2.1. Forging process design

The first step in designing forging dies is selection of parting line. Selection of parting line affects the load, effective stress and grain flow in the component. In the current study three different parting lines are selected. The three different paring lines are named as top parting line, bottom parting line and central parting line which are shown in the Figure 1.

After selection of parting line various allowances like finish allowance, draft allowance, fillet and corner radii are added to the component on basis of IS-3469[11]. A three stage forging process which includes an upsetter, blocker and finisher is incorporated. Now the flash is designed based on Neuberger & Mockel's [1] equations 1 and 2. The thickness and width of flash are 2.4mm and 7.8mm respectively.

$$\text{Flash thickness} \quad t\,(mm) = 0.89\sqrt{W} - 0.017W + 1.13 \tag{1}$$

$$\text{Flash width} \quad w\,(mm) = t\,(3 + 1.2 \times e^{-1.09}W) \tag{2}$$

Where W is weight of forging in kg.

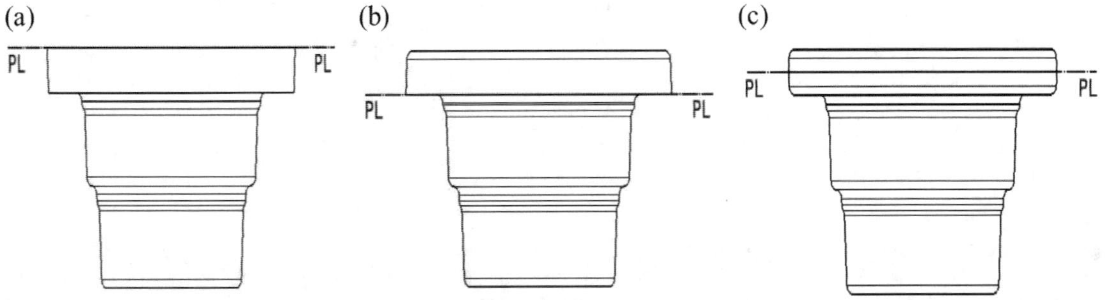

Figure 1. Coupling flange with various parting lines (a) Top parting line, (b) Bottom parting line,(c) Central parting line

The finisher is obtained by scaling the forging by 1.05 times to compensate the contraction which occurs during the cooling of the component from forging temperature to room temperature. The blocker is designed from finisher by reducing each side along the width by 0.3mm and increasing the depth, keeping the volume constant. A cup type upsetter is designed, so that the upsetted billet will sit in the blocker bottom die. For this the cup radius of the upsetter bottom die is kept 0.3mm less than the minimum radius of the blocker. The cross sectional views of upsetter, blocker and finisher die sets with top parting line are shown in the figure 2. Similarly two die sets with bottom parting line and central parting line are designed.

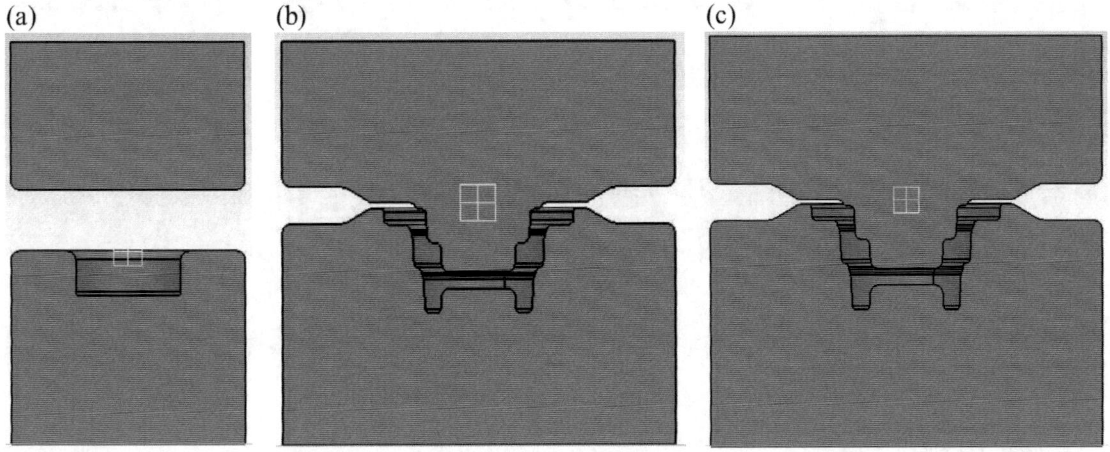

Figure 2. Die sets (a)Upsetter, (b) Blocker with top parting line, (c) Finisher with top parting line

For inverted dies a different approach is used as the placement of the billet will be difficult in such type of dies. Blocker and finisher dies are designed in the same fashion with a difference that the impression is cut completely in the top die instead of bottom dies. The bottom die of the upsetter is kept same as that in the previous case but the top die is given the shape of the blocker bottom die, and the width vise dimensions are reduced by 0.6mm.The cross sectional views of upsetter, blocker and finisher die sets with inverted dies are shown in the Figure 3.

2.2. Selection of equipment

In the current study two equipments – mechanical press and hammer, are considered for studying the effect of parting on the load, stress and grain flow. Initially required equipment capacity for mechanical press is calculated by using equation 3 and the mass of falling part for drop hammer is calculated by using equation 4.

$$\text{Capacity of mechanical press P (kg)} = 8\,(1-0.001\,D)\,(1.1 + 20/D)^2 \,.\, \sigma \,.A \qquad (3)$$

$$\text{Mass of falling part in hammer M (kg)} = 10\,(1-0.005\,D)\,(1.1 + 2/D)^2\,(0.75+0.001D) \times D \times \sigma \qquad (4)$$

Where D is maximum diameter including flash width, A is plan projected area including flash and σ is tensile strength at forging temperature. For medium carbon steel it is 7-8 kg/mm^2. In the present work 7 kg/mm^2 is considered which resulted in a press capacity of 1041.88 Tons. Hence a standard mechanical press of 1500 Tons is considered. Mass of falling part of hammer is 3445.7 kg. Hence a standard drop hammer of 10000lb of falling mass is considered.

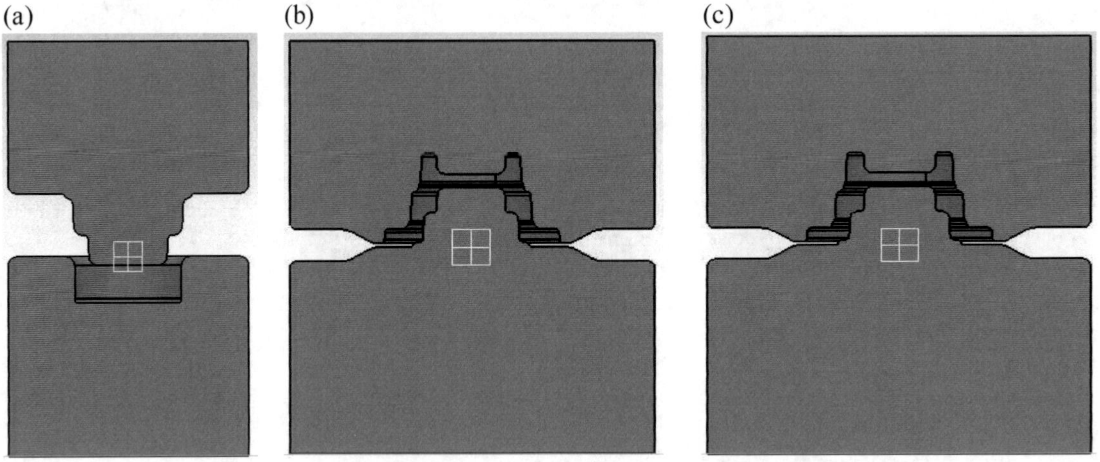

Figure 3. Inverted Die set (a)Upsetter, (b) Blocker, (c) Finisher

2.3. Numerical analysis

2.3.1. *Rigid-viscoplastic material.* The analysis of the material is carried out considering the material to be rigid-viscoplastic material. This is the formulation most commonly used for analysis of a forging process and the finite element based software DEFORM-3D also uses the same formulation in default. The main assumption of this formulation is that there is a linear increase in stress with strain rate till a threshold is reached, this level is referred as the limiting strain rate. The formulation of these materials can be expressed as the following equation.

$$\emptyset = \int_v E\,(\dot{\varepsilon})\,dV - \int_{S_i} f_i v_i\,ds + K^* \int_v \frac{1}{2}(\dot{\epsilon}_v)^2\,dV \qquad (5)$$

Where, K* is a very big penalty constant.

2.3.2. *Heat Transfer Equation.* The temperature distribution of the component and the dies is governed by the energy balance equation which is shown in the equation 6.

$$\int_v k\,T_i \delta T_i dV - \int_v \rho c \dot{T} \delta T dV - \int_v k^* \bar{\sigma} \dot{\bar{\varepsilon}}\,dV - \int_{S_q} q_n \delta T\,ds = 0 \qquad (6)$$

The equation can be written in matrix from as follows using numerical discretization,

$$KT + C\dot{T} - Q = 0 \quad (7)$$

Lagrangian Incremental method of analysis is adopted for calculation of heat transfer and amount of deformation at each time-step.

2.3.3. *Meshing*. The size of the element greatly defines the accuracy of the analysis, so it is very important to decide the mess size before the start of the analysis. Coarser mesh in the places of minute details like fillet and corner radii will result in degradation of mesh, loss in volume and remeshing problems. Finer mesh results in excessive computational time. Mesh size also depends on the output results of concern. If our concern is about die-filling, approximate load, stress and temperature distributions then a moderate mess size is sufficient. If emphasis is more on folds then the minute details have to be meshed properly. For such cases it is established by the software provider that the size ratio of the element must be 1/3rd of the minimum detailed size. Following these rules the, billet initial elements are 89020, number of nodes are 19377and surface polygons 16416,for top die - initial elements are 31885, number of nodes are 7247 and surface polygons 6058, for bottom die - initial elements are 52983, number of nodes are 11839 and surface polygons 10832. Further, during simulation if there is any mesh degradation the software has a capability of remeshing. The volume loss permitted while remeshing is 2 percent.

2.3.4. *Initial condition.* The work piece material considered is AISI 1045 with dimensions ø55mm x 120mm. The top and bottom dies are considered to be of H13 die steel as it is the most commonly used material. The forging range of AISI 1045 is 850°C to 1100°C. So, in all the cases, the maximum forging temperature i.e., 1100°C is considered as the initial billet temperature. The temperature of the dies is considered to be 300°C. Hot forging under lubrication is considered for which the coefficient of friction is 0.3.

The numerical analysis is carried out in DEFORM-3D with the die sets for all the three parting line and inverted die set in both mechanical press and Hammer.

3. Results and discussion

3.1. Load distribution

The load vs stroke curve for the component with bottom parting line when mechanical press is used is shown in the figure 4 and 5 shows the load distribution in mechanical press and hammer with top parting line. The table 5, shows the maximum load in various cases,

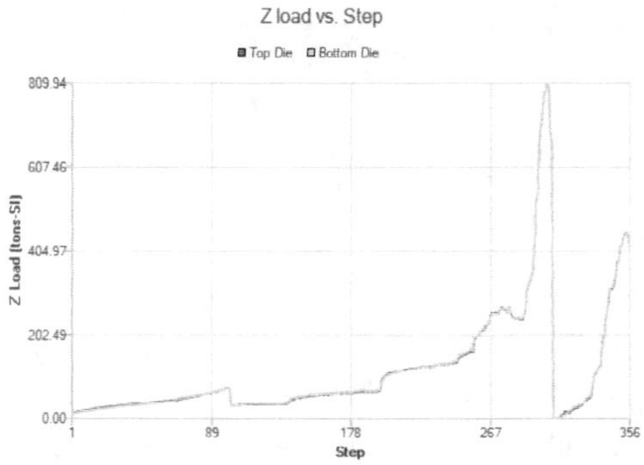

Figure 4. Load distribution in mechanical press with top parting line

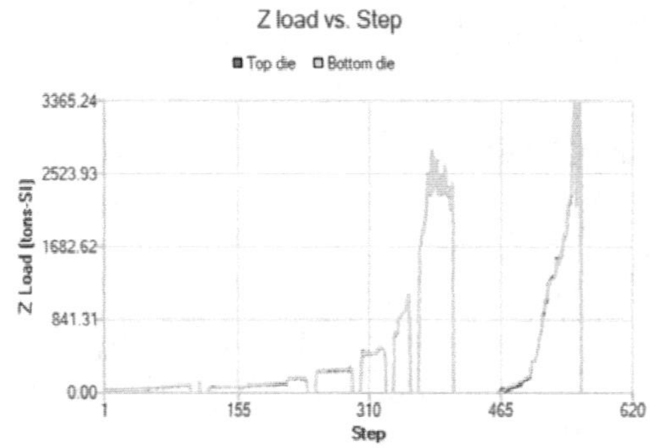

Figure 5. Load distribution in hammer with top

parting line

From the Table 1, it can be seen that the load is minimum when top parting line is used in mechanical press so it is more suitable to forge coupling flange in mechanical press with a top parting line. All the remaining conditions yielded in a higher load.

While using a hammer, bottom parting line should not be considered as the load in both the blocker and finisher is very high. By comparing blocker loads in top parting and inverted die-set, it can be seen that the top parting line yielded in less load compared to central parting line, but looking into finisher loads it is evident that inverted dies resulted in lesser load. In general as the finisher operation is critical operation, the amount of load coming on to the finisher dies should be kept as minimum as possible. With respect to load, inverted die-set is more suitable for forging a coupling flange in hammer. In a mechanical press the metal flows very easily by forward extrusion process whereas in hammer metal flows by reverse extrusion process easily.

Table 1. Maximum load in various conditions

	Mechanical press blocker load (tons)	Mechanical press finisher load (tons)	Hammer blocker load (tons)	Hammer finisher load(tons)
Top Parting Line	810	440	2578	3365
Bottom Parting Line	1225	683	3710	4680
Central Parting Line	1256	628	3260	2910
Inverted dies	684	628	3020	2560

3.2. Stress distribution.

The effective stress distribution for blocker and finisher with top parting line are shown in the figure 6 and 7. And the table 2, shows the variation in maximum stress in various cases. From the table 2, it can be seen that the stresses developed in the component is least when top parting line is used in mechanical press so it is more suitable to forge coupling flange in mechanical press with a top parting line. All the other conditions yielded in a higher stress. This is in well agreement with load criteria which is mentioned in the previous section 3.1.

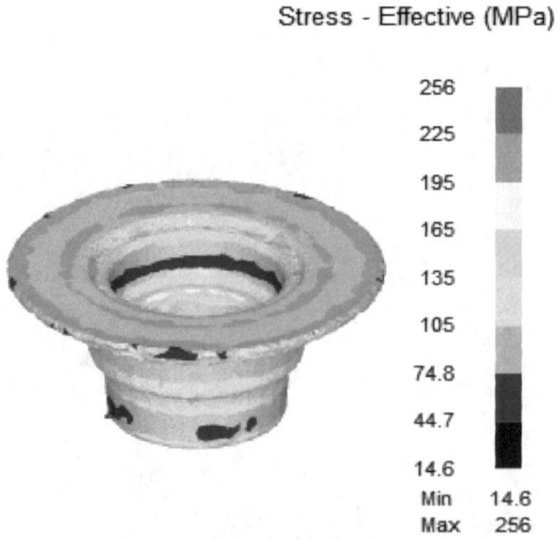

Figure 6. Blocker effective-Stress distribution in mechanical press with top parting line

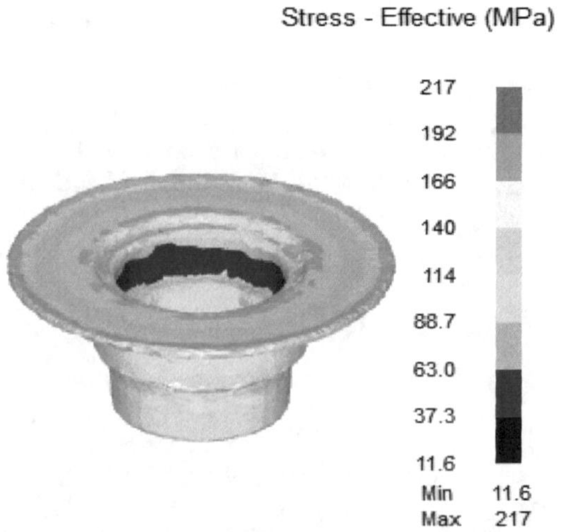

Figure 7. Finisher effective-Stress distribution in mechanical press with top parting line

Table 2. Maximum stress in various conditions

	Mechanical press blocker stress (MPa)	Mechanical press finisher stress (MPa)	Hammer blocker stress (MPa)	Hammer finisher stress (MPa)
Top Parting Line	256	217	709	718
Bottom Parting Line	574	511	794	849
Central Parting Line	390	341	622	763
Inverted dies	323	338	640	657

While using a hammer, bottom parting line should not be considered as the effective-stress in both blocker and finisher is very high. The effective stress developed is almost same in the central parting line and inverted die set. But while comparing the stresses developed in the finisher, it can be seen that the stresses in the component forged in inverted dies-set is less compared to that in case of central parting line. Hence according to minimum effective-stress criteria, top parting line has to be selected for forging coupling in hammer. This is in well agreement with the load criteria which is mentioned in the section 3.1. Hence for hammer forging, an inverted die design where the impression is cut completely in the top die should be consider.

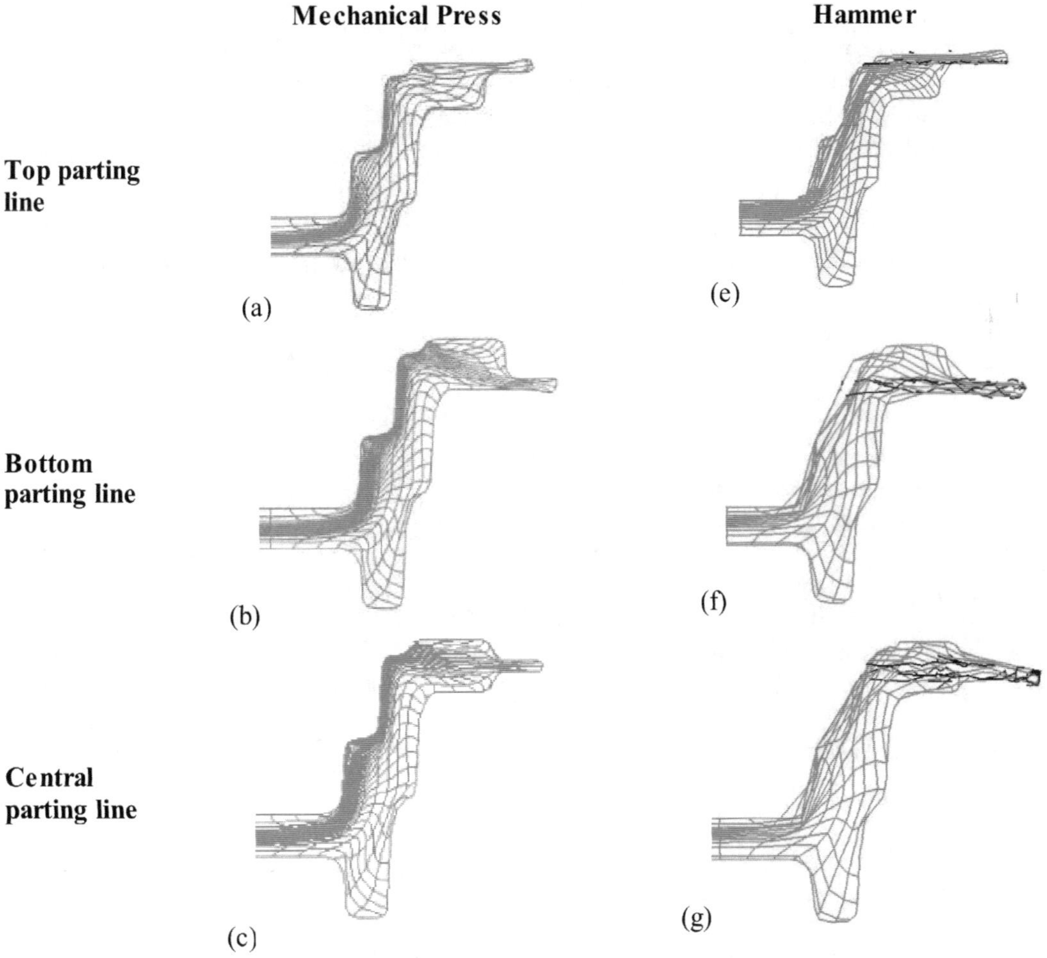

Inverted dies

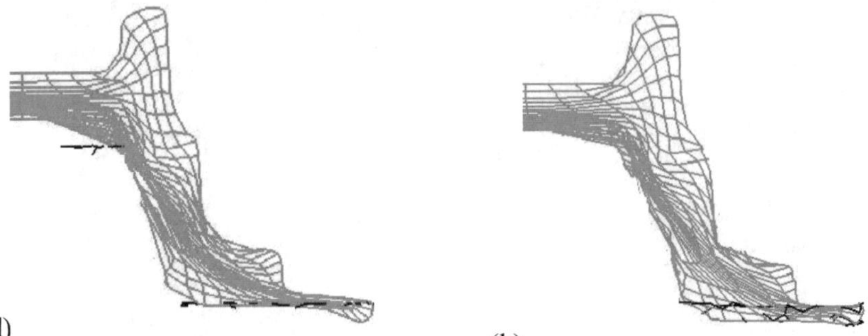

(d) (h)

Figure 8. Grain flow in various conditions

3.3. Grain flow

From the figure 8, it can be observed that the grain flow in various cases. There most favourable grain from can be observed in mechanical press with top parting line. We can observe that thers is no back flow of flow next hence there is less chance of formations of defects like folds or labs. In the cases of bottom paritng line and central paritng line the grain flow reversal is obsered. Flow reveral is more predominent in bottom paritng line hence it should not be selected in any equipment. For hammer, top paring line and inverted dies have similar flow pattern but there is more dencification of flow net in flange region for top parting line. Hence for hammer inverted die set can be selected.
Based on the discussions in sections 3.1, 3.2 and 3.3 the most suitable parting line for both mechanical press and hammer is top paritn line.

4. Conclusion

- Numerical analysis is carried out for section of optimal parting line in case of mechanical press and hammer forgings.
- The paper describes the effect of selection and location of parting line on load, effective-stress and grain flow.
- The paper also presents most suitable equipment and suitable parting line location for forging of coupling flange.
- Three different parting lines and inverted die design are selected for forging of a coupling flange in mechanical press and hammer.
- The component with top parting line (that is when impression is completely accommodated in bottom die) when forged in mechanical press resulted in minimum load, minimum effective stress and favourable grain flow.
- The most suitable die design for hammer forging is to accommodate coupling flange completely in the top die and it results in minimum load, minimum effective stress and favourable grain flow.
- Bottom parting line should not be selected in any case, as it resulted in high load and high-effective stresses.
- As far as grain flow is concerned top parting line and inverted dies resulted in favourable grain flow.

References

[1] Thomas. A., Die Design, Forging Handbook. Drop Forging Research Association, Sheffield, UK. 1995.
[2] ASM Metals Handbook, Forming and Forging, April 1996 Vol 14.
[3] Poli, C., Knight, W.A. Design for Forging Handbook, University of Massachusetts, Amherst, USA, 1981.
[4] S.N. Prasad, R. Sharan, N.P. Saksena, Forging Die Design and Practices, 1982.

[5] Zhenchao Qi, Xingxing Wang, Wenliang Chen, A new forming method of straight bevel gear using a specific die with a flash, The International Journal of Advanced Manufacturing Technology, 100, 2019, 3167–3183.
[6] Doriana M. D' Addona, Dario Antonelli, Neural Network Multiobjective Optimization of Hot Forging, Procedia CIRP 67, 2018, 498 – 503.
[7] F. Fereshteh-Saniee, A. H. Hosseini, The effect of flash allowances and bar size on forming load and metal flow in closed die forging, Journal of Materials Processing Technology, 177, 2016, 261-265.
[8] Yahui Liu, Jun Wang, Donghong Wang, Numerical optimization on hot forging process of a connecting rods based on RSA with experimental verification. The International Journal of Advanced Manufacturing Technology, volume 90, 2017, 3129–3135
[9] Xinhai Zhao, Guoqun Zhao, Xiahui Huang and Yiguo Luan, Multiple objective preform die shape optimization design fir controlling forging's shape and deforming force during forging process, Advanced Materials Research, Vols 306-307, 2011, 1504-1507.
[10] DEFORM V12.0.2 , User's Manual, 2020.
[11] Indian Standards IS: 3469 (Part I to III) – 1974.
[12] Mathala Prithvi Raj, Manoj Kumar, Ajit Kumar Pramanick, Yield improvement in hot forging of differential spider, materials today:PROCEEDINGS, Volume 26, Part 2, 2020, 3107-3115.
[13] Ibrahim Khoury, Laurence Giraud-Moreau, Pascal Lafon, Carl Labergere, Towards an optimisation of forging processes using geometric parameters, Journal of Materials Processing Technology, 177, 2006, 224–227.
[14] J. S. Chung, S. M. Hung, Optimal process design in non-isothermal, Non-steady metal forming by the Finite element method, Int. J. Numer. Meth. Engng. 42, 1998, 1343-1390.

Response surface methodology approach for optimization of methylene blue adsorption using activated carbon from rice husk

Lam Van Tan[1,2,*], **Hong-Tham T. Nguyen**[1,2]

[1]NTT Hi-Tech Institute, Nguyen Tat Thanh University, Ho Chi Minh City, Vietnam.
[2]Center of Excellence for Green Energy and Environmental Nanomaterials, Nguyen Tat Thanh University, Ho Chi Minh City, Vietnam

Corresponding author: labasm2013@gmail.com; lvtan@ntt.edu.vn

Abstract. In this study, an activated carbon (AC) material was derived from rice husk and tested for adsorption efficiency against methylene blue (MB). Response surface methodology (RSM) and central composite design (CCD) were adopted as the optimization techniques. The affordable, non-toxic and easily obtainable rice husk was carbonized and thereafter activated by KOH to produce porous AC, which was then characterized by XRD, FTIR, SEM and BET analyses. In the optimization study, the independent and collaborative effects of three experimental parameters, including dye concentration, pH solution and contact time, were taken into account. The adsorption capacities showed up to follow the direction that was well congruent with the confirmation experiments, which disclosed the robustness and viability of the optimization technique.

Keywords: adsorption, response surface metholodogy, methylene blue, activated carbon, rice husk

1. Introduction

One of the most contemporary issues that are causing paramount concerns in various parts of the world is the alarming rate of environmental pollution. Among many types of contaminants, colorants, notably those categorized as aggressive, prompt, basic, and acidic nature, have been regarded as a persistent threat to the water ecosystem [1]. They express higher water solubility, presenting difficulty in removing them from the water using the conventional methods [2,3]. Therefore, removal of industrial dyes has been an issue that attracts extensive attention from many fields.

Methylene Blue (MB) is a cationic dye whose usage is prevalent among garment and paper industries and has been known for the ability to reduce water transparency, hinder the activity of photosynthesis, and increase the demand for chemical oxygen, which interferes with the bacterial metabolism of waterlogged bacteria [4]. Due to the complicated structure and synthetic origin, The removal of MB dyes could not be easily realized by conventional wastewater treatment techniques, hence calling for further studies dealing with treatment of dye-containing wastewater [5,6].

It is a fact that porous materials play an important role in treating wastewater owing to a number of advantages including large contact area and reactivity of the material surface, fine porosity and ability to be tailored to adapt to different pollutants and media [7]. In this sense, the present work aims to investigate the optimal factors to the efficiency adsorption process of methylene blue dye by using response surface methodology.

2. Material and methods

2.1. Materials

The activated carbon was synthesized from rice husk source materials according to the previous report. Firstly, rice husk was hand-distributed upon the floor to be sun dried. This drying process takes from some days to a week depending on the amount available sun irradiation. Then, the rice husk was crushed into powder to prepare for the next activated stage. In this stage, solid KOH was used to activate dry powder according to ratio 1:1 (g/g) for 24 hours. The impregnated sample was dried at 105°C for 24 hours prior to the physical carbonization phase with 99.9 % N_2 gas flow at 500°C. Residual KOH was removed from activated carbon by washing with distilled water and HCl solution. The obtained carbon was then dried in an oven for 24 hours. The final product renders an adsorbent suitable for wastewater treatment, which was abbreviated AC_RH.

2.2. Instrumentation

Powder X-ray diffraction (PXRD) patterns were recorded with a Jasco-4700 diffractometer equipped with an X'Celerator detector and using Cu Kα radiation (λ=1.54060). FT-IR spectra were collected on a Jasco-4700 spectrometer (Japan). The specific area and pore distributions of as-prepared -samples were calculated from the N_2 isotherm adsorption branch by BET (TriStar 3000 V6.07 A). Field Emission Scanning electron microscopy (FESEM) studies were carried out by Hitachi FE-SEM S4800 (Japan). The concentration of MB dyes was determined by using UV−Vis spectra at 664 nm wavelength (UV-Vis Ligent Cary 60).

2.3. Batch adsorption studies

The adsorption efficiency of materials is assessed based on the efficient removal of methylene blue (MB). Briefly, the amount of AC_RH and 100 mL of MB solution were mixed into a 250 ml glass beaker. Then, the solution was stirred at 200 rpm by a shaker in 180 minutes, 4 ml solution was withdrawn in the regulated time period, and centrifuged to separate the solid from the dye solution. Finally, the concentration of the balance dye was determined in wavelength 664 nm by spectrophotometry. The following equation used to calculate the removal efficiency was calculated based on concentration before and after the adsorption process:

$$q_e = (C_o - C_e)/m.V \qquad (1)$$

where, C_o and C_e are the respectively initial and equilibrium dye concentrations (mg/L).

3. Results and discussion

3.1. Material characteristics

Figure 1 depicts the XRD pattern and FT−IR spectra of AC_RH. According to **Figure 1.A**, the board range at 20-30 could be indexed to the amorphous carbon structure [8]. In addition, the weak and broad C (101) diffraction peak (2θ = 40-50°) is due to the a axis of the graphite structure [2]. Besides, FT−IR spectra was used to overcome the spectral broadenings and band overlaps **(Figure 1.B)**. A broad spectrum is found in areas of surface functional groups such as phenols, carboxylic acids and carboxylic acid derivatives as well as in the presence of the -OH group on the material's surface in the range 3554-2947 cm^{-1} [3,4]. These bands are formed by the prolonged oscillation of the OH functional group and become wider in the case of coal oxidation. Covalent fluctuations of C−H bonding of groups $-CH_2$, $-CH_3$ were observed at peaks 2898 cm^{-1} and 2810 cm^{-1} [5,6]. In addition, the typical positions for C−O and C = O oscillations of the ester, ketone, aldehyde, lactone or carboxylic groups appear in the range of 1486-1728 cm^{-1}. The adsorption of the carbonyl ester group C = O of the lactone and phenol groups was observed in the range 919-1438 cm^{-1} [7,8]. These functional groups contribute greatly to improving the adsorption efficiency.

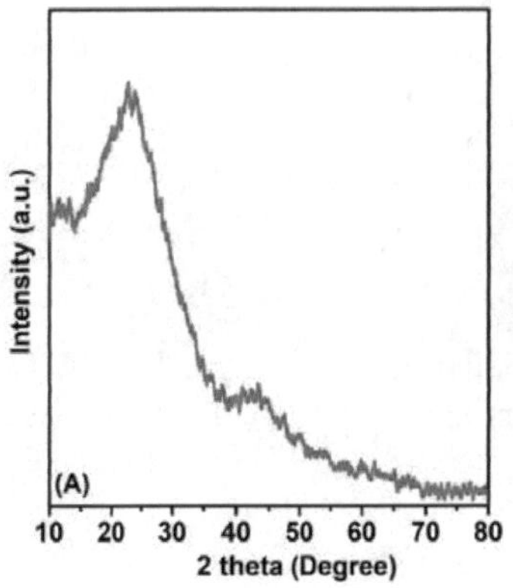

Figure 1. XRD pattern (A) and FT–IR spectra (B) of AC_RH

Figure 2 illustrate the surface morphology of AC_RH. Visually, the surface of the adsorbent exhibited good porosity and irregular shape with even cavities and fine open pores, indicating relatively high surface area.

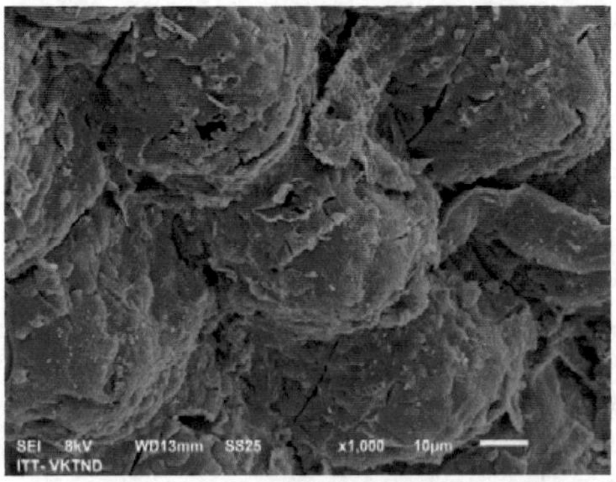

Figure 2. SEM image of AC_RH

To make sure the porous material evenly, N_2 isotherm adsorption / desorption was analyzed **(Figure 3)**. AC_RH exhibit the type I isotherm according to definitions of International Union of Pure and Applied Chemistry (IUPAC). In detail, the surface area and pore diameter of AC_RH are 493.46 m^2/g and 2.45 nm, respectively. With the presence of large pores, it is very likely that methylene blue molecules were trapped and adsorbed into the pores.

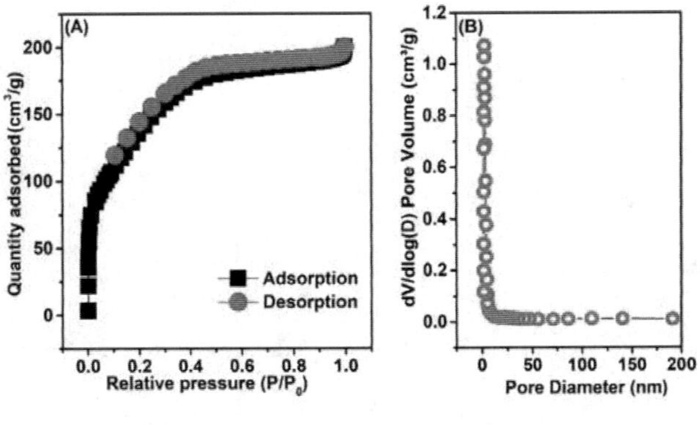

Figure 3. N_2 isotherm adsorption/desorption of AC_RH

3.2. RSM (CCD) study

To realize the optimization process of the MB adsorption process with AC_RH, Design Expert sofware was utilized. The software was configured to adopt the central composite design (CCD) with design inputs taken from the preliminary investigations. A total of 20 experiments were devised as given in Table 1. Those experiments would be attempted to produce data for model estimation.

Table 1. Ranges, levels and thresholds of parameters in the experimental design with Central Composite Design

Factors	**Units**	**Levels**				
		Low (-)	**Central (0)**	**High (+)**	**-α**	**+α**
Initial MB Concentration	mg/L	40	50	60	33.18	66.82
pH	-	7	8	9	6.32	9.68
Contact time	min	170	180	190	163.18	196.82

Table 2 summarized the detailed experiment runs with the corresponding actual and estimated responses. In this model, the diference between the predicted efficiency removal and the experiment efficiency removal is negligible.

Table 2. Generated parameters for RSM experiments and their corresponding responses

Standard Runs	**Coded Variables**			**Tested**	**Predicted**
	Concentration (mg/g)	**pH**	**Time (min)**	**%**	**%**
1	40	7	170	36.05	35.90
2	60	7	170	32.66	33.34
3	40	9	170	36.53	37.40
4	60	9	170	32.52	32.63
5	40	7	190	36.11	36.61
6	60	7	190	35.81	35.55
7	40	9	190	38.11	38.05
8	60	9	190	34.01	34.78
9	33.18	8	180	35.10	34.71
10	66.82	9	180	30.28	29.80
11	50	6.32	180	36.4	36.30
12	50	9.68	180	37.64	36.93

13	50	8	163.18	37.67	37.07
14	50	8	196.82	39.72	39.46
15	50	8	180	38.24	38.70
16	50	8	180	38.47	38.70
17	50	8	180	38.66	38.70
18	50	8	180	39.10	38.70
19	50	8	180	38.63	38.70
20	50	8	180	38.96	38.70

Table 3. Data ANOVA for MB removal efficiency on the composite adsorbent.

Source	Sum of Squares	Degree of Freedom	Mean Square	Value F	Prob. > F	Comment
Model	118.85	9	13.21	32.93	< 0.0001	$R^2 = 0.9674$
A-Concentration	29.01	1	29.01	72.35	< 0.0001	$R_{adj}^2 =$ 0.9380
B-pH	0.4647	1	0.4647	1.16	0.3070	
C-Time	6.93	1	6.93	17.29	0.0020	
AB	2.44	1	2.44	6.07	0.0334	
AC	1.13	1	1.13	2.81	0.1244	
BC	0.0022	1	0.0022	0.0055	0.9425	
A^2	74.86	1	74.86	186.66	< 0.0001	
B^2	7.85	1	7.85	19.57	0.0013	
C^2	0.3429	1	0.3429	0.8550	0.3769	

The ANOVA results of methylene blue removal using AC_RH are given in Table 3. It is evident from the data in **Table 3** that the predicted R^2 value (0.9674) and the adjusted R^2 value (0.9380) are closely related to each other and thus ideally aligned with the quadratic model. Given the results of ANOVA **(Table 3)**, the model F-value of 32.93 and the p-value of <0.0001 % implies that the system is clinically meaningful. The above demonstrates that the current mathematical formula fits well into the experimental results demonstrate the feasibility of an inconsequential lack of fit **(Table 3)**.

The plot between both the experimental (actual) and the predicted MB removal values is shown in **Fig. 4**. Evidently, the difference between predicted and experimental responses was small across observations, at around less than 0.1. This suggests high explanatory power of the estimated model and the well-fitness of the data to the quadratic function. Those results suggest that it is reasonable to assume that the validated numerical vibration.

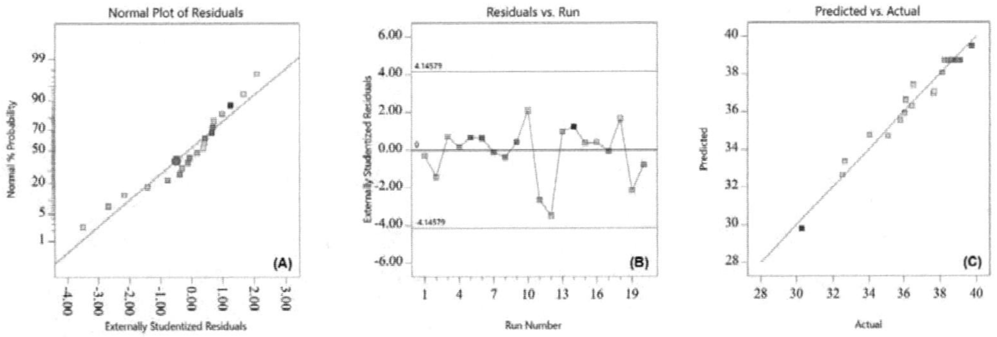

Figure 4. Normal plot of residuals (A), Residuals versus run plot (B), Predicted versus actual plot (C)

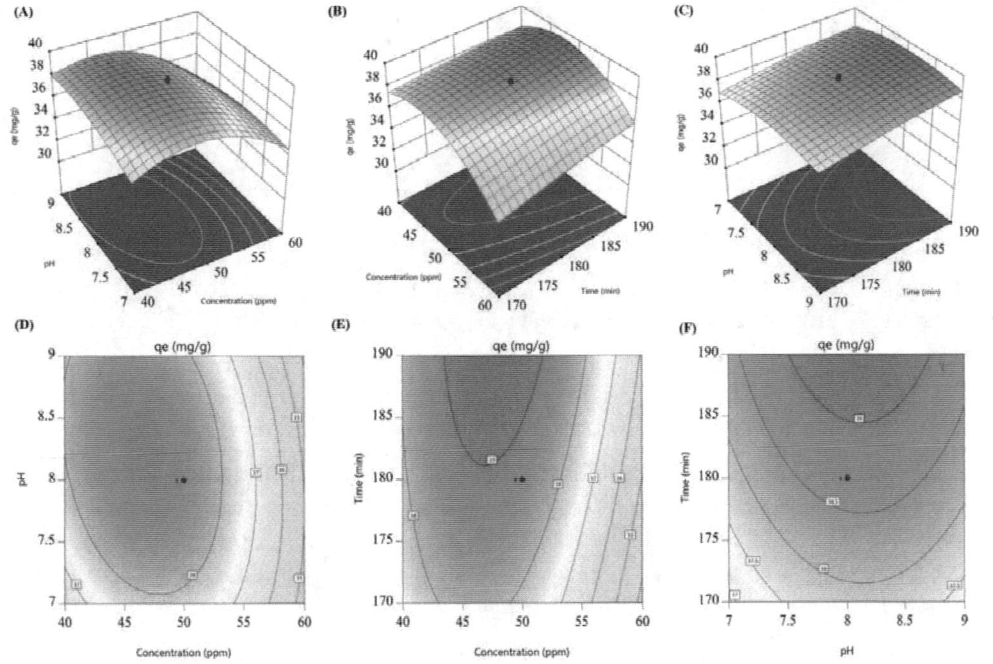

Figure 5. Response surface plots a, b concentration (mg/L) and pH solution; c, d concentration (mg/L) and contact time (min); e, f, contact time (min) and pH solution

Figure 5 show the the 2D-contour and 3D-response surface plots, which were characterized for the adsorption capacity (mg.g^{-1}). The pertinence of dichotomous relationships between three selected variables (pH, contact time, and initial concentration of MB) can be clearly deduced from these graphs. In detail, both two variables including pH solution and contact time affected remarkably the removal capacity of Methylene Blue from aqueous solution. Further addition of initial concentration dye led to decline in adsorption efficiency. From the figure, it is also evident that improvements in adsorption efficiency are due to decreased dye concentration in the media and that decreased contact time could also cause better interaction between dye and adsorbent, which is likely responsible for enhanced adsorption.. At low pH values (2–6.0), the removal of dye seemed to be inefficient and irresponsive to contact time. However, at higher pH range (8.0–9.0) and long contact time, pH value was found to associate with noticeable changes in the response. Current results revealed the dependence of optimal pH on the pH range, which in turn remarkably affect dye removal capacity.

4. Conclusion

In the present study, activated carbon was produced from rice husk source, which was evaluated as a porous material with high surface area (493.46 mg^2/g). By using response surface method (RSM), three variables (initial MB concentration, pH solution, and contact time) were adopted to explore the individual and combined effects thereof on methylene blue adsorption. The results revealed that each factor individually and significantly affected the adsorption process. The reliability of the developed has been guaranteed by the high magnitude of the coefficient of determination between the predicted experimental and model values. Among all the process variables, the initial dye concentration was found as the most significant factor. Current results support the use of optimization techniques in wastewater treatment processes to find out most suitable conditions and materials for specific contaminants.

ACKNOWLEDGMENT

This research is funded by Nguyen Tat Thanh University, Ho Chi Minh City, Vietnam.

References

[1] Ait Ahsaine H, Anfar Z, Zbair M, Ezahri M and El Alem N 2018 Adsorptive Removal of Methylene Blue and Crystal Violet onto Micro-Mesoporous Zr 3 O/Activated Carbon Composite: A Joint Experimental and Statistical Modeling Considerations *J. Chem.* **2018** 1–14

[2] Zong M-H, Duan Z-Q, Lou W-Y, Smith T J and Wu H 2007 Preparation of a sugar catalyst and its use for highly efficient production of biodiesel *Green Chem.* **9** 434

[3] Tran T Van, Thi D, Nguyen C, Thi H and Le N 2019 Comptes Rendus Chimie Amino-functionalized MIL-88B (Fe) -based porous carbon for enhanced adsorption toward cipro fl oxacin pharmaceutical from aquatic solutions *Comptes rendus - Chim.*

[4] Sarma G K, SenGupta S and Bhattacharyya K G 2011 Methylene Blue Adsorption on Natural and Modified Clays *Sep. Sci. Technol.* **46** 1602–14

[5] Thuan T Van 2018 Preparation of activated carbon from sugarcane bagasse using $ZnCl_2$ for the removal of Cu (II) ion from aqueous solution: application of response surface methodology (RSM) *Vietnam J. Sci. Technol.* **54** 277

[6] Huong P T L, Huy L T, Phan V N, Huy T Q, Nam M H, Lam V D and Le A T 2016 Application of Graphene Oxide-MnFe2O4Magnetic Nanohybrids as Magnetically Separable Adsorbent for Highly Efficient Removal of Arsenic from Water *J. Electron. Mater.* **45** 2372–80

[7] Bansal R C and Goyal M 2005 *Activated Carbon Adsorption* (CRC Press)

[8] Liu X-Y, Huang M, Ma H-L, Zhang Z-Q, Gao J-M, Zhu Y-L, Han X-J and Guo X-Y 2010 Preparation of a Carbon-Based Solid Acid Catalyst by Sulfonating Activated Carbon in a Chemical Reduction Process *Molecules* **15** 7188–96

[9] Hai F I, Yamamoto K and Fukushi K 2007 Hybrid Treatment Systems for Dye Wastewater *Crit. Rev. Environ. Sci. Technol.* **37** 315–77

[10] Arslan-Alaton I, Tureli G and Olmez-Hanci T 2009 Treatment of azo dye production wastewaters

using Photo-Fenton-like advanced oxidation processes: Optimization by response surface methodology *J. Photochem. Photobiol. A Chem.* **202** 142–53

[11] Gogate P R and Pandit A B 2004 A review of imperative technologies for wastewater treatment I: oxidation technologies at ambient conditions *Adv. Environ. Res.* **8** 501–51

[12] Kumar P, Prasad B, Mishra I M and Chand S 2008 Decolorization and COD reduction of dyeing wastewater from a cotton textile mill using thermolysis and coagulation *J. Hazard. Mater.* **153** 635–45

[13] Lu X, Yang B, Chen J and Sun R 2009 Treatment of wastewater containing azo dye reactive brilliant red X-3B using sequential ozonation and upflow biological aerated filter process *J. Hazard. Mater.* **161** 241–5

[14] de Souza S M de A G U, Bonilla K A S and de Souza A A U 2010 Removal of COD and color from hydrolyzed textile azo dye by combined ozonation and biological treatment *J. Hazard. Mater.* **179** 35–42

An approach to eliminate shell cracking problem in fused deposition modeling pattern based investment casting process

Simranjeet Singh[1], Parlad Kumar[2*], Jagmeet Singh[3]

[1]Design Engineer, Shiva Enterprises, Sonepat, Haryana, India
[2,3]Department of Mechanical Engineering, Punjabi University, Patiala, Punjab, India

*Corresponding author email: pkgarg4050@gmail.com

Abstract. Fused deposition modeling process can be used to make complex plastic parts directly from the three dimensional CAD drawings without any requirement of tooling. For low volume production, these plastic parts can be used instead of wax patterns in investment casting process to reduce high cost of tooling required for making wax patterns. But with plastic patterns there is a problem of ceramic shell cracking during the pattern burnout process. This shell cracking is mainly due to thermal expansion of the plastic sprue and plastic patterns. In this paper, a method has been proposed to eliminate the problem of shell cracking by using a wax sprue instead of plastic sprue. A joint has been designed to attach plastic patterns with wax sprue. It has been found that the proposed method is successful in eliminated the shell cracking problem. In addition to this, the proposed method is expected to reduce the process cycle time and cost.

1. Introduction

Investment casting (IC) is an old process and it is used to make complex metal parts with good surface detailing. In IC process, metallic dies are required for making wax patterns. The manufacturing of die is costly and time consuming process. Therefore, the IC process is cost effective only for high volume of production. But in additive manufacturing, objects are directly produced from 3D CAD drawings and no hard tooling is required [1]. Therefore, it is considered cost effective for low volume of production also. For high volume of production additive manufacturing is expensive due to its costly build material and slow speed. Different types of additive manufacturing techniques are available with different types of material processing capabilities. A suitable additive manufacturing technique and its material can be selected as per the required properties of the end product [2]. Fused deposition modeling (FDM) is an additive manufacturing technique that can be used to make plastic parts. Common FDM machines can print different types of plastic materials but ABS (Acrylonitrile- Butadiene-Styrene) plastic is widely used for different applications. Studies show that ABS plastic patterns can be used as a substitute to wax patterns in IC process [3-5]. For this purpose, the plastic patterns are prepared directly from 3D CAD drawings on FDM machine. These plastic patterns are used instead of wax patterns in the conventional IC process with very little change in the process. There are two main advantages of this method. Firstly, the change in design is quick and easily possible as no hard tooling is involved in the process and only drawing is to be changed. Secondly, the time and cost involved in hard tooling is saved in this method.

This method has been successfully used by many researchers to reduce the cost and lead time of the low volume production. However, there are some problems associated with this method. The main problems associated with plastic patterns and plastic sprues are poor surface finish, poor dimensional accuracy, ash accumulation and ceramic shell cracking. The solutions of these problems have been suggested by different researchers. The problem of poor surface finish is due to poor surface finish of pattern prepared by FDM machine due to layer by layer manufacturing. This problem can be eliminated by using various techniques as reported by many researchers [6-8]. The dimensional inaccuracy of the FDM pattern based IC is due to the fact that FDM parts shrink after manufacturing. The shrinkage of casted parts during solidification is also responsible for some dimensional inaccuracy. The dimensional accuracy can be improved to a great extent by using the techniques suggest by some researchers [9-12]. The problem of ash accumulation in cavity is due to ash produced after the burnout of ABS plastic pattern. Pressurized air or water can be used to remove this ash from the cavity [4,13]. The problem of ceramic shell cracking is more predominant and in this problem the ceramic shell (mold) breaks during the dewaxing/burnout process of IC. The main reason behind the shell cracking is the thermal expansion of the plastic patterns at high temperature during burnout process. The shell cracks when the coefficient of expansion of pattern is more than coefficient of expansion of the shell.

Many studies have been conducted by different researchers to address this problem of shell cracking. Harun et al. [14] found that the hollow patterns of ABS material offers better results than solid patterns in terms of mold cracking and internal mold cleanliness at temperature 300-500°C. Chen et al. [15] reported that the temperature range 200-250°C in burnout process is dangerous because the strength of ceramic shell decreased rapidly after 200°C. An addition of high temperature polymers can increase the strength of ceramic shell and prevents the shell cracking. Sivadasan et al. [16] found that the thermal expansion of the plastic patterns produced by FDM process can be reduced by using different build styles to generate distinct internal web structure in the pattern. Marwah et al. [17] reported that the maximum hoop stress is higher on square internal structure as compared to the hollow internal structure. The hollow pattern experienced 40% less strain as compared to the square pattern. Kumar et al. [12] suggested that a thin layer of wax on FDM pattern significantly reduces the chances of shell cracking during pattern burnout stage. Yao and Leu [18] found that the collapsibility of pattern and shell cracking are strongly related to the span length of the internal web of plastic pattern, cross-sectional dimensions and the shell thickness. The shell cracking can be prevented by buckling of epoxy webbed pattern in early stages of burnout process. Bansode et al. [19] suggested that the shell cracking can be reduced by adjusting silica concentration, ratio of filler to binder and the number of secondary layers. It has also been reported by many authors that the addition of different types of fibers in the slurry to make fiber composite slurry can also enhance the shell strength. Different types of fibers have been tried by different researchers. The fibers of ceramic, nylon, polypropylene, aluminum silicate, glass or carbon can be used with variable success rate [20-22].

The problem of ceramic shell cracking becomes more severe if both pattern as well as sprue are manufactured by using plastic. The use of plastic sprue becomes necessary especially in the large and heavy castings because there is no set procedure to join plastic pattern with wax sprue. In the present study, it has been proposed that even for heavy castings, only patterns should be prepared with plastic whereas the sprue should be prepared with wax by using the conventional method. If it happens, the wax sprue would melt at a relatively low temperature and it would leave some space for expansion of the plastic pattern. Thus the force on the ceramic shell would decrease and chances of ceramic shell cracking would be reduced. For this purpose, there is a need to join the plastic patterns with the wax sprue. Therefore, a method has been proposed to make a strong joint between plastic patterns and wax sprue. This joint is supposed to carry the load of pattern as well as slurry coated on it. For designing the joint, different design approaches have been considered and the best approach has been selected on the basis of joint strength. The designed joint has been checked for its effectiveness for reducing the problem of shell cracking during IC process.

2. Design Approaches

A schematic diagram for arrangement of sprue, gate and pattern in an IC tree is represented in Figure 1. In this figure only one pattern of any random geometry has been shown attached to the sprue. In actual practice the IC tree is formed by attaching a number of patterns around the sprue.

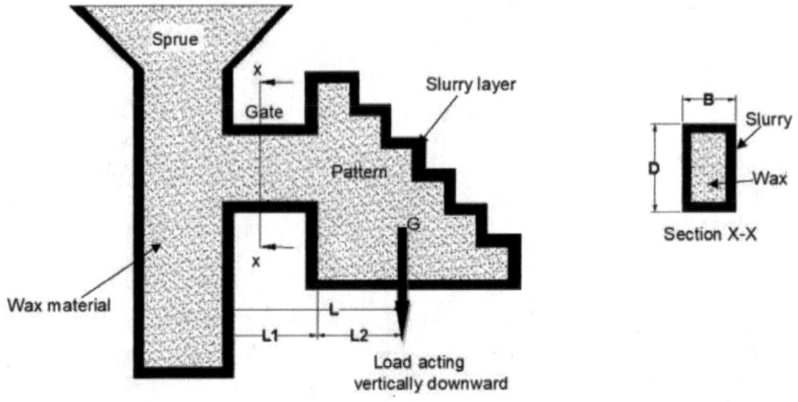

Figure 1. Arrangement of sprue, gate and pattern (not to scale).

This tree is coated with many layers of slurry to make a ceramic shell around the tree. In conventional method the wax patterns are attached to wax sprue after local heating of the surfaces. But in the present study the problem is to attach the plastic pattern with wax sprue and wax gate. No set procedure is available to make a strong joint between plastic and wax. Therefore, an approach of embedding one material in other has been tried. There may be different possible arrangements to produce a locking effect between sprue and patterns. Some of the possible arrangements have been worked out and represented in Figure 2 (a-f). In order to represent the details, these figures are not as per the actual scale. The proposed wax-plastic joint is in the gate area. All of the joint design approaches shown in Figure 2 (a-f) can be used to produce a locking effect between a plastic pattern and wax gate. In all of these methods, plastic patterns have been designed with some plastic extensions. The required plastic extensions can be made in 3D drawing before printing the patterns so that this projected extension become an integral part of the pattern after printing. These plastic extensions can either hold the wax or can be embedded into the wax as per their design to make a joint. Now the aim is to find the best approach out of these different approaches. For this purpose, comparative load carrying capacity of each joint was estimated. It was found that the design approach shown in Figure 2(e) has maximum load carrying capacity as compared to other approaches. Therefore, this method has been selected for making a joint of plastic pattern with wax gate. The detailed design of this method has been explained in the next section.

3. Detailed Design

In this section the selected joint type has been designed in detail to find each and every dimension of the required plastic extension with the pattern. The required dimensions of the joint depends upon the total load on the joint due to weight of pattern and weight of slurry covered on the pattern. In this study, the joint has been designed to carry a plastic pattern of volume 300 cm^3. This size of pattern can be used to make castings up to mass of 2.5 kg, depending upon the density of material to be casted. It can be seen from the arrangement that in a casting tree the gate acts as a cantilever beam. The bending moment on this cantilever depends upon the load as well as the distance equal to the sum of gate length L1 and distance of center of gravity L2 from the gate as shown in Figure 2(e). The size of gate cannot be same for different sizes and geometries of the patterns. The location, type, and size of the gate is designed on the basis of experience and as per the requirement.

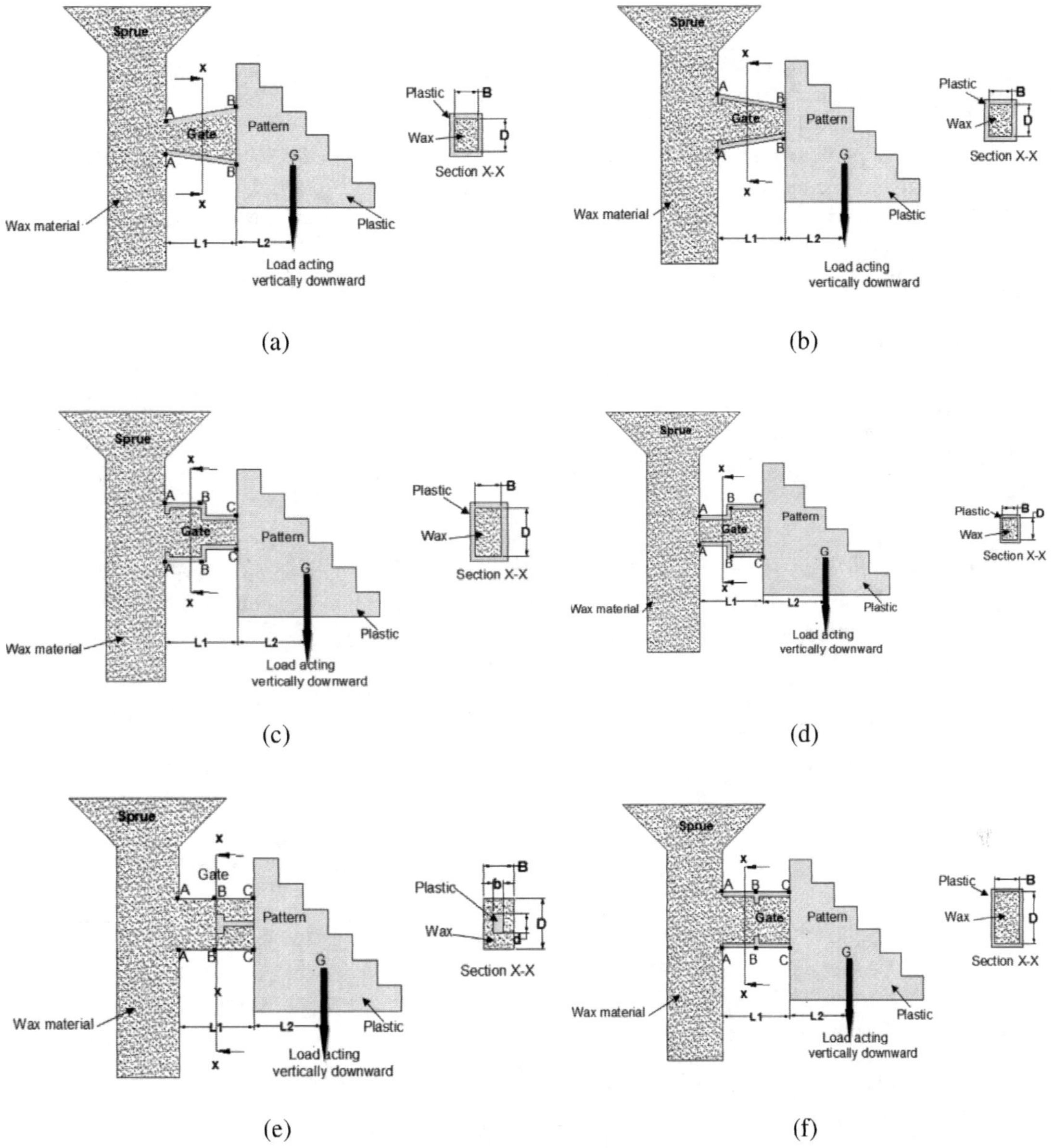

Figure 2. (a-f) Different approaches for joining plastic pattern with wax sprue (figures not to scale).

There may be more than one gate, if required. For the present geometry, a rectangular gate of length 30 mm and cross-section 25mm x 10mm has been used, as suggested by foundry. The center of gravity, volume of the pattern and volume of slurry have been obtained from the 3D CAD model of the pattern. The volume of this pattern is 300 cm^3 and the volume of 8 mm thick layer of slurry for this geometry is 334 cm^3. The density of the ABS plastic supplied by Stratasys is 1.04 g/cm^3 [23] and the density of the slurry is 2.2 g/cm^3 [24]. The total mass of the pattern and slurry can be calculated as given below.

IOP Conf. Series: Materials Science and Engineering 1091 (2021) 012035 doi:10.1088/1757-899X/1091/1/012035

Mass of pattern $m_p = V_p . \rho_p = 300 \times 1.04 = 312$ g
Mass of slurry $m_s = V_s . \rho_s = 334 \times 2.2 = 735$ g
Total mass $m = m_p + m_s = 312 + 735 = 1.047$ kg

The total load (P) due to pattern material and slurry material is

$P = m \times a = 1.047 \times 9.81 = 10.27$ N

The maximum permissible stress of wax and plastic can be calculated by dividing the failure stress by factor of safety. The value of tensile strength of wax material is 1.38 MPa [25]. The tensile strength of ABS parts printed by FDM depends upon various process parameters of FDM but a suitable selection of these parameters gives the tensile strength of about 30 MPa [26]. The values of maximum permissible stress for wax and ABS plastic material by taking a factor of safety equals to 2 are as following:

Maximum permissible stress of wax material $\sigma_{wax} = 0.69$ MPa
Maximum permissible stress of ABS material $\sigma_{abs} = 15$ MPa

The factor of safety has been selected judiciously to consider the effect of load due to any possible unconsidered parameter. The weakest cross-section gives the value of load that can be carried safely by the joint. The bending moment (M) can be calculated using the equations (1) and (2).

$$M = \sigma \times Z \ ; \ Z = \frac{BD^2}{6} \text{ (For rectangular section)} \quad (1)$$

$$M = P \times L \ ; \ L = \text{Length of beam} \quad (2)$$

There are various possible cross-sections where the design can fail. The design should be safe on all the cross-sections. Figure 3 shows the schematic diagram representing various possible failure cross-sections in the gate area. The cross-sections X-X, M-M, Y-Y and Z-Z have been considered as possible failure cross-sections for designing the joint.

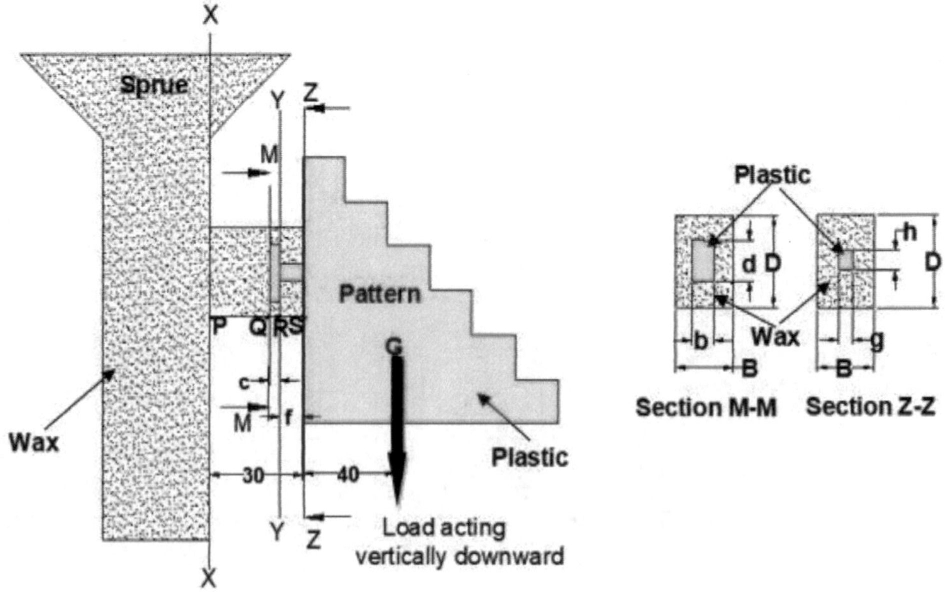

Figure 3. Schematic diagram of gating arrangement (Figure not to scale, dimensions are in mm).

The next step is to find the dimensions of the different parts of the joint. The calculations for different failure cross-sections and the obtained design values have been represented in Table 1.

IOP Conf. Series: Materials Science and Engineering 1091 (2021) 012035 doi:10.1088/1757-899X/1091/1/012035

Table 1. Calculation of design values at various cross-sections.

S. No.	Failure Cross-section	Equations used	Calculations	Calculated Design values
1	Calculation of B and D (Considering wax failure at cross-section X-X)	M = P×L M = σ × Z $M = \sigma \times \frac{BD^2}{6}$ Where P= 10.27 N L = 70 mm σ = 0.69 MPa D= 2.5B	M = 10.27×70 M =718.9 Nmm $718.9 = 0.69\times \frac{BD^2}{6}$ BD^2= 6251.30 B^3 = 1000.21 B = 10 mm D = 2.5×B D = 25 mm	B = 10 mm D = 25 mm (Already selected values of wax gate cross-section are 25mm x 10mm, therefore it is safe)
2	i) Calculation of *g* and *h* (Considering plastic failure at cross-section Z-Z)	M = P×L M = σ × Z $M = \sigma \times \frac{gh^2}{6}$ Where P= 10.27 N L = 40 mm σ = 15 MPa h= 2.5g	M = 10.27×40 M = 410.8 Nmm $410.8 = 15\times \frac{gh^2}{6}$ gh^2= 164.32 mm^4 g^3 = 26.29 mm^3 *g* = 2.97 mm h = 2.5×2.97 h = 7.43 mm	*g* = 2.97 mm h = 7.43 mm
	ii) Calculation of *g* and *h* (Considering plastic failure at cross-section Y-Y)	M = P×L M = σ × Z $M = \sigma \times \frac{gh^2}{6}$ Where P= 10.27 N L = 50 mm (taking *f*=10) σ = 15 MPa h= 2.5g	M = 10.27×50 M = 513.5 Nmm $513.5 = 15\times \frac{gh^2}{6}$ gh^2= 205.4 mm^3 g^3 = 32.86 mm^3 *g* = 3.20 mm h = 2.5×2.86 h = 8.00 mm	*g* = 3.20 mm h = 8.0 mm The larger of the two values should be selected, therefore *g* = 3.20 mm and h= 8.00 mm
3	Calculation of *c* (Considering plastic failure between cross-section Y-Y and M-M)	2(g + h) × c = Area $Area = \frac{P}{\sigma}$ Where P= 10.27 N *g* = 3.20 mm *h* = 8.00 mm σ = 15 MPa	2(3.20 +8.00) × c = 0.68 *c* = 0.03 mm Practically this value is very small to make by FDM. So it can be taken as minimum 2 mm.	*c* = 2 mm
4	Calculation of *b* and *d* (Considering wax failure at cross-section M-M)	M = P × L M = σ × Z $M = \sigma \times \frac{BD^3-bd^3}{6D}$ Where P= 10.27 N L = 52 mm σ = 0.69 MPa B=10 mm D=25 mm	M = 10.27×52 M =534.04 Nmm $534.048 = 0.69\times \frac{10\times25^3-bd^3}{6\times25}$ bd^3 = 40154.35 mm^4 b^4 = 2569.87 mm^4 *b* = 7.12 mm *d* = 2.5×b *d* = 2.5×7.12 *d* = 17.80 mm	*b* = 7.12 mm *d* = 17.80 mm

The various calculated dimensions of the joint for the component are shown in Table 2. These dimensions can be directly used for making a joint for carrying any pattern up to size 300 cm^3 and for the components whose center of gravity is up to 40 mm from the gate. For making this joint, a T-shaped projected portion of plastic with dimensions shown in Table 2 has to be made with the plastic pattern. This projected portion will be embedded in the wax in the gate area to get the necessary joint.

Table 2. Designed values for various dimensions of the joint.

S. No.	Dimension (in Figure 3)	Design value (in mm)	Rounded value of dimension (in mm)
1	B	10	10.0
2	D	25	25.0
3	*g*	3.20	3.5
4	*h*	8.00	8.0
5	*c*	2.00	2.0
6	*f*	10.00	10.0
7	*b*	7.12	7.5
8	*d*	17.80	18.0

4. Case Study

In order to check the effectiveness of this method, castings of a component shown in Figure 4 were produced. This drawing is of a biomedical implant used for total hip replacement in human body. Many other types of customised biomedical implants can also be manufactured by using FDM patterns and IC process. Castings were prepared by using two different methods. In first method plastic patterns were attached to plastic sprues through plastic gates as shown in Figure 5 whereas in second method the plastic patterns are attached to a wax sprue with the help of proposed joint as shown in Figure 6. Two patterns were attached to one sprue in both the methods.

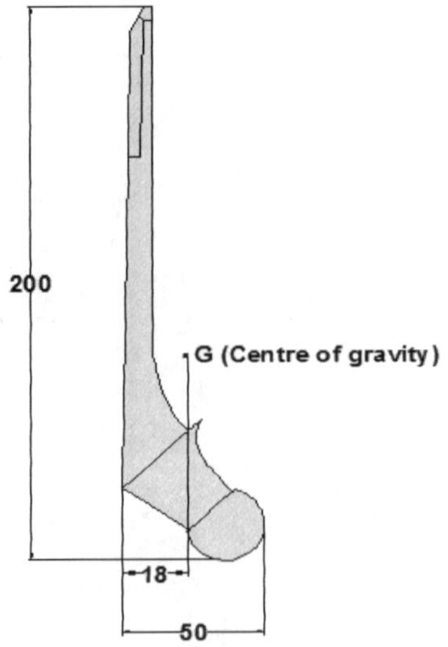

Figure 4. Component to be casted (All dimensions in mm).

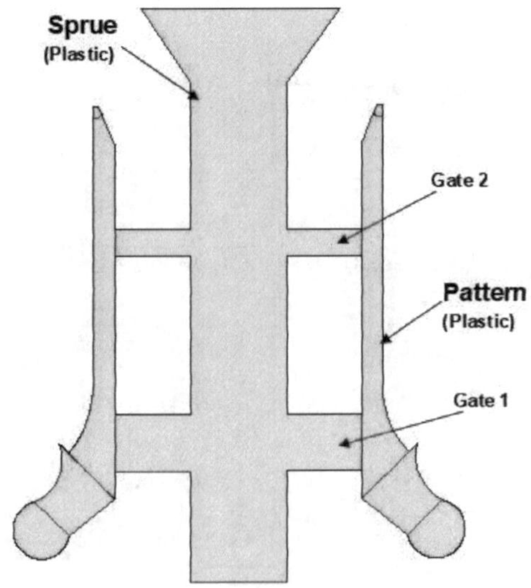

Figure 5. Schematic diagram of casting tree with method 1.

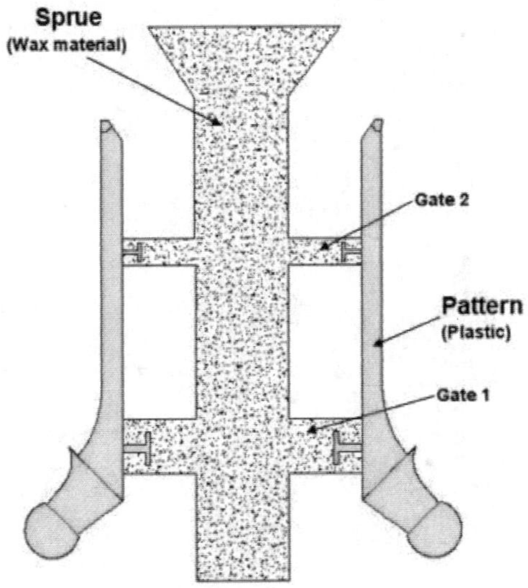

Figure 6. Schematic diagram of casting tree with method 2.

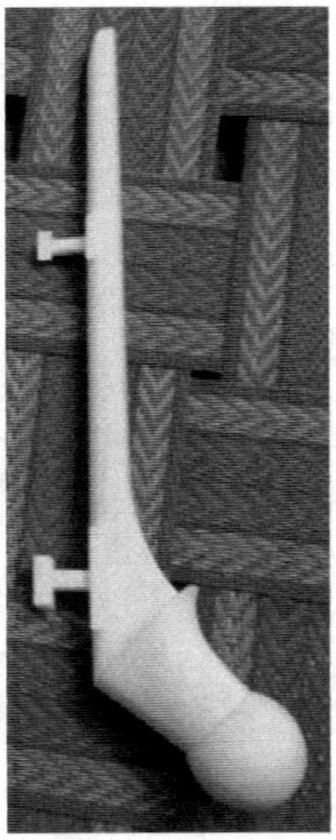

Figure 7. FDM pattern with extensions.

The design dimensions calculated in the previous section were used directly. The length of the part to be casted was 200 mm, therefore two gates were suggested by the foundry for its better stability and easy filling of the mold. However, the load of patterns and slurry could be carried by the main joint in gate 1 only. For ease, all the dimensions of the second gate were kept as half of the first gate. The size of both the sprues was 30 mm x 30 mm in cross-section.

A 3D CAD model of the selected component was prepared by using Solidworks software. An FDM technology based 3D printer (make: Stratasys uPrint SE) was used to prepare ABS patterns for the castings. The patterns were prepared hollow from inside and a wall thickness of 2 mm was used for the outer profile. The internal support filling of the pattern was kept at lowest possible density as per the machine setting available on the machine. The T-shaped projected portion on pattern required for making joint was made solid to give it strength. Total eight patterns were prepared and out of eight, four patterns were used for method 1 and other four were used for method 2.

In method 1, four patterns and two sprues were prepared by using ABS material on the FDM machine. The gates were prepared as an integral part of sprue. The patterns were easily attached with the gates by using ABS glue as per the arrangement shown in Figure 5. In method 2, two wax sprues were prepared by using the conventional method of foundry and four patterns were prepared by using ABS plastic on FDM machine. These patterns were produced with T- shapes plastic projections as shown in Figure 7.

A suitable arrangement was made to prepare the gates for embedding the projections of these FDM patterns into wax. A plastic sheet of thickness 0.5mm was taken and folded into U-shape as shown in Figure 8. The patterns were kept in the contour of U-shaped folded sheet as shown in Figure 9. Then this contour was filled with molten wax for embedding the plastic projections in wax for both the gates. After solidification of wax the gates were cut from the center to get two different FDM patterns attached with wax gates as shown in Figure 10. The wax in this figure appears dark in color as this is recycled runner wax taken from foundry. In this way the gates were prepared for all the four patterns used in method 2. These wax gates were locally heated and attached to wax sprues to prepare the casting trees as per the arrangement shown in Figure 6.

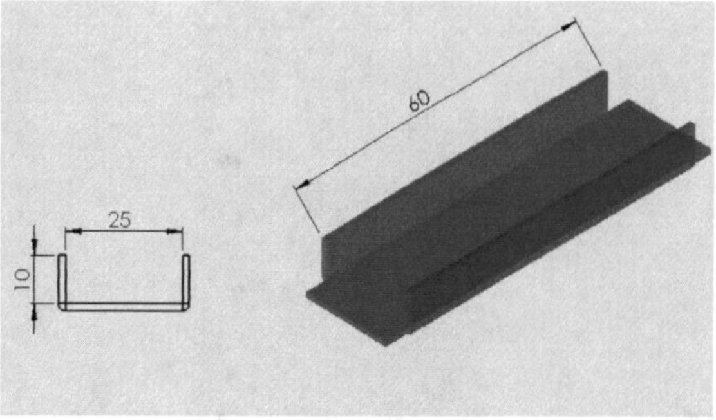

Figure 8. Drawing of folded plastic sheet for gate preparation.

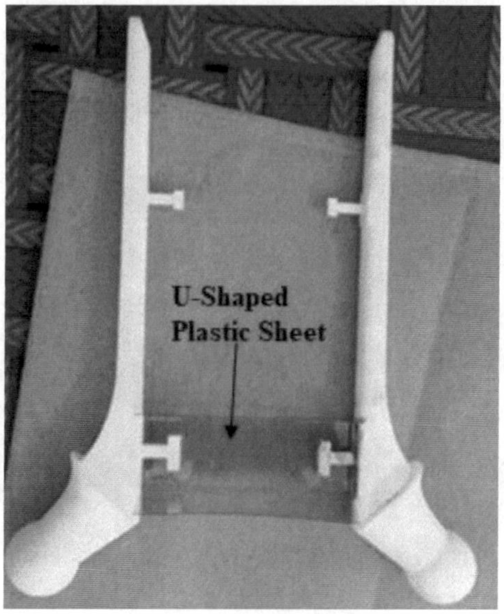

Figure 9. Gate preparation setup.

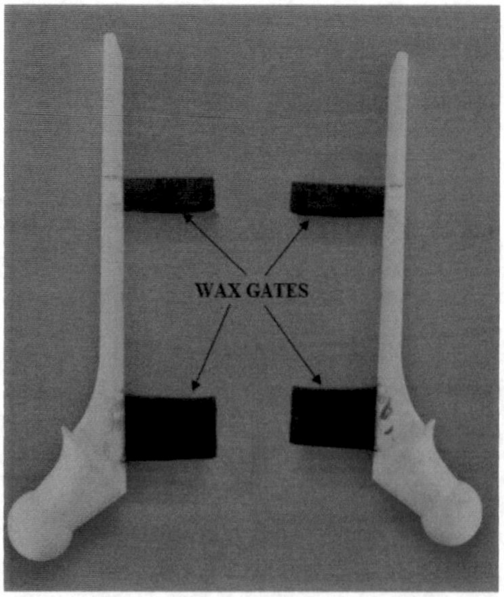

Figure 10. Patterns attached with wax gates.

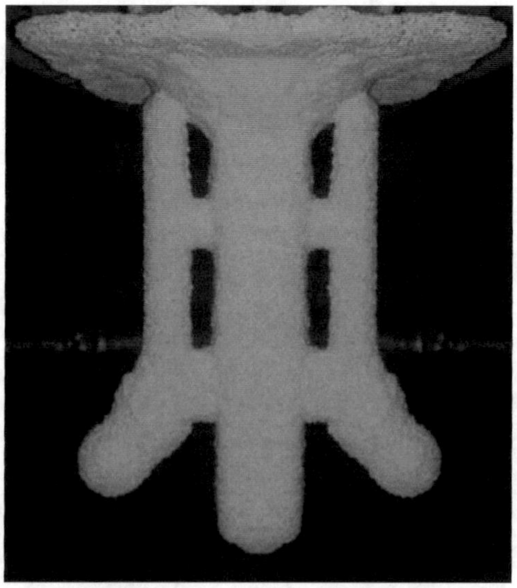

Figure 11. Dried ceramic shell.

The prepared IC trees were repeatedly dipped into the ceramic slurry to prepare a hard ceramic shell around the patterns. The ceramic shells were built through the successive stages of dipping and stuccoing. After each coat the shells were dried in air for minimum 4 hours at room temperature of 30°C. This procedure was repeated till eight layers of slurry material were coated on a tree. A casting tree after deposition of slurry layers is shown in Figure 11.

All the four casting trees prepared with both the methods were put in a furnace for dewaxing/burning out of the ABS patterns and to get the cavities. The ABS patterns become softer at much lower temperature but for complete burning of the ABS patterns, a temperature of 1020°C is required [6].

However, the temperature of furnace was increased up to 1120°C for complete removal of the pattern and for sintering of the shell. During the heating process all the shells made by using method 1 were broken beyond repair and could not be used for pouring the molten metal. Whereas all the shells prepared by using method 2 sustained the heating process and the cavities were obtained. The observations of shell cracking have been shown in Table 3. The obtained cavities were used for pouring the molten stainless steel 316L to get the castings as shown in Figure 12.

Table 3. Results of shell cracking.

S. No.	Method 1 (Plastic sprue)	Method 2(Wax sprue)
1	Shell cracked. No casting obtained.	No crack appeared and casting obtained
2	Shell cracked. No casting obtained.	No crack appeared and casting obtained
3	Shell cracked. No casting obtained.	No crack appeared and casting obtained
4	Shell cracked. No casting obtained.	No crack appeared and casting obtained

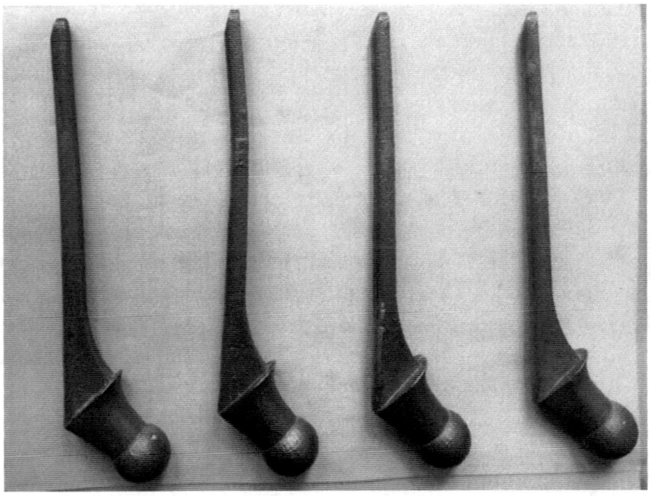

Figure 12. Castings obtained with method 2.

Table 3 shows that all of the four shells having plastic sprue were cracked during the burnout process. It was observed that these ceramic shells cracked before reaching temperature of 200°C. It is interesting to note that the coefficient of thermal expansion of wax is higher than ABS. Even then the casting trees with more plastic content were broken and the trees having more wax content sustained. To understand its reason, we have to know about the behaviour of ABS during temperature increase. The ABS is an amorphous polymer and it does not have any specific melting point. The melting point of such material is represented as glass transition temperature (*Tg*). The Tg is a temperature range in which the polymer structure turns to a viscous liquid. The *Tg* for ABS plastic is around 108°C and on heating, its specific volume increases relatively linearly below *Tg*, but around *Tg* the thermal expansion coefficient increases about three times [27]. Therefore, it may be believed that the ABS material creates maximum stress in the ceramic shell mainly around the *Tg* and after this temperature it becomes soft and the value of stress on the wall of the shell decreases.

In this study, both types of casting trees were loaded at room temperature in a furnace and after loading, the temperature was increased gradually up to 1020°C. In method 1, the expanded volume of ABS at *Tg* was much higher than the expansion of mold material, therefore all of the shells cracked. Whereas in method 2, all of the four shells sustained during the burnout process. It may be due to the fact that the melting point of wax is 65°C which is much lower than the *Tg* of ABS. The wax gates and sprue melted at lower temperature. Some molten wax is flowed out of the shell and some was absorbed by the dried ceramic shell. This created some space for the expansion of pattern material at a higher temperature which resulted in decrease in stress on the shell. However, the stress in the shell could not be eliminated completely and deformation of one shell was observed as shown by a slight bend in piece number 2 in Figure 12.

It has been observed that the dewaxing is an important step and during dewaxing, the ceramic shell may fail even without any plastic part in it. In conventional IC process, when every part of casting tree is of wax, the dewaxing is done in a closed chamber by using steam at a temperature of 180°C-220°C and at 8 bar (0.8 MPa) pressure [28]. Actually, the green ceramic shell is weak and it cannot bear the stress due to thermal expansion of the wax. The high temperature steam quickly enters the pores of the ceramic shell and outer layer of the wax patterns melts quickly. This molten wax is absorbed by the ceramic shell, thus leaving some space for further expansion of the wax pattern. When ABS patterns attached with the wax sprue are tried to be dewaxed at 180°C with conventional method, the ABS suddenly expands and the shell breaks. Therefore, if ABS patterns are to be used then the dewaxing method has to be modified and the temperature of the dewaxing unit should be increased slowly. Because initially at low temperatures there is thermal expansion of both wax and ABS. Therefore, if plastic patterns are used along with wax sprues then the step of steam dewaxing should be skipped and the casting trees should be placed in the furnace that can reach up to a temperature of 1120°C. This high temperature serves dual purpose as it removes the ABS pattern as well as strengthen the ceramic shell by sintering it. After loading the furnace, the temperature should be increased gradually. Enough time should be given to melt the wax at lowest possible temperature to leave space for expansion of ABS pattern. If the shell remains unbroken up to 200°C, there are very less chances of their cracking.

In this way, the developed joint was successful to hold the patterns throughout the process. Therefore, the problem of shell cracking can be successfully eliminated by using the wax sprues instead of ABS sprues with appropriate joints. It would lead to saving in cost and time as the material cost and manufacturing time for making plastic sprues would be saved. However, it is a matter of further study that whether this method is successful for large and complex castings or not. For complex and large castings, a combination of two or three different methods to prevent the shell cracking may be used.

5. Conclusions

This study represents a new approach for eliminating the problem of ceramic shell cracking in FDM pattern based IC. In this study, a joint has been designed to attach ABS plastic patterns with wax sprue for making castings. For making this joint, a T-shaped projected portion of plastic has been proposed as an integral part of the plastic pattern. This projected portion is embedded in the wax in the gate area to get the necessary joint. In order to check the success of the design, castings of a biomedical implant have been prepared by using both old as well as proposed methods. The designed joint was capable to carry the total load of pattern and slurry material. All the ceramic shells containing plastic sprue cracked during the burnout process but no shell cracking was observed with the proposed method. The proposed method can also reduce the cycle time and the cost as well.

References

[1] Rosochowski A and Mutusuzak A 2000 Rapid tooling: the state of the art. *Journal of Materials Processing Technology* **106** 191-8

[2] Pham DT and Gault RS 1998 A comparison of rapid prototyping technologies. *International Journal of Machine Tools and Manufacture* **38** 1257-87

[3] Lee CW, Chua CK, Cheah CM, Tan LH and Feng C 2004 Rapid investment casting: direct and indirect approaches via fused deposition modeling. *International Journal of Advanced Manufacturing and Technology* **23** 93-101

[4] Chua CK, Feng C, Lee CW and Ang GQ 2005 Rapid investment casting: direct and indirect approaches via model maker II. *The International Journal of Advanced Manufacturing Technology* **25** 26-32

[5] Kumar P, Ahuja IPS and Singh R (2012) Application of fusion deposition modeling for rapid investment casting – a review. *International Journal of Materials Engineering Innovation* **3** 204-26

[6] Galantucci LM, Lavecchia F and Percoco G 2010 Experimental study aiming to enhance the surface finish of fused deposition modeling parts. *CIRP Annals Manufacturing Technology* **58** 189-92

[7] Boschetto A and Bottini L 2015 Surface improvement of fused deposition modeling parts by barrel finishing. *Rapid Prototyping Journal* **21** 686-96

[8] Kumar P, Ahuja IPS and Singh R 2016 Effect of process parameters on surface roughness of hybrid investment casting. *Progress in Additive Manufacturing* **1** 45–53

[9] Sood AK, Ohdar RK and Mahapatra SS 2009 Improving dimensional accuracy of fused deposition modelling processed part using grey Taguchi method. *Materials and Design* **30** 4243–52

[10] Noriega A, Blanco D, Alvarez BJ and Garcia A 2013 Dimensional accuracy improvement of FDM square cross-section parts using artificial neural networks and an optimization algorithm. *International Journal of Manufacturing Technology* **69** 2301–13

[11] Guo H and Zhou YT 2014 Research on the impacting factors of dimensional accuracy with silicone mold casting products based on FDM. *Advanced Material Research* **912–914** 309-13

[12] Kumar P, Ahuja IPS and Singh R 2015 Investigations on dimensional accuracy of the components prepared by hybrid investment casting. *Journal of Manufacturing Processes* **20** 525-33

[13] Omar MFM, Sharif S, Ibrahim M, Hehsan H, Busari MNM and Hafsa MN 2012 Evaluation of direct rapid prototyping patterns for investment casting. *Advanced Material Research* **463-464** 226-33

[14] Harun WSW, Sharif S, Idris MH and Kadirgama K 2009 Characteristic studies of collapsibility of ABS patterns produced from FDM for investment casting. *Material Research Innovation* **13** 340-43.

[15] Chen X, Li D, Wu H, Tang Y and Zhao L 2011 Analysis of ceramic shell cracking in stereo-lithography based rapid casting of turbine blade. *International Journal of Advanced Manufacturing and Techn*ology **55** 447-55

[16] Sivadasan M, Singh NK and Sood AK 2012 Investment casting -Rapid tooling approach. *International Journal of Mechanical and Industrial Engineering* **1** 99-103

[17] Marwah OMF, Sharif S, Ibrahim M, Shukri MS and Mohamad EJ 2016 An experiment analysis on strain and stress for ABS pattern developed by portable FDM machine used in direct investment casting. *ARPN Journal of Engineering and Applied Sciences* **11** 6496-501

[18] Yao WL and Leu MC 1999 Analysis of shell cracking in investment casting with laser stereolithography in investment casting patterns. *Rapid Prototyping Journal* **5** 12-20

[19] Bansode SN, Phalle VM and Mantha SS 2020 Influence of slurry composition on mould properties and shrinkage of investment casting. *The Indian Institute of Metals* **73** 763-73

[20] Huang P, Lu G, Yan Q and Mao P 2019 Effect of ceramic and nylon fiber content on composite silica sol slurry properties and bending strength of investment casting shell. *Materials* **12** 2788

[21] Lü K, Liu X, Lu Y and Du Z 2017 The interfacial characteristics and action mechanism of fibre-reinforced shell for investment casting. *The International Journal of Advanced Manufacturing Technology* **93** 2895-902

[22] Li Z, Liu X and Lv K 2020 Study on the strength of short carbon fiber-reinforced silicon sol shells for the investment casting process. *International Journal of Metalcasting* **14** 432-41

[23] Stratasys Datasheet for ABSplus, retrieved from https://www.stratasys.co.in/materials/search/absplus [accessed 04 April 2020]

[24] Jebaraj PM and Reddy AC 1998 Prediction of thermal shock of ceramic shells using fused silica as reinforcing filler at casting conditions. *National Conf. on Advances in Production Technology*, Feb. 7-9, (Bangalore, India) p 52-6

[25] Desain JD, Brady BB, Metzler KM, Curtiss TJ and Albright TV 2009 Tensile tests of paraffin wax for hybrid rocket fuel gains. 45th AIAA/ASME/SAE/ASEE Joint Propulsion Conference and Exhibit; Aug 2-5 (AIAA, Denver, Colorado, U.S.A.) p 2009–5115

[26] Samykano M, Selvamani SK, Kadirgama K, Ngui WK, Kanagaraj G and Sudhakar K 2019 Mechanical property of FDM printed ABS: influence of printing parameters *The International Journal of Advanced Manufacturing Technology* **102** 2779-96

[27] Hale A 2002 Thermosets. *Handbook of Thermal Analysis and Calorimetry, vol.3* (Elsevier) p 317

[28] Jones S, Jolly MR, Blackburn S, Gebelin JC, Cendrowicz A and Lewis K 2003 Effect of moisture upon mechanical properties of ceramic moulds during high pressure steam dewaxing *Material Science and Technology* **19** 907-14

Construction automation

Akula Prakash[1], Kandlagunta Mounika[2], Anand Goud K[3] and Vittalaiah A[4]

[1,4]Assistant Professor, Department of Civil Engineering, GRIET, Hyderabad, India.
[2] Assistant Professor, Department of Civil Engineering, CMRIT, Hyderabad, India.
[3] Assistant Professor, Department of Civil Engineering, IARE, Hyderabad, India.

Corresponding Author: akulprakash93@gmail.com

Abstract. Construction industry is creating a lot of enthusiasm for the construction network throughout the most recent couple of decades. This incorporates with skilled work power, adequate cash, and time when it is finished by ordinarily. As these necessities are not constantly accessible in all circumstances, the industry is confronting issues like consistently declining profitability, low work productivity in construction, a high mishap rate, low quality, deficient control of the building site, and the evaporating of the talented work power. To conquer the previously mentioned issues the terms ROBOTICS and AUTOMATION presented. The robots utilized in the construction have followed a similar idea as those utilized in assembling. Building robots have been utilized in different errands including material taking care of, different inside and outside completing assignments and quality control. Be that as it may, this use of mechanical autonomy in the construction business is not completely embraced. The purposes behind this circumstance are distinguished as existing robots are not all around embraced to building construction, issues related with the routinely planned structure, hard to legitimize robot business financially and administrative hindrances. The construction business requests a profitability increment that cannot be achieved by methods for customary techniques that have just arrived at their framework confines and cannot add to advance viability. Further justification must be accomplished by the decrease of work and construction times. Maybe if no critical research advances, temporary workers will have the option to tackle their issues by bringing in mechanical technology and procedure-controlled machines from abroad. Consequently, there is a need to sustain the advancement of effective innovative work programs in construction mechanization and apply autonomy. This paper makes an endeavour to utilize robots in the construction business, recorded to diminish risks at work destinations, improve quality, and decrease the expense of construction. A contextual investigation has been considered to show the appropriateness of robots in the construction business.

1. Introduction

Directly from the presence of the humankind, the fundamental needs – nourishment, apparel, and sanctuary lead to progress. To fulfil these fundamental needs, construction of building, advancement of transportation, water system ventures and so forth., has occurred in the construction field as a piece of human advancement. The improvement in construction field utilizing innovative advancement made ready to construction industry. "Yesterday's innovation isn't the specific answer for tomorrow's concern". Along these lines, in the construction business mechanization through mechanical technology is presented as an answer. Improvement of automated frameworks for construction applications has progressed significantly in recent years

Construction mechanization and mechanical autonomy have been producing a lot of enthusiasm for the construction network during the most recent two decades. The way toward dispersing the early outcomes from innovative work of construction robotization and mechanical autonomy into industry practice is presently gradually occurring and the quantity of automated models have been planned and assembled.

The construction industry requires a different kind of robots. Typically, in the manufacturing field, robots are stationary, and the product moves along the assembly line. Automation is easier to incorporate because each product is identical with repetitive tasks done over and over. However, construction robots must move about the site because buildings are stationary and of a large size. They require engines, batteries, or motors and drive themselves. Construction robots are also faced with the changing sites and must be reprogrammed with each new condition. Therefore, it requires digital control with manipulators using coordinate systems to direct three-dimensional motion. They also must be able to function under adverse weather conditions including variations in humidity and temperature. Additionally, they are constantly exposed to dust and dirt on the site

Goal of Automation

Reports of the U.N. are demonstrating a deficiency of 600 million dwelling units all through the world. Remorsefully, there is no proof that the world has this limit at present, since assessments of lack continue expanding, India produces 2 million homes according to a main Indian financial expert, while United Nations reports that India produces 3.3 million homes for every year, which is far not exactly the prerequisite in India's tenth plan. Initially, we must find or design a framework that can create houses at the ideal speed. Next, we must deliver them at reasonable expenses on the off chance that we are to address the average workers and others.

Objective of Automation

- Identifying the workplaces where the robots can be actualized in the building site, to bring mindfulness among the individuals about computerization and mechanical autonomy and to demonstrate the recipients through a contextual analysis.
- Identify the parameters relying upon the idea of building including the idea of the undertakings to be performed.
- Calculate the expense of robot utilizing parameters, and
- Develop the mechanical framework for full scale experimentation for practical appraisal of efficiency.

Competitive Advantages

As found in different enterprises, robots ordinarily work quicker than people do. Many accept that robots are additionally increasingly trustworthy. With machines, there is less worry about fluctuating efficiency due to the high points and low points of life. Since these robots are viewed as quicker than people at work tasks, temporary workers ought to hope to finish extends sooner. At the point when construction work is done in front of calendar, manufacturers set aside cash because of decreased place of work overhead. Also, they have a chance to offer and work different tasks sooner than anticipated. Also, building proprietors are satisfied to increase early inhabitants to structures and a snappier profit for their speculation. As noted, safety is a significant perspective to any construction ventures. Be that as it may, the expense of security is noteworthy. Putting resources into fall assurance, individual defensive gear, and safety the board can be exorbitant. A portion of these expenses can be diminished if robots are used instead of people.

Inherent Barriers

Construction robots can dispense with laborers from being presented to risky, filthy, and overwhelming work. In any case, they set aside extensive effort for set-up, altering and tidy up. Skilled staff are expected to work and consistently screen the robots. As noted, single errand robots are constrained since they are not equipped for distinguishing and fixing issues continuously. Accordingly, skilled specialists must be available during work activities.

In addition, robot administrators need preparing. Unique preparing in PC use will be compulsory. Laborers will require a more grounded foundation in scholastic territories, for example, perusing and issues tackling to be beneficial. Indeed, even with robots, temporary workers should in any case put vigorously in preparing and instructing the specialist. The principal goal of this postulation is to acquaint a pathway with automated application in construction industry. A contextual investigation has been considered to show the relevance of robots for a specific exercise and the resultant points of interest are talked about.

Parameters Deciding the Expense:

The significant parameters deciding the expense of the robotized work can be in this way isolated into three principal gatherings.

- Parameters reliant on the mechanical framework incorporate its cost, its work envelope (and afterward its secured territory), its speed of development and its method of activity (degree of human control).
- Parameters reliant on the idea of the structure site incorporate the idea of the undertakings to be played out, their quality, the quantity of moves between work areas and the areas/circulation of work in each segment.
- Parameters reliant on the assignment to be performed incorporate the yield every hour, the materials and the assistant works required.

The expense per every hour is recognized utilizing the accompanying

$$C = \frac{\{P * pr\,(i,n)\} + C_m}{H} + C_o$$

Where

C = Cost per hour

P = Investment on the robot (including cost of carriage, effectors, sensors etc.,)

pr (i ,n) = Capital recovery factor (depreciation and interest factor, assuming annual interest (I) and economic life (n) in years.

C_m= Cost of repairs and high-level maintenance of the robot per year

C_o = Operating costs (including some wear effected parts) per hour

H = Number of robot employment hours / year

2. Research Methodology

The strategy for the use of robots comprises of following advances:

- Identify the practicality for utilizing the robots.
- Identify the different innovative and authoritative issue of execution.
- Select the robot setup. (for example, reach, pay load limit, detecting characteristics and control framework for each errand)
- Identify the adjustment legs, beginning work point and the area of work material.
- Develop the pre-modified way to work the robot by remote control.
- Equip sensor gadgets and control instruments to interface with nature.
- Identify the spots where robots can recreate the components.
- Identify the parameters relying upon the idea of building including the idea of the errands to be performed, amount, number of moves between workstations and area of work in each segment.
- Identify the parameters relying upon the assignment to be performed including yield/hour, materials, and the helper work.
- Calculate the expense of the robot utilizing parameters including devaluation, enthusiasm on venture, upkeep and working costs.
- Analyze the expense every hour 'C' for robot.
- Develop the mechanical framework for full scale experimentation for sensible appraisal of efficiency.

The above technique is shown utilizing the accompanying contextual investigation

3. Case Study and Results

A Residential Building in Kukatpally, Hyderabad, Telangana, India. has been considered for the application of robotics in building construction. In this case study the robotics is applied only for plastering. The study has been done by comparing the cost and time of the work done (plastering) by conventional method with the robot projected values of cost and time. the building has the following data regarding plastering.

For walls:

Plastering 20mm thick in CM (1:6 proportion) and CM (1:4 proportion) for 1st and 2nd coat respectively with dubara sponge finishing including making 12mm grooves on either side of the column and beams at the function of panel walls including cost and conveyance of all materials to side curing, scaffolding charges, and all incidental and operational charges etc., complete.

For ceiling:

Ceiling plastering 12mm thick in CM (1:5 proportions) and CM (1:3 proportions) for 1st and 2nd coat respectively with dubara sponge finishing including cost and conveyance of all materials to site, curing, scaffolding charges, and all incidental and operational charges etc., complete.

Details of plastering manually (conventionally):

For walls:

- The total estimated area of the plastering for the walls **7246.32 Sq. m.**
- The above-mentioned plastering for walls was done by **30 Masons** and **56 M.M & W.M** in **56days**
- The total cost for the plastering walls manually including all the charges are **Rs. 6,28,000**

For ceiling:

- The total estimated area of the plastering for the ceiling 2473.09 Sq. m.
- The above-mentioned plastering for ceiling was done by 30 Masons and 43 M.M & W.M in 21days
- The total cost for the plastering ceiling manually including all the charges are Rs. 1,85,000

Calculations:

The total cost for the plastering both walls and ceilings
= Cost for plastering of walls + cost for plastering of ceilings
= Rs. 6,28,000 + Rs. 1,85,000
= **Rs. 8,13,000**

The total time for the plastering both walls and ceilings
= 56 + 21 = **77 days**

Cost and Time estimation of robotic systems for plastering works:

Time Estimation:

The total time for plastering the whole building can be calculated as follows

- Plastering capacity of a robot = 45 Sq. m per hour
- Total area to be plastered = 7246.32 + 2473.09 = **9719.41 Sq. m.**

$$\text{Time to be taken} = \frac{\text{Total Area}}{\text{Plastering Capacity}} = \frac{9719.41}{45} = 215.986 \text{ hrs}$$

= 8.999 days = **9 days** (approx.)

Cost estimation:

- Plastering cost of a robot = **Rs. 1450 per hour**
- Total time for plastering job = **216 hours**

- The cost of plastering for whole building = 1450 * 216 = **Rs 3,13,200**

Type of system	Plastering area (Sq. m.)	Cost (Rs)	Time (Days)
Conventional System	9719.41	8,13,000	77
Robotic systems	9719.41	3,13,200	9
Savings		***499800***	***68***

- Percentage of Costs saved = $\frac{savings\ costs}{actual\ costs} * 100 = \frac{499800}{813000} * 100 = \mathbf{61.476 = 62\%}$ (Approx.)
- Percentage of Time saved = $\frac{savings\ time}{actual\ time} * 100 = \frac{68}{77} * 100 = \mathbf{88.311 = 88\ \%}$ (Approx.)
- From the above case study values displayed in the table, we can infer the following points.
  - ✓ Costs can be saved up to **62 percentage**
  - ✓ Time can be saved up to **88 percentage**

4. Conclusions

The accompanying ends are made dependent on the theory work.

- The contextual analysis uncovers that a measure of Rs. 4,99,800 can be made as sparing sum properly utilizing the robot for putting work, in correlation with the ordinary work i.e., 62%
- The timeframe likewise can be chopped down from 77 days to 9 days, therefore contributing or significant sparing in labour costs without any mishaps or wounds to the work power.
- The Construction industry requests an efficiency increment by methods for computerization and advanced mechanics
- Robot, Safety, strength, and subjective components ought to be given need in the visionary thoughts.
- Architects and Engineers need to advocate more normalized plans that consider simpler computerization.
- In request for Construction computerization with robots to be completely operational, Designers must attempt to improve extends however much as could reasonably be expected identifying with repetitiveness and normalization of construction components

Hindrances:

- We have a few issues distinguished prompting the low business utilization of Construction Robots:
- Many improvements are made for explicit employments locales and are not appropriate for other people.
- No further advancement of beginning frameworks is embraced because of absence of input from various building destinations
- Insufficient advancement of development mechanical models
- Insufficient regard for building plan to the imperatives of robotized development.
- Insufficient monetary avocation for mechanical autonomy in building.
- Difficult administrative condition.

The reasons distinguished for the failure of robots are

- Lack of suitable structure locales
- Need for additional improvement/adjustments
- Lack of monetary avocation.
- Lack of enthusiasm for organization's administration.
- Availability of better hardware for same capacity.

- Problems with material stock.

The purposes behind the current circumstances are identified as follows:

- Existing robots are not very much evolved to building development
- There are issues related with the ordinarily planned structures.
- It is hard to legitimize robot work financially

Future Scope

Further research should be encouraged to develop the prototypes suiting to the requirements of the construction industry. A certain innovative leap is required by a systems approach that combines the existing construction technologies with new information, automaton, and robot technologies. Research needs to see as an investment not as an expenditure in the industry.

5. References

[1] D. Castro-Lacouture Construction automation Springer Handbook of Automation, Part G, Springer (2009), pp. 1063-1078
[2] P. Keerthanaa, K. Jeevitha, V. Navina, G. Indira, S. Jayamani (2013), Automatic wall painting robot, IJIRSET vol 2, issue 7, ISSN: 2319-8753
[3] Sorour Md. A. Abdellatif (2011), Development of Roller-Based Interior Wall Painting Robot Int Journal of Mechanical and Mechatronics Engg. Vol 5, no 11
[4] Journal of Automation in Construction – Special Issue :The best of ISARC 2002, Vol 13, Issue 1, January 2004, pp 5-19.
[5] Robotics and Automation in Construction, Published online 01, October 2008 ISBN 978-953-7619-13-8
[6] C. Balaguer, M. Abderrahim, S. Boudjabeur, P. Aromaa, K. Kahkonen, S. Slavenburg, D. Seward, T.Bock, R. Wing, & B. Atkin, Future Home: An Integrated Construction Automation Approach, IEEE Robotics & Automation Magazine (2002) 55-66.
[7] Balaguer, C. (2003). EU Future Home project results (keynote paper), 20th International Symposium on Robotics and Automation in Construction (ISARC'03), The Netherlands, September 2003, Eindhoven
[8] M. De Grassi, B. Naticchia, A. Giretti An automatic four color spraying device carried by a robot," ISARC, December 17-19, Sendai, Japan, 2007.
[9] B. Naticchia, A. Giretti, A. Carbonari: "Set up of a robotized system for interior wall painting," Proceedings of the 23 Tokyo, Japan, 2006. Journal of Construction 132-151, 2004
[10] Crawford, F.S. (1988) "Culvert Whistlers Revisited", Journal of Physics. 56(8), 752-754
[11] Amarjit Singh. (2004) "High Quality, Low Cost Architectural Flexible and Quick Turnaround Mass Housing for the World's Billion". Proc of International Conference on Advances in Concrete and Construction ICACC- 2004 Hyderabad. India pp.851-868.
[12] Dereck Seward and Khaled Zied (2004), "Graphical Programming and Development of Construction Robots", Journal of Construction Engineering and Management, ASCE. 65.
[13] V. S. S Kumar and Balaji Narsimhulu (1999), "Robotics in Construction Industry" . 1-3.
[14] Everret, J.G (1994), "Automation and Robotics Opportunities Construction Versus Manufacture", Journal of Construction Engineering and Management, ASCE. 120(2), 443- 452.
[15] Farid, F. (1993), Journal of Construction Engineering and Management, ASCE. 119(2), 193-195.
[16] Krom (1997), "Industrialization and Robotics in Construction", Journal of Construction Engineering and Management, ASCE. 111(3), 260-280
[17] Taylor, M., Wamuziri, S., Smith, I. Automated construction in Japan. Proceedings of the Institution of Civil Engineers-Civil Engineering, 156(1): 34-41, 2003.
[18] Kim, T., Lee, U.-K., Yoo, W.S., An, S.-H., Cho, H., Lee, Y., Doh, N. Benefit/cost analysis of a robot-based construction automation system. In International Conference on Control, Automation and Systems, Gyeonggi-do, Korea, pages 616-621, 2010.

Evaluation of optimum dosage in blended self curing concrete on the impact of supplementary cementitious materials (scm's) and durability studies on different proportions

Vivek Kumar C[1] Mohammed Aiman Parvez[2], Dr.T.Srinivas[3]

[1]Assistant Professor, Civil Engineering Department, GRIET, Hyderabad, India. Corresponding author : vivekkumarcr@gmail.com
[2]M Tech Student Scholar, Structural Engineering, GRIET, Hyderabad, India.
[3]Professor, Civil Engineering Department, GRIET, Hyderabad, India.

Corresponding Author: vivekkumarcr@gmail.com

Abstract. Now a days, the shortage of potable water increases, for domestic and commercial purposes due to environmental conditions. Due to this scarcity, curing of concrete is affected and in this point of view water assets are more valuable every day and becomes more economical. Traditional Curing used for retaining moisture within the concrete elements during the early days and beyond to develop the desired properties in terms of strength & durability. To develop the modern era curing, a few of self-curing agents like Super Absorbent Polymers, Polyethylene Glycol (PEG 400) has been used for internal curing of concrete used for reducing the hydration process. In this experimental study, Supplementary Cementitious Materials like Class F Flyash, Silica Fume and Ground Granulated Blast Furnace Slag (GGBFS), have been casted for 65%+25%+15% (M2), 65%+20%+15% (M6), 65%+25%+10% (M10) and 65%+20%+15% (M11) of optimal mix proportions. The sorptivity assessments have being carried out on all blended mixes of Self Cured Concrete with size of 15 cm x 15cm x15cm. The pH value of the acidic medium was at 0.45 by submerging in acid, the specimen samples are brought out and were cleaned in repeatedly with water and kept in atmospheric condition for two days for constant weight and then the specimens are balanced using weighing scale and loss in weight and hence the percentage in loss of weight was premeditated within and after 28 days with different mix proportion of M30 grade of concrete supplementary cementitious materials like Flyash, Ground Granulated Blast Furnace Slag and Silica Fume with PEG 400 as 1% by its weight.

1. Introduction

Plain Cement concrete (PCC) is very weak in tension comprises several micro cracks which propagates within the concrete matrix underneath the constantly applied load. Subsequently, PCC members cannot endure tensile stresses created due to the applied force without the addition of reinforcing elements which are capable to resist these stresses. The addition of Supplementary Cementitious Materials (SCM's) like Flyash, Ground Granulated furnace Slag and Silica Fume to the structural concrete enhances its stiffness, ductility, and load bearing capacity, while at the identical time reduced crack enhancement and propagation. According to the composite material theory, positive synergy of distinct admixtures may be supplementing with each other to make a new composite material with high structural implementation and excellent economic benefits.

Ternary blended concrete demonstrates the adjunct of numerous pozzolanic materials to the concrete with cement behaving as the most important binding material. Fly ash from power plants and metakaolin have being the both essential in modern day concrete practice. Widening the extent of SCM's such as Fly ash, Silica fume, GGBFS, Rice Husk Ash and Metakaolin in the consumption of concrete making, leads to the concept of blended cements and concretes. Thus, due to the influence for the point of early hydration was decline in generation of excess $Ca(OH)_2$, control on heat of hydration and mitigation of uninterrupted bleed channels. By reorientating the mix design parameters for superior structural performance of concrete with thrust on controlling OPC content but increasing the whole cementitious material and controlling water content, consuming chemical admixture, for amended workability, for offsetting deliberate hydration with SCM's.

The response of Fly ash versus Calcium hydroxide has been released during the cement hydration method which have being affected by that oxidation procedure, mineralogical and structure level as well as grain size distribution. Several amendments remained provided the information regarding the chemical activation of the Fly ash versus $Ca(OH)_2$ reaction are also called pozzolanic reaction, in order to consent the escalation of the fly ash content in the blended Portland cements exclusive of a major negative impact on the early strength in the effort to create clinker free binders. Concrete incorporating self-curing agents will imply a new set of drift in the construction in the new millennium, due to the intensified use of performance built concrete. Newer procedures may be used for the inclusion of internal or self-curing of water in concrete. Most of the researchers proposed to the use of soaked aggregates to deliver internal curing for concrete and polyethylene glycol PEG 400 in concrete mixes as self-curing agent.

2. Research Significance

The primary objective of this section is to assess the outcomes of distinct kinds and percentage of self-curing agents on the mechanical properties of concrete such as Compressive and Flexural strength of reinforced concrete beam with M30 grade mix. According the Compressive strength of concrete of M30 Grade with water-binder ratio as 0.40 and mix proportion of 1 : 2.12 : 3.75 with addition of self curing agents PEG as 1% .

The self-curing agent Polyethylene Glycol 400 which utilized in this paper have been added as constant, where the proportioning of blending proportions have been made with the results have been concluded with the compressive strength and split tensile results. The optimal self-cured concrete is have been derived as 65%+25%+15% (M2), 65%+20%+15% (M6), 65%+25%+10% (M10) and 65%+20%+15% (M11) and with the same combination of blending proportioning is made for performance of blended concrete, which was testing under loading. Also, these results offer more knowledge about the determination of self-curing agent ratios and the greatest category to optimize the mechanical properties of self cured concrete built upon their performance of structural behaviour under loading conditions.

In this investigation, an attempt has been made to investigate the performance of strength parameters of casted control samples with incorporation of Polyethylene Glycol, PEG 400 as 1.0% in Blended Self Curing Concrete and controlled self-curing concrete with M30 Grade concrete mix. Ordinary Portland Cement, Fine and Coarse Aggregate, Potable Water, Class F Fly Ash, Ground Granulated Blast Furnace Slag, Silica Fume, Poly-Carboxylate Super Plasticizer and PEG Plasticizer were the materials used in the recent findings.

3. Materials and Proportioning of M30 Blended Mix

Ordinary Portland cement, Class F Flyash, GGBFS and silica fume, commonly available in India for modern-day concrete, silicious river sand with a fine aggregate fineness modulus of 2.73 and a nominal maximum size of 20 mm, were used to take this study. The coarse aggregate consumed cooperates with the Indian Standard IS 383-1970 code and is evaluated for its physical characteristics, its specific gravity 2.6 as an angular coarse aggregate, in conjunction with the Indian Standard IS 2386-1963.Bulk density with free state as 1455 kg/m3 and compressed state as 1529 kg/m3. Potable water have been used in the investigations for both mixing of SCM specimens.

Class F Flyash were acquired with a specific gravity of 2.172 from Mettur Thermal Power Plant. Though with the chemical configuration consisting of 63.98 % silica content, 1.721 % calcium oxide, 1.0 % magnesium oxide, 10 % pH value and 2.12 % ignition loss. GGBFS is more importantly resistant to the ingress of ions found in chloride and within concrete away with each other in the concrete mixtures, associated with decreased permeability.

Ground Granulated Blast Furnace Slag have been obtained in Jindal South West (JSW)Steel Ltd, Karnataka mainly having Silica content of 35.25%, Alumina as 19.2 % and Calcium Oxide content as 34.90 % was found. Consequently, the silica fume used has a 92.5 % SiO_2 content and was in fine and powdered form. In the present study, low - molecular weight polyethylene glycols (PEGs) (400) have been used, whereas these chemical compounds were systematically mixed with water in PEG as a curing agent subsequent to the addition or mixing of water into the concrete.

As super plasticizer (polycarboxylate-type), a new-generation high-variety water mitigating chemical admixture corroborating to American Society for Testing and Materials (ASTM) C494 has been used to improve the workability of self-curing blended mix with reduced water-cement ratio.

Several number of dosages of Class F Flyash, Ground Granulated Blast Furnace Slag and Silica fume proportion has been provided to the optimal content arrived from the previous investigation made as per literature study providing the blended proportioning of self cured concrete as 65%+25%+15% (M2), 65%+20%+15% (M6), 65%+25%+10% (M10) and 65%+20%+15% (M11) has derived in Table 2 [17], of numerous mix proportions have been determined, in which the performing the optimum level of dosage of blended mix elements were arrived for the given mix M30 grade of concrete with polycarboxylate-type super plasticizer and PEG 400 as 1% to the weight of the cement content. Though the specimen's samples of M30 grade blended concrete beams made with optimum amount of Flyash, GGBFS and Silica fume replacement were casted. Tests for workability was obtained by Slump cone test and defined in Table 2 [17].

Table 1. Proportioning of Standard Self Cured mix for M30 Concrete

Materials	Cement	Fine Aggregate	Coarse Aggregate	PEG 400	Super Plasticizer	Water	Water-Cement Ratio
Quantity	350	744.25	1314.18	2.28	3.5	175	0.4

Table 2. Mix proportions of Blended Self Curing Concrete for M30 Grade **[17]**

Mix (%)	Cement	Flyash	GGBFS	Silica fume	Fine Aggregate	Coarse Aggregate	S.P	Water	Slump (mm)
M1 (100)	350	--	--	--					244
M2 (65+25+15)	227.50	87.50	35.0	--	1314.2	744.3	3.50	140.0	255
M3 (65+25+15)		--	87.50	35.0					255
M4 (65+25+15)		87.50	--	35.0					256
M5 (65+20+15)		70.0	52.50	--					251
M6 (65+20+15)		--	70.0	52.50					251
M7 (65+20+15)		70.0	--	52.50					254
M8 (65+25+10)		87.50	--	35.0					252
M9 (65+25+10)		87.50	35.0	--					255
M10 (65+25+10)		--	87.50	35.0					252
M11 (65+20+15)		70.0	--	52.50					248
M12 (65+20+15)		70.0	52.0	--					251
M13 (65+20+15)		70.0	--	52.50					253

**** The Slump Cone values for mixes of SCM's with cement from M1 to M13 with addition of PEG 4000 as 1.0 %.*

4. Mechanical Behaviour of Blended Self Curing Concrete with SCM's

The mixing both of materials were controlled for a period of two minutes throughout this blended mix of self-curing concrete associated with SCMs. In order to achieve the desired workability without even any segregation resistant over fresh condition, the amount of polycarboxylate-type superplasticizer was altered for each appropriate mix. Numerous specimens were cast using compressive cubic moulds and split-tensile cylindrical moulds accompanying the mixing of concrete.

After the castings were filled with a consolidated mix of self-curing concrete that were well compacted, the surface of the concrete was levelled and the specimens were kept in the moulds for 24 hours or one day in experimental conditions, while the mould surface were protected by plastic wrap and afterwards, and during experimental study, the demoulded specimens were incubated at room temperature. For cubic specimens, compressive strengths and split tensile strengths were carried out, whereas strengths were properly evaluated at 7, 14 and 28 days, as well as for split tensile tests for blended concrete grade M30 mixtures.

Self Curing Concrete mixtures with their strength parameters blended with Supplementary Cementitious components in various proportions of different percentages of Class F Flyash, Ground Granulated Blast Furnace Slag including Silica Fume of various mixed percentages of M1 (Conventional mix) and M2 to M13 has been tabularized in Table no.3.

Table 3. Strength parameters of Self Curing Concrete blended with SCM's in various proportions

Mix (%)	Compressive Strength (N/mm^2)			Split Tensile Strength (N/mm^2)		
	7 days	14 days	28 days	7 days	14 days	28 days
M1 (100)	12.48	20.15	31.21	1.34	2.1	3.4
M2 (65+25+15)	16.04	26.67	41.12	1.48	3.11	4.21
M3 (65+25+15)	16.79	25.6	38.16	1.54	3	4.18
M4 (65+25+15)	15.52	25.89	37.85	1.47	2.99	4.13
M5 (65+20+15)	13.06	24.52	35.21	1.42	2.6	4.03
M6 (65+20+15)	13.97	22.52	36.76	1.39	2.48	4.1
M7 (65+20+15)	14.03	20.78	35.07	1.35	2.56	4.19
M8 (65+25+10)	13.36	21.42	34.25	1.47	2.49	3.94
M9 (65+25+10)	14.08	20.76	33.53	1.49	2.44	3.91
M10 (65+25+10)	14.64	23.78	34.86	1.45	2.5	3.86
M11 (65+20+15)	15.97	25.29	38.95	1.33	2.24	3.54
M12 (65+20+15)	15.17	23.9	35.29	1.29	2.19	3.68
M13 (65+20+15)	14.25	21.19	33.93	1.3	2.28	3.79

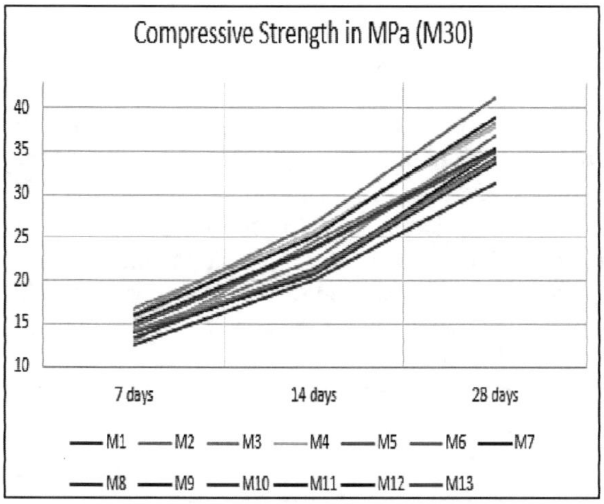

Fig 1. Impact of SCM's % on Compressive strength of Blended Self-curing concrete

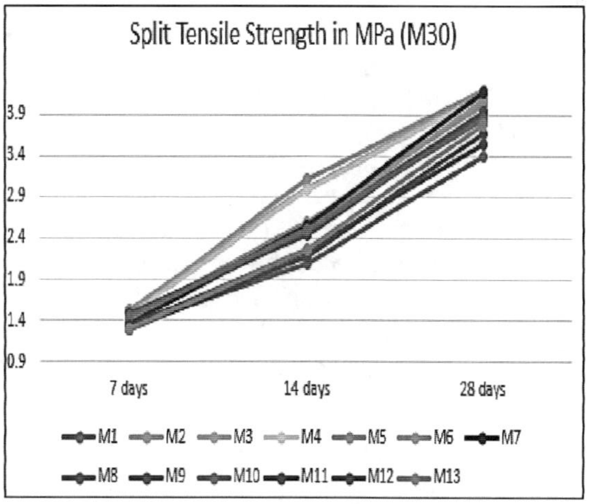

Fig 2. Impact of SCM's % on Split Tensile strength of Blended Self Curing concrete

4.1. pH Value

The pH values of all studied concretes of various proportions 65%+25%+15% (M2), 65%+20%+15% (M6), 65%+25%+10% (M10) and 65%+20%+15% (M11) for the specified mix of M30 grade blended concrete with Flyash, GGBFS and Silica fume with Cement have been derived in Table 2 [17]. As the blended self curing concrete mix which exhibits the amount of heat of hydration have been increased steadily with time due to the production of Calcium Hydroxide ($Ca(OH)_2$) as the result of continuance of hydration of cement content and definite higher values compared to their corresponding concretes with Ground granulated blast furnace slag and Silica fume after 28 days. Fig.3. demonstrates the difference in the Ph values of the concrete mixes studied due to the exposure to blended mixes of Class F FA, Ground Granulated Blast Furnace Slag and Silica fume content.

Comparing the results, these blended SCC mix incorporates the difference in pH values, Where the pH value have been lowered in the with percentage with 25 % of Class F Flyash and 15% of Ground granulated blast furnace slag, meanwhile there is a gradual increase in the mix proportion of 25 % of GGBFS and 10 % of Silica Fume and it has been lowered 3 % over the mix of 25 % Flyash and 15 % Silica fume. All the concrete mixes which exhibits the continuous increase and sudden decrease in the pH value with distinct rates corresponding to the concrete type.

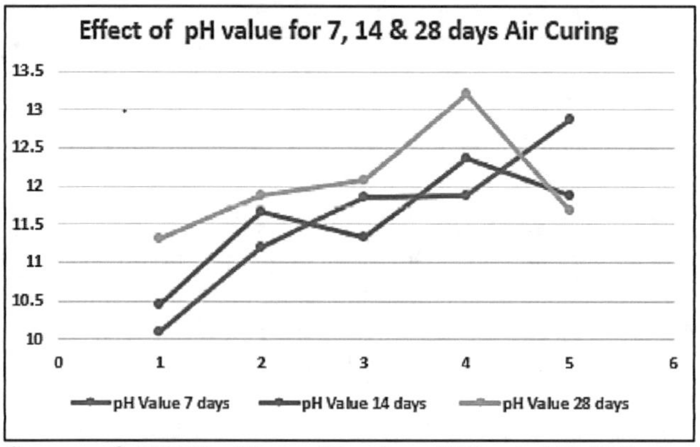

Fig 3. Influence of SCM's % on pH value of Blended Self Curing concrete

4.2. Sorptivity Test

Sorptivity experiments were run on all Self Curing Concrete specimens with dimensions of 15x15 x15cm. Water impermeability of their adjacent faces was also included in the preparation of samples, decreasing the concentration of water evaporation. The experiment started with the observation of the weight of the samples and then they were positioned in contact with the water depth capable of drenching them from around 5 mm in the recipient. The samples were withdrawn from the recipient after a predetermined period of to proceed to weight identification. These samples of surface water with a damp cloth were removed prior to weighing. The samples were replaced in the container immediately after weighting until the following measurement period was reached. The procedure was repeated consecutively at different intervals, including 15 minutes, 30 minutes, 1 hour, 2 hours, 4 hours, 6 hours, 24 hours, 48 hours, 72 hours, 7 days, 14 and 28 days.

Table 4. Sorptivity values at early and consequent stages ($mm/s^{0.5}$)

Mix %	Early absorption	Resultant absorption
M1 (100)	0.0288	0.0081
M2 (65+25+15)	0.0395	0.0069
M3 (65+25+15)	0.0412	0.0088
M4 (65+25+15)	0.0325	0.0099
M5 (65+20+15)	0.0255	0.0072
M6 (65+20+15)	0.0274	0.0075

M7 (65+20+15)	0.0287	0.0068
M8 (65+25+10)	0.0271	0.0088
M9 (65+25+10)	0.0277	0.0096
M10 (65+25+10)	0.0286	0.0092
M11 (65+20+15)	0.0415	0.0103
M12 (65+20+15)	0.0396	0.0098
M13 (65+20+15)	0.0310	0.0083

4.3. Resistance to Acid

In accordance with ASTM C2677 (2001), the relative acid resistance has been calculated . (2001). (2001). Using 5% pH sulfuric acid (H2SO4) of 0.50. pH, the aggressive acid environmental condition was replicated. The 28 days of cured mortar cubes were immersed for 28 days in an aggressive acid environment. In order to mitigate evaporation, the glass containers with submerged mortar specimens were kept covered during the test phase, as shown in Figure 5. The blended SCC mortar specimens were washed with distilled water at 2, 7, 14,28, 56 and 90 days of exposure; the acid resistance was then measured by calculating the weight loss of the specimens as follows: where is the weight (grams) of the specimens before immersion and the weight Wt (grams) of the specimens cleaned after day immersion. Average weight losses were reported in Fig 4 for each mortar specimen.

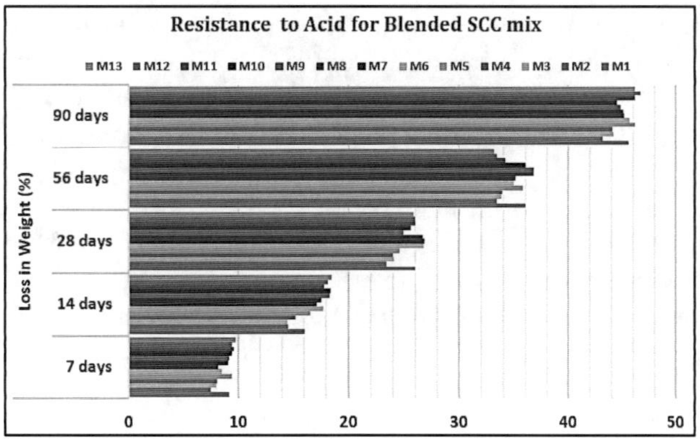

Fig 4. Loss of Weight (%) due to the influence of SCM's % of Blended Self-curing concrete

4.4. Mass loss

It can be seen from the test results that, the mass loss of all self-curing and conventional concretes increased gradually with time under air curing due to water evaporation from concrete as shown in Fig. 7. The presence of silica fume and self-curing agents in concretes caused an additional reduction in mass loss relative to conventional concrete during the experiment to confirm that a better water retention occurred. It could be seen on the test results that, due to water evaporation from concrete, the mass loss of both self-curing and traditional concretes increased substantially with time under air curing, which is shown in Fig. 7. Also during experiment, the development of silica fume and self-curing agents in concrete caused a modification actions in mass loss compared to conventional concrete to demonstrate a stronger preservation of water.

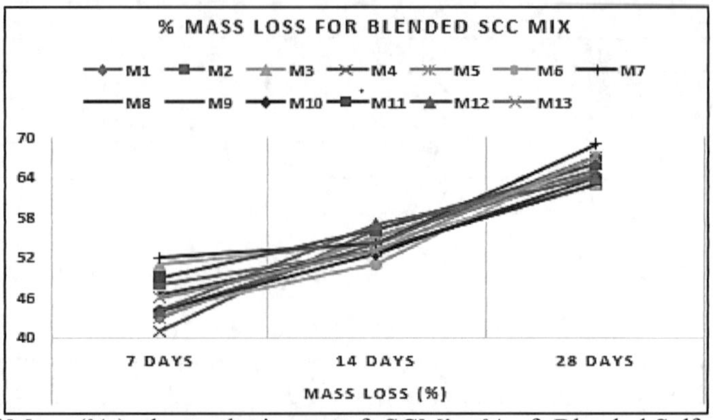

Fig 5. Loss of Mass (%) due to the impact of SCM's % of Blended Self-curing concrete

5. Discussions

The concrete was formed according to Mix Design M30 grade (IS 10262:2009) and blended concrete was examined for strength properties, which include compression test and split tensile test with different ratios, such as 65% +25 % +15%, 65 % +20 % +15 %, 65 % +25 % +10 % and 65 % +20 % +15 % from M1 as Traditional SCC mix. The concentrations of Polyethylene Glycol 400 were adjusted from 1 percent by concrete weight as a self-relieving agent with different concentrations of SNF rapidly changing from 1 percent by solid weight through one self-renovating solid blend and distinguishing the test result and the triple mixed self-curing concrete with various percentages replacements of FA, SF, GGBFS and their combinations of concrete from these final quantities continued to optimize the quantities for various blended SCC mixes of grades considered. The percentage adjustments, correlating to their desired strengths, are shown in table 3. The final optimal quantities of various grades of Cementitious materials are listed in Table-4, along with their total powder content and water/powder ratios. In Table-5, compressive strengths are tabulated for different binary, ternary and quaternary blended optimal SCC blends, the self-curing performance of PEG 400 applied to concrete is the maximum if 45 kg/m3 PEG 400 incorporated water through ensures with 1 kg/m3.

Table 4. Parameters of optimally blended M30 grade concrete mixes for 28 days

Mix	Mix Designation (% by weight)	Compressive Strength in MPa	Split Tensile Strength in MPa	Resultant absorption mm/s $^{0.5}$	pH Value	Resistance to Acid %	Mass Loss %
M2	65%+25%+15%	41.12	4.21	0.0069	11.89	23.41	63
M6	(65%+20%+15%	36.76	4.19	0.0075	12.08	26.77	67
M10	65%+25%+10%	34.86	3.94	0.0093	13.20	25.60	66
M11	65%+20%+15%	38.95	3.79	0.0103	11.69	26.00	64

6. Conclusion

The following conclusions are documented as below, based on the investigational studies and test findings;

The addition of polyethylene-glycol into concrete reduces water evaporation, contributing to the increase in the concrete's water retention capacity, subsequently resulting in increased mechanical properties. Although 45 kg/m3 of water is used in practices for 1 kg/m3 of self-curing concrete, the prevalence of internal self-curing by PEG 400 procedures used for concrete is higher. The PEG 400 seems to be more effective than traditional concrete curing methodologies. The concrete component and the water to cement ratio remain significantly determined by the implementation of the self-curing

agent. Micro silica use has been important and mandatory for the development of high-strength concrete mixes (M30) due to the enhanced reactive and micro-filler characteristics. Either GGBS and Silica fume will be incorporated to optimize the opportunities of micro-filler functionality in M30 grade fly ash integrated concrete mixes. Due to its greater sensitivity, the addition of silica fume (SF) to blended concrete mixes facilitates accelerated hydration.
M30 concrete mixes blended with 65 % OPC +25 % FA +10 % GGBFS optimally blended mix grade tends to produce both necessary workability and predicted compressive strengths. From that same observation, GGBFS transmits high strength in blended SCC mixtures, although both Fly Ash and GGBFS are expected in blended concrete mixes developed directly proportional with lower water powder. As the Flyash and GGBS content with M2, M3 and M4 getting a higher compressive strength for blends having M5, M6 and M7, M8, M9 and M10 increased to an optimum point in M11, M12 and M13, the compressive strength properties of conventional concrete. And therefore it was determined that, again for efficient use of GGBFS content, the optimum level yields the highest strength in the proportions of M2, M3 and M4 at 65 percent + 25 percent + 15 percent. To increase its strength by about 30 % of the total binder content, that results the optimum level of GGBS content.

As the Flyash and GGBFS content has increased significantly in Concrete mixture proportions of M2, M3 and M4 with a higher compressive strength for mixes with a marginal decrease of M5, M6, M7, M8, M9 and M10, the split tensile strength characteristics of concrete increase to an optimal level in M11, M12 and M13. It can therefore be hypothesized that, for the optimized use of GGBFS material, there is an acceptable amount that delivers the maximum intensity in M2, M3 and M4.

Comparison of these findings, the differential in pH values is included with the blended SCC mixture, where the pH value has been decreased in the percentage of 25 % of Flyash and 15 % of GGBFS, although the mix ratio of 25 % of GGBFS and 10 % of Silica Fume has increased dramatically and 3 % has been decreased over the combination of 25 % Flyash and 15 % Silica Fume. The resultant absorption in the sorptivity test is significantly greater in M11 with 65 percent OPC+20 percent FA +15 percent SF with 0.0103 mm/s 0.5 compared to the mixing proportions of M2,M6 and M10 compared to the mixing proportions of M2,M6 and M10. This indicates that even more absorption is transferred by the combination of Flyash and Silica Fume comparable to GGBFS.
Within that combination M2 with 65 percent OPC+25 percent FA+15 percent GGBFS compared to M6, M10 and M11 mix proportions, acid resistance is much lower, suggesting that the aggressive acid ambient environment was associated with 5 percent sulfuric acid (H2SO4) which is much more resistant to GGBFS blends. It is observed that there is a gradual increase in mass loss between 7 days to 28 days ranging from M2 to M13 blends, implying that the mass loss significantly increased by 60% of its weight continuously with air curing period caused by the absorption of water from concrete consisting of all blended supplementary cement materials.

7. Discussions

[1] Bentz D P and Stutzman P E (2006), "Curing, Hydration, and Microstructure of Cement Paste," *ACI Materials Journal*, 103, **5**,pp.348-356.
[2] Ruhal Pervez Memon , Abdul Rahman Bin Mohd. Sam et.al, (2018)A Review: Mechanism, Materials and Properties of Self-Curing Concrete, *ARPN Journal of Engineering and Applied Sciences*, **13**, pp. 9398 -9409
[3] J. Castro, L. Keiser, M. Golias, J. Weiss, (2011), Absorption and desorption properties of fine lightweight aggregate for application to internally cured concrete mixtures, *Cement Concrete Composites.*, **33**, pp. 1001-100.
[4] Chand M. S. R., Giri P. S. N. R., Kumar G. R. & Kumar P. R. (2014). Paraffin wax as an internal curing agent in ordinary concrete. *Magazine of Concrete Research*. **67**, 82-88.
[5] Di Bella, C., Villani, C., Phares, N., Hausheer, E. & WEISS, J, (2012), Chloride transport and service life in internally cured concrete. *Structures Congress*. 686-698.
[6] Bentz D. P. & Stutzman P. E. 2008. Internal curing and microstructure of high performance mortars. *ACI SP-256, Internal Curing of High Performance Concretes: Laboratory and Field Experiences,* pp 81-90.

[7] Kohler K., & Jensen O. M. (2007). Internal curing of concrete, state-of-the-art Report of RILEM Technical Committee 196-ICC. *RILEM Publications S.A.R.L., RILEM Report* 41, Lausanne, Switzerland.

[8] IS 4031-1996 Part I, "Methods of Physical Tests for Hydraulic Cement", *Bureau of Indian Standards,* New Delhi

[9] IS: 12269- 1987, "Specification for 53 Grade Ordinary Portland Cement, *Bureau of Indian Standards* New Delhi.

[10] IS 10262-2009, "Guidelines for Concrete Mix Design Proportioning" *Bureau of Indian Standards* New Delhi

[11] IS 2386-1963 Part III "Methods of Test for Aggregates for Concrete" *Bureau of Indian Standards* New Delhi

[12] IS 516-1959, "Methods of Tests for Strength of Concrete", *Bureau of Indian Standards,* New Delhi.

[13] IS 5816-1999, "Split Tensile Strength of Concrete - Method of Test*", Bureau of Indian Standards* New Delhi

[14] Dhir R K, Hewlett P C, Dyer T D, (1998)"Mechanisms of water retention in cement pastes containing a Self curing agent" , Magazine of Concrete Research, **50**, pp. 85-90.

[15] Pamnani, N J, Verma, A and Bhatt D (2013) Comparison of compressive strength of medium strength self-compacted concrete by different curing techniques, *International Journal of Engineering Trends and Technology,* **4**, 1451-1457.

[16] Zhutovsky S and Kovler K, (2012), Effect of internal curing on durability-related properties of high performance concrete, *Cement and Concrete Research,* **42,** pp. 20-26.

[17] Vivek Kumar C, et.al, (2020), "Strength Behaviour of Blended Self Cured Concrete made with Cement by partial replacement of Supplementary Cementitious Materials (SCM's), *E3S Web of Conferences,* 184, 01100.

[18] ASTM, C1585-04: 2004 Standard Test Method for Measurement of Rate of Absorption of Water by Hydraulic-Cement Concretes, ASTM International.

[19] BS 1881-208: 1996 Testing concrete. Recommendations for the determination of the initial surface absorption of concrete.

IOP Conf. Series: Materials Science and Engineering 1091 (2021) 012038 doi:10.1088/1757-899X/1091/1/012038

Schedule delay analysis of a multi-storied residential project using primavera

Balakrishna Ch[1], Anand Goud K[2], Akula Prakash[3] and Kandlagunta Mounika[4]

[1,2] Assistant Professor, Department of Civil Engineering, IARE, Hyderabad, India.
[3] Assistant Professor, Department of Civil Engineering, GRIET, Hyderabad, India.
[4] Assistant Professor, Department of Civil Engineering, CMRIT, Hyderabad, India.

Corresponding Author: balakrishnaace@outlook.com

Abstract. : Procrastination in schedule is a major problem for any construction project. Completing the project on time is a very hard task as the delay in construction occurs due to various factors. Loss of time and money are the outcomes of the delay in the project. The delay may occur through various factors, which are mostly unpredictable. These delays cause losses to the contractor; sometimes this may also lead to discontinuation of the contract and affect the possession date of the project. It is necessary to determine the factors that are responsible for the delay. In this study, schedule delay analysis is conducted on the construction of a multi-storied residential project and the factors and parties responsible for the delays are found out. Many unforeseen factors and ignored issues that lead to the delay are identified. Using various features and tools of Primavera software, the schedule delay analysis is conducted, and an alternate schedule is created with increased resources and time so that it reduces the effect of delays, which would be worthwhile to the contractor and all other parties involved in the project.

1. Introduction

Planning is the most important factor which decides the quality of any project. A project which is well planned and executed will save a lot of time and money and minimizes the losses for the parties involved. An effective plan includes the proper usage of money, time and provided resources. If the plan is done without proper usage of resources, the quality of the project will be affected. Hence it is necessary that planning of a project is done with an objective of getting quality results. Before the start of a project, a schedule of activities is created by the project planner, which includes the list of all the activities involved in the construction, duration of each activity, parties involved in the activity. This schedule is included in the tender documents during the tendering process. After the tender is accepted, the construction process begins as per the schedule. If all the activities are done on time without any obstacles, the possession will be on time and all the parties will be in profits. But, due to many unpredicted factors by various parties result in the delay of construction.
Schedule delays are common in the construction projects as the factors may not be unpredictable. Sometimes, various unexpected factors may lead to the delaying of the project. It is not sure that a project might be completed as per the schedule and many factors may lead to the postponement of completion. Even if a project is planned and is being executed as per schedule, some obstacles may interrupt the project resulting in the delay. In recent times, delays are most common in the construction projects. Even with advancements in technology and reduced manual work, delays occur in the projects. Even if the planning of the project is done in an organized manner, the project is being

pushed beyond the deadline. There is no such method which can guarantee the completion of a project without any delays. Even some projects which are profitable to the contractors are turned into loss ventures due to the delays and time overrun. Equation).

Various factors and obstacles affect the project in several ways. Some of these are excusable as they are uncontrollable, while some are non-excusable and cost the contractor and other parties involved. Below are listed some of the factors that are responsible for construction delays

- The conditions of the site are not always the same. They change from time to time and sometimes can be difficult to complete the work.
- The owner may sometimes change his tastes and hence it is needed to change some designs and therefore delay occurs.
- The weather conditions are not always the same. It is unpredictable in countries like India. Differing weather conditions result in delays
- The labor, materials or equipment may not be available sometimes and this shortage may lead to delay in some activities.
- Failure in some plans may occur and some activities have to be rescheduled and delay occurs due to these activities.

2. Schedule Delay Analysis

Schedule delay analysis is conducted to identify the factors that are responsible for delays in the project. The impact of delays can be reduced if the remaining project is done at a faster pace. The activities which are delayed are first identified and then the cause of delay for these activities. The remaining schedule is modified based on the impact of delays. Primavera is used to conduct the delay analysis. Various tools available in the software help us to identify factors responsible for the delays, impact of delays and lag beyond the finish date. This helps to rectify and resolve the issue at an early stage and to think of an alternate situation to complete the project as soon as possible and reduce the impact of delays. This helps to reduce the lag and extra cost to be spent on the project.

Al-Momani stated that "Construction technology has a great potential to improve productivity and decrease project duration. Delay happens in many construction projects, although the priority of delay causes is different in various countries due to environmental effects. Delays can lead to considerable negative effects such as lawsuits between owners and contractors, loss of productivity and revenue, and contract termination

Megha desai and Rajeev Bhatt stated that "Delay could be defined as the time overrun either beyond completion date specified in a contract or beyond the date that the parties agreed upon for delivery of a project. It is a project slipping over its planned schedule and is considered as common problem in construction projects. To the owner, delay means loss of revenue through lack of production facilities and rentable space or a dependence on present facilities. In some cases, to the contractor, delay means higher overhead costs because of longer work period, higher material costs through inflation, and due to labor cost increases.

Aziz and Remon Fayek stated that "Time is money; delay in a certain construction project affects time and thus money, which is the lifeblood of any economy. The timely completion of highway construction projects is considered one of the most important factors referring to the project success, as well as the quality and the safety. All around the world many construction projects face one of the biggest construction problems which is the delay, delays differ from a country to another, from a construction project to another and from construction type or cost to another due to every project circumstances Sullivan and Harris.

Tsegay Gebrehiwet and Hanbin Luo stated that "A construction project is commonly admitted as successful when it complete on time, with budget, according the specifications, and stakeholder satisfaction. However, most of the projects did not finish as the expected timetable. Instead, they completed before or after the schedule due to uncertainties of events and its uniqueness [1]. Construction projects experienced 70% of time overruns and 76% of contractors and 56% of consultants have indicated that they have been facing average time overrun of 10 to 30% from the original duration that causes 50% cost overrun.

IOP Conf. Series: Materials Science and Engineering 1091 (2021) 012038 doi:10.1088/1757-899X/1091/1/012038

Mohammed M. Marzouk and Tarek I. El Rasas stated that "Construction delay means a time overrun either beyond the contract date or beyond the date that the parties have agreed upon for the delivery of the project. In both cases, a delay is usually a costly situation [1]. Delay was also defined as an act or event which extends required time to perform or complete work of the contract manifests itself as additional days of work [2]. Poor site management can cause project delay and affect productivity.
Ogunlana et al. Identified 26 delay causes affecting construction industry in a fast-growing economy in Thailand categorized them into 6 groups, and data were collected by visiting sites and mailing to 17 contractors, 18 consultants and design firms and one project owner. 8 contractors and 6consultants gave approval of which only 12 projects were selected for visits. (a) problems of shortages or inadequacies in industry infrastructure (mainly supply of resources); (b) problems caused by clients and consultants and (c) problems caused by contractor incompetence/inadequacies.
Venu Malagavelli, Balakrishna Chetlapally, Prasad Bollini stated that "Delay analysis is a well-developed technique to find the delays in the project using Primavera. Delay analysis is used to find the delays and its effects on the project cost and factors affecting the project. The delays analysis also focuses on the precautionary measurement to complete the project within the time. Indian Building industry in the recent times is encountering a numerous issue resulting in the loss of a billion ringgit. The principle factor which produces the unsteadiness to the building profession is price and span."
Odeh and Battanieh identified 28 delay causes affecting construction projects with traditional type of contracts in Jordan; first, a survey questionnaire was developed to assess perceptions of contractors and consultants of the relative importance of construction delay causes. Second, the questionnaire was distributed to a random sample of contractors and consultants working on large projects in Jordan. The Spear-man s rank correlation coefficient was then used to test association between the contractors and consultants ranking.
Yang and Wei identified 35 delay causes, 15 causes in the planning phase and 20 causes in the design phase for construction projects in Taiwan by sending a structured questionnaire to engineers at the A/E companies for public construction projects resulting in 95 valid responses identifying the delay causes. This study used the Likert scale in questionnaire design to plot the importance frequency matrix and ranked the factors by the importance and frequency of delays using the relative importance index then calculated the severity index.

INVESTIGATION:
A detailed analysis of the sequence of activities (dependent or independent) executed from the commencement of the construction to its completion.

- Collection of ideas
- Literature review
- Site Visit
- Collection of raw information from site engineer and contractor
- Data Collection, which is the basis to prepare manual schedule
- Preparation of WBS and corresponding activities
- Preparation of the plan and schedule by using the different elements of primavera
- Comparing the updated schedule with original schedule to perform delay analysis

The multi storied residential project is located at Hyderabad. The residential project is a 6.5-acre gated community spread across 8 towers, of 12 floors each. The duration of the project was 623 days, when it was scheduled by the project planner. It took 705 days during execution phase. Delay analysis is conducted to identify the delay factors and the effect on the project completion. A schedule is prepared which helps to complete the project in planned duration.

3. Methodology
There are five major processes in the project planning management which include initiation, planning, execution, forecasting and controlling, close. Initiation is the process in which all the actions required at the start of a project take place. Proper planning is necessary to complete a project in a given time and limited budget. This process involves two stages namely scheduling and costing. Scheduling is

further divided into various groups: calendar creation, WBS (work breakdown structure) creation, activity creation, setting duration for each activity, assigning constraints and baseline creation. Execution process involves day to day updating of the project. The update can be done in two methods. If the project is being executed without any delays, spotlight updating is done. If there are any delays manual updating is done using duration and units. Monitoring is done to check the status of the project at any point of time and find out if any delays present or not. Controlling is the steps taken to reduce the further delays and reduce the impact. Monitoring is done by a tracking process where the project can be analyzed at any instant to find out any obstacles. After the completion of execution of the project, reports are generated to know the delays, their impact on the project and the loss occurring to various parties involved in the project. Then, these reports are exported to the contractor and owner and analysis is done on the delays and their impact on the project.

A. Critical Path Method

Calculating the free float and total float using the relationships between two or more activities and some of the float calculating formulae is done in this process. By this method, whether the project gets completed on time or gets delayed can be determined. For calculating the float of an activity, the float of its predecessor or successor activity is also necessary.

B. Constraints

In the as-planned schedule, various constraints are assigned to activities and most of the activities are given flexibility which was an error, and this led to some delays in the construction stage. In the proposed schedule, the critical activities are assigned with mandatory start and mandatory finish constraints. The non-critical activities are assigned with start on and finish on constraints. This reduces the flexibility and the resources involved will have to complete the activities on time. This minimizes the delays to zero. As the delays are identified, we can conclude that there was no perfection in the planning phase and the constraints were not set properly by the planner. This plays a major role in delaying of certain activities. In the updated schedule, constraints were set to critical as well as non-critical activities. The assigned constraints include start on or before, finish on or before, mandatory start and mandatory finish.

C. Adjusting Baselines

A baseline is a copy of the project data, which is used to determine the delay of activities. It can be used to measure the project progress. An as-planned schedule is used to create a baseline for the project. Using this baseline, the performance of as-built schedule is measured, whether it is following the as-planned schedule is identified. For a project, many baselines can be created upon the request of the management. The activity bars that cross the baseline not only get delayed, but also may delay the other activities related to them and even the finish date of the project. Hence, monitoring the baselines and controlling is necessary to minimize the effect of delayed activities on the entire project. Hence, baselines should be adjusted by setting constraints, increasing work hours and resources to reduce the impact of delays on the successive activities.

D. Time Impact Analysis

The process of finding out the activities that are being delayed, the causes are found out in the time impact analysis. This analysis is helpful to find out the impact of delays and the parties responsible for the delays. Depending upon the impact of delays on the project, some modifications are done to Necessary steps are taken after the analysis to minimize the effect of the delays.

E. Duration of Project

The duration of the project was 705 days with a delay of 82 days constituted by 15 parameters. The proposed schedule is created with utmost care and resolving the errors of the as-built schedule and hence the duration of this schedule is less than the as-built one. The duration of the proposed schedule from the start to finish of the project is 657 days. When this schedule is compared with the as-built schedule, it is 45 days ahead and hence if the proposed schedule is used, the project can be completed earlier. The resources assigned for as-planned schedule is 100% but in case of the proposed schedule, 120% resources are assigned. There would be no extra cost as the activities will be completed the earliest. The working hours for both as planned and proposed schedules are 48 hours per week. The proposed schedule is created referring to the errors that happened in the as-built schedule and the factors that caused delays are taken into consideration. The proposed schedule takes more time to build compared to the as planned schedule considering the weather factors that may disturb the pace of the project progress. In this case, the overall cost of the project increases, but it is less compared to the extra cost incurred due to contractor

4. Results and Discussions

The process of finding out the activities that are being delayed, the causes are found out in the time impact analysis. This analysis is helpful to find out the impact of delays and the parties responsible for the delays. Depending upon the impact of delays on the project, some modifications are done to Necessary steps are taken after the analysis to minimize the effect of the delays.

A. Calculation of float

To find out the impact of delayed activities on the remaining activities, float for the delayed activities is found out. For calculating the float, early start, early finish, late start, late finish of an activity are required. Table I shows the calculation of total float and free float for delayed activities. It is necessary that the activities that caused delays are identified to minimize delays. The activities that caused delays and their causes are shown in Table II.

Table-I: Calculation of total flow and free flow for critical activities

Activity	Duration	Start (planned)	Finish (planned)	Start (as-built)	Finish (as- built)	Total float	Free float
1-2	20	0	20	0	20	0	0
1-3	25	0	25	5	30	5	5
2-3	10	20	30	20	30	0	0
2-4	12	20	32	23	35	3	3
3-4	5	30	35	30	35	0	0

Table-II: Causes and factors for the delays

Activity	Delay (days)	Causes for delay	Parties responsible
Backfilling & consolidation	4	Labor shortage	Contractor
Plinth beam concrete	4	Shuttering delay	Contractor
Supply of steel at site	2	Late delivery	Supplier
Slab shuttering	10	Labor shortage	Contractor
Reinforcement	8	Less equipment	Contractor
Conduit laying	2	Plumber	Contractor
Slab concreting	2	Equipment fail	Equipment
Conduits &plumbing	3	Plumber	Contractor
Terrace waterproofing	4	Less equipment	Contractor
External plastering	10	Scaffold delay	Labor
External painting	8	Shortage	Contractor

B. Comparison of Durations

After the creation of updated schedule, both the planned and proposed schedules are compared. This is done to know the number of days that are saved and that can be reduced by using the proposed schedule. After comparison, it is found out that the proposed schedule reduces 45days of duration. Figure 1 shows the pictorial results of comparing both the schedules.

C. Resource usage

The usage of resources in the proposed schedule is Increased so that the delay can be reduced. The resources used is increased by 20% for the proposed schedule. This might increase the cost of the project, but it is less than the extra loss incurred on the contractor. The comparison of resource usage for planned and proposed schedule is shown in pictorial representation

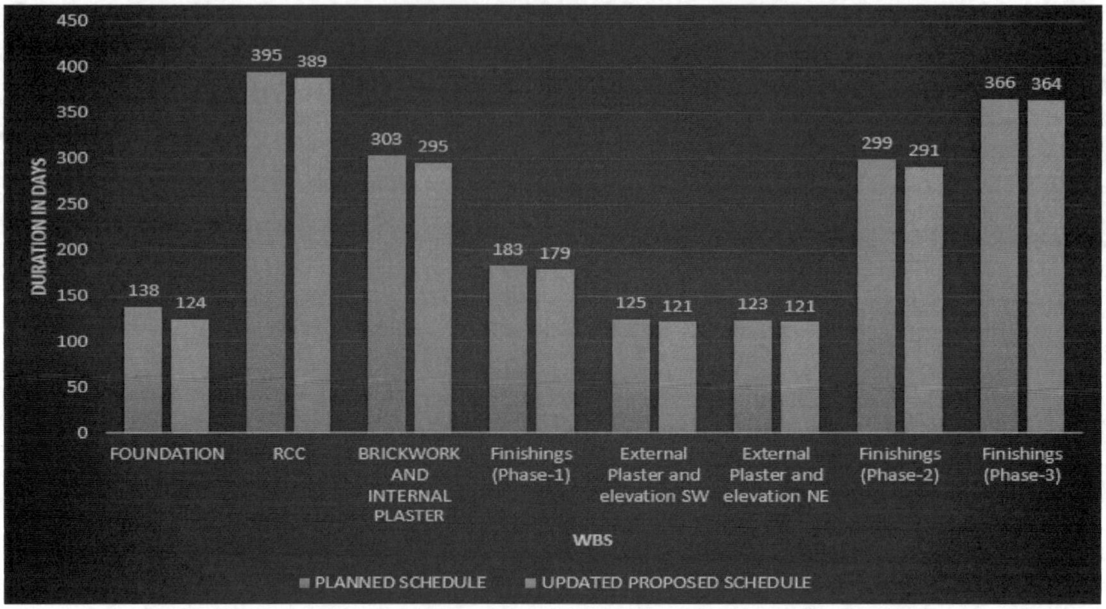

Fig.1 Comparison of durations (planned and proposed schedule)

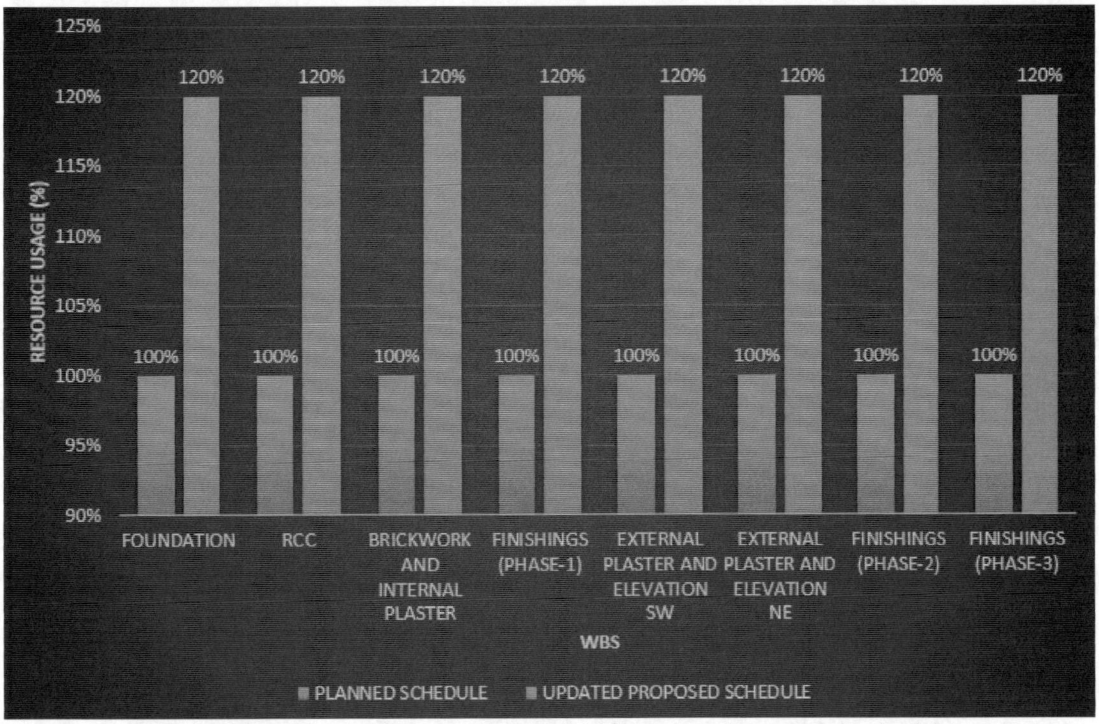

Fig.2 Comparison of resources used (planned and proposed schedule)

5. Conclusions

Delay analysis is done on the planned schedule of the multi-storied residential project and the following were the observations:

- The updated duration after completion of the project is 705 days.
- Constraints are not properly assigned to the critical activities, which led to delay of several activities.
- The contractor was observed to be the cause of delay for various factors.
- Out of the 15 parameters, 4 were observed to be excusable and the other parameters lead to the delay in critical activities.
- After the proposed schedule, the duration of the project is reduced by 45 days.
- The actual duration of the project is 705 days but the duration of the proposed schedule is 660 days.

6. References

[1] Ar. Meena. V, Ar. K. Suresh Babu; "Study on Time Delay Analysis for Construction Project Delay Analysis"; International Journal of Engineering Research & Technology Vol. 4, March-2015.

[2] Megha Desai and Rajeev Bhatt; "Critical Causes of Delay in Residential Construction Projects"; International Journal of Engineering Trends and Technology; V4 (4):762-768 Apr 2013.

[3] Aziz, Remon Fayek (2013); "A Critical Literature Review on Main Cause of Delay in Construction Projects"; International Journal of Engineering Research & Technology, Volume: 05.

[4] Tsegay Gebrehiwet and Hanbin Luo; "Analysis of Delay Impact on Construction Project Based on RII and Correlation Coefficient"; Creative Construction Conference 2017.

[5] Mohamed M. Marzouk, Tarek I. ElRasas; "Analyzing delay causes in Egyptian construction projects"; Journal of advanced research (2014) 5, 49-55.

[6] Al-Momani; "The importance of new technology for delay mitigation in construction projects" American Journal of Civil Engineering and Architecture. 2015, Vol. 3, No. 1, 15-20.

[7] El-Razek et al.; "Studying the suitability of different data mining methods for delay analysis in construction projects"; Journal of Applied Research on Industrial Engineering Vol. 2, No. 1 (2015) 15-33.

[8] L Muhwezi, J. Acai, and G. Otim; "An Assessment of the Factors Causing Delays on Building Construction Projects in Uganda" International Journal of Construction Engineering and Management 2014; 3(1): 13-23.

[9] Mr. I. Michael Raj, Ms. M. Panimalar; "Schedule delay analysis in construction projects using Primavera" International Journal of Engineering Research & Technology e-ISSN: 2395-0056 Volume: 06 Issue: 05.

[10] G. Sweis, R. Sweis, A. Abu Hammad, A. Shboul; "Delays in construction projects: the case of Jordan"; International Journal of Project Management 26 (2008) 665-674.

[11] Venu Malagavelli, Bala Krishna Chetlapally, Prasad Bollini; "Delay analysis of raw water reservoir and pumping station", International Journal of Engineering and Advanced Technology (IJEAT) ISSN: 2249- 8958, Volume-9, February, 2020.

[12] S.O. Ogunlana, K. Promkuntong, V. Jearkjirm, "Construction delays in a fast- growing economy: comparing Thailand with other economies", Int. J. Project Manage. 14 (1) (1996) 37-45.

[13] Odeh, H.T. Battaineh, "Causes of construction delay traditional contracts", Int. J. Project Manage. 20 (1) (2002), 67-73

Studies on strength parameters of blended self cured concrete with natural and manufactured sand on the impact of fly ash and ground granulated blast furnace slag

Vivek Kumar C[1], Dr. A.N.Swaminathen[2], I.Chandana[3], Dr.J. Selwyn Babu[4]

[1,3]Assistant Professor, Civil Engineering Department, GRIET, Hyderabad, India.
Corresponding Author : vivekkumarcr@gmail.com.
[2]Professor, Civil Engineering Department, Sree Sakthi Engg. College, Coimbatore, TN, India.
[4]Professor, Civil Engineering Department, Malla Reddy Engg. College (A), Hyderabad, India.

Abstract. Construction companies now use manufactured sand instead of natural river sand in concrete for a day, where the exploration limitation applies to river sand along with the implementation rule. Water, meanwhile, is also the most prevalent aspect of the concrete manufacturing curing process, and it is now scarce for a few days due to a greater rise in human population growth. In contrast to the traditional curing process, self-curing concrete minimizes the use of potable water. In addition to chemical admixture with the mixture, this paper deals with the various studies of fresh and hardened concrete specimens and findings with the self-curing agent mixed with PEG 400 with Fly Ash and GGBFS in different proportions. The results for the various combinations of blended proportions are correlated with the substitution of manufactured sand (M-Sand) with river sand with both ordinary self-curing concrete and mixed self – curing concrete with M30 grade. Attempts have been made to study the properties of blended self – curing concrete with Manufactured Sand.

1. Introduction

Because of the high compressive strength of the most used building material in the world of hard concrete. To gain greater durability, conditional curing needs to be achieved one day after fresh concrete. Hydration of cement occurs after levelling the concrete in a specified position. Internal concrete curing decreases the demand for water than traditional curing. In addition to compaction, the curing process plays a major role in achieving improved strength, durability, water tightness, abrasion resistance, volume stability and thawing resistance. Curing is a very important component that remains important for the hydration process, especially in the early ages of concrete strength. Though, acceptable curing is not continuously probable by conventional curing methods and it is suggested to make the entrenched water available for curing to overcome this problem.

To minimize water evaporation from the concrete surface and thereby increase the water retaining ability of the concrete relative to traditional concrete, self-curing agents should be added. This problem also is not solved by conventional curing methods, but internal curing involving materials such as hydrophilic compounds, lightweight aggregates (LWA) and super-absorbent polymers (SAPs) contributes to specific concretes with superior performance (Zhutovsky et al., 2002). Practically good curing is not possible in many cases due to practical difficulties. There the benefit of self-curing admixture is more significant.

The process of self-curing agents relates to the atmospheric state of surface moisture. The hydrogen bond between polymerized fresh concrete water molecules decreases the evaporation of surface water as time contributes to concrete reinforcement. In case chemical admixture does not react

readily with cement paste, causes the concrete to shrink with open pores and cracks at early ages in chemical shrinkage. After the potable water, which is not recurrently available, substantial autogenous deformation and early-age cracking may result due to hydration and other processes.
Empty pores are formed within in the cement paste due to its chemical shrinkage hitting during cement hydration, prominent to a reduction in its internal relative humidity and shrinkage influencing the cracking of the early age. For this purpose, the significance of self- curing agents is largely effective in reducing the water content in concrete as well as internal curing is also possible.

Internal curing (IC) is a way of providing the water to hydrate all the cement, achieving what the mixing water by itself does not do. Internal curing provides water to maintain higher relative humidity and prevent the occurrence of self-desiccation. It also eliminates autogenous shrinkage and retains mortar/concrete early age strengths, i.e. from 12 to 72 hours above the level, where cracking can be generated by internal & externally induced strains. The main extent of the paper is experimentally to examine the effect of polyethylene glycol (PEG 400) on strength characteristics of Self-curing concrete.

The purpose of the experimental study was to understand the strength characteristics of internal cured concrete and to acquire the optimal percentage of compounds which are self-curing. Also, the durability properties of hardened concrete should be studied. Despite improved knowledge of concrete curing methods, research reports suggest that even the most essential curing requirements in practical engineering are not met or are completely ignored. Attempts have thus been made to facilitate internal curing in concrete through the addition of self-curing admixtures, which may produce concretes that are equal or superior to traditional forms of curing

The aim of this research is to vary the mechanical characteristics of concrete such as compressive strength, split tensile strength and rupture modulus by varying the % PEG proportion by cement weight for both ordinary self-curing and blended self-curing concrete with partial replacement of M Sand in Fine Aggregate for grade concrete M30.

2. Materials

All the constituents of concrete making materials are properly tested under laboratory condition. The materials are listed as follows.

Polyethylene Glycol is an ethylene oxide group, general formula H(OCH2CH2)nOH. Where n represents the average number molecular weight in ethylene oxide group ranges from 400 to 4000. PEG is non-toxic, odourless, lubricating, neutral pH value & non-irritating and used in many medical applications. Jagannadha Kumar et al. studied the impact of admixture (PEG 400) on compressive, split tensile strength and modulus of rupture by differing the proportion of Polyethylene Glycol 400 by weight of cement varies from 0% to 2% for M20 and M40 mixes. It was found that 1% of PEG 400 by weight of cement was optimum for M20, while 0.5 % for M40 grade concrete for achieving maximum strength without compromising workability. Joseph et al., experimented on self-cured concrete with PEG400 for M20 grade of concrete also suggested 1% of PEG400 by weight of cement gives optimum strength for M20 grade of concrete and gets increased the workability. Shikha Tyagi et al., (5) made M25 and M40 grade of concrete and found the optimum dosage of PEG-400 to be 1% for M25 and 0.5% for M40 grades of concrete.

As per previous investigations, it is founded that, Polyethylene Glycol (PEG) is widely used self-curing compound, which is taken to be chemical admixture in this experimental work having molecular weight as 400, pH value as 6.5 and specific gravity of PEG 400 as 1.12 respectively.

To conform with the IS 1489-1991 standard, ordinary Portland (OPC 53) cement and a specific gravity of 3.05 were used. With sieve analysis, locally available aggregates of sizes from 4.75mm and 20.0mm were used. Relevant gravity, water absorption and fineness modulus are measured according to IS: 383-1970, 2,74, 0,25 % and 3,05 respectively.

The manufactured sand (M-Sand) used in this experimental work conforms to the 4.75 mm sieve passage. The other properties were also been checked and it is having the specific gravity as 2.6, Water Absorption as 1.3% and fineness modulus as 3.10, respectively. Normally employed quality drinking water conforming to IS:456-2000. In this study, it was acquired from the National Thermal Power Plant, Dadri, where fly ash Class F was used. The characteristics of this fly ash are light grey and the specific gravity ranges from 1.90-2.55 if the uniformity is 3.1-10.7. Apart from diminished permeability, GGBFS blended concrete is considerably more resistant to the ingress of chloride ions through concrete.

In Jindal South West Steel Ltd, Karnataka, GGBFS was acquired predominantly with SiO2 at 35.20 percent, Al2O3 at 19 percent and CaO at 34.90 percent. GGBFS is also an aluminosilicate with major components such as SiO2, Al2O3, CaO, and MgO, and it has been discovered that the incorporation of GGBFS develops workability and prolonged compressive strength.. The integration of Flyash and granulated blast furnace slag was found to deliver lower early age strength to coral aggregate-concrete although superior late age strength due to late production of the pozzolanic reaction. As a super plasticizer, a polycarboxylate-type, new-generation high-range water reducing admixture confirming ASTM C494 was used to enhance the flow or workability of blends with a decreased water-cement ratio.

Table 1. Chemical Composition of Class F Fly Ash

Chemical properties	SiO_2	Al_2O_3	Fe_2O_3	CaO	MgO	Na_2O	K_2O	SO_3	*Loss of Ignition*	*Cl*
Compounds %	45.98	23.55	4.91	18.67	1.54	0.24	1.80	1.47	2.31	0.0053

3. Mix Proportioning of M30 Mix

The mix proportioning was based on the Indian Code for Mix design approach (IS 10262: 2019 for mixing concrete M30 of binary and ternary mixed concretes rendered with optimum mixtures of Fly ash (FA) and Ground Granulated Blast furnace Slag (GGBFS). The proportioning of a blend was made with reference of the Indian Code for Design Mix (IS 10262: 2019 for blending M30 concrete of binary and ternary combined concretes with optimal mixtures of Fly ash (FA) and Ground Granulated Blast Furnace Slag (GGBS). From the above literatures and investigations, Slump Cone test was conducted for Workability and variation in dosage of 0.5%, 1% and 1.5% , was taken for both the natural sand and Manufactured sand as shown in Table 2.

Table 1. Proportions of Self Cured Concrete with M30 grade mix

Material	**Quantity (kg/m^3)**
Cement	350
Fine aggregate (River Sand)	744.25
Fine aggregate (M Sand)	744.25
Coarse aggregate (Angular)	1314.18
Poly Ethylene Glycol (PEG 400)	2.28
Super Plasticizer	3.5
Water	175
Water Binder ratio	0.4

As per IS 456-2000, clause 8.2.4.2, the maximum cement content is limited to 450 kilograms per cubic meter of concrete. Revised quantities in kg per cubic metre for grade (M30) mixed concrete mix are reached in subsequent trail mixes without compromising the necessary strength property. The estimated quantity of OPC is 350 kg and pozzolanic materials such as Flyash (FA) and Ground Granulated Blast Furnace Slag (GGBFS) have been segregated from the literature studies on different dosages formulated in Table 3.

Slump cone test was the common way of finding workability in both construction site and laboratory. In the desired design mix, as per code provision IS:1199-1959, the percentage of PEG 400 kept constant as 1 % for the slump value also increases consequentially in different replacements of Flyash and Ground Granulated Blast Furnace Slag with the blended mixing parameters of River sand and Manufactured Sand (M-Sand) as shown in Table 2.

Table 2. Slump values for various PEG 400 percentages for Natural & M-Sand

Percentage of PEG	Slump in Natural Sand	Slump in Manufactured Rock fines
0.0%	74	55
0.5%	92	78
1.0%	**112**	**106**
1.5%	140	130

Table 3. Mix Proportion of M30 grade Blended mix for various percentages for Flyash & GGBFS

Mix (%)	Cement	Flyash	GGBFS	C.A	River Sand	M sand	S.P	Water
M1 (100)	350	--	--	1314.18	744.255	--	3.5	140
M2 (65+25+15)	227.5	87.5	35.0		744.255	--		
M3 (65+15+25)		35.0	87.5		744.255	--		
M3 (65+25+15)		87.5	35.0		--	744.255		
M4 (65+15+25)		35.0	87.5		--	744.255		
M5 (65+20+15)		70.0	52.5		744.255	--		
M6 (65+15+20)		52.5	70.0		744.255	--		
M7 (65+20+15)		70.0	52.5		--	744.255		
M8 (65+15+20)		52.5	70.0		--	744.255		
M9 (65+10+25)		87.5	35.0		744.255	--		
M10 (65+25+10)		87.5	35.0		744.255	--		
M11 (65+10+25)		87.5	35.0		--	744.255		
M12 (60+20+20)		35.0	87.5		--	744.255		
M13 (60+20+20)	210.0	70.0	70.0		744.255	--		
M14 (70+10+20)		70.0	70.0		--	744.255		
M15 (70+20+10)	245.0	35.0	70.0		744.255	--		
M16 (70+10+20)		70.0	35		744.255	--		
M17 (70+20+10)		35.0	70.0		--	744.255		
M18 (70+20+10)		70.0	35		--	744.255		

**All the materials proportioned in this table is having the unit of kg/m^3*

*** Slump values with different SCM's on cement mixes between M1 to M13 only with incorporation of PEG 400 as 1.0% .*

4. Mechanical Properties of Blended SCC mix with M-Sand replacements

The mixing both of materials were controlled for a period of two minutes throughout this blended mix of self-curing concrete associated with SCMs. To achieve the desired workability without even any segregation resistant over fresh condition, the amount of polycarboxylate-type superplasticizer was altered for each appropriate mix with Natural sand and M Sand replacement levels.

Numerous specimens were cast with accurate blending of Flyash and Ground Granulated Blast Furnace Slag using compressive cubic moulds and split-tensile cylindrical moulds accompanying the mixing of concrete. After the castings were filled with a consolidated mix of self-curing concrete that were well compacted, the surface of the concrete was levelled and the specimens were kept in the moulds for 24 hours or one day in experimental conditions, while the mould surface were protected by plastic wrap and afterwards, and during experimental study, the demoulded specimens were incubated at room temperature.

Self Curing Concrete mixtures with their strength parameters blended with Supplementary Cementitious components in various proportions of different percentages of Class F Flyash and Ground Granulated Blast Furnace Slag of various mixed percentages of M1 (Conventional mix) and M2 to M18 has been tabularized in Table no.4. The cube and cylinder specimens are casted in accordance to table 3 for 7 days and 28 days curing of conventional full curing and self-curing concrete specimens kept at shadow of temperature 300c ± 30c. The different curing was carried out till the respective concrete specimens were tested after 7 & 28 days for compressive and split tensile strength.

Table 3. Strength parameters of M30 grade Blended SCC for various ratios for Flyash & GGBFS

Mix (%)	Cement	Flyash	GGBFS	River Sand	M sand	Compressive Strength (MPa)		Split Tensile Strength (MPa)	
						7 Days	28 Days	7 Days	28 Days
M1 (100)	350	--	--	744.255	--	10.87	35.1	1.98	4.89
M2 (60+25+15)	227.5	87.5	35.0	744.255	--	9.87	33.2	1.68	3.56
M3 (60+15+25)		35.0	87.5	744.255	--	9.9	34.1	1.49	4.28
M4 (60+25+15)		87.5	35.0	--	744.255	9.4	34.6	1.59	3.6
M5 (60+15+25)		35.0	87.5	--	744.255	9.5	33.8	1.68	3.49
M6 (65+20+15)		70.0	52.5	744.255	--	9.49	33.1	1.77	3.48
M7 (65+15+20)		52.5	70.0	744.255	--	9.66	33.5	1.89	4.26
M8 (65+20+15)		70.0	52.5	--	744.255	9.4	32.8	2.1	4.15
M9 (65+15+20)		52.5	70.0	--	744.255	8.98	33.41	2.01	3.75
M10 (65+10+25)		87.5	35.0	744.255	--	8.69	32.1	1.98	4.21
M11 (65+25+10)		87.5	35.0	744.255	--	8.48	32.14	2	4.1
M12 (65+10+25)		87.5	35.0	--	744.255	8.36	31.21	2.12	3.99
M13 (60+20+20)		35.0	87.5	--	744.255	8.29	32.12	1.89	3.87
M14 (60+20+20)	210.0	70.0	70.0	744.255	--	9.3	33.21	1.87	3.86
M15 (70+10+20)		70.0	70.0	--	744.255	9.12	31.4	1.78	3.73
M16 (70+20+10)	245.0	35.0	70.0	744.255	--	9.14	32.6	1.65	4.25
M17 (70+10+20)		70.0	35	744.255	--	8.96	33.2	1.88	3.65
M18 (70+20+10)		35.0	70.0	--	744.255	8.88	34.1	2.16	4.01
M19 (70+20+10)		70.0	35	--	744.255	9.15	33.6	2.01	3.77

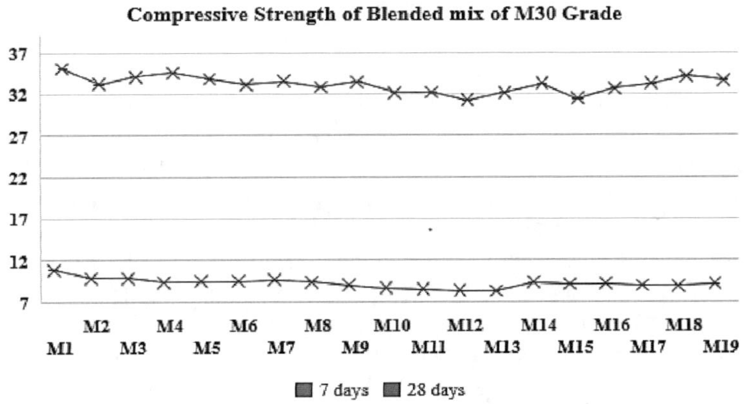

Fig 1. Compressive Strength of M30 grade Blended SCC for various ratios for Flyash & GGBFS

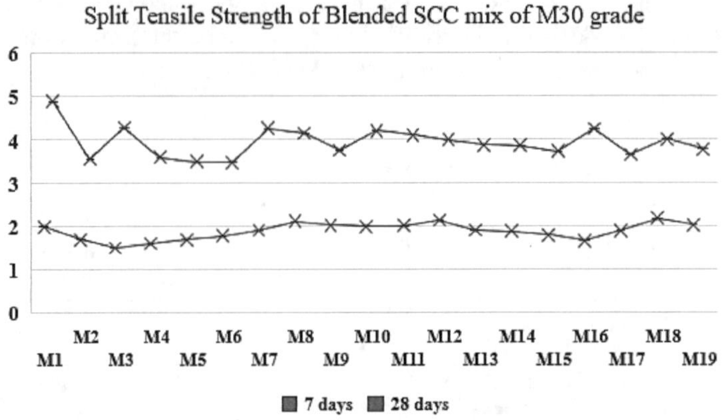

Fig 2. Split Tensile Strength of M30 grade Blended SCC for various ratios for Flyash & GGBFS

5. Results and Discussions

The mechanical properties like workability, compressive strength, split tensile strength and acid resistance which been improved with binary and ternary blended with the Flyash and GGBFS in the concrete mix of grade M30 with several mixes of River Sand and Manufactured Sand. After this percentage of replacement the mechanical properties decreased and very similar for Manufactured sand to the river sand . Therefore, an optimum value of 1% of Polyethylene glycol is considered and added to the mix.

For all replacements of Flyash and GGBFS when M-Sand was incorporated there was an increase in mechanical properties in the strength parameter in M4 (60% OPC +25% FA+15% GGBFS), M9 (65% OPC +15% FA+20% GGBFS) M13 (60% OPC +20% FA+20% GGBFS)and M18 (70% OPC +20% FA+10% GGBFS) provides very slight variation in the compressive strength. More over with the blended mix of 20 % of Ground Granulated Blast Furnace Slag provides greater strength in all the mixes of Self Cured Concrete, which did not affects the other parameters of concrete strength. The mix of 15% to 20 % of Class F Flyash with M-Sand achieved a maximum of 15 % increase in compressive strength when compared to control concrete mix at the age of 28 days.

The split tensile strength of mix 10% Flyash and GGBFS with M-Sand was found to decrease to a maximum of 10 % when compared to the control mix. While for the mixes M3 (60% OPC +25% FA+15% GGBFS), M7 (65% OPC +15% FA+20% GGBFS) M10 (60% OPC +20% FA+20% GGBFS)and M16 (70% OPC +20% FA+10% GGBFS) incorporating M-Sand varying with 15 % to 25% of Flyash and 10% to 20 % of the slight increase and decrease in split tensile strength at the age of 28 days was in the range of 8 to 15%.

In this experimental investigation, it is found that use of manufactured sand doesn't really influence the compressive strength of concrete rather than river sand Replacing cement with high volume fly ash enhances the compressive strength marginally by up to 10 percent, after which the rise in quantity of fly ash gradually decreases its compressive strength.

In the slump test, a workable slump of 112 mm for natural sand and 106 mm for Polyethylene Glycol 400 manufactured sand was obtained for a water-cement ratio of 0.45, which is taken in mixed design in accordance with IS 10262:2009. It is noted from the test result that the reduction in strength of concrete cubes is more due to the rise in the days of specimen immersed in chemicals, so that in the case of an acid resistance test, the reduction in compressive strength was maximum.

In the initial 7 days, there was a substantial loss of weight. After the percentage of weight loss has been steadily increased up to 28 days with a minimum percentage of weight loss at 3.5 kg/m3, it can therefore be concluded that M Sand can be used as an efficient 4 kg/m3 natural sand replacement in terms of strength, workability and durability properties.

6. Conclusions

As the percentage replacement of natural sand through manufactured sand is increased, it is observed that there is a decrease in workability properties. Characteristics for minimizing workability are the angular aggregate form and surface structure of processed sand, which imparts greater internal friction through reducing its flow properties of the concrete.

It is noted throughout this experiment conducted that use of the M sand instead of natural river sand doesn't really affect its concrete's strength development. Replacing cement for high volume fly ash up to 15 percent marginally raises the compressive strength, and the rise in fly ash quantity steadily reduces the compressive strength and also the split tensile strength. Whereas the blending of GGBFS provides the initial early strength and increased workability of M30 grade self cured mix with PEG 400 as 1% ranging percentage varies from 10 % to 20 % of blended mix.

The prospect of manufactured sand and cement substituting natural fine aggregate with industrial by-products such as fly ash, silica fume, metakaolin, etc. The GGBFS provides technological, environmental, and economic advantages that would be of great importance in the current development circumstance in the construction industry.

It concluded that high fines, compared to concrete made from natural sand, whereas the concrete had higher flexural strength, improved strength and ductility, and higher unit weight and lower permeability due to micro fines filling the pores. Compared to river sand, there is no significant difference in dry shrinkage in concrete made with manufactured sand.

7. References

[1] Plunge W.H (2001) Vision 2030,"A vision for the U.S. Concrete Industry" *Concrete International,* **23** 25-34

[2] Kumar Mehta P (2007) Cements and Concrete Mixtures for Sustainability *Proceedings of Structural Engineering* World Congress (Bangalore, India,)

[3] Ruhal Pervez Memon, Abdul Rahman Bin Mohd. Sam, et.al, (2018) A Review: Mechanism, Materials and properties of Self-Curing Concrete, ARPN Journal of Engineering and Applied Sciences, **13** (24) ISSN 1819-6608.

[4] Specification and Guidelines for Self-Compacting Concrete, EFNARC Association House, 99 West Street, Farnham, Surrey GU9 7 EN, UK

[5] Nimitha Vijayaraghavan and Vayal A S 2013 Effect of manufactured sand on durability properties of concrete *American journal of Engineering Research (AJER)* **2** 437-440

[6] Praveen Kumar K and Radhakrishna (2015) Strength and workability of cement mortar with manufactured sand *International Journal of Research in Engineering and Technology* 4 186-189

[7] Donza H., Cabrera O. and Irassar E.F. 2002 High-strength concrete with different fine aggregates *Journal of Cement and Concrete Research* **32** 755-1761

[8] Magudeaswaran P,R Jagadeesh. M, Bhuvaneswari. R, (2017) Experimental Study on Self Compacting Concrete Contains Partially Manufactured Sand and Recycled Clay Roof Tile, *International Journal of Civil Engineering and Technology* **8** (3), 599–608

[9] IS 383:1970 Indian Standard Specification for coarse and fine aggregates from natural sources for concrete (New Delhi, India: Bureau of Indian Standards)

[10] Ganeshwaran P A and Suji P (2013) Behavioral studies on potential aspects of self compacting concrete with M-sand and Fly Ash *International Journal of Emerging Technology and Advanced Engineering* **3** 192-197 2250-2459.

[11] Nagendra R and Sharada Bai H (2013) Study on use of manufactured sand in self compacting concrete, Seminar document on alternatives to river sand- a sustainable approach.

[12] Nanthagopalan Prakash and Santhanam Manu 2011 Fresh and hardened properties of self compacting concrete produced with manufactured sand, *Cement & Concrete Composites* **33** 353-358

[13] Dr.M.Palanisamy P.Sudheer kumar, (2019), Aggregate replacing with Marbles in M30 grade concrete, *International Journal of Management, Technology and Engineering*, 9 583-588
[14] Bilodeau A and Malhotra V. 2008 High- Volume Fly Ash System: Concrete Solution for Sustainable Development ACI Materials Journal 97-116 January-February
[15] Pillai Anil Kumar (2013) Sustainability and Durability in Concrete, *Bulletin of ACCE(India)* **12** 40-43
[16] Deo Shirish V. and Pofale Arun D. (2015) Parametric Study for Replacement of Sand by Fly Ash for Better Packing and Internal Curing *Open Journal of Civil Engineering* **5** 118-130
[17] IS:12269-2013 Ordinary Portland cement,53 grade-specification (New Delhi, India: Bureau of Indian standards)
[18] IS 456-2000 Code of practice for plain and reinforced cement concrete (New Delhi, India: Bureau of Indian standards)
[19] IS 9103-1999 Concrete admixtures- specification (New Delhi, India: Bureau of Indian standards)
[20] BS: 5075- 3:1985," Concrete admixtures. Specification for super plasticizing admixtures"
[21] IS 516:1959 Reaffirmed in 2004,"Methods of tests for strength of concrete" (New Delhi, India: Bureau of Indian standards).

Soil strength enhancement by adding of fly ash

P.V.V.S.S.R KRISHNA[1], G.Swetha[2], Chandana Sukesh[3]

[1,2]Assistant Professor, Civil Engineering Department, GRIET, Hyderabad, India.
[3] Assistant Professor, Civil Engineering Department, Vasireddy Venkatadri Institute of Technology, Guntur, India.

Corresponding Author: siva.polinas@gmail.com

Abstract. Fly ash is an industry development and also freely available so we intend to apply it. This paper deals with the experimental have a look at on development of homes of expansive soil via fly ash replacement. Soil pattern used on this task is an expansive soil that is collected from Bachupally, By engaging in diverse assessments like plastic restriction, liquid limit, free swell index, unconfined compressive test (UCC) and California bearing ratio (CBR) we came to know that the soil sample is weak in shear, bearing and plasticity hence we repeated those experiments by replacing the soil sample with fly ash in various percentages like 10%, 20%, 30% and 40% and came to know that the soil properties are improved at 30% replacement like max dry density increased by 5.5%,plasticity index decreased by 25.4% and CBR value increased by 200%.

1. Introduction

In India, Nearly 20% of land is covered by using expansive soils.
With the rapid development of industrialization and urbanization, the lack of land appears to be an imminent challenge.
So buiiding civil engineering structures on expansive soils, but, pose a primary danger to the shape in itself, due to the extra diploma of instability in such kinds of soils. Quantified in billions of dollars consistent with yr is the loss in belongings every 12 months globally thanks to the uncertainity inside the expansive soils. Dumping of fly ash has ended up turning out to be an emerging difficulty. India, as a growing us of a, is exceedingly depends on thermal flora based on coal for production of electricity, and this dependency isn't going to falter every time quickly. Coal pulvarization in these power palnts generate many waste substances, consisting of fly ash. As of the year 2012, the generation of fly ash rose to one hundred thirty MT/12 months. However, best 56% of this generated fly ash wastes had been simplest applied. The residual fly ash is disposed off in locations, and this leads hazard to health, and the land region reduces which can be in any other case applied for functions apart from the disposal of fly ash. The following targets are studied. To determine the Atterberg limits of the clay pattern for exceptional probabilities of fly ash replacements (0%, 10%, 20%, 30%, and forty%), CBR cost of the clay sample for special probabilities of fly ash replacements the use of CBR take a look at, shear power fee of the clay sample through UCC check, swelling of the clay sample by means of free swell index test

2. Methodology

Various tests were conducted on expansive soils to assess the effect of the fly ash as a stabilizing agent, in which the fly ash in the expansive soil was varied in values of 10% to 40% (multiples of 10) by weight of the total quantity taken. The following standard Indian codes were taken as reference for conducting experiments

1. Standard proctor compaction test IS : 2720 (Part 7) - 1980
2. Unconfined compressive strength (UCS) test IS : 2720 (Part 10) - 1991
3. California bearing ratio (CBR) test IS : 2720 (Part 16) - 1987 20
4. Free swell index test IS 2720 (Part 40) - 1977
5. Liquid & Plastic limit test IS 2720 (Part 5).

Table 1. Soil properties for various fly ash percentages

PROPERTIES	0% Fly ash	10% Fly ash	20% Fly ash	30% Fly ash	40% Fly ash
Liquid limit (%)	73	69.5	61	59	54
Plastic limit (%)	29	28	26	24.2	17.3
Plasticity index (%)	44	41.5	35	32.8	36.6
Max dry density(g/cc)	1.62	1.662	1.69	1.71	1.675
Optimum moisture content (%)	25.6	22.7	18.4	14.3	12.6
UCC(N/mm^2)	0.197	0.195	0.206	0.176	0.170
CBR (%)	2.53	3.80	5.83	7.61	6.85
Free swell index (%)	100	50	40	20	10

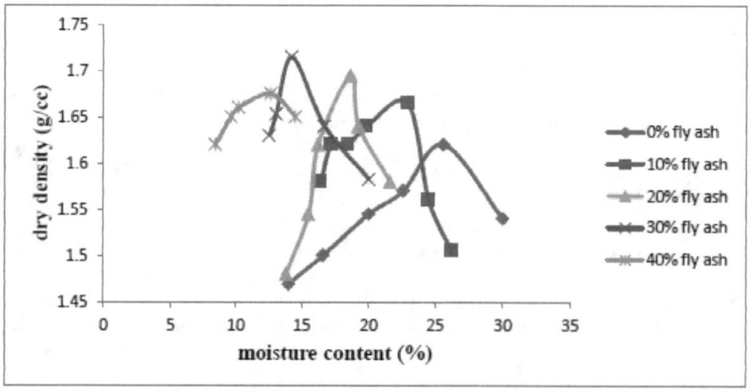

Fig 1 comparison curves of compaction for various fly ash percentages

III. Results and Discussions

1.From the graphs it is clear that dry density increases up to 30% and then decreases, whereas optimum moisture content goes on decreasing. It is also observed that OMC value goes on decreasing upon fly ash replacement and maximum dry density value is attained at 30% fly ash replacement

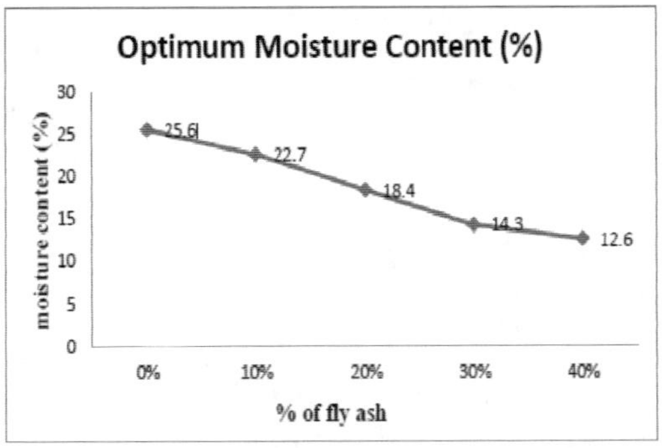

Fig 2: variation of optimum moisture content for increase in fly ash content:

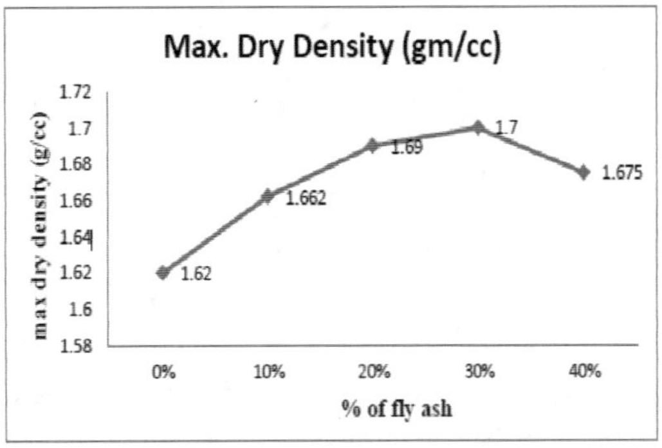

Fig 3 : variation of max dry density for increase in fly ash content

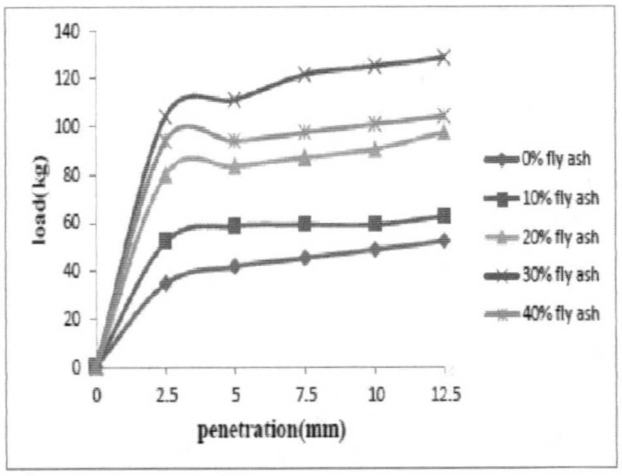

Fig 4 : variation of CBR curves for increase in fly ash content

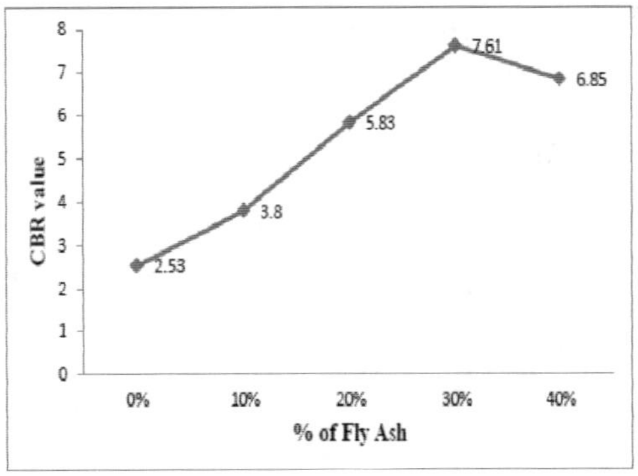

Fig 5 : variation of CBR values for increase in fly ash content

2.Soil can take greater load up to 30% replacement and maximum CBR value is attained at 30% fly ash replacement from the above graphical representations

IOP Conf. Series: Materials Science and Engineering 1091 (2021) 012040 doi:10.1088/1757-899X/1091/1/012040

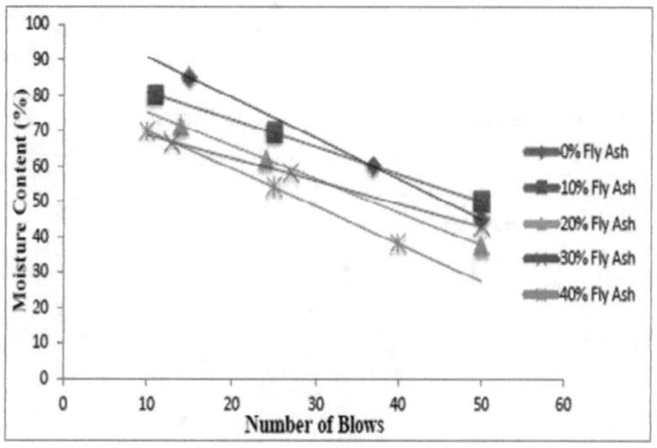

Fig 6 : variation of liquid limit curves for increase in fly ash content

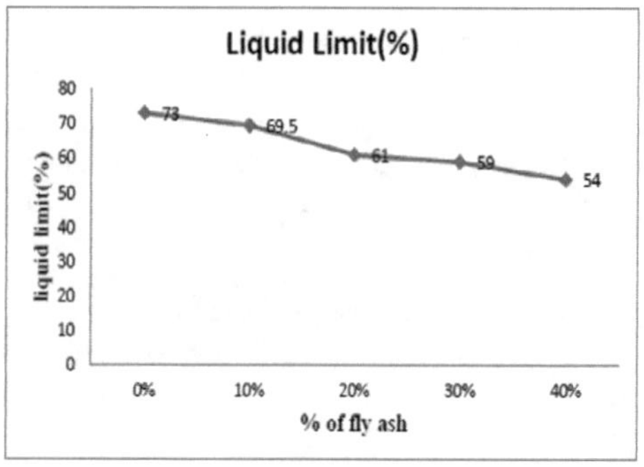

Fig 7 : variation of liquid limit values for increase in fly ash content

3.The above curve shows that liquid limit goes on decreasing upon fly ash replacement and the above graph shows that liquid limit goes on decreasing upon fly ash replacement.

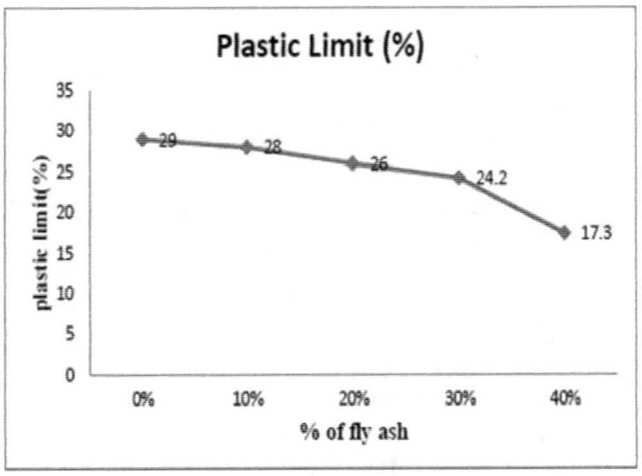

Fig 8 : variation of plastic limit for increase in fly ash content

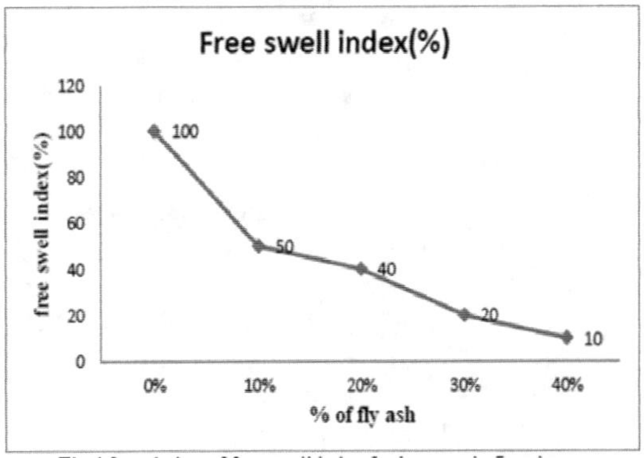

Fig 9 : variation of free swell index for increase in fly ash content

4.Plastic limit goes on decreasing upon fly ash replacement and FSI goes on decreasing upon fly ash replacement from the above charts

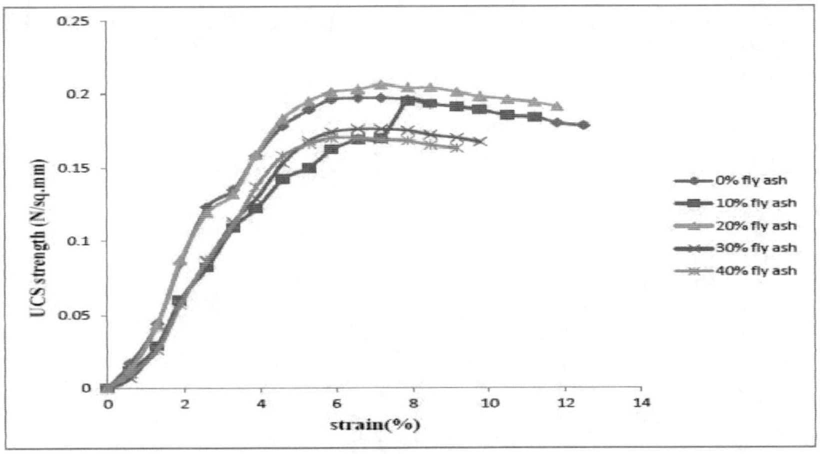

Fig 10 : UCS variation in expansive soil with fly ash content

From the above graph it is observed that the peak UCC value is attained at 20% replacement

IV. Conclusions

On Basis of the results obtained and comparisons made in this research paper , it is determined that the soil pattern is properly graded clay. The most dry density (MDD) cost of the expansive soil expanded until 30% substitute with fly ash. Then MDD values continuously decreased thereafter. The share of boom of mdd at 30% is 5 %.

The superior moisture content material (OMC) goes on reducing with growth in fly ash content in soil. It reached the favored value at 30% alternative. The percent of lower in OMC at 30% is 44%.

Soil unconfined compressive energy (UCC) with variant of fly ash material decrease to begin with then it reaches the peak price for 20% substitute of soil by fly ash then it diminishes. Within the california bearing ratio (CBR) test conducted on soil performed with various fly ash content, the cbr price expanded gradually with the growth in fly ash content material until 30% substitute then it decreased thereafter. The percentage of boom of CBR at 30% is 2 hundred%.

The liquid limit and plastic limit of the soil is going on reducing with the increase in fly ash content material. It reached the preferred cost at 30% substitute.

The free swell index of the expansive soil decreases swiftly for 10% and 30% replacements. This expansive soil sample can be changed by using fly ash up to 30%.

V.References

[1]. Bose, B. (2012), "Geo-engineering Properties of Expansive Soil Stabilized with Fly Ash", Electronic Journal of Geotechnical Engineering, Vol. 17, pp. 1339-1353.

[2]. Cokca, E. (2001), "Use of Class C fly ash for the stabilization of an expansive soil", ASCE Journal of Geotechnical and Geoenvironmental Engineering. Vol. 127, Issue 7.

[3]. Holtz, W. G., and Gibbs, H. J. (1956), "Engineering properties of expansive clays" Transactions ASCE. Vol. 121, pp. 641–677.

[4]. Kumar, A., Walia, B.S., and Bajaj, A. (2007), "Influence of Fly Ash, Lime and Polyester Fibers on Compaction and Strength Properties of Expansive Soil", Journal of Materials in Civil Engineering, Vol. 19, Issue three, pp. 242-248.

[5]. Malik, A., and Thapliyal, A. (2009), "Eco-pleasant Fly Ash Utilization: Potential for Land Application", Critical Reviews in Environmental Science and Technology, Vol. 39, Issue four, pp. 333-366.

[6]. Misra, A., Biswas, D., and Upadhyaya, S. (2005), "Physical-mechanical behavior of self-cementing magnificence C fly ash-clay minerals", J. Fuel, Vol. Eighty four, pp. 1410-1422.

[7]. Nath, P., and Sarker, P. (2011), "Effect of Fly Ash on the Durability Properties of High Strength Concrete", Procedia Engineering, Vol. 14, pp. 1149-1156.

[8]. Phanikumar, B.R. And Sharma, R. S. (2004), "Effect of Fly Ash on Engineering Properties of Expansive Soil", Journal of Geotechnical and Geoenvironmental Engineering. Vol. 130, Issue 7, pp. 764-767.

[9]. Consoli, N.C., Johann, D.R.A. And Saldanha, R.B. (2011). Variables governing electricity of compacted soil-fly ash-lime mixtures. ASCE Journal of Materials in Civil Engineering, 23: 432-440.

[10]. Deb, T. And Pal, S.K. (2014). Effect of fly ash on geotechnical residences of nearby soil-fly ash mixed samples. IJRET: International Journal of Research in Engineering and Technology, three(5): 507–516.

[11]. Elsharief, A.M., Elhassan, A.A.M and Mohamed, A.E.M. (2013). Lime stabilization of tropical soils from Sudan for street creation. International Journal of GEOMATE, four(2): 533–538.

Replacement of opc by zirconium nano particle in concrete

M Manikanta Sai Swaroop [1*]

[1] M.Tech Student, GRIET, Hyderabad, Telangana, India.
*Corresponding Author: kantaswarop@gmail.com

Abstract. It has been shown nanoparticles can act as heterogeneous nuclei for cement pastes, further accelerating cement hydration because of their high reactivity. As a nano-filler, it densifying the microstructure, thereby, leading to a reduced porosity. Because of ultrafine size, nanoparticles show unique physical and chemical properties different from those of the conventional materials. Because of their unique properties, nanoparticles have gained good attention in construction fields. In this study, 1% cement was placed with ZrO_2 nanoparticle of 33 nm to improve the mechanical (compression strength and split tensile strength), durability (Rapid chloride permeability and carbonation test) and microstructural (FESEM) properties in concrete structures and results are compared with normal concrete.

***Key words*:** ZrO_2 nanoparticle; FESEM; Compressive Strength test; Split tensile strength test; Rapid chloride permeability test; Carbonation Test.

1. Introduction

Concrete is a combination of cement, water, and aggregates that are distinct materials. This plays an important part in the field of building. It is possible to prepare concrete with various materials which can replace the original ingredients. In this plant, cement with a nanoparticle of ZrO2 is put. To change the properties of concrete for particular applications, chemical properties such as super plasticizers may be applied in small quantities. Furthermore, ZrO2 nanoparticles are able to act as nanofillers and, by decreasing pores, recover the pore structure of specimens. If nanoparticles are mixed with conventional building materials, the new materials may have outstanding properties for the Building super-high-rise, long-term or smart civil infrastructure systems.

In the related Concrete Structures experiment, we used commonplace cement. With totally different quantities of ZrO2 Nps with a median particle size of twenty nm, the experimental or E series (E1–E4) mixtures were ready. 0.125, 0.25, 0.5 and 2.0 percent ZrO2 Nps/cement by weight is ready for the experimental mixtures. With split longevity, flexural strength and setting time strategies, the altered cement was studied with ZrO2 nanoparticles. Final results showed that zirconium dioxide Nps as Additives Concrete Structures could be used for its meliorative functions. 1]. [1]. Increased compression, split tensile & flexural strength in the presence of ZrO2 nanoparticles due to more hydrated product formation. The water absorption coefficient is improved at 2 days of curing by increasing the content of nanoparticles up to 4.0 wt. (a percentage) [2]. The addition of ZrO2-TiO2 nanoparticles to orthodontic adhesives improves in vitro compressive strength, tensile strength and shear bond strength, but in vivo studies and randomized clinical trials are required to confirm current findings[3]. Cement could be beneficially supplemented with nano-ZrO2 particles up to a maximum limit of 2.0% with a mean particle size of 15 nm when the specimens are healed in saturated lime water[4]. Bulking strength of nanocomposite zirconium/epoxy before and after application of damage. Applying harm with two kinds of Drilling methods for cutting: flat and ball-end. Studying the study of buckling power and failure [5].

In modern times, concrete is as critical as computers or electricity. In fact, it is widely used throughout the industry and consumed annually at a rate of half a dozen billion tons. Concrete may be a blend of a hardened paste of cement and water to control stone and sand. A area of research that is currently gaining an excellent amount of interest is nanoscience and nanotechnology. Whereas small amounts of nanoparticles (less than 1 cement weight chronicles) are usually decent in square measurement to increase the efficiency of the use of composites in concrete requires that contain a large amount of those particles. Therefore, having the capacity to supply nanoparticles in giant amounts at a coffee price is predominant. Many technologies for the development of nanoparticles are common.

2. Significance of the study

To Synthesis of ZrO_2 nanoparticles and its characterization by X-ray diffraction (XRD) and Field Emission Scanning Electron Microscope (FESEM). Fabrication of two types of concrete (a) normal concrete and (b) modified concrete with modification concrete with ZrO_2 nanoparticles. Evaluation of concrete properties such as mechanical (compressive and split tensile strength), durability (RCPT, carbonation), microstructural (XRD and SEM) and thermal (TG/DTA) properties.

3. Experimental procedure

The experimental procedure implemented for this project has been adopted from the available literature. In this work we compare the strength properties of Concrete replaced with Zro_2 Nanoparticle and Control concrete.

3.1 Materials

3.1.1. Fine aggregate

Sand was taken from the storage yard with the help of shovel and it is spread over the open platform for dying. Sand was allowed to dry until all moisture content in the sand gets evaporated. At the end of the day, the dried and was sieved by 4.76mm sieve a shown in Figure 3.1. Then the sieved sand was collected in empty bags and kept ready for making concrete. According to the requirement of sand bags were weighed and batched for casting.

FIGURE 3.1 SEIVING OF SAND

3.1.2 Coarse aggregate

Coarse aggregate consists of broken stone of 20mm size was sieved through 20mm sieve. The sieved aggregate was collected in empty bags and weighed and batched for casting.

IOP Conf. Series: Materials Science and Engineering 1091 (2021) 012041 doi:10.1088/1757-899X/1091/1/012041

Figure 3.2 coarse aggregate

3.1.3 Nanoparticle

The nanoparticle used is Zirconium(ZrO_2) nanoparticle. Zirconium oxide nanoparticles appear in the form of a white powder. It is bought in commercial grade and then it is converted to nanoparticle by using BALL MILLING UNIT. Then it is stored in dry place for casting. The powdered zirconium nanoparticle is shown in figure 3.3.

Figure 3.3 Zirconium Nanoparticle

FESEM Analysis of ZrO_2 nanoparticle

A scanning electron microscope (SEM) is a type of electron_microscope that produces images of a sample by scanning the surface with a focused beam of electrons

Figure 3.4 FESEM image of ZrO_2 Nps

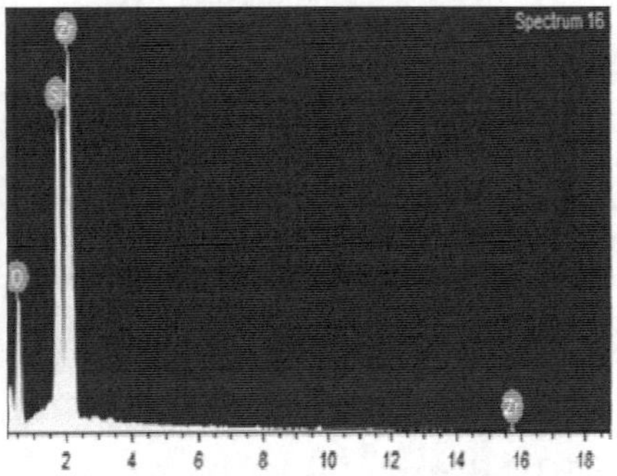

Figure 3.5 EDAX image of ZrO_2 Nps

Table 3.1 Weight % ratio of chemical composition of ZrO_2 (EDAX analysis)

Element	Weight%	Atomic%
O K	45.94	76.01
Si K	12.72	11.99
Zr L	41.34	12.00
Totals	100.00	

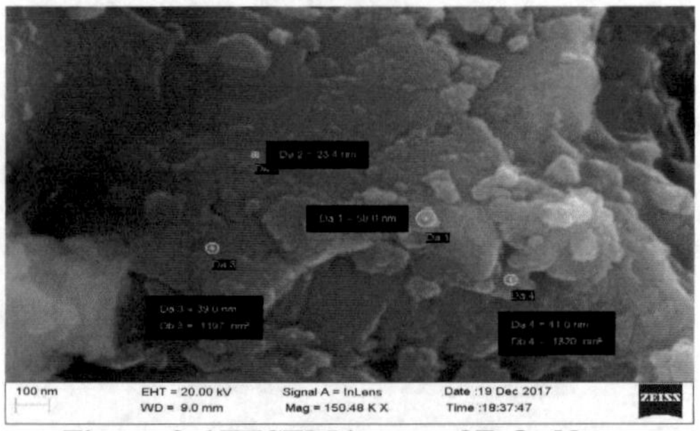

Figure 3.6 FESEM image of ZrO_2 Nps

3.2 Casting of specimens

3.2.1 Preparation of concrete

Concrete is prepared by the mixture of cement, aggregates and water. Coarse aggregates are added to the mixer and the mixer is turned on. then fine aggregate and cement is added to the mixer and leave it for few minutes then add water to the mixer with nanoparticle and admixture and leave it for few minutes. Then the prepared concrete is taken in trolley for casting of specimen.

Figure 3.7 Concrete Mixer

Figure 3.8 Adding nanoparticle

3.2.2 *Casting and curing of specimen*

The concrete was mixed using mixer machine. Mixed concrete was brought near the moulds from mixer machine by trolley. Wet concrete is filled in the moulds. Compacting of concrete was done using compaction rod. Compaction of concrete using compaction rod. Finishing was done in two stages. The first one was done immediately after placing and compacting concrete in such a way that the concrete fills the full mould and leveled. The second one which is final finishing, done after two hours such that top surface is smoothened and leveled. The next day of casting cubes and cylinders were demoulded and placed inside the curing tank. Then the cube and cylinders was allowed to cure for 7, 14, 28 and 56 days respectively. Testing of specimens were done after curing for required days.

4. Experimental Investigation

4.1 Compressive strength Test

Test specimens of concrete cubes having 150 mm x150 mm x150 mm were used for compressive strength. This test is carried out for 7 days, 14 days, 28 days and 56 days respectively.

Figure 4.1 Compressive strength of concrete cube

Table 4.1 Compressive strength test results of cube specimen

Specimens type	Compressive strength (N/mm^2)			
	Age of specimens			
	7	14	28	56
Normal concrete	42.31	44.98	49.84	61.53
Concrete with ZrO_2	44.73	51.24	61.59	68.86

4.2 Split tensile strength Test

Test specimens of concrete cylinders having 200 mm x 100 mm were used for split tensile strength test. This test is carried out for 28 days and 56 days respectively. The split tensile strength test of concrete cylinder is shown in the figure 4.2.

Figure 4.2 Split tensile strength test of concrete cylinder

Table 4.2 split tensile strength test results of concrete cylinders

Specimens type	Split tensile strength (N/mm^2)	
	Age of specimens	
	28 days	56 days
Normal Con concrete	2.656	3.676
Concrete with ZrO_2	3.341	4.679

4.3 Rapid chloride permeability test

Test specimens of concrete cylinders having 100 mm x 50 mm were used for Rapid chloride permeability test. This test is carried out for 28 days. The rapid chloride permeability test consists of moulds with electrodes, connecting wires and RCPT machine. The rapid chloride permeability test apparatus is shown in figure 3.14 and 3.15.

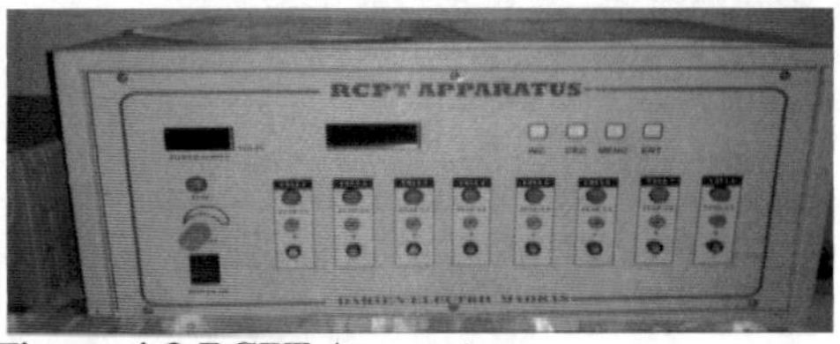

Figure 4.3 RCPT Apparatus

Figure 4.4 Rcpt Moulds

Table 4.3 Rapid chloride permeability test results

Age of sample / exposure days of sample (days)	Specimen type	
	NC	ZrO_2
28	1258	829

Table-4.4 Chloride ion penetrability (ASTM C 1202–09)

Chloride ion penetrability based on Charge passed (ASTM C 1202–09)	
Charge passed (Coulombs)	Chloride ion penetrability
> 4000	High
2000-4000	Moderate
1000-2000	Low
100-1000	Very low
< 100	Negligible

4.4 Carbonation test

Carbonation is a chemical reaction in which calcium hydroxide reacts and forms insoluble calcium car bonate with carbon dioxide:

$$Ca(OH)_2 + CO_2 \rightarrow CaCO_3 + H_2O$$

The process of forming a carbonate is sometimes referred to as "carbonation", although this term usually refers to the process of dissolving carbon dioxide in water.

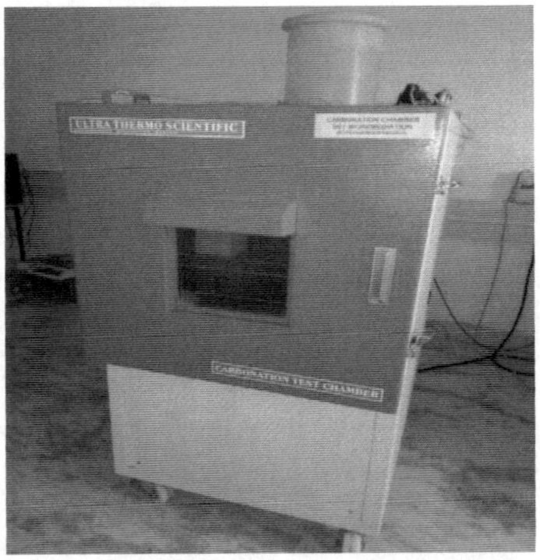

Figure 4.5 Accelerated carbonation chamber

5. Results and Discussion

5.1 *FESEM results*

FESEM studies was performed to evaluate the morphological and microstructure of concrete.

Figure 5.1 FESEM images of 28 days cured (a) normal concrete (b) concrete modified with ZrO_2 nanoparticles.

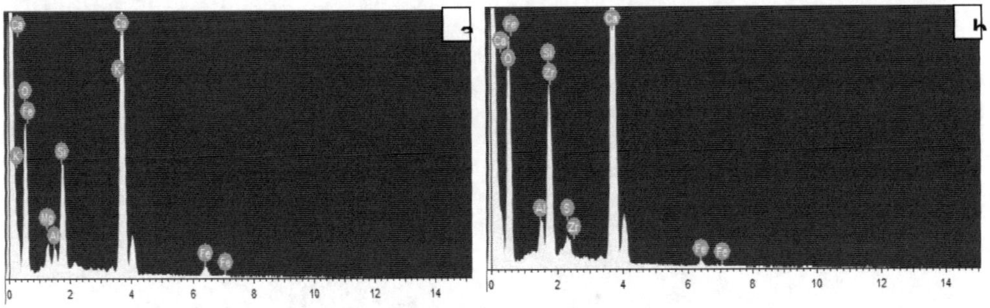

Figure 5.2 EDAX analysis of 28 days cured (a) normal concrete (b) concrete modified with ZrO_2 nanoparticles

IOP Conf. Series: Materials Science and Engineering 1091 (2021) 012041 doi:10.1088/1757-899X/1091/1/012041

Table-5.1 EDAX analysis of 28 days cured concrete specimens (Weight%)

Specimens	Age of the specimen (days)	C	O	S	Mg	Al	Si	K	Ca	Fe	Zr
Normal concrete	28 days	-	62.97	-	1.88	1.27	6.36	0.48	24.76	2.27	-
Concrete modified with ZrO_2		-	65.91	0.68	-	1.49	7.04	-	23.95	0.88	0.05

Figure 5.1 shows the FESEM images of 28 day cured normal concrete and concrete modified with ZrO_2 specimens. Both the specimens showed compact morphology. However, compared to normal concrete (fig.1a) concrete modified with ZrO_2 (fig.1b) showed early stage needle-shaped ettringite ($C_6A_3H_{32}$). This ettringite formation happens when, calcium aluminates react with gypsum ($CaSO_4$) which is the reason for more strength in concrete structures. It will contribute to the early strength development of concrete since the needle-shaped crystals can work as reinforcement for the surrounding C–S–H. Energy dispersive X-ray analysis (EDX) was used to analyze the chemical composition of the all the specimens (Figure 5.2 a & b). It is elementally composed of C, O, S, Mg, Al, Si, K, Ca, Fe and Zr. The presence of Zr was confirmed in concrete modified with ZrO_2 specimens (Table-5.1).

5.2 Compressive strength test

Table 4.1 shows the results of compressive strength test of normal concrete and concrete modified with zirconium nanoparticles. From this we can observe that the strength of concrete modified with zirconium nanoparticles is greater than normal concrete.

5.3 Split tensile strength test

Table 4.2 shows the results of Split tensile strength test of normal concrete and concrete modified with zirconium nanoparticles. From this we can observe that the tensile strength of concrete modified with zirconium nanoparticles is greater than normal concrete.

5.4 Rapid chloride Permeability test

Table 4.3 shows the results of Rapid chloride permeability test of normal concrete and concrete modified with zirconium nanoparticle This shows that the chloride ions passed through concrete modified with zirconium. From this we can conclude that zirconium nanoparticle can reduce the pores in concrete.

5.5 Carbonation test

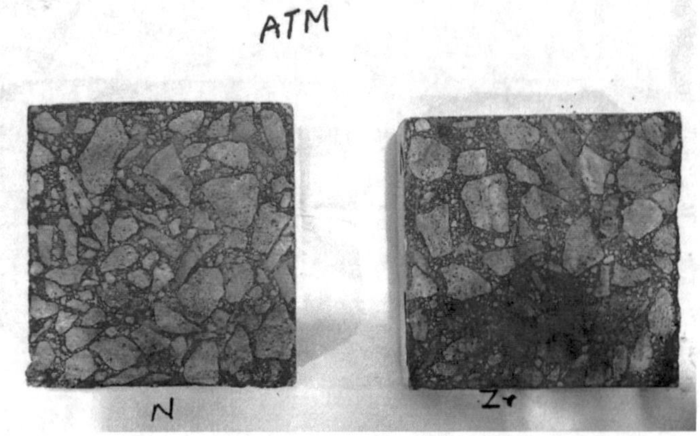

Figure 4.3 Concrete subjected to atmospheric condition

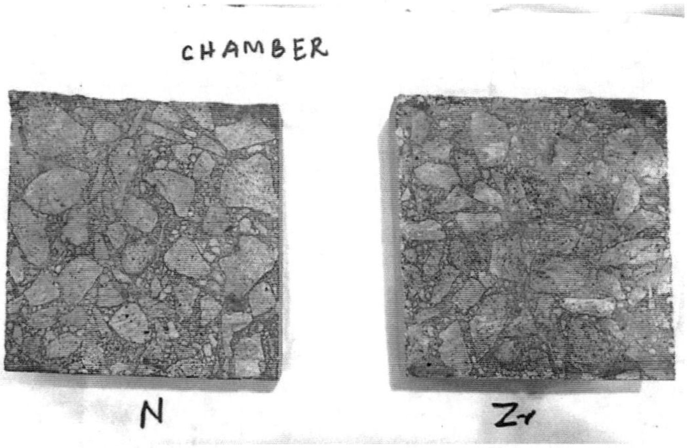

Figure 4.4 Concrete subjected to Accelerated carbonation chamber

From figure 4.3 and 4.4 we can conclude that carbonation has not occurred for both specimens which were subjected to natural atmospheric conditions and accelerated carbonation chamber. The carbonation test was carried out as per RILEM CPC-18. The reduction of the pH value could be made visible by applying a phenolphthalein solution to a freshly fractured surface of concrete. The non-carbonated areas turn red or purple while carbonated areas remain colorless. All the specimens were cured for 28 days and were subjected to natural atmospheric conditions and carbonation chamber (20^oC temp., 65% relative humidity and 1% CO_2).

6. Conclusion

The applications of engineering science have appeared today in all the fields of science mentioned. In various sciences, if engineering science is often used as purposeful, useful, and constructive, its role is often determined. The main goal of engineering these days is to have firm buildings and concrete structures. As long as hydraulic cement plays a key role in most buildings and concrete trade, an important objective is cement production with a replacement combination that can reinforce and have high strength. We tend to get motivation from engineering science here. In order to extend quality and tolerance in concrete structures, we tend to use ZrO2 nanoparticle during this study. Our findings This issue confirmed that ZrO2 nanoparticle may well be correlated with the role of improvement within the hydraulic cement framework that is the primary element of concrete structures. Hydraulic cement and concrete structures can well be affected by the use of nanoparticles in particular amounts, and ultimately this application results in a price saving. On the other hand, the study of various nanoparticles in the sector is suggested.

7. Reference

[1] Masoud Negahdary, Amir Habibi-Tamijani, Asadollah Asadi, and Saeid Ayati, 2012 "Synthesis of zirconium nanoparticles (Nps) and their ameliorative roles as additives concrete structures". Research article, pp 1-3.
[2] Ali Nazari, and Shadi Riahi, 2010 "The effects of the ZrO_2 nanoparticles on physical and mechanical properties of high strength self-compacting concrete". Journal of Experimental Nanoscience, Vol. 8, No. 4, 413–433.
[3] Nayef H.Felemban , Aravind Mohan , Ajith Johny, 2017 "The influence of adding modified Zirconium oxide-titanium dioxide nano-particles on mechanical properties of orthodontic adhesive: an in vitro study". Vol 146, pages 8-19.
[4] Shadi Riahi, 2012 "The effects of ZrO_2 nanoparticles on strength assessments and water permeability". Vol 5, pages 45-50.

[5] Yasser Rostamiyan, and Amir Habibi-Tamijani, 2017" The effect of nano Zirconium dioxide and drilling on the buckling strength of epoxy-based nanocomposites". International journal of Engineering research and Applications, Vol 2, pp 230-233.
[6] C Vivek Kumar, Patam Manisha, Pooja Sadula, Behavior of Monotonic Loading for Glass Fibre based High Performance Concrete in External Beam-Column Joint using ANSYS. Analysis, E3S Web of Conferences, 184 01088 (2020), DOI: 10.1051/e3sconf/202018401088, (2020).
[7] C Vivek Kumar, Mohammed Aiman Parvez, T Srinivas, Strength Behaviour of Blended Self Cured Concrete made with Cement by partial replacement of Supplementary Cementitious Materials (SCM's), E3S Web of Conferences, 184 01100 (2020), DOI:10.1051/e3sconf/ 202018401100, 2020
[9] K. B. Girish, 2018 "Synthesis and Mechanical Properties of Zirconium Nano Reinforced with Aluminium Alloy Matrix Composites".
[10] Jan Sieniawski, Maryana Zagula Yavorska, Jerzy Morgiel, Jolanta Romanowska 2015" Nanoparticles in zirconium-doped aluminide coatings". Vol. 39, pp 50-54.
[11] Aysegul Kutluay Baytak, 2016 "A sensitive determination of terbutaline in pharmaceuticals and urine samples using a composite electrode based on zirconium oxide nanoparticles".

Health state of hussain sagar

Akula Prakash[1], Rathod Ravinder[2] , Anand Goud K[3] and Kandlagunta Mounika[4]

[1,2] Assistant Professor, Department of Civil Engineering, GRIET, Hyderabad, India.
[3] Assistant Professor, Department of Civil Engineering, IARE, Hyderabad, India.
[4] Assistant Professor, Department of Civil Engineering, CMRIT, Hyderabad, India.

Corresponding Author: akulprakash93@gmail.com

Abstract. In the heart of Hyderabad City, India, Hussain Sagar Lake, a historical lake constructed in 1952 by Hussain Shaw Wali, is situated. Earlier until the 1930s, for both twin cities of Hyderabad and secunderabad, it was the primary supply of water to satisfy drinking and irrigation requirements. Now this lake became a mere garbage bin without proper treatment for wastewater loads coming from point and non-point sources. The lake then lacks the capacity to self-purify a mechanism that has an important element that relies on the overall ambient oxygen uptake. Several schemes have been initiated to develop HUSSAIN SAGAR Lake and its surroundings by the authorities concerned. HMDA, a regulatory body set up by the government of Andhra Pradesh, recently launched 'Hussain sagar lake and catchment area development project' in 2004 with reference to the lake health status and set up STPS and wastewater interception and diversion systems. But all such steps are unsuccessful. Ultimately, by gathering samples based on time intervals and various places, we expect findings based on our test results.

1. Introduction

Lakes are one of the reservoirs of fresh water for human uses and they are most valuable for construction practices in and near the regions where they live, acting as a means of groundwater penetration. During the reign of the Qutub Shah dynasty at Golconda fort, Hussain Sagar Lake was constructed in 1562. It is an artificial lake built on the Musi River, a tributary. In addition to adding an artistic charm to the twin cities, the lake enters the cities of Hyderabad and Secunderabad. From 1884 to 1930, lake water was used for agriculture and drinking water purposes. The approximate lake catchment area is 240 square kilometers (93Sq mi). The lake was the primary source of water supply for main feeder nalas called picket nalla, Kukatpally nalla, Banjara nalla and Balkapur nalla, excess water from the catchment area reaching the lake until the 1930s. The overwhelming number of Ganesh idols in the lake has worsened last year.

Hussain Sagar Lake is an important part of Hyderabad, but it has now become the largest sewage collection region for the twin towns of Hyderabad and Secunderabad. The entire Hussain Sagar ecology has changed because of strong anthropogenic stresses, such as unplanned urbanization. Over the last three decades, the quality of water has declined considerably. Owing to siltation and plant litter deposition, the lake has become shallow. A analysis of water and sediment lakes in 2008 showed that it surpassed its eutrophic status, achieving super eutrophic status. Many adverse changes have occurred in the composition of biological populations, and certain major organisms have either declined or vanished entirely. A project was created to understand the significance of the protection of Hussain Sagar Lake. APPCB, Hyderabad, India, will carry out this initiative. In this regard, a project feasibility report for acceptance and funding has been prepared. A significant volume of untreated waste and industrial effluents continue to pour into the lake after the commissioning of the sewage treatment system on the western shore.

Table 1 location of Hussain Sagar Lake

Location	Hyderabad, Telangana, India
Coordinates	17.45N, 78.5E
Type	Artificial lake
Basin countries	India
Max length	3.2 KM
Max width	2.8 km
Surface area	4.4sq.km
Max depth	32 ft
Surface elevation	17,559,990 ft
Islands	Gibraltar rock
Settlements	Greater Hyderabad

Massive volumes of water used in Hyderabad by residents and industrial corridors go to shop in Hussain Sagar (tank bund). Our analysis is that the water from Hussain Sagar Lake is obtained with this water for 3 samples in a comparable sequential time. We have the pH water sample, conductivity, turbidity, chlorine demand, chemical oxygen demand, bio-chemical oxygen demand, potassium and nitrogen content, iron content, jar test in our laboratory. Either way, the obtained water is not ideal for drinking purposes. With the separate experimental observations, we can say that the water should be used for secondary uses such as industrial use, i.e., cooling water, coal mines, etc.

Massive amounts of water are lost either by not recycling or using the water for secondary applications. As water shortage exists mainly in the summer season in the Hyderabad environment, we can maintain the water through this experimental result. One of the most frequent commercial uses of recycled treated waste is the reuse of cooling water.

The project Hussain Sagar Lake and Catchment Area Development Project, launched in March 2006, intends to increase the quality of lake water by stopping pollutants from entering the lake from both point and non-point contamination sources, in addition to eliminating nutrient-rich sediments. As part of the scheme, in September 2011, seven fountains were built in the lake to increase the dissolved oxygen level in the water and to support marine life and to bring charm to the lake. In addition, the execution of the project included the new 30 MLD sewage treatment plant (STP) on the Picket nalla and 20 MLD STP adjacent to Khairatabad Flyover at Balkapur nalla, the installation of ring sewers across the lake and a 5 MLD small treatment plant at Rangadhamini chervu. Additionally, dredging and removal of sediment is expected for this undertaking. Dredging from Picket nalla, Balkapur nalla and Banjara nalla has begun to produce seven lakh cu.m of non-hazardous and nutrient-rich sediment. In addition, several more initiatives, such as the A Clean Hussain sagar initiative and Canadian technologies to rejuvenate Hussain sagar lake, are operating today to clean the lake. As seen in Table 2, several projects are being undertaken for the creation of Hussain Sagar Lake and its catchment area.

Table 2: Interventions undertaken for Hussain sagar lake improvement

S. No	Name of the Project	Scheme	Budget, Rupees	Year of Sanctioned	Status
1.	Abatement of pollution of Hussain sagar lake project	MCH	40 Crore	1998	Completed
2.	Green Hyderabad Environment Program	HUDA	Not available	2002	Completed
3.	Musi River Action Plan	NRCD	344 Crore	2004	Completed
4.	Hussain sagar lake and Catchment area Improvement Project	HMDA	370 Crore	2006	Completed

2. Methodology and Test Results

Table 3: pH of the samples at different locations

pH of the Samples	Sample 1	Sample 2
Boating point	6.6	6.7
Nimarjanam point	7.4	7.4
Necklace road	7.5	7.5

Table 4: Turbidity of the samples at different locations

Turbidity of Samples (NTU)	Sample 1 (NTU)	Sample 2 (NTU)
Boating point	27	91
Nimarjanam point	25	124
Necklace road	53	96

Table 5: Conductivity of samples at different locations

Conductivity of Samples (micro-Siemens/cm) at different Sample locations	Sample 1 (micro-Siemens/cm)	Sample 2 (micro-Siemens/cm)
Boating point	0.28	0.29
Nimarjanam point	0.29	0.30
Necklace road	0.29	0.30

Total dissolved salts of different samples
Let us consider the factor as 0.8, and then the total dissolved salts in samples are:
Total dissolved salts = conductivity* factor
From Table 5.
Total dissolved salts in the Boating point sample 1 = 0.28 * 0.8 = 0.224
Total dissolved salts in the Nimarjanam point sample 1 =0.29 * 0.8 = 0.232
Total dissolved salts in the Necklace road sample 1 = 0.29 * 0.8 = 0.232
Similarly, the total dissolved salts of all the samples are:

Table 6: Total Dissolved Salts in the samples at different locations

Total dissolved salts (micro-Siemens/cm) at different sample locations	Sample 1 (micro-Siemens /cm)	Sample 2 (micro-Siemens /cm)
Boating point	0.224	0.232
Nimarjanam point	0.232	0.24
Necklace road	0.232	0.24

Table 7: Determination of Total Solids of the sample at different locations

Sample collecting locations	Sample 1 (gm/lit)	Sample 2 (gm/lit)
Boating point	47.725	83.275
Nimarjanam point	62.7	14.225
Necklace road	35.825	39.35

Table 8: Determination of Chlorides of the sample at different locations

Sample collecting locations	Sample 1 (mg/lit)	Sample 2 (mg/lit)
Boating point	106.89	90.09
Nimarjanam point	138.25	87.143

Necklace road	109.186	89.50

Table 9: Determination of Iron Content of the sample at different locations

Sample collecting locations	Sample 1 (mg/lit)	Sample 2 (mg/lit)
Boating point	0.0135	0.0133
Nimarjanam point	0.023	0.0143
Necklace road	0.025	0.0116

Table 10: Determination of Alkalinity of the sample at different locations

Sample collecting locations	Sample 1 (mg/lit)	Sample 2 (mg/lit)
Boating point	952	216.67
Nimarjanam point	1008.5	40
Necklace road	1547	53.3

Table 11: Determination of Acidity of the sample at different locations

Sample collecting locations	Sample 1 (mg/lit)	Sample 2 (mg/lit)
Boating point	30	112.99
Nimarjanam point	11.531	36.892
Necklace road	55	33.013

Table 12: Determination of Dissolved Oxygen of the sample at different locations

Sample collecting locations	Sample 1 (mg/lit)	Sample 2 (mg/lit)
Boating point	1.18	2.167
Nimarjanam point	7.19	1.280
Necklace road	2.561	0.591

Table 13: Determination of Alum Dosage of the sample at different locations

Sample collecting locations	Sample 1 (mg/lit)	Sample 2 (mg/lit)
Boating point	37.5	37.5
Nimarjanam point	37.5	25
Necklace road	37.5	12.5

Table 14: Determination of Chlorine Demand of the sample at different locations

Sample collecting locations	Sample 1 (mg/lit)	Sample 2 (mg/lit)
Boating point	2.5	2.5
Nimarjanam point	5	2.5
Necklace road	2.5	5

Table 15: Determination of 5 Day Biochemical Oxygen Demand (BOD) of the sample

Sample collecting locations	Sample 1 (mg/lit)	Sample 2 (mg/lit)
Boating point	0.606	0.809
Nimarjanam point	0.77	0.68
Necklace road	0.37	0.46

3. Conclusions

The health state of Hussain sagar is determined by conducting several experiments as prescribed in the IS code 3025-1964. However, several cleaning programs were developed by the government of Telangana; with the conclusion that the water of Hussain sagar cannot be used for drinking purpose. Based on results obtained, following conclusions can be drawn:

- The pH, conductivity, and total dissolved salts of the samples 1 and 2 at 3 different locations are within limit as per code IS 10500-2012
- Iron content, Total solids, conductivity, and acidity of all the samples are below the limit(acceptable) as per 10500-2012
- The amount of oxygen required for organisms to live is very less than required as per IS 10500-2012, hence there will not be any chance of presence of living organisms in Hussain sagar lake.
- The alkalinity values of sample 1 at all different locations are very high indicates that water has capability to neutralize high amount of acid, whereas the second sample has the alkalinity value within the limit as per IS 10500-2012.
- By over test results we have concluded that due to acid rains occurred in March just a day before the sample collection resulted in acid precipitation which lowered the pH value of lake.
- The DO level of water at Nimarjanam point is more when compared to other locations due to immersion of idols.
- The turbidity value of second sample is more than the first sample due to acid rain

4. References

[1] Jayesh Ranjan, Reddy KS (2008) "Conservation of Hussain Sagar Lake Hyderabad" (Andhra Pradesh, India). Proceedings of Taal 2007: *The 12th World Lake Conference,* pp.1753-1756.

[2] M Vikram Reddy, K Sagar Babu, V Balaram, M Satyanarayanan (2012) "Assessment of the effects of municipal sewage, immersed idols and boating on the heavy metal and other elemental pollution of surface water of the eutrophic Hussain sagar Lake", Hyderabad, India. Environmental Monitoring and Assessment 184(4):1999-2000

[3] VVS Gurunadha Rao, Mitsuo Yoshida, BA Prakash, SVN Chandrasekhar, K Mahesh Kumar (2004) "Environmental Impact of Human activities to Urban Lake Sediments: Potentially Toxic Elements (PTEs) Contamination in Hussain sagar Lake", Hyderabad. The 11th National Symposium on Hydrology, National Institute of Hydrology, Roorkee (India), p.22-23.

[4] S Umamaheswara Reddy, P Venugopal Rao (2013) "Analysis of Sediment and Pore Water samples of a Contaminated Lake - A Case Study of Hussain Sagar". *International Journal of Emerging Technology and Advanced Engineering* 3(3):841

[5] R Srikanth, AM Rao, CS Kumar, A Khanum (1993) "Lead Cadmium Nickel, and Zinc Contamination of Ground Water around Hussain Sagar Lake", Hyderabad, India. Bulletin of Environmental Contamination and Toxicology 50(1):138-143.

[6] Bob Pears, CVV Prasad, M Chandra Sekhar (2016) "Assessment of Water Quality Status of Musi River-A Tributary of River Krishna in India". *International Journal of Environmental Science and Computing* 6(11):3415-3422.

[7] Basmah Muner Mahmood (2016) "Sewage Treatment Plant of Hussain Sagar Lake, Hyderabad A Case Study". *International Journal of Research*3 (14):515-534.

[8] Sridhar kumar (2014) "Assessment of water quality in Hussain Sagar Lake and its inlet channels using multivariate statistical techniques". *IJSER* 5(9):327-333.

[9] APHA, AWWA, WEF (1995) "Standard Methods for Examination of water &wastewater". 19th edition, APHA, AWWA, WEF, Washington, DC, USA.

[10] The Hindu, News Article –vision lab report, Hyderabad, 2017.

[11] Bureau of Indian Standard (BIS) Indian standard Drinking water Specification (Second Revision) Indian Standard.
[12] Usharan iK (2010) "Physico chemical and bacteriological characteristics of Noyyal River and ground water quality of perur". India Journal of Applies Sci Environ Management 14(2):29-35.

Studies on the seismic behaviour of setback building

J. Selwyn Babu[1], J. Rex[2], K. Bhanu Prakash[3], S P Raju V[4]

[1]Professor, [2]Associate Professor, [3]M.Tech Student
Civil Engineering Department, Malla Reddy Engg College (A), Hyderabad, India.
[4]Assistant Professor, Department of Civil Engineering, GRIET (A), Hyderabad, India.

Corresponding Author: selwynbabu@gmail.com

Abstract Setback is kind of anomaly (irregularity) which can inductee various kinds irregularities for example, mathematical, mass or firmness anomalies and this along vertical side. The current examination is restricted to fortified cement outlined structures intended for seismic burdens (DL, LL and EL). The seismic conduct of G+10, G+15, G+20 and G+25 structures with and without misfortunes were contemplated. The models were examined utilizing Response Spectrum technique in ETABS. The impact of Setback is contemplated considering the boundaries, for example, storey displacements, storey floats & storey shears are correlated with the structure without a setback.
Keywords: Storey displacements, storey drifts, storey shear, setback, Response spectrum

1. Introduction

Tremors are the most capricious and decimating of every cataclysmic event, which are hard to spare over designing properties and life, against it. Consequently, so as to defeat these issues we have to distinguish the seismic exhibition of the assembled climate through the advancement of different diagnostic systems, which guarantee the structures to withstand during continuous minor quakes and produce enough alert at whatever point exposed to serious tremor functions so it can spare however many lives as would be prudent. The investigation system evaluating the tremor powers and its interest contingent upon the significance and cost, the technique for dissecting the structure fluctuates from direct to non-straight. The conduct of a structure during a quake relies upon a few elements, solidness, and sufficient parallel quality, and malleability, straightforward and customary arrangements. The structures with customary math and consistently conveyed mass and firmness in plan just as in height endure substantially less harm contrasted with unpredictable setups. Regardless, nowadays need and solicitation of the latest age and creating people has made the modelers or masters unavoidable towards masterminding of inconsistent plans. From now on quake planning has developed the primary concerns of enthusiasm for understanding the capacity of building setups. Most late seismic tremors have indicated that the anomalies in plan, height, conveyance of mass, solidness and qualities may cause genuine harm in basic frameworks. Nonetheless, an exact assessment of the seismic conduct of unpredictable structures is very troublesome and a convoluted issue. There are various models in the harm report of past quakes in which the reason for disappointment of multi-celebrated strengthened solid structures is anomalies in designs.

Figure 1. Irregular shape structure disaster

2. Scope of the study

- The system of this study is limited to supporting the design of concrete block for seismic loads (D.L, L.L & E.L).
- The tremor behaviour of G+10, G+15, G+20 and G+25 models with & without setbacks were premeditated.
- The models were analysed utilizing Response Spectrum Method
- The effects of rebound can be studied by considering the parameters related to non-rebound buildings (such as floor displacement, floor drift, and floor shear).

3. Methodology

3.1. Response spectrum:

This technique was also well known as the modal or mode overlap technique. This procedure is relevant for individual models if more than the essential modes have a particular impact on the response of the model. This study is generally associated with investigating the dynamic response of models that are asymmetric or that have areas of discontinuity or irregularity in the linear range of motion. In particular, it can be applied to the force and deformation analysis of multi-story buildings due to moderately strong floor oscillation causing a moderately large but essentially linear response to the structure. For a specific type of damping, this method can calculate the response in any natural vibration mode independently of other modes, which is an effective model for many buildings, and the modal response can be combined to determine the overall response. It is based on facts. Each mode has its own deformation mode (mode shape), its own frequency (modal frequency) and its own modal damping to respond.

4. Modelling

- Structure – Ordinary Moment Resisting Frame
- Number of stories - G+10, G+15, G+20 and G+25
- Category of building - Regular & Symmetrical in top view
- Top view area - 12 met. x 8 met.
- Elevation of the model - 30 m, 45 m, 60 m and 75 m
- Storey elevation- Bottom floor height - 3.0 m, Typical floor ht. - 3.0 m
- Support type - Fixed
- Seismic zone - V
- Concrete grade type - M30

IOP Conf. Series: Materials Science and Engineering 1091 (2021) 012043 doi:10.1088/1757-899X/1091/1/012043

- Steel grade type - Fe415
- Reinforced concrete density - 25 kN/m^3
- E_c - 27386127.87 kN/m^2
- E_s - 2 x 10^8kN/m^2
- Floor finish value- 1.5 kN/m^2
- Live load burden on floors - 3 kN/m^2
- Wall. burden on beams - 3.9 kN/m^2
- Parapet. wall burden - 1 kN/ m^2
- Zones V - 0.36
- Importance. factor (I) – 1
- R.R.F (R) – 5%
- Soil category - II
- Damping. ratio value – 0.05
- Column – 900 mm x 600mm
- Beam – 600 mm x 300 mm
- Slab – 175

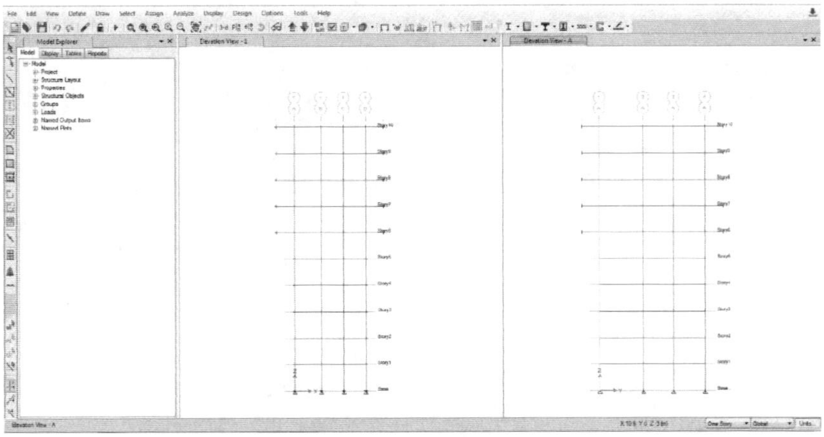

Figure 2 Elevation of G+10 building with setbacks

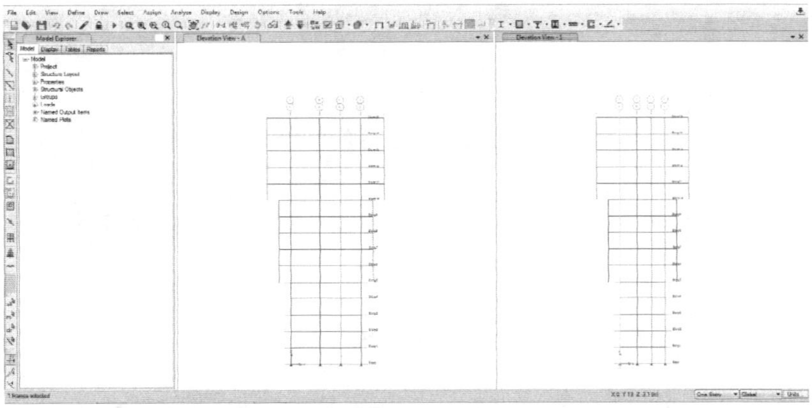

Figure 3 Elevation of G+15 building with setbacks

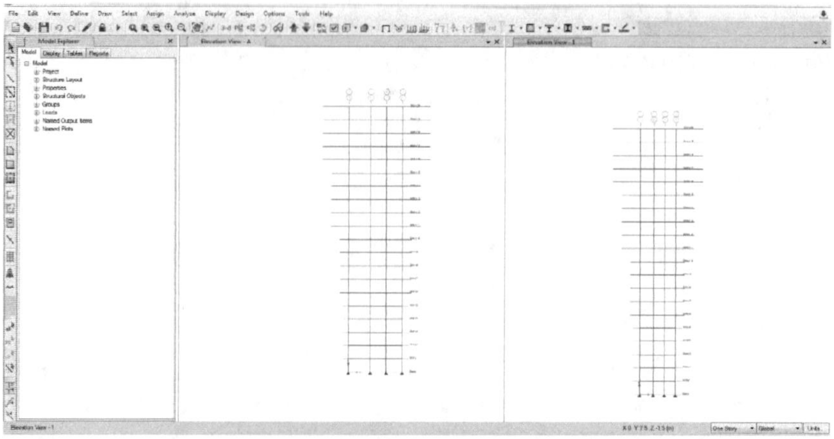

Figure 4. Elevation of G+20 building with setback

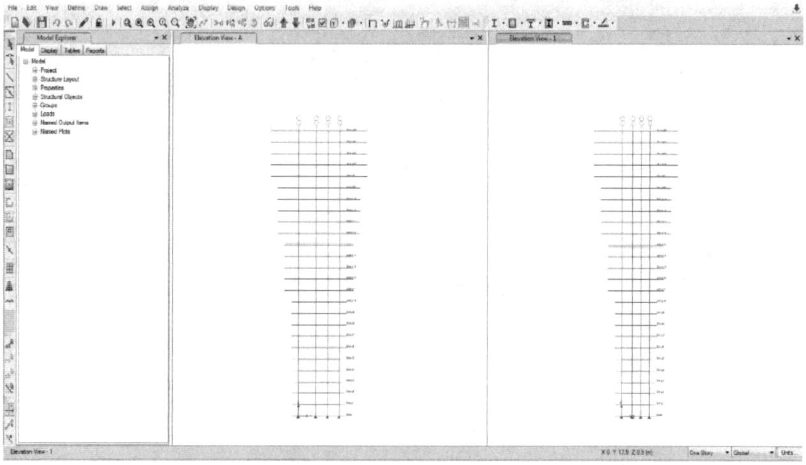

Figure 5. Elevation of G+25 building with setbacks

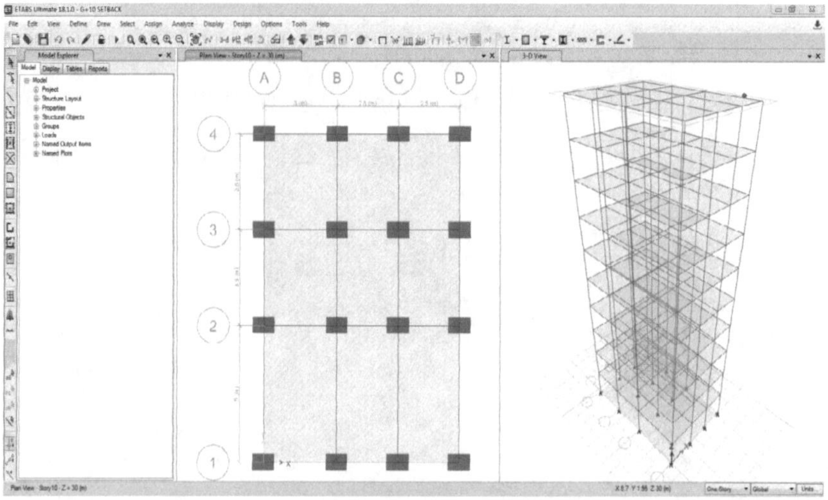

Figure 6. Plan and 3D view of G+10 building without setbacks

5. Results and Discussions

5.1 Storey displacements

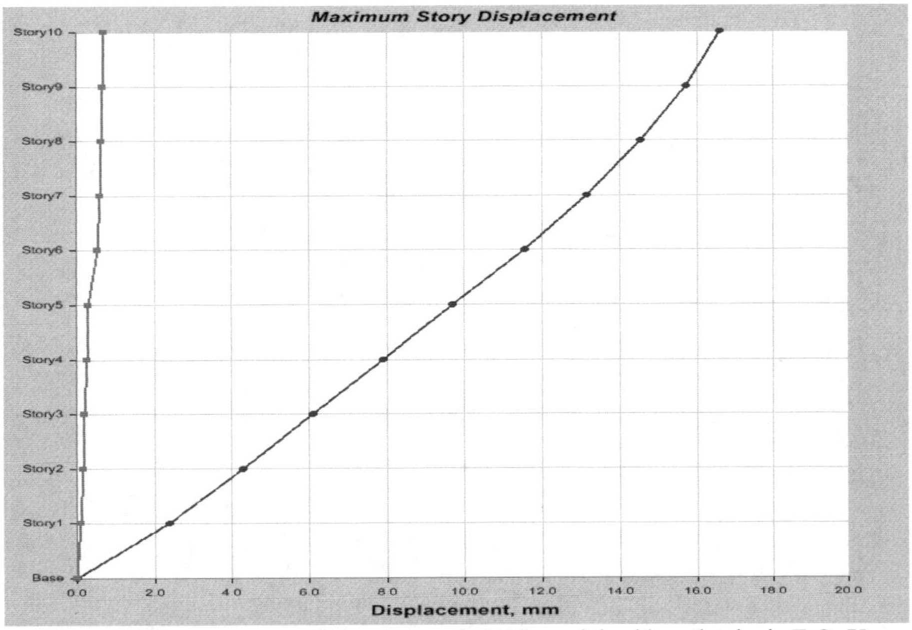

Figure 7. Extreme storey dislocations of G+10 model with setbacks in E.Q. X

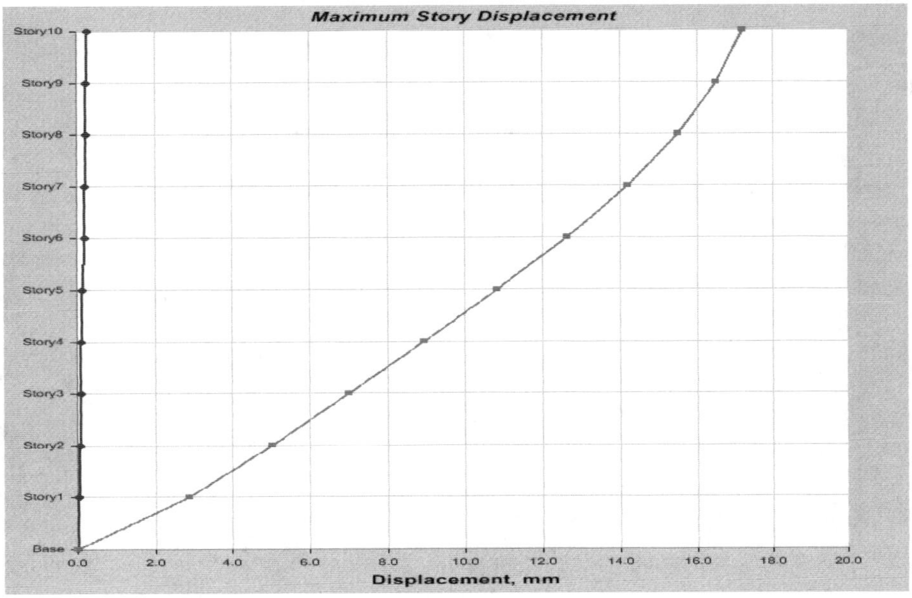

Figure 8. Extreme storey dislocations of G+10 model with setbacks in E.Q. Y

Storey dislocation is the entire movement of any floor w.r.t ground. Storey displacements for the G+10 model with setbacks have higher displacement in quake Y direction with.respect.to X path as shown in fig. 7 & 8.

5.2 Storey drifts

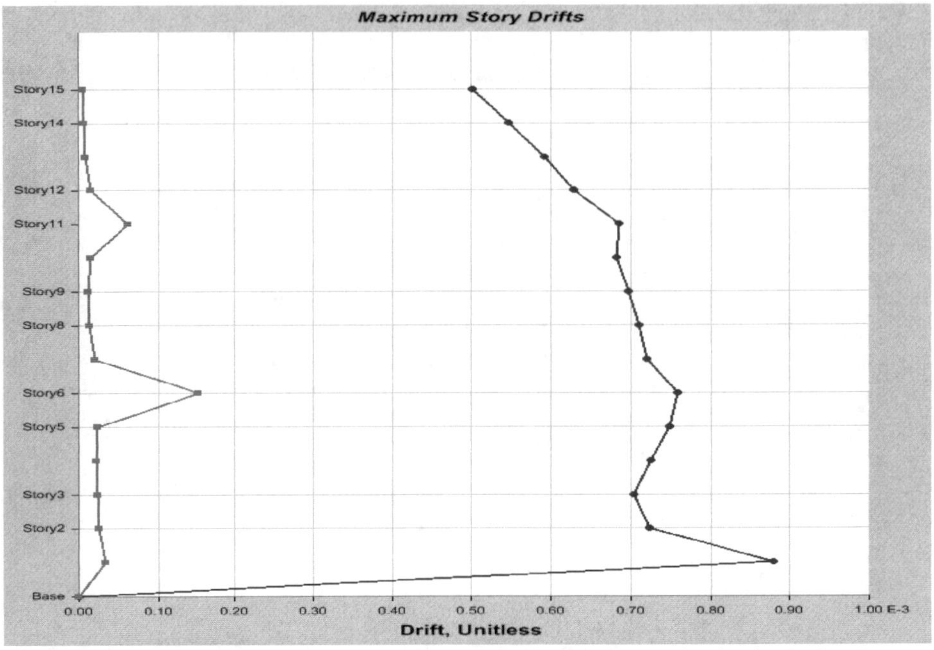

Figure 9. Extreme storey drift of G+15 model having setbacks for E.Q. X

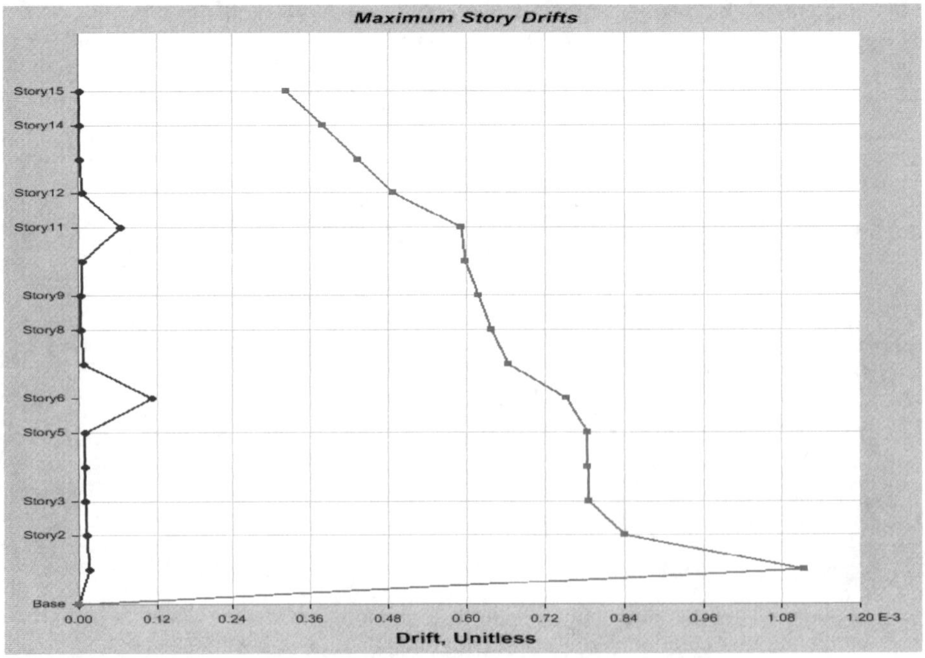

Figure 10 Extreme storey drift of G+15 model having setbacks for E.Q. Y

Storey drift is the horizontal dislocation of one storey relative to the storey above or below. For G+15 model with setbacks it was observed that drifts are more in X dir. w.r.t Y dir. as shown in fig. 9 & 10.

5.3 Storey shear:

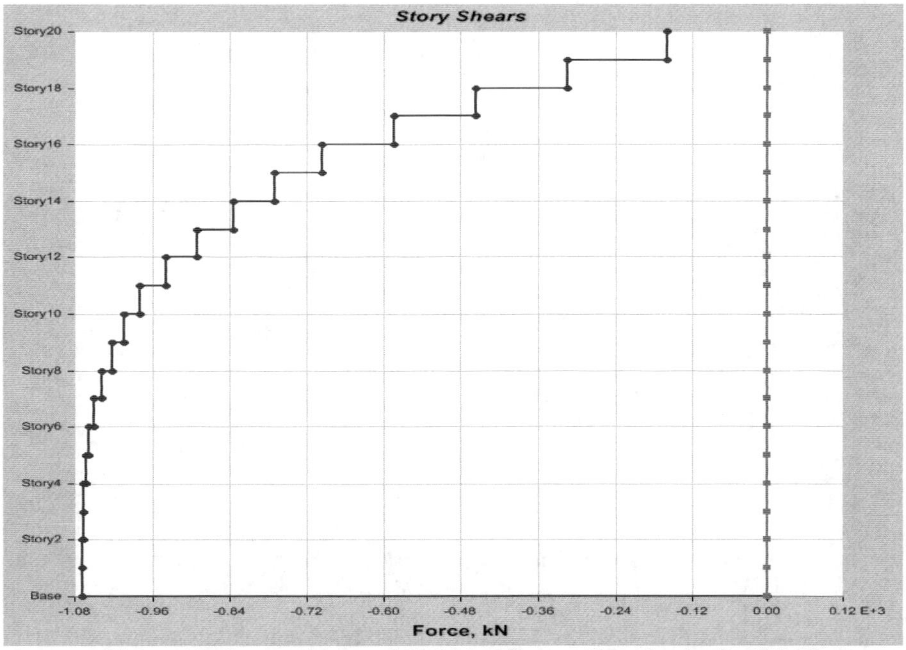

Figure 11. Storey shears of G+20 building with setbacks for EQ X

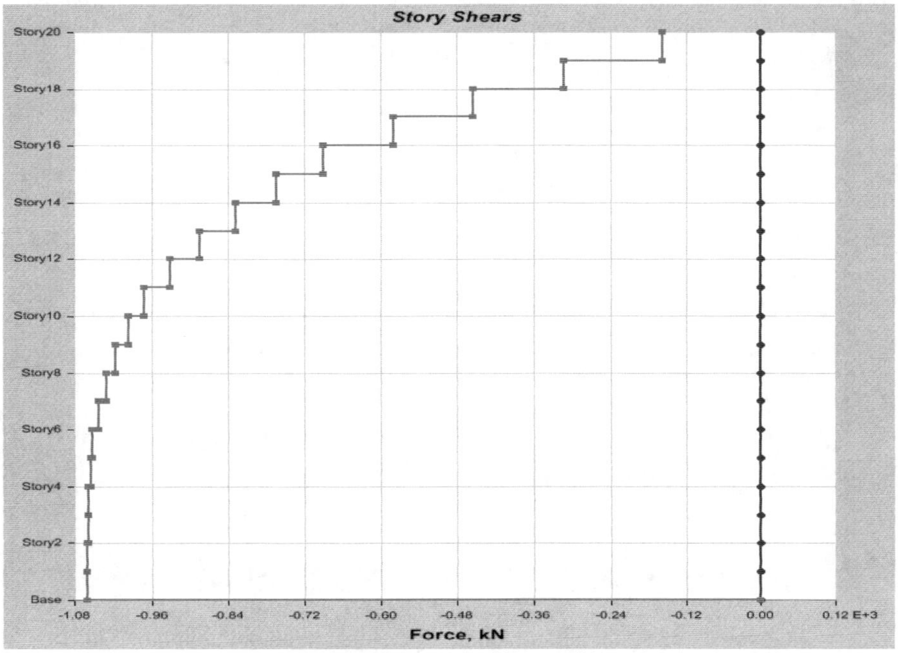

Figure 12. Storey shears of G+20 building with setbacks for EQ Y

Story shear factor is the proportion of the story shear power when story breakdown happens to the story shear power when complete breakdown happens. Storey shear for G+20 model was observed that shear was increased in Y direction w.r.t X direction as shown in figure 11 & 12

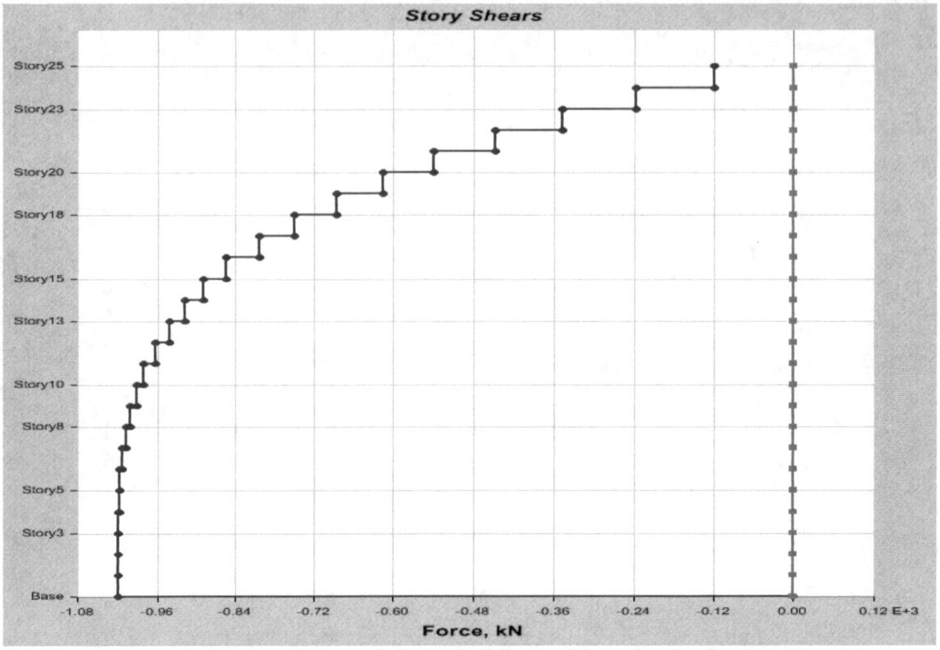

Figure 13. Storey shears of G+25 building with setbacks for EQ X

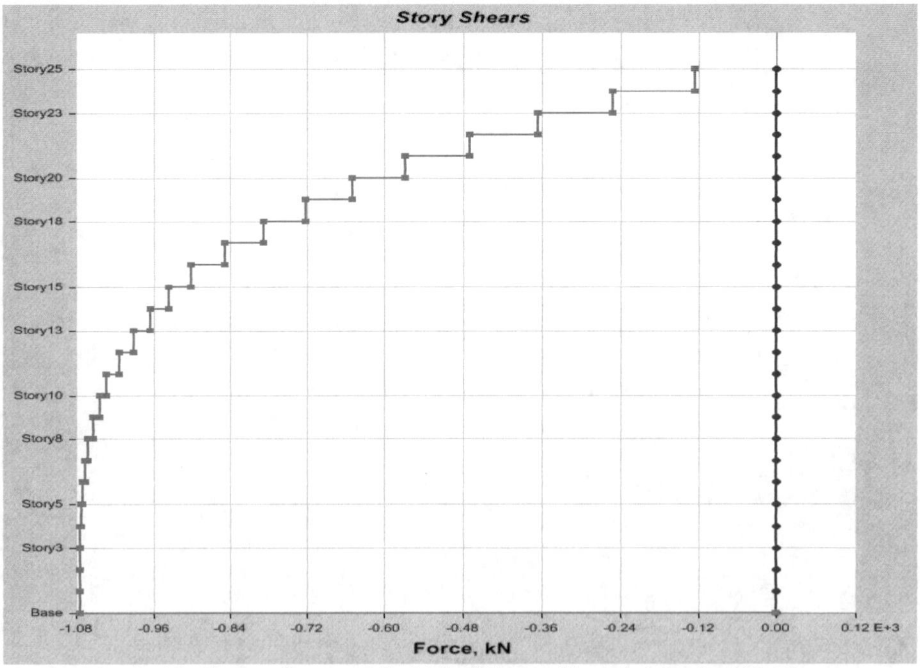

Figure 14. Storey shears of G+25 building with setbacks for EQ Y

Storey shear for G+25 model was observed that it was increased in X direction more w.r.t Y direction as shown in figures 13 & 14

6. Conclusion

Conclusions drawn from the analysis results of G+10, G+15, G+20 and G+25 buildings with and without setbacks are as follows.

- As the storey increases the displacements and drifts increases and the storey shears decreases.
- For the plan area of 12 m X 8 m the G+25 building without setbacks is unstable and the displacements and drifts are very high. As the column and beam sizes considered are maximum but G+25 building is unstable without setbacks.
- The displacements and drifts are more for the buildings with setbacks as compared to buildings without setbacks. But the buildings with setbacks are having less shears and they are more stable than buildings without setbacks
- The maximum storey displacements of G+10 building with setbacks are 11% and 10% more than the G+10 building without setbacks in X & Y dir. respectively.
- The extreme storey displacements of G+15 building with setbacks are 18.5% and 14.5% more than the G+15 building without setbacks in X & Y dir. respectively.
- The Extreme storey displacements of G+20 building with setbacks are 25% and 23% more than the G+20 building without setbacks in X & Y dir. respectively.
- The maximum storey displacements of G+25 building with setbacks are 80.2 mm and 67 mm in X and Y directions respectively.
- As the storey is higher the maximum story displacements of a higher storey building are increasing 40% approximately than the lower storey building.
- The storey drifts are higher for the G+15 building with setbacks as compared to other buildings with and without setbacks. The maximum values are 0.000879 and 0.001113 in X and Y directions respectively.
- The storey shears are higher for the G+15 building with setbacks as compared to other buildings with and without setbacks. The maximum values are 1156.68 KN and 1074.92 KN in X and Y directions respectively.

7. References

[1] Fabio, "Nonlinear seismic examination of R.c. encircled structures with incidents retrofitted by damped underpins", https://doi.org/10.1016/j.engstruct.2016.07.055
[2] Beskos & Bazeos, "Seismic reaction of plane steel MRF with misfortunes: Estimation of inelastic distortion requests", https://doi.org/10.1016/j.jcsr.2007.12.002
[3] Sebastian, Leonardo & Ignacio, "Trial cyclic reaction of RC dividers with difficulty discontinuities", https://doi.org/10.1016/j.engstruct.2018.10.054
[4] Achilleas & Triantafyllos Makarios, "Surmised seismic examination of multi-story structures with mass and firmness abnormalities",
https://doi.org/10.1016/j.proeng.2015.11.147
[5] Zeris & Giannakouras, "Seismic execution of unpredictable RC outlines structured by the DDBD approach", https://doi.org/10.1016/j.engstruct.2018.12.058
[6] Yexian, kaoshan & Solomon, "Seismic execution-based structure and hazard investigation of warm force plant working with thought of vertical and mass abnormalities"
https://doi.org/10.1016/j.engstruct.2018.03.001
[7] Hema Chandra & Bharath, "Seismic flexible execution of L-formed structure outlines through arrangement abnormalities", https://doi.org/10.1016/j.istruc.2020.05.017
[8] Hamzeh & Maedeh Zakersalehi, "Mathematical reproduction of seismic breakdown systems of vertically unpredictable steel elevated structures",
https://doi.org/10.1016/j.jcsr.2019.105914

[9] I.S 1893, (2002) part one, code book Proposals for quake safe plan of structures.
[10] Anil.K. chopra, “Dynamics of structures”

Flexural behaviour of high strength basalt FRP rebar concrete

[1] E Giri Prasad Goud, [2] Marimuthu Kannimuthu, [3] V Srinivasa Reddy, [4]V Ram Singh

[1]Research Scholar of Civil Engineering, K L University, Andhra Pradesh, India. Email id: giriprasadgoud7@gmail.com
[2]Department of Civil Engineering, K L University, Andhra Pradesh, India
[3]Professor of Civil Engineering, GRIET, Hyderabad, Telangana, India. Email: vempada@gmail.com
[4]M.Tech (Structural Engineering), Department of Civil Engineering, GRIET, Hyderabad, India

Abstract. In this paper, the flexural behaviour of under-reinforced M70 grade basalt FRP rebar concrete beams in terms of first crack load, pre-cracking and post-cracking behavior, deflection pattern, crack development pattern and ultimate load carrying capacity is evaluated. Loads at first crack in basalt FRP rebars beams are higher than the normal steel rebar concrete beams confirming that basalt FRP rebars beams offer greater resistance to cracking. The ultimate flexural strength of basalt FRP rebar concrete beams have improved when compared to normal steel rebar concrete beams. First crack formation was delayed in basalt FRP rebar concrete beams due to which fatigue strength is increased which in turn increases the time taken for first crack occurrence and thereby increasing the load carrying capacity. The deflection at the mid span decreased in basalt FRP rebar concrete beams which indicates the improved flexural stiffness of the elements thereby reducing the structural member's deformability, increasing strength and hence controlling deflection. Mid-span deflections at peak load reduced in basalt FRP rebar concrete beams. The ultimate deflection in case of basalt FRP rebar concrete beams is less than that of normal steel rebar concrete beams.

Keywords: Basalt fibres, BFRP rebars, flexural behaviour, load carrying capacity, reinforced beams

1. Introduction

Ductility is the deformation capacity of a member or structure after the first yield. It gives a measure of energy dissipation capacity of a member or structure. Work done on a structure after yield cannot be stored in the material so it is converted into heat energy and is dissipated to the environment. Energy dissipation is desirable for structures because the heavy energy imparted by the ground motions, etc., are to be released.

Up to application of load at first crack the curve is linear and on further application of load multiple cracks are caused and the deviation in curve is observed. For under reinforced beams with the increase in load the multiple cracks increased. After the multiple cracking stages, it is found that there is yield in steel so the P-δ curve become more or less flat till the ultimate load is reached. On further increase in load a drop in the load is observed with propagation of cracks. All the beams failed by compression of concrete and the load deformation curves are plotted up to failure stage.

IOP Conf. Series: Materials Science and Engineering 1091 (2021) 012044 doi:10.1088/1757-899X/1091/1/012044

In beams, curvature increases gradually with increase in moment up to the multiple cracking stages and beyond, the curvature increased drastically at constant moment or with small variation in the moment. The M-Φ curve becomes more or less flat till the ultimate moment is reached. Finally all the beams failed by compression of concrete and the moment curvature plots are drawn up to failure stage.

2. Objectives

To evaluate the flexural behaviour of under-reinforced M70 grade basalt FRP rebar concrete beams in terms of first crack load, pre-cracking and post-cracking behavior, deflection pattern, crack development pattern and ultimate load carrying capacity

3. Experimental Investigations

A M70 grade basalt FRP rebar concrete, under-reinforced beams of size 100 x 150 x 1200 mm were cast and tested under two point loading as per IS: 9399 – 1979 and their behaviour in flexure was investigated to understand the effect of Basalt FRP rebar on its deflection characteristics and cracking behavior. The beams are tested under symmetrical two-point flexural loading. From the above investigations, the parameters observed are first crack load, pre-cracking and post-cracking behavior, deflection pattern, crack development pattern and ultimate load carrying capacity of basalt FRP rebar agent incorporated beams. The beam sizes and length were chosen to ensure that the beams would fail in flexure. The beam dimensions were also sufficiently large to simulate a real structural element. The beam details are shown in Table 1 and 2 and Figure 1 shows reinforcement arrangement. The beam specimen is simply supported over an effective span of 1000 mm. The clear cover of the beam was 20 mm.

Table 1. Rebar details of M70 grades beams

Grade of the concrete beam	Steel on compression side	Steel on tension side	Shear reinforcement
M70	2-10 mm Dia.	2-16 mm Dia.	8 mm Dia. @90 mm c/c

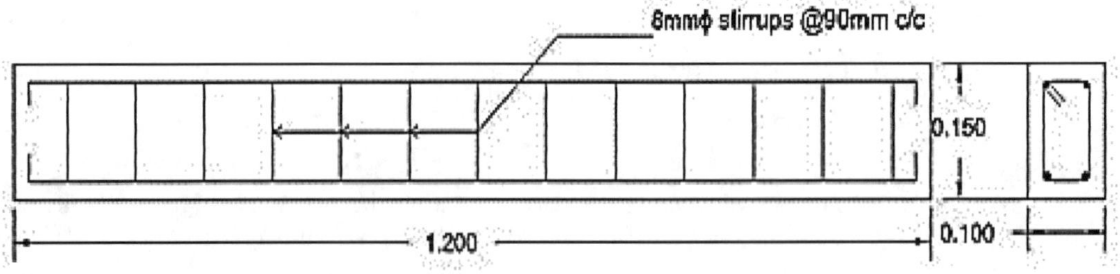

Figure 1. Geometry of beam specimens (All measurements are in meters)

Table 2. Beam descriptions

Beam Description	Grade	Size of Beam L x B x H	Type of rebar	Type of Section
M70N-UR	M70	1200 mm x 100 mm 150	Steel	Under-reinforced

M70BFRP-UR		mm	Basalt FRP	Under-reinforced

High yield strength deformed steel bars of diameter 16 mm and 10 mm were used as the longitudinal reinforcement in the specimens. The reinforcement details are given in the table 2 for both compression steel and tension steel. Two legged vertical stirrups of 8 mm diameter at spacing of 90 mm c/c were provided as shear reinforcement.

One day before the testing, the cured beams are white washed and the locations of supports; load points and the central deflection gauge are marked with a pencil. The test specimen was mounted in a UTM of 1000 kN capacity. The supports of the beam rested on an adjustable steel rollers. The effective span of the beam is 1000 mm. The load was applied on two points each 333 mm away from the support. Dial gauges of 0.01 mm least count were used for measuring the deflections under the load points and at mid span for measuring the deflections. Beams are subjected to symmetrical two-point flexural loading under strain rate control. The dial gauge readings were recorded at different loads until failure. The load was applied at the rate of 2.5 kN/sec until the first crack was observed. Subsequently, the load was applied at the rate of 5kN/sec. The behavior of the beam was observed carefully and the first crack was identified using a hand held microscope. The failure mode of the beams was also recorded. Deflections at the central point and the ultimate load carrying capacity (Peak load) are measured for M70N-UR and M70BFRP-UR beams considered. The test is continued until the load drops to 15-20% of peak load recorded, in the descending portion of load deflection curves. At that stage the testing is stopped by gradually unloading. The experimental test set-up is shown in Figure 2.

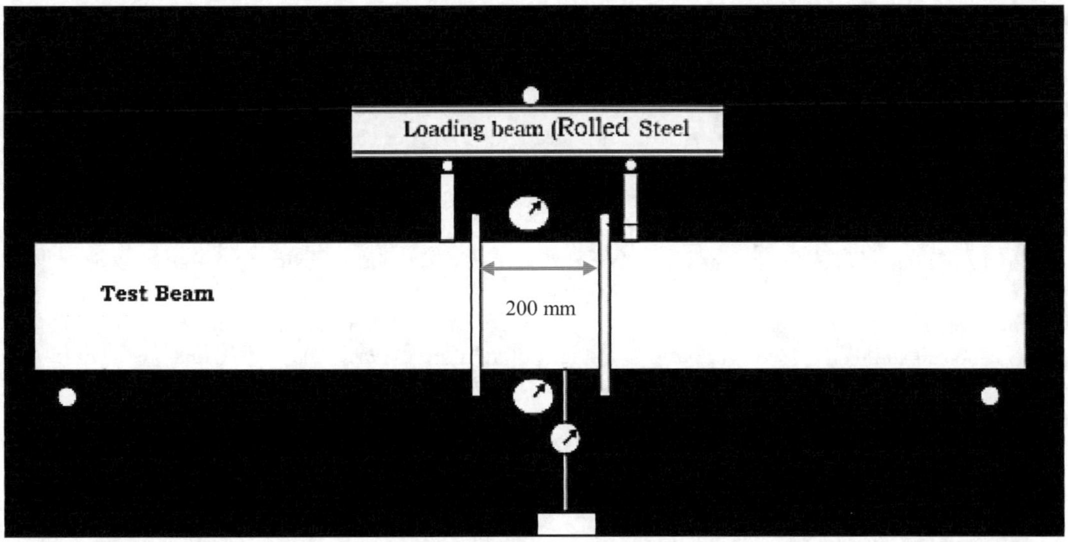

Figure 2. Flexural Test Set up

The strain in beam was measured using vernier calipers. The beam was instrumented to record the strain profile across a length of 200 mm at midspan at the extreme top for strain in compression fibre and at the bottom for strain in tensile steel as shown in figure 2. All measurements were recorded manually. The change in length of the portion under observation is taken at every interval of load using digital vernier calipers. The behaviour of the beam was observed carefully and the first crack was identified with magnifying glass. The deflections and strain values were recorded for respective load increments until failure. The failure mode of the beams were also recorded.

4. Load-Deflection and Moment –Curvature Relations

Deflections were measured at the central point and under the load using the deflection meters to plot load-deflection curves. The values of moments and curvatures are also obtained. A sample calculation for curvature and moment is presented below. Load-deflection behavior of M70 grade of Basalt FRP rebar beams were studied using load –deflection plots. Moment curvature relationships are very important to assess out ductility of the structure and the amount of possible redistribution of stresses.

The deformations measured are divided by the gauge length (200 mm) to obtain the strains at the particular level. From the top and bottom strains, the average curvatures were calculated. From these results, M-Φ diagrams are plotted for under reinforced Basalt FRP rebar concrete beams.

Test results regarding load at first crack, ultimate load carrying capacity and deflection at maximum load of normal and basalt FRP rebar concrete beams are tabulated in Table 3 to 6

No horizontal cracks were noticed at the level of the reinforcement indicating that there was no occurrence of bond failure. Vertical flexural cracks were observed in the middle third region (constant moment) and ultimate failure occurred due to crushing of the compression concrete with significant amount of deflection. The flexure cracks were the first to initiate in the Constant Bending moment Zone (middle third) as expected. Loads at first crack in basalt FRP rebars beams are higher than the normal steel rebar concrete beams confirming that basalt FRP rebars beams offer greater resistance to cracking. The ultimate flexural strength of basalt FRP rebar concrete beams have improved when compared to normal steel rebar concrete beams. First crack formation was delayed in basalt FRP rebar concrete beams due to which fatigue strength is increased which in turn increases the time taken for first crack occurrence and thereby increasing the load carrying capacity. The deflection at the mid span decreased in basalt FRP rebar concrete beams which indicates the improved flexural stiffness of the elements thereby reducing the structural member's deformability, increasing strength and hence controlling deflection. Mid-span deflections at peak load reduced in basalt FRP rebar concrete beams. The ultimate deflection in case of basalt FRP rebar concrete beams is less than that of normal steel rebar concrete beams.

From the above investigations, it can be concluded that, the ultimate flexural strength of basalt FRP rebar concrete beams have improved significantly when compared with normal steel reinforced beams due to better bond strength at the interfacial zones. Mid-span deflections at ultimate load in beams incorporated with basalt FRP rebar are reduced significantly. The recovery of deflection after removal of the load was more in case of basalt FRP rebar concrete beam indicating more elastic behaviour. Thus basalt FRP rebar concrete beam are more suitable to take fatigue loads than normal steel reinforced beams. From the above investigations it is evident that the load at first crack, cracking behavior, deflection pattern, crack development pattern and ultimate load carrying capacity of basalt FRP rebar concrete beams have improved enormously.

In normal concrete beams the visible flexural cracks developed at 70% to 80% of the ultimate load of each beam and in basalt FRP rebar concrete beams they developed at 75% to 85% of the ultimate load of each beam. All the beams exhibited a tension failure which is a ductile failure. The cracks are accompanied by pronounced bulging. When the load is further increased, cracks propagated towards the top of the beam. As the beams are forced to deform further, the cracks became more pronounced and the concrete crushed at one or both the ends. With further increase in beam deflection, the load decreased, accompanied by concrete spalling. This crack pattern is observed to be same for all the normal and basalt FRP rebar concrete beams.

The differences noticed in the moment-curvature behaviour of basalt FRP rebar concrete beams are, the increase in the horizontal plateau of the moment-curvature plots and increase in ultimate moment in beams with basalt FRP rebar beams than that in normal steel rebar beams.

In basalt FRP rebar concrete beams, the bondage is more which arrests the micro cracks developed in the matrix which results in requirement of more energy. This lead to an improvement in load at first crack. In under-reinforced beams after the multiple cracking stages, the yielding of steel was found to be more, and ultimate load corresponds to the yielding of steel and crushing of concrete. The deflection at service load is determined from load deflection plot corresponding to a load of Pu/1.5. Deflections observed from the load-deflection curves for all the specimens at service loads (P_u/1.5) are less than the maximum permissible deflection of 4mm i.e span/250 specified by IS 456-2000.

Table 3. Observations of flexural test on steel rebar concrete beam

M70N-UR											
Load kN	Dial gauge reading		Deflection mm		Strain gauge reading		Ec	Et	1/R	d mm	Moment kN.m
	Centre	Under load	Centre	Under load	Top	Bottom					
0	0	0	0	0	199.91	199.32	0	0	0	122	0.000
5	11	6	0.11	0.06	199.9	199.34	5E-05	0.0001	1.21E-06	122	0.833
10	23	14	0.23	0.14	199.9	199.35	5E-05	0.000151	1.62E-06	122	1.667
15	35	23	0.35	0.23	199.89	199.36	0.0001	0.000201	2.43E-06	122	2.500
20	47	32	0.47	0.32	199.89	199.37	0.0001	0.000251	2.83E-06	122	3.333
25	59	41	0.59	0.41	199.88	199.38	0.00015	0.000301	3.64E-06	122	4.167
30	71	51	0.71	0.51	199.88	199.38	0.00015	0.000301	3.64E-06	122	5.000
35	83	61	0.83	0.61	199.87	199.39	0.0002	0.000351	4.45E-06	122	5.833
40	95	72	0.95	0.72	199.86	199.41	0.00025	0.000452	5.66E-06	122	6.667
45	108	83	1.08	0.83	199.85	199.42	0.0003	0.000502	6.47E-06	122	7.500
50	121	94	1.21	0.94	199.84	199.44	0.00035	0.000602	7.68E-06	122	8.333
55	134	105	1.34	1.05	199.83	199.45	0.0004	0.000652	8.49E-06	122	9.167
60	147	116	1.47	1.16	199.82	199.46	0.00045	0.000702	9.3E-06	122	10.000
65	162	128	1.62	1.28	199.8	199.48	0.00055	0.000803	1.09E-05	122	10.833
70	177	140	1.77	1.4	199.79	199.5	0.0006	0.000903	1.21E-05	122	11.667
75	192	153	1.92	1.53	199.78	199.51	0.00065	0.000953	1.29E-05	122	12.500
80	207	166	2.07	1.66	199.77	199.52	0.0007	0.001003	1.37E-05	122	13.333
85	222	179	2.22	1.79	199.76	199.53	0.00075	0.001054	1.45E-05	122	14.167
90	237	193	2.37	1.93	199.75	199.54	0.0008	0.001104	1.54E-05	122	15.000
95	253	206	2.53	2.06	199.75	199.56	0.0008	0.001204	1.62E-05	122	15.833
100	269	220	2.69	2.2	199.74	199.58	0.00085	0.001304	1.74E-05	122	16.667
105	285	234	2.85	2.34	199.73	199.59	0.0009	0.001355	1.82E-05	122	17.500
110	302	248	3.02	2.48	199.72	199.6	0.00095	0.001405	1.9E-05	122	18.333
115	319	263	3.19	2.63	199.71	199.62	0.001	0.001505	2.02E-05	122	19.167
120	337	278	3.37	2.78	199.7	199.64	0.00105	0.001605	2.14E-05	122	20.000
125	355	293	3.55	2.93	199.69	199.66	0.0011	0.001706	2.26E-05	122	20.833
130	377	308	3.77	3.08	199.68	199.68	0.001151	0.001806	2.38E-05	122	21.667
135	403	328	4.03	3.28	199.67	199.73	0.001201	0.002057	2.63E-05	122	22.500
140	429	354	4.29	3.54	199.66	199.76	0.001251	0.002208	2.79E-05	122	23.333
145	455	380	4.55	3.8	199.65	199.79	0.001301	0.002358	2.95E-05	122	24.167
150	481	406	4.81	4.06	199.63	199.84	0.001401	0.002609	3.23E-05	122	25.000
155	507	432	5.07	4.32	199.61	199.89	0.001501	0.00286	3.52E-05	122	25.833
160	551	476	5.51	4.76	199.59	199.94	0.001601	0.003111	3.8E-05	122	26.667
165	605	530	6.05	5.3	199.56	200.01	0.001751	0.003462	4.2E-05	122	27.500
170	734	659	7.34	6.59	199.45	201.13	0.002301	0.009081	9.18E-05	122	29.833

IOP Conf. Series: Materials Science and Engineering 1091 (2021) 012044 doi:10.1088/1757-899X/1091/1/012044

Table 4. Observations of flexural test on basalt FRP rebar concrete beam

M70BFRP-UR											
Load kN	Dial gauge reading		Deflection mm		Strain gauge reading		Ec	Et	1/R	d mm	Moment kN.m
	Centre	Under load	Centre	Under load	Top	Bottom					
0	0	0	0	0	206.19	205.97	0	0	0	122	0.000
5	10	7	0.1	0.07	206.19	205.99	0	9.71E-05	7.83E-07	122	0.833
10	21	15	0.21	0.15	206.18	206	4.85E-05	0.000146	1.57E-06	122	1.667
15	33	23	0.33	0.23	206.17	206.01	9.7E-05	0.000194	2.35E-06	122	2.500
20	45	32	0.45	0.32	206.16	206.02	0.000145	0.000243	3.13E-06	122	3.333
25	57	40	0.57	0.4	206.15	206.03	0.000194	0.000291	3.91E-06	122	4.167
30	69	51	0.69	0.51	206.15	206.04	0.000194	0.00034	4.31E-06	122	5.000
35	81	61	0.81	0.61	206.14	206.05	0.000242	0.000388	5.09E-06	122	5.833
40	93	72	0.93	0.72	206.13	206.07	0.000291	0.000486	6.26E-06	122	6.667
45	105	83	1.05	0.83	206.12	206.08	0.000339	0.000534	7.04E-06	122	7.500
50	118	94	1.18	0.94	206.11	206.09	0.000388	0.000583	7.83E-06	122	8.333
55	131	105	1.31	1.05	206.1	206.11	0.000436	0.00068	9E-06	122	9.167
60	144	116	1.44	1.16	206.1	206.13	0.000436	0.000777	9.78E-06	122	10.000
65	157	127	1.57	1.27	206.09	206.14	0.000485	0.000825	1.06E-05	122	10.833
70	171	141	1.71	1.41	206.08	206.16	0.000533	0.000922	1.17E-05	122	11.667
75	185	154	1.85	1.54	206.07	206.18	0.000582	0.00102	1.29E-05	122	12.500
80	199	166	1.99	1.66	206.07	206.19	0.000582	0.001068	1.33E-05	122	13.333
85	213	179	2.13	1.79	206.06	206.2	0.00063	0.001117	1.41E-05	122	14.167
90	227	193	2.27	1.93	206.05	206.21	0.000679	0.001165	1.49E-05	122	15.000
95	241	206	2.41	2.06	206.04	206.22	0.000727	0.001214	1.57E-05	122	15.833
100	256	220	2.56	2.2	206.03	206.23	0.000776	0.001262	1.64E-05	122	16.667
105	271	233	2.71	2.33	206.02	206.24	0.000824	0.001311	1.72E-05	122	17.500
110	286	247	2.86	2.47	206.01	206.25	0.000873	0.001359	1.8E-05	122	18.333
115	301	262	3.01	2.62	206	206.26	0.000921	0.001408	1.88E-05	122	19.167
120	316	276	3.16	2.76	205.99	206.28	0.00097	0.001505	2E-05	122	20.000
125	332	292	3.32	2.92	205.98	206.3	0.001018	0.001602	2.11E-05	122	20.833
130	348	307	3.48	3.07	205.97	206.32	0.001067	0.001699	2.23E-05	122	21.667
135	364	326	3.64	3.26	205.96	206.33	0.001115	0.001748	2.31E-05	122	22.500
140	380	342	3.8	3.42	205.95	206.35	0.001164	0.001845	2.43E-05	122	23.333
145	398	360	3.98	3.6	205.94	206.39	0.001212	0.002039	2.62E-05	122	24.167
150	424	386	4.24	3.86	205.92	206.43	0.001309	0.002233	2.86E-05	122	25.000
155	455	417	4.55	4.17	205.9	206.48	0.001406	0.002476	3.13E-05	122	25.833
160	493	455	4.93	4.55	205.87	206.53	0.001552	0.002719	3.44E-05	122	26.667
165	550	512	5.5	5.12	205.83	206.59	0.001746	0.00301	3.84E-05	122	27.500
170	634	596	6.34	5.96	205.79	206.71	0.00194	0.003593	4.46E-05	122	28.333
175	781	743	7.81	7.43	205.73	207.63	0.002231	0.008059	8.3E-05	122	29.167

4.1 Model calculation of Deflection, E_c, E_t, Curvature and Moment

Deflection in mm= dial gauge reading X L.C (0.01)
Ec = (201.46- top strain gauge reading)/201.46
Et = (bottom strain gauge reading -200.3)/200.3
1/R = (Ec + Et)/d
Support reaction = load/2
Distance of support from application of load L/3 where L is 1m.
So Maximum Moment in kN-m = (load/2)*(L/3)= (load x L)/6 = load x 1m/6

Table 5. Flexural features of reinforced beam made with steel rebars

M70 grade reinforced beam made with steel rebars (M70N-UR)						
Service load kN	Load at first crack (kN)	Ultimate Flexural Strength (kN)	Deflection at service load mm	Central Deflection at Failure (mm)	Deflection Under load at failure (mm)	Maximum Crack width at failure (mm)
49	65	170	3.72	7.34	6.59	1.14

Table 6. Flexural features of reinforced beam made with basalt FRP rebars

M70 grade reinforced beam made with basalt FRP rebars (M70BFRP-UR)						
Service load kN	Load at first crack (kN)	Ultimate Flexural Strength (kN)	Deflection at service load mm	Central Deflection at Failure (mm)	Deflection Under load at failure (mm)	Maximum Crack width at failure (mm)
49	70	175	3.69	7.81	7.43	1.09

5. Conclusions

Based on the results reported in this research work and key findings during the experimental investigations, the following conclusions are drawn:

1. The load carrying capacity of beams was found to be increasing with the usage of basalt FRP rebars.
2. The maximum deflections of basalt FRP rebar concrete beams were more when compared to conventional steel rebar beams. This indicates increase in ductility of basalt FRP rebar beam specimens.
3. The load-deflection and moment-curvature behaviour of under reinforced basalt FRP rebar beams shows increased values of loads at ultimate and at first crack.
4. Basalt FRP rebar concrete beams control the initiation of micro cracks, improve the first crack load, the ultimate load and ductility of concrete specimens under flexure. They are also effective in resisting deformation at all stages of loading from first crack to failure.
5. The decrease in crack width and reduction in deflections at service loads is observed in under-reinforced basalt FRP rebar beams.

6. References

[1] ACI, ACI 440.1R-06 Guide for the design and construction of structural concrete reinforced with FRP bars, American Concrete Institute, 2006.
[2] Adhikari, S., 2009, "Mechanical Properties and Flexural Applications of Basalt Fiber Reinforced Polymer (BFRP) Bars", *MS thesis, Department of Civil Engineering*, University of Akron, Ohio, USA.
[3] Banibayat P, Patnaik A. Variability of mechanical properties of basalt fibre reinforced polymer bars manufactured by wet-layup method. Mat Des 2014; 56: 898-906. Garbacz A., Łapko A., Urbański M., *Investigation on concrete beams reinforced with basalt rebars as an effective alternative of conventional R/C structures*, Procedia Engineering 57/2013, 1183-1191.
[4] Dawid Pawłowski and Maciej Szumigała, "Flexural behaviour of full-scale basalt FRP RC beams – experimental and numerical studies" Procedia Engineering 108 (2015) 518 – 525
[5] GangaRao, H. V. S., and Vijay, P. V., 1997, "Design of Concrete Members Reinforced with GFRP Bars," *Proceedings of the Third International Symposium on Non- Metallic (FRP) Reinforcement for Concrete Structures (FRPRCS-3)*, V.1, Japan Concrete Institute, Tokyo, Japan, pp. 143-150.
[6] Li, Lu, Fang, Liu, Li "Flexural study of concrete beams with basalt fibre polymer bars" Structures and Buildings – ICE publishing
[7] Padmanabhan Iyer, Sara Y. Kenno and Sreekanta Das "Mechanical Properties of Fiber-Reinforced Concrete Made with Basalt Filament Fibers" DOI: 10.1061/(ASCE)MT.1943-5533.0001272
[8] Thilan Ovitigala, "Structural Behaviour of Concrete Beams Reinforced with Basalt Fibre Reinforced Polymer (BFRP) bars". University of Illinois at Chicago, 2012.
[9] Tomlinson, D., and Fam, A. (2014). "Performance of concrete beams reinforced with basalt FRP for flexure and shear" J. Compos. Constr., 10.1061/(ASCE)CC.1943-5614.0000491, 04014036.
[10] Urbanski M, Lapko A, Garbacz A. Investigation on concrete beams reinforced with basalt rebars as an effective alternative of conventional R/C structures. Procedia Eng 2013; 57: 1183-91.

Influence of heat treatment on interfacial microstructure of cast Babbitt /carbon steel bimetallic materials

Mohamed Ramadan[1,2*], Naglaa Fathy[3]

[1] Central Metallurgical Research and Development Institute (CMRDI), P.O. Box 87, Helwan 11421, Egypt
[2] College of Engineering, University of Ha'il, P.O. Box 2440 Ha'il, Saudi Arabia
[3] College of Science, University of Ha'il, P.O. Box 2440 Ha'il, Saudi Arabia

*Corresponding author: mrnais3@yahoo.com

Abstract. Bimetallic material has been extensively employed as an advanced functional material in different industrial applications for its unique physical and mechanical properties. Bi-metallic material can be fabricated by bonding of similar and dissimilar materials. Variety of physical and chemical properties, such as wettability, reactivity, phase transformation, as well as thermal properties must be compatible to bond each other. The results showed that bonding interface area of heat treated Babbitt/steel bimetal using tinning mixture of flux+ Sn was improved by 9%. Otherwise, bonding interface area of heat treated Babbitt/steel bimetal using tinning mixture of flux + Sn + glycerol and, flux +Sn + petroleum jelly was improved by 93% and 70% respectively. For all used tinning conditions, the interface thickness layer increase with heat treatment process of Babbitt/steel bimetal.

1. Introduction

Sn- based Babbitt alloys mainly contain variety weights of Sb and Cu elements. They have low wear, high corrosion resist, good run-in properties and good behaviour in the absence of sufficient lubrication. For their good corrosion resistant in addition to previous mentioned properties, Sn- base Babbitt alloys are commonly used in the components of the steam turbines, engine, and compressor bearings. Sn- based Babbitt alloys are easily cast and retain reasonable mechanical properties at elevated temperatures. Sn- based Babbitt alloys are used as a lining metal bearing surface. The Sn-based Babbitt metal can be bonded to the metal bearing shell using chemically or mechanically routes. Commonly, there are three standard fabrication techniques for applying the Sn-based Babbitt to a substrate shell: static casting, centrifugal casting and arc spraying [1-5].

Among all other bi-metal bearings fabrication techniques, liquid-solid static casting technology considered the most economical one that allows the production of bearing layer elements directly. For the developing liquid-solid, bi-metal casting techniques much more attention on improving the interface bond of the bearing metal and substrate shell should be considered for the higher bond strength, durability and performance of the bi-metal bearings [6-10]. For the most liquid – solid bi-metal casting fabrication techniques, low carbon steel, stainless steel and/ or Cu alloys were used as solid substrate shells [2, 6-10].

Powder tin with flux mixture was considered as one of direct tinning process that is successfully used for Babbitt-steel bimetallic materials fabrication. Direct tinning process is a promising tinning process for solid steel substrate used for flat and horizontal bearings applications [11]. Recently, Glycerol and petroleum jelly were added to tin/flux mixture in order to facilitate the mixture for curvature surface solid steel shells application (ex. Journal Bearings) [11]. The addition of glycerol or petroleum jelly to tinning mixture significantly increases the voids and forming unbonded interface layer areas. The aim of current work is to improve the interfacial bonding of Babbitt/steel bimetal composite that fabricated using tinning mixtures contain glycerol or petroleum jelly.

2. Experimental Work

Babbitt/ carbon steel bimetal specimens were fabricated by using liquid-solid bimetal casting technique. Tinning process of Cu alloy solid substrate involves metallic tin / flux mixture deposition on the surface of Cu alloy solid substrate was explained elsewhere [12-13] The tinning mixtures using tin powder, flux, glycerol or petroleum jelly were reported in detail in previous work [11]. The chemical compositions of Babbitt alloy and carbon steel substrate used for bimetal are given in Table 1. Specimens of approximate 40x20x18 mm^3 were cut from three as-cast Babbitt/ carbon bimetal for heat treatment and microstructure investigation as well as hardness measurements. Specimens were heated to 210 ° C for 120 min in an electrical heating furnace with heating rate of 10 ° C/ min. The samples were air cooled into room temperature.

Table 1. Chemical compositions of Babbitt alloy and carbon steel substrate used for bimetal.

Materials	Chemical Compositions (wt.%)								
	Sb	Cu	Pb	C	Si	Mn	Cr	Sn	Fe
Babbitt Alloy	6.5–7.5	2.5–3.5	0.35	-	-	-	-	Bal.	
Carbon Steel	-	0.21	-	0.18	0.32	0.45	0.14	-	Bal.

The Babbitt/ steel bimetal specimens were cut for microstructure investigation, ground, polished and etched with a 4% nital. Optical microscope and scanning electron microscope (SEM) fitted with an energy-dispersive X-ray spectroscopy (EDS) were used to investigate the Babbitt/ steel bimetal specimens. The hardness of Babbitt/ steel bimetal specimens was measured by Vickers microhardness testing machine. the hardness average value of five measurements of each of bimetal specimen were taken.

3. Results and Discussions

Microstructures of tin-based Babbitt alloy and carbon steel substrate used for bimetallic composite fabrication are shown in Figure 1. SEM microstructures (Figure 2a) and EDS analysis (Figure 2b&c) of Cu_6Sn_5 and SnSb hard phases of tin-based Babbitt alloy. Needles and asterisks shape of Cu_6Sn_5 phase are observed. It was reported [14–16] that the matrix of solid solution contains Cu_6Sn_5 and SnSb hard embedded phases produce a Babbitt matrix has fatigue resistant properties.

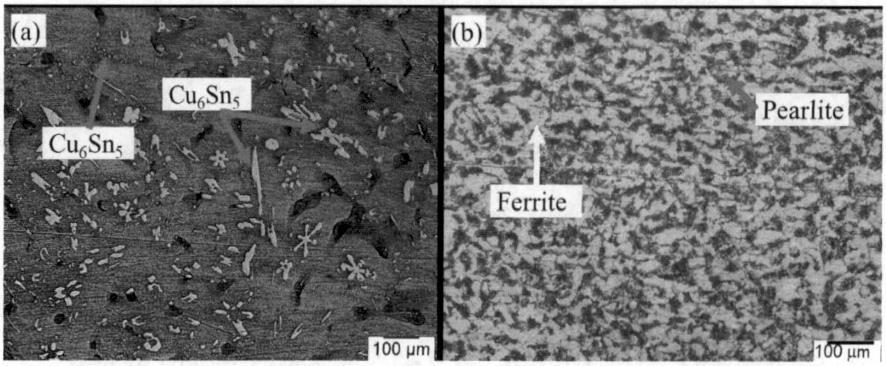

Figure 1. Microstructure of (a) Babbitt alloy and (b) carbon steel substrate.

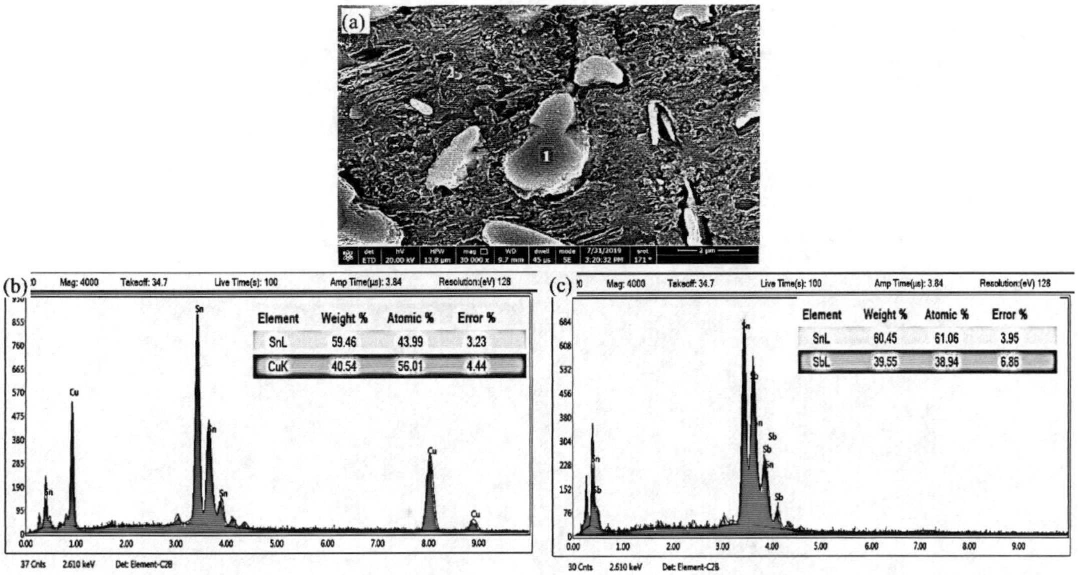

Figure 2. SEM micrographs and EDS analysis from the as-cast Babbitt alloy, (b) and (c) EDS analysis corresponding to the areas 1 and 2 indicated in (a).

Figure 3 a, c & e shows the interfacial microstructures of Babbitt/steel bimetal for as-cast using tinning mixture of flux+ Sn, flux + Sn + glycerol and, flux + Sn + petroleum jelly respectively. Figure 3 b, d & f shows of interfacial microstructures of Babbitt/steel bimetal for heat treated bimetal of (a), (c) and (e) respectively. It is reported [11] that morphology of the interfacial layer of Babbitt /steel bimetal are a significantly affected by addition of bath glycerol and petroleum jelly to the mixture of tinning mixture. Using flux +tin mixture produces a discontinuous interfacial layer (Figure 3 a), otherwise the addition of glycerol or petroleum jelly to the tinning mixture produce a continuous interfacial layers (Figure c&e).

Although, the addition of petroleum jelly and glycerol to the tinning mixture stabilized the interfacial tin layer, much more interfacial unbonded area were produced especially with glycerol addition (see Fig. 3 c& e). Otherwise, the heat treated as-cast bimetal show significant improvements for interfacial bonding areas and their structure morphology (Figure 3 b, d&f). The influence of heat treatment especially heating time on the microstructure of tin-based alloy have been studied [17]. The precipitates morphology is changed by increasing the heating time, whereas the SbSn phase with cuboids shape is turned to sphere shape. In current research work the changes in interface structure and the bonding area could be due to surface energy considerations. At the higher heating temperatures of 210 ° C, which is near to solidus temperature of Babbitt (220-230 ° C) alloy, the effect of surface energy on Babbitt matrix is highly considered. Therefore, low melting point matrix area that could contains some impurities and in addition to the metal flow at relatively high temperature fill the voids on the interface resulted in increasing the interfacial bonding area.

Figure 4 shows the bonded interface area of as- cast and heat treated Babbitt/steel bimetal using tinning mixture of flux+ Sn, flux + Sn + glycerol and, flux +Sn + petroleum jelly. Bonding interface area of heat treated Babbitt/steel bimetal using tinning mixture of flux+ Sn is improved by 9%. Otherwise, bonding interface area of heat treated Babbitt/steel bimetal using tinning mixture of flux + Sn + glycerol and, flux +Sn + petroleum jelly is improved by 93% and 70% respectively.

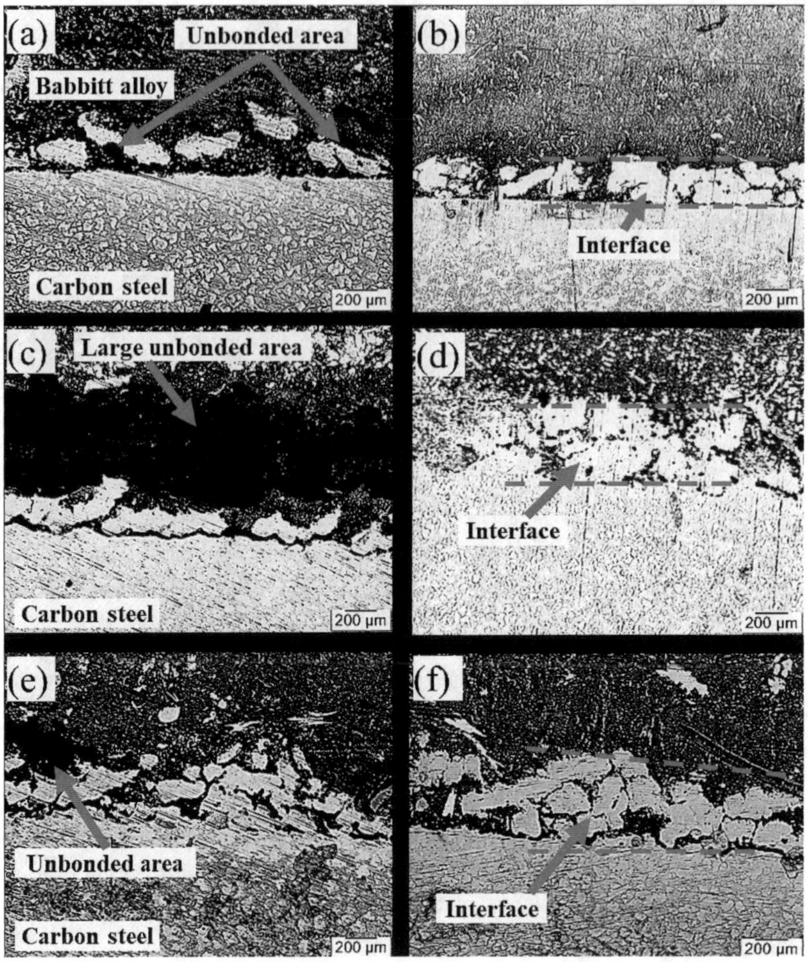

Figure 3. Interfacial microstructures of Babbitt/steel bimetal for (a),(c) and (e) as-cast bimetalusing tinning mixture of , flux+ Sn, flux+Sn+glycerol and , flux+Sn+petroleum jelly respectively and (b),(d) and (f) heat treated bimetal of (a),(c) and (e) respectively.

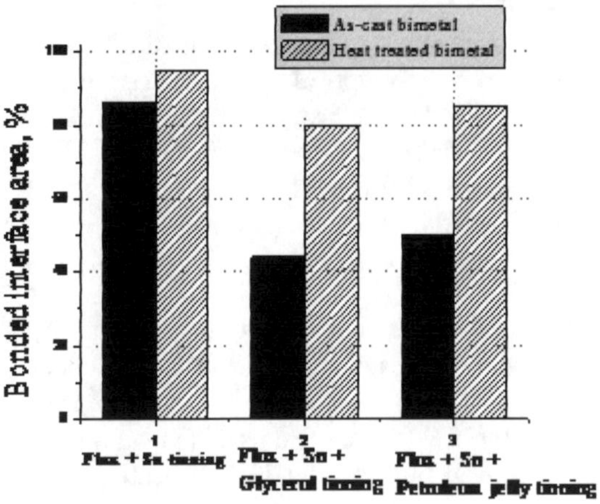

Figure 4. Bonded interface area of as- cast and heat treated Babbitt/steel bimetal using tinning mixture of, flux+ Sn, flux+Sn+glycerol and , flux+Sn+petroleum jelly respectively.

Figure 5 shows the interface layer thickness of as- cast and heat treated Babbitt/steel bimetal using tinning mixture of flux + Sn, flux + Sn + glycerol and flux + Sn + petroleum jelly. For all tinning conditions, the interface thickness layer increase with heat treatment process of Babbitt/steel bimetal. The increasing of interface layer thickness is mostly due to the higher diffusion rate of elements and the formation of FeSn intermetallics.

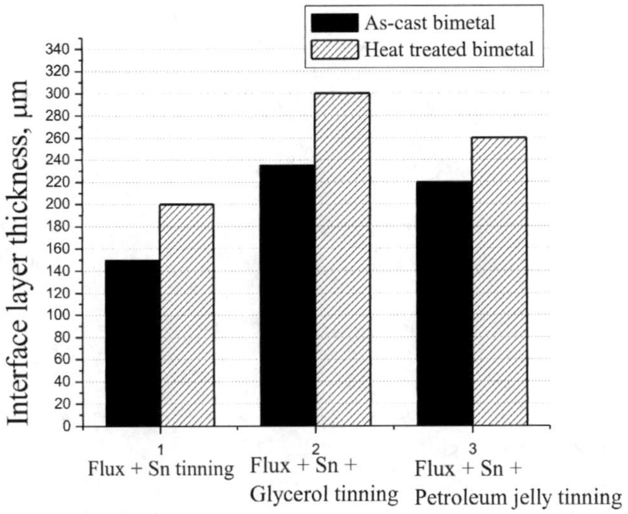

Figure 5. Interface layer thickness of as- cast and heat treated Babbitt/steel bimetal using tinning mixture of flux+ Sn, flux+Sn+glycerol and , flux+Sn+petroleum jelly respectively.

The heat treatment process applied in present study has improved the diffusion bond of Babbitt/steel bimetal using tinning mixture of flux + Sn, flux + Sn + glycerol and, flux + Sn + petroleum jelly. Whereas, higher diffusion at relatively higher near solidus temperature promotes the aimed acceptable bonding area between Babbitt bearing alloy and carbon steel shell. Therefore, based on the above analysis, the heat treatment for bimetal liquid – solid compound cast could be an effective tool for improve interfacial structure and bonding areas of sound and defected Babbitt/steel bimetal materials.

4. Conclusions

A significant improvement of interfacial bonding area of interfacial microstructure of heat treated cast Babbitt /carbon steel bimetallic materials prepared with flux+ Sn, flux+Sn+glycerol and , flux+Sn+petroleum jelly was achieved. Bonding interface area of heat treated Babbitt/steel bimetal using tinning mixture of flux+ Sn is improved by 9%. Otherwise, bonding interface area of heat treated Babbitt/steel bimetal using tinning mixture of flux + Sn + glycerol and, flux +Sn + petroleum jelly is improved by 93% and 70% respectively. For all used tinning conditions, the interface thickness layer increase with heat treatment process of Babbitt/steel bimetal.

Acknowledgment
The authors gratefully acknowledge technical Staff of Foundry Technology Lab of Central Metallurgical R&D Institute for their support of their research program.

References

[1] Diouf P and Jones A 2010 Metallurgical and Materials Transactions A Investigation of Bond Strength in Centrifugal Lining of Babbitt on Cast Iron **41A** 603-609.
[2] Fathy N and Ramadan M 2018 AIP Conference Proceedings Influence of volume ratio of liquid to solid and low pouring temperature on interface structure of cast Babbitt-steel bimetal composite, **1966** 020028; doi: 10.1063/1.5038707..
[3] Valeeva A K, Valeev I S and Fazlyakhmetov R F 2014 J Frict. Wear. Effect of Structure of B83 Babbit on Its Wear **35** 311–315.
[4] Sadykov F A, Barykin N P Valeev I Sh, and Danilenko V N 2003 J. Mater.Eng. Perform. Influence of the Structural State on Mechanical Behavior of Tin Babbit **12** (1) 29–36.
[5] Ji X and Chen Y 2016 J. Mater.Eng. Perform. Tribological Behavior of Babbitt Alloy Rubbing Against Si3N4 and Steel Under Dry Friction Condition **25** (3) 750-755.
[6] Ramadan M, Fathy N, Abdel Halim K S and Alghamdi A S 2019 International Journal of Advanced and Applied Sciences New trends and advances in bi-metal casting technologies **6** (2) 75-80.
[7] Gawronski J, Szajnar J, Wróbel P 2004 Journal of Materials Processing Technology Study on theoretical bases of receiving composite alloy layers on surface of cast steel castings **157-158** 679-682.
[8] M. Ramadan, B. Ayadi, W. Rajhi and A.S. Alghamdi 2020 Key Engineering Materials Influence of Tinning Material on Interfacial Microstructures and Mechanical Properties of Al12Sn4Si1Cu /Carbon Steel Bimetallic Castings for Bearing Applications **835** 108-114.
[9] Cholewa M, Wróbel T and Tenerowicz S 2010 Journal of Achievements in Materials and Manufacturing Engineering Bimetallic layer castings,. **43**: 1 385-391.
[10] Ramadan M 2015 Advanced Materials Research Interface characterization of bimetallic casting with a 304 stainless steel surface layer and a gray cast iron base **1120-1121** 993-998.
[11] Fathy N 2018 Engineering, Technology & Applied Science Research Interface Microstructure Investigation of Babbitt -Carbon steel Composite Using Flux With Glycerol and Petroleum Jelly Additives **8** (3) 3028-3031.
[12] Ramadan M, Alghamdi A S, Subhani T and Abdel Halim K S 2020 Materials Fabrication and Characterization of Sn-Based Babbitt Alloy Nanocomposite Reinforced with Al2O3 Nanoparticles/Carbon Steel Bimetallic Material **13** 2759; doi:10.3390/ma13122759.
[13] Ramadan M, Alghamdi A S, Hafez K M, Subhani T and Abdel Halim K S 2020 Materials Development and Optimization of Tin/Flux Mixture for Direct Tinning and Interfacial Bonding in Aluminum/Steel Bimetallic Compound Casting **13** 5642.
[14] Koutský J and Veselá J 2004 J. Mater. Process. Technol. Evaluation of white metal adhesion (conventional casting and thermal wire arc spraying) by ultrasonic non-destructive method **157–158** 724–728.
[15] Barykin N, Sadykov F and Aslanian I 2000 J. Mater. Eng. Perform. Wear and failure of Babbitt bushes in steam turbine sliding bearings. **9** 10–115.
[16] Sadykov FA, Barykin NP, Valeev I S and Danilenko V N 2003 J. Mater. Eng. Perform. Influence of the structural state on mechanical behaviour of tin Babbitt **12** 29–36.
[17] Moazami Goudarzi M, Jenabali Jahromi S A and Nazarboland A 2009 Materials and Design Investigation of characteristics of tin-based white metals as a bearing material, **30** 2283–2288.

Near-Field Fractal RFID Reader Antenna for Item Level Tagging

Chitra Varadhan[1,2,3], **Dr Arulselvi**[2,4,5]

[1] Research Scholar
[2] Department of Electronics and Communication Engineering, B.H.I.E.R, Chennai, Tamilnadu, India - 600 073
[3] Email - chitralaya@gmail.com
[4] Associate Professor
[5] Email - arulselvi2003@gmail.com

Abstract: This paper proposes a novel methodology for designing near-field fractal antenna operating at European Ultra-High Frequency (UHF) for Radio Frequency Identification (RFID) systems. The principle of Oppositely Directed Current (ODC) technique is implemented to generate homogeneous magnetic field above the surface area meant for near-field RFID application. The conceptual theoretical framework of the design based on the fractal structure can be applied. The fabricated RFID reader antenna operates at 867 MHz with reflection coefficient S_{11} -30 dB, thus the reader antenna covering UHF RFID band range. The reading capabilities of the antenna are determined by the interrogation area of, 190mm x 243mm. It is adequate for object tracking and detection systems in pharmaceutical management and chemical industries.

Keywords: *Near-field analysis, Radio Frequency Identification (RFID) reader antenna, Ultra-High Frequency (UHF), Item level tagging, Fractal antenna.*

1. INTRODUCTION

In tracking and detection system, most researchers investigating that the effective functioning of near-field radio frequency identification antenna (RFID) requires optimum magnetic field strength to be provided within specific interrogation zone [1]. Achieving intense and constant magnetic field dissemination becomes more challenging, during operating frequencies upsurge to ultra-high frequencies (UHF) band [2].

RFIDs find important applications in today's technology driven society and it is a promising method in contactless tracking and identification. The applications can be library book management, employee attendance management, factory inventory management, vehicle identification and traffic toll collection. As frequency is high, the signal can travel in short time, resulting in fast tracking however, this initiates signal distortion due to obstructions such as liquids and metals [3]. Here, the permanent tags are spotted more persistently on numerous target objects specifically bottles containing liquids like drugs, acids, oil, water, clothes and small items, causing communication barrier between reader and tag [4]. Especially, tracking the items in drug and pharma industries which has glitches due to impediments in near field.

Recent reports in UHF RFID system (usually 900 MHz) proves that near field communications have been broadly pertinent for item-level tagging. Ordinary RFID reader antennas for far-zone reading lags on performance efficiency in near-zone RFID applications [5 - 7]. The near field communication is an emerging communication technology, for which the application which demands fast data transfer in short distance with low power.

In near-field analysis, the RFID reader antenna has to produce uniform strong magnetic field. To improve the reading stability, the amount of magnetic field should be distributed uniformly in the reading zone [8]. Commonly, loop antennas having capacity to make strong magnetic field. Amid the several forms of loop antenna, single or multi-turn antennas and conventional solid line loop antennas are used extensively in near-field reader antennas. On the other hand, smaller circumference loops produce intense magnetic field at the loop core centre, but there is rapid decrease in

IOP Conf. Series: Materials Science and Engineering 1091 (2021) 012046 doi:10.1088/1757-899X/1091/1/012046

magnetic field strength with distance from the centre, thereby obtaining limited reading area [9&10]. A Modified Fractal Pinwheel tiling is proposed here, to improvise the electrical performance with compact structure.

2. ANTENNA DESIGN & OPERATION

2.1 Opposite Direction Current

Highly strong magnetic field can be generated by Opposite Directed Currents (ODCs) procedure. ODCs are generated over a board surface area by duel fed dipoles with current direction of 180^0 out of phase [11]. However, the duel feed structure improves the complexity in the RFID system. Hence, single feed ODC can be implemented in the reader antenna. It consists of triangular slot and gradual increase of the distance between ODC pair is achieved by triangular slot at the middle of the antenna; thereby producing high intensity magnetic field on surface of the antenna. The figure 1 shows the direction of surface current.

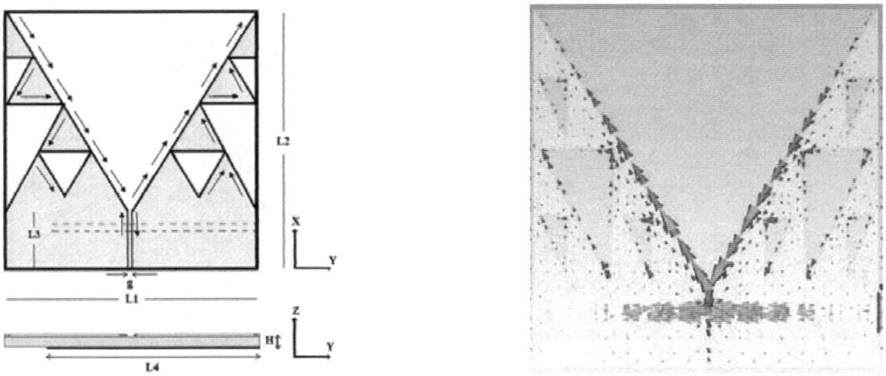

Figure 1. Simulated Result of the Proposed Antenna Indicates Surface Current Direction

Modified fractal Pinwheel tiling structure helps to increase the electrical length and application of ODC principle is implemented at the edges of the tapered slot opening. The tapered slots were used for high directivity and high frequency applications.

2.2 Antenna Geometry

In this, the reader antenna was designed on FR-4 substrate (dielectric constant ε_r=4.3, loss tangent δ = 0.02 and thickness H = 1.5 mm). It has triangular shaped opening, and the slot line is fed by a very narrow micro-strip line, carefully placed at antenna's posterior side. The slot width is around 0.35m which is very small in comparable to its wavelength. Complete dimension of the antenna is 90 x 90 mm. Modified fractal pinwheel tiling is implemented as a space filling structure in this paper. Fractals are self-similar and space filling patterns. The antenna operates at 867MHz with S_{11} of -39dB at second iteration.

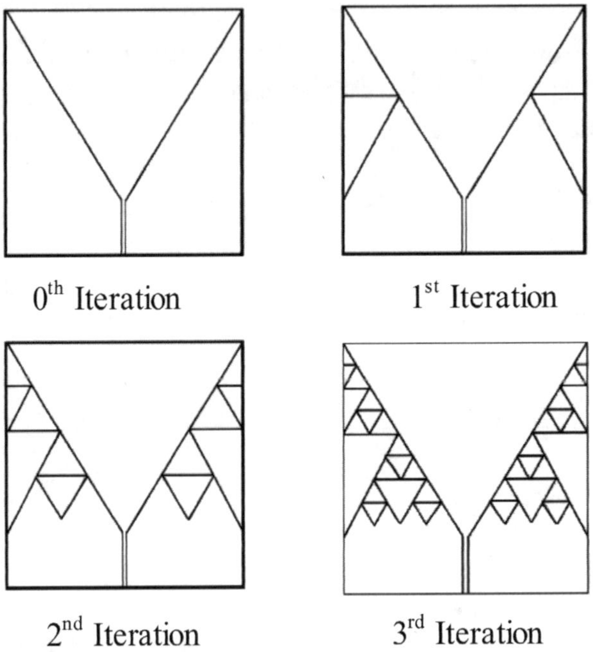

Figure 2. Different Iterations of Modified Fractal Pinwheel Tiling

The fractal geometry is based on the Modified pinwheel tiling is shown in figure 2. Let T be the right triangle. That T can be divided in four copies of its image. The iterations can be obtained by suitable rescaling and translating/rotating of triangles by infinite increasing sequence, all made of isometric copies of T.

3 RESULT AND DISCUSSION

3.1 S_{11} Characteristics

Figure 3 indicates the measured and simulated results of RFID reader antenna. The measured impedance bandwidth is 190MHz (780-970 MHz) after the optimization, simulated result is (at 10 dB level) 823-914 MHz, the obtained bandwidth is 91 Megahertz. The results are found to be in good correlation with simulated results.

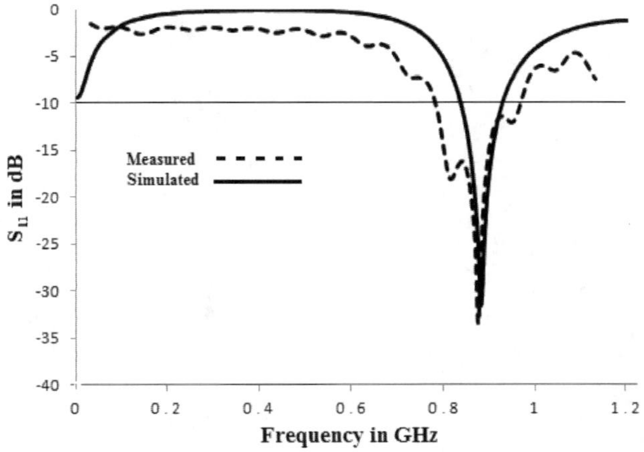

Figure 3. Measured and Simulated S_{11} vs Frequency Graph

3.2 VSWR Bandwidth

VSWR calculation was carried out by taking 2 as a reference line. The bandwidth of VSWR is 93 MHz i.e., from 824 to 917 MHz. Least VSWR of 1.06 is attained at 867 MHz, obviously at operating frequency.

3.3 Surface Current Distribution

In near-field analysis, surface current flow is very important. The direction and flow of the surface current distribution is depicted in figure 1 and observation of opposite direction currents is attained on this RFID reader antenna plane.

The minimum and optimum power detection by a viable RFID near-field tag is 20dBA/m. After the simulation, taking 20dBA/m as a reference, the obtained interrogation area is 190mm x 243mm. It is significant that there are no nulls transpiring at centre of the antenna. The interrogation area is more than two times greater than the proposed antenna size. Also found that the maximum field strength is -7.20dBA/m by a distance of 50 mm from reader antenna.

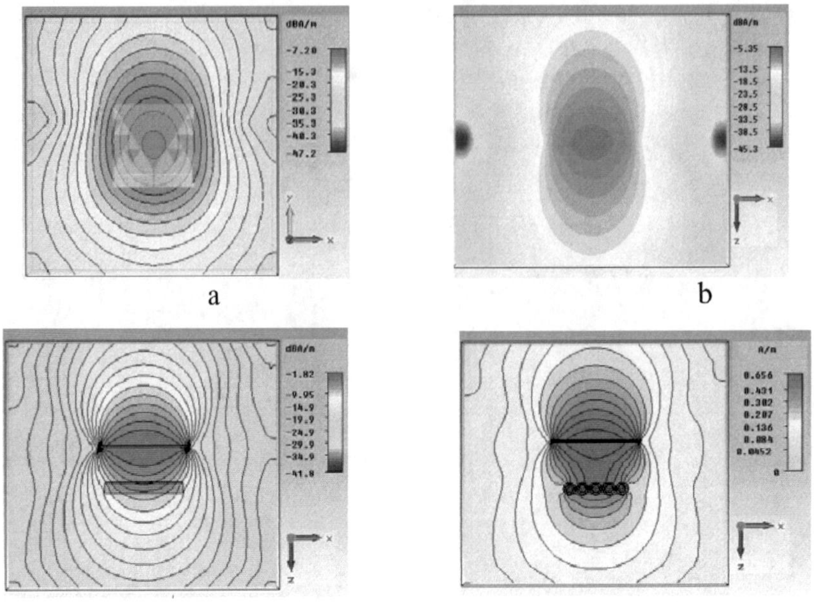

a b

c d

Figure 4. RFID Reader Antenna of magnetic field at 50 mm.

Figure 4 a: x-y cut plane in z axis

Figure 4 b: x-z cut plane in y axis.

Figure 4 c: x-z cut plane in y axis with Bakelite.

Figure 4 d: x-z cut plane in y axis with Glycerine in test tube

The Figure 4(a) shows the RFID Reader Antenna of magnetic field at distance 50mm in the x-y plane. Figure 4(b) presents x-z cut plane at a height 50mm from the antenna centre. The obtained magnetic field is roughly -23 dBA/m in the near field which gratifies the near field detection necessities. Thus, good read distance is comparatively achieved.

Figure 4(c) shows the simulation result which is obtained in real time application. Taking Bakelite as obstruction placed in front of the antenna with 45 mm distance. The dimension of the Bakelite is 40 x 50 x 10 mm length, height and thickness respectively. The simulation result gives slight degradation around -9.9 dBA/m.

Figure 4(d) depicts simulation results of different real time application. It is the x-z cut plane of magnetic field with glycerine contained in a Plexiglas test tube as obstruction. The dimensions of each test tube are approximately 1.4 cm diameter and 14 cm height. The antenna is positioned at 50mm distance from the test tube and kept at a spacing of 2 mm. The simulated result indicates a real time scenario similar to chemical tracking, and reduction of around -0.3 dBA/m is obtained after the obstruction, and it is tolerant.

4 FABRICATION AND MEASUREMENT

4.1 Fabricated Prototype

The antenna is developed on FR-4 (lossy) substrate and the substrate thickness is 1.5mm having sufficient mechanical stiffness. At the back side, the strip-line is deposited with PEC material with the thickness of 0.35 m. With the same thickness, front side also PEC panel deposited with V shape opening and fractal design is etched out carefully. The prototype of fractal RFID Reader antenna is shown in figure 5.

Figure 5. Fabricated Prototype of Fractal RFID Reader Antenna

4.2 Measurement Setup

The RFID reader antenna is developed on FR-4 substrate and the fabrication cost is reduced due to the availability of FR-4 material. In this design of antenna, no active elements were used so that low profile and simplicity were maintained. The micro-strip feed was soldered to sub-miniature version-A (SMA) connector of 50Ω impedance on the prototype.

Measurements have been carried out using Agilent E8363B vector network analyser (VNA) to trace the magnetic field strength in the Z direction of the proposed antenna. The antenna was fed by a microwave source, with a power of 20 dBm. The measurement setup is shown in figure 6. A fixture was made using foam with dielectric constant 1 and the antenna was placed over it with a set of five test tubes with a distance of 2mm each, containing liquid as obstruction which will ensure the performance of the antenna. A field meter with field probe provision is used to measure the magnetic field. The position marks are made on the probe kept at the fixture to uphold precision and the procedure was repetitive for the field measurements at different points on the z axis.

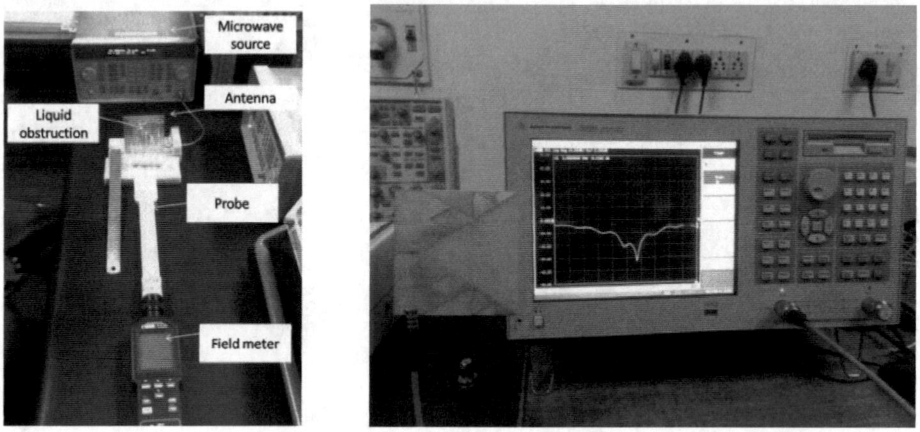

Figure 6. Measurement Setup: *Near-Field Measurement*

Table 1 compares the interrogation area with other literatures and the proposed antenna, results in larger interrogation area which is good enough for RFID applications.

Proposed design has minimum input power 20dBm with good read range. The antenna prototype is small as well as simple compared to the existing antennas. Dealing with the RFID applications, this antenna will be useful in several applications mainly chemical and pharmaceutical item tracking, as a compact and cost efficient alternative.

Table 1. Comparison Result with Other Antennas

Sl.No.	Reference No.	Interrogation Area (mm)	Antenna Size (mm)
01	12	200 x 190	95 x 200
02	13	154 x 154	175 x 189
03	Proposed Antenna	190 x 243	90 x 90

The Table 2 shows the performance analysis of the novel RFID reader for pharmaceutical and chemical applications. This includes the results from the measured and simulated antenna.

Table 2 Performance Analysis of the Fractal RFID Reader Antenna

Sl.No.	Parameter	Value
01	Operating frequency	867MHz
02	Impedance Bandwidth	190 MHz (measured)
03	S_{11} at 867 MHz	-29dB
04	VSWR Band width	93MHz (simulated)
05	Interrogation area at 50 mm from plane in Z direction	190 mm x 243 mm
06	Size of the Antenna	90 mm x 90 mm
07	Read distance	9.47 cm
08	Read distance with Bakelite	8.7 cm
09	Read distance with Glycerine as Obstruction	8.9 cm
10	Antenna input power	20dBm

5. CONCLUSION

It will be always challenging to design a simple and low profile antenna for UHF near field RFID applications. The proposed modified fractal pinwheel tiling antenna has substantiated the capability of producing a uniform intense magnetic field in and around the antenna. The complete size of the antenna is approximately 90 x 90 mm only. The designed antenna operates at 867 MHz, and possesses appropriate bandwidth of 190 MHz which is sufficient to cover USA and Canadian UHF RFID bandwidth of 902 - 928 MHz and European UHF RFID bandwidth of 865 - 868 MHz. The prototype is fabricated and measured. Taking Bakelite and Glycerine as obstruction, small acceptable degradation in the simulated results was found for tracking it in chemical and pharmaceutical industries. The simulated results are validated using experimental measurements and results are in consistent association with simulated results. Thus proposed work is an acceptable cost efficient and compact alternative for item level tagging.

References

1. F. K. Byondi, C. Sul and Y. Chung, 2018 "Super Long Range Small Empty Cavity UHF RFID Tag Antenna Design for a Metal Cart," *IEEE Region Ten Symposium (Tensymp)*, Sydney, Australia, pp. 145-148.
2. C. Cho, J. Ryoo, I. Park, and H. Choo, 2010, "Design of a novel ultra-high frequency radio-frequency identification reader antenna for near-field communications using oppositely directed currents," *Microwave. Antennas Propagation*, vol. 4, no. 10, pp. 1543–1548.
3. C. Varadhan, J. K. Pakkathillam, M. Kanagasabai, R. Sivasamy, R. Natarajan and S. K. Palaniswamy, 2013, "Triband Antenna Structures for RFID Systems Deploying Fractal Geometry," in *IEEE Antennas and Wireless Propagation Letters*, vol. 12,pp.437-440.
4. V. Laquerbe, G. Le Fur, D. Belot, L. Feat and R. Contreres, 2020, "Simple and Robust Probes for Near-Field Antenna Measurements at low UHF bands," *2020 14th European Conference on Antennas and Propagation (EuCAP)*, Copenhagen, Denmark, pp. 1-4.
5. M. Lalkota, G. Gupta and A. R. Harish, 2018 "UHF reader antenna for near and far field RFID operation," *2018 IEEE International Conference on RFID (RFID)*, Orlando, FL, , pp. 1-7.
6. M. Shirazi and X. Gong, *2016,* "Singly-polarized reconfigurable slot-ring antenna/array with fractal shapes," *IEEE International Symposium on Antennas and Propagation (APSURSI)*, Fajardo, 2016, pp. 843-844.

7. Gorai, M. Pal and R. Ghatak, 2017, "A Compact Fractal-Shaped Antenna for Ultrawideband and Bluetooth Wireless Systems with WLAN Rejection Functionality," in *IEEE Antennas and Wireless Propagation Letters*, vol. 16, pp. 2163-2166.
8. J. R. Smith and H. H. Sigmarsson, 2015, "Dual-band second order microstrip filter using dual-mode perturbed meander loop resonators," *2015 IEEE 16th Annual Wireless and Microwave Technology Conference (WAMICON)*, Cocoa Beach, FL, pp. 1-3.
9. J. Anguera, A. Andújar, J. Jayasinghe, D. Uduwawala, M. K. Khattak and S. Kahng, 2016, "Nature-Inspired High-Directivity Microstrip Antennas: Fractals and Genetics," *2016 8th International Conference on Computational Intelligence and Communication Networks (CICN)*, Tehri, pp. 204-207.
10. Z. Liang, J. Liu, Y. Li and Y. Long, 2016, "A Dual-Frequency Broadband Design of Coupled-Fed Stacked Microstrip Monopolar Patch Antenna for WLAN Applications," in *IEEE Antennas and Wireless Propagation Letters*, vol. 15, pp. 1289-1292.
11. J. Shi, X. Qing, Z. N. Chen, and C. K. Goh, Mar. 2013 "Electrically large dualloop antenna for UHF near-field RFID reader," IEEE Trans. Antennas Propag., vol. 61, no. 3, pp. 1019–1025.
12. H. Liu, C. Weng, and C. Yang, 2011, "Design of near-field edge-shorted slot microstrip antenna for RFID handheld reader applications," IEEE Antennas Wireless Propag. Lett., no. 2010, pp. 1135–1138.
13. L. Yan, Y. Zhang, L. T. Yang, and H. Ning, 2008 "The Internet of Things: From RFID to the Next-Generation Pervasive Networked Systems." Boca Raton, FL, USA: CRC Press, pp. 11–12.
14. Constantine A. Balanis, "Antenna Theory", Second Edition, Wiley student edition.

Acknowledgments

Mrs. Chitra Varadhan, received Bachelor degree in Electronics and Communication Engineering from Regional Engineering College (NIT-affiliated with Bharathidhasan University) Trichy in 1996 and completed Master of Engineering in College of Engineering, Anna University, Guindy, Chennai in 2008. Currently pursuing PhD in Fractal Antennas in BIHER [Bharath Institute of Higher Education & Research] Chennai, Tamil Nadu, India.

Dr S. Arulselvi, working as an Associate Professor, in BIHER [Bharath Institute of Higher Education & Research] Chennai, Tamil Nadu, India. Areas of specialization include Networking and communications. Currently guiding Mrs Chitra Varadhan in the area of Fractal Antennas.

Some investigative studies on dissimilar metals by friction stir welding

Vuppula Prasanna[1*], Dr. P. Ramesh Babu[2]

Abstract. The aim of the current work is to develop weldments and to investigate the microstructure and mechanical properties of the aluminium alloy 6061-T6 and magnesium alloy AE42 plates (120x100x3 mm) which are welded through FSW. H13 conical thread tool pin was selected to prepare the welded joints. Two plates that were co-ordinated vertically according to the welding directions were well joined. In this study, two parameters such as weld speed at 60,150 mm / min and tool rotation speed at 800, 1200 RPM are taken into account as defect free. By changing these parameters similar and dissimilar alloys are welded together. Tensile specimens (according to ASTM standards) were developed to estimate mechanical properties including tensile strength, % elongation, yield strength and joint efficiency. It was found that an increase in the tool rotational speed or welding speed of the tool led to an increase in the tensile strength, which reached a maximum value and then decreased. (LOM) Light optical Microscopy research was accustomed examine and study the weld zone properties. Dynamic recrystallization was discovered within the weld region in addition as within the thermo-mechanical heat-affected zone (TMAZ). There's a transparent reduction within the quantity of precipitate through the TMAZ and from the BM (Base Material) into the weld zone. Observed precipitates from similar and dissimilar welded joints - Mg_2Si, Al_2O_3, SiC, $Mg_{17}Al_{12}$, Mg_2Al_3, $Al_{11}Ce_3$ and Al_2Ce. Welds are without porosity. XRD with the EDS characterization was executed in the NZ (Nugget Zone) demonstrated the existence of SEM intermetallic phases and their weight in percent. Rapid phase structures of welded joints were detected using X-ray diffraction (XRD), however, the SEM fractograph indicated a flexible ductile fracture mechanism. Vickers micro-hardness test was performed on the thickness of the plate within the weld space to review and perceive the variation of hardness with thickness. An interrelationship between size of precipitate and micro hardness was ascertained. The corrosion behaviour of the base alloy and welded joints was measured in 3.5% NaCl solution using the salt spray test and by investigating the corrosion parameters (corrosion rate, exposure time and weight reduction). Corroded products were analysed and classified using optical microscopy and SEM. The corrosion rate decreases with increasing exposure time, but remains uniform as the corrosion time increases. Finally, a three-dimensional finite element phases was developed for temperature evolution in the friction-stir welded joint plunge depth, dwell, moving phases, and the heat conduction effect is arbitrary in the LAGARAGIAN-EULERIAN formulation in the ABAQUS / EXPLICIT, Johnson-Cook (JC) elastic-plastic model Will be used. The JC model defines the strength of the material as a function of three parameters, i.e. the strength of the material depends on the strain hardening effects, strain rate effects and temperature and verifies the simulated results with the experimental results.

1. Introduction

After two decades of development, Friction Stir Welding (FSW) has become a viable and important manufacturing alternative, especially in the automotive industry (alloy wheels) with aluminium alloys and magnesium alloys. Although FSW was started as an alternative to dissimilar materials, it is now advancing to higher temperature metals, including aluminium alloys and magnesium alloys. Interest has also increased for FSW of dissimilar metals and alloys, especially when welding by conventional welding [1-5] is difficult or impossible.

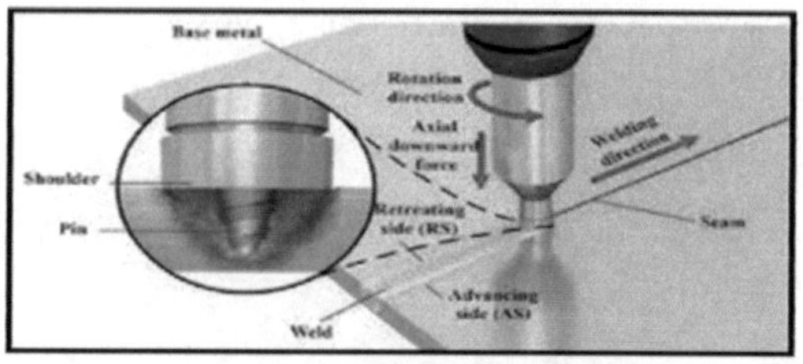

Fig 1: A Schematic FSW

1.1 Problems in traditional welding dissimilar metals – Problem Identification

Although magnesium alloys are lightweight and do not have sufficient strength to use in structural material, the combination of aluminium alloy and magnesium alloy is a good construction. However, conventional method on unequal joints is not appropriate because the large coefficient of expansion of the Mg12Al17 intermetallic compound structure is in the stir zone. The use of arc welding is not possible because it produces large HAZ by melting the base metal. It produces error due to high heat input and slow cooling rate and is difficult to joint uneven material due to different chemical composition such as melting point. To overcome the disadvantages, friction stir welding is an alternative method and this process takes place below the melting point of the alloy. The intermetallic compound layer in the stir zone is shorter than the conventional welding process. Improving efficiency by enhancing the intermetallic compound by using filler rod is not required in the FSW. There is a less study in joining asymmetric AE42 magnesium alloy and aluminium AA6061-T6.

Table 1: Literature Review of Dissimilar Metals

S. No	Author/Year	Process Parameters	Type of Tool Used	Material Used	Remarks
1.	Edward D Nicholas / 1998	T: 6mm T.W.S: 1.2 m/min	AISI H13 tool steel	7075-T7351 alloy	UTS: increased by 6% to 25% compare to base metal
2.	K. Colligan / 1999	T.W.S: 30 mm /min T: 6.4 mm	H13 Threaded surface	Al 6061 & 7075	Aim of this study was to perform tests to reveal the material movement patterns that take place in friction stir welds.
3.	Joseph R. Pickens/ 2000	T: 3mm	Base metals	Al alloys 2090 & 8090	UTS:272 Mpa , Y.S: 137 Mpa, El%: 19.6
4.	Chao Y.J et al./ 2001	T: 9mm T.S: 113mm/min T.R.S: 215 rpm	Steel tool threaded pin	AA2024-T3 and AA7075-T735	Yield stress and strain rate effects of dissimilar metals.
5.	Wayne M Thomas / 2002	T: 6mm T.S: 120mm/min	Whorl type probes	5083-O & 6082-T6	Lap welds of 190% of the plate thickness, improvement in weld integrity, reduction in upper plate thinning and an increased welding speed.
6.	Per Bakke /2003	UTS: 280 Mpa % El: 10-12%	Base metals	Mg alloys	Higher RE contents the ductility decreases due to formation of a continuous skeleton of intermetallic decorating the grain borders.

7.	Zhili Feng et al / 2004	T: 6mm T.R.S: 375,450 rpm. T.W.S: 152.4 mm/min	H13 steel tool	Al 319 & Al 356	Comparisons of tensile properties of the base metal and Al 319 and Al 356 weldments.
8.	W H Jiang et al / 2004	T: 6mm T.R.S: 914r/min T.W.S: 140mm/ min.	H13 steel tool	6061-T6 & AISI 1018 steel	The tensile failure occurred at the boundary between the NZ and the TMAZ of the base metal has high joining strength.
9.	Erdinc Kaluc. E et al. / 2005	T: 6.45 mm T.R.S:1600-1650rpm T.W.S:50, 62, 70 74 mm/min. T.A: 2°	High carbon high chromium steel M5 threaded	5083 H321-5086 H32 aluminium alloys	It has been observed that friction stir welded joints have so good mechanical properties and metallurgical properties as/ than TIG and/or MIG welded joints.
10.	Cavaliere. P et al / 2005	T.R.S: 700 rpm T.W.S: 2.67 mm/s T.T.A: 3° T: 2. 5mm	H13 tool	2024-7075 aluminium alloys	FS weldments decrease in fatigue life respect to the FSW 7075-T6 ones
11.	Peel. M. J et al./ 2006	T.R.S:280, 560, 840 rpm T.W.S:100, 200, 300 mm/min	Cylindrical pin	AA5083-AA6082	As a result, such metrics may not be suitable for characterizing the conditions under which welds are produced
12.	R. Sathish et al /2014	T.R.S: 1000-1250rpm Friction pressure: 18 to 32 MPa	HSS tool	Al 7075-T6 & Al 6061-T6	U.T.S:81-224 Mpa Eln: 13.78%
13.	Vuppula Prasanna et.al /2016	T: 5mm A.F: 4KN T.T.A.2°	H13, HSS	Mg alloys	Review paper on Magnesium alloys and tools used for FSW.
14.	Vuppula Prasanna et.al /2016	T: 5mm A.F: 4KN T.T.A.2°	H13, HSS	Similar and dissimilar alloy metals	Review paper on FSW and its tools used for FSW.
15.	Vuppula Prasanna et.al /2018	T: 3mm T.R.S : 600-1400 rpm T.W.S: 40-120m/min A.F: 4KN T.T.A.2°	H13 tool	Mg alloy AE42	HV is max at FSZ with 600 rpm and 40mm/min is 68.9 HV. Tensile strength is max at 190 Mpa at 1400 rpm.
16.	Gianluca D Urso et al/2017	T: 4mm T.R.S: 1000,1500,2000 rpm T.W.S: 10,35,60mm/ min	H13 steel	Al7075, 6060 & 2024 alloys	The best conditions in terms of mechanical strength were obtained using the "intermediate" values of rotational speed, and, in general, when the process parameters result in low values feed rate per unit revolution (F/S), that corresponds to the higher thermal contribution to the joint region
17.	Z. Hou et al/ 2018	T: 2.5mm T.R.S: 600 rpm T.W.S: 30mm/min T.T.A: 2°	H13 steel	Al 2024 & Mg alloy A Z31	Lower residual stresses and better weld quality were obtained with tool offset to the aluminium side

18.	Vuppula Prasanna et.al /2018	T: 3mm T.R.S : 400,600,800, 1200 rpm T.W.S:35, 40 80 & 100m/min A.F: 4KN T.T.A.2°	H13 tool	Mg alloy AE42	Impact energy value max at FSZ at 34°C is 4 at 1200 RPM with 100mm/min.
19.	Vuppula Prasanna et.al /2018	T: 3mm T.R.S : 400,600,800, 1200 rpm T.W.S:35, 40 80 & 100m/min A.F: 4KN T.T.A.2°	H13 tool	Mg alloy AE42	Corrosion rate of the corroded Magnesium AE42 specimen exposed in 3.5% of NaCl solution for different exposure time of 6.5, 7, 7.5 days for salt spray testing. It was observed that with the increase in exposure time, the corrosion rate decreases.
20.	Vuppula Prasanna et.al /2019	T: 3mm T.R.S : 800, 1200 rpm T.W.S:40-150m/sec A.F: 40KN T.T.A.2°	H13 tool	AA6061-T6 and Mg AE4	UTS is max at 800 rpm, HV is 98Hv at weld zone, Joint Efficiency: 80%
21.	Vuppula Prasanna et.al /2020	T: 3mm T.R.S : 800, 1200 rpm T.W.S:60 & 150m/min A.F: 40KN T.T.A.2°	H13 tool	AA6061-T6	Max HV at 1200 rom is 117 Max UTS is 215 Mpa J.E= 89.5%.

2. Experimental Setup

Material and Specimen preparations

Mg AE42 is collected in 3mm thick slabs. AA6061-T6 belongs to the heat-treatable 6xxx series, of that aluminium, magnesium and silicon are the main composition. Dimensions for weldments are ($120x100x3mm^3$). Table 1 &2 and Table 3 & 4 shows composition and properties of dissimilar metals.

Table 2: Chemical Composition of Base metal (Wt. %) of Aluminium alloy AA6061-T6

6061-T6	Cu	Mg	Mn	Fe	Cr	Si	Al
Standard	0.31	0.99	0.08	0.25	0.16	0.66	Balance
Observed	0.30	1.00	0.06	0.19	0.13	0.67	Balance

Table 3: Mechanical properties of Aluminium Base metal AA6061-T6 alloy

Alloy AA6061-T6	0.2% Yield Strength (MPa)	UTS (MPa)	Elongation (%)	Hardness (VHN)
Standard	198	240	26	105
Observed	183	236	22	103

Table 4: Chemical Composition of the base metals (Wt. %) of Magnesium AE42 alloy

Wt. %	Al	Zn	RE(Ce+ Nd+ La+ Th+ Pr)	Mn	Mg
Standard	4	-	2(1.2+0.4+0.6+0.1+0.1)	0.3	Balance
Observed	3.9	0.08	2.2	0.31	Balance

Table 5: Mechanical properties of Magnesium Base Metal AE42 alloy

Alloy AE42	Yield Strength (MPa)	UTS (MPa)	Elongation (%)	Hardness (VHN)
Standard	130	225	6	75
Observed	124	198	3	72

Table 6: FS welding Tool Design

H13 Tool Conical tapered threaded	Tool shoulder diameter, D (mm)	Tool shoulder length, L (mm)	Pin diameter, d (mm)	D/d ratio of tool	Pin length, L (mm)	Pitch (mm) of threaded pin
	12	16	2.5	4.8	2.9	1

FSW depends on design of tool, process parameters. Design of tool and geometry of it's made. For current research on aluminium alloy AA6061-T6 and magnesium alloy conical tapered threaded tool pin profile is selected. The determined material for high speed steel material is hardened 60HRc [6-8].

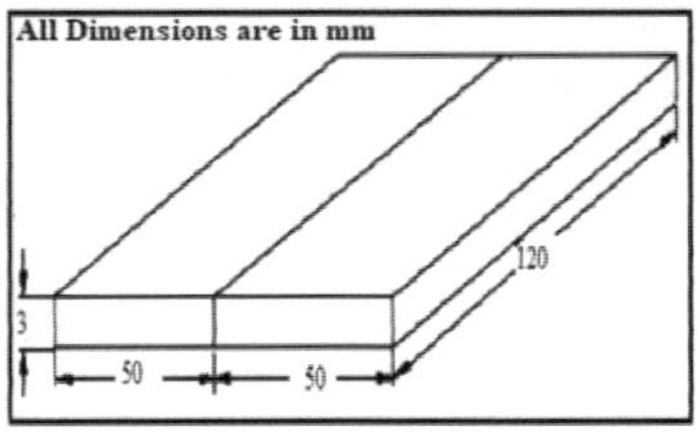

Fig 2: Metal plates used for Friction Stir Welding

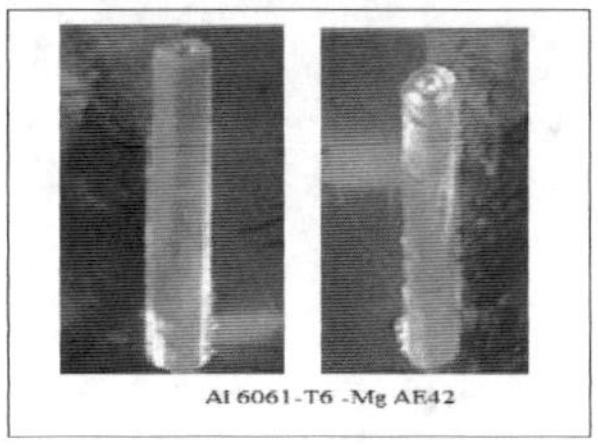

Fig 3: Dissimilar FSW joints

Infra-red technique could be a smart activity methodology for temperature. However, the device is extremely expensive and also the temperature below the shoulder of the tool can't be measured at once. K-type grounded thermocouples with a sheath diameter of one metric linear unit were used. The digital measuring system, the TM-747D, was wont to connect four thermocouples to a private computer, with an information acquisition system put in to record temperature histories throughout FSW. Tiny holes with a diameter of one metric linear unit were trained on either side of the work in accordance with the thermocouples. The positions of the thermocouples within the work area unit shown in Fig. 4. Every drilling hole is formed to stop the thermometer from being crushed by a little indented clamp at the doorway. The sensing head of the thermometers is one metric linear unit long and also the holes wont to place the thermocouple area unit eleven metric linear unit deep. Therefore, thermocouples are securely embedded in the holes and the temperature can be measured accurately without external disturbances. The direction of rotation and the direction of movement of the tool are shown in Fig.4. As further described, two of the four thermocouples (TC1 and TC3) are placed on the advancing side, and the remaining two (TC2 and TC4) are placed on the reverse side. The distances between the thermocouples and the tips of the joint line are all 6 mm. This type of layout has been used to measure temperature histories of developing and backward sides [13-18].

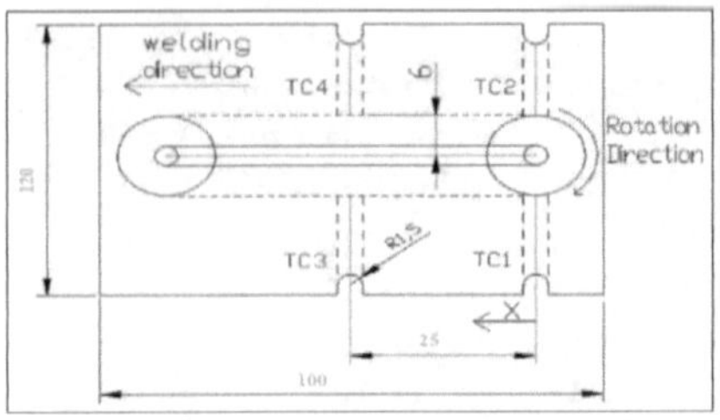

Fig 4: Layout of Thermocouples inside the workpiece

Table 6: FS Welding Parameters

S. No.	Similar/Dissimilar metals	Rotational speed (rpm)	Travel Speed (mm/min)	Thickness (mm)	Tool depth (mm)	Initial Heat Time (Sec)	Weld Length (mm)
1	Al-Mg	800	150	3	2.9	2	110
2	Al-Mg	1200	150	3	2.9	2	110
3	Al-Mg	800	60	3	2.9	2	110
4	Al-Mg	1200	60	3	2.9	2	110

Results and Discussions

MICROHARDNESS

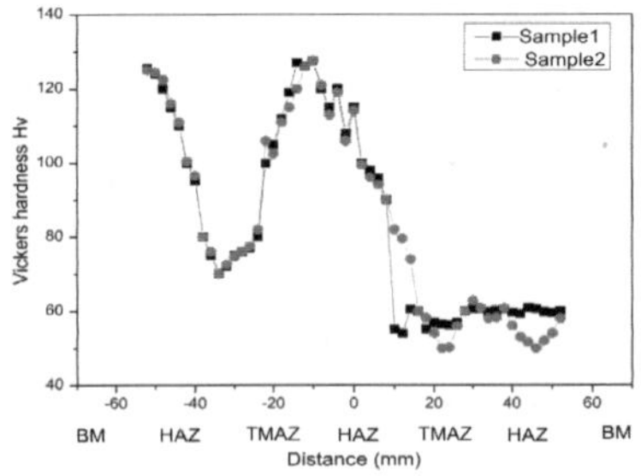

Fig 5: Vickers hardness graph for Sample 1 and Sample 2

Vickers of uneven weldments were measured with the dotted lines shown in Fig.5. Microhardness provides uneven distribution. BM average hardness values of dissimilar metals were 105HV and 75HV. NZ microhardness values supported phase elements and local recrystallization. The higher NZ, values will be high because of the presence of hard and brittle intermetallic compounds [19-26]. NZ grains are recrystallized and smaller when compared to base material. NZ hardness will increase within the FSZ and depends on the strength of the grain size hardening metal alloys. There's no report on careful analysis of precipitate distribution across the Al / MG weld joint. However, the FSW study on Al-MG-Si compounds are studied [27-29] indicates that the end of sharp needle deposits to a lower hardness within the stir zone.

X-RAY DIFFRACTION

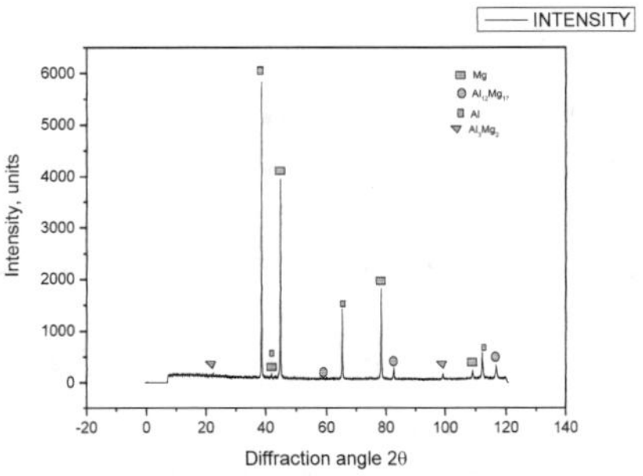

Fig 6: XRD pattern of dissimilar metals

Figure 6 contains some of the peaks obtained from the XRD analysis on the specimen. Large peaks of the intermetallic compound $Al_{12}Mg_{17}$ are found. $Al_{12}Mg_{17}$ and Al_3Mg_2 are intermetallic compounds near the intermetallic layer of NZ found by XRD samples. Intermetallic structure of 2 sorts of compounds ($Al_{12}Mg_{17}$ and Al_3Mg_2) has conjointly been recorded in Al-MG welds [30-36]. It is observed, intermetallic compounds like Mg, $Al_{12}Mg_{17}$, Al, Al_3Mg_2 at different diffraction angles with number of peaks of each intermetallic compound are Mg peaks 4 at angles of 42°, 45°, 78°, 108° out of which has maximum intensity at 4000 units at an angle of 45° and minimum intensity at 150 units at an angle of 42°, similarly $Al_{12}Mg_{17}$ has peaks 3 and angles of 58°, 82°, 118° out of which has maximum intensity at 300 units at an angle of 118° and minimum intensity at 120 units at an angle of 58°. Al has peaks 4 and angles of 38°, 42°, 65°, 113° out of which has maximum intensity at 5800 units at an angle of 38° and minimum intensity at 150 units at an angle of 42°, similarly Al_3Mg_2 has peaks 2 and angles of 22°, 100° out of which has maximum intensity at 150 units at an angle of 100° and minimum intensity at 120 units at an angle of 22° [37-39].

MICROSTRUCTURE

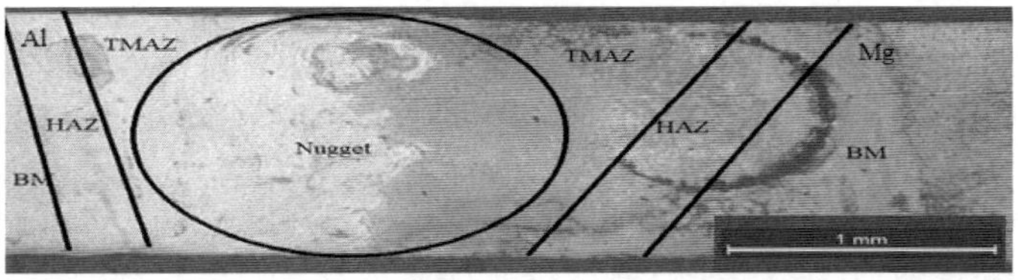

Fig 7: Macrostructure of a FSW joint; cross section showing the HAZ TMAZ and nugget zone

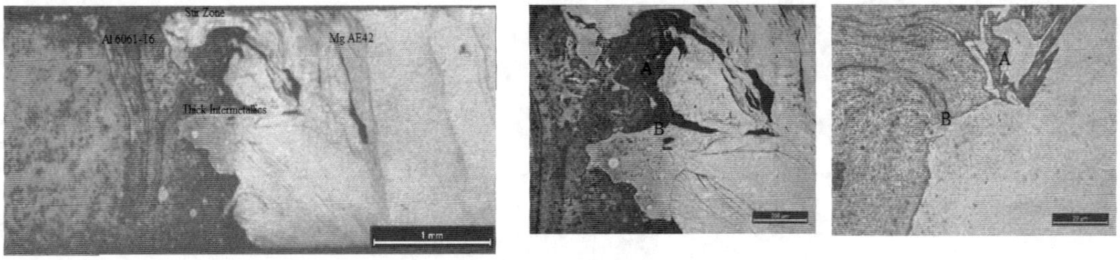

Fig 8: EPMA of dissimilar metals

Detailed examination of the asymmetric microstructure shown in Fig. 7 geometrical section at joint is distinguished in a few different areas. The parent material heat effect zone (HAZ), also known as material deformation, significantly modified the microcirculation and mechanical properties, modifying the thermo-mechanically affected zone (TMAZ), in which the material was deformed into plastic and also had some thermal inhalation. Finally, welding zone is called the nugget; NZ is a redesigned fine, identical grains containing a few micrometres of original grains and sub-grain borders.

BM contains a larger grains because of more temperature when compared to 1.3 micro meter of base material. The grain size is 25 to 30 micrometre and is stable. Silicon precipitate also can be observed they are 10-15 micrometres long. Thermal and mechanical effect add each other so that the structure is remarkably distorted and the grain starts smashing in TMAZ and the morphology of the grain does not change but the dimension is reduced up to 10 micron. The silicon precipitates rise and are spread in the structure. In addition to recrystallization, both Al-enrichment and β phase dissolve in the interior of α-Mg grains in the nugget area, leading to an increase in Al concentration and corrosion resistance improvement has been noted in the nugget area.

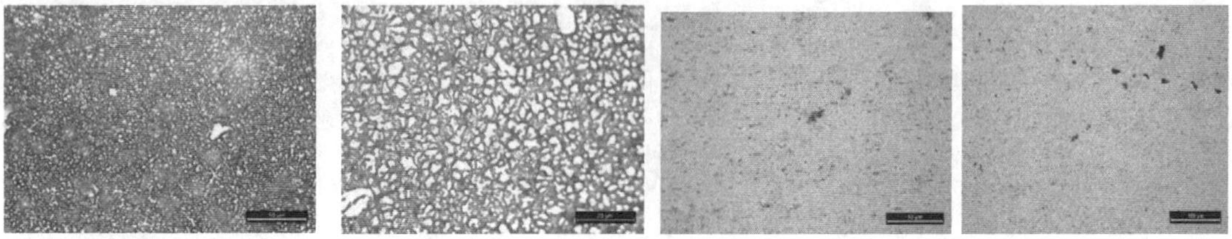

Fig 9: Microstructure of FS welded joints on the Mg side nugget zone clockwise from (a) at 800 rpm with 60mm/min and 1200rpm at different magnifications.

Scanning Electron Microscopy:

SEM with EDS analysis performed at middle of NZ on weldments. Fig 9 discuss about spot 1, 2 & 3 in different rotational speeds on dissimilar joints. Figure 10 shows images of areas of different patterns with percentage of compositions in tables. Mg (19.26% Mg and 80.74% Al) is abundant in spot 1 and spot 2 has a composition of 91% Mg and 5.58% Al. Similarly Spot 3 contains 2.78% Al and 17.32% Mg and Zone A and Zone B contain 38–54% Al and 46–62% Mg. The Al-Mg binary phase diagram shows two equilibrium intermetallic phases; Phases $Mg_{17}Al_{12}$ (39.5–55% Al) and Mg_2Al_3 (59.7–61.46% Al) [39]. The composition of Zone B on each the Al and Mg aspect intermetallic layer corresponds to the $Mg_{17}Al_{12}$ part and therefore the Mg aspect zone A closely corresponds to the Mg_2Al_3 phase composition vary. Elemental analysis by energy scattering X-ray analysis indicates the presence of semi-quantitative intermetallic compounds. Sensible proof for the existence of intermetallic compounds is shown in figure.

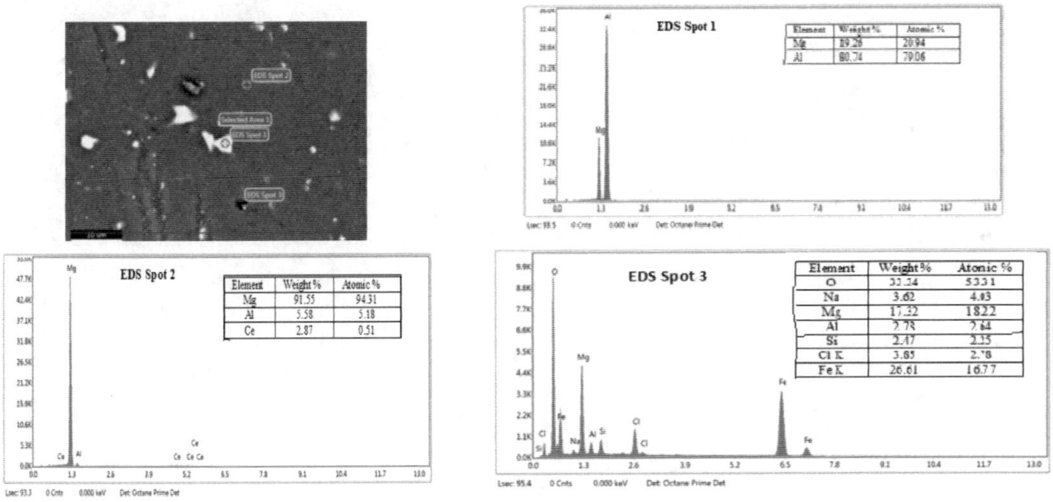

Fig 10: Results of EDS analysis of the dissimilar metal

Tensile Strength

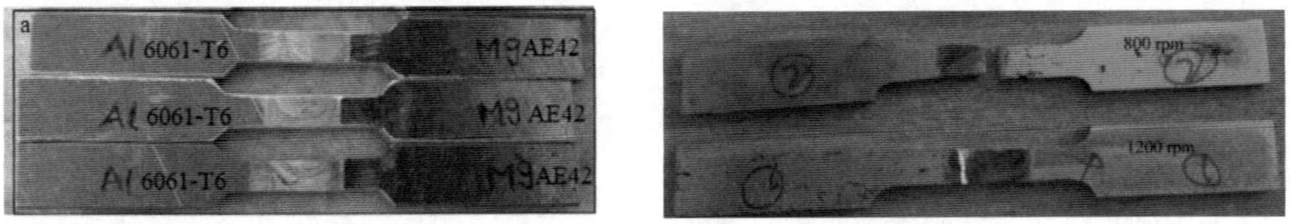

Fig 11: a) Tensile testing specimens

Table 7: Process Parameters

S. No	Rotational Speed (rpm)	Travel speed (mm/min)	Width (mm)	Thickness (mm)	Sp. Gauge length (mm)	0.2 Yield Strength (MPa)	Ultimate Tensile Strength (MPa)	Elongation (%)	Efficiency (η) %
1	800	150	5	3	25	140	190	3.0	84.4
2	1200	150	5	3	25	165	187	3.5	83.1
3	800	60	5	3	25	124	167	3.0	74.2
4	1200	60	5	3	25	123	155	3.0	68.8

Tool rotation speed is enforced to keep up a forced relationship on the joint tension tests. The strength of the bottom metals is 195 MPa. In each joints, 1200 revolutions per minute tool rotation speed, 60 millimetre / min travel speed, 5 KN axial force of 187 MPa with 72% joint. Employing a rotational speed of 1200 revolutions per minute, it's created to indicate low T.S (Tensile Strength) correlated to the weldment, having joint efficiency of 68% at 800 revolutions per minute discussed in table 7 [40-43].

FRACTOGRAPHY

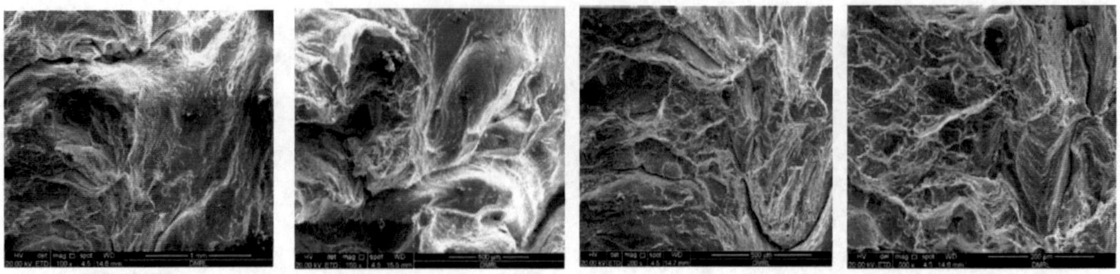

Fig 12: Effect of tool rotational speed of 1200 rpm on fracture surface morphology

Figure 13 shows the crack surface morphology demonstrating the effect of the rotation speed of the joint. Tension failure of weld joints is caused only by fracture along the central continuous vertical but slightly curved intermetallic metallic (IM) layer. Due to the brittle nature of the joint minimum cross section and IM along the weld center line at the minimum cross section it can be classified that all weld joints failed along the IM layer during tensile testing. Alloy and mg Alloy Cross section of weld joint, tensile fracture pattern. Failure of welded joint specimens is caused by a very narrow IM membrane, leaving IM compounds on the al and mg sides of the fractured surfaces sides of the fracture surfaces [44-45].

Transgranular cracking occurs in the brittle IM phases and the phase-like characteristics similar to cleavage facets. Fig shows SEM fracture morphology from the general direction on the AS to the fracture surface, i.e. towards the AA6061-T6 Al alloy. Cleavage like crack can be observed on the fracture surface, indicating that the asymmetric weld has completely failed through the brittle fracture mode.

CORROSION STUDIES

Table 8 shows the corrosion attack on specimens with a rotational speed of 800 and 1200 rpm. These data shows the estimated weight reduction with a solution of NaCl from the salt spray test. The corrosion attack on the magnesium alloy side is much higher compared to the aluminium alloy side. Corrosion testing is performed at 30 h and 90 h intervals. The surface for corrosion evaluation (for salt mist) is mechanically ground with 1200 grit SiC paper, rinsed with distilled water and dried with warm running air [45-48].

Salt spray corrosion testing involves exposing samples in the salt spray as per ASTM B117 standards and methodically evaluating the corrosion-tested sample according to ASTM G1-03. The weight loss is more due to the high rotational speed and high corrosion time as shown in fig 14. The difference in weight loss before and after conducting the experiments are shown in table 8.

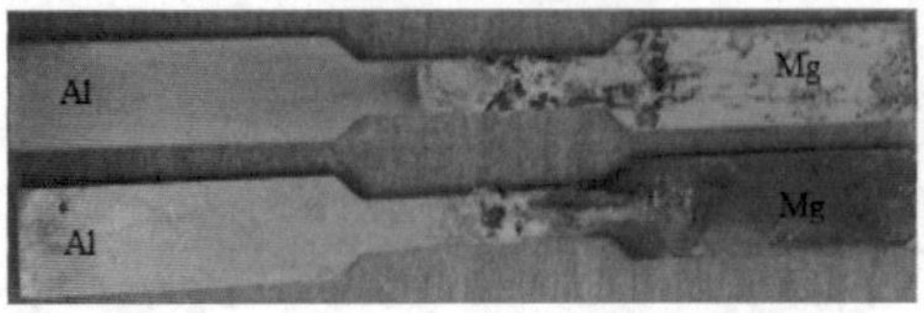

Fig 13: Corroded samples of dissimilar metals with corrosion time 30 hrs and 70hrs

Table 8: Results of salt spray test on weight reduction

S. No	RPM	PD	TS	Corrosion Time (hrs)	Weights (g) Before	Weights (g) After	Change in weights (g)
1	800	3	60	30	5.15	4.87	0.28 grams

2	1200	3	150	70	5.18	4.86	0.32 grams

Corrosion morphology

Fig. 15 shows the SEM image of the polished sample after 30 and 70 hours of salt spray corrosion testing. The exposed sample surface was prepared for microscopic examination with a small test. Corrosion test specimens for scratch-free surfaces were polished on a disc-polishing machine. To determine the depth and diameter of the pits, the exposed specimens were cut in the cross-sectional direction, the corrosion products were removed, and then the specimens were covered with hot setting resin and the surface was observed at 200 X magnification. Here, it can be seen that the weld joint is attacked towards the magnesium and the other part of the region is related to the distribution of the β -phase performance in Fig. 15. The aluminium in the weld joint is not attacked which shows corrosion is higher for aluminium alloys and lower for magnesium alloys [44].

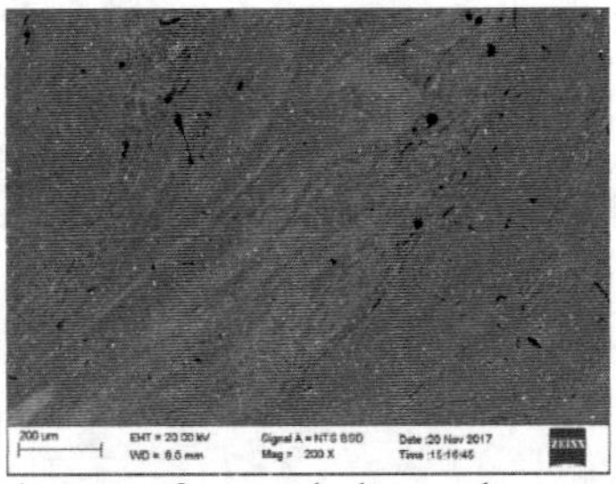

Fig 14: SEM image of corroded sample at weld nugget

NUMERICAL STUDIES

Numerical simulation of FSW is highly complex due to non-linear contact interactions between tool and work piece.

3.1 Geometry, boundary conditions and the FE mesh

The 3-D numerical model is based on the C3D8RT element type, which is a thermo-mechanically coupled hexahedral element with 8 nodes, each with a trilinear displacement. Plate dimensions 120mmx100mmx3mm in numerical pattern in plunging stage. The mesh contains 23608 nodes and 20972 elements. The numerical pattern of the welding plate, the tool is shown in the figure 6 [40, 41].

3.2 Thermal model

FSW heat generation has two sources: friction heating at the tool work piece interface and plastic energy dissipation due to shear deformation in the nugget zone. When heat is dissipated by circulation into the workpiece, tool and backing plate, and also by convection and radiation from the surfaces. Traditional losses are considered in this model are lower because of lower temperatures. They can be combined with convection from the top surface of the plate by using a slightly higher heat transfer coefficient, the governing equation for heat transfer is

$$\frac{\partial}{\partial x}\left(k\frac{\partial T}{\partial x}\right)+\frac{\partial}{\partial y}\left(k\frac{\partial T}{\partial y}\right)+\frac{\partial}{\partial z}\left(k\frac{\partial T}{\partial z}\right)+q_V=\rho c_p\frac{\partial T}{\partial t}$$

Where, k is the materials conductivity W/m K

q_v is the rate at which energy is generated per unit volume of the medium W/m^3

ρ is the density kg/m^3

c_p is the specific heat capacity J/ (Kg °C)

$$\sigma_{JC} = [A + B\varepsilon^n][1 + Cln\frac{\dot{\varepsilon}}{\dot{\varepsilon}_o}][1 - (\frac{T - T_r}{T_m - T_r})^m]$$

.

Fig 15: Numerical model of the welding plate, and the tool

Table 10: Mechanical Characteristics of parent material Al alloy 6061-T6 and Mg alloy AE42

Material Properties	Value of Al 6061-T6	Value of Mg AE42
Young's Modulus of Elastic (Gpa)	198	120
Poisson's ratio	0.33	0.35
Thermal Conductivity (W/m-K)	167	139
Coefficient of Thermal Expansion (°C^{-1})	23.4x10^{-6}	26.1x10^{-6}
Density (kg/m^3)	2.7	1.80
Specific Heat Capacity J/(Kg °C))	0.896	1.45
Latent Heat (J/g)	400	373
Temperature of melt (°C)	412	435

Table 11: Johnson cook elastic plastic constants for dissimilar metals

Dissimilar Metals	A	B	n	m	Melting temp T_m (°C)	Ambient temp T_r (°C)
AA6061-T6	285	94	0.41	1.34	588	25
Mg AE42	172	360.7	0.45	0.95	650	25

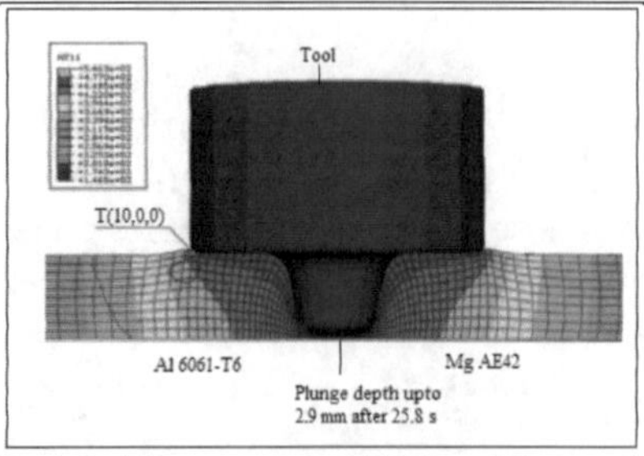

Fig 16. Temperature fields of the tool

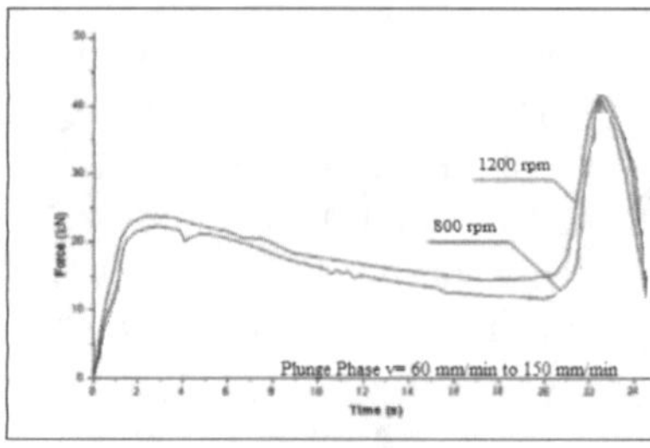

Fig 17. Force dependence of the time during the plunge stage

Study of temperature fields and the plunge force of Al alloy 6061-T6 to magnesium alloy AE42 was developed by coupled thermomechanical model, under different rotating speeds: (800, 1200) rpm with travel speeds of 150mm/min and 60mm/min. Figure 8 shows the coordinates of point T (10, 0, 0) used for measuring the temperature dependence of the time. The heat transfer coefficient through the welding plate is 687 W/m^2 K, A constant coefficient of friction is 0.3 assumed between the tool and the welding with corresponding heat convective coefficients on the surface of the welding plates are h=15 W/m^{2}K with the ambient temperature of 210 ℃ at 30 s. when the plunge speed is 20mm/min and the rotation speed is 800mm/min.

CONCLUSION

The temperature fields, plunge force associated plastic deformations of dissimilar weldments completely different rotating speeds throughout the FSW method were analysed employing a couple thermo-mechanical model and therefore the observations associated with an intense plastic deformation influence of the warmth generated because of surface friction plastic deformation. Observations of macrostructure and microstructure shows different zones of weldments. Coarse grains were found to be in HAZ and fine grains within the nugget. Tensile strength of the parent material found was 84%. This behavior is said to the plastic deformation result of warmth generated by surface friction plastic deformation. Maximum temperature made by weldments will vary from 85% - 95% to the melting temperature of the BM. Speed of rotation will increase, the temperature will increase. The temperature field is symmetrical. The contact between the rotating pin and therefore the fastening plate begins to extend, and reaches maximum value at 1200 rev and 800 rev to the fabric malleability and is softened and the plunge force may be reduced. The results show that software ABAQUS is particularly important in simulating mechanical properties in this process. Of the two rotational speeds and welding speeds used to fabricate dissimilar joints, the joint fabricated at a tool rotational speed of 800 rpm and welding speed of 60mm/min yielded superior tensile properties. The above joint showed a maximum joint efficiency of 76% compared to other joints. During high speeds, the high heat generation causes the grain growth in the stir zone. Moreover, a higher tool rotational speed causes excessive release of stirred materials to the upper surface, leads to voids in the stir zone. On the other side, at lower speeds the area of the stir zone decreases with decreasing and affects the temperature distribution in the stir zone. Macrostructure observations have shown that joints formed at low welding speeds (60 mm / min) have defects such as piping defect in the tile or stair zone and as a result have low tensile properties. Low welding speed (60 mm / min) leads to high temperature and slow cooling rate in the weld zone, leading to high grain growth, which in turn can lead to tensile properties of the joints. The joint made at a welding speed of 150 mm / min exhibited high tensile strength and this may be due to insufficient heat output, which in this condition allows the material to flow plastic with appropriate mechanical work.

Acknowledgments
Thanks to DMRL, Head, Department of Metallurgy, Hyderabad, for permission to use the equipment in connection with this project.

References:
[1] Edward D Nicholas “Developments in the Friction Stir Welding of Metals” Aluminium Alloys Vol- 1 Proceedings of ICAA-6 / 1998
[2] K. Colligan “Material Flow behaviour during Friction stir welding of Aluminium” Supplement to The welding journal, July 1999
[3] Joseph R. Pickens “Review Recent developments in the weldability of lithium-containing Aluminium alloys” Journal of Materials Science 25, 2000, 3035-3047.
[4] Chao Y.J et al. “Effect of Friction Stir Welding on Dynamic Properties of AA2024-T3 and AA7075-T7351” Welding Research Supplement 197-s - 200-s August 2001.
[5] Wayne M Thomas, et.al, “Friction Stir Welding-Tools and Developments” FSW seminar, IST-Porto, Portugal, 2002.
[6] Per Bakke, et al. “Enhanced Ductility and Strength Through RE Addition to Magnesium Die Casting Alloys” Magnesium Technology (The Minerals, Metals & Materials Society) 2003
[7] Zhili Feng et al. “G. Friction Stir Welding and Processing of Advanced Materials” High Strength Weight Reduction Materials, Progress Report, FY 2004.
[8] W H Jiang et al, “Feasibility study of friction stir welding of 6061-T6 aluminium alloy with AISI

1018 steel" Proceedings of the Institution of Mechanical Engineers, Part B: Journal of Engineering Manufacture 2004 218: 1323.

[9] Erdinc Kaluc. E et al. "Investigation on Mechanical and Microstructural Properties of TIG, MIG And FS Welded 5083-5086 Aluminium Alloys. Conference Paper 2005.

[10] Cavaliere. P et al. "Mechanical response of 2024-7075 aluminium alloys joined by Friction stir Welding" Journal of Materials Science 40 (2005) 3669-3676.

[11] Peel. M. J et al. "Dissimilar Friction Stir Welds in AA5083-AA6082 Part I: Process Parameter Effects on Thermal History and Weld Properties" Metallurgical and Materials Transactions A, Volume 37A, July 2006-2183

[12] Huang Deming et al, "Indentation creep behaviour of AE42 and Ca-containing AE41 alloys" Materials letters, Elsevier 2007, 1015-1019.

[13] Taiki Morishige et al "Dissimilar Welding of Al and Mg Alloys by FSW" Materials Transactions, vol.49, No.5 (2008) pp. 1129 to 1131, Special Issue on platform Science and Technology for Advanced Magnesium Alloys, IV The Japan Institute of Metals.

[19] Vivekanandan. P et al. "The Experimental Analysis of Friction Stir Welding on Aluminium Composites" International Journal of Metallurgical Engineering 2012, 1(4); 60-65.

[20] H Mizushima et al "Weldability in friction stir welding between dissimilar metals with circular Weld line" Proceedings of 1St International Joint Symposium on Joining and Welding, 2013.

[21] Hazman Seli et al "Evaluation of Properties and FEM Model of the Friction Welded Mild Steel-Al6061- Alumina, Materials Research. 2013; 16 (2); 453-467.

[22] Raguraman. D et al "Corrosion Study in Friction Stir Welded Plates of AA6061 and AA7075" International Journal of ChemTech Research CODEN (USA): IJCRGG ISSN; 0974-4290 Vol, No.4, pp 2577-2582, July-Aug 2014.

[23] R. Satish et al "Corrosion Studies on Friction Welded Dissimilar Aluminium Alloys of AA7075-T6 and AA6061-T6, Int. J. Electrochem. Sci., 9(2014) 4104-4113

[24] J. Mohammadi et al "Friction stir welding joint of dissimilar materials between AZ31B Magnesium and 6061 aluminium alloys: microstructure studies and mechanical Characterizations" Materials Characterization, 2104

[25] Paul Kah et al "Investigation of weld defects in friction –stir welding and fusion welding of Aluminium alloys" International Journal of Mechanical and Materials Engineering (2015).

[26] Mohammad Amin Bozorgzadeh et al "Friction stir Welding" International Journal of Review in Life Science, ISSN: 2231-2935, Volume 5 (2015), Issue (Jul-Sep), Pages 72-75.

[27] Ho-Sung Lee et al "Friction Stir Welding Process of Aluminium- Lithium Alloy 2195" International Conference on Manufacturing Engineering and Materials, ICMEM 2016.

[28] Kamaljit Singh et al "Friction Stir Welding of Similar and Dissimilar Materials: An Overview" Int. J. New. Inn, Procedia Engineering 149 (2016) 62-66.

[29] V. Prasanna, Dr. A. Seshu Kumar, Dr. P. Ramesh Babu, "Friction Stir Welding of Magnesium Alloys- A Review" International Journal of Advanced Engineering Research and Science, Vol-3, Issue-12, Dec-2016.

[30] V. Prasanna, Dr. A. Seshu Kumar, Dr. P. Ramesh Babu, "Analogy of Conventional Welding Methods with Solid-State Welding Techniques" Imperial Journal of Interdisciplinary Research, Vol-2, Issue-9, 2016.

[31] Sabitha Jannet et al. "Microstructure Analysis of friction Stir Welded AA5083 and AA6061-T6 Alloys "International Journal of Mechanical Engineering and Technology (IJMET), Volume 8, Issue 7, July 2017, pp.328-333, 2017.

[32] Nilesh V. Desai et al "A Methodology for Friction Stir Welding of Aluminium 6061-T6 and 304L Stainless Steel" International Journal of Innovative Research in Science, engineering and Technology, Vol.6, Issue 12, December 2017.

[33] Kuntala Goutham Krishna et al "Study on mechanical Properties of Friction Stir Welded Dissimilar AA2024 and AA7075 Aluminium Alloy joints" International Journal of Nanotechnology and Applications, ISSN 0973-631X Volume 11, Number 3 (2017), pp. 285-291.

[34] M.A Unnkrishnan et al "A Survey on Friction Stir Welding of Dissimilar Magnesium Alloys" IOP Conf. series: Materials Science and Engineering 247, ICRAAE-2017.

[35] Gianluca D Urso et al "The Influence of process parameters on mechanical properties and Corrosion behaviour of friction stir welded aluminium joints" International Conference on the Technology of Plasticity, ICTP 2017, 17-22 September 2017.
[36] V. Prasanna, A. Seshu Kumar, and P. Ramesh Babu, "Aspects of Friction Stir Weldments of rare earth AE42 Magnesium alloy", International Journal of Applied Engineering Research, vol. 12, no.17, pp. 6647-6652, 2017.
[42] V. Prasanna, Dr. A. Seshu Kumar, Dr. P. Ramesh Babu, "pivotal Study on CVN value of Friction Welded Magnesium AE42 Alloy at Different tool Rotation and Transverse Speed" International Journal of Research in Advent Technology, special Issue (ICOMACE-2018).
[43] V. Prasanna, A. Seshu Kumar, and P. Ramesh Babu, "An Investigation on Corrosion Behaviour of Similar Friction Stir Welded Joints", Journal of Mechanical Engineering Research and Developments ISSN: 1024-1752 Website: http://www.jmerd.org Vol. 41, No. 1, 2018, pp. 136-141.
[44] Chaitanya Sharma et al "Friction Stir Welding of Dissimilar Aluminium Alloys AA5086 and AA7039, 2019
[46] Ahmed El-Keran et al "Mechanical Behaviours of Joining AL-Alloys based FSW parameters and Welding tool design" International journal of Scientific & Engineering Research, Volume 10, Issue 6, June-2019.
[47] Vuppula Prasanna, Dr. A. Seshu Kumar, and Dr. P. Ramesh Babu, "Experimental Studies od Dissimilar Metals Welded Joint by F.S.W", Elsevier Materials Today: Proceedings 18 (2019) 3276–3285.
[48] Vuppula. Prasanna, Dr. P. Ramesh Babu, "The Effect of process Parameters on Microstructural And Mechanical Properties of Friction Stir Welded Age-Hardenable Aluminium Alloy' International Journal of Engineering and Advanced Technology (IJEAT), Volume-9, Issue-3, February-2020.

IOP Conf. Series: Materials Science and Engineering 1091 (2021) 012048 doi:10.1088/1757-899X/1091/1/012048

Experimental validation of laser powder bed fusion simulation

P Minetola[1], V Stiuso[1], F Calignano[1], M Galati[1], M S Khandpur[1], and L Fontana[1]

[1] Dept. Management and Production Engineering (DIGEP), Politecnico di Torino, Corso Duca degli Abruzzi 24, 10129 - Torino, Italy

paolo.minetola@polito.it

Abstract. In many industrial sectors, laser powder bed fusion (L-PBF) is the main additive manufacturing technology for producing end-usable metal parts. Although L-PBF technique has been developed in the last twenty years, ensuring process feasibility and achieving maximum product quality at the first building session is still a difficult goal to pursue. Simulation software packages are available in the market for the prediction of induced stresses and deformation in L-PBF products to help the user getting the part right at the first time. In this paper, Amphyon software by Additive Works is tested and experimentally validated for the production of Ti6Al4V parts in an EOSINT M270 Dual Mode machine. First, the sensitivity of the software is evaluated by changing the main process parameters by +/- 20% with respect to Ti64 reference values. After calibration, the software is validated by comparison of the predicted deformed shape of a reference part with the real geometry using 3D scanning. Experimental results show that Amphyon software is able to predict the deformed shape for L-PBF parts correctly. The deviations from the real geometry depend on a simplified simulation model that considers a limited set of parameters for the L-PBF process.

1. Introduction

Among Additive Manufacturing (AM) technologies for metals, powder bed fusion (PBF) is the most widespread process for the production of end-usable part directly from a CAD model and without the need for specific tools or dies [1]. In laser powder bed fusion (L-PBF), which is also renowned as Selective Laser Melting (SLM), a laser beam is used to selectively melt metal particles layer after layer until part completion [2].

When compared to subtractive manufacturing, apart from advantages deriving from the higher design freedom, AM makes mass customization viable because of the absence of specific moulds or dies. In AM, the size of the economic lot is pushed down to the single unit of a product and getting a part right at the first time is of paramount importance [3, 4]. This need holds for most AM applications in the biomedical or racing sectors, where a unique part is requested with a specific design and shape that are different from those of other parts [5]. Therefore, especially for parts with a complex shape, optimization of the L-PBF process is fundamental to reduce the defect rate and associated costs, while maximizing part quality and profit.

Unlike the expensive experimental trial and error approach, numerical methods can be used for the optimization of the L-PBF process by means of virtual simulations. Several different approaches have been presented in the literature for the simulation of the L-PBF process [6-18]. The simulation aims to predict the behaviour of the metal powder in the interaction with the laser source during the melting process and the subsequent cooling phase in each layer of the build. The computational methods evaluate the effect of the L-BPF process on the material properties and its impact on the final part geometry and quality [19]. Therefore, virtual simulation is an aid for the optimization of powder bed fusion through the definition of part orientation, support structures, and optimal machine parameters so that residual stresses in the part and the associated geometrical distortions can be minimized [20, 21].

Distortions are the inevitable result of the thermal gradient between the different areas of the powder bed. During the layerwise process, the bed area invested by the spot of the laser beam will have a temperature close to the material melting point. This area is generally named heat-affected zone (HAZ) [22, 23]. Conversely, the temperature of the surrounding areas will be much lower.

The material in the HAZ undergoes thermal expansion and forces the surrounding colder material to bend for accommodating the expansion. When the action of the laser is over, the material in the HAZ cools and its contraction imposes tensile stressing in the underlying layers. The stresses due to the expansion and contraction are so high as to cause permanent plastic deformation. Excessive distortions during the build cause the delamination between the layers with consequent upward bending of the upper part layer. Due to this deformation, the reacoter might collide with the upper layer of the part while depositing the next layer of powder on the bed. Depending on the type of deformation, the collision can lead to the safety stop of the machine with consequent interruption of the construction of the piece or even of the entire job [24]. The remaining parts are incomplete and the corresponding material is a production waste.

To prevent distortions during the L-PBF process, constant heating of the build plate is used to reduce thermal gradients. Moreover, support structures can be added to the part to conduct heat away from the hottest areas of the build. Thus, the addition of supports prevents part distortion and warping as well as cracks induced by thermal stresses. The other main roles of support structures are to fix the part on the build platform and to support overhanging geometries. The number and volume of supports depend primarily on the orientation of the part in the build volume of the AM machine [25, 26]. Since the support structures are made of the same metal as the part, further post-processing and metal cutting operations are needed with an increase of manufacturing times and costs. Hence, the simulation of the L-PBF process allows considering different orientation alternatives, supporting strategies, process parameters and resulting part distortions resulting from the thermal history of the layers in the powder bed.

The general manufacturing route of a metal product by L-PBF involves several different steps and operations (figure 1).

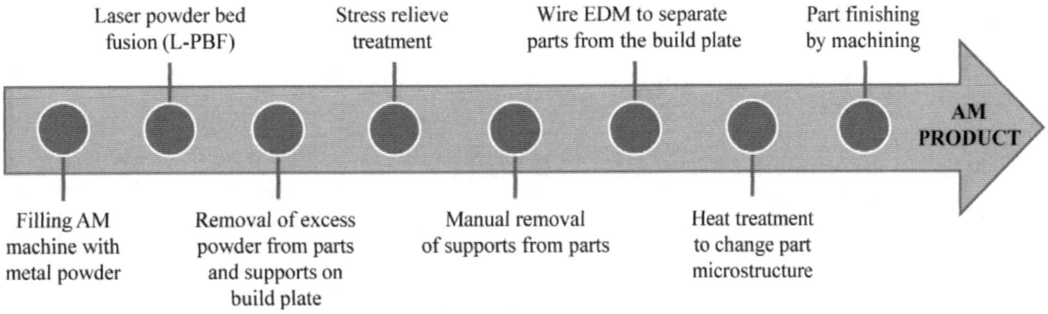

Figure 1. Manufacturing route of a metal part by L-PBF.

First the metal powder is loaded into the feeding the AM machine, then the parts are built layerwise by the L-PBF process. Once the build is complete, the unmelted powder is removed from the build plate to clean the parts that are still fixed by the supports. Before separating the part from the plate, a stress relieve treatment is normally applied to reduce the residual stresses and related part distortions. After this first thermal treatment, supports are manually removed to avoid further part distortions at the separation from the build platform. At this stage, the removal of the support structures reduces the time and cost of the final finishing step. After support removal, the parts are separated from the build plate by wire electro-discharge machining (WEDM) not to cause additional stresses by a mechanical cutting action. Final finishing is needed to remove the material allowance and get parts to their final shape in compliance with dimensional and geometric tolerances. The final machining step might be preceded by a second thermal treatment for the modification of part microstructure to meet required material properties.

Nowadays most complete simulation methods also consider the effect of stress relieving treatment and the minimization of part warpage by compensation of the geometric distortions [27]. Internal stresses caused by the thermal gradient in the L-PBF process are normally set equal to zero when the simulation includes the stress relieving. Based on the distortion information, software packages for L-PBF simulation compute the compensated shape of the part. If simulation results are reliable, by producing the compensated geometry under the same conditions and inputs of the simulated L-PBF process, the real distorted shape of the AM part should be very similar to the one of the nominal CAD model, i.e. near-net-shaped.

Several software packages for the simulation of the L-PBF process are available in the market [27]. In this paper, Amphyon software is tested for the optimization of powder bed fusion for Ti6Al4V using an EOSINT M270 Dual Mode machine. The experimental equipment and methodology are described in the following section, while experimental results of the calibration and validation of the L-PBF simulations are presented in the third section. Conclusions and future developments are discussed in the last section.

2. Materials and methods

Amphyon software is developed and sold by the German company Additive Works GmbH and is included in the Altair Partner Alliance (APA) program. In this work, the 2020 version of Amphyon, which is the fifth major version since the beta release of 2016, was used. The user-friendly framework consists of five modules to be used in L-PBF chains from the CAD model to the build job, including the stress-relieving treatment and the compensation of geometric distortions.

First of all, the sensitivity of Amphyon software was evaluated by changing the simulation value of the process parameters by +/- 20% with respect to the standard values adopted by the authors for Ti6Al4V powder with an EOSINT M270 Dual Mode machine [28]. The main process parameters are resumed in Table 1 for a layer thickness of 30 µ. The EOSINT M270 has a Ytterbium (Yb) fiber laser source with a power of 200 W for a laser spot of 100 µm. The working volume of the machine is 250 x 250 x 215 mm and the build chamber is filled with argon during the L-PBF process, limiting the oxygen content to 0.1% to reduce the reactivity of the titanium powder.

Table 1. Process parameters of the EOSINT M270 machine for Ti6Al4V.

Parameter	Core	Skin	Contour
Laser power (W)	170	150	120
Laser spot size (µm)	100	100	100
Scan speed (mm/s)	1250	1000	1250
Hatching distance (mm)	0.10	0.10	0.10

Amphyon software uses a unique set of parameters for the simulation of the L-PBF process without distinguishing between the core or skin or contour regions. However, while the skin and contour parameters are set for specific and limited zones of the part in the build layer, the core covers most of the part area in the powder bed if the cross section is not too small. Therefore, the melting of the metal powder and its thermal history are mainly influenced by the laser parameters for the core region. For this reason, the core parameters from Table 1 were used as input for Amphyon simulations together with a build platform temperature of 100 °C. The Ti6Al4V material properties were set as Young modulus of 110 GPa, Yield stress of 1060 MPa and Poisson coefficient of 0.3. These values were extracted from the EOS datasheet for Ti6Al4V powder processes with an EOSINT M270 machine.

A reference geometry provided by Additive Works and named "AW Box" was used for the sensitivity analysis and the subsequent experimental validation. This geometry a hollow thin square tube that is built with the faces oriented at 45 degrees to the build direction and with a vertical edge to support the tube and connect it to the build platform. The tube is designed to be produced without supports. It is 1.67 mm thick and 100 mm long and its overall dimensions are shown in Figure 2.

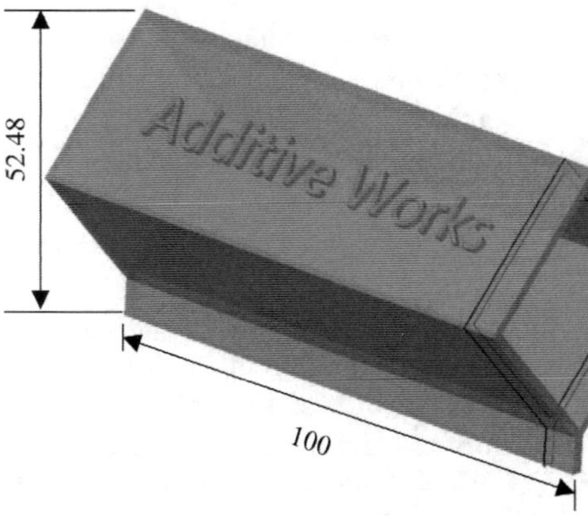

Figure 2. Reference geometry of the hollow square tube with the outline of its cross section.

Figure 3. Structure of the mesh for the simulation model in Amphyon software.

Table 2. Settings of the simulation model and solver of Amphyon software.

Parameter	Mechanical simulation	Thermal simulation
Mesh type	Adaptive	Adaptive
Mesh resolution	2	2
Solver accuracy	0.5	0.2

At the end of the sensitivity analysis, the influence of the variation of +/- 20% of the process parameters is considered over the following results of Amphyon simulation: the temperature at the end of the process, maximum residual stress, maximum deformation and build time. In order to avoid differences due to the mesh structure, the properties of the simulation model were not changed in Amphyon software during the analysis. The mesh resolution setting ranges from 0 (coarse) to 10 (fine), whereas the solver accuracy can be set from 0.0 (fast) to 1.0 (accurate). The adopted values for the simulation and solver are listed in Table 2 because the mechanical and thermal simulations are run separately. A mesh resolution equal to 2 and a solver accuracy of 0.5 for the mechanical simulation and 0.2 for the thermal one were selected to limit the computational load and time in a standard notebook. The mesh structure of the simulation model is shown in figure 3.

To account for the specific material and machine, the calibration of the simulation parameters is implemented in Amphyon through an experiment-based approach that requires the use of cantilevers. For precise calibration based on the anisotropy of the material, three samples of the cantilever must be fabricated using a different hatching strategy of the laser beam. The cantilever measures 180 x 7.2 x 8 mm and the three hatching strategies to be considered are the one parallel to the longitudinal direction of the cantilever, the one orthogonal to the first, and the average one, which combines the other two strategies (figure 4). In the specific case of the EOS machine, the main scanning pattern of the laser is rotated by a hatch angle of 67° between consecutive layers.

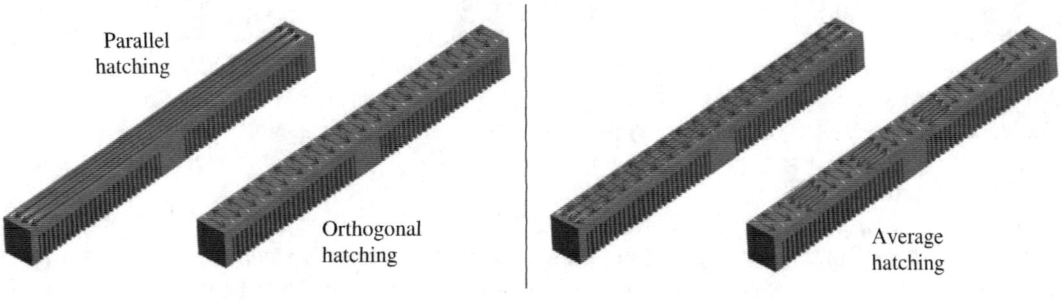

Figure 4. Different hatching strategy of the laser beam to account for material anisotropy.

IOP Conf. Series: Materials Science and Engineering 1091 (2021) 012048 doi:10.1088/1757-899X/1091/1/012048

The three strategies generate a different thermal history in the cantilever with different results in terms of induced residual stresses and consequent bending of the specimen. For the evaluation of the cantilever deformation, it is essential not to apply the stress-relieving treatment to avoid excessive stress relaxation and reduced part distortions. To detach the specimen from the build platform, WEDM cutting should be applied in the longitudinal direction and the maximum deflection of each test specimen along the main axis of the cantilever should be measured.

One replica of the three cantilever specimens and one of the reference geometry (figure 5) were produced with EOS Ti6Al4V powder with 30 μm layer thickness using the EOSINT M270 Dual Mode machine. After manufacturing, dimensional inspection with a structured light 3D Scanner Atos Compact Scan 2M by GOM GbmH was carried out to capture the real geometry of the replicas. The ATOS Compact Scan (figure 6) has two 2 megapixel cameras for stereoscopic vision and a working volume of 125 x 90 x 90 mm. The ATOS scanner uses fringe projection with blue LED light to measure the 3D coordinates of points on the surface of the inspected object through the triangulation principle. According to the acceptance test of VDI/VDE 2634 guideline part 3, the length measurement error of the scanner is smaller than 20 μm. The real geometry of the specimens was acquired before and after separation from the build platform and consequent deformation.

Figure 5. Ti6Al4V replica of the reference square tube by Additive Works.

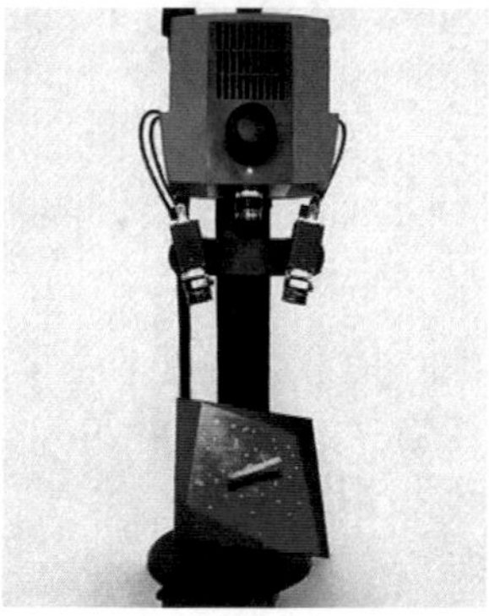

Figure 6. Inspection of a cantilever using a structured light Atos Compact Scan 2M.

An ECUT32D CNC Wire Cut EDM machine by Suzhou Baoma was used to separate the replicas from the build platform. The cantilever specimens were cut in the longitudinal direction following the indications provided by Additive Works in Amphyon reference guide. The maximum deflection of the specimens was measured using the scan data from the ATOS scanner and the measurement results were used for the calibration of Amphyon software.

After calibration, the production of the reference geometry (figure 2) was simulated using Amphyon. The deformed shape resulting from the simulation was compared to the scan data of the real part for final validation of Amphyon accuracy in predicting L-PBF part distortions.

3. Experimental results

The results of the experimental activities developed following the methodology described in the previous section are described hereafter. The software calibration and all simulations were run on a notebook with Windows 10 Home 64-bit operative system. The PC has an HP 820D motherboard with an Intel(R) Core(TM) i7-6700HQ CPU @ 2.60 GHz, 16 GB RAM, an NVIDIA GeForce GTX960M graphic card and a SATA SSD for data storage.

3.1. Sensitivity analysis

The results of the sensitivity analysis are presented for steps of 10% change of the four simulation parameters (Table 1). These parameters are varied one at a time, so sixteen simulations are run to complete the sensitivity analysis with four levels (-20%, -10%, +10%, +20%) of parameter change. The results of the simulation are displayed in the graphs of figures 7, 8, and 11. The red dashed line in the graphs shows the reference value that is obtained with the parameters listed in Table 1 for the core region. The temperature at the end of the process (figure 7) increases by almost 10 °C if the laser power is increased by 20%. A reduction of the scan speed or the hatching distance has the effect of delivering more energy per unit area on the powder bed. Therefore, when the scan speed or the hatching distance is reduced by 20%, the temperature at the end of the process is 10 °C higher than with standard parameters. A change of the layer thickness does not influence the temperature at the end of the process.

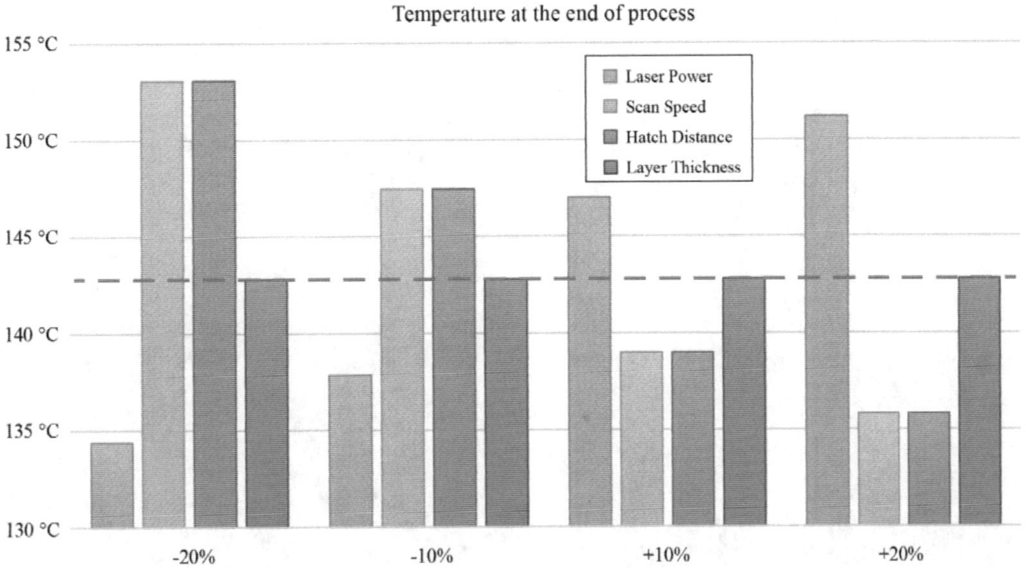

Figure 7. Results of the sensitivity analysis for the temperature at the end of the process.

The influence of the variation of the simulation parameters on the maximum residual stress (figure 8) is similar to one on the temperature at the end of the process. As the laser power increases, the temperature gradient is higher and induced residual stresses grow.

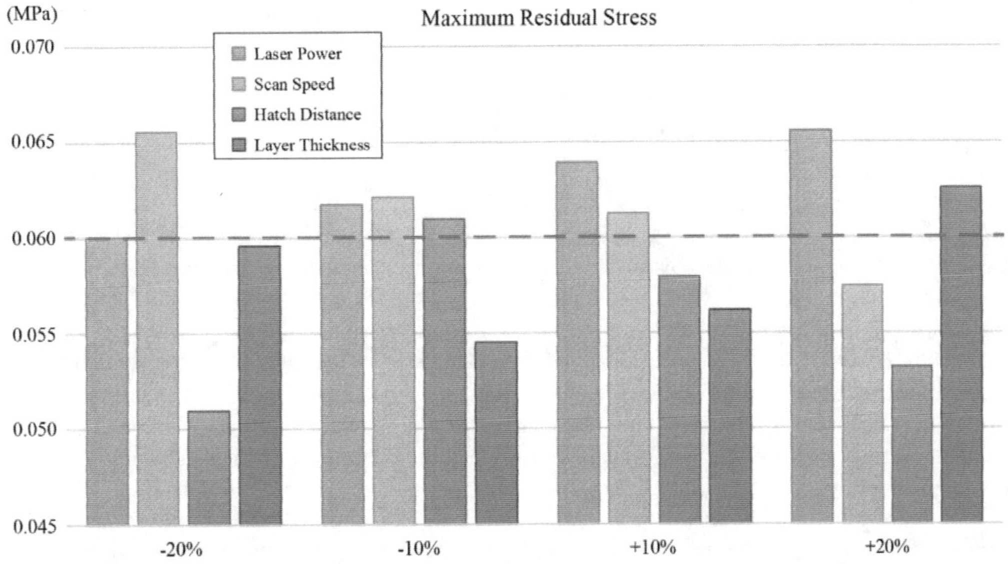

Figure 8. Results of the sensitivity analysis for the maximum residual stress.

The residual stresses for the square cube are concentrated in the red and yellow areas of figure 3, which also shows the structure of the mesh used for the simulation model. The maximum stress is located on the sharp edges that are usually the location where cracks originate due to thermal gradients.

The value of the maximum residual stress ranges from 0.060 MPa for -20% of the laser power to 0.066 MPa when the power of the laser is raised by 20%. Conversely, when the scan speed is increased by 20%, less heat is delivered to the powder bed in the unit time and the maximum residual stress value is 0.057 MPa. By decreasing the scan speed by 20%, the maximum residual stress raises to 0.066 MPa. The variation of the other two parameters of hatching distance and layer thickness from -20% to +20% does not show a clear trend, because the value of the maximum residual stress is oscillating.

To investigate the effect of the simulation parameters on the maximum part deformation, dimensional inspection and comparison of the deformed shape of the square tube are carried out using GOM Inspect software. The deformed shape resulting from the simulation with the standard set of parameters in Table 1 is used as a reference for the comparison. The deformed shape resulting from each simulation of the sensitivity analysis is then aligned to the reference geometry and the comparison results are shown with coloured maps by GOM Inspect (figure 9). The maximum deviation is then measured in the position of the sharp edge where part deformation is maximum, as already shown in figure 3.

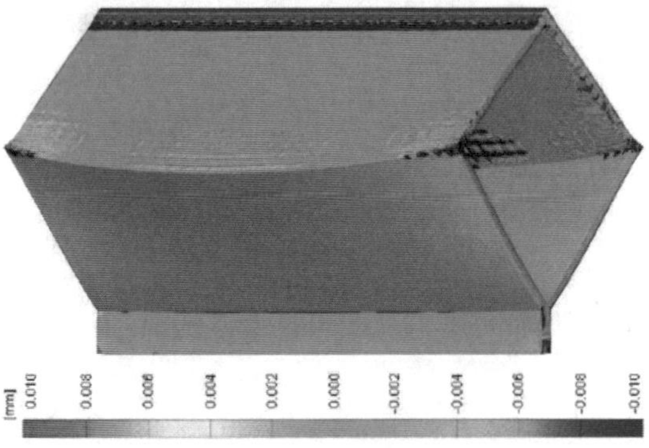

Figure 9. Results of the comparison of maximum deformation for +20% laser power.

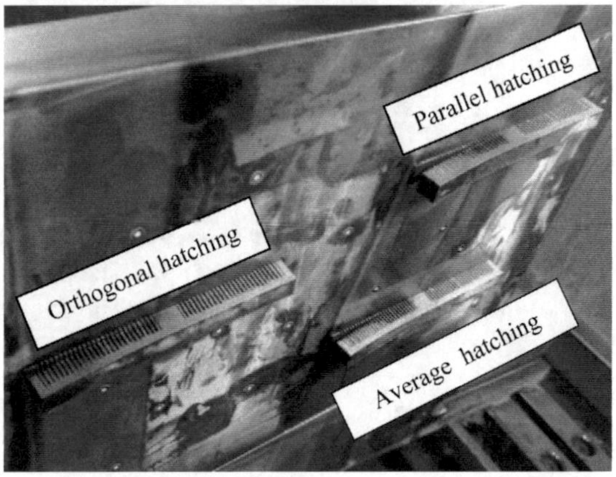

Figure 10. Deformation of the three cantilevers specimens during WEDM.

For the sake of conciseness, the graph with the results of the comparisons about the maximum deformation is omitted. However, since the distortions are a direct consequence of the residual stresses, the trends are similar to those in figure 8. The maximum difference to the deformation with standard parameters is obtained for +20% of the layer thickness. The corresponding maximum distortion is always located on the sharp edge of the square tube and is about 80 μm larger.

Like in all additive manufacturing processes, the build time depends on the part height in the build direction and on the layer thickness, which defines how many layers of material are to be deposited or processed. In the case of the L-PBF process, the hatching distance and scan speed also affect the build time. Accordingly, the results of Amphyon simulations do not show differences in the build time when laser power is varied (figure 11). The increase of the scan speed of the laser provides a small reduction of the build time as the metal powder is melted in less time in every single layer. The same result is obtained when the hatching distance is increased because the exposure to the laser source of the part cross section corresponding to the single layer is completed in a shorter time. The parameter that mainly affects the build time is the layer thickness.

This correct result is in good agreement with the practice. The smaller the layer thickness, the higher the number of layers within the build. When the build volume of the machine is not saturated and a unique small part is produced as in the case of the square tube the recoating operation for depositing the metal powder on the bed takes more time than the exposure and melting of the part cross section. The variation of the scan speed or hatching distance from -20% to +20% shortens the build time from 7 hours to 6 hours and a half. The modification of the layer thickness from -20% to +20% makes the build time decrease from more than 8 hours to 5 hours and a half. This result is consistent with the real manufacturing time since the build job with the three cantilevers and the reference part was fabricated in about 8 hours.

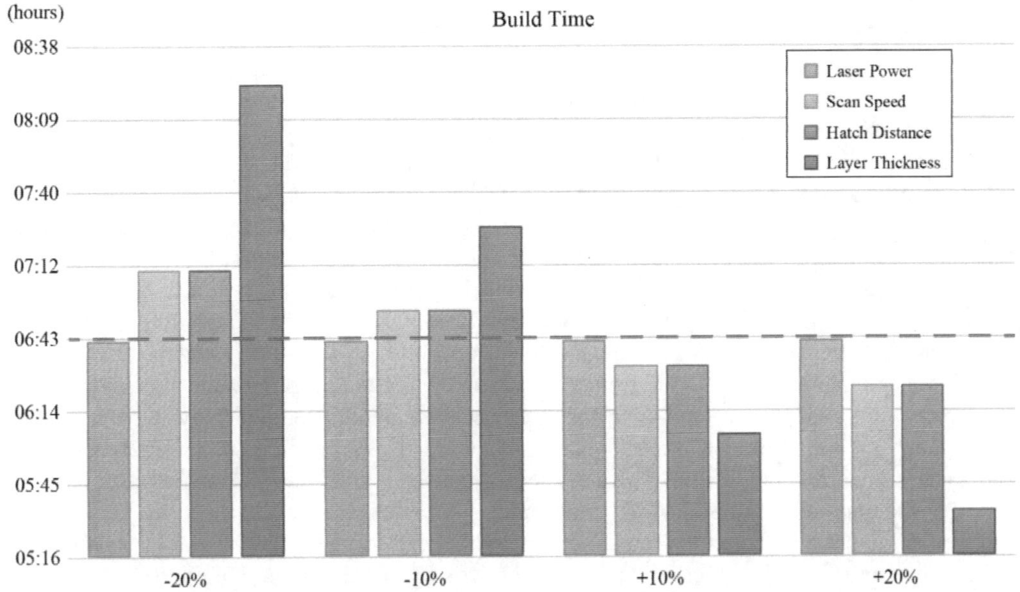

Figure 11. Results of the sensitivity analysis for the build time.

3.2. Calibration of Amphyon software

After the fabrication, the real geometry of the three cantilevers and the one of the square tube are captured using the ATOS Compact Scan before and after the WEDM cut. For the calibration of Amphyon software, the maximum deflection of the three cantilever specimens after separation from the build platform by WEDM is measured. The specimen deformation was visible during the WEDM process as shown in figure 10. The measurements are carried out with GOM Inspect software using the scan data of the ATOS Compact Scan. A reference plane is first defined by best fitting (least-squares method) two sets of points at the two ends of the top surface of each cantilever. The maximum deflection of each specimen is then measured by selecting different points in the middle of the top surface of the cantilever, wherein the distortion is maximum. For each point, the projected distance from the best-fit plane is computed (figure 12). The maximum distances measured for the three cantilevers (Table 3) are used for the calibration of Amphyon software, which takes about 1 hour and a half to be completed.

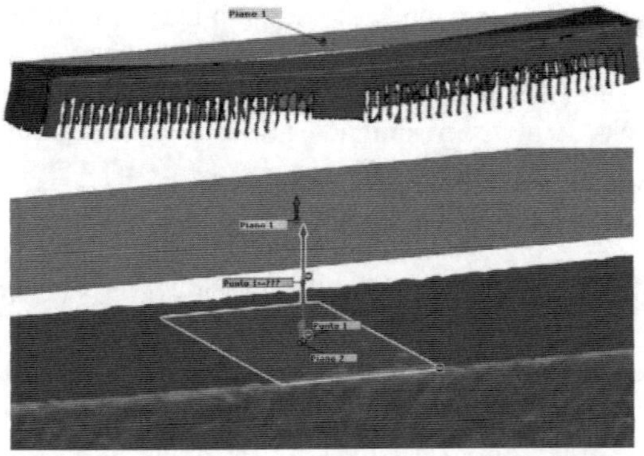

Figure 12. Best fit plane (above) and maximum deflection by the projected point (below).

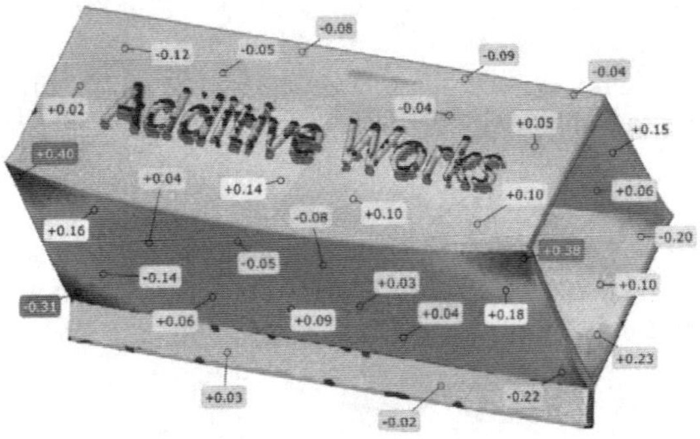

Figure 13. Deviations (in mm) between the real geometry and the deformed shape by Amphyon.

Table 3. Calibration values of Amphyon software from the three cantilever specimens.

Cantilever	Parallel hatching	Orthogonal hatching	Average hatching
Maximum deflection	2.74 mm	2.09 mm	2.28 mm

3.3. Validation

The validation of the simulation results of Amphyon software is carried out by running one last simulation for the square tube after the calibration using the reference process parameters for the core region of Table 1. The deformed shape of the part that was calculated by Amphyon is then exported into the STL format and imported into GOM Inspect software.

In GOM Inspect the simulated geometry is then aligned and compared to the scan data of the square tube after separation from the build platform by WEDM cut. The result of the comparison is shown in figure 13 as a coloured map representing the deviations between the compared shapes. The labels in figure 13 show the local deviations. The statistics provided by GOM Inspect software for the coloured map include an average of 0.05 mm and a standard deviation of 0.30 mm.

The area of the marking of the part with "Additive Works" text is not considered in the following because it is not significant for this study. Generally, the absolute difference between the predicted shape from Amphyon and the real shape of the square tube is below 0.10 mm. In the most stressed areas of the reference part (figure 3), where the deformation is higher, the difference between the compared geometries reaches the maximum of about 0.40 mm. It should also be considered that sharp edges cannot be correctly manufactured by the L-PBF process because the spot of the laser beam has a finite size, that is 0.10 mm in the case of the EOSINT M270 machine.

Therefore, it can be concluded that using the embedded experimental calibration procedure for the material-parameters pair of the specific machine, Amphyon software has good prediction capabilities of the distortion of the as-built part after the L-PBF process.

4. Conclusions

In this paper, the commercial software package Amphyon by Additive Works is tested and validated for the simulation of the L-PBF process using EOS Ti6Al4V powder in an EOSINT M270 Dual Mode machine. As a case study for the analysis, the reference geometry of the "AW Box" by Additive Works was used.

First, a sensitivity analysis is carried out to investigate how a small variation of the main process parameters is taken into consideration by Amphyon solver and related simulation results. Then the software for experimentally calibrated through the production of three cantilever specimens to account for the anisotropy of the material.

After calibration, the production of the reference part was simulated with Amphyon and the resulting deformed shape was compared with the real geometry of the as-built part, which was digitized by 3D scanning. The comparison shows that Amphyon software can correctly predict the distortion induced by the thermal history of the material during the L-PBF process. For the case study, the average difference between the predicted shape from the simulation and the real shape was 0.05 mm, while the maximum deviation was smaller than 0.50 mm.

However, one important feature of Amplyon software was not tested in this work. With its pre-deformation (PRE) module, the software can compensate for the residual distortions of the L-PBF process and can compute a pre-compensated geometry. This geometry can be exported as an STL file and used as the input in the Additive Manufacturing route to get the as-build part with the desired geometry equal to the one of the nominal CAD model. The analysis of the PRE module is ongoing at the Centre of Integrated Additive Manufacturing (IAM@PoliTO) of the Politecnico di Torino and the authors will present the results of their study in a future paper.

Acknowledgments

The authors gratefully acknowledge Mr. Nils Keller, CEO of Additive Works GmbH, for granting a free educational license of Amphyon software to be tested at the Centre of Integrated Additive Manufacturing (IAM@PoliTO) of the Politecnico di Torino.

References

[1] Atzeni E and Salmi A 2012 *Int. J. Adv. Manuf. Tech.* **62**(9-12) 1147-55
[2] Calignano F *et al.* 2017 *Proc. IEEE*, **105**(4) 593-612
[3] Atzeni E, Iuliano L, Marchiandi G, Minetola P, Salmi A, Bassoli E, Denti L and Gatto A 2013 Additive manufacturing as a cost-effective way to produce metal parts *High Value Manufacturing* ed PJ Bartolo et al. (London: CRC Press) pp 3-8
[4] Salmi A, Calignano F, Galati M and Atzeni E 2018 *Virtual Phys. Prototyp.* **13**(3) 191-202
[5] Calignano F, Galati M, Iuliano L and Minetola P 2019 *J. Healthc. Eng.* **2019** 9748212
[6] Papadakis L, Loizou A, Risse J and Schrage J 2014 *Proc. CIRP* **18** 90-5
[7] King W, Anderson AT, Ferencz RM, Hodge NE, Kamath C and Khairallah SA 2015 *Mater. Sci. Tech.* **31**(8) 957-68
[8] King WE, Anderson AT, Ferencz RM, Hodge NE, Kamath C, Khairallah SA and Rubenchik AM 2015 Applied Physics Reviews **2**(4) 041304
[9] Li C, Fu CH, Guo YB and Fang FZ 2016 *J. Mater. Process. Tech.* **229** 703-12
[10] Dunbar AJ, Denlinger ER, Gouge MF and Michaleris P 2016 *Addit. Manuf.* **12** 108-20
[11] Bugatti M and Semeraro Q 2018 *Addit. Manuf.* **23** 329-46
[12] Koepf JA, Gotterbarm MR, Markl M and Körner C 2018 *Acta Mater.* **152** 119-26
[13] Moges T, Ameta G and Witherell P 2019 *J. Manuf. Sci. E.-T. ASME* **141**(4) 040801
[14] Mayer T, Brändle G, Schönenberger A and Eberlein R 2020 *Heliyon* **6**(5) e03987
[15] Olleak A and Xi Z 2020 *Manuf. Lett.* **24** 140-4
[16] Cook PS and Murphy AB 2020 *Addit. Manuf.* **31** 100909
[17] Ninpetch P, Kowitwarangkul P, Mahathanabodee S, Chalermkarnnon P and Ratanadecho P 2020 *AIP Conf. Proc.* **2279**(1) 050002
[18] Hodge NE 2020 *Addit. Manuf.* **37** 101600
[19] Borovkov AI, Maslov LB, Ivanov KS, Kovaleva EN, Tarasenko FD and Zhmaylo MA 2020 *IOP Conf. Ser.: Mater. Sci. Eng.* **986** 012033
[20] Salmi A, Piscopo G, Atzeni E, Minetola P and Iuliano L 2018 *Proc. CIRP* **67** 191-6
[21] Ramesh Sagar V, Lorin S, Wärmefjord K and Söderberg R 2020 *J. Manuf. Sci. E.-T. ASME* **2020** 1-35
[22] Luo Z and Zhao Y 2018 *Addit. Manuf.* **21** 318-32
[23] Roy M and Wodo O 2020 *Addit. Manuf.* **32** 101017
[24] Kayacan MY, Özsoy K, Duman B, Yilmaz N and Kayacan MC 2019 *Mater. Manuf. Process.* **34**(13) 1467-75
[25] Jiang J, Xu X and Stringer J 2018 *J. Manuf. Mater. Process.* **2**(4) 64
[26] Bartsch K, Lange F, Gralow M and Emmelmann C 2019 *J. Laser Appl.* **31** 022302
[27] Peter N, Pitts Z, Thompson S and Saharan A 2020 *Addit. Manuf.* **36** 101531
[28] Minetola P, Galati M, Calignano F, Iuliano L, Rizza G and Fontana L 2020 *Proc. CIRP* **88** 399-404

Multiferroic Bismuth ferrite nanocomposites as a potential photovoltaic material

Naimur R. Niloy*, M.I. Chowdhury, M.A.H. Shanto, J. Islam and M.M. Rhaman

Department of Electrical and Electronic Engineering, Ahsanullah University of Science and Technology, Dhaka, Bangladesh.

*Corresponding author: naimurniloy@gmail.com

Abstract. Bismuth ferrite (BFO) is one of the few materials which is multiferroic (antiferromagnetic and ferroelectric) at 25°C. In this work, we synthesized multiferroic BFO via sol-gel method and observed its structural and optical properties. While synthesizing BFO, the annealing temperature was at 550°C. X-ray diffraction (XRD) analysis exhibited a rhombohedral crystal structure. Rietveld refinement via MAUD software has been employed to obtain the structural parameters of BFO. Williamson-Hall (W-H) plot method is used to analyze the XRD data. We obtained crystallite size of 37.61 nm through W-H plot analysis. Using the crystallographic information file, we measured the bond angle and bond length of BFO nanoparticles. BFO nanoparticles with both uniform distribution and spherical shape was observed in Field Emission Scanning Electron Microscopy (FESEM). The average particle size of 157 nm was measured from FESEM analysis. The atomic % of Bismuth (Bi), Oxygen (O) and Iron (Fe) elements in BFO was obtained from Energy dispersive spectroscopy. The optical bandgap energy (E_g) of BFO was calculated using UV-Vis-NIR, and we found the bandgap energy of 2.04 eV for BFO annealed at 550°C. This bandgap energy makes BFO a potential photovoltaic material.

1. Introduction

We are now heavily dependent on non-renewable energy resources such as fossil fuel, coal, gas, etc. The population on our earth is rapidly increasing. This increasing population needs more energy. Energy consumption will grow by nearly 50% between 2018 and 2050. In 2016, the total world energy came from 5% nuclear, 80% fossil fuels, 5% renewable (solar, hydro, geothermal), and 10% biofuels. When we burn fossil fuels, they release CO_2 as well as other greenhouse gases. These increase the heat in our atmosphere. Moreover, these non-renewable energy sources are also limited. To face energy challenges and reduce the impact of fossil fuels, we need to rely more on renewable energy sources. Silicon has a bandgap limitation of 1.1 eV. But the optimum bandgap energy for PV cells is 1.4 eV [1]. That's why photovoltaic cells based on silicon can produce maximum efficiency of 33% only. Therefore, for improving an energy source that is both renewable and an effective alternative to our existing energy resources, researchers have begun to look over a broad range of possible materials for thin-film solar cells [2, 3]. BFO is one of the most eligible materials for this purpose. BFO shows Ferroelectric and Ferromagnetic properties simultaneously and hence is a multiferroic material. Multiferroic materials possess more than one property of ferroics simultaneously in the same phase [4]. Bandgap energy (E_g) of BFO is 2.0 to 2.7 eV. That's why it offers a rousing prospect to research novel materials for solar cells [5, 6]. Oxide materials can not only be cost-friendly but also widely available. Their properties like electrical energy, ferromagnetic and ferroelectric effects, and bandgap energy are in a favorable zone. These properties can also be changed through doping. Thus, it makes BFO a credible candidate for photovoltaic applications. Furthermore, for the separation of charge, the ferroelectric property of BFO is a distinct path of photo-excited careers. The ferroelectric materials like $LiNbO_3$, $Pb(Zr/Ti)O_3$, and $BaTiO_3$ were investigated 30 years ago [7]. Back then the efficiencies of PV were limited because of the small current density and large E_g of the investigated materials. As a result, the power conversion efficiency (PCE) was inadequate. Those also had some structural limitations. As BFO has lower E_g, it provides the opportunity to discover novel materials [8].

Compared to other ABO_3 (ferroelectric perovskite), BFO has a lower E_g and a higher maximum polarization of $90C/cm^2$. Among single-phase ABO_3 (perovskite-type) multiferroics $BiFeO_3$ (T_{CE} = 1110K, T_N = 670K) [9], which can be considered as high regulating temperatures. Again, the optical properties of BFO remain less explored yet. Optical characteristics of BFO materials are coming to light recently through academic analyses which reveal that the E_g of BFO is between the visible light range (380-740 nm). This is also the reason why BFO is suitable for photovoltaic cells. In this work, the structural and optical properties of BFO nanocomposites were measured. While observing the optical properties through UV-Vis-NIR, we focused on its potential as a future photovoltaic material.

2. Materials and Methods

2.1. Material Synthesis

BFO nanomaterial was prepared using sol-gel technique. For preparing BFO nanomaterial, we used bismuth nitrate pentahydrate, glacial acetic acid, iron(III) nitrate nonahydrate, citric acid, iron(III) nitrate nonahydrate, and ethylene glycol. To synthesize 1 gm of bismuth ferrite nanoparticles, 1.55 gm of $Bi(NO_3)_3.5H_2O$ and 1.29 gm of $Fe(NO_3)_3.9H_2O$ and had initially been dissolved in 400 ml deionized water. At the beginning of the heating process, we used $C_6H_8O_7$ as a chelating agent. When the gel began to form, $C_2H_6O_2$ was used as a polymerization mediator. Adding NH_4OH, we controlled the pH to the level of 2.0. The detailed procedure can be found in the other work [10]. The prepared gel had been dried at 100-120°C in an oven for 20-24h for getting xerogel powder. After obtaining the xerogel, we annealed it for 2-2.5h at 550°C with a furnace. We maintained the heating rate at 3°C per min and then finally grinded the xerogel again to obtain nanostructured materials. Now, the 1 gm nanocomposites are ready for characterization.

2.2. Methods

X-ray diffractometry (PANalytical EMPYREAN) was used to obtain X-ray diffraction patterns. Using FESEM (JEOL JSM-7600F), the morphology and elemental analysis were studied. With the help of UV–VIS-NIR spectrophotometry (PerkinElmer UV–Vis–NIR Spectrometer Lambda 1050), the optical bandgap energy was calculated.

3. Result and Discussion

3.1. X-ray Diffraction

Structural characterization of nanoparticles was done employing X-ray diffraction (XRD) at around 25°C. On the nanoparticle samples, X-ray beams were diffracted which had a high intensity. The scanning limit had been kept from 10°–80°. Rietveld refinement of XRD data was done using the MAUD software. The lattice space group and structural parameters of BFO nanoparticles were obtained after Rietveld refinement. Figure 1. shows the XRD patterns for BFO nanoparticles after the refinement at 550°C. BFO nanoparticles were washed with CH_3COOH [11]. Generally, Nitric acid (HNO_3) is used to eliminate the secondary phase ($Bi_{25}FeO_{40}$). That phase was composed during the synthesis of BFO nanocomposites. Small trails of the secondary phase ($Bi_{25}FeO_{40}$) were found at 27°. But this HNO_3 may also dissolve the primary strong phase of Bismuth ferrite. Considering this issue, a weak acid (CH_3COOH) had been used to wash the samples. The X-ray diffraction pattern confirms the crystalline nanocomposites of BFO. The secondary phase ($Bi_{25}FeO_{40}$) in BFO is a major concern. BFO containing particles Bi^{3+} and/or Fe^{3+} participate in creating this type of secondary phase which also consists of Bi^{3+} and/or Fe^{3+} ions. The off-stoichiometric and metastable nature of BFO are primary reasons for forming the secondary phase. Considering the application point of view secondary phase increases leakage current and decreases ferroelectric property. So, this secondary phase must be as minimum as possible. BFO nanoparticles annealed at 500°C contains less secondary phase [10]. But its primary peak intensity is comparatively low. So, in this work, we considered to prepare BFO nanocomposites at 550°C of annealing temperature. Crystallite size was calculated from the Scherrer equation:

$$D = \frac{k\lambda}{\beta cos\theta} \quad (1)$$

Here, β = Full width at half maximum intensity (FWHM),

k = 0.9,

λ = Wavelength of Cu-Kα radiation. (~1.542 Å)

θ = Bragg angle of diffraction.

D = Crystallite size.

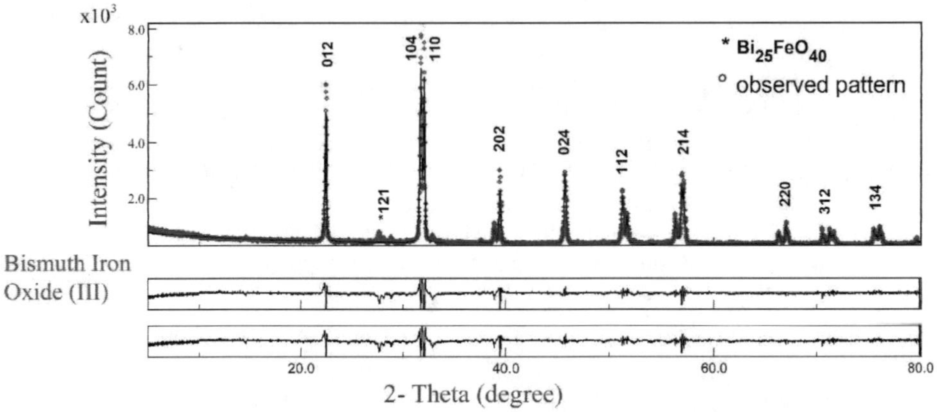

Figure 1. X-ray diffraction pattern of BFO nanocomposites.

The structural and optical parameters of BFO was measured and listed in Table 1. Scherrer formula does not consider the lattice strain. So, Williamson Hall (W-H) plot method was also used to analyze the XRD data. For Williamson Hall (W-H) plot method analysis uniform deformation model (UDM) was used. Uniform lattice strain through the crystal was considered in the UDM. The inconsistent FWHM (β) values in (312) and (134) diffraction peaks were not considered for the W-H plot method. Here at equation (2), ε denotes for lattice strain. Considering lattice strain, we obtain for FWHM (β):

$$\beta = \frac{k\lambda}{D cos\theta} + 4\varepsilon\ tan\theta \quad (2)$$

Rearranging the equation (2), we get:

$$\beta cos\theta = \frac{k\lambda}{D} + 4\varepsilon\ sin\theta \quad (3)$$

In Figure 2, W-H plot analysis from XRD pattern is shown. Crystallite size (D) is calculated from the Y-intercept of linear fit and the lattice strain can be denoted as the slope of linear line. The lattice strain can also be calculated using Bragg's law:

$$d = \frac{n\lambda}{2sin\theta} \quad (4)$$

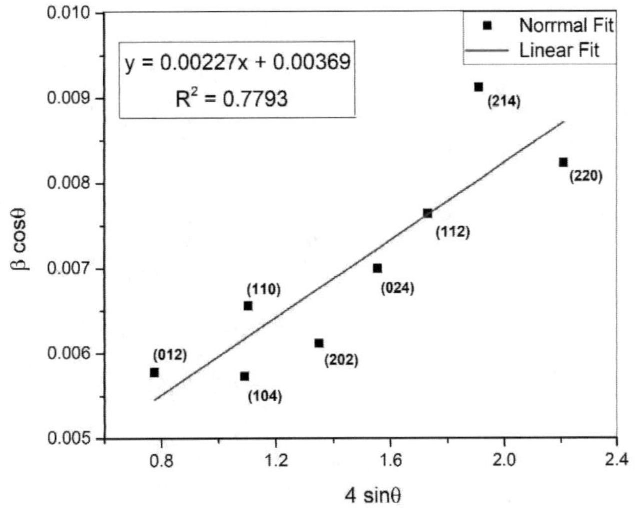

Figure 2. Williamson-Hall plot analysis from X-ray diffraction of BFO nanoparticles

Table 1. Structural and optical parameters of BFO

Parameters	Value
Lattice Space group	R3c
Lattice Parameters (Å)	a = b = 5.57875 c = 13.86691
Volume (Å)^3	373.7414
Density (g/cm^3)	8.3387
R (expected) (%)	3.96786
R (profile) (%)	3.74782
Goodness of fit	1.19087
Bond length of Fe-O (Å)	1.9567 (3); 2.1077 (3)
Average bond length (Å)	2.0322
Bond angle (degree)	154.7642
Crystallite size (nm) (Scherrer formula)	34.3
Crystallite size (nm) (W-H plot)	37.61
Lattice strain (%) (Bragg's law)	0.2641
Lattice strain (%) (W-H plot)	0.2270
Particle size (nm)	157
Optical bandgap (eV)	2.04

Here, d is lattice spacing and n is order of reflection. Now, lattice strain,

$$\varepsilon = \frac{\Delta d}{d} \tag{5}$$

Where,

$$\Delta d = d_1 - d_2 \tag{6}$$

IOP Conf. Series: Materials Science and Engineering 1091 (2021) 012049 doi:10.1088/1757-899X/1091/1/012049

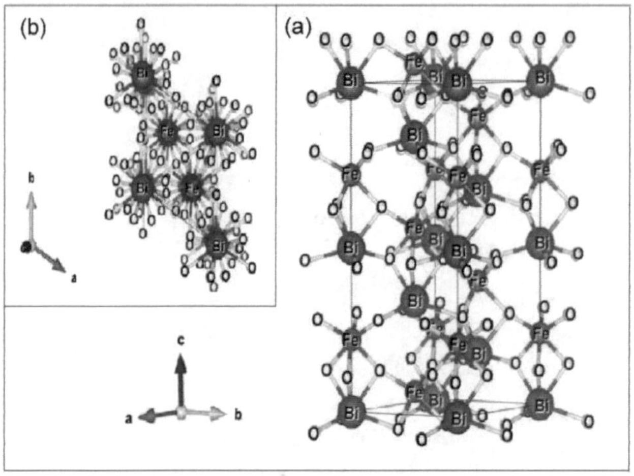

Figure 3. (a) Unit cell structure along [u v w] (b) Rhombohedral unit cell along the normal to (h k l)

A unit cell of Bismuth ferrite nanocomposites is presented in Figure 3. Using the crystal information file (CIF) in VESTA software, BFO unit cell structure was made. That CIF file was obtained after the Rietveld refinement of XRD data via MAUD software. The Fe-O-Fe (bond angles) and Fe-O (bond length) were determined in VESTA software.

3.2. Field emission scanning electron microscopy (FESEM)

In Figure 4, FESEM micrograph of BFO powder annealed at 550°C is shown. At high temperature, vapor pressure is also high on smaller particles than larger particles. So small particles vaporize and condense on larger particles. In this mechanism attainment of uniform particle size distribution is quite difficult. The sizes of the particles were measured from the micrographs of FESEM using ImageJ software. Average particle size was determined through the particle size distribution histogram. It was found that the particle size from the FESEM micrographs was larger than the size of the crystallite which was calculated from XRD analysis through Scherrer formula and W-H plot method. The reason behind this larger crystallite size is the ratio of surface area to volume of the nanoparticles is higher. This higher ratio leads to agglomeration and it results in assembling several crystallites into a particle [12].

Figure 4. FESEM image of BFO nanoparticles

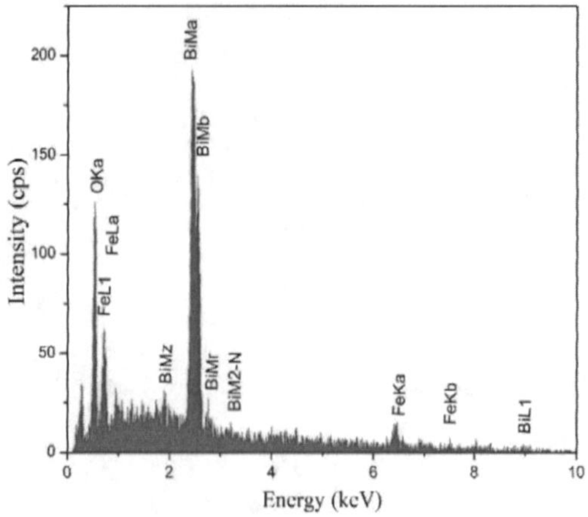

Figure 5. EDS analysis of BFO nanoparticles.

IOP Conf. Series: Materials Science and Engineering 1091 (2021) 012049 doi:10.1088/1757-899X/1091/1/012049

Table 2. EDS weight and atomic ratio of BFO nanocomposites

$BiFeO_3$ Composite	Bismuth (Bi)		Iron (Fe)		Oxygen (O)	
	Weight (%)	Atomic (%)	Weight (%)	Atomic (%)	Weight (%)	Atomic (%)
	74.34	26.53	14.68	19.46	10.98	54.01

From Table 1, it is observable that crystallite size of BFO annealed at 550°C is 37.61 nm and particle size is 157 nm. In Figure 5, Energy dispersive spectroscopy (EDS) shows the existence of Fe, O and Bi. Table 2 shows the measurements of the elements at atomic (%) and also weight (%).

3.3. Optical Bandgap of BFO

BFO is very efficacious as an irradiator of visible-light because its bandgap energy is small. The photovoltaic absorption capability of BFO is immediately related to its structure of electrons. In order to calculate the bandgap energy, photovoltaic absorption phenomenon is important [13]. With the help of the diffuse reflectance spectra, the optical bandgap (E_g) of BFO nanoparticles was measured using a UV–Vis–NIR spectrophotometer. For computing the E_g of all BFO, the Kubelka–Munk equation was used. The diffuse reflectance values had been converted to the Kubelka–Munk function:

$$F(R) = \frac{(1-R)^2}{2R} \tag{7}$$

Here at equation (7), the diffuse reflectance is expressed by R. Figure 6 represents the $[F(R)h\nu]^2$ Vs. energy (hν, eV) plots for all nanoparticles, it mostly plots the violet to red solar radiation which has wavelengths ranging from 380-750 nm that can hit the substance with sufficient energy to remove electrons from their infirm bonds and produce electric current [3]. The energy of the light is calculated by $E = hc/\lambda$. Infrared rays do not have notable energy to displace the electrons because the energy of light (E) is inversely proportional to the wavelength (λ). Thus, we only considered the range of wavelengths from 300–800 nm for recording the values.

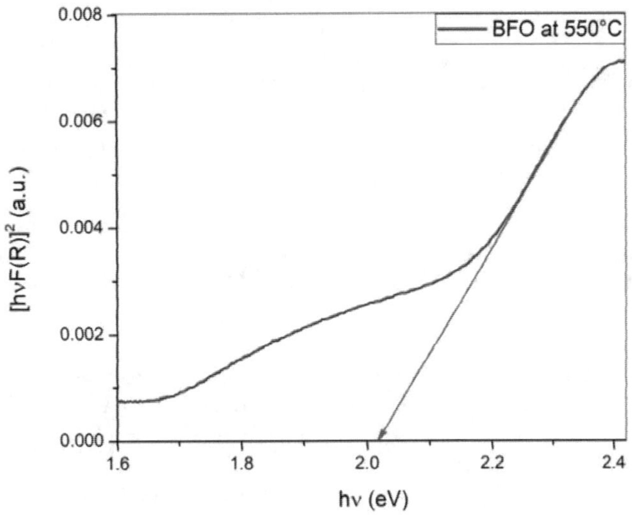

Figure 6. $[h\nu F(R)]^2$ Vs energy (hν) for BFO nanocomposites

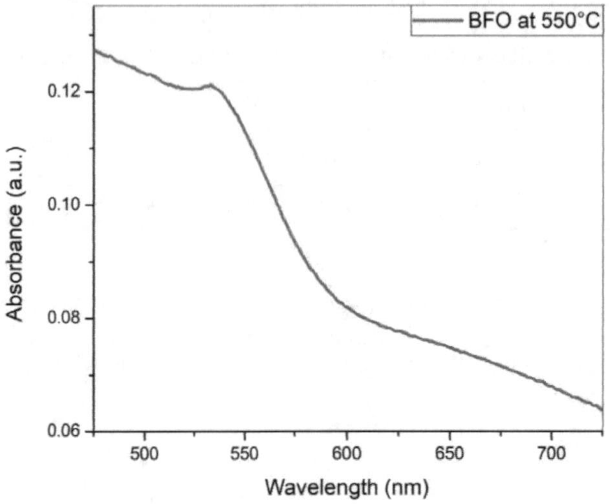

Figure 7. Absorbance Vs Wavelength for BFO nanoparticles

The bandgap energy (eV) is obtained from the intersection point of the tangent line at $[F(R)h\nu]^2 = 0$ [14]. In Table 1, the value of Eg for BFO nanoparticles annealed at 550°C is shown. Since the absorbance is an important phenomenon, we also plotted that for the wavelength of 475–725 nm in Figure 7. It has been observed from that figure that for the wavelength of 520–595 nm, BFO exhibits a higher rate of absorbance.

4. Conclusion

BFO samples had been synthesized using the sol-gel method. The structural properties had been analyzed through X-ray diffraction and FESEM micrograph. Optical properties were measured by UV-Vis-NIR. Rietveld refinement via MAUD software provided a number of structural parameters of BFO nanoparticles. It had been found that the crystallite size calculated by the W-H plot method is slightly higher than the crystallite size measured by Scherrer formula. The average particle size measured from FESEM is significantly larger than the crystallite size. Lattice strain of 0.2641% was measured by Bragg's law. A high absorption rate of BFO nanoparticles was observed at visible wavelength. We measured the optical bandgap energy of 2.04 eV for BFO nanoparticles via UV-Vis-NIR. This bandgap is nearly close to the ideal value of 1.50 eV (the maximum power conversion efficiencies for a semiconductor-based photovoltaic cell). So these BFO nanoparticles formed at 550°C may discover potential applications in the photovoltaic solar cells.

Acknowledgements

The authors are grateful to the Department of Glass and Ceramic Engineering, Bangladesh University of Engineering and Technology, Dhaka, Bangladesh for experimental support.

References

[1] Miller O D, Yablonovitch E, and Kurtz S R 2012 Strong Internal and External Luminescence as Solar Cells Approach the Shockley–Queisser Limit *IEEE J. Photovolt.* **2** 303-311.

[2] Rhaman M M, Matin M A, Hossain M N, Mozahid F A, Hakim M A, and Islam M F 2019 Bandgap engineering of cobalt-doped bismuth ferrite nanoparticles for photovoltaic applications *Bull. Mater. Sci.* **42** 190.

[3] Rhaman M M, Matin M A, Hossain M N, Mozahid F A, Hakim M A, Rizvi M H, and Islam M F 2018 Bandgap Tuning of Sm and Co Co-doped BFO Nanoparticles for Photovoltaic Application *J. Electron. Mater* **47** 6954-6958.

[4] Schmid H 2008 Some symmetry aspects of ferroics and single phase multiferroics. *J. Phys. Condens. Matter* **20(43)** 434201.
[5] Rhaman M M, Matin M A, Hossain N M, Hakim M A and Islam M F 2019 Prospect and Barrier of Multiferroic Bismuth Ferrite for Application in Solar Cell *2nd Int. Conf on Innovation in Engineering and Tech. ICIET 2019* 1-5. DOI: 10.1109/ICIET48527.2019.9290733
[6] Rhaman M M, Matin M A, Al Mamun M A, Hussain A, Hossain M N, Das B C, Hakim M A and Islam M F 2020 Enhanced electrical conductivity and multiferroic property of cobalt-doped bismuth ferrite nanoparticles *J. Mater. Sci. Mater. Electron.* **31** 8727.
[7] Fridkin V M and Popov B N 1978 The anomalous photovoltaic effect and photoconductivity in ferroelectrics Ferroelectrics **21** 611-613.
[8] Matin M A, Rhaman M M, Hakim M A and Islam M F 2020 $Bi_{(1-y)}Sm_yFeO_3$ as prospective photovoltaic materials *Bull. Mater. Sci.* **43** 1-9. DOI: 10.1007/s12034-020-02118-2
[9] Kurian M and Kunjachan C 2014 Investigation of size dependency on lattice strain of nanoceria particles synthesised by wet chemical methods *Int. Nano Lett.* **4(4)** 73.
[10] Niloy N R, Chowdhury M I, Anowar S, Islam J and Rhaman M M 2020 Structural and Optical Characterization of Multiferroic $BiFeO_3$ Nanoparticles Synthesized at Different Annealing Temperatures *J. Nano- Electron. Phys.* **12(5)** 05015. DOI: 10.21272/jnep.12(5).05015
[11] Xu X, Guoqiang T, Huijun R and Ao X 2013 *Ceram. Int.* **39(6)** 6223.
[12] Gupta S, Tomar M, James A R and Gupta V 2014 Ce-doped bismuth ferrite thin films with improved electrical and functional properties *J. Mater. Sci.* **49(15)** 5355-5364.
[13] Irfan S, Shen Y, Rizwan S, Wang H C, Khan S B and Nan C W 2017 Band- Gap Engineering and Enhanced Photocatalytic Activity of Sm and Mn Doped $BiFeO_3$ Nanoparticles *J. Am. Ceram. Soc.* **100(1)** 31-40.
[14] Mocherla P S, Karthik C, Ubic R 2013 Tunable bandgap in $BiFeO_3$ nanoparticles: the role of microstrain and oxygen defects *Appl. Phys. Lett.* **103(2)** 022910.

Development of Teff-Starch based Edible Film: Mechanical and Optical Properties

Venkatesa Prabhu. S[1*], Hemalatha. P[1], Belachew Zegale Tizazu[1] and Amsalu Gosu[1]

College of Biological and Chemical Engineering, Addis Ababa Science and Technology University, Addis Ababa, Ethiopia.

Corresponding author: venkatchemdata@gmail.com

Abstract. Teff is a unique staple food in Ethiopia. Teff has not been investigated in detail with respect to biofilm base material. In this study, a biofilm was prepared using teff-starch. Further, water barrier, mechanical, and optical properties of the teff-starch based films were investigated. These properties were compared with the cassava-starch based biofilm that was developed with same combination of plasticizer, surfactant and agar. The result showed that teff-starch based edible film is good compactable for food packaging material. Also, the mechanical and optical properties compared with standard cassava-starch based biofilm showed teff-starch films are better appropriate as for a packaging material.

1. Introduction

Synthetic plastics are known to be good for different desirable properties such as mechanical and barrier properties. They can be produced in mass production at comparatively low cost [1]. Though having these advantages, most of the synthetic plastics are handled by incineration or landfill. Particularly, the synthetic plastic materials after use as food packaging may cause environmental pollution. The food packaging is mainly focusing to reduce food quality deterioration that occurs by biological and chemical parameters [2]. Oxidation of lipids from the food is one of the principal reasons deterioration of the foods. So, food industries use synthetic antioxidants for reducing lipids oxidation [3].

Due to consumer concerns with respect to safeness of synthetic antioxidants, food industries are focusing towards biofilms for food packaging materials since they are biodegradable and environmentally friendly. Biopolymers such as gum, wax, protein, and starch that are derived from raw food resources can be used as biofilm base materials [4]. Such environmentally-friendly materials are rapidly biodegraded when they are landfilling and harmless. Recently, starch-based biofilms are gaining huge interest among the researchers for developing bio-based food packaging materials due to its easy processing, and colorless and odorless nature [5]. Even though the biofilms obtained from starch have been widely investigated recently by several researchers, their mechanical and physical properties still need to be improved.

Teff is known to be a unique grain in Ethiopia. It is identified as a gluten-free grain. Even though several studies using teff starch have been carried out for substituting wheat in baking [6], starch from teff has not been subjected to investigate in detail as a biofilm base material. In specific, the blends concerning of teff-starch with major ingredients such as sorbitol as plasticizer, agar and tween-80 as surfactant have not been studied yet. Teff has been characterized as 73% starch content that contains 18%–27% amylose. The starch content makes teff is suitable for preparation of biofilms [7]. In this study, biofilm was prepared using the starch that isolated from the teff with predetermined amounts of sorbitol as plasticizer, agar and tween-80 as surfactant. Further, the physical and barrier properties of the developed film were examined. The developed film was further compared with biofilm prepared with standard cassava starch that contain 82% by weight of amylopectin 18% by weight of amylose.

2. Materials and Methods

2.1. Materials

Teff powder was purchased from city market of Addis Ababa, Ethiopia. Food grade-Sorbitol, food grade-tween-80, agar, glass-ware and other required chemicals were procured from reputed chemicals at Addis Ababa.

2.2. Isolation of Teff-Starch

Starch from the teff was extracted using the procedure given by Bultosa et al. (2002) with some modifications. After proper purification, teff grains were grounded by grinder. Further, the powder was sieved with 600-mesh sieve. 100g of teff powder was taken 1000mL beaker and mixed with 500mL distilled deionized water. The blend was well-stirred at 30 °C for 1 hr. The resultant slurry was grounded well using wet milling. Further, it was filtered twice with a 300-mesh sieve. Using test-tube centrifuge, the filtrate was subjected to centrifuge at 900 × g, 30 °C, for 6 min. The gray layer was separated using a spatula after the supernatant was rejected. The resulted sludge was further well-dispersed using distilled water. These steps were repeated until the gray layer not to be visible. The resultant product was spread using deionized water and dried at 55 °C for 30 hr. The isolated starch was grounded by a mortar than it was filtered by passing through sieve (300-mesh). Then, it was stored at 4 °C and used for the preparation of biofilm.

2.3. Determination of chemical composition of isolated teff-starch

Chemical composition of isolated teff-starch such as moisture content (%), ash content (%), starch content (%), amylose (%), and amylopectin (%) were determined using AOAC methods (1995).

2.4. Standard cassava starch

Cassava starch that contains amylopectin-82% (by weight) and amylose-18% was obtained from Sigma-Aldrich as analytical grade.

2.5. Preparation of Biofilms

Solutions with starches (3g), plasticizer (0.8 ml), agar (0.8g) and surfactant (0.8 ml) were considered to prepare films by casting method. The clear-homogenous solution, used for film forming purpose, was first arranged by dissolving starch and agar in deionized water without plasticizer. Further, the surfactant and plasticizer were added to the solution. This solution was exposed to heat slowly up to 75 ± 5 °C with mild stirring (150 rpm). Further, the mixture was saved back for 35 min to develop a clear-homogenous film forming solution. This solution was subjected to degas using vacuum environment. The solution was carefully transferred into the Petri dishes placed with well-leveled surface and allowed to dry at 30 °C in a clean chamber with controlled temperature for 50 h. After the determined time, films were cautiously detached from the Petri dishes. The detached films were equilibrated at 30 °C for 80 h before they subject to analysis.

2.6. Characterization of the developed films

2.6.1. Mechanical Properties. The tensile properties such as, elongation (%), tensile strength (MPa), and Young's modulus (MPa) were determined by a proper material testing equipment with appropriate load cell based on standard technique of ASTM, D 882-91(1996). Sample of film strips of appropriate size were cut before analysis. Film thickness of the prepared samples were determined by a digital micrometre.

2.6.2. Barrier Properties. The film thickness was determined by an appropriate digital micrometer. 30 trails for measurements of thickness were done at different locations randomly. The mean values were calculated and recorded. The examination for moisture content present in the film was carried out by gravimetric method. The important property, water vapor permeability for the prepared films was assessed based on the procedure given by Mali et al (2004) [8]. Thickness of the film was determined after tests, and Water vapor permeability was determined by the Eq. 1.

Water vapor permeability = (Transmission rate of the water vapor)/S(R_1- R_2)D (g $m^{-1}s^{-1}Pa^{-1}$) …… (1)

R_1 refers to relative humidity of the desiccators, R_2 refers saturation vapor pressure of water (Pa) at the room temperature, R_2 refers the relative humidity of the permeation cell. D refers to the thickness of the film (m). Film solubility (SOL) was examined using the method marked by Romero-Bastida et al (2005). Using Eq. 2, the total soluble matter of the film sample was determined.

$$\text{SOL}(\%) = \left(\frac{M_0 - M_1}{M_1}\right) \times 100 \quad ..... (2)$$

Where, M_1 and M_0 refer to weight of the dry sample after and before the test, respectively. Another important property, swelling capacity of acquired films in water was determined based on the procedure explained by Hu et al (2009). The films with predetermined size were taken in the containers having 100mL (0.1 mol l^{-1}) NaOH and 100mL (0.1 mol l^{-1}) HCl solution, respectively. Then, the vessels were perfectly sealed and kept at 30 $^{\circ}$C.

They were slightly shaken at regular time interval. The changes of sample appearance were noted to examine the stability of films at acid and alkali medium. Transparency of the film (TR) was examined using the procedure explained by Ozdemir & Floros (2008) [9, 10]. The rectangular predetermined-shapes of the films were positioned inside a spectrophotometer cell. The transparency of each film was determined at 560nm using Eq. 3.

$$\text{TR}(\%) = \left(\frac{I_r}{I_0}\right) \times 100 \quad ...... (3)$$

I_r refers to light intensity with the specimen in the beam and I_o refers to light intensity with no specimen in the beam. Sample of film strips of appropriate size were cut before testing and the thickness of the samples were determined by a digital micrometre [11].

3. Results and Discussions

3.1. Mechanical properties of developed films

Mechanical strength is needed to keep the structural integrity as well as barrier properties. The investigation on the mechanical properties having much importance because of their influence on performance and also the acceptance of the consumers. Edible films with acceptable mechanical properties are great potential and ecological substitute for replacing synthetic packaging in various applications. This work focused starch-based biodegradable biofilms were developed and the mechanical properties, Young's modulus, percentage elongation, tensile strength of the starch-based biofilms were studied.

Tensile strength is defined as the ability of any material that withstands against tensile (pulling) force. Tensile strength can be determined in units of force per cross-sectional area. It was observed that the teff-starch joined with agar molecule which induced the formation of biofilm with elevated tensile strength (4.8 MPa) [12]. The observed value is good for the development of dense structure due to establishment of inter-molecular H bonds between starch and agar. This molecular linkage directly improved the tensile strength of developed films. However, the film developed from starch derived from cassava-starch exhibited higher tensile strength (5.46 MPa) than the teff- starch films. Elongation break point is widely used to explain the structural character of materials. The developed films were plasticized with sorbitol that exhibited lower value of elongation (33.12%). This value little high compared with cassava-starch biofilm (31.3 %) because of its interference with amylose packing in starch chains and its further effect on chain attraction. Young's modulus explains material's stiffness property, specifically for elastic materials. It can be defined as the ratio of the stress to strain on the linear part of the stress–strain curve.

The results from teff-starch biofilm characterization towards determination of Young's modulus showed that the teff-starch biofilm has reduced the Young's modulus (146 MPa). It showed the hydrophilic nature of the teff-starch. However, the cassava- starch biofilm had higher modulus value (158 MPa). Barrier and optical properties are important for the materials when they focused for permeable object transmits from one side to the other [13].

3.2. Optical and barrier properties of developed films

The knowledge on optical and barrier properties will be much helpful for packing as well as designing the structure of packing materials quite reasonably and efficiently. So, the optical and barrier properties such as thickness, water vapor permeability, moisture content, swelling capacity, anti-permeability against oil, solubility, stability in acid/alkali solution, and transparency were examined. The developed films were good transparent with no apparent bubbles, also they were easily separated from the Petri plates without apparent exudation of the surfactant on the film surface. They were easy to handle for both starch, surfactant, plasticizer, and agar utilized in this work.

Determination of moisture content provides the evidence with regards to the amount of water present in the films. The observed results provided that moisture content of developed films differed pointedly. Teff-starch based biofilm showed that the moisture content 18.24 %, but for cassava-starch films exhibited 14.35 %. Irrespective of the used plasticizer, the moisture content of the teff-starch based film was showed higher value than cassava-starch based film. Since the starch was more readily disintegrated by absorbing water and having poor barrier properties due to its low hygroscopic property. Water vapor permeability of the developed films is an important property. This property has great impact on utility of film in different applications. It was observed that the teff-starch in the film formulation showed 0.67×10^{-10} g/msPa and cassava-starch showed 0.68×10^{-10} g/msPa. These values were nearly same that could be the fact that, the hydrophilic nature of the starches favouring the adsorption of the water.

Water solubility and swelling capacity are important factors while choosing the films for specific uses. The properties of solubility and swelling capacity are much anticipated properties in various cases, in specific, the encapsulation. Solubility and swelling capacity of biofilms explain the pertinence of film to pack water-rich foods like unwrapped fruits. They are also much related to the biodegradability. So, the solubility and swelling capacity of films in water was studied [14]. It was observed that the predetermined film-formulation showed 31%-solubility and swelling capcity-82%. It was accredited to the occurrence of higher hydrophilic groups in the films that can attracted more water. Cassava-starch based films showed little lower value of solubility (29%) and swelling capacity (81%) than teff-starch films.

The edible starch-films prepared in this study for packing material to pack slight acid or alkali materials. Both materials could influence the stability of starch-based films. The determination of prepared starch-films in acid/alkali solution was examined. Starch-films were cut into 4 × 4 cm and submerged in 0.1 mol/L hydrochloric acid solution. The size of the films was increased little bit but such tendency did not go more beyond 3 days. After 35 days, the films still hold reliability. This kind of property showed that the starch films developed on both teff and cassava could withstand acidic condition. But, the teff-starch film exhibited a noticeable swelling at 1st day itself in 0.1 mol/L NaOH. The films start to break at end of 2nd day. It was fully collapsed in 4 days. The same kind of results were obtained for the cassava-starch biofilm. This was because of the OH-groups of starch molecules joined with NaOH, partially demolish the H- bonds and reduce the contact among the starch macromolecules that make the observed swelling and gelatinization property of starch. Transparency of films is much important while the biofilms are used for food packaging. The prepared films were observed to be transparent and homogenous. In addition, it was observed that, teff-starch based films showed higher transparency (up to 85%) than cassava-starch films (up to 81%). The higher transparency could be possibly related to its relatively higher amylopectin and lower amylose content.

References

[1] Gontard, N, Guilbert, S & Cuq, JL 1992, 'Edible wheat gluten films: Influence of the main process variables on film properties using response surface methodology', Journal of Food Science, vol. 57, pp. 190-195.

[2] Mali, S, Grossmann, MVE, García, MV, Martino, MN & Zaritzky, NE 2004, 'Barrier, mechanical and optical properties of plasticized yam starch films', Carbohydrate Polymers, vol. 56, pp. 129-135.

[3] Hu, G, Chen, J & Gao, J 2009, 'Preparation and characteristics of oxidized potato starch film', Carbohydrate Polymers, vol. 76, pp. 291-298.

[4] Lu, YS, Tighzerta, L, Dole, P & Erre, D 2005, 'Preparation and properties of starch thermoplastics modified with waterborne polyurethane from renewable resources', Polymer, vol. 46, pp. 9863–9870.

[5] Rafiemanzelat, F, Zonouz, AF & Emtiazi, G 2012, 'Synthesis and characterization of poly (ether-urethane) derived from 3,6-diisobutyl2,5-diketopiperazine and PTMG and study of their degradability in environment', Polymer Degradation and Stability, vol. 97, pp. 72-80.

[6] Orhan, Y, Hrenoviü, J & Büyükgüngör, H 2004, 'Biodegradation of plastic compost bags under controlled soil conditions', Acta Chimica Slovencia, vol. 51, pp. 579í588.

[7] Zhong, F, Li, Y, Ibanz, AM, Oh, MH, Mckenzie, KS & Shoemaker, C 2009, 'The effect of rice variety and starch isolation method on the pasting and rheological properties of rice starch pastes', Food Hydrocolloids, vol. 23, pp. 406–414.

[8] Singh, B & Sharma, N 2008, 'Mechanistic implications of plastic degradation', Polymer Degradation and Stability, vol. 93, pp. 561-584.

[9] Talja, RA, Helén, H, Roos, YH & Jouppila, K 2007, 'Effect of various polyols and polyol contents on physical and mechanical properties of potato starch-based films', Carbohydrate Polymers, vol. 67, no. 3, pp. 288–295.

[10] Zhong, F, Li, Y, Ibanz, AM, Oh, MH, Mckenzie, KS & Shoemaker, C 2009, 'The effect of rice variety and starch isolation method on the pasting and rheological properties of rice starch pastes', Food Hydrocolloids, vol. 23, pp. 406–414.

[11] Ozdemir, M & Flores JD 2008, 'Optimization of edible whey protein films containing preservatives for water vapor permeability, water solubility and sensory charateristics', Journal of Food Engineering, vol. 84, pp. 116-123.

[12] Di Franco, CR, Cyras, VP, Busalmen, JP, Ruseckaite, RA & Vazquez, A 2004, 'Degradation of Polycaprolactone/starch blends and composites with sisal fiber', Polymer Degradation and Stability, vol. 86, pp. 95-103.

[13] Avella, M, De Vlieger, JJ, Errico, ME, Fischer, S, Vacca, P & Volpe, MG 2005, 'Biodegradable starch/clay nanocomposite films for food packaging applications', Food Chemistry, vol. 93, pp. 467–474.

[14] Abrham W, Fantahun M, Subas CD, Naod GS, Ebisa T, 2014, 'Literature review on Enset starch: physico-chemical properties and pharmaceutical applications', Journal of Drug Delivery and Therapeutics, vol.4, no. 3, pp. 1-6.

Experimental and Simulation Study of Heat Transfer During Metal Melting

B P Hari Chandra[1*], Dinesh P A[2], Sumukh K S3

1,3 Department of Mechanical Engineering, M S Ramaiah Institute of Technology, Bengaluru,
2 Department of Mathematics, M S Ramaiah Institute of Technology, Bengaluru

* Corresponding Author's email: bpharichandra@msrit.edu

Abstract. Melting of metals, during a casting process involves an enormous amount of heat transfer from source to the raw material. Study of heat transfer, although helps in optimum utilization of power, practically it is a very difficult task owing to high temperatures. This paper reports an attempt to perform computer aided analysis of heat transfer/temperature distribution in an Electrical Resistance Furnace (ERF) and experimental validation of the same. Several novel techniques were used to serve the purpose. Lead, with a melting point of 327 °C was used for simulation and experimental validation. The experimentation involves the placement of thermocouples in planned locations in the furnace, in the crucible and within the metal during the melting process. On similar lines, monitoring points were considered during the computational technique. The simulation was carried out using ANSYS's Icepack module. The results indicate the suitability of the software-based computational/simulation technique for the melting process with an error of up to 20%, depending on the location of the thermocouples/monitoring points. The correlation of experimental vs. simulated values was very high in metal (R2=0.990 to 0.995) and lower (R2=0.635 to 0.666) in air.

Keywords: Metal melting, computational technique, heat transfer, electric resistance furnace, crucible, lead.

1. Introduction

Melting is an inevitable step in any casting process which requires an enormous amount of heat transfer from source to the raw material. In an electric resistance heating process, the heat transfer happens from the resistance coils to the raw material placed in a crucible. Study of heat transfer helps in optimum utilization of power, and also to understand the temperature profiles during melting. In practical terms, measuring and observing heat transfer/temperature distribution in the furnace during melting of metals is a challenging task owing to very high temperature ranges. Off late, the computational techniques have become prominent tools to determine solutions to heat transfer problems.

Successful studies have been conducted on the usage of computation techniques in heat transfer during various processes. Earlier investigators try to simulate phase change of paraffin wax using a prismatic block. The authors try to simulate the heat temperature profile and liquid fraction. However authors do not validate the same experimentally and limited the study to low melting point non-metals [1] Another study tries to simulate the melting point of aluminium using ANSYS software; based on experimental and simulation analysis on a prismatic block and determines the melting temperature. However, this simulation study is limited only to determine the melting point of Aluminium [2]. One study has discussed the use of CFD in determining convection heat loss in a furnace used in forging application when the furnace door was opened [3]. Yet another study tries to use ANSYS-CFX module to simulate the transient heat distribution in various locations of a gas-fired furnace as applied to heating of metal block for its heat treatment, and have validated the same using experimental method and report an error of ±5%. The study does not include determining temperature distribution within the metal block [4]. Transient exergy transfer in Aluminium holding furnace is also carried out using simulation techniques. This study does not consider any kind of phase change in the material but only tries to work out on exergy during the transfer of molten Aluminium from furnace to mould [5]. Attempts have also been made to numerically simulate the transient heat transfer in electric steel tempering furnace, which essentially has metal heated only in solid state [6]. However, study on heat transfer in a given metal during the melting operation in a furnace showing the transient temperature change within the metal, and when the metal is being heated from solid state till melting temperature with experimental validation has seldom been documented. This paper is a report of a study which aims at determining the transient temperature distribution in a furnace; both within the furnace and within the metal at typical locations. The paper reports the experimental determination of the transient temperature distribution and compares the same with CFD analysis adopting several novel approaches. The primary objectives of the study that is being carried out are i) To determine the transient temperature profiles in electric resistance furnace with the help of thermocouples placed at typical positions in the furnace and within the metal. ii) To perform simulation of transient temperature profile in an electric resistance furnace, including the metal during a melting process using computer-aided computational heat transfer techniques at different time intervals, at typical locations in the furnace and within the metal. iii) To compare the experimental and simulation results.

2. Experimentation

For ease of temperature measurement and financial constraints, a metal with a low melting point was to be selected. Hence lead, with a melting point of 327 ºC was selected. The overall setup includes an electrical resistance furnace with resistance coil for the heat source, a graphite crucible seven K-type thermocouples, two temperature indicators (Figure 1) with four channels each.

Fig. 1. Overall experimental setup

A custom designed mild steel stand for facilitating placement of thermocouples (Figure 2), raw material blocks in random shapes and sizes and video camera.

Fig. 2. Custom designed stand to support the thermocouples

A novel approach was followed in experimentation. The experimental procedure had following stages (i) Pre melting and solidification of raw material (ii) Facilitating placement of thermocouples within the raw material (iii) Setting up of the experimental setup (iv) Heating (melting) and temperature measurement.

Pre-melting of the raw material was the preliminary stage of the experimentation. This was carried out to serve two main purposes. (i) To melt lead and allow to solidify to the shape of the crucible: One of the unique strategy adopted during the experimentation was in connection to the raw material shape used for melting. Normally during any melting process, the raw material would invariably be irregular in shapes and sizes. However, to simplify the modelling and simulation, in the current research the raw material was pre-melted and allowed to solidify in the crucible to facilitate the solidified lead to take the form of the crucible. For pre-melting, 2 kg lead of irregular shape was placed in the crucible of known dimensions, which in turn was kept in the ERF for heating. The furnace temperature was set to 385 ℃, a temperature of ~60 degrees above the melting point of lead. (ii) To determine the transient temperature of the coil during the melting process: Another important purpose the pre-melting served was to determine the coil temperature during the melting process. This is required since coil temperature is not fixed and keeps chaning with time. To determine the coil temperature, two calibrated thermocouples were hooked to the resistance coils inside the furnace (Figure 3)**Error! Reference source not found.** such that the sensing point of the thermocouple at the coil

Fig. 3. Thermocouple placements during pre-melting

The thermocouples wwere connected to two digital temperature display units. To record the temperature, a video camera was used. Further, another thermocouple and temperature display unit which was a part of the ERF, with the sensing point located at the bottom of the furnace indicated the air temperature at the location.

Once the set-up was complete, the furnace was switched on. The video camera was also switched on to record the temperatures. Once the pre melting process was over, the video feed was used to record the temperatures for every 10 seconds. The data thus obtained is used in the simulation process to input the data into software during providing the boundary conditions.

Once the lead melted, the crucible was moved out of the furnace and was allowed to solidify naturally. After solidification, the solidified lead was removed, which had the inside shape of the crucible (Figure 4).

Fig. 4. Lead after pre-solidification

To place the thermocouples at different positions in the pre solidified raw material (lead) three holes (of diameter 2mm) were drilled at different positions and to different depths in the raw material.
Preparation of the experimental setup for the final experimentation involved the following steps: The pre-melted and solidified raw material was placed inside the crucible. Altogether seven pre-calibrated thermocouples were used for recording the temperatures at seven typical locations within the furnace and the raw material. Of these, three thermocouples were inserted into the holes drilled into the raw material, till the sensing tip of the thermocouples touches the bottom of the holes. The thermocouples were supported and made to be in a position with the help of a custom designed stand. The fourth thermocouple was placed on the surface of the raw material at the centre. The fifth thermocouple was placed inside the crucible, in air, above the surface of the raw material. Sixth and seventh thermocouples were used to sense the temperature of the air between the crucible and the furnace coils the stand was used to position the thermocouples (Figure 5 - Schematic). The thermocouples were tied to the stand with the help of GI wires so that they stay in position.

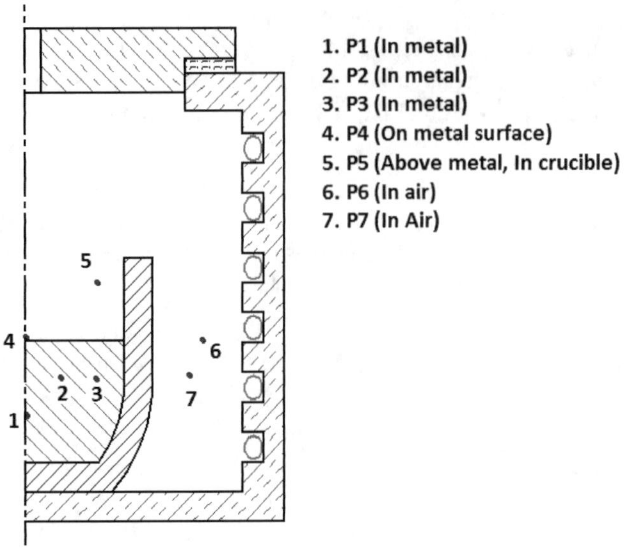

Fig. 5. Placement of seven thermocouples

The crucible along with the thermocouples mounted with the help of the stand was placed inside the ERF. The leads of the thermocouples were connected to the display devices placed outside the furnace. It was quite challenging to keep the complete setup in the furnace and make sure that every thermocouple is placed at the specified points, the glass wool was used to close the furnace completely so that no heat loss should occur while experimentation during the melting process. The lid of the furnace had a sufficient amount of glass wool to avoid leakage of heat from the furnace to the surroundings.

For melting, the furnace cut off temperature was set to 385 ºC (measured at the bottom inside of the furnace), similar to the pre-melting process. Just before the switching ON the furnace, all the seven temperatures were noted and corrected. The furnace was switched ON and temperatures at all the seven positions were noted, for every five minutes in the beginning 10 minutes, and later on for every 3 minutes, manually. The temperatures were noted from all the seven thermocouples till the melting temperature reached, and beyond (up to 2820 seconds and metal temperature stabilized at 396 ºC)

3. Simulation

The computational assessment included geometric modelling, defining various geometries, meshing, defining boundary conditions, computations and simulation. CATIA was used for 3D geometric modelling. Modelling and analysis were carried out using ANSYS Icepak version 16.0.

3.1 Geometric modelling:

Axial half section of the furnace, crucible and raw material were modelled as three different bodies as per their dimensions. A section thickness of 10 mm was considered while modelling the furnace and slots were provided for placement of the coils. The crucible, which was cylindrical in the top half and arc section at the bottom half was approximated as a cylindrical section. The raw material (lead) which was used for melting, obtained to the shape of the crucible was also modelled with similar cylindrical section approximated from that of the crucible.

3.2 Computations:

The geometric model created in CATIA was imported into Icepack. as a surface model. As part of the procedure to conduct the analysis, the various surfaces generated were required to be identified and named. Figure 6 shows the naming of the various surfaces of the furnace, crucible and raw material.

Furnace walls were named with wall numbers. The crucible and raw materials were also named appropriately.

The heating coils were modelled directly in Icepack according to the slots modelled during furnace modelling. Finally, all surfaces and volumes were verified in the "computational domain".

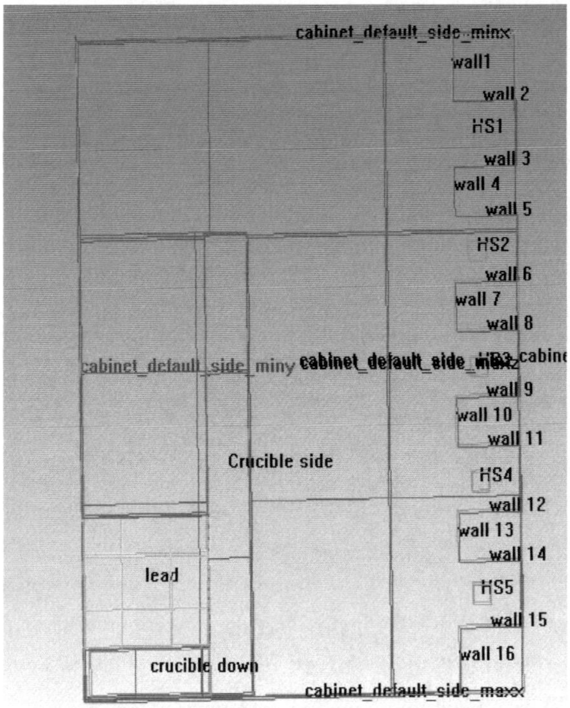

Fig. 6. CFD Model in Icepak

3.3 Meshing:

Meshing was carried out using Hexa Cartesian structured mesh. A volume mesh was created for the airspace (inside furnace), crucible and raw material.

Apart from volume mesh of the air space, crucible and raw material, the side, top, bottom walls and coil are surface meshed. There are 50872 elements and 56222 number of nodes in the model. The complete domain of the analysis includes coil, air crucible and lead. The cut plane of the domain is shown in Figure 7.

IOP Conf. Series: Materials Science and Engineering 1091 (2021) 012051 doi:10.1088/1757-899X/1091/1/012051

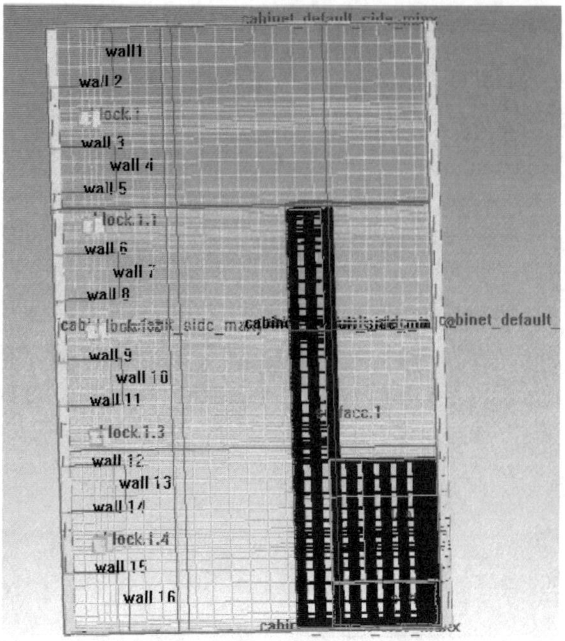

Fig. 7. Cut Plane of Computational Domain.

The mesh was checked for its quality using the quality check option, and the quality index (QI) obtained was 0.9959. Thus in the current case, since the quality is high, the running time can be expected to be minimum.

3.4 Boundary Conditions:

The boundary conditions (BCs) required to be given were the properties of the materials, furnace properties, temperature inputs and other inputs/assumptions.

3.4.1 Properties of materials:

The material property inputs to be given were the raw material (lead), furnace and crucible properties. For raw material properties, the thermal conductivity, specific heat and density were to be assigned. Since the thermal conductivity and the specific heat change with temperature (Table 1) [7]; but does not have a well defined pattern, 'piecewise linear' option available in the software was used. Although density also changes with temperature, it was assumed to be constant the software does not provide an option for varying-density input.

Table 1. Properties of lead [7]

Thermal conductivity W/(m^2.k)	35.1 at 0 ºC
	34.8 at 25 ºC
	30.3 at 260 ºC
	17 at 400 ºC
	20 at 600 ºC
Density kg/m^3	11,350 at 25 ºC

IOP Conf. Series: Materials Science and Engineering 1091 (2021) 012051 doi:10.1088/1757-899X/1091/1/012051

Specific Heat(J/g-C)	0.12 at 0 °C
	0.1287 at 25 °C
	0.132 at 125 °C
	0.368 at 227 °C
	0.449 at 527 °C
Emissivity	0.63
Melting point	327 °C

The crucible used being made of graphite, the properties of graphite available in the 'materials library' of the software were used. Further, transparency of raw material and crucible were input as 'opaque' since they were so.

3.4.2 Furnace properties:
The properties of the furnace wall to be input were its mobility (stationary or mobile), heat flux at walls, wall initial temperature. All walls (top, bottom and near coil) forming the surface of the furnace were considered as 'stationary'. The furnace walls were assumed to be adiabatic, and hence the heat flux through the wall was assumed as 0 W/m2. Further, the initial wall temperature was noted to be 30 °C, and the same was input.

3.4.3 Temperature inputs:
The various temperature inputs to be given are the Coil temperature (profile), crucible initial temperature, raw material initial temperature. The coil temperatures that ware noted during the pre-melting was input while providing the coil temperature profile using a piece-wise linear model (Figure 8). The initial temperature of the crucible was measured to be 30 °C, and the same was input. On similar lines, the raw material initial temperature was also inputted as 30 °C.

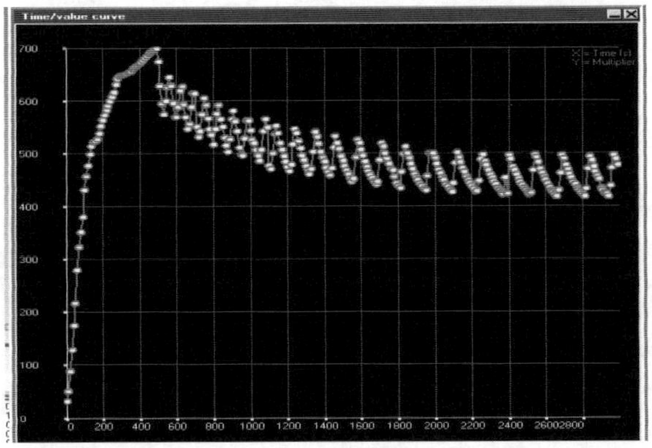

Fig. 8. Piece-wise linear model of coil Temperature in CFD

The other inputs and assumptions made were as follows. The flow of air in the airspace was assumed to be as laminar. A normal downward gravitational acceleration of 9.81m/s2 was considered in the analysis. Since majorly, the heat transfer in the current work largely involves high temperature radiation, discrete ordinance radiation model was assumed.

3.5 Simulation:
The analysis being transient, time steps and the iteration for each step plays an important role. In the current work, the time step for the analysis is considered as 1 second, and analysis was performed for ~50 minutes, which also was the time during the pre-melting experiment. During the simulation,

monitoring of the temperature was at seven discrete points, called the "monitoring points". The monitoring points were defined and their coordinates were noted. The monitoring points were in accordance with the thermocouple placement during experimentation.

4. Results and discussion

The results of the simulation are presented and discussed in two parts: (i) Contours (heat maps) at typical time elapsation (ii) Temperature profiles (at different monitoring points; both for experimental and computational techniques and their comparison)

4.1 Contours (heat maps):

The simulation results at several time intervals were generated. From the heat maps it could be noted that temperatures of the airspace, crucible and raw material are continuously increasing with the time, reaches the melting point of lead and further increases to stabilize. A typical heat map (Figure 9) at 2400 sec is discussed here. The temperatures are highest at the coils (496 C), and lowest at the base of the crucible (396.5 C).

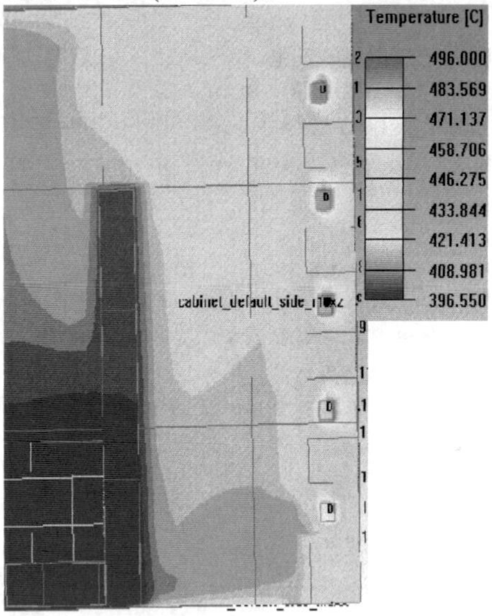

Fig. 9. Heat map at 2400sec

4.2 Comparison of results:

The results of experimentation and simulation at the monitoring points and corresponding thermocouple locations were compared. It can be noted that the temperature at the monitoring point P1 is increasing with time (Figure 10). The temperature stabilizes at ~385 ºC. The observation is true for both simulation and experimental methods. There is an error (simulation – experimental) of 0 to 35 ºC (up to 18.3%) (Table 2.). Correlation analysis indicates and R2 value of 0.995 indicating a very high correlation between the experimental and simulation methods. Regression equation, $y = 1.015x + 10.527$ (y=simulated temperature, x = experimental temperature) indicates that the simulated values are on an average 10.5 ºC higher than the experimental values.

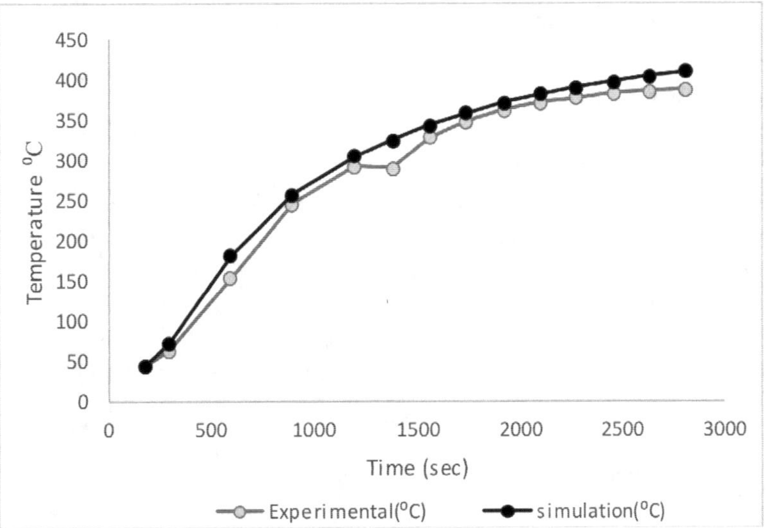

Fig. 10. Temperature profiles at P1

On similar lines, the profiles of temperatures at monitoring point P2 to P7 were compared with the corresponding thermocouple values, and the results of the same are summarized (Table 2.). The temperature profiles at points P2 to P4 were similar to that of P1. Although the temperature profile at P5 which is the point just above the metal is similar to the previous points and had a high correlation (R2=0.995) the error was highest (27.7%).

It can be noted that at points P6 & P7, which were air temperatures, the correlation between the experimental and computational values were poor, with R2 being 0.666 and 0.635 respectively.

Table 2. Summary of differences between Simulated and Experimental values

Point	Max. Error (°C)	Max. % Error	Regression Equation	R^2
P1	35	18.3	y = 1.015x + 10.527	0.995
P2	31	13.4	y = 1.0209x + 4.6428	0.994
P3	38	12.9	y = 1.0341x + 3.6903	0.996
P4	23	14.3	y = 0.9666x + 11.638	0.990
P5	32	27.7	y = 0.9582x + 26.631	0.995
P6	62	20.1	y = 0.5289x + 190.44	0.666
P7	45	12.6	y = 0.5242x + 196.19	0.635

4.3 Discussion:

From the results of the temperatures from locations which were inside the raw material, and on the surface of raw material i.e., from P1 to P4 the deviation between experimental and simulation temperatures were between 23 to 38 ℃, which amounts to an error between 13 to 19%. The differences could be due to assumption in the boundary condition that the furnace walls are adiabatic; while in actual, there is heat loss from the furnace walls. Also, the absorption of heat to the mild steel stand which was placed in the furnace to support the thermocouples would have contributed to the difference. Further, the crucible shape in the computational model was slightly simplified one, and not the exact shape.

The thermocouple/monitoring point P5 placed in the air the raw material which was inside the crucible, the maximum % error was 27.7 ℃ and the maximum deviation was 32 ℃; while on an average was higher than the experimental values (by ~ 27 ℃). Here again, there was a good correlation between the simulated and experimental values as in cases of P1 to P4.

The thermocouples/monitoring points P6 and P7 were in the air-space between the crucible and furnace wall. The error between the simulated and experimental values was quite significant although the overall pattern of the changes in temperatures was the same. The regression equation also does not show a slope near 1.0, but at ~0.5. There is no correlation between the experimental and simulated values as in the case of points P1 to P5, although the percentage of error is ~12 to 20%.

Further, there was no repetition in the experimentation. Repetition of the experiments could further help in understanding the periodicity property of the present experiment. Overall it was inferred that a good correlation existed between the experimental and simulated values in the metal than in the air space.

5. Conclusions

Computational method of determining heat transfer in electric resistance furnace is quite suitable with and an error of ~15% for temperatures in metal, and ~20% for temperatures in air.

Computational technique is a promising method to determine heat transfer in an electric resistance furnace, which would help in reducing experimental conduction of the melting process, during research; which will in turn help reduce time, money, energy and effort.

6. References

[1] A Yadav, S Soni, "Simulation of Melting Process of a Phase Change Material (PCM) using ANSYS (Fluent)," Int Res J Eng Technol, 2017; **4**, 3289-94
[2] Y Y Wang, P Guo, Y Wu, Z L Zhang, and S M Jiang "Experiment and ANSYS simulation analysis for metal aluminum solid and fluid conversion," IOP Conference Series: Earth and Environmental Science, 2017; **93**, 1-6
[3] M Filipponi, F Rossi, A Presciutti, S D Ciantis, B Castellani, and A Carpinelli, "Thermal analysis of an industrial furnace" Energies 2016; **9**, 1-13
[4] V N Chougule, O F Daud, G R Gupta, V H Shelke, and R V Ugale, "Numerical Analysis of Combustion Furnace Performance by CFD," IOSR J Mech Civ Eng, 2013; **55**, 72–9
[5] L Acevedo, S Usón, and J Uche, "Exergy transfer analysis of an aluminum holding furnace," Energy Convers Manag, 2015 **89**, 484–96
[6] C I D Palacio, P N Alvarado-Torres, and L F Cardona-Sepúlveda, "Numerical Simulation of the Flow and Heat Transfer in an Electric Steel Tempering Furnace," Energies, 2020; **13**, 3655
[7] V Sobolev, "Thermophysical properties of lead and lead–bismuth eutectic," J Nucl Mater, 2007; **362**, 235–47

Tribological and mechanical investigation of coir-silk reinforced polypropylene based matrix composites.

Vinay Kumar.D[1,4]*, C.Nanjundaswamy[2], Sachit.T.S[3], Mohan.N[4]

[1,3]Department Of Mechanical Engineering, Symbiosis Institute of Technology (SIT), Symbiosis International (Deemed) University (SIU), Lavale, Pune, Maharashtra, India

[2] Department of Civil Engineering Research Centre, Dr Ambedkar Institute of Technology, Bengaluru, Karnataka, India. And affiliated to Visvesvaraya Technological University, Belagavi, Karnataka, India

[4] Department of Industrial Engineering and Management Research Center, Dr Ambedkar Institute of Technology, Bengaluru, Karnataka, India. And affiliated to Visvesvaraya Technological University, Belagavi, Karnataka, India

Abstract: Silane treated natural fibers such as silk, coconut coir is reinforced into the polypropylene (PP) with a varying weight fraction of fibers and matrixes are developed in twin screw extruder at a constant temperature followed by the hot compression molding. Totally three different weight percentage of the polymer composites are developed. Such as silk fiber reinforced composites (S-PP), coir fiber reinforced composites (C-PP) and silk along with coir fiber reinforced composites (S-C-PP). The developed composites are then subjected to various tests. The water absorption capabilities of the all three composites are investigated. The composites are tending to abrasive wear test as per the ASTM standards. Design of experiments L-9 orthogonal array was introduced to identify the optimal wear condition of the composites developed. From the results it is observed that the tricholo vinyl silane chemical treatment to the coir and silk fibers enhances the tribo-mechanical properties of the fiber reinforced polymer composites. Also, the wear life of the composites was increased by 35% and water absorption capability increased by 42%. The scanning electron microscopy helped in identifying the impact of wear test over the surface of the composites at higher magnifications. The detailed surface morphology was executed in order to identify the behavior of the composites upon application of load.

Keywords: Silane treatment, Design of experiments, Twin screw extruder, Tribology, Polymer matrix composites.

1. Introduction

Natural fiber reinforced composites manufacturing is becoming one of the important development in the recent decades. These natural fibers are easily available and less in cost. The mass production of natural fibre reinforced composites helps in economical production and recycling of the materials[1-4]. Polymer matrix composites are having multiple engineering applications as they possess promising tribo-mechanical properties. As polymer composites are built with variety of continuous and short fibres which helps in transferring the loads between Matrix and the fibres[5-8]. In the recent years lot of research work is done to develop polymer composites in order to fulfill the industrial needs. In fact many researchers have proved that natural fibre reinforced polymer matrix composites are having potential to replace many industrial materials. The tribological and the mechanical characteristics of these composites exhibited promising results. Also, when these natural fibres are chemically treated they will release the moisture [9-12]. And when moisture content is reduced in the fibres this will help in proper bonding with the matrix material. Coconut coir comes under the category of natural fibers. They are drawn out from the outer husk of coconut and are used in making several daily use products. Coir has fibrous properties found between the hard-inner shell and the external coat of a coconut. Coir has the most upraised cluster of lignin, resulting in stronger yet less flexible than cotton and undesirable for coloring [13-15]. The elasticity of coir is less as compared to abaca. However, it has considerable protection from "microbial action and saltwater harm". A coarse, short fiber is found in ropes, sleeping mattress, brushes, geotextiles and vehicle seats. It is a natural fiber which have a biomechanical property obligatory to initially stabilize the linkage, its fiber is a nonstop string of extraordinary rigidity estimating from 500 to 1500 meters lengthy, having a diameter varying from approximately (10 to 13) microns. In woven silk, the fiber's three-sided structure goes about like a prism which refracts light, resulting in silk fabric and is exceptionally valued "normal shimmer". It also has great permeability, less conductivity and dyes without any problem [16-18]. In this experimental investigation different weight fraction of natural fibers were reinforced with the polypropylene matrix materials. The natural fibers were treated chemically in order to reduce the moisture content and for the better reinforcement in the matrix material. This paper shows an assessment of the different utilization of coconut coir and silk natural cellulosic fibers for the plan & assembling of composite materials [19-24].

2. Materials and method:

Commercially available polypropylene is utilized as matrix material and naturally procured coconut coir and silk fibers are used as reinforcements. The matrix polypropylene granules are fed in the twin-screw extruder at a temperature of 160 degree Celsius, followed by the compression molding process. The silane treated natural fibers are chopped for the length of 1mm – 2mm and are fed into the twin screw extruder which is followed by the compression molding. The below table 1.1 shows the weight fraction of various materials used in this work. The fibers are distributed randomly all over the matrix and for the uniform mixing fibers are fed into the twin screw extruder.

Table 1.1: weight fraction of the composites used in the investigation.

Composite	Coconut coir fiber weight percentage	Silk fiber weight percentage	Polypropylene matrix weight percentage	Total
C1	20		80	100
C2		20	80	100
C3	20	20	60	100

2.1. Twin-screw extruder

This is a process where two or more components are mixed together made to form a compound. This is more suitable for extruding reactive polymer materials. This process has two identical screws which are kept engaged and are mounted on shafts and are made to rotate in the same direction inside a closed container. These screws rotate with or against each other and this motion provide a special mixing form such as kneading block, forward and reverse capabilities, etc. Their movement indicates the characteristic of mixing.

3. Results & Discussion

3.1. Design of experiments

To identify the optimal wear rate of the polymer matrix composite samples prepared, L-9 orthogonal array was implemented. As shown in the table 1.2 it is a 3-factor and 3-level investigation. From the Taguchi's approach L-9 orthogonal array was derived and it is having 9 rows and 3 columns. The column 1designated for weight fraction of the composites, column 2 designated for the load in N and it is varied from 5N to 15N, column 3 is designated for speed of disc rotation varying from 200RPM to 600RPM. And from the MINITAB software the analysis of the experiment was done.

Table 1.2: L9 orthogonal array

Sl No.	Composites	Load in N	Speed of the disc in RPM	Wear Rate * 10^3	SNRA2	MEAN2
1	C1	5N	200RPM	935	-59.4162	935
2	C1	10N	400RPM	1150	-61.214	1150
3	C1	15N	600RPM	1246	-61.9104	1246
4	C2	5N	400RPM	1144	-61.1685	1144
5	C2	10N	600RPM	1280	-62.1442	1280
6	C2	15N	200RPM	1285	-62.1781	1285
7	C3	5N	600RPM	801	-58.0727	801
8	C3	10N	200RPM	867	-58.7604	867
9	C3	15N	400RPM	950	-59.5545	950

Table 1.3: Linear Model Analysis: SN ratios versus Composites, Load on the pin, Speed of the disc rotation. Estimated Model Coefficients for SN ratios.

Term	Coef SE	Coef	T	P
Constant	-60.491	0.06972	-867.687	0
Composite C1	-0.3559	0.09859	-3.609	0.069
Composite C2	-1.3393	0.09859	-13.584	0.005
Load on 5N	0.9385	0.09859	9.519	0.011
Load on 10N	-0.2152	0.09859	-2.183	0.161
Speed of 100RPM	0.3728	0.09859	3.781	0.063
Speed of 200RPM	-0.1547	0.09859	-1.569	0.257

IOP Conf. Series: Materials Science and Engineering 1091 (2021) 012052 doi:10.1088/1757-899X/1091/1/012052

S = 0.2091 R-Sq = 99.6% R-Sq(adj) = 98.2%.

Table 1.4: Analysis of Variance for SN ratios:

Source	DF	Seq SS	Adj SS	Adj MS	F	P
Composites	2	14.3815	14.3815	7.19075	164.39	0.006
Load on the Pin	2	4.3509	4.3509	2.17546	49.73	0.02
Speed of the disc rotation	2	0.6313	0.6313	0.31565	7.22	0.122
Residual Error	2	0.0875	0.0875	0.04374		
Total	8	19.4512				

Table 1.5: Response Table for Signal to Noise Ratios: Smaller is better.

Level	Composites	Load On the Pin	Speed of the disc rotation
1	-60.85	-59.55	-60.12
2	-61.83	-60.71	-60.65
3	-58.80	-61.21	-60.71
Delta	3.03	1.66	0.59
Rank	1	2	3

Table 1.5 is displaying the final results of the Taguchi's design of experiment L-9 orthogonal array. To get the statistically significant reading 3 replicates were used for running each test. And the average wear rate is noted upon completion of each experiment. After this ANOVA used to know the significant factors affecting the major wear rate of the composites. From the table it is observed that the percentage of weight fraction of fibers in the composites is the most influencing factor followed by load and disc rotation speed respectively. Also, residual error of the entire experiment is only 0.8763%. From the figure it is observed that composites C2 are the least affected with respect to wear rates are concerned.

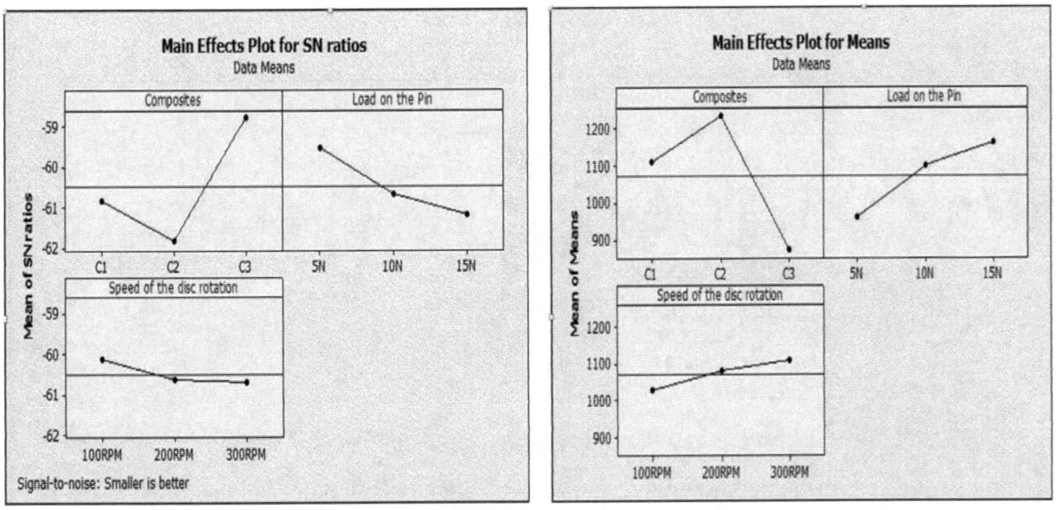

Figure 1.1: Main effects plot for SN rations and means

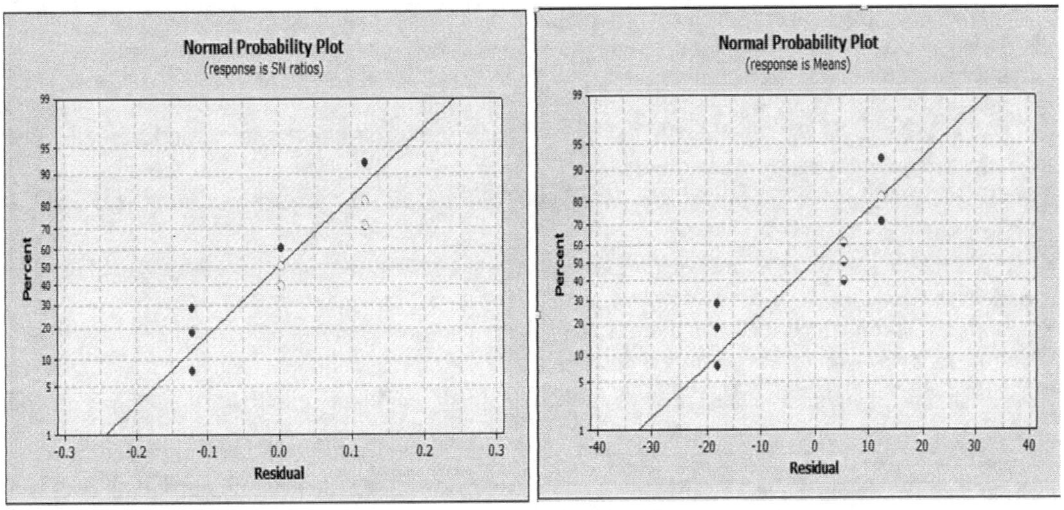

Figure 1.2: Normal probability plot for SN ratios and means

From the figure 1.1 and 1.2 it is observed that composites C2, silk fiber reinforced composites are tend to lose more when they are let to dry body abrasion wear test. Whereas the C3 composite, which is a combination of coir and silk fibers are tend to wear less upon higher load and distance.

3.2. Water absorption

Figure 1.3 shows the percentage of water absorption by each composite. From the figure it is seen that the composites C3 are good at absorbing higher amount of water content. This property is due to the chemical treatment given to the fibers before manufacturing. C3 composites are able to absorb 1.55% of water followed by C2 1.13% and C1 0.88%.

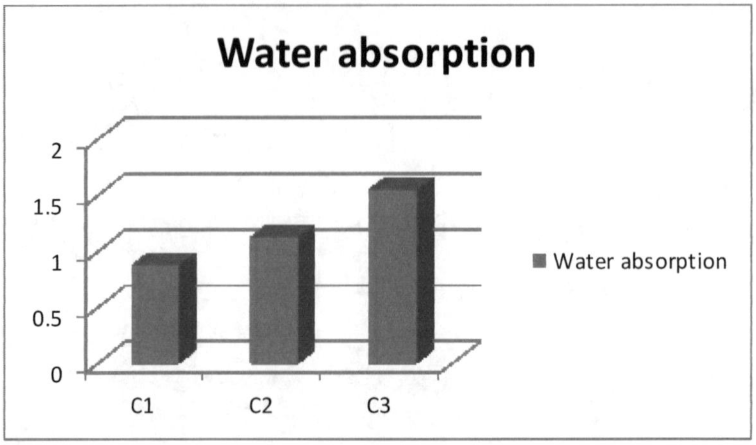

Figure 1.3: Water absorption of the composites

In this project, fiber reinforced polypropylene composites were used to study their tribo-mechanical properties. Coir fiber was chemically treated with alkali solution. Hot compression method was used in manufacturing the composite followed by twin-screw extrusion. The composites were prepared with different weight fraction compositions of coir and silk fiber. The composites were then let for the tribo-mechanical tests. Following are the conclusions drawn from the results:

- Twin screw extrusion method helps in producing good surface finish composites and this manufacturing process found effective.
- The wear life of the C3 composites was increased by 35% in comparison with the C1 and C2 composites.
- From the design of experiments it was observed that weight fraction of the composites are the major influencing factor followed by the load on pin and disc rotation speed.
- Water absorption capability of the C3 composites was increased by 42%. This is due to the tri chloro vinyl silane treatment to the composites.

References:

[1]. Leong Y W, Ishak Z M and Ariffin A 2004 Mechanical and thermal properties of talc and calcium carbonate filled polypropylene hybrid compositesJ. Appl. Polym. Sci. 91 3327–36

[2]. Ahmed S and Jones F R 1990 A review of particulate reinforcement theories for polymer compositesJ. Mater. Sci. 25 4933–42

[3]. Fu S Y, Feng X Q, Lauke B and Mai Y W 2008 Effects of particle size, particle/matrix interface adhesion and particle loading on mechanical properties of particulate–polymer composites Composites Part B: Engineering 39 933–61

[4]. Matykiewicz D, Lewandowski K and Dudziec B 2017 Evaluation of thermomechanical properties of epoxy–basalt fibre composites modified with zeolite and silsesquioxane Compos. Interfaces 24 489–98

[5]. Kasgoz A, Akın D and Durmus A 2012 Rheological and mechanical properties of cycloolefin copolymer/organoclay nanocomposites J. Reinf. Plast. Compos. 31 1329–41

[6]. Oleksy M, Heneczkowski M, Oliwa R, Budzik G, Dziubek T, Markowska O and Jesionowski T 2014 Hybrid composites with epoxy resin matrix manufactured with vacuum casting technology (Rapid Communication) Polimery 59 977

[7]. Rudawska A, Jakubowska P and Kloziński A 2017 Surface free energy of composite materials with high calcium carbonate filler content Polimery 62 434–40

[8]. Sancaktar E andWalker E 2004 Effects of calcium carbonate, talc, mica, and glass- fiber fillers on the ultrasonic weld strength of polypropylene J. Appl. Polym. Sci. 94 1986–98

[9]. Suresha B, Kumar B R, Venkataramareddy M and Jayaraju T 2010 Role of micro/nano fillers on mechanical and tribological properties of polyamide66/polypropylene composites Mater. Des. 31 1993–2000

[10]. Vaia R A, Ishii H and Giannelis E P 1993 Synthesis and properties of two-dimensional nanostructures by direct intercalation of polymer melts in layered silicates Chem. Mater. 5 1694–6

[11]. Eslami-Farsani R, Khalili S M R, Hedayatnasab Z and Soleimani N 2014 Influence of thermal conditions on the tensile properties of basalt fiber reinforced polypropylene–clay nanocomposites Mater. Des. 53 540–9

[12]. Anne R 2006 A ross basalt fiber; alternative to glass Compos Technol. Publisher: Gardner Publications Incorporated 12 ISSN: 1083-4117

[13]. Bhat T, Chevali V, Liu X, Feih S and Mouritz A P 2015 Fire structutal resistance of basalt fiber composite Compos. Part A Appl. Sci. Manuf. 71 107

[14]. Kumar B R, Suresha B and Venkataramareddy M 2009 Effect of particulate fillers on mechanical and abrasive wear behaviour of polyamide 66/polypropylene nanocomposites Mater. Des. 30 3852–8

[15]. Basavaraj E and Siddaramaiah 2010 Sliding wear behavior of nanoclay filled polypropylene/ultra-high molecular weight polyethylene/ carbon short fiber nanocompositesJournal of Macromolecular Science, Part A: Pure and Applied Chemistry 47 558–63

[16]. Shi G, Zhang M Q, Rong M Z, Wetzel B and Friedrich K 2003 Friction and wear of low nanometer Si3N4 filled epoxy composites Wear 254 784–96

[17]. Chen W X, Li F, Han G, Xia J B, Wang L Y, Tu J P and Xu Z D 2003 Tribological behavior of carbon-nanotube-filled PTFE composites Tribol. Lett. 15 275–8

[18]. Li F, Hu K A, Li J L and Zhao B Y 2001 The friction and wear characteristics of nanometer ZnO filled polytetrafluoroethylene Wear 249 877–82 [19] Vijay R, Si

[19]. Kumar, V., Mohan, N., Bongale, A., & Khedkar, N. (2019). Impact of silane treated basalt fibers and montmorillonite nano-clay on polypropylene composites. Materials Research Express, 6(12), 125325.

[20]. Kumar, D. V., & Mohan, N. Tribological investigation of coconut coir/banana fibers/glass fiber reinforced hybrid polymer matrix composites.

[21]. Senthamaraikannan P and Kathiresan M 2018 Characterization of raw and alkali-treated new natural cellulosic fiber from Cocciniagrandis.L. Carbohydrate Polym. 186 332–43

[22]. Vinod A, Vijay R, Singaravelu D L, Sanjay M R, Siengchin S and Moure M M 2019 Characterization of untreated and alkali-treated natural fibers extracted from the stem of Catharanthusroseus Mater. Res. Express 6 085406

[23]. Kathirselvam M, Kumaravel A, Arthanarieswaran V P and Saravanakumar S S 2019 Characterization of cellulose fibers in Thespesiapopulnea barks: influence of alkali treatment Carbohydrate-Polymers 217 178–89

[24]. Vijay R, Lenin Singaravelu D, Vinod A, Sanjay M R and Siengchin S 2019 Characterization of alkali-treated and untreated natural fibers from the stem of parthenium hysterophorusJ. Nat. Fibers (https://doi.org/10.1080/15440478.2019.1612308)

Effect of process parameters on material removal rate in magnetic abrasive flow machining of Al/SiC/B_4C metal matrix composites

Anil Jindal[1*], Parlad Kumar[2], Sushil Mittal[3]

[1,2]Department of Mechanical Engineering, Punjabi University, Patiala, Punjab, India
[3]Department of Mechanical Engineering, Chandigarh University, Gharuan, Punjab, India

*Corresponding author E-mail: aniljindal11@gmail.com

Abstract. In the recent times, the necessity for magnetic abrasive flow machining (MAFM) process in the industry arises as it is superior to conventional machining processes. MAFM is used to obtain high surface finishing and materials removal rate for the cylindrical, complex and intricate profiles. In this paper, the effect of process parameters has been studied on the material removal rate in MAFM process of metal matrix composites of Al/SiC/B_4C prepared by stir casting process. The process parameters are magnetic field, extrusion pressure, no. of cycles, mesh size of abrasives, workpiece material and concentration of abrasives.Taguchi design of experiments has been used for design of experiments and the analysis of variance technique has been used to check the significance level of each input parameter. The results show that the intensity of magnetic field has a dominant effect on the material removal rate.

1. Introduction

Metal Matrix Composites (MMCs) have excellent mechanical properties and are widely used in aerospace, structural and automotive fields [1-2]. To meet the standard industrial requirements, various processes such as Magnetic Abrasive Flow Machining (MAFM), grinding, laser polishing, lapping etc. are used for surface finishing operations [3-4]. In MAFM process, the abrasive particles are used in combination with iron powder and lubricant solution to obtain machining of inner surface and provide lubrication [5]. AFM process can provide the significant improvement in the surface finishing of the aerospace, automobile and rail locomotive components in just few minutes which show that this process is suitable for advance manufacturing [6]. MAFM process uses magnetic field to enhance the cutting force and obtain superior surface finishing as compared with conventional machining processes [7]. Therefore, MAFM process provides accuracy and significant surface finishing of the materials [8]. Shabgard et al. [9] reported the MAFM of H13 tool steel. It was found that the MAFM process provides good surface finishing of hard materials. Mittal et al. [10] found that AFM process increases the surface finishing of Al/SiC MMCs.

Using the magnetic field in the abrasive flow machining process, the surface finishing greatly improved in difficult to machine work pieces [11]. The reinforcement of SiC in the hybrid composites showed significant improvement in the wear resistance [12]. Research study showed that the surface roughness was significantly reduced using MAFM process for Inconel 718 work piece [13].The magnetic field helped in improving the surface finishing of the Al-6061 hybrid composites using MAFM process [14].The wear rate and coefficient of friction of Al/SiC/B_4C MMCs are lower as compared to Al/SiC MMCs.SEM images study showed uniform distribution of materials and no defect due to SiC and B_4C reinforcement in the aluminium [15]. MAFM process may not be suitable for drilling holes in the work pieces. However, the in some cases, it also offers little high cost [16]. The metallurgical and hardness properties of aluminium are greatly improved with reinforcement of SiC and B_4C particles [17]. MAFM process improves the surface finishing unto 95 nm for achieving nano-finishing in case of Ti-6Al-4V flat discs [18]. MAFM process helps in improving the tool life by 50-60% in case of Ti-6Al-4V work piece [19-20]. MAFM process is also suitable for achieving nano-finishing of bio-titanium alloys [21-22]. It was reported that the machining efficiency can be increased by 40% using MAFM process assisted with ultrasonic vibration [23].

In the field of precision engineering, various processes such as magnetic float polishing [24], magnetorheological abrasive flow machining [25], magnetic abrasive finishing [26], magnetorheological finishing [27] are available for achieving better surface finishing with good flexibility and controllability [28-29]. Al/SiC/B_4C MMCs are used in aerospace, industries, rail locomotives, automobile and medical sector. The magnetic field, work piece materials, no. of cycles and extrusion pressure has the significant effect on the MRR in magnetic field abrasive flow machining. The magnetic field beyond certain value showed decrease in the MRR for the work piece as the rough peaks vanishes after certain level of machining [30].

2. Materials and Methods

In the present paper, Al/SiC/B_4C MMCs is prepared using stir casting method. The main advantage of preparing the Al/SiC/B_4C MMCs with stir casting is the lower cost as compared to other available methods and stir casting allows uniform distribution of the materials. The Al/SiC/B_4C MMCs are fabricated using micro-EDM process and then machined using MAFM process.

The various process parameters taken in the present research are magnetic field density, extrusion pressure, no. of cycles, mesh size of abrasives, workpiece material and concentration of abrasives. The response parameter taken is MRR. The effect of process parameters on response parameter was investigated. The magnetic field distribution is investigated using simulation software. The XRD analysis used to check the uniform distribution of materials and SEM used to compare the machined surfaces before and after machining.

2.1. Experimental Setup

The experimental setup for MAFM process consists of the electromagnets, media cylinders, pistons, workpiece fixtures and hydraulic unit. The electromagnets provide the magnetic field using the electric current. The cylinders used to guide the reciprocating pistons. Nylon fixtures with a hole used to hold the workpiece in the right position. Hydraulic unit with the capacity of withstanding pressure of 10 MPa was used for the experimentation. Hydraulic unit consists of direction, pressure control valves, hydraulic cylinders, tank, pressure gauges and gear pumps.

Figure 1. Experimentalsetup for MAFM process

2.2. Experimentation

The work pieces of Al/SiC/B_4C MMCs were prepared using the stir casting process. The process parameters used are magnetic field, extrusion pressure, no. of cycles, mesh size of abrasives, workpiece material, concentration of abrasives as given in Table 1. The response parameter is material removal rate (MRR). The hydraulic oil was used in addition to the abrasive mixture and iron powder to provide the lubrication to the inner surface. The abrasive mixture with varying concentration was made up of Al, SiC and B_4C particles for machining the work pieces. The experiments were carried out by varying the process parameters to achieve the optimal material removal rate. The response parameter MRR was calculated using the following expression:

$$MRR = \frac{\text{Initialweight} - \text{Finalweight}}{\text{Time}}$$

Table 1. Level of Input Parameters

Symbol	Factors	Level 1	Level 2	Level 3
A	Magnetic Field (T)	0.2	0.4	0.6
B	Workpiece Material (percentage of SiC in Al/SiC/B_4C)	10	20	30
	Workpiece Material (percentage of B_4C in Al/SiC/B_4C)	3	5	7
C	Mesh Number	100	150	200
D	Abrasives concentration	50	55	60
E	Extrusion pressure (MPa)	3	5	7
F	No. of Cycles	100	200	300

The specimens were fabricated using the micro-EDM process. After fabrication, the specimens were machined using MAFM process. Carbonyl iron particles used with the abrasive mixtures and hydraulic oil to finish the inner surface of Al/SiC/B_4C hybrid MMCs. Hydraulic oil acts as lubricant and binder for the abrasive mixture passing through the hollow inner surface of the work piece.

Figure 2. Machined work pieces

The significance of each process parameter was analyzed using the ANOVA technique. The distribution of the magnetic field in the machining region was analyzed using the simulation software as shown in the Figure 3.

IOP Conf. Series: Materials Science and Engineering 1091 (2021) 012053 doi:10.1088/1757-899X/1091/1/012053

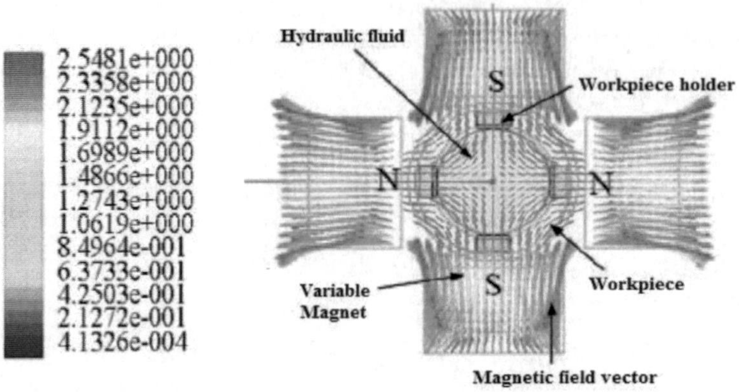

Figure 3. Distribution of Magnetic field lines in the machining region

3. Results and Discussion

Experiments were performed using Taguchi's L27 Orthogonal array. Three repetitions of the experiments were done. The obtained values of MRR are given in Table 2. The experiments were performed using different values of input parameters as given by Taguchi L27 array.

Table 2. Experimental Observations for MRR

Exp. No.	MRR1 (in µg/s)	MRR2 (in µg/s)	MRR3 (in µg/s)	Mean MRR (in µg/s)
1	2.10	1.98	1.66	1.91
2	2.31	3.16	2.84	2.77
3	5.02	4.77	5.14	4.98
4	2.12	1.85	2.64	2.20
5	2.53	4.02	3.09	3.21
6	5.00	4.10	4.05	4.38
7	2.53	2.05	2.76	2.45
8	2.54	3.23	2.81	2.86
9	4.22	5.01	4.38	4.54
10	2.55	2.96	2.58	2.70
11	6.65	5.13	5.00	5.59
12	7.96	4.44	5.35	5.92
13	3.24	3.38	2.83	3.15
14	7.99	6.99	6.35	7.11
15	4.55	4.62	4.81	4.66
16	3.44	4.02	3.37	3.61
17	6.47	6.88	7.09	6.81
18	4.52	5.49	5.00	5.00
19	9.78	9.77	8.46	9.34
20	3.99	4.34	3.36	3.9
21	8.69	8.67	8.89	8.75
22	9.13	9.97	9.66	9.59
23	4.14	4.78	4.38	4.43
24	8.19	8.63	8.27	8.36
25	9.02	9.24	9.32	9.19
26	4.33	3.99	4.37	4.23
27	8.35	8.23	8.20	8.26

The values of MRR obtained in Table 2 have been used for calculation signal to noise ratio. For this purpose, 'higher the best' approach has been used. The graphs for signal to noise ratio have been shown in Figure 4 and response values for means are given in Table 3. To get the percentage contribution of each input parameter the analysis of variance has been shown in Table 4.

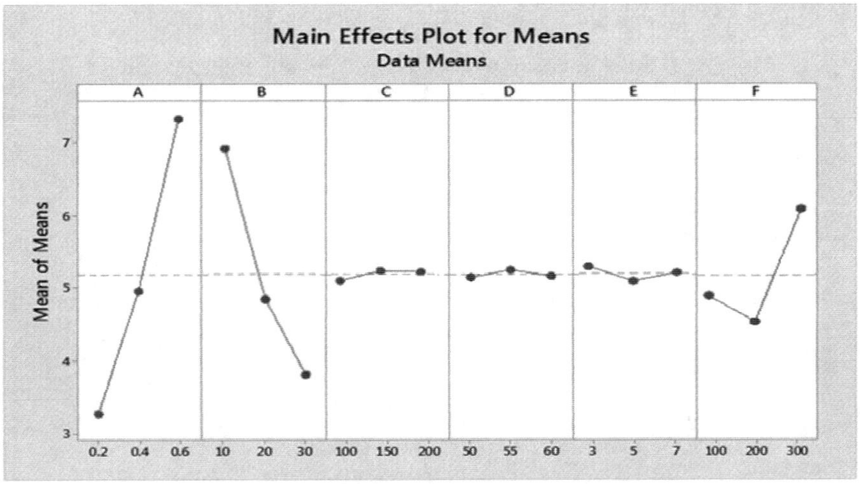

Figure 4. S/N ratio plots for process parameters

Table 3. Response Table for Means

Level	A	B	C	D	E	F
1	3.256	3.856	5.14	5.096	5.273	4.904
2	4.95	4.852	5.244	5.232	5.076	4.546
3	7.339	6.537	5.16	5.217	5.196	6.094
Delta	4.083	2.881	0.104	0.137	0.198	1.549
Rank	1	2	6	5	4	3

Table 4.ANOVA for MRR

Source	D.F.	Seq. SS	Contribution	Adj. SS	Adj. MS	F-Value	P-Value
A	2	75.755	49.97%	75.7546	37.8773	23.86	0.001
B	2	41.455	27.34%	41.4552	20.7276	13.05	0.007
C	2	0.179	0.12%	0.1787	0.0893	0.06	0.945
D	2	0.055	0.04%	0.0533	0.0247	2.76	0.041
E	2	11.832	7.80%	11.8319	5.9159	4.55	0.030
F	2	0.101	0.07%	0.1008	0.0504	7.64	0.013
Error	4	22.228	14.66%	22.2281	1.5877		
Total	16	151.605	100.00%				
R-Sq = 99.59%	R-Sq(adj) = 98.66%						

Table 4 represents that the mesh size of abrasives has little effect on the MRR. It can be clearly observed that the magnetic field density is the major process parameter having most significant effect on the MRR of the Al/SiC/B_4C hybrid MMCs in case of MAFM process. The percentage contribution shows that the Magnetic field density, work piece materials, extrusion pressure and no. of cycles has a significant effect on MRR. The surface finishing was visibly improved after the machining.

3.1. Optimization for MRR using Taguchi Method

Overall MRR (M) = 5.18 μ g/s

Optimum MRR considering most significant factors as per Taguchi combination

$= A_3 + B_3 + F_3 - 2M$
= 7.33 + 6.53 + 6.09 – 2 x 5.18
= 9.59μ g/s

Now, Confidence Interval, $CI = \pm\sqrt{\frac{F\ (1,n,\alpha)Ve}{ne}}$

F = F – ratio, α = risk

n = degree of freedom for error

V_e = Variance

n_e = no. of replications

At 95%, Confidence Interval

$F_{(1,\ n,\ \alpha)} = F_{(1,\ 2,\ 0.05)} = 18.50$

$$n_e = \frac{N}{1+ \text{dof of all the factors}}$$

$$n_e = \frac{27}{(1+6)} = 3.80$$

$V_e = 0.051$

$$CI = \pm\sqrt{\frac{18.50\ X\ 0.051}{3.80}} = \pm 0.494$$

Mean experimental value after confirmation experiments = 9.81 μg/s

$$\%\ \text{error} = \frac{9.81-9.59}{9.81} \text{x}100 = 2.24\ \%$$

3.2 XRD and SEM analysis

The XRD graph in Figure 5 shows that there are no unwanted compounds present in the Al/SiC/B_4C MMCs. Most of the peaks are of Al, SiC and B_4C only. So, the uniform distribution of the materials achieved using the stir casting process. The surface defects, cracks and irregularities were significantly improved from the work piece using MAFM. It was observed that the magnetic field around the work piece in the machining region enhanced the MRR of the Al/SiC/B_4C MMCs.

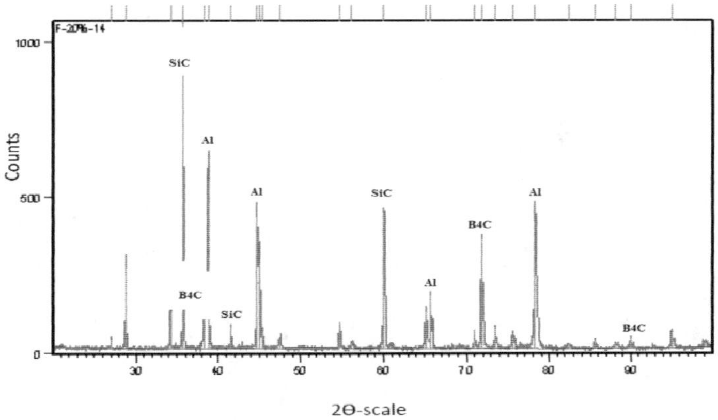

Figure 5. XRD graph for the Al/SiC/B_4C hybrid MMCs

SEM images were captured before and after machining of MAFM process. The surface defects, cracks and irregularities were significantly removed from the workpiece using MAFM. Figure 6 shows SEM surface before machining and Figure 7 shows a work piece after machining.

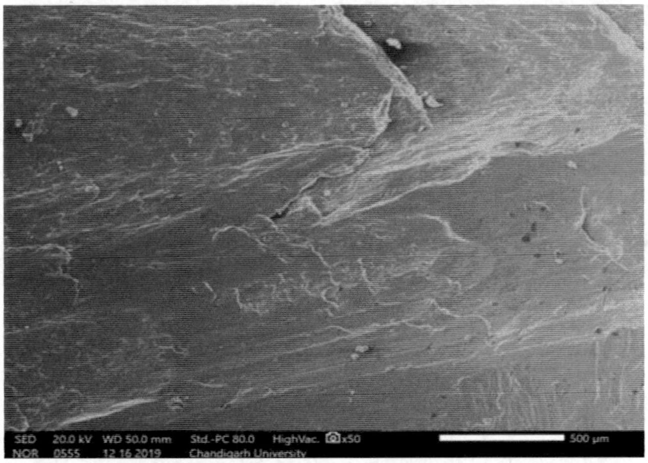

Figure 6. SEM before MAFM process

Figure 7. SEM after MAFM process

4. Conclusions

1. In the present study it has been found that the significant factors in MAFM process are: magnetic field, work piece materials, no. of cycles and extrusion pressure.
2. It has been found that MRR increases with the increase in the magnetic field, no. of cycles and extrusion pressure.
3. The intensity of magnetic field has strong correlation with MRR. The MRR increases with increased magnetic field intensity. Also, the MRR decreased with increasing the hardness of the material.
4. The SEM and XRD images confirm the results of increased surface finish and removed material.

References

[1] Mohapatra J, Nayak S, Mohapatra M 2020 Mechanical and Tribology properties of Al 4.5% Cu 5% TiC Metal Matrix Composites for light weight structures *International Journal of Lightweight Materials and Manufacture***3** 120-126.

[2] Rawal S2001Metal Matrix Composites for space applications *JOM The Member Journal of the Minerals, Metals and Materials Society***53** 14-17.

[3] Amnieh SK, P Mosaddegh, Tehrani AF 2017Study on Magnetic Abrasive Finishing of spiral grooves inside of aluminium cylinders *International Journal of Advanced Manufacturing Technology***91** 1-10.

[4] Mingareev I, Bonhoff T, El-Sherif AF, Richardson M 2013 Femtosecond laser post-processing of metal parts produced by laser additive manufacturing *Journal of Laser Applications***25** 1-4.

[5] Sadiq A, Shunmugam MS 2009Investigation into magnetorheological abrasive honing MRAH *International Journal of Machine Tools and Manufacture***49** 554-60.

[6] Groeger, Segel F, Uhlmann E, Robkamp S 2018 Definition of edges in correlation to abrasive flow machining as a finishing process *Surface Topography Metrology and Properties***6** 1 – 11.

[7] Jain VK 2002Advanced Machining Processes *Allied Publishers Pvt. Ltd.*

[8] Shinmura T, Takazawa K and Hatano E 1985 Study on magnetic abrasive process-application to plane finishing *Bulletin of Japan Society of Precision Engineering***19** 289-91.

[9] Shabgard M, Tabriz F, Gholipoor A 2016 Experimental study of the effects of abrasive particle size and workpiece hardness in magnetic abrasive flow machining *Modares Mechanical Engineering* **16** 131-138.

[10] Mittal S, Kumar V, Kansal H 2016 Multi objective optimization of process parameters involved in micro-finishing of Al/SiC MMCs by abrasive flow machining process *Journal of Materials: Design & Applications,* **232** 1-14.

[11] Zou Y, Xie H, Dong C, Wu J 2018 Study on complex micro surface finishing of alumina ceramic by the magnetic abrasive finishing process using alternating magnetic field *International Journal of Advanced Manufacturing Technology***97** 2193–2202.

[12] Prasad BK 2007 Investigation into sliding wear performance of zinc-based alloy reinforced with SiC particles in dry and lubricated conditions *Wear***262** 262–273

[13] Guo J, Tan ZE, Au KH, Liu K 2017 Experimental investigation into the effect of abrasive and force conditions in magnetic field assisted finishing *International Journal of Advanced Manufacturing Technology***90** 1881–1888.

[14] Sharma VK, Kumar V, Joshi RS 2018 Investigation on Surface Roughness During Finishing of Al6061 Hybrid Composites Tube with Traces of Rare Earth Metals Using Magnetic Abrasive Flow Machining *Advances in Unconventional Machining and CompositesProceedings of AIMTDR* 493-501.

[15] Sekar K, Ananda Rao DV 2020 Investigation of Hybrid Composite A7075/SiC/B_4C by Stir and Squeeze Casting method *Materials Today: Proceedings***22** 1398–1408.

[16] Singh S, Shan HS 2002 Development of magneto abrasive flow machining process *International Journal of Machine Tools and Manufacturing* 953-959.

[17] Pandiyarajan R, Prabakaran MP, Rajkumar T, Vetrivel Kumar K, Manikandan R 2020 Metallurgical and mechanical properties of SiC/ B4C reinforced with aluminum composites synthesized by mechanical alloying *Materials Today: Proceedings,* 1-5.

[18] Parameswari G, Jain VK, Ramkumar J, Nagdeve L 2019 Experimental investigations into nano-finishing of Ti6Al4V flat disc using magnetorheological finishing process *International Journal of Advances in Manufacturing Technology***100** 1055–1065.

[19] Yamaguchi H, Hendershot P, Pavel R, Iverson JC 2016 Polishing of uncoated cutting tool surfaces for extended tool life in turning of Ti–6Al–4V *Journal of Manufacturing Process***24** 355–360.

[20] H Yamaguchi, Srivastava AK 2012 Magnetic abrasive finishing of cutting tools for machining of titanium alloys *CIRP Annals Manufacturing Technology***61** 311–314.

[21] Barman A, Das M 2017 Design and fabrication of a novel polishing tool for finishing freeform surfaces in magnetic field assisted finishing (MFAF) process *Precision Engineering* **49** 61–68.

[22] Barman A, Das M 2019 Tool path generation and finishing of bio-titanium alloy using novel polishing tool in MFAF process *International Journal of Advanced Manufacturing Technology***100** 1123–1135.

[23] Zhou K, Chen Y, Du ZW, Niu FL 2015 Surface integrity of titanium part by ultrasonic magnetic abrasive finishing *International Journal of Advances in Manufacturing Technology***80** 997–1005.

[24] Umehara N, Kirtane T, Gerlick R, Jain VK, Komanduri R 2006 A new apparatus for finishing large size/large batch silicon nitride Si_3N_4 balls for hybrid bearing applications by magnetic float polishing (MFP) *International Journal of Machine Tools Manufacturing***46** 151–169.

[25] Das M, Jain VK, Ghoshdastidar PS 2012 Nanofinishing of flat workpieces using rotational–magnetorheological abrasive flow finishing (R-MRAFF) process *International Journal of Advances in Manufacturing Technology***62** 405–420.

[26] Kheelraj P, Pulak MP 2018 Use of chemical oxidizers with alumina slurry in double disk magnetic abrasive finishing for improving surface finish of Si *Journal of Manufacturing Process***32** 138–150.

[27] Singh AK, Jha S, Pandey PM 2012 Nano-finishing of a typical 3D ferromagnetic workpiece using ball end magnetorheological finishing process *International Journal of Machine Tools Manufacturing***63** 21–31.

[28] Guo J, Tan ZE, Au KH, Liu K 2017 Experimental investigation into the effect of abrasive and force conditions in magnetic field assisted finishing *International Journal of Advances in Manufacturing Technology* **90** 1881–1888.

[29] Zou Y, Xie H, Dong C, Wu J 2018 Study on complex micro surface finishing of alumina ceramic by the magnetic abrasive finishing process using alternating magnetic field *International Journal of Advances in Manufacturing Technology* **97** 2193–2202.

[30] Singh R, Walia RS 2012 Hybrid magnetic force assistant abrasive flow machining process study for optimum material removal *International Journal of Applied Engineering Research* **7** 2121-2124.

Mix design challenges of porous concrete pavements modified with mineral admixtures

D.Tarangini [1*], B.Radha Kiranmaye [2], Dr.P.Sravana [3], Dr.P.Srinivasa Rao [4]

1 Research Scholar, Department of Civil Engineering, JNTUH, Hyderabad, India
3 Professor & Head, Department of Civil Engineering, JNTUH, Hyderabad, India
* corresponding author e-mail: dtarangini05@gmail.com

Abstract: Pervious concrete has been gaining importance in pavement constructions and low volume road applications. It has applications in parking lots, footpaths and can be used with other materials to prevent erosion. The main problem of porous concrete pavement is its strength. This mainly depends on its pore structure, cement matrix, aggregate gradation, cement content, admixtures and water/cement ratio. The research aims at determining the compressive strength and permeability of pervious concrete with W/C of 0.34.Variations in aggregate sizes is considered and admixtures like fly ash, micro silica, nano silica and superplasticizer are used. The test results show significant difference in compressive strength with variations in admixtures addition. The results were in good agreement in the case of mixtures with W/C of 0.34 and 3% nano silica with respect to compressive strengths and permeability.

Keywords: pervious concrete, porous concrete, permeability.

1. Introduction

In recent years, porous concrete has its wide application in pavement construction as an effective stormwater management tool. With rapid growth in population and continous urbanization, the percolation of rainfall through the ground has decreased. Porous concrete pavements have become a tool in conservation of water resources. The conventional concrete cannot capture rain water and does not support in percolating into the ground. Moreover, normal concrete in pavement construction utilizes more land for retention ponds and other stormwater management techniques.The major advantage of pervious concrete over conventional concrete is the presence of voids which allows the water to percolate. Pervious concrete also has structural, economic and road - user benefits in pavement systems[1],[2]. However, there is a gap in research and use of pervious concrete as a pavement material. One of the major resons being lack of standardized technique for material preparation and testing [3],[4]. Various studies have been carried out on various parameters like aggregate grading,type of aggregate, cement paste content , water-binder ratios etc [3]. The mechanical properties such as compressive strength, permeability and void content have been studies through these parameters [3], [5],[6],[7],[8],[9],[10],[11].

Few researchers have emphasized on partial replacement of Cement with rice husk ash (RHA) for testing the properties of pervious concrete [3] [12] and further the strength and abrasion characteristics have been improved by addition of super plasticizer, silica fume and polymers [13],[14],[15],[16]. Kevern et al used fibers to improve the freeze and thaw action [17]. Few researchers have introduced rubber in pervious concrete [18]. Recycled aggregates and municipal solid waste incinerated bottom ash have also been introduced into pervious concrete as aggregates [7], [19], [11],[20],[21],[22].

The minimum cementitious material used per volume of pervious concrete varies for about 350 kg/m^3 and permeability ranges from 100 to 900 L/min/ m^2 of surface [23],[27]. The literature on pervious concrete confirms that the void content and the water percolation rate are directly related. The target compressive strength achieved by most reserchers varied between 3.5 to 28 MPa [27]. The selection of superplasticizer should be such that it allows high reduction of water content without disturbing the consistency of the mix [27]. Also, improvenment in the mechanical properties such as compressive and flexural strength can be achieched by adding mineral admixtures. The w/c ratio used is around 0.35 to 0.45 with a void content of 15 to 25% [27]

2. Materials and Methods

The work was carried out to find compressive strength, split tensile strength and permeability of pervious concrete by using locally available materials which are given below

2.1 Cement

The cement used was 53 grade Ordinary Portland Cement (Ultra Tech) confirming to properties like specific gravity, initial and final setting time and fineness.

2.2 Fine Aggregates

Natural river sand is used as fine aggregate in the study. The sand is tested for various properties like specific gravity, bulk density and water absorption. In this present study coarse aggregate is partially replaced with 0% , 5 % and 10% fine aggregates.

2.3 Coarse Aggregates

Crushed granite of 10 mm and 20 mm nominal size are used as coarse aggregate. A different proportion of (20 mm: 10 mm) in ratios of 50:50 was used.

2.4 Water

Locally available water has been used for mixing and curing which is potable and free from impurities like oils, salts, sugar, organic materials etc.

2.5 Admixtures

Commercially available fly ash, micro silica and nano silica are used as admixtures . The fly ash used was class F from Hyderabad industries Ltd., Its specific gravity is 1.97. Micro silica used in this work was obtained from Oriental Texxim pvt ltd). Its specific gravity is 2.2. Nano silica used was from Elkem industries, with its specific gravity as 2.12 and particle size of 17nm.

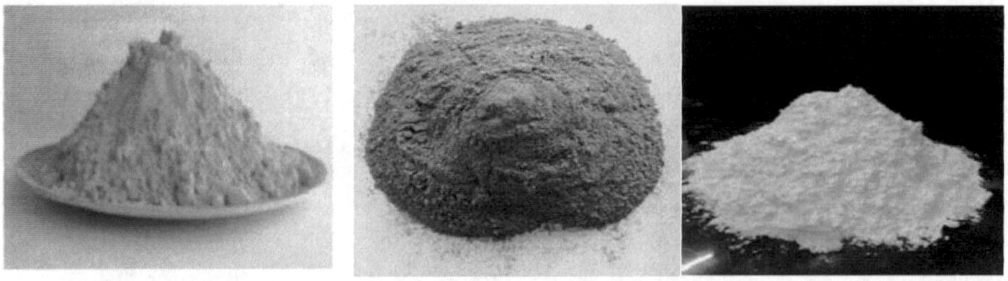

Figure 1 (a)Fly ash (b) Micro silica (c) Nano silica

Conplast SP 430– superplasticizer is used in this research. It is mixed with 0.7% by the weight of cement and then mixed with water.

2.6 Mixture Proportion

In each of the control mixes, 30% (by weight) of cement was partially replaced by fly ash ,10 % by micro silica and 3% by nano silica. The coarse aggregates were partially replaced by fine aggregates by 10% (by weight). About 50% of the total coarse aggregates were taken in size range of 19–9.5 mm and 50% in the size range of 9.5–4.75 mm were used. Based on trails conducted, the desired dosage of plasticizer of about 0.7% by weight of cement was taken. A constant water-binder (w/b) ratio of 0.34 was taken for all the mixes . The mixes were designed according to ACI 522 R-10 . The mechanical properties such as compressive strength, permeability and void characteristics of pervious concrete have been studied. The details of various mixes used are given in Table 1.

Table 1: Pervious concrete mix designations with quantities of ingredients used in mixtures

S.No	Mix Designation	Cement Kg/m^3	Sand Kg/m^3	Coarse aggregate Kg/m^3	Water Kg/m^3	Fly ash Kg/m^3	Micro silica Kg/m^3	Nano silica Kg/m^3
1.	PCF0	350	0	1698	119	0	0	0
2	PCF5	350	85	1613	119	0	0	0
3	PCF10	350	169	1529	119	0	0	0
4	FPCF0	266	0	1843	129	114	0	0
5	FPCF5	266	92	1751	129	114	0	0
6	FPCF10	266	184	1659	129	114	0	0
7	MPCF0	342	0	1843	129	0	38	0
8	MPCF5	342	92	1751	129	0	38	0
9.	MPCF10	342	184	1659	129	0	38	0
10	NPCF0	388	0	1940	136	0	0	12
11	NPCF5	388	97	1843	136	0	0	12
12	NPCF10	388	194	1746	136	0	0	12

2.7 Casting of pervious concrete

Casting process of pervious concrete is as follows.

1. The materials are tested initially and weighed properly as per the mix design.
2. First aggregates and admixture including cement are mixed for 1 min. Then 50% water is added to the mix
3. Remaining water with superplasticizer is then added to mix and mixed for 2 min.
4. It is then compacted with tamping rod .
5. The curing is done at a temperature of 20 ± 1 °C, and relative humidity of 95% . After one day, the specimen is removed from the mould and kept into a curing tank for specific age [28].

Figure 2. (a) casting of pervious concrete

3. Testing of specimens

3.1. Compressive strength

Standard cubes of size 150 × 150 × 150 mm were casted for each mix and curing was done. Compressive strength was tested at 7,28 and 56 days in accordance with the Indian standard IS:516-1959 [3].

3.2. Split tensile strength

Standard cylindrical specimens of size 150 mm diameter and 300 mm height were casted for each mix and curing was done. Split tensile strength was tested at 7, 28, and 56 days in accordance with the Indian standard IS:5816-1999 [3].

3.3. Permeability

In order to check the permeability of porous concrete, a special experimental setup has been fabricated based on falling head permeability method (figure 2). This was initiated by Neithalath, Weiss and Olek [24]. Specimens of dimensions 100 mm dia and 200 mm height were casted and checked for permeability after 28 days of curing [3].

3.4. Porosity

Pervious Concrete is related to porosity and permeability. The porosity was measured for cylinder specimens based on ASTM C1754 guidelines. The equation for calculating it is as follows:

$$P= (1-(W2-W1)/pw *V)) *100 \%$$

Where P = Porosity, %

W1= Weight under water, Kg

W2= Air dried weight, Kg

V= Volume of sample, m3

pw = Density of water, Kg/m3 [29].

Figure 3. (a) Porosity Test (b) Permeability Test

4. Results and Discussions

4.1 Influence on Compressive strength

The compressive strength of pervious concretes with different percentages of mineral admixtures (with 30% fly ash,10 % micro silica and 3% nano silica replacement of cement) are shown in figure 4 and figure 5. It is noticed that the compressive strength of pervious concrete increased with increase in percentages of fine aggregate. The compressive strength varied from 15.45 to 25.45 MPa (at 28 days) . The addition of mineral admixtures has greatly influenced the strength of pervious concrete. The strength of pervious concrete along with admixture has been highest with 10% replacement of coarse aggregate with sand. This trend has been followed by all three mineral admixtures (with 30% fly ash,10 % micro silica and 3% nano silica replacement of cement) at 28 days of curing. The highest strength of 35.56 MPa for 28 days and 40.42 MPa for 56 days is obtained by 3% nano silica replacement of cement and 10% replacement of coarse aggregate with sand. According to Indian standards, the strength requirement for lean cement concrete (LCC) for sub base of flexible pavement should lie in the range of 3.7-7.2 MPa (28 days) and dry lean concrete (DLC) for sub base of rigid pavements should lie in the range of 7.-10 MPa (28 days) [3]. The results obtained satisfy the Indian standards and hence can be recommended for the use as sub-base material in pavements.

Nano silica has shown a better performance in pervious concrete when compared with flyash and microsilica. Therefore nano silica is highly recommended as sub base in pavements. However different percentages of nano silica as replacement of cement should be checked in pervious concrete. A variation in percentages of flyash and microsilica should also be done to check the performance of pervious concrete.

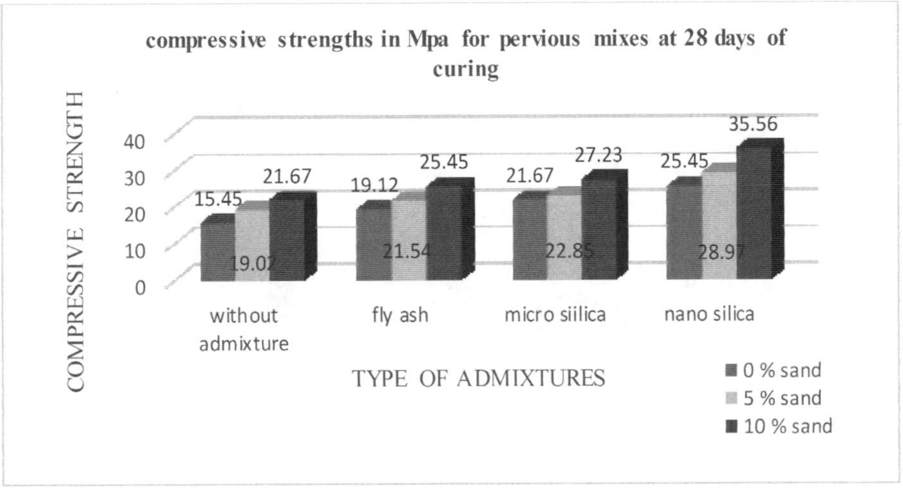

Figure 4. Graph representing compressive strengths with different admixtures and percentages of fine aggregates for 28 days curing

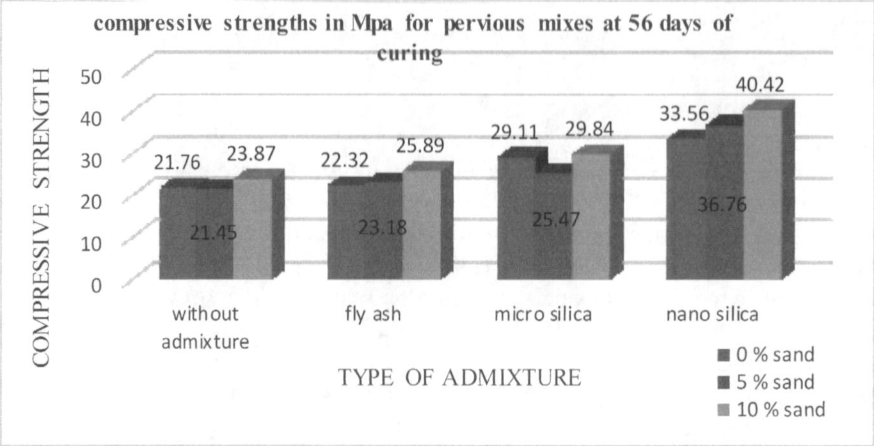

Figure 5. Graph representing compressive strengths with different admixtures and percentages of fine aggregates for 56 days curing

4.2 Influence on split-tensile strength

The influence of different admixtures and varying percentages of fine aggregate in pervious concrete on split tensile strength is grapically represented in figure 6. It is observed that this graph also follows a similar pattern as that of compressive strength. A highest strength of 3.91 MPa is obtained for pervious concrete with nano silica as replacement of cement and 10% sand as replacement of coarse aggregate. There was a noticeable increase of 17 % with fly ash and 10% fines when compared with without fash and fines. Similarly, an increase of 31 % and 78% with the addition of micro silica and nano silica is seen.

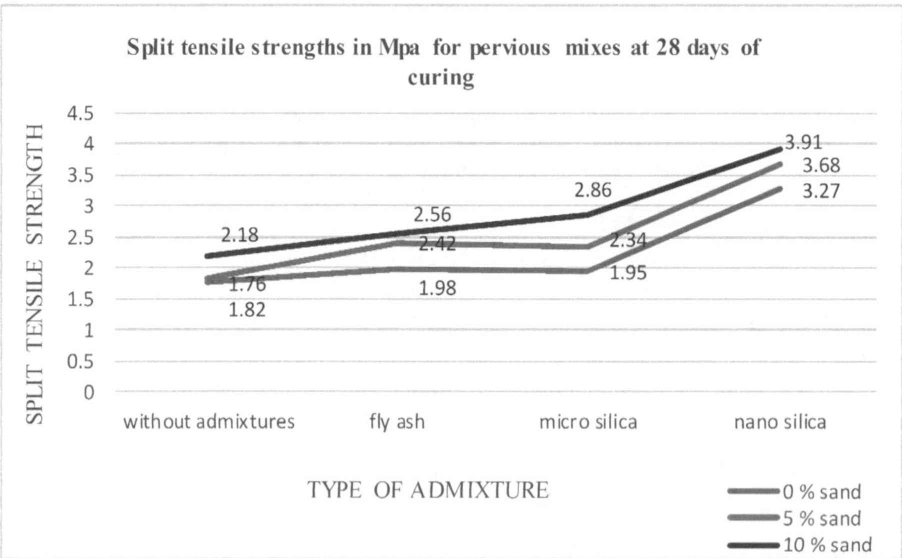

Figure 6. Graph representing split tensile strength with different admixtures and percentages of fine aggregates for 28 days curing

4.3 Influence on Permeability

The effect of different admixtures along with different percentages of fine aggregate on permeability are graphically represented in figure 6. It is observed that the permeability decreases with addition of admixtures. The highest permeability of 1.48 cm/s is obtained with no admixture and no fines pervious concrete. When admixtures are added,a highest permeability of 0.956 cm/s is obtained with cement replaced with 30% flyash. Among all the mixes, the lowest value obtained is 0.401cm/s for nanosilica with 10% sand. However, the results obtained fall in the range of 0.1 to 2 cm/s as reported by various reserchers [25],[26].

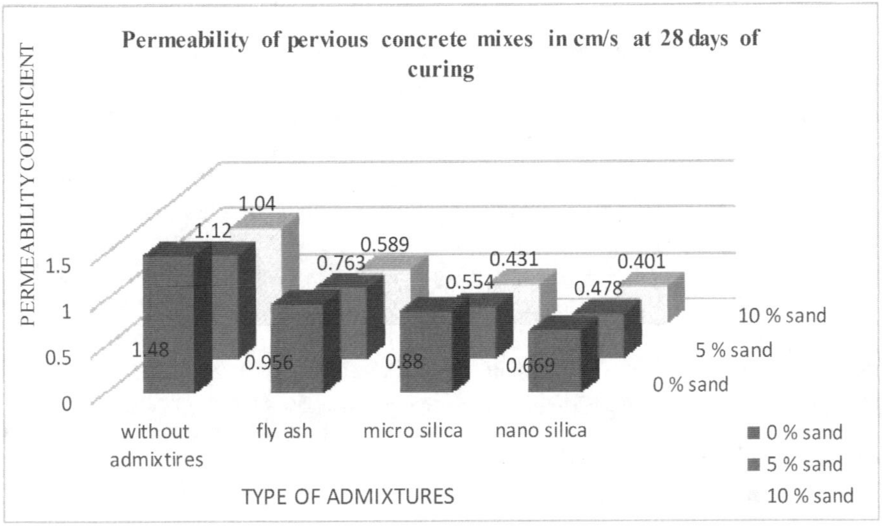

Figure 6. Graph representing permeability coefficients with different admixtures and percentages of fine aggregates for 28 days curing

4.4 Influence of voids on compressive strength characteristics

The voids percentage with no admixtures and different percentages of sand are in the range of 21.21 to 26.23 % . Similarly, 17.34 to 23.24 % (for 30% fly ash content), 16.43 to 21.63 % (for 10% micro silica content) and 16.11 to 20.98 % (for 3% nano silica content).The compressive strength ranges from 15.45 - 35.56 MPa from zero admixtures to 3 % nano silica (Figure 7).

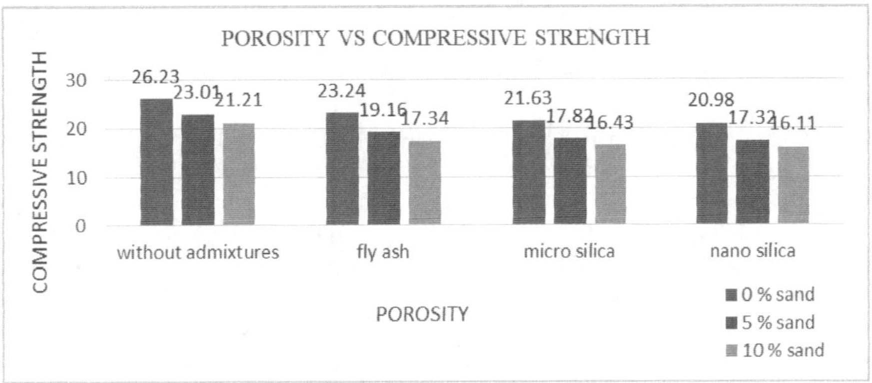

Figure 7. Graph representing permeability coefficients with different admixtures and percentages of fine aggregates for 28 days curing

From the results, it is noiced that the voids percentage is less with nano silica when comared with flyash and micro silica. This is because of more fine particle size. This causes more filler effect and lesser is the void content. Hence the nano silica replacement has shown lesser void content when compared to fly ash and micro silica.The reduction in voids has lead to the compactness of the concrete and increased strength.

5. Conclusions

From the present study, the combined effect of different types of admixtures along with different percentages of fine aggregate on pervious concrete is studied. The compressive strength results obtained with various combinations of fine aggregate percentages along with admixtures recommend the use of admixture based pervious concrete as a sub base material in pavement constructions. Replacement of cement up to 3% by nano silica over 30 % fly ash and 10 % micro silica has significantly increased the compressive strength. However further variation in admixture percentages should be checked. The split- tensile strength results follow the similar trend as that of the compressive strength results. The permeability values obtained through various combinations fall within the common range of 0.1 to 2 cm/s as obtained by many reserchers. The voids content obtained are also in the permissible range. Finally, it can be concluded that the presence of mineral admixtures enchances the performance of pervious concrete. Among the three admixtures considered for the study, the performance of nanosilica was better. This is because of the filling effect caused by the increased surface area of nano silica particles.The addition of superplasticizer has improved the fluidity of the mix with suitable consistency.It has no specific change with respect to strength rather maintaining proper binding between all the powder content and the aggregates.

References

[1] McCain, G.N. and Dewoolkar, M.M., 2010. Porous concrete pavements: mechanical and hydraulic properties. Transportation research record, 2164(1), pp.66-75.
[2] Nguyen, D.H., Sebaibi, N., Boutouil, M., Leleyter, L. and Baraud, F., 2014. A modified method for the design of pervious concrete mix. Construction and Building Materials, 73, pp.271-282.
[3] Uma Maguesvari, M. and Sundararajan, T., 2017. Influence of fly ash and fine aggregates on the characteristics of pervious concrete. International Journal of Applied Engineering Research, 12(8), pp.1598-1609.
[4] Chandrappa, A.K. and Biligiri, K.P., 2016. Pervious concrete as a sustainable pavement material–Research findings and future prospects: A state-of-the-art review. Construction and Building Materials, 111, pp.262-274.
[5] Lee, M.G., Huang, Y.S., Chang, T.K. and Pao, C.H., 2011. Experimental study of pervious concrete pavement. In Emerging Technologies for Material, Design, Rehabilitation, and Inspection of Roadway Pavements (pp. 93-99).
[6] Ibrahim, A., Mahmoud, E., Yamin, M. and Patibandla, V.C., 2014. Experimental study on Portland cement pervious concrete mechanical and hydrological properties. Construction and Building Materials, 50, pp.524-529.
[7] Cheng, A., Hsu, H.M., Chao, S.J. and Lin, K.L., 2011. Experimental study on properties of pervious concrete made with recycled aggregate. International journal of pavement research and technology, 4(2), p.104.
[8] Lian, C. and Zhuge, Y., 2010. Optimum mix design of enhanced permeable concrete–an experimental investigation. Construction and Building Materials, 24(12), pp.2664-2671.
[9] Girish, G. and Manjunath Rao, R., 2011. A step towards mix proportioning guidelines for pervious concrete. International Journal of Earth Sciences and Engineering, 4(6), pp.768-771.

[10] Kevern, J., Wang, K., Suleiman, M.T. and Schaefer, V.R., 2006, May. Pervious concrete construction: Methods and quality control. In Submitted to Concrete Technology Forum-Focus on Pervious Concrete, National Ready Mix Concrete Association, Nashville, TN, May (pp. 23-25).
[11] Bhutta, M.A.R., Hasanah, N., Farhayu, N., Hussin, M.W., bin Md Tahir, M. and Mirza, J., 2013. Properties of porous concrete from waste crushed concrete (recycled aggregate). Construction and building materials, 47, pp.1243-1248.
[12] Hesami, S., Ahmadi, S. and Nematzadeh, M., 2014. Effects of rice husk ash and fiber on mechanical properties of pervious concrete pavement. Construction and Building Materials, 53, pp.680-691.
[13] Huang, B., Wu, H., Shu, X. and Burdette, E.G., 2010. Laboratory evaluation of permeability and strength of polymer-modified pervious concrete. Construction and Building Materials, 24(5), pp.818-823.
[14] Lian, C. and Zhuge, Y., 2010. Optimum mix design of enhanced permeable concrete–an experimental investigation. Construction and Building Materials, 24(12), pp.2664-2671.
[15] Wu, H., Huang, B., Shu, X. and Dong, Q., 2011. Laboratory evaluation of abrasion resistance of portland cement pervious concrete. Journal of materials in civil engineering, 23(5), pp.697-702.
[16] Dong, Q., Wu, H., Huang, B., Shu, X. and Wang, K., 2013. Investigation into laboratory abrasion test methods for pervious concrete. Journal of Materials in Civil Engineering, 25(7), pp.886-892.
[17] Kevern, J.T., Biddle, D. and Cao, Q., 2015. Effects of macrosynthetic fibers on pervious concrete properties. Journal of Materials in Civil Engineering, 27(9), p.06014031.
[18] Senin, M.S., Shahidan, S., Abdullah, S.R., Guntor, N.A. and Leman, A.S., 2017. A review on the suitability of rubberized concrete for concrete bridge decks. In IOP Conf. Series: Materials Science and Engineering 271 (2017) 012074 doi: 10.1088/1757-899X/271/1 (Vol. 12074).
[19] Nguyen, D.H., Boutouil, M., Sebaibi, N., Leleyter, L. and Baraud, F., 2013. Valorization of seashell by-products in pervious concrete pavers. Construction and Building Materials, 49, pp.151-160.
[20] Güneyisi, E., Gesoğlu, M., Kareem, Q. and İpek, S., 2016. Effect of different substitution of natural aggregate by recycled aggregate on performance characteristics of pervious concrete. Materials and Structures, 49(1-2), pp.521-536.
[21] Hossain, T., Salam, M.A. and Kader, M.A., 2012. Pervious concrete using brick chips as coarse aggregate: An experimental study. Journal of Civil Engineering (IEB), 40(2), pp.125-137.
[22] Zhu, Y.G., Kou, S.C., Poon, C.S., Dai, J.G. and Li, Q.Y., 2013. Influence of silane-based water repellent on the durability properties of recycled aggregate concrete. Cement and Concrete Composites, 35(1), pp.32-38.
[23] Mulyono, T., 2019, November. Laboratory Experiment: Pervious Concrete for Permeable Pavement, Focus in Compressive Strength and Permeability. In IOP Conference Series: Earth and Environmental Science (Vol. 366, No. 1, p. 012019). IOP Publishing.
[24] Neithalath, N., Weiss, J. and Olek, J., 2006. Predicting the permeability of pervious concrete (enhanced porosity concrete) from non-destructive electrical measurements. United States: Purdue University.
[25] Kevern, J.T., Schaefer, V.R., Wang, K. and Suleiman, M.T., 2008. Pervious concrete mixture proportions for improved freeze-thaw durability. Journal of ASTM International, 5(2), pp.1-12.
[26] Martin III, W.D., Kaye, N.B. and Putman, B.J., 2014. Impact of vertical porosity distribution on the permeability of pervious concrete. Construction and Building Materials, 59, pp.78-84.
[27] Mulyono, T., 2019, November. Laboratory Experiment: Pervious Concrete for Permeable Pavement, Focus in Compressive Strength and Permeability. In IOP Conference Series: Earth and Environmental Science (Vol. 366, No. 1, p. 012019). IOP Publishing.
[28] Liu, H., Liu, R., Yang, H., Ma, C. and Zhou, H., 2018, February. Experimental study on the performance of pervious concrete. In IOP Conference Series: Earth and Environmental Science (Vol. 113, No. 1, p. 012126). IOP Publishing.
[29] ACI 2008 Building Code Requirements for Structural Concrete (ACI 318M-08) and Commentary. (United States of America: ACI Committee

Experimental study on metakaolin & nano alumina based concrete

Kunchala Ashok[1], B Kameswara Rao[2] and B Sarath Chandra Kumar[3]

[1]Asisstant Professor ©, Department of Civil Engineering, JNTUH college of Engineering Hyderabad, Kukatpally, TS, India.
[2]Professor, Department of Civil Engineering, Koneru Lakshmaiah Education Foundation, Vaddeswaram, AP, India.
[3]Associate Professor, Department of Civil Engineering, Kallam Haranadhareddy Institute of Technology, Chowdavaram, Guntur, AP, India.

ABSTRACT
The experimental work carries around the impact of investigation of Nano Particle like Nano Alumina (NA) and further more one of the beneficial cementitious (pozzolanic) material in cement to improve the mechanical properties of the concrete. Nano Particle and pozzolanic material is partially replaced by the weight of cementitious material into concrete. The experimental work examine was done on Nano replaced concrete were cured for 28, 56 and 90 days of water for various concrete mixes like M30, M40 and M50 to calibrate the mechanical properties of concrete like compressive strength, flexural strength and split tensile strength of concrete and also Non Destructive tests like UPV and Rebound hammer test on the hardened concrete. The workability and mechanical properties were also studied for the conventional and Nano replaced Concrete. Based on the experimental test results obtained the influence of Nano particle in concrete can be examined in both fresh and hardened state.
KEYWORDS: Nano Alumina, Metakaolin, Workability, NanoTechnology, Mechanical property, NDT.

1. INTRODUCTION

The normal cement influences the ordinary Portland cement which will act as a fillers, for example, sand, coarse, admixtures and water. This blend of ingredients permits concrete to be produce in a liquid structure that can be siphoned and formed. The measure of CO2 radiated from the overall creation of OPC relates to roughly 8% of the outflows into the world's environment [2] [13]. To reduce the utilization of natural resources, for example, lime stone, shale, clay, characteristic stream sand, shakes that are being devoured for the improvement of human humankind that are not rewarded the earth and utilization of waste materials in solid that likewise forestalls tremendous zone of land that is utilized for the capacity of waste materials that outcomes in the land, air and water contamination [3] [12] [16 – 23]. The concrete should bring about the manageable advancement without decimation of natural resources.

Nano innovation can be utilized for development cycles and plan in numerous regions since the Nano innovation produced materials have numerous special qualities [4] [15]. The utilization of Nano particles in solid will lessen huge measure of outflows and the utilization of warm protections the exhibition will bring about effective utilization of energy for cooling [5]. As the solid material is generally usable in the construction industry it has been needed to improve its quality and execution over conventional concrete [6]. Nano substituted cement is characterized as the material made by making up for the short comings and pores in Normal concrete utilizing Nano particles of lesser size than 30 Nano meters [7]. The diverse Nanoparticles are accessible like Nano silica, alumina, iron, carbon cylinders and titanium dioxide [22]. In the current age Nano innovation was created with perceptible rate and furthermore the imaginative expected employments of Nano particles there is a worldwide premium in examination of consolidating of Nano particles in development materials in concrete mortar and cement [8]. The utilization of Nano particles has as of late been investigated to defeat the insufficiency of low early age compressive strength in concretes. The Nano particles have a high molecule surface zone to volume proportion and will go about as a filler impact by possessing up the voids between concrete grains with the correct arrangement brings about higher packing density [9].

In the exploration work the impact of Nano particle like Nano Alumina and also one of the strengthening cementitious (pozzolanic) material like metakaolin in concrete to improve the mechanical properties of ordinary standard and high strength concrete. The Nano particle like Nano Alumina and pozzolanic material like metakaolin were incompletely supplanted with concrete by 1% and 15% separately in the concrete blends for various evaluations of concrete.
The Nano particle will bring about more prominent execution in strength and manageability. The fractional substitution of concrete by a mineral admixture called metakaolin will shows better execution in both better and economy. The concrete presented with a blend of Nano particle and mineral admixture improves the miniature structure just as decline the calcium hydroxide fixation by devouring it through a pozzolanic response. The resulting change of the miniature structure concrete composites improves the mechanical properties, functionality properties and increment the administration life properties adding to economically assembled climate.

2. MATERIAL PROPERTIES
Nano replaced concrete comprises of cement, metakaolin, Nano alumina, fine aggregate, coarse aggregate, water and chemical admixture.

2.1 *Cement*
The binding material utilized in the current research work is standard Portland cement of 53 grade and the essential properties of concrete like fineness, specific gravity, consistency, initial setting time and final setting time have been researched according to BIS 4031: 1996 and is introduced in table 1. [11]

Table 1: Properties of 53 grade of Ordinary Portland Cement (KCP)

S.No	Property	Test results
1.	Fineness	3.5%
2.	Specific gravity	3.1
3.	Standard consistency	29%
4.	Initial setting time	45 minutes
5.	Final setting time	450 minutes
6.	Compressive strength at 28 days	51.5 N/mm^2

2.2 *Metakaolin*
White shaded metakaolin of explicit surface region of 150000 to 180000 cm^2/gm having explicit gravity of 2.5 and mean grain size of under 1 micron and having a fineness of 700 to 900 kg/m^2. The metakaolin is an exceptionally responsive material which can contend silica [5] [10].

2.3 Nano particle (Nano Alumina)
The properties of Nano particles are particle size, purity, specific gravity, density and colour are appeared in beneath tables. The actual properties of Nano particle like Nano alumina is presented in table 2 and the chemical composition of Nano particle like Nano alumina was referred to in the table 3 [22].

Table 2 : Properties of Nano Alumina

Properties	Nano Alumina
Morphology	spherical
Colour	White
Purity	99.9%
Particle size	30 – 50 nm
Specific gravity	3.9
Density	90 (g/l)

* Properties of Nano Alumina is provided by the supplier.

Table 3 : Chemical Composition of Nano Alumina (Al_2O_3)

Al_2O_3	Cao	Fe_2O_3	Mgo	Ca
>99.5%	<0.017%	<0.035%	<0.01%	<0.05%

* Chemical Composition of Nano Alumina is provided by the supplier.

2.4 Aggregates

Fine aggregate passing through 4.75 mm IS sieve with fineness modulus 2.47 and specific gravity 2.67 was used for the present work. Coarse aggregate less than 20 mm size with fineness modulus 7.05 and specific gravity 2.62 were used after testing in the research work. The testing was conducted as per BIS 2386:1963 [12].

2.5 Water

The water used in the present research work is potable water which having a p^H value is 6.5 to 8. The limits for permissible solids in water as mentioned in IS 456:2000 [22].

2.6 Super plasticizer

Fosroc Conplast SP 430 which appears in blackish brown colored fluid based super having plasticizer was utilized in the investigation which having a specific gravity is 1.18. To provide excellent acceleration of strength gain at early ages and significant expansions in strength at all ages by essentially reducing water interest in a concrete blend. Especially suitable for precast concrete and

IOP Conf. Series: Materials Science and Engineering 1091 (2021) 012055 doi:10.1088/1757-899X/1091/1/012055

other high early strength necessities. To essentially improve the functionality of site blended and precast concrete without increasing water demand. The pH estimation of SP 430 is 7 to 8 [5].

3. EXPERIMENTAL INVESTIGATION

3.1 Proportioning of Mix Design

The mix proportioning for traditional concrete specimens was designed as per code IS 10262:2019 with Ordinary Portland Cement, fine aggregate, coarse aggregate, and water [13]. The Nano replaced concrete was prepared by the partial substitution of normal concrete by Nano alumina and metakaolin by 1% and 15% of material is substituted by the mass of cement. The water cement ratios used for different grades are 0.48, 0.45 and 0.4.The different mixes used in the experimental work for different grades and their mix proportions are represented in the table 4. The sample designation for different concrete grades like M30, M40 and M50 with and without Nano Alumina are represented like NC30, NA30, NC40, NA40, NC50 and NA50.

- NC – Normal Concrete
- NA – Nano Alumina Concrete

Table 4: Mix proportions for specimens

Grade of concrete	Type of concrete	Cement (kg/m^3)	Metakaolin (kg/m^3)	Nano Alumina (kg/m^3)	Fine Aggregate (kg/m^3)	Coarse aggregate (kg/m^3)	Water (kg/m^3)	SP (kg/m^3)
M 30	NC30	365	-	-	672	1194	175	-
	NA30	307	54.75	3.65	672	1194	175	3.65
M 40	NC40	378	-	-	662	1206	170	-
	NA40	317.5	56.70	3.78	662	1206	170	3.78
M 50	NC50	390	-	-	652	1242	156	1.95
	NA50	327.1	59.0	3.90	652	1242	156	5.85

The supplementary cementitious material in each sample was substituted by Nano alumina and metakaolin by 1% and 15% of the weight of cement. The super plasticizer solution with a dosage of 1 and 1.5% is incorporated to Nano particle concrete mix proportions for improving the workability and flow characteristics of concrete.

3.2 Preparation of concrete specimens

The concrete samples are casted containing Nano particle for various evaluations were set up by beginning drying blending of cement, sand and metakaolin for around 2 minutes in container blender. The chemical compound admixture was broken down in 30% of fluids and Nano particles were added and mixed at a pace of 350 RPM for around 5 minutes for uniform scattering with no coagulation. At that point this arrangement was added to the dry combination with the excess 70% of water followed by ceaseless blending for about an additional 3 minutes in pan blender. The control samples was set up in the comparable manner as depicted above, however with no expansion of Nano particle. The cubes of standard in measurements and cylinder of 150 mm in diameter and length of 300 mm were casted in steel form and compacted. The prism molds of size 100 mm x100 mm x500 mm in measurements are likewise projected in steel molds. The samples were demolded from the molds 24 hours in the wake of projecting, and were permitted to cure in water for 28, 56 and 90 days. After the curing time the samples with standard dimensions of cubes were utilized for the assurance of compressive strength, 150 x 300 mm cylinders for tensile strength and 100 x 100 x 500 mm prisms for flexural strength were determined according to BIS codes. The Non destructive tests like ultra sonic pulse velocity test and rebound hammer test on hardened concrete for 90 days.

4. POZZOLANIC REACTION

The pozzolanic reaction describes that the aluminous and siliceous materials that itself has a practically no cementitious material worth anyway that will, in finely isolated structure inside seeing moistness falsely react with calcium hydroxide at typical temperatures to outline compound having cementitious properties. The hydration of concrete improvement the calcium hydroxide and calcium silicate hydrate are released inside the hydration of two guideline parts of cement specifically tricalcium silicate and dicalcium silicate. The occupation of Nano particles will go about as fillers in the voids or void spaces and all around dissipated Nano particles will go probably as a nucleation or crystallization focal points of the hydrated items, in this way extending the hydration rate, i.e., Nano particles will helped towards the improvement of littler size CH valuable stones and homogeneous gatherings of C-S-H structure. They improved the structure of the advancement zone among aggregates and stick and the effect of Nano particles on the mechanical quality of cementitious materials, the development of Nano particles to solid pastes was found to extend the axial strength to some extent that was dependent on the Nano atom content, water-to-cover weight extent (w/b), and alleviating time. The Nano particles can improve the filler sway and besides the high pozzolanic reactivity of fine particles increases impressively the measure of C-S-H gel.

It can adjust the microstructure in the ITZ progress zones and therefore the assessment of C-S-H gel achieves reducing the water vulnerability. The purpose of the utilization of super fine added substances (Nano particles) in cementitious structures is to improve the characteristics of the plastic and hardened material. The Nano particles have a filler sway by garnish off the pores between the solid grains with the right creation; the higher squeezing thickness achieves a lower water solicitation of the mix and it. It moreover adds to quality improvement as a result of the diminished fine porosity. Moreover speaking to its solidness examines, results show that extension of Nano particles shows better strength execution of the solid. Making of Nano changed solid will achieve better with respect to quality, strength and viability. The fragmented replacement of cement by a mineral admixture called Metakaolin will gain both better and economy. Solid mixes are changed with extension of admixtures and Nano particles, which improve the microstructure similarly as decrease the calcium hydroxide center by burning-through it through a pozzolanic response. The subsequent difference in the microstructure of solid composites improves the mechanical properties, strength and constructs the existence administration properties adding to reasonably fabricated climate.

5. RESULTS AND DISCUSSIONS

5.1 WORKABILITY

The workability property of Nano altered cement represents to that the level of Nano particles are expanded beyond the permissible value the slump values are diminishing for various evaluations for various water concrete proportions. The workability property like slump test is led on concrete by supplanting the Nano particles. The expansion of Nano particles and mix of mineral admixtures additionally diminishing the water concrete proportions. The slump values are represented underneath table 5.

Table 5: Workability for different Mixes and its density

Grade of concrete	**Type of concrete**	**Slump (mm)**	**Density (kg/m^3)**
M 30	NC30	65	2350
	NA30	95	2360

M 40	NC40	60	2380
	NA40	85	2390
M 50	NC50	60	2410
	NA50	80	2420

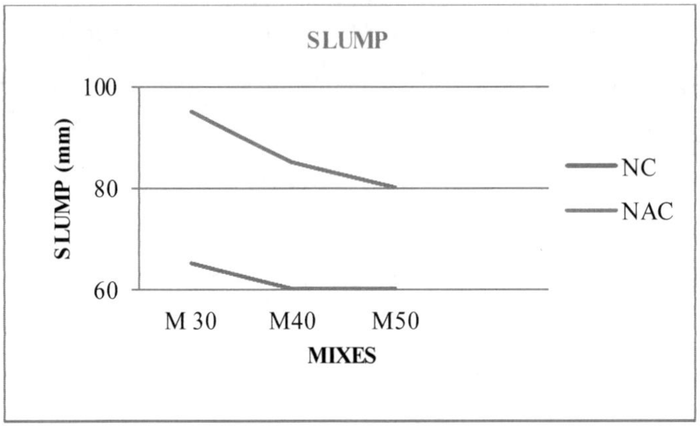

Figure 1: Slump for different mixes

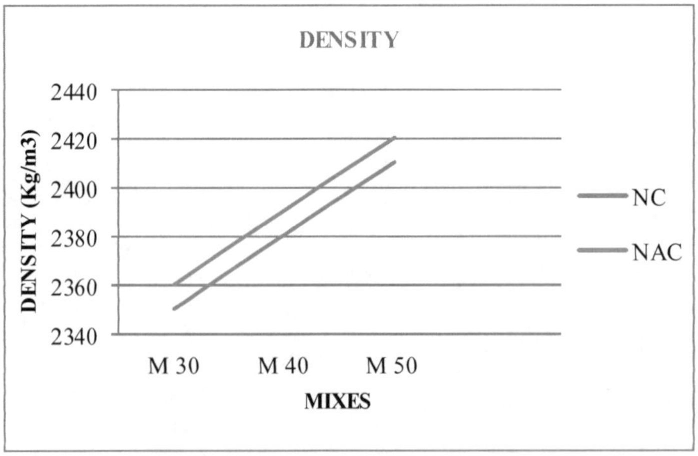

Figure 2: Density for different mixes

From the above representation in graphically from figures 1 and 2 the workability and density of Normal and Nano Alumina concrete are represented to in graphically and the functionality of Nano Alumina solid abatements contrasted with Normal cement because of the Nano particles and the mineral admixture acquainted in cement due with the water fastener proportion diminished and furthermore the Nano particles have a surface to volume proportion the Nano Alumina molecule is under 20 Nano meters in width. The particle have high responsive and the conduct of such materials is mostly impacted by substance responses at the interface. The higher surface zone is to be wetted, it diminishes the free dispersant water in watery framework accessible in blend. The density of Nano Alumina concrete expanded contrasted with Normal cement for various grades like M30, M40 and M50 because of the particles shows enormous surface area to volume proportion.

5.2 COMPRESSIVE STRENGTH

The compressive test was performed by the codal arrangements i:e IS 516:1959 on samples of blocks 150 mm X 150 mm X 150 mm for 28, 56 and 90 days of curing. The test results consequences of compressive strength are appeared in table 6 which was normal of average three samples. The results outcomes indicated that the samples containing Nano particles demonstrated an expansion in concrete compressive strength contrasted with Normal concrete.

Table 6: Compressive Strength of Concrete

Grade of concrete	Type of concrete	28 Days (N/mm^2)	56 Days (N/mm^2)	90 Days (N/mm^2)
M 30	NC30	39	42	47
	NA30	46	53	59
M 40	NC40	48	51	56
	NA40	55	60	67
M 50	NC50	58	61	69
	NA50	66	70	75

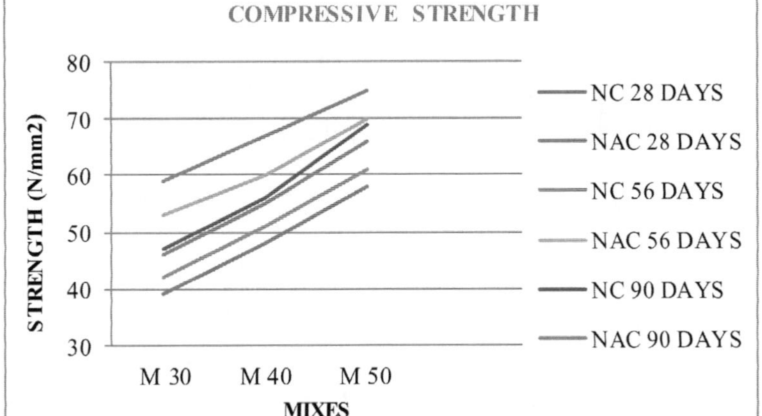

Figure 3: Compressive strength

The importance of above test results obtained from the average of three samples as represented in figure 3 the results displays that the samples containing Nano particles shows an expansion in compressive strength when contrasted and the Normal concrete. The results outcomes shows that the concrete altered by Nano Alumina of 1% respectively of 18%, 26%, 26% & 15%, 18%, 15% and 14%, 20%, 10% are improved in compressive strength with that of Normal cement for 28, 56 and 90 days for various grades like M30, M40 and M50. The upgrade of compressive strength of concrete can be primarily because of that Nano particles go about as a cores in advancing the concrete hydration and topping off of pores to augment in the compressive strength of concrete. The impact of Nano Alumina particle in concrete improves the low early age compressive strength in concrete.

5.3 SPLIT TENSILE STRENGTH

The split tensile strength test was performed by the codal arrangement i:e IS 516:1959 on cylindrical specimens of 150 mm diameter and length of 300 mm for 28, 56 and 90 days of curing water. The Nano particle and other pozzolanic material mixtures are shows preferable results over the traditional concrete. The test results of split tensile strength are exhibits in table 7 which was average of three casted samples. The results shows that the casted specimens containing Nano particles shows an increase in split tensile strength compared to traditional concrete.

Table 7: Split Tensile Strength of Concrete

Grade of concrete	**Type of concrete**	**28 Days (N/mm^2)**	**56 Days (N/mm^2)**	**90 Days (N/mm^2)**
M 30	NC30	3.28	3.96	4.13
	NA30	4.2	4.5	4.78
M 40	NC40	3.65	4.06	4.2
	NA40	4.4	4.78	5
M 50	NC50	4.62	4.8	4.9
	NA50	5.0	5.3	5.5

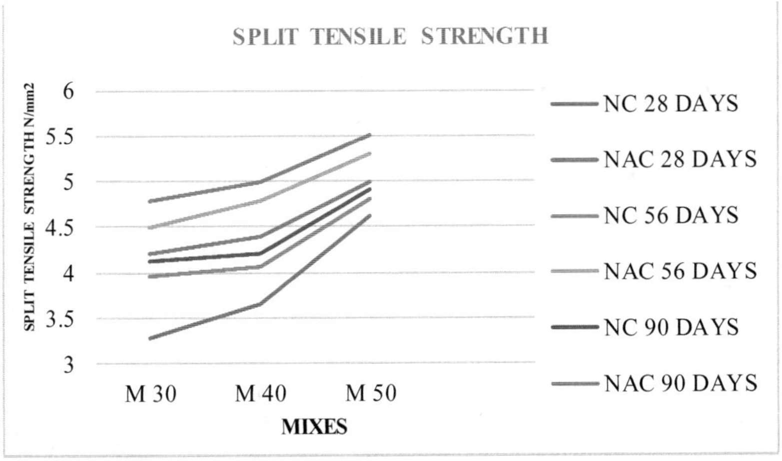

Figure 4: Split tensile strength

The above test results of split tensile strength obtained from the above average of three samples represented as shown in figure 4 the results represents that the specimens incorporated Nano particles shows an increase in strength when compared with the Normal concrete. The results exhibits that the concrete replaced by Nano Alumina of 1% respectively of 30%, 14%, 16% & 21%, 18%, 19% and 8%, 10%, 12% enhanced in indirect tensile strength with that of traditional concrete for 28, 56 and 90 days for different grades like M30, M40 and M50. The enhancement of split tensile strength of concrete can be due to that Nano particles incorporated in concrete which could be more C-S-H gel formation in the presence of Nano particle in concrete and the denser packing of material at their interfacial transition zone.

5.4 FLEXURAL STRENGTH

Based on the obtained results of Flexural strength was performed by the IS codal provisions i:e IS 516:1959 on casted specimens of prisms of size 100mm X 100mm X 500 mm for 28, 56 and 90 days of curing. The test results are shown in table 8 which was average of three casted samples. The results obtained showed that the casted specimens incorporating of Nano particles showed an increase in concrete flexural strength compared to traditional concrete. The Nano particles and other mineral admixtures mixes shows very preferable outcomes results over the normal concrete.The results showed that the specimens containing Nano particles showed an increase in concrete flexural strength compared to Normal concrete.

Table 8: Flexural Strength Results

Grade of concrete	Type of concrete	28 Days (N/mm^2)	56 Days (N/mm^2)	90 Days (N/mm^2)
M 30	NC30	3.76	4.8	4.97
	NA30	4.61	5.07	5.6
	NC40	4.12	5.22	5.3

M 40	NA40	5.08	5.46	5.91
M 50	NC50	5.38	5.85	6..05
	NA50	5.72	6.15	6.3

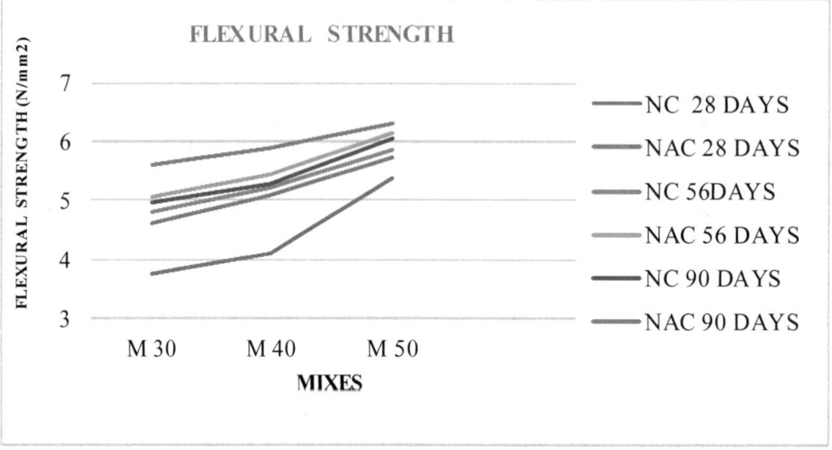

Figure 5: Flexural strength

The above test results obtained of flexural strength of concrete obtained from the average of three casted samples as shown in figure 5 the results obtained shows that the specimens incorporated Nano particles shows an increase in the flexural strength of concrete when compared with the traditional concrete. The test results shows that the concrete replaced by Nano Alumina of 1% respectively of 23%,5.6%, 13% & 23%, 4%, 5% and 6%, 6.3% ,4% enhanced in flexural strength with that of traditional concrete for 28, 56 and 90 days for different grades like M30, M40 and M50. The enhancement of flexural strength of concrete can be mainly due to that Nano particles induced in concrete in promoting the cement hydration and filling up of pores to increase the flexural strength of concrete.

5.5 CYLINDER COMPRESSIVE STRENGTH

The cylinder compressive test was performed by the IS codal arrangements i:e IS 516:1959 on samples of diameter 150 mm and length of 300m mmfor 28, 56 and 90 days of curing. The test consequences of cylinder compressive strength are appeared in table 9 which was normal of three casted samples. The result outcomes indicated that the samples containing Nano particles demonstrated an expansion in concrete cylinder compressive strength contrasted with traditional concrete.

Table 9: Cylinder Compressive Strength of Concrete

Grade of concrete	Type of concrete	28 Days (N/mm^2)	56 Days (N/mm^2)	90 Days (N/mm^2)
M 30	NC30	27	34	39
	NA30	38	45	53
M 40	NC40	36	41	46
	NA40	45	52	56
M 50	NC50	46	50	57
	NA50	53	60	63

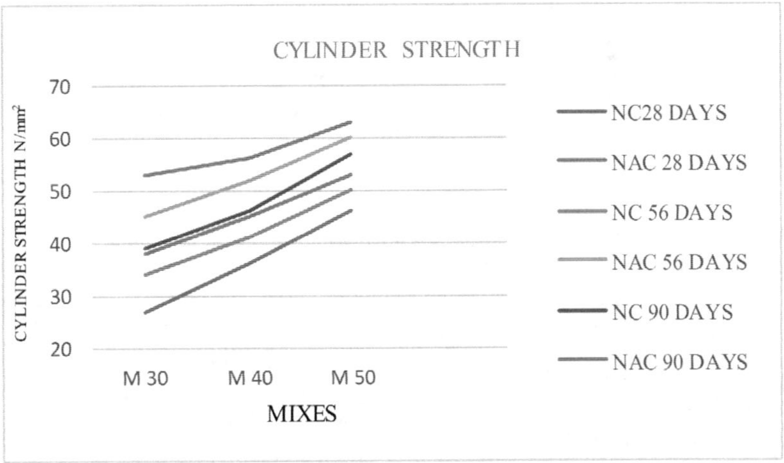

Figure 6: Cylinder Compressive strength

The importance of above test results obtained from the average of three casted samples as represented in figure 6 the results displays that the samples containing Nano particles shows an expansion in cylinder compressive strength when contrasted and the Normal concrete. The outcomes shows that the concrete altered by Nano Alumina of 1% respectively of 41%, 32%, 36% & 25%, 27%, 27% and 15%, 22%, 11% are improved in cylinder compressive strength with that of Normal concrete for 28, 56 and 90 days for various grades like M30, M40 and M50. The upgrade of cylinder compressive strength of concrete can be primarily because of that Nano particles go about as a cores in advancing the concrete hydration and topping off of pores to augment in the cylinder compressive strength of concrete. The impact of Nano Alumina particle in concrete improves the low early age compressive strength in concrete.

5.6 NON DESTRUCTIVE TESTS

5.6.1 Ultra Sonic Pulse Velocity

The ultra sonic pulse velocity test criterion for concrete quality grading based on IS 13311:Part 1. The Nano alumina concrete has high velocity compared to Normal concrete. This is due to the contribution of Nano alumina due to dense packing the ultrasonic pulse pass through the cube with high velocity. The Nano material in concrete results in less porous and impermeable concrete. The estimated strength may vary from the actual strength by +/- 20 percent.

Table 10: Velocity ranges for concrete quality grading from IS 13311: Part 1.

S.no	Pulse Velocity (Km/sec)	Concrete Quality Grading
1.	Above 4.5	Excellent
2.	3.5 – 4.5	Good
3.	3.0 – 3.5	Medium
4.	Below 3.0	Doubtful

5.6.2 Rebound Hammer Test

The rebound hammer test is conducted by IS 13311:Part 2 on concrete hardened surface. When the rebound hammer plunger is pressed against the concrete surface the spring controlled mass rebounds such rebounds depends upon concrete surface hardness. The surface hardness and therefore rebound is considered here to be related to compressive strength of concrete. The rebound is read off along a graduated scale and is designated as the rebound number or rebound index. The strength of the concrete by rebound hammer is varying by +/- 25% with respect to actual strength of concrete.

Table 11: Ultra Sonic Pulse Velocity Test and Rebound Hammer Test at 90 Days

Grade of concrete	Type of concrete	UPV (90 Days)			Rebound Number(90 Days)	
		Velocity (km/s)	Quality	Compressive Strength (N/mm^2)	Rebound No	Compressive Strength (N/mm^2)
M 30	NC30	3.7	Good	55	41	44
	NA30	4.6	Excellent	71	44	51

M 40	NC40	3.9	Good	66	45	53
	NA40	4.8	Excellent	79	48	60
M 50	NC50	4.0	Good	78	49	64
	NA50	5.0	Excellent	89	54	74

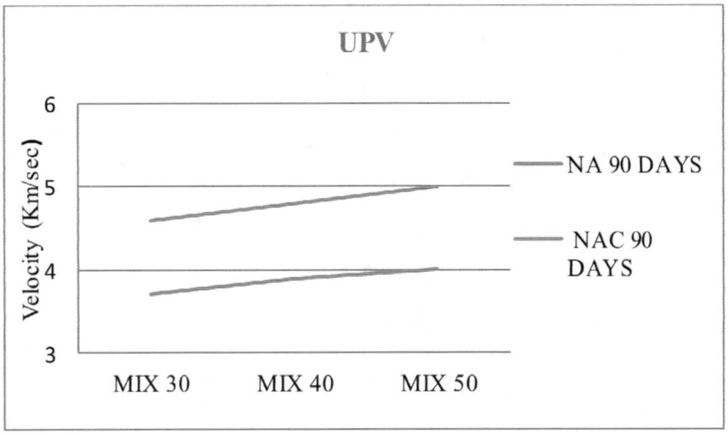

Figure 7: Ultra Pulse Velocity at 90 days

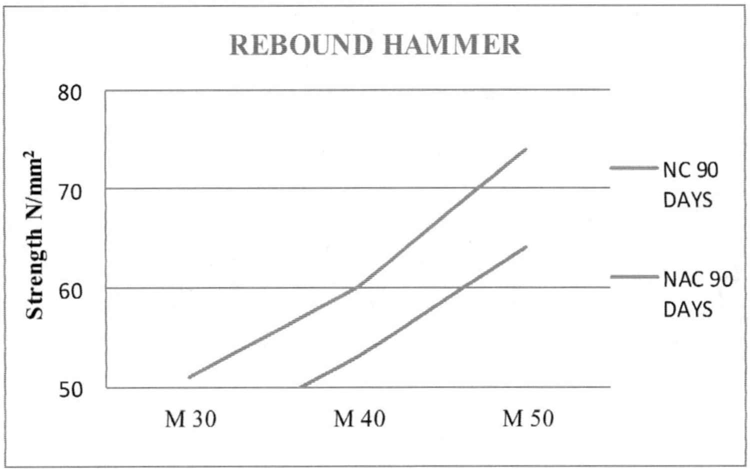

Figure 8: Rebound Hammer test at 90 days

The test results of Rebound number and ultra-sonic pulse velocity obtained from table 11 shows that the quality of Nano Alumina concrete is increases compared to traditional concrete. The quality of Nano Alumina concrete is above 4.0 km/sec and the quality of Normal concrete is below 4.0 km/sec due to dense particle packing done by the influence of Nano Alumina particle in concrete which have huge surface area to volume ratio. The results derived from the rebound number of Nano Alumina concrete is more compared to Normal concrete due to more surface hardness due to the influence of Nano Alumina in mixes of concrete which improves more compactness.

5.7 WATER ABSORPTION

Water absorption test was done according to ASTM C 642-06 on sample of size 100 mm X 100 mm following 90 days of curing water. The test results are shown in table 12 and based on the test results evident that NA has greater reduction in water permeation amongst addition of individual Nano particle due to lesser size of particles in effective filling of gel pores present in the concrete specimen.

Table 12: Water Absorption Test at 90 Days

Grade of concrete	Type of concrete	Water Absorption (%)
M 30	NC30	2.2
	NAC30	1.6
M 40	NC40	2
	NAC40	1.3
M 50	NC50	1.8
	NAC50	1.1

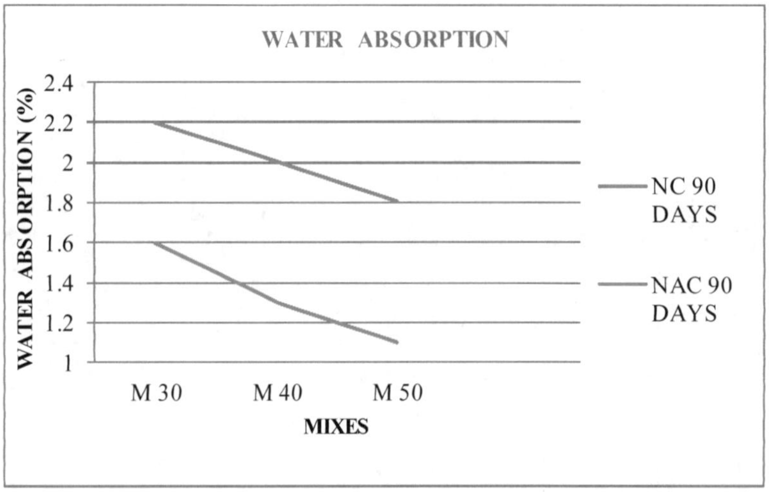

Figure 9: Water absorption test results at 90 days

The above test results obtained from the figure 9 represents in a graphically manner such that the influence of Nano particle in concrete reduces the water permeation amongst the addition of Nano particle in concrete. This is mainly due to the lesser size of particles less than 20 Nm in effective occupying of the voids and the gel pores present in the concrete specimen. From the test result obtained the values are 27%, 35% and 39% of water absorption is decreased compared to Normal concrete due to the reason filling of pores at Nano scale and densification results in reduction of permeable pores present in the concrete specimen at 90 days.

6. CONCLUSIONS

The following observations and conclusions are made on the basis of the experiments conducted on the normal concrete and Nano alumina concrete specimens are

1. The compressive strength of Nano replaced concrete by 1% and 15% met kaolin is found to be increased to a maximum values of 18%, 26%, 26% & 15%, 18%, 15% and 14%, 20%, 10% with the partial substitution of Nano alumina with respect to traditional concrete specimen at 28, 56 and 90 days for M30, M40 and M50.
2. The split tensile strength of Nano replaced concrete by 1% and 15% met kaolin is found to be increased to a maximum values of 30%, 14%, 16% & 21%, 18%, 19% and 8%, 10%, 12% with the partial substitution of Nano alumina with respect to Normal concrete specimen at 28, 56 and 90 days for M30, M40 and M50.
3. The flexural strength of Nano replaced concrete by 1% and 15% met kaolin is found to be increased to a maximum values of 23%, 56%, 13% & 23%, 4%, 5% and 6%, 6.3% ,4% with the partial substitution of Nano alumina with respect to Normal concrete specimen at 28, 56 and 90 days for M30, M40 and M50.
4. The cylinder compressive strength of Nano replaced concrete by 1% and 15% met kaolin is found to be increased to a maximum values of 41%, 32%, 36% & 25%, 27%, 27% and 15%, 22%, 11% with the partial substitution of Nano alumina with respect to Normal concrete specimen at 28, 56 and 90 days for M30, M40 and M50.
5. The water absorption of Nano replaced concrete by 1% and 15% met kaolin is found to be decreased to a value of 27%, 35% and 39% with the partial substitution of Nano alumina with respect to Traditional concrete specimen at 90 days for M30, M40 and M50.
6. The influence of Nano particle by 1% and 15% met kaolin was enhanced in improvement of concrete which could be because of high explicit zone of 20 Nm size of particle. The utilization of Nano Alumina in cement improves the early age advancement of concrete strength.
7. The partial substitution of Nano particle by 1% and 15% met kaolin in concrete exhibits high velocity which is very excellent compared to Normal concrete. The setting time of concrete is decreased by the contribution of Nano particle in concrete and also it modify the structure of ITZ between cement paste and aggregate.
8. Based on the above preliminary mixes the increase of Nano particle in concrete decreases the workability and setting time of concrete is decreases. Nano particle decreases the water absorption of concrete and the Nano particles in cementitious systems is to improve the characteristics of the plastic nature and hardened concrete.
9. The contribution of metakaolin and Nano Alumina by 1% and 15% met kaolin will enhance in the improvement of the micro structure as well as reduce the free calcium hydroxide concentration by utilizing it through a pozzolanic reaction.
10. The modification of the micro structure of concrete by 1% and 15% met kaolin will improves the strength properties of the Nano particle concrete and increase the service life characteristics which will contributing to sustainable built environment.
11. The utilization of Nano particles by 1% and 15% met kaolin in concrete have overcome the deficiency of of low early age compressive strength in concrete.

REFERENCES

[1]. PuvvadaVeerendra and B. Sarath Chandra Kumar, "A Critical Review of Experimental Studies on Strength and Durability Properties of Concrete Incorporating Mineral Admixtures and Fibres, International Journal of Advanced Research in Engineering and Technology, Volume 11, Issue 07, 2020, pp. 49-79.

[2]. Jugalmistry, "Study on effect of nano materials on various properties of concrete", Journal of Civil Engineering and Environment Technology, Vol. 1, number 4, 2014, pp. 36 – 39.

[3]. Abdul Rahim and Sandanu. R. Nair, "Influence of Nano-Materials in High Strength Concrete", Journal of Chemical and Pharmaceutical Sciences, Special Issue 3: 2016, pp.15-22.

[4]. A. H. Shekaria, M. S. Razzaghib, "Influence of Nano Particles on Durability and Mechanical Properties of High Performance Concrete", The Twelfth East Asia-Pacific Conference on Structural Engineering and Construction, Procedia Engineering 14, 2011, pp. 3036–3041.

[5]. S Sanju, S. Sharadha, J. Revathy, "Performance on the Study of Nano Materials for the Development of Sustainable Concrete" International Journal of Earth Sciences and Engineering, Volume 09, No. 03, 2016, pp.294-300.

[6]. A. M. Said, M. S. Zeidan, M. T. Bassuoni, Y. Tian, "Properties of Concrete incorporating Nano-silica", Construction and building materials 36, 2012, pp. 838-844.

[7]. Mohamed Amin, Khaled Abu el-hassan, "Effect of using different types of Nano materials on mechanical properties of high strength concrete", Construction and Building Materials, 80, 2015, pp. 116-124.

[8]. Sujing Zhao, Wei Sun, "Nano mechanical behavior of a green ultra high performance concrete", Construction and Building Materials, 63, 2014, pp. 150-160.

[9]. Rahini, Kannayiramoorthi, NoorulHakkeemsubuhudeen, Manivannan, Venkatesan, "Experimental Study on Behaviour of Nano Concrete", International Journal of Innovative Research in Science, Engineering and Technology, Vol. 6, Issue 3, 2017, pp. 4438-4443.

[10]. B. Sarath Chandra Kumar and P. Saha, "Contribution of Flyash to the properties of Mortar and Concrete" participated in "International Conference on Advances in Civil Engineering" in International Journal of Earth Sciences and Engineering, Vol. 04, Issue 06, 2011, pp 1017-1023.

[11]. B. Sarath Chandra Kumar and K. Ramesh, "Experimental Study on Metakaolin and GGBS based Geopolymer Concrete", International Journal of Engineering and Technology, Vol. 9, Issue 02, 2017, pp. 341-349.

[12]. B. Sarath Chandra Kumar, K. Ramesh and P. Poluraju, "An Experimental Investigation on Flexural Behavior of GGBS and Metakaolin based Geopolymer Concrete", ARPN Journal of Engineering and Applied Sciences, Vol.12, Issue 07, 2017, pp. 2052-2062.

[13]. B.Sarath Chandra Kumar, V. Vamsi Krishna Varanasi and P.Saha, "Sustainable Development Using Supplementary Cementitious Materials and Recycled Aggregate", International Journal of Modern Engineering Research, Vol.2, Issue 01,2012, pp-165-171.

[14]. B. Sai Doondi, B. Sarath Chandra Kumar and P. Saha, "Green Buildings – on the Move", "International Journal of Modern Engineering Research, Vol.2, Issue 01, 2012,pp-349-352.

[15]. B. Sarath Chandra Kumar and K. Ramesh (2018), "Analytical Study on Flexural Behaviour of Reinforced Geopolymer Concrete Beams by ANSYS", IOP Conf. Ser.: Mater. Sci. Eng., Vol. 455, pp. 01- 09.

[16]. DoddaSravanthi, Y. Himath Kumar and B. Sarath Chandra Kumar, "Comparative Study on Flow Characteristics, Strength and Durability of GGBS Based Geopolymer Concrete" IOP Conf. Ser.: Mater. Sci. Eng., Vol. 912, pp. 01- 11.

[17]. ARaziyaTasneem, B. Sarath Chandra Kumar and Y Himath Kumar, "Service Life Prediction of Geopolymer Concrete with Respective to Chloride Ion Penetration", IOP Conf. Ser.: Mater. Sci. Eng., Vol. 912, pp. 01- 10.

[18]. Pulikonda Suman, B. Sarath Chandra Kumar and K. Ramesh "Strengthening of RC Beams with M-Sand Using External Bonding of Steel Plates" IOP Conf. Ser.: Mater. Sci. Eng., Vol. 912, pp. 01-12.

[19]. Vaddeswaram Sangeetha, B. Sarath Chandra Kumar and Y. Himath Kumar "Effect of Ferrocement Wrapping System on Strength and Behaviour of Geopolymer Concrete Beam" IOP Conf. Ser.: Mater. Sci. Eng., Vol. 912, pp. 01- 13.
[20]. ParuchuriVenkata Rao, B. Sarath Chandra Kumar and K. Ramesh "Influence of Carbon Steel Fibers on the Flexural Crack width of RC Beam" IOP Conf. Ser.: Mater. Sci. Eng., Vol. 912, pp. 01-13.
[21]. N. V. Lakshmana Rao, V. Vamsi, J. Chetan Anand and B. Sarath Chandra Kumar, "Experimental Study on Reinforced Carbon Steel Fibers Beams with GGBS as Partial Replacement of Cement", Journal of Critical Reviews, (ISSN 2394- 5125), Volume 07, Issue 12, 2020, pp. 1171-1179.
[22]. Kunchala Ashok and B. Sarath Chandra Kumar, "Influence of Nano Particles in Concrete – A Review", International Journal of Advanced Research in Engineering and Technology, Vol. 11, Issue 06, 2020, pp. 248-263.
[23]. B. Sarath Chandra Kumar, Sadasivan Karuppusamy and K. Ramesh, "Correlation between Compressive Strength and Split Tensile Strength of GGBS and MK Based Geopolymer Concrete using Regression Analysis", Journal of Mechanics of Continua and Mathematical Sciences, Volume 14, Issue 1, 2019, pp. 21-36.

Evaluation of correlation between compressive and splitting tensile strength of concrete using alccofine and nano silica

K Ashwini and P Srinivasa Rao

K Ashwini, Research scholar, Department of civil engineering, Jawaharlal Nehru Technological University Hyderabad, (Telangana State) - 500085, India

P Srinivasa Rao, Professor, Department of civil engineering, Jawaharlal Nehru Technological University Hyderabad, (Telangana State) - 500085, India

Abstract. In the present study correlation between compressive and splitting tensile strength of concrete was studied using both experimental and analytically predicted results. Three grades of concrete M40, M50, and M60 were considered by replacing its cement content with alccofine and nano-silica. Experimental results of compressive and splitting tensile strength were evaluated using 150mm cubes and cylinders of diameter 150mm and height 300mm for 28 days of curing in water. An equation was proposed for correlation between compressive and splitting tensile strength from experimental data using regression analysis and analytical results were predicted. Accuracy of empirical equations reported by researchers and the proposed equation was analyzed employing Root Mean Square Error, Integral Absolute Error, Normal Efficiency and Mean Absolute Error statistical parameters. Good compliance between experimental and analytical results was observed.

Key words: alccofine, cement, concrete, mechanical properties, nano silica,

1. Introduction

Concrete is one of the utmost versatile and durable construction materials as it is strong and economical. High-performance concrete(HPC) is found to be cost-effective, regardless of its high initial cost as compared to conventional concrete as it increases the service life of the structure and thereby reduces overall costs. HPC produced by replacing cement with one or more additives like GGBS, flyash, microsilica, etc. improves the fresh, mechanical, and durability properties to a large extent. Nowadays researchers taking a keen interest in producing high-performance concrete using micro and nano particular size additives.

Alccofine is an ultrafine material showing high reactivity and controlled particle size distribution [11] is being used as an additive to produce high-performance concrete. Use of alccofine in concrete results in denser pore structure and high strength because maximum SiO_2 and CaO content in it enables it to react both in the pozzolanic and hydraulic manner [10,12]. It enhances the mechanical properties like split tensile strength, compressive strength, and flexural strength [13] and durable against acidic environment [14] and also chloride penetration, chances of corrosion [15]. Nano particular sized nano-silica is progressively attracting the consideration of researchers to boost the strength and durability of high-performance concrete. The compressive strength and bond strength of concrete using nano-silica is much greater than control concrete mixes [16]. Nano silica fills up the pores and increases the hydration process [18].

Compressive and split tensile strength is the most essential properties of concrete. As the direct tensile test is difficult to execute, therefore indirect tensile strength tests namely split and flexural strength are performed [7]. Split tensile strength is one of the crucial properties of concrete and is used to design structural elements of concrete subjected to transverse shear, torsion, shrinkage, etc. [21]. As a matter of comport, square root functions are used mostly to find tensile strength from compressive strength(f_{ck}) [2]. But many research studies were conducted to determine the correlation between compressive and splitting tensile strength for conventional concrete, geopolymer concrete, self-compacting concrete, roller-compacted concrete, etc. using power-type equations and found that the power of compressive strength varies [1-6]. Error analysis using statistical parameters of Root Mean Square Error, Integral Absolute Error, Normal Efficiency, and Mean Absolute Error are used to determine the accuracy of these power-type equations developed from experimental results [2, 9].

Hence in this research, an effort is made to determine a correlation between the compressive and splitting tensile strength of alccofine and nano silica-based concrete.

2. Research significance

Mostly many researches investigated the behavior of concrete and very few researches are there on the correlation between split tensile and compressive strength of concrete. Therefore the present study investigates the correlation between split tensile and compressive strength of concrete using alccofine and nano-silica by considering analytical and experimental results for 28 days of curing.

3. Experimental Programme

Alccofine and Nano silica was procured from Ambuja cement Pvt. Ltd. and Astrra Chemicals Chennai. The particular size, fineness, and specific gravity of Alccofine and Nano silica are 4-6μm, 12000 cm2 /gm, 2.86 and 17nm, 202 m2/g, 2.2-2.4 as provided by suppliers. KPC cement grade 53 conforming to IS 12269-1987 used for the study was obtained from local suppliers. Locally available sand having the density of 1520 Kg/m3, confirming to Zone-II, having fineness modulus 3.18 was used. Coarse aggregate of density 1416 Kg/m3 having fineness modulus 7.6 and size 20mm was used. Locally available tap water was used.

All the concrete mixes of grade M40, M50, and M60 were designed as per IS10262:2019 and IS 456:2005. In the present research, all the grades of concrete were partially replaced by 15% Alccofine(Al) and 15+3% (Al) Alccofine and nano-silica (Ns) by weight of cement. Mix proportions for all the grades of concrete are given in table 1.

Table 1. Mix proportions for all the grades of concrete (Kg/m3)

Concrete grades	Cement	Fine aggregate (FA)	Coarse aggregate (CA)	w/c
M40	400	667	1248	0.4
M50	440	642	1243	0.36
M60	527	596	1218	0.3

Evaluation of experimental results of compressive strength was done using 150mm cubes and split tensile strength using cylinders of diameter 150mm and height 300mm for 28 days of curing in water using standard test methods confirming to IS:516-1959 and IS:5816-1999. Regression analysis was used to develop a proposed power-type equation to determine the correlation between compressive and splitting tensile strength. The predicted results of split tensile strength from the proposed power-type equation are compared with experimental results. A proposed power-type equation was compared with the equations developed by other researchers and the accuracy of the equations was measured using statistical parameters of Root Mean Square Error, Integral Absolute Error, Normal Efficiency, and Mean Absolute Error.

Table 2. Tests results of the compressive strength and splitting tensile strength(Mpa).

Concrete grade	Additives	Compressive strength (fck)	Split tensile strength (ft)
	0	51.2	4.18
M40	Al	62.7	5.26
	Al+Ns	66	5.65
	0	58.7	4.91
M50	Al	67.6	6.05
	Al+Ns	70.7	6.43
	0	68.5	5.61
M60	Al	80.1	6.68
	Al+Ns	82.69	7.08

4. Results and discussions

Experimental results of compressive and splitting tensile strength of concrete using alccofine and alccofine + nano-silica as additives for concrete grades of M40, M50, and M60 are given in table 2. Experimental results given in the table are average values of three specimens tested. It was observed that compressive and splitting tensile results using (Al+Ns) were greater than control mixes for all the grades of concrete. Using regression analysis a power-type equation was developed to determine the correlation between compressive and splitting tensile strength considering all the grades of concrete. From figure 1 it was observed that split tensile strength for all grades of concrete increases with an increase in compressive strength.

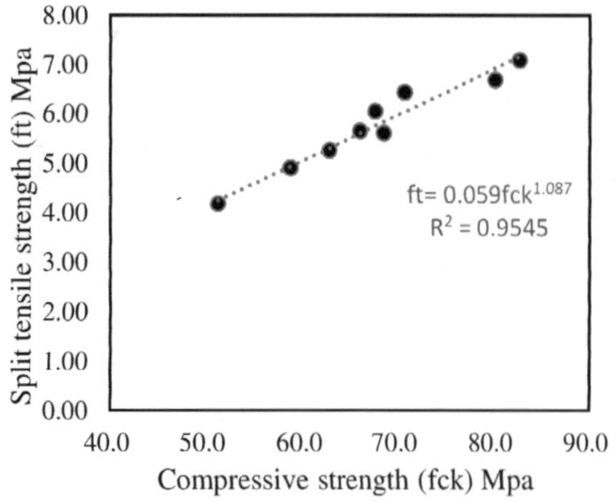

Figure 1 Correlation between compressive and splitting tensile strength

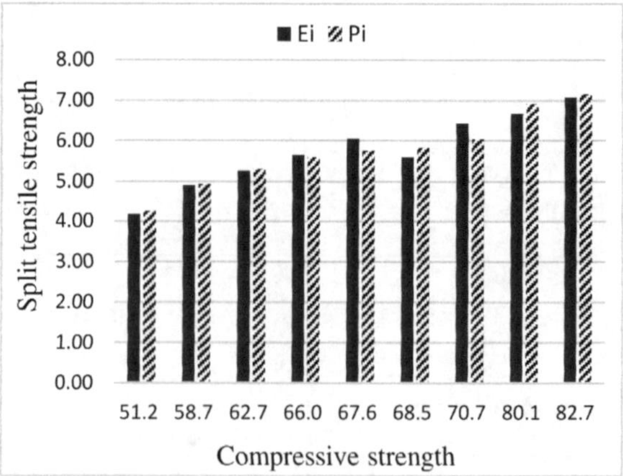

Figure 2 Compressive strength Vs Experimental & Predicted results of split tensile strength

Equation 1 given below is the proposed power-type equation adopted for the correlation between compressive and splitting tensile strength of concrete considering all the grades of concrete at 28 days of curing obtained from figure 1 using regression analysis. R^2=0.9545 acquired from experimental results and power-type regression equation justifies that 95.45% of experimental results were correlated to the power-type regression equation and a good relation between compressive and splitting tensile strength was observed.

$$f_t = 0.059 f_{ck}^{1.087} \quad \ldots\ldots 1$$

f_t = Split tensile strength
f_{ck} = Compressive strength

Table 3. Split tensile strength calculated from different equations

Experimental results		Predicted values						
		Proposed Equation	ACI318-14[1]	Selim et al. [6]	Lavanya et al. [2]	Smadi et al. [5]	Carino et al. [3]	Oluokun et al. [4]
Comp. strength (fck)	Split tensile strength (ft)	$0.059f_{ck}^{1.087}$	$0.56\sqrt{f_{ck}}$	$0.106f_{ck}^{0.948}$	$0.249f_{ck}^{0.772}$	$0.46\sqrt{f_{ck}}$	$0.272f_{ck}^{0.71}$	$0.294f_{ck}^{0.69}$
	Ei	Pi	E1	E2	E3	E4	E5	E6
51.2	4.18	4.25	4.01	4.42	5.20	3.29	4.45	4.44
58.7	4.91	4.94	4.29	5.03	5.78	3.52	4.90	4.88
62.7	5.26	5.30	4.43	5.36	6.08	3.64	5.14	5.11
66.0	5.65	5.61	4.55	5.63	6.32	3.74	5.33	5.29
67.6	6.05	5.75	4.60	5.76	6.44	3.78	5.42	5.38
68.5	5.61	5.84	4.63	5.83	6.51	3.81	5.47	5.43
70.70	6.43	6.04	4.71	6.01	6.67	3.87	5.59	5.55
80.1	6.68	6.92	5.01	6.76	7.34	4.12	6.11	6.05
82.69	7.08	7.16	5.09	6.97	7.52	4.18	6.25	6.19

Split tensile strength values calculated from the proposed power-type equation and other equations are given in table 3. From figure 2 it can be concluded that experimental results and results predicted from the proposed equation are almost similar. Not much information regarding the validity and accuracy of the equations utilized for determining split tensile strength from compressive strength is available in the literature [2]. Error analysis was done using statistical parameters of Root Mean Square Error, Integral Absolute Error, Normal Efficiency, and Mean Absolute Error [2, 9, 19] to analyze the accuracy of the proposed power-type equation as compared to the equations developed by other researchers are given below.

Root Mean Square Error (RMSE) = $\sqrt{\frac{1}{n}\sum_{i=1}^{n}(E_i - P_i)^2}$ 2

Integral Absolute Error (IAE) = $\sum_{i=0}^{n}\frac{\sqrt{(E_i-P_i)^2}}{\sum E_i} \times 100$..3

Normal Efficiency (NEF) = $\left(1 - \frac{1}{n}\sum_{i=1}^{n}\frac{|P_i-E_i|}{E_i}\right) \times 100$..4

Mean Absolute Error (MAE) = $\frac{1}{n}\sum_{i=1}^{n}|P_i - E_i|$...5

E_i= Experimental results
P_i=Predicted results from proposed equations

Table 4: Results of error analysis from statistical parameters

Source	Equations	RMSE	IAE	NEF	MAE
Proposed Equation	$0.059f_{ck}^{1.087}$	0.201	2.74	99.77	0.158
ACI318-14[1]	$0.56\sqrt{fck}$	1.29	20.28	98.35	0.158
Selim et al. [6]	$0.106f_{ck}^{0.948}$	0.215	3.12	99.75	0.18
Lavanya et al. [2]	$0.249f_{ck}^{0.772}$	0.711	11.57	98.96	0.667
Smadi et al. [5]	$0.46\sqrt{fck}$	2.07	34.51	97.12	1.988
Carino et al. [3]	$0.272f_{ck}^{0.71}$	0.507	7.196	99.42	0.414
Oluokun et al. [4]	$0.294f_{ck}^{0.69}$	0.544	7.79	99.37	0.449

From the error analysis values shown in table 4 split tensile strength calculated from the proposed equation based on experimental results is much accurate compared to those calculated using design codes and other equations obtained by researchers. The split tensile results calculated from the equation obtained from Selim et al. [6] were almost similar to predicted results. It can be concluded that the equation having power other than square root can also be used to calculate split tensile strength from compressive strength and this is in agreement with many other researchers [2-7]. As the regression equation is acceptable if the IAE value ranges within the limit of 0 to 10% [20], the proposed equation showed the lowest value of IAE i.e. 2.74 as compared to other equations. Equations showing the least value of Root Mean Square Error, Integral Absolute Error, Normal Efficiency, and Mean Absolute Error can be considered as best [2, 9, 19]. From table 4 RMSE, IAE, NEF, MAE values of proposed equation are 0.201, 2.74, 99.77, 0.158. Hence proposed equation can be used to predicted split tensile results from compressive strength results.

5. Conclusion

- Good compliance between experimental and analytical results was observed.
- R^2=0.9545 acquired from experimental results and power-type regression equation justifies that 95.45% of experimental results were correlated to the power-type regression equation
- The split tensile results calculated from the equation obtained from Selim Pul [6] were almost similar to predicted results.
- Equation having power other than square root can also be used to calculate split tensile strength from compressive strength

- From error analysis using statistical parameters the proposed equation $f_t = 0.059f_{ck}^{1.087}$ is accurate, as it showed least value as compared to other equations and can be used to calculate split tensile results from compressive strength results.

Acknowledgement
The authors are grateful to the Jawaharlal Nehru Technological University Hyderabad (India) for providing the necessary laboratory facilities to carry out the research work discussed in the present paper.

Conflict of Interest
The authors declare that there is no conflict of interest regarding the publication of this paper.

References
[1] ACI Committee 318, 2014. Building Code Requirements for Structural Concrete and Commentary. *American Concrete Institute,* Farmington Hills, MI, pp. 107.
[2] Lavanya, G and Jegan, J., 2015. Evaluation of relationship between split tensile strength and compressive strength for geopolymer concrete of varying grades and molarity. *Int. J. Appl. Eng. Res.* **Vol. 10 (15),** pp 35523–27.
[3] Carino, N.J., Lew, H.S., 1982. Re-examination of the relation between splitting tensile and compressive strength of normal weight concrete. *ACI Mater. J.* **Vol.79 (3)**, pp 214–219.
[4] Oluokun, F.A., Burdette, E.G., Deatherage, J.H., 1991. Splitting tensile strength and compressive strength relationships at early ages. *ACI Mater. J.* **Vol.88 (2)**, pp 115–121.
[5] Mohammad Smadi and Ezzeddeen Migdady 1991 Properties of high strength tuff light weight aggregate concrete. *Cement & concrete composites.* **Vol. 13**, pp 129-135.
[6] Selim Pul. 2008 Experimental investigation of tensile behaviour of high strength concrete. *Indian Journal of Engineering & Materials Sciences.* **Vol. 15,** pp. 467-472.
[7] Chamroeun Chhorn, Seong Jae Hong and Seung Woo Lee, 2018. Relationship between compressive and tensile strengths of roller-compacted concrete. *J. Traffic Transp. Eng. (Engl. Ed.).* **Vol. 5 (3)**, pp 215-223.
[8] Mutiu A. Akinpelu, Samson O. Odeyemi, Oladipupo S. Olafusi, and Fatimah Z. Muhammed, 2019 Evaluation of splitting tensile and compressive strength relationship of self-compacting concrete *Journal of King Saud University – Engineering Sciences.* **Vol.31**, pp 19–25.
[9] Richard Campos, Matias M. Mendez Larrain, Musharraf Zaman and Victor Pozadas, 2020 Relationships between compressive and flexural strengths of concrete based on fresh field properties, *International Journal of Pavement Research and Technology* (article in press). DOI: https://doi.org/ 10.1007/s42947-020-1074-0
[10] Dr. Narayan V. Nayak and Yatin Joshi, 2019 Fine And Ultrafine Materials For Concrete Durability, *Civil Engineering & Construction Review,* **Vol.32**(9), pp. 48-56.
[11] Rajesh Kumar S, Amiya K Samanta, Dilip K. Singha Roy, 2015 An experimental study on the mechanical properties of alccofine based high-grade concrete, *International Journal of Multidisciplinary Research and Development*, **Vol.2(10)**, pp 218-224.
[12] Shamshad Ahmad, Mohammad Shakeel, Sirajuddin Ahmed. 2016 Use of Ultrafine Slag in High Strength Concrete, *International Journal of Science and Research.* **Vol.5(10)**, pp1341-44.
[13] P. R. Kalyana Chakravarthy and R. Rathan Raj. 2017 Analysis on compressive strength of concrete with partial replacement of cement with Alccofine. *ARPN Journal of Engineering and Applied Sciences,* **Vol.12(8),** pp 2392-95.
[14] K. Gayathri, Dr. K. Ravichandran, Dr. J. Saravanan. 2016 Durability and Cementing Efficiency of Alccofine in Concretes. *IJERT*, Vol.5(5), pp 460-467.
[15] Susanto Teng , Tze Yang Darren Lim, Bahador Sabet Divsholi, 2013 Durability and mechanical properties of high strength concrete incorporating ultra-fine Ground Granulated Blast-furnace Slag, *Construction and Building Materials*, Vol. 40, pp 875-881.
[16] YE Qing, Zhang Zenall, Sheng Li and Chen Rongshen, 2006 A Comparative Study on the Pozzolanic Activity between Nano-SiO2 and Silica Fume, *Journal of Wuhan University of Technology - Mater. Sci. Ed.*, Vol. 21(3), pp 153-157.

[17] Ye Qing, Zhang Zenan, Kong Deyu and Chen Rongshen, 2007 Influence of nano-SiO2 addition on properties of hardened cement paste as compared with silica fume, *Construction and Building Materials.* Vol.21, pp 539–545.
[18] D.VivekK., S.Elango, R.Saravanakumar, B.MohamedRafek, P.Ragavendra, S.Kaviarasan and E.Raguram, 2020 Effect of Nano-Silica in high performance concrete, *Materials today proceedings* (article in press)
[19] Ali Jaber, Iqbal Gorgis and Maan Hassan, 2018 Relationship between splitting tensile and compressive strengths for self-compacting concrete containing nano- and micro silica, *MATEC Web of Conferences* **162**, pp 1-8.
[20] Nihal Arioglu, Z. Canan Girgin, and Ergin Arioglu, 2006 Evaluation of Ratio between Splitting Tensile strength and Compressive Strength for Concretes up to 120 MPa and its Application in Strength Criterion, *ACI Materials Journal.*, **vol.103 -91**, pp19 - 24.
[21] Bureau of Indian standards IS 5816:1999, Splitting tensile strength of concrete-method of test, New Delhi, India.

AUTHORS PROFILE

K Ashwini is currently pursuing Ph.D in Structural Engineering from Jawaharlal Nehru Technological University Hyderabad, Telangana, India. She completed her M.E in Structural Engineering from AISSMSCOE Pune. She has published three research articles in international journals.

Dr. P Srinivasa Rao is currently working as a Professor in the department of civil engineering at JNTUH college of engineering Hyderabad, Telangana, India. He has been associated with number of design projects, for numbers of organizations and involved as a key person in quality control and mix designs. He has guided 8 Ph.D students, guiding 10 Ph.D students and guided more than 200 M. tech projects. He is the member of ISTE, ICI and Institute of Engineers and has delivered invite lectures in many organizations and institutions. He was recipient of best teacher award of 2015 govt. of Telangana and also outstanding concrete engineer of Telangana-2015 (ICI). He specialized in structural engineering and his research interest are concrete technology, structural design, high performance concrete, prefabricating structures, special concretes and use of micro silica, fly ash building materials.

Slag resistant and refractory ceramic compositions in MgO – Al_2O_3 – SiO_2 system on the base of local raw materials of Uzbekistan

Z A Babakhanova, M Kh Aripova

Department of silicate materials and rare noble metals, Tashkent Chemical-Technological Institute, 32 Navoi street, Tashkent 10011, Uzbekistan

Corresponding author e-mail: zebo.babakhanova@gmail.com

Abstract. Forsterite and spinel-corundum ceramic materials were obtained in the MgO-Al_2O_3-SiO_2 system based on local raw materials of Uzbekistan - talc-magnesite of Zinelbulak deposit, kaolin of Angren deposit and inorganic additives (fused periclase PPE-87, electric fused corundum). The formations of various phases were studied using XRD and SEM-EDS. Study of chemical-mechanical properties of ceramic samples showed that spinel-corundum composition have density from 2520 to 2770 kg/m^3, water absorption rate from 10.52 to 11.71%, open porosity from 27 to 31%, apparent density from 2420 to 2750 kg/m^3, compressive strength from 100 to 120 MPa, characterized by increased refractoriness (it was 1750 °C for forsterite ceramic materials, for spinel-corundum more than 1800 °C) and slag resistance under metallurgy slags. The synthesis of spinel in optimal compositions occurs at 300 °C lower than the traditional sintering temperature of spinel products, which contributes to significant energy savings in the synthesis of highly refractory spinel-corundum materials.

1. Introduction

The synthesis of refractories based on local raw materials is an urgent task for an effective development of the metallurgical industry of the Republic of Uzbekistan. More than 900 metallurgical enterprises are operating in Uzbekistan, widely using refractory ceramic materials in the processes of casting liquid metals and linings of metallurgical furnaces. For instant, ferrous and non-ferrous metallurgy consumes up to 70 % of the refractory ceramic materials produced around the world [1-2].

There are fireclay and acid-resistant refractory materials produced at several ceramic enterprises of Uzbekistan. Due to the limited availability of suitable raw materials, highly needed high-alumina, magnesite, magnesia-chrome, chromite refractories are not produced in Uzbekistan. In this regard, scientists are currently working intensively on development of methods of enrichment of raw materials and comprehensive study of new types of raw materials [3]. To increase the refractoriness and slag resistance of fireclay refractories, the new compositions of composite materials are being developed, in particular, special attention is paid to the selection of certain binders and special additives [4].

Currently, compositions of new generation refractory materials are obtained on the basis of fine and highly purified oxides [5], nano- and other advanced materials [6-7].

S. Pagliosa et al. [8] have developed refractory products for metallurgical industry based on periclase-alumina-carbon system, as a carbon component nanographite was used to increase the efficiency of the synthesized masses. Compared to natural graphite, the oxidation resistance of nanographite is 50% higher due to the presence of nanosized particles. At the same time, the MgO – Al_2O_3 – SiO_2 system is of considerable interest for obtaining ceramic materials with high refractoriness, heat and slag resistance [9]. The study of the phase diagram of the MgO – Al_2O_3 – SiO_2 system show the presence of such refractory compounds as corundum, periclase, quartz, forsterite (melting temperature 1890 °C), mullite (1850 °C), spinel (2100 °C) [10].

Spinel-corundum, mullite and corundum products, including those with carbon content, are used in metallurgy for preparation of refractory crucibles, melting tanks and furnace lining. The effect of spinel-corundum refractories on the environment is less than that of periclase-chromite refractories, since they do not contain chromium compounds [11-12]. In this regard, spinel corundum products are also widely used in rotary kilns in the production of cement and lime.

In this research work the crystallization regions of forsterite, corundum and spinel in MgO-Al_2O_3-SiO_2 system were chosen in order to obtain ceramic materials with increased refractoriness and slag resistance.

Forsterite crystallization area. Forsterite compositions (42 w.% SiO_2) cannot be fired prior to sintering up to very high temperatures, because the liquid phase does not form in a system at temperatures below 1850 °C. In the compositions lying in forsterite-enstatite region, containing mainly forsterite, the liquid is formed at 1557 °C. Since the liquidus curve in this case has a steep slope, the amount of liquid present in the system changes insignificantly with temperature. Therefore, such compositions have a wide sintering range and can be easily sintered. In contrast, in compositions containing mainly enstatite (55, 60, and 65 % SiO_2), even at low temperatures, a large amount of liquid is formed, and its content rapidly increases with increasing of temperature. These compositions have a narrow sintering range.

In the MgO-Al_2O_3-SiO_2 system, MgO, $MgSiO_3$, $MgAl_2O_4$ (spinel) and $Mg_2Al_4Si_5O_{18}$ (cordierite) coexist in equilibrium with forsterite at 1350 °C. Spinel forms a simple binary euthetic system with forsterite, t_m = 1710 °C. The resulting melt makes it difficult to achieve equilibrium and refractories of the spinel-forsterite composition are practically not obtained. In the forsterite - cordierite system, the lowest eutectic temperature is 1370 °C. Cordierite is not refractory (incongruent decomposition temperature is 1465 °C), but it has a relatively low thermal expansion coefficient of $1.5 \cdot 10^{-6}$. Thus, cordierite is included in the composition of some ceramic refractories as a binder to increase their heat resistance [13].

Spinel crystallization area. In the technology for the production of refractory ceramics, spinel compositions take an important place, while two types of spinels are distinguished: normal spinel AB_2O_4, where A^{2+} ions are located in tetrahedral regions, and B^{3+} ions are located in octahedral regions. This structure was observed for $ZnFe_2O_4$, $CdFe_2O_4$, $MgAl_2O_4$, $FeAl_2O_4$, $CoAl_2O_4$, $NiAl_2O_4$, $MnAl_2O_4$, and $ZnAl_2O_4$. In reverse spinels, the A^{2+} ions and half of the B^{3+} ions are located in octahedral regions; the other half of B^{3+} is at tetrahedral sites, $B(AB)O_4$. This is a more common structure and is observed for $FeMgFeO_4$, Fe_3O_4, $ZnSnZnO_4$, $FeNiFeO_4$, and many other ferrites that are important for obtaining ceramics with magnetic properties [14].

Magnesium aluminate spinel is of great importance as a structural ceramic due to its use in refractories. It provides a combination of the desired physical, chemical and thermal properties at both normal and elevated temperatures. Spinel melts at 2135 °C, exhibits high resistance to most acids and alkalis and has low electrical losses. Due to these properties, spinel and ceramics with spinel structure have a wide range of applications in the chemical, optical, metallurgical and electrical industries. They are used as refractory lining in steel furnaces, transition and sintering zones of cement rotary kilns [15], glass furnace regenerators [16], side walls and bottom of steel ladles in glass furnaces and melting tanks.

Well known methods for the synthesis of spinel $MgO \cdot Al_2O_3$ are the traditional solid-phase reaction (SSR) [17], joint processing [18] and the process of gelation. Ping et al [19] reported that the traditional SSR method is the most used in spinel preparation. However, it has several disadvantages such as longer processing times, several calcination steps, the need for very high sintering temperatures, accompanied by uneven and abnormal grain growth, and residual porosity.

Periclase and spinel form a certain range of solid solutions; spinel forms with Al_2O_3 an extensive region of spinel solid solutions, which reaches a maximum at 85 wt. % Al_2O_3 at 1860 ° C, and a more limited region of MgO solid solution in spinel, reaching 62 wt.% Al_2O_3 at 1995 °C. There is also a solid solution of Al_2O_3 in periclase at elevated temperatures. On the other hand, in the corundum field for MgO, the solid solution field is not shown, since the solubility of MgO in corundum is no more than a few hundred parts per million.

Mixtures of magnesia spinel with periclase or corundum are used for production of highly refractory spinel, spinel-periclase and spinel-corundum materials with a melting point of at least 1925 °C [20]. The tendency of spinels to form solid solutions is a positive factor due to ability of flux oxides (Fe_2O_3, K_2O, Na_2O), common in the composition of local raw materials to form solid substitution solutions. As a result, the negative effect of flux oxides on the physical-chemical characteristics of refractory materials could be significantly reduced.

Magnesia-alumina spinel was synthesized by Krasny B. L. [21], using a finely dispersed mixture as an alumina-containing component, in wt. %: talc - 5-20, electrocorundum - 80-95, as a binder - lignosulfonate. The firing of products was carried out at a temperature of 1600 ± 50 ° C with isothermal holding for the time required for the formation of an equilibrium amount of magnesium-alumina spinel. However, the material obtained was characterized by low compressive strength (45-50 MPa). This disadvantage is due to the fact that the products were manufactured in a one-stage method, in which there is a linear and volumetric expansion of the material due to the spinel reaction. Authors reported that in this way it is impossible to obtain products based on an alumina-magnesia spinel with zero open porosity, density close to theoretical, and a high level of strength.

The objective of this research work was development of high refractory and slag resistant ceramics with forsterite and spinel-corundum structure based on the MgO-Al_2O_3-SiO_2 system using raw materials of Uzbekistan.

2. Research methods

For the synthesis of slag-resistant refractory materials in the MgO-Al_2O_3-SiO_2 system, regions of crystallization of forsterite, spinel, and corundum were selected. Kaolin of Angren Region and talc-magnesite of Zinelbulak deposit were used as local raw materials of Uzbekistan.

The compositions of raw materials and calculated ceramic masses were reduced to a three-component system to ensure the mathematical planning of the properties of refractory compositions. Analysis of the regions in state diagram of MgO-Al_2O_3-SiO_2 make it possible to determine the main crystalline phases and the amount of the liquid phase at the refractory operating temperature.

To identify crystalline phases in the synthesized materials, a Bruker AXS D8 Advance diffractometer, Bruker, Germany was used (Cu-Kα - cathode, step - 0.05, shooting speed 2 sec). Identification of crystalline phases was accomplished by means of a Match! program package (Crystal Impact GbR, Bonn, Germany).

The structure of materials was studied by scanning electron microscopy on an electron microscope (Carl Zeiss, Germany) with an energy-dispersive elemental analyzer (Oxford Instruments, Great Britain).

An oxidation level and thermal stability of graphite-containing ceramic materials was determined by heat treatment of a sample at a temperature of 1000-1500 °C in an oxidizing environment, sawing along the axis with measurement of the decarbonized zone. For a qualitative assessment of the slag resistance of ceramic compositions in relation to a specific type of slag, a static method for assessing slag resistance at 1500-1550 °C was chosen. Slag corrosion was evaluated by the depth and area of the refractory corrosion. This method also takes into account the influence of the porosity of the product on its slag resistance. Additionally, the chemistry of the interaction of slag with refractories was studied by electron microscopy to detect new phases in the material.

Refractoriness of materials was determined by the method of pyrometric cones in conditions of trade production company “Ogneupor” (Tashkent city). Standard methods were used according to GOST 2409-2014, ASTM C20-00, C356-03, C773-88 to determine the apparent porosity, water absorption, specific gravity, density, linear shrinkage and mechanical compressive strength of ceramic samples.

Analysis of the given compositions obtained on the basis of two main raw materials - kaolin AKF-78 of the Angren deposit and talc-magnesite from the Zinelbulak deposits showed that all compositions (compositions No. 40-49 in figure 1) are situated on the state diagram of the MgO-Al_2O_3-SiO_2 system in the region of the initial crystallization of non-refractory cordierite and sapphirine.

In this regard, for synthesis of refractory and slag-resistant ceramic materials in the MgO-Al_2O_3-SiO_2 system, the compositions were calculated in crystallization region of forsterite (compositions No. 51-58 in figure 1), as well as in crystallization area of corundum and spinel (No. 70-79 in figure 1).

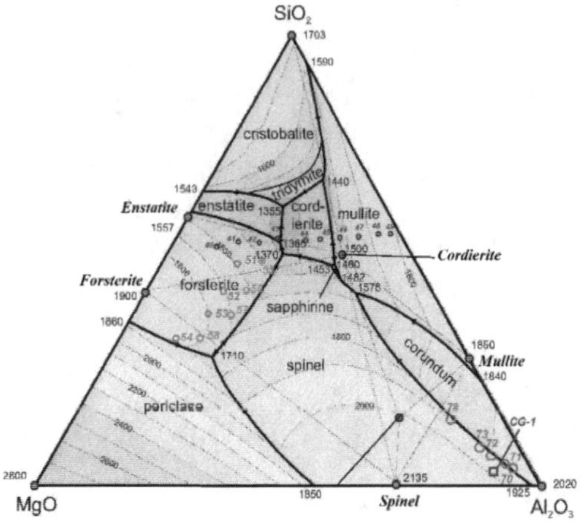

Figure 1. Arrangement of calculated compositions on MgO-Al_2O_3-SiO_2 system state diagram.

To obtain refractories of forsterite composition (No. 51-58), enriched talc-magnesite of the Zinelbulak deposit (Uzbekistan), AKF-78 kaolin (Uzbekistan) and fused periclase of the PPE-87 grade (Russian Federation) were used as raw materials. The samples were fired in a laboratory furnace with silit rods at 1450 °C for 2 hours.

To obtain high-alumina compositions, area of interest were found in the crystallization field of $MgAl_2O_4$ spinel with a melting point of 2105 °C. Ceramic masses were synthesized with an electrocorundum content (Kazogneupor) from 90 to 70 wt.%, Zinelbulak talc-magnesite from 5 to 30 wt. % and enriched graphite concentrate of Zakhchakhona (Uzbekistan) from 5 to 10 wt. % (compositions No. 70-82).

The chemical and mineralogical composition of the initial local raw materials, as well as methods of enrichment of local graphite and the high-temperature processes of phase formation were studied earlier and are presented in our previous works [22]. The samples were fired in a silit furnace at 1500-1520 °C for 2 hours.

3. Results and its discussion

Mineralogical composition of calcined ceramic samples of forsterite and spinel composition have been studied by semi-quantitative XRD- and SEM-EDS analysis.

3.1. Study of phase formation in forsterite-enstatite region of MgO-Al_2O_3-SiO_2 system

The results of studying the mineralogical composition of fired samples of forsterite composition (compositions No. 51-58) by semi-quantitative XRD-analysis (figure 2) showed that the main crystalline phase in composition No. 51-52 is represented by enstatite; in compositions No. 53-54 and No. 57-58 by forsterite; and compositions No. 55-56 contain a significant amount of non-refractory cordierite.

In compositions No. 51-53, lying in the forsterite-enstatite region, containing mainly forsterite, the liquid is formed at 1557 °C. In this regard, compositions No. 51-53 are characterized by formation of a large amount of glass phase. Composition No. 54 is characterized by a predominant amount of forsterite in the crystalline phase; according to the state diagram of the system, the liquidus curve in this case changes slightly with temperature: composition No. 54 has a fairly wide sintering range, is easily sintered and promising composition for production. Physical and mechanical properties of ceramic samples are given in table 1.

Table 1. Physical and mechanical properties of ceramic samples synthesized in forsterite-enstatite region.

No	Open porosity, %	Water adsorption, %	Specific gravity, kg/m^3	Main crystalline phases (according to XRD analysis)
51	5.35	3.34	2920	Enstatite, forsterite, periclase
52	5.57	3.87	2900	Enstatite, forsterite, periclase
53	8.80	5.88	2850	Forsterite, enstatite, periclase
54	9.20	6.05	2840	Forsterite, periclase, enstatite, spinel
55	3.21	2.35	2950	Forsterite, cordierite, periclase, spinel
56	3.52	2.78	2920	Forsterite, cordierite, periclase, spinel
57	8.51	6.45	2830	Forsterite, periclase, cordierite, spinel
58	8.92	7.20	2800	Forsterite, enstatite, periclase

Analysis of the properties, structure and phase composition of the ceramic samples synthesized in forsterite-enstatite region showed that the optimal composition for obtaining refractory materials is composition No. 54, the phase composition is represented by forsterite - 87, periclase - 11 and spinel - 2% (figure 2). Refractoriness of the optimal composition was 1750 °C, which refers these compositions to high refractory materials, the temperature of the onset of deformation under load (0.02 kN/cm) - 1600 °C; heat resistance - 7 thermal cycles; residual change in dimensions when heated at 1350 °C - 0.6 cm; porosity – 7 %. A technological scheme and technological regulations for the production of refractory forsterite products were developed, synthesized forsterite refractory materials recommended to use at heating units, regenerators, open-hearth furnaces, etc.

3.2. Phase formation in the area of spinel crystallization

Study showed that in compositions No. 70-82 calcined at 1000 °C the microstructure of materials is loose and contains forsterite, enstatite and corundum phases; at 1200 °C in MgO-Al_2O_3-SiO_2 system in the presence of a small amount of impurities, fusible compounds is formed and crystallization of spinel accelerated. According to the results of semi-quantitative XRD-analysis, the mineralogical composition of materials synthesized at 1500-1520 °C is represented by corundum Al_2O_3 - from 60 to 70% and spinel $MgAl_2O_4$ - from 30 to 40% (figure 3). The parameters of the crystal cell of the cubic spinel are a=0.8078 nm (8.0775 Å), which is in good agreement with the reference standard No. 00-075-1796 (Match, a = 8.078 Å, space group Fd3m).

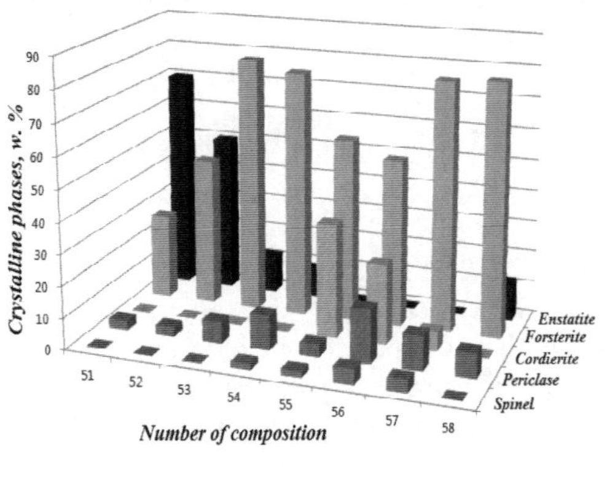

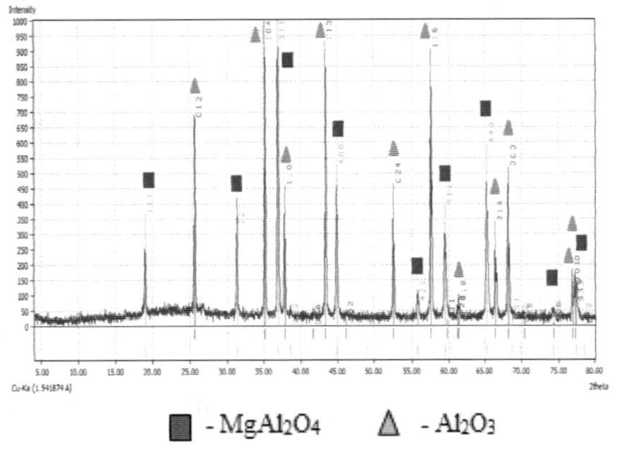

IOP Conf. Series: Materials Science and Engineering 1091 (2021) 012057 doi:10.1088/1757-899X/1091/1/012057

Figure 2. Mineralogical composition of ceramic samples No. 51-58 synthesized in forsterite-enstatite region at 1450 °C.

Figure 3. Diffraction pattern of the optimal composition of ceramic sample No. 77 synthesized in spinel crystallization region at 1500 °C.

The crystalline phase in calcined at 1500 °C ceramic composition No. 77 presented by corundum (65 wt.%) and spinel (35 wt.%) (figure 3), the binder is represented by a high-alumina-silicate phase. The pores in the ceramic sample are generally isolated, rounded, ranging in size from 10 to 50 μm. The formation of a spinel phase in compositions No. 75-78 leads to an increase in the slag resistance of refractory compositions, in contrast to graphite-chamotte refractories. At the same time, the synthesis of spinel in compositions No. 75-78 occurs at much lower temperatures (at 1500 °C) than the traditional sintering temperature of spinel products (1750 °C) [23].

Formation of binding in ceramic materials requires much more reactions than the crystalline phase formation processes. The binder phase in spinel-corundum refractories is formed during the decomposition of talc and magnesite, followed by the interaction of periclase with the silicate phase. As a result, a magnesium-containing alumina silicate phase is formed (spectrum 6 in figure 4), which binds individual crystals of corundum and spinel increasing mechanical strength of the samples. A binder in refractories is the most important factor that determines the processing methods and properties of the synthesized material; it is represented in the material by round, irregularly shaped grains with a size of <25 μm (figure 4).

Figure 4. Electron microscopic image and EDS of a ceramic sample No. 77 synthesized at 1500 °C: formation of spinel phase.

The main compaction of materials is performed at the stage of pressing; therefore, when synthesizing refractories, it is extremely important to obtain materials with a higher density and lower values of water absorption and porosity. An increase in the density of the material contributes to an increase in the chemical and thermal resistance of the refractory material. In this regard, when determining the optimal compositions, the main attention was paid to the analysis of the density and porosity of ceramic samples.

The study of the physical and mechanical properties of the synthesized materials showed that ceramic materials of spinel-corundum composition No. 70-82 have a density from 2520 to 2770 kg/m^3, water absorption from 10.52 to 11.71%, an apparent specific gravity from 2420 to 2750. The compressive strength of the samples was from 100 to 120 MPa. The temperature of the beginning of deformation under load (0.02 kN / cm) - more than 1800 °C; heat resistance - 9 thermal cycles; residual change in dimensions when heated at 1350 °C - 0.2 cm. The synthesized materials are high refractory with a melting point of more than 1800 °C. Compared with the ceramic materials given in [21], the synthesized spinel-corundum materials were distinguished by a higher mechanical strength: 100-120 MPa versus 45-50 MPa. This indicates the possibility of using the synthesized materials for the production of high refractory ceramic products with the required mechanical strength.

Determination of slag resistance of materials in relation to metallurgical slag of the Almalyk Mining and Metallurgical Complex (AMMK) by the static method for 2 hours at 1550 °C showed that compositions No. 75-77 are demonstrated a higher resistance to molten metals.

Chemical composition of metallurgical slags from copper-smelting production of AMMK,%: Cu - 0.95; Pb-0.2; Zn 0.77; S - 1.19; Fe - 35.15; SiO_2 - 34.24 [24-25]. To determine the zones of penetration of metallurgical slag, studies were carried out using an electron microscope on the distribution of elements over the surface of a ceramic sample. The results are shown in figure 5.

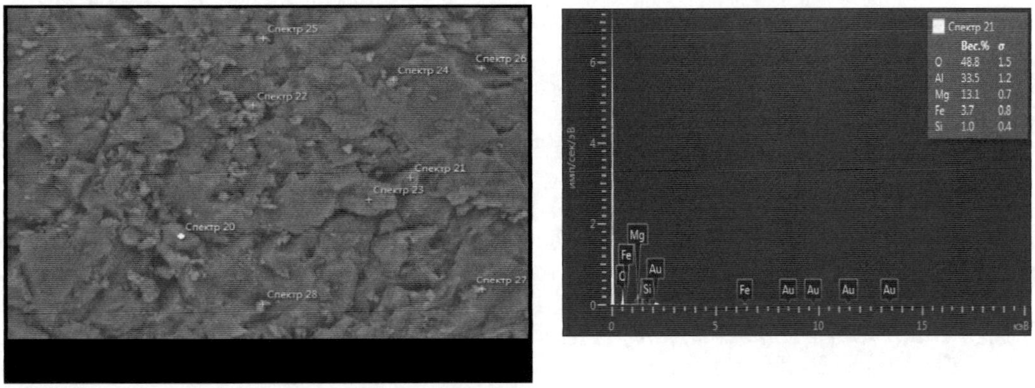

a b

Figure 5. Electron microscopic image and spectral analysis of sample No. 77 after slag resistance test: (a)- general view, (b) – spinel composition with Si and Fe content.

The main crystalline phases in a ceramic specimen after testing for slag resistance are represented by crystals of corundum (crystal sizes 10-25 mk), fine-crystalline spinel phase (crystals sizes 1-5 mk) (figure 5). The amount of alumina silicate binder increased, and penetration of metallurgical slag occurs in small amounts along the phase boundary and pores of the sample. The area of penetration of slag in specimens No. 75-77 was 0.25-0.3 cm, under the same conditions the depth of penetration of slag in fireclay refractories was 1.5-2.2 cm. To determine the oxidation of graphite-containing ceramic materials, fired specimens were sawn and the inner sections were examined microscopically to determine the decarbonized area. Studies have shown that the optimal compositions No. 75-77 with a carbon content of 7 % are distinguished by the best indicators of resistance to oxidation (the smallest thickness of the decarbonized layer).

4. Conclusion

In the MgO-Al_2O_3-SiO_2 system based on local raw materials and inorganic additives, refractory materials of forsterite (compositions No. 51-58) and spinel-corundum composition (compositions No. 70-79) with high refractoriness (1750-1800 °C) and slag resistance in relation to metallurgical slag were synthesized. The formation of a spinel bond in corundum refractories at a temperature of 1500 °C, which is lower than the traditional sintering temperature of spinel products (1750 °C), has been established. This contributes to significant energy savings in the synthesis of highly refractory materials of spinel-corundum composition. A higher resistance to metallurgical slag of refractories of spinel-corundum composition, which is also distinguished by higher refractoriness (1800 °C), has been established. In this connection, spinel-corundum compositions No. 75-77 with a carbon content of 7% can be recommended for use in crucial processes of metallurgical production, for operation in direct contact with metallurgical slag.

References

[1] Akselrod L M 2017 *Novye ogneupory* **11** 3-14. doi: 10.17073/1683-4518-2017-11-3-14
[2] Raja B V R 2007 Recent Developments in Secondary Steel Making Processes *IIM Metal News*

10 **[1]** 19-26
[3] Ruzibaev B and Saloxidinova S 2020 The possibilities of using loess clay of kattabog deposit in the production of ceramic products *Chemistry and chemical engineering* **2** 27-30 https://uzjournals.edu.uz/cce/vol2020/iss2/5 (accessed 07.01.2021)
[4] Mkrtchyan R and Aripova M 2020 Influence of the aluminum oxide additive on the phase composition of mullit-forming aluminosilicate minerals *Chemistry and chemical engineering* **2** 22-26 https://uzjournals.edu.uz/cce/vol2020/iss2/4 (accessed 07.01.2021)
[5] Roy I, Kujur M, Singh R et al. 2020 Development of Castables with High Thermal Shock Resistance and Volume Stability *Interceram* **69** 38-43 https://doi.org/10.1007/s42411-020-0113-9
[6] Mendonça Spera N, Fernandes L, Sakihama J. et al. 2020 Designing Colloidal Silica-Bonded Porous Structures of In-situ Mullite for Thermal Insulation *Interceram. - Int. Ceram. Rev.* **69** 54–63 https://doi.org/10.1007/s42411-020-0120-x
[7] Singh A K and Sarkar R 2016 Nano mullite bonded refractory castable composition for high temperature applications *Cer. Inter.* **42** 11 12937-45
[8] Pagliosa S, Kampos A and Frejere N 2013 *Ogneupory i tehnicheskaja keramika* **10** 24-28
[9] Lee W E, Zhang S and Karakus M 2004 Refractories: Controlled microstructure composites for extreme environments *J.Mater.Sc.***39** 6675-85
[10] Bobkova N M 2007 *Fizicheskaja himija tugoplavkih nemetallicheskih i silikatnyh materialov* (Minsk: Vysshaja shkola) p 301
[11] Calbo J, Sorlí S, Llusar M, Tena M A and Monrós G 2004 *British Ceram.Trans.* **103** 1 3-9 doi: 10.1179/096797804225012729
[12] Nath Mithun, Song Shengqiang, Li Yawei, and Xu Yibiao 2017 *Ceram.Int.* **44** 2383-89 https://doi.org/10.1016/j.ceramint.2017.10.207
[13] Mei S Yang J and Ferreira J M 2001 Cordierite-based glass-ceramics processed by slip casting *J. Casting. J.Eur.Cer.Soc.***21**185-193
[14] Braulio M A L, Rigaud M, Buhr A, Parr C and Pandolfelli V 2011 Spinel-containing alumina-based refractory castables *Ceram.Int.* **37** 1705-17 doi:10.1016/j.ceramint.2011.03.049
[15] Bartha P and Södje J 2001 *CN Refractories* **5** 62–71
[16] Cemail A, Brain R, Frank L R and Paul D W 2002 *J.Eur.Cer.Soc.* **22** 745-754
[17] Sarkar R, Das S K and Banerjee G 2000 *Ceram.Int.* **26** 25-28
[18] Raveron H, Campos D G, Rodriguez R M and Bonassin J C 2002 *Materials Letters* **56** 97-101
[19] Ping L R, Azad A M, and Dung T W 2001 *Materials Research Bulletin* **36** 1417-30
[20] Resende W S, Stoll R M, Pallone E, Cunha F N and Bradt R C 2001 *Miner.and Metallurg. Processing* **18** 68-74 doi:10.1007/BF03402874
[21] Krasnyj B L e.a. 2003 *Sposoby izgotovlenija izdelij iz korundovoj keramiki* [Methods of manufacturing products from corundum ceramics] Patent RF no. 2198860
[22] Babakhanova Z A and Aripova M Kh 2019 *Refractories and Industrial Ceramic* **59** 5 454-458
[23] Kashheev I D, Zemljanoj K G 2017 *Novye ogneupory* **3** 127-133
[24] Matkarimov S T, Yusupkhodjayev A A, Berdiyarov B T, Nosirkhujayev S Q U and Matkarimov Z T 2020 *International Journal of Advanced Science and Technology* **29** 3 5633-39 Retrieved from http://sersc.org/journals/index.php/IJAST/article/view/6189
[25] Matkarimov S T, Yusupkhodjaev A A, Khojiev Sh T, Berdiyarov B T and Matkarimov Z T 2020 *Journal of Critical Reviews* **7** 5 214-230

IOP Conf. Series: Materials Science and Engineering 1091 (2021) 012058 doi:10.1088/1757-899X/1091/1/012058

Numerical investigation of convective heat transfer in fluid flow past a tandem of triangular and square cylinders in channel

M. A. Moussaoui[1*], **Y. Admi**[1*], **E. B. Lahmer**[1], **A. Mezrhab**[1]
[1] Laboratory of Mechanics & Energetics, Faculty of Sciences, Department of Physics 60000 Oujda, Morocco

Email: ma.moussaoui@ump.ac.ma admi_youssef1718@ump.ac.ma

Abstract. A numerical study of a two-dimensional laminar flow and heat transfer characteristics in a horizontal channel containing two in-line obstacles of different shapes, namely an upstream triangular cylinder and a downstream square cylinder, was carried out. It is based on a coupling between the Lattice Boltzmann Method (LBM) and the Finite Difference Method (FDM). The airflow (Pr = 0.71) is assumed to be laminar and incompressible. All physical properties of the fluid, expect its density, are supposed to be constant. The two cylinders are kept at a constant hot temperature, while the incoming flow is at a cold temperature. The Reynolds number and the horizontal separation distance between the cylinders are varied to investigate their influences on fluid flow and heat transfer.

1. Introduction

Flow around cylinders has been a subject of research for many scientists for a long time due to its practical applications. The majority of numerical or experimental investigation has focused on the effects of the form, number, and position of cylinders, Reynolds number, and the boundary of the computational domain. Many applications can be found in heat exchangers, buildings, bridges, mechanical systems, and electronic devices. Looking at the literature, one can conclude that the majority of numerical papers are about the flow passes on bluff bodies with the same shape [1-15] such as circular cylinders [1-4] square cylinders [5-8] rectangular cylinders [9-12], and triangular cylinders [12-16]. All these works are done with a variety of numerical methods. From the above, it can be concluded that very limited attention has been focused on problems associated with the flow around two cylinders composed by the square cylinder and triangular cylinders or used triangular cylinders as disturber cylinders in front of square cylinders. For this reason, this study concentrates on the effects of Reynolds number and horizontal distance separation in a channel containing two in-line obstacles of different shapes, namely a triangular cylinder upstream and a square cylinder downstream.

2. Problem and numerical details

2.1. Physical model

Fig. 1 illustrates the physical model of the plane channel with the two obstacles. The air flow (Pr = 0.71) is considered laminar and incompressible. With the exception of its density, all the physical properties of the fluid are considered invariable. The L/d ratio is fixed at 30.75 and the blockage ratio is d/H=1/4. The triangular cylinder is located at a distance Xin/d = 8 downstream of the inlet section of the channel. Both cylinders are maintained at a constant hot temperature h = 0.5, whereas the incoming flow is at a cold temperature θc = - 0.5. The input velocity profile is parabolic. The boundary conditions implemented on the top and bottom walls of the channel are u = v = 0, for velocity, and adiabatic for temperature. For the boundary conditions, the rebound scheme and a spatial quadratic interpolation [17, 18] are used.

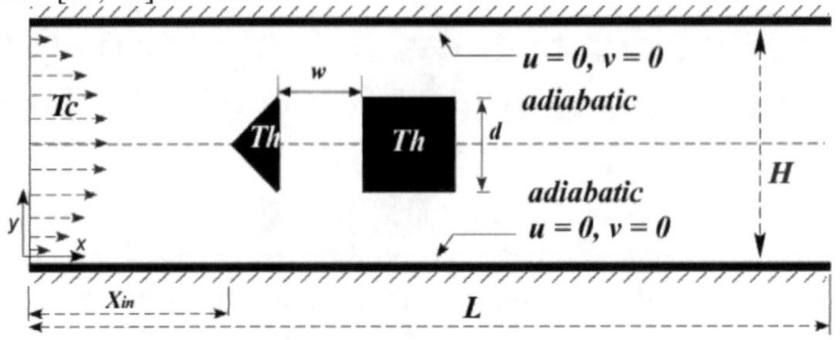

Figure. 1. Schema of computational domain.

3. Numerical simulation

To simulate the fluid flow, the numerical method used is the multi relaxation time (MRT) lattice Boltzmann method [19] which can be formulated as follows:

$$\boldsymbol{f}(\boldsymbol{x}+\boldsymbol{e},t+1)-\boldsymbol{f}(\boldsymbol{x},t)=-M^{-1}.S.\left(\boldsymbol{m}(\boldsymbol{x},t)-\boldsymbol{m}^{eq}(\boldsymbol{x},t)\right) \tag{1}$$

Where M is the 9×9 transformation matrix such as $\boldsymbol{m} = \boldsymbol{M.f}$ and S is the relaxation matrix where is a diagonal matrix, i.e., $S = diag(0,s_1,s_2,0,s_4,0,s_6,s_7,s_8)$.

We consider a D2Q9 model and the particle speed $\vec{e_i}$ are given by:

$$\vec{e_i}=\begin{cases}(0,0), & i=0\\ \left(\cos[(i-1)\pi/2],\sin([(i-1)\pi/2])\right)c, & 1\le i\le 4\\ \sqrt{2}\left(\cos[(2i-9)\pi/4],\sin([(2i-9)\pi/4])\right)c, & 5\le i\le 8\end{cases} \tag{2}$$

Where $c= dx/dt$ is the lattice speed, and dx and dt are the lattice width and time step, respectively. Here, dt is chosen to be equal to dx, thus $c=1$.

The nine moments with the associated distribution functions are classified in this order:

$$m=(\rho,e,\varepsilon,j_x,q_x,j_y,q_y,p_{xx},p_{xy})^T$$

$$f=(f_0,f_1,f_2,f_3,f_4,f_5,f_6,f_7,f_8)^T$$

$$\underbrace{\begin{pmatrix}\rho\\ e\\ \varepsilon\\ jx\\ qx\\ jy\\ qy\\ p_{xx}\\ \vdots\end{pmatrix}}_{m}=\underbrace{\begin{pmatrix}1&1&1&1&1&1&1&1&1\\ -4&-1&-1&-1&-1&2&2&2&2\\ 4&-2&-2&-2&-2&1&1&1&1\\ 0&1&0&-1&0&1&-1&-1&1\\ 0&-2&0&2&0&1&-1&-1&1\\ 0&0&1&0&-1&1&1&-1&-1\\ 0&0&-2&0&2&1&1&-1&-1\\ 0&1&-1&1&-1&0&0&0&0\\ \vdots\end{pmatrix}}_{M}\cdot\underbrace{\begin{pmatrix}f_0\\ f_1\\ f_2\\ f_3\\ f_4\\ f_5\\ f_6\\ f_7\\ \vdots\end{pmatrix}}_{f} \tag{3}$$

The equilibria of the moments m^{eq} at the equilibrium are:

$$m^{eq} = (0, e^{eq}, \varepsilon^{eq}, 0, q_x^{eq}, 0, q_y^{eq}, p_{xx}^{eq}, p_{xy}^{eq})^T \tag{4}$$

Where:

$$\begin{aligned} e^{eq} &= -2\rho + 3(j_x^2 + j_y^2) \\ \varepsilon^{eq} &= \rho - 3(j_x^2 + j_y^2)/\rho \\ q_x^{eq} &= -j_x \\ q_y^{eq} &= -j_y \\ p_{xx}^{eq} &= (j_x^2 - j_y^2)/\rho \\ p_{xy}^{eq} &= j_x j_y / \rho \end{aligned} \tag{5}$$

The macroscopic fluid density and moment flux are calculated by:

$$\begin{aligned} \rho(\boldsymbol{x},t) &= \sum_i f_i(\boldsymbol{x},t) \\ J(j_x, j_y) &= \sum_i f_i(\boldsymbol{x},t)\boldsymbol{e}_i \end{aligned} \tag{6}$$

The finite difference method [19, 20] is used to solve the macroscopic temperature. An explicit coupling between both schemes MRT and FDM is carried out. The energy equation is given by:

$$\partial_t \theta + U \cdot \nabla \theta = \alpha_f \Delta \theta \tag{7}$$

The above equation is resolved explicitly by using first-order direct time difference scheme and the second-order central difference scheme for space discretization, with the same grid points as for the LBM scheme.

4. Results and discussion

The numerical code (HTLBM) has been validated in the previous work (see [21-25]). A preliminary work was performed to identify the optimum grid. Various simulations were carried out for w= 4d, Re = 100, and for different uniform grids (Nx×Ny) in order to examine the grid independence. To optimize grid refinement with simulations efficiency, the grid 1567×203 was selected for all the further calculations.

The local Nusselt number (Nu) based on square cylinder height is expressed by the following relationship:

$$Nu = -d \frac{\partial \theta}{\partial n} \Big|_{\text{block surface}} \tag{8}$$

Figure 2 depicts the instantaneous velocity plots for w= 4d, Pr = 0.71, and for four values of the Reynolds number (Re=20, 40, 80, and 100). In general, when a cylinder is positioned in the wake of another cylinder in the transverse flow, its unsteady charge becoming dependent not only on the flow characteristics in its wake but also on those in the wake of the upstream cylinder. For Re=20, It is apparent that behind each cylinder, two symmetrical vortex zones appear on either side of the wake which turns in opposite senses. The velocities vectors present a perfect symmetry in the channel. The increase of the Reynolds number leads to an increase in the magnitude and strength of these recirculation zones. For Re = 80, the symmetrical behaviour observed behind each cylinder disappeared. The pattern of the fluid flow changes and the wake loses its original symmetry. An unsteady separation flow over each cylinder is observed where the fluid flow oscillates with a repetitive state of swirling vortices. The von Karman vortex streets are visualized behind the rear face of the triangular and square cylinder. The oscillations in the wake grow in magnitude for Re=100. The unsteadiness in the flow increases with increasing Reynolds number.

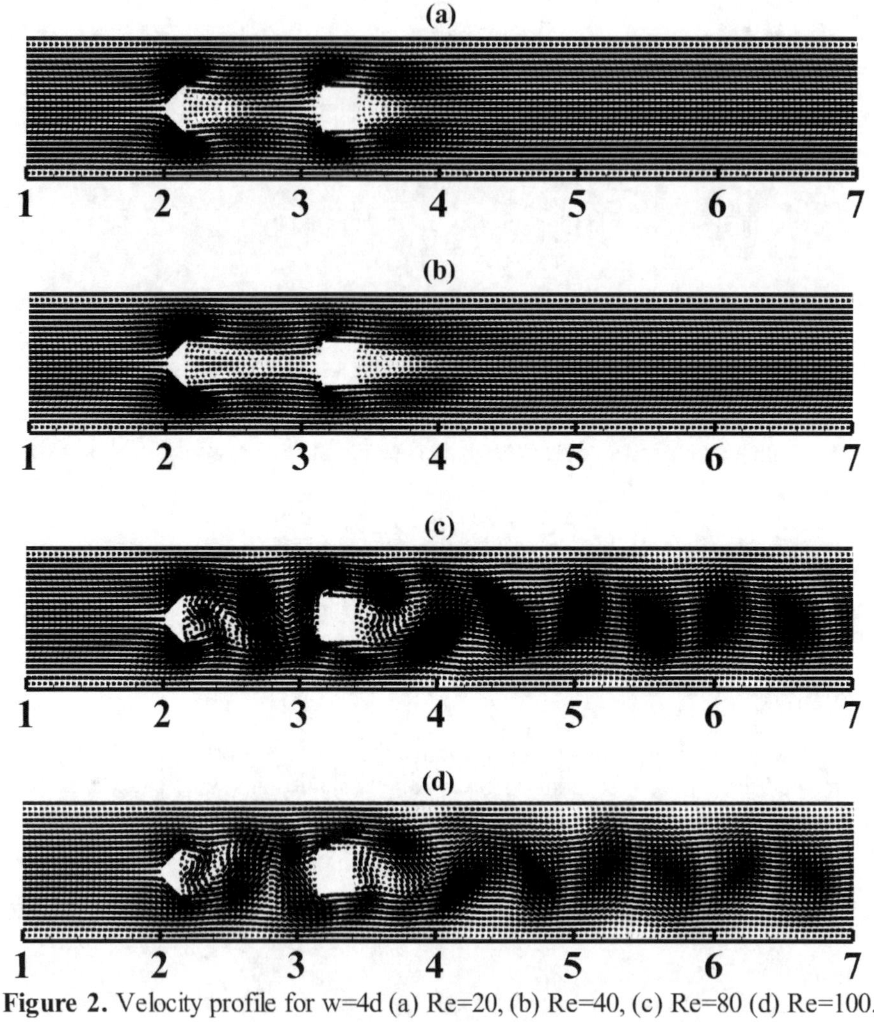

Figure 2. Velocity profile for w=4d (a) Re=20, (b) Re=40, (c) Re=80 (d) Re=100.

Figure 3 presents the isotherms behaviours for various Reynolds numbers. From major observations, it can be deduced that the increase of Re affects favourably the heat transfer from the hot cylinders to the flow. The increase of Re from 20 to 40 leads to the reduction of the thickness of the thermal layer near to the cylinders walls. For these values of Re, the flow is symmetrical and characterized by a vertical gradient of temperature through the channel.

For Re=80 and 100, the flow structure is changed to be asymmetric. The shape of the temperature fields reflects the fluid motion. An indicative temperature gradient is generated around the cylinders reflecting the presence of a significant heat transfer in these zones.

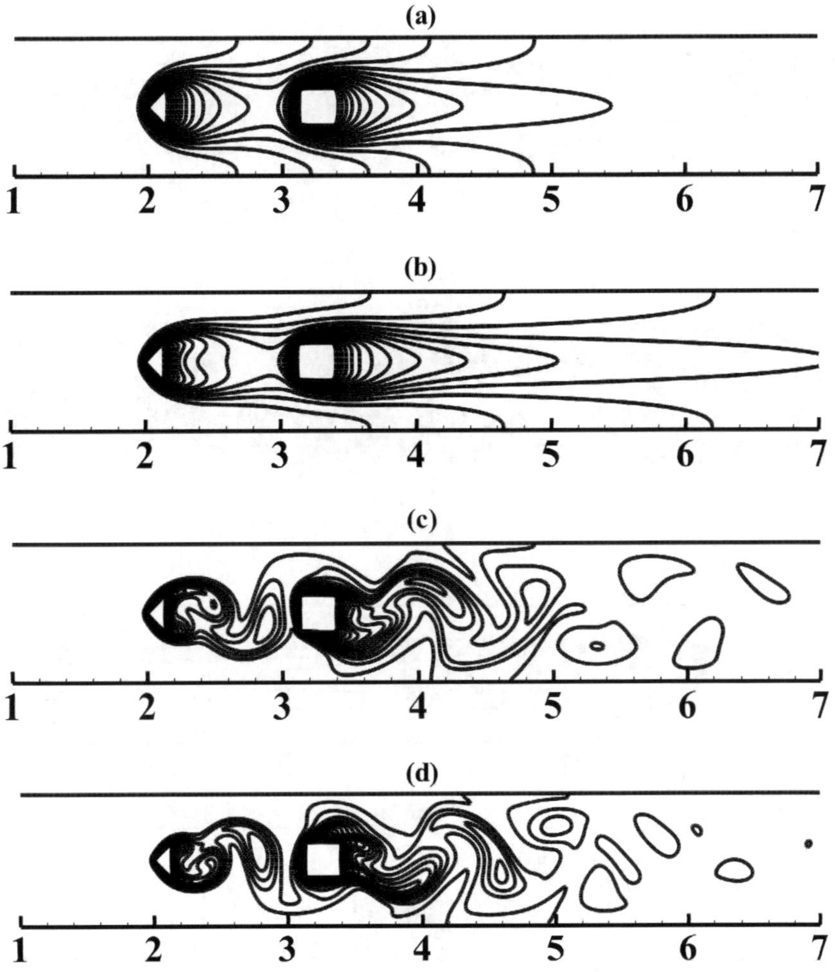

Figure 3. Isotherms for w=4d (a) Re=20, (b) Re=40, (c) Re=80 (d) Re=100.

In order to to understand the influence of varying gap spacing on fluid flow and heat transfer, Figure 4 and 5 illustrate the instantaneous velocity profiles and isotherms at Re= 100 and for various spacing ratios W=2d, 4d and 6d. Let us note that at low Reynolds number (Re=20), the flow present a constant symmetry and the gap spacing don't affect the behaviors of the fluid flow and the heat transfer. When Re=100, two modes are observed depending on the gap space between the two cylinders. When w=2d, the square cylinder is very close to the triangular; which prevents the appearance of vortex in the space between them. As w increases to w=4d and 6d, the flow the flow pattern is qualitatively changed. An alternate vortex are created between the cylinders and in the downstream of the second cylinder. At w=2d there is no flow separation and the isotherms have diffusion-type profiles while they become more non-uniform at w=4d and 6d because of the flow separation.

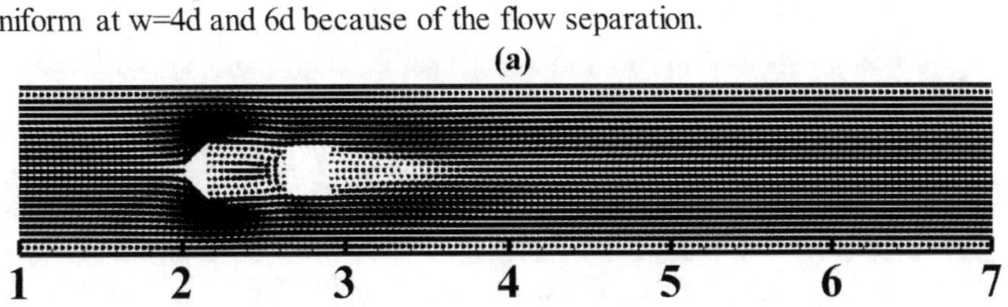

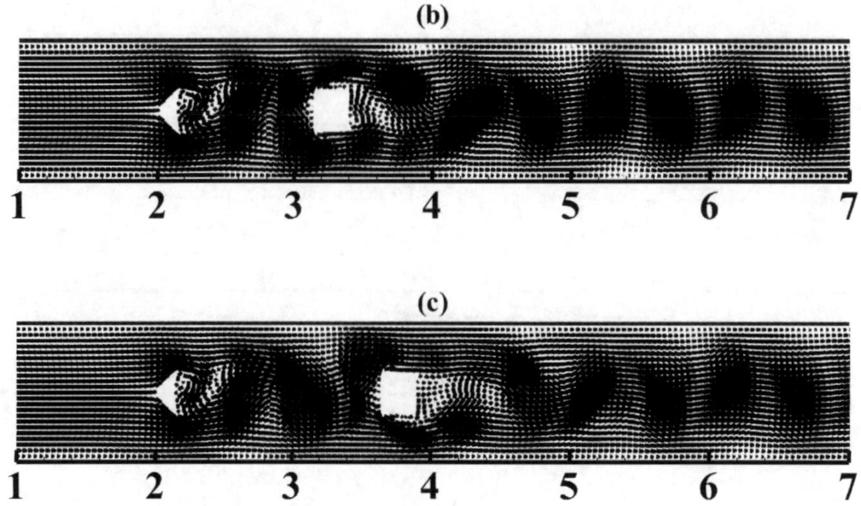

Figure 4. Velocity profile for Re=100 (a) w=2d, (b) w=4d, (c) w=6d.

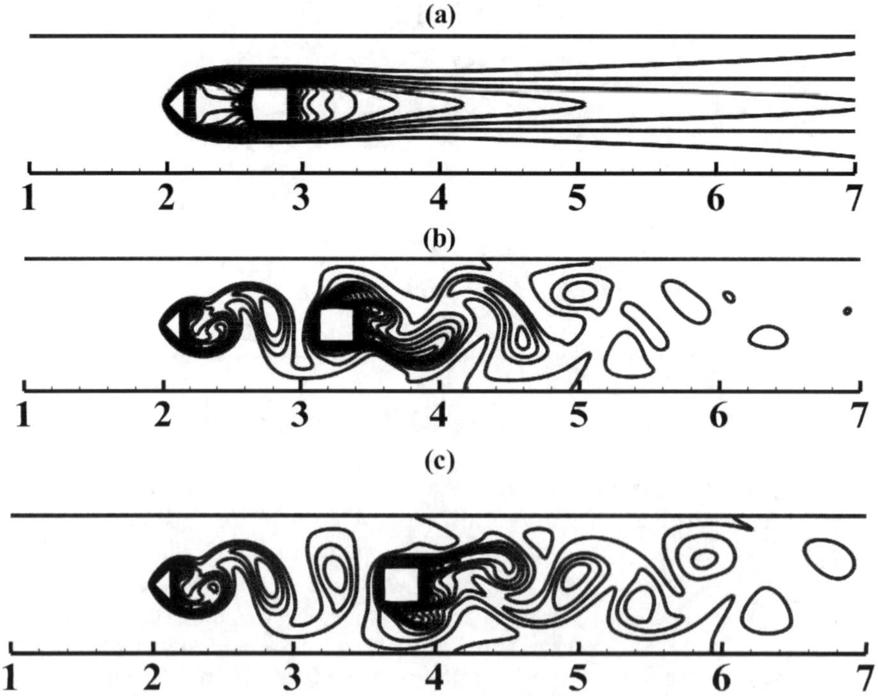

Figure 5. Isotherms for Re=100 (a) w=2d, (b) w=4d, (c) w=6d

Figure 6 shows the variation of the local Nusselt number (Nu) along the surfaces of the square cylinder for w=4d and different Re. An increase of Nu is observed on all faces of the cylinder with the increase of Re. Similarly, we observed that the Nusselt number on the frontal face, where the isotherms are tightly wrapped, is the highest of all cylinder faces; which signifies greater heat transfer. In the front face of the square cylinder, the Nusselt number increases from the stagnation point to achieve a maximum value on the borders. On the upper and lower faces of the square cylinder, the heat transfer decreases along the direction of flow owing the advection of the heat flow downstream by the fluid.. The lowest value of the Nusselt number is found behind the obstacle where it is almost constant.

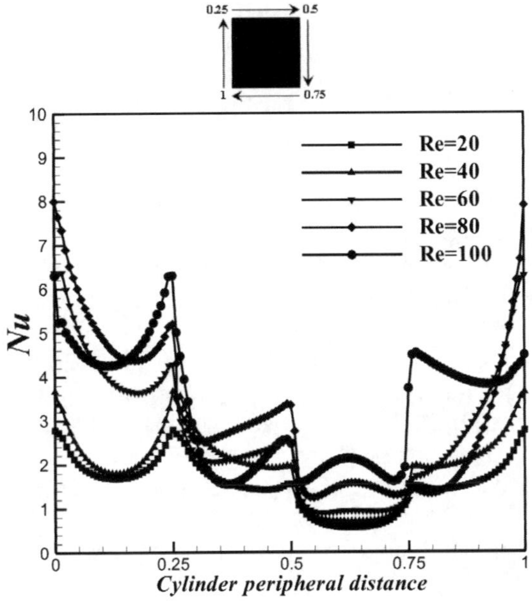

Figure 6. Variation of Nu in square cylinder with different Re for w=4d.

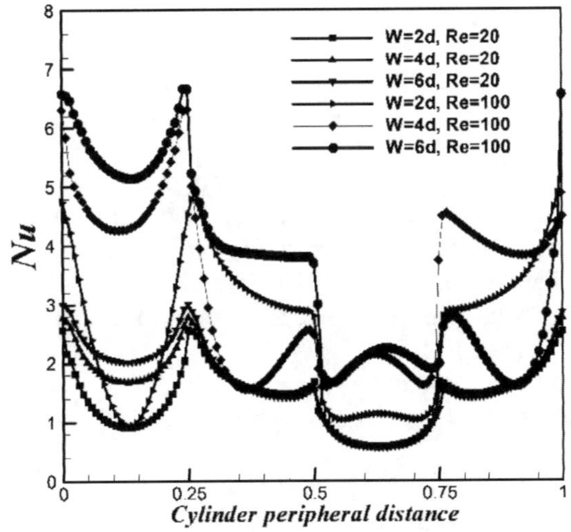

Figure 7. Variation of Nu in square cylinder with different w for Re=20 and Re=100.

Figure 7 illustrates the variation of the local Nusselt number (Nu) along the surfaces of the square cylinder for various values of gap spacing (w=2d, 4d, 6d) and for Re=20 and Re=100. For Re=20, the local Nusselt number present the same comportment for all the gap spaces studied. Whereas, when

Re=100, a higher fluid velocity results in a stronger temperature gradient, which means an increased heat transfer from the cylinder surface. On the front face, Nu increases as d increases, but decreases on both the side faces comparing w=2d and w=4d.

5. Conclusion

The 2D HTLBM was implemented to laminar flow (air: Pr=0.71) and the heat transfer in a channel containing a tandem of triangular cylinder and square cylinder for different Reynolds number and gap spacing. Based on the results, one can conclude that the flow structure is strongly influenced by the Reynolds number and the gap space. The heat transfer is clearly affected by varying gap spaces between the two cylinders, particularly for large Reynolds numbers.

References

[1] Carmo BS, Meneghini JR.: Numerical investigation of the flow around two circular cylinders in tandem. J. of Fluids and Structures, **22** pp. 979–988(2006).
[2] Deng J, Ren AL, Zou JF, Shao XM.: Three-dimensional flow around two circular cylinders in tandem arrangement. Fluid Dynamics Research, **38**, pp. 386–404(2006).
[3] Mahbub Alam Md., Zhou Y.: Strouhal numbers, forces and flow structures around two tandem cylinders of different diameters, J. of Fluids and Structures, **24**, pp. 505–526(2008).
[4] Dehkordi B.G., Moghaddam H.S., Jafari H.H.: Numerical simulation of flow over two circular cylinders in tandem arrangement, J. of Hydrodynamics, **23** (1), pp.114-126(2011).
[5] Pratish P.P. & Tiwari, S.: Numerical Investigation of Laminar Unsteady Wakes Behind Two Inline Square Cylinders Confined in a Channel, Engineering Applications of Computational Fluid Mechanics, **3**, No. 3, pp. 369–385(2009).
[6] Bao, Y., Wua, Q., Zhou, D.: Numerical investigation of flow around an inline square cylinder array with different spacing ratios, Int. J. of computers & fluids, **55**, pp. 118–131(2012).
[7] Moussaoui, M.A., Jami, M., Mezrhab, A., Naji, H.: Computation of heat transfer and fluid flow in an obstructed channel using lattice Boltzmann method, Engineering Computations: Inter. J. for Computer aided Engineering and Software, **27**, (1), pp. 106-116(2010).
[8] Sohankar, A., Mohagheghian S., Dehghan A.A., Manshadi M.D.: A smoke visualization study of the flow over a square A cylinder at incidence and tandem square cylinders, J. of Visualization, **18** (4), pp. 687-703(2015).
[9] Davis, R.W., Moore, E.F., Purtell, L.P.: Anumerical-experimental study of confined flow around rectangular cylinders, Phys. Fluids, **27**, pp. 46–59(1984).
[10] Maiti, D.K.: Dependence of flow characteristics of rectangular cylinders near a wall on the incident velocity, Acta Mech., 222, pp. 273–286(2011).
[11] Maiti, D.K.: Numerical study on aerodynamic characteristics of rectangular cylinders, Ocean Eng, 54, pp. 251–260(2012).
[12] Maiti D.K., Bhatt R.: Vortex shedding suppression and aerodynamic characteristics of square cylinder due to offsetting of rectangular cylinders towards a plane, Ocean Engineering, **82** pp. 91–104(2014).
[13] Zhai Q., Wang H.K. and Yu G.L.: Numerical Study of Flow Past Two Transversely Oscillating Triangular Cylinders in Tandem at Low Reynolds Number, J. of Applied Fluid Mechanics, 10, No. 5, pp. 1247-1260(2017).
[14] Sahin B., Manay E. and Ozceyhan V.: Over all heat transfer enhancement of triangular obstacles, Int. J. of Automotive and Mechanical Engineering IJAME, **8**, pp. 1278-1291(2013).
[15] Wang, S., Zhu, L., Zhang, X., He G.: Flow Past Two Freely Rotatable Triangular Cylinders in Tandem Arrangement, J. of Fluids Engineering, 133, pp. 081202-1(2011).

[16] Zhi. Y. Ng, Vo T., Hussam W.K. and Sheard G.J.: Two-dimensional wake dynamics behind cylinders with triangular cross-section under incidence angle variation. J. of Fluids and Structures, 63, pp. 302-324(2016).
[17] Bouzidi, M., Firdaouss, M., Lallemand P.: Momentum transfer of a Boltzmann-lattice fluid with boundaries, Phys. of Fluids, **13**, pp. 3452-3459(2001).
[18] Lallemand P., Luo L-S.: Lattice Boltzmann method for moving boundaries, J. Comput. Phys., **184**, pp. 406-426(2003).
[19] Mezrhab A., Bouzidi M., Lallemand P.: Hybrid lattice Boltzmann finite-difference simulation of convective flows, Comput. and Fluids, **33**, pp. 623-641(2004).
[20] Lallemand P. and luo L.-S.: Hybrid finite-difference thermal lattice Boltzmann equation, Int. J. Modern Phys. B, **17**, pp. 41-47(2003).
[21] Moussaoui M.A., Jami M., Mezrhab A., Naji H., Bouzidi M.: Multiple relaxation-time lattice Boltzmann computation of channel flow past a square cylinder with an upstream control bi-partition, Int. J. for Numer. Methods in Fluids, **64**, pp. 591-608(2010).
[22] Moussaoui, M.A., Jami, M., Mezrhab, A., Naji H.: Convective heat transfer of laminar flow around two square blocks arbitrary located in a plane channel using hybrid LBM-FDM, Heat and Mass Transf., **45**, pp. 1373-1381(2009).
[23] Moussaoui M.A., Jami M., Mezrhab A., Naji H.: MRT-Lattice Boltzmann simulation of forced convection in a plane channel with an inclined square cylinder, Int. J. of Therm. Sci., **49** pp. 131–142(2010).
[24] Derfoufi S., Moussaoui M.A., Mezrhab A., Fontaine J.P., Bouzidi M.: Natural convection in Γ-shaped cavity using HTLBM, IEEE (2017).
[25] Moussaoui M.A., Mezrhab A., Fontaine J.P., and Bouzidi M.: Study of heat transfer in an enclosure with a square cylinder using Lattice Boltzmann method, Eur. Phys. J. Appl. Phys. **78**, pp(2017).

PID Controller Design with Cuckoo Search Algorithm for Stable and Unstable SOPDT Processes

Durga Vivek[1], **Dr.P.Bharath kumar**[2]
[1] PG student, Dept. Of EEE, JNTUA, Ananthapuram, AP, India
[2]Assistent Professor, Dept. Of EEE, JNTUA, Ananthapuram, AP, India

[1]E-mail:-durga.vivek10@gmail.com
[2]E-mail:- pbk.eee@jntua.ac.in

Abstract. The design of Proportional–Integral–Derivative (PID) controller for the various types of stable and unstable processes with the dynamic model having Second-Order Plus-Dead-Time (SOPDT) is focused in this work. This work initiated with the rejection of load disturbance by connecting PID controller to the system through feedback and then a set-point filter is connected in forward path to achieve proper tracking of reference. The direct synthesis method is applied to get the ranges of PID parameter values and the sharp optimum parameter values which suits for the process can be achieved by introducing Cuckoo Search algorithm which is taken from a class of meta-heuristic algorithms.

KEY WORDS: SOPDT, PID Controller, Direct Synthesis, Cuckoo Search Algorithm.

1. Introduction

In process industries, the equipment like heating boilers, liquid storage tanks, liquid level system and chemical reactors having some transportation delays, analysis loops for composition and recycle loops, etc. due to this, the equipment dynamic response become slow with large time constant [1]. To control such type of systems (FOPDT, SOPDT and integrating systems), some authors proposed variety of design techniques of controllers [2-5].

Using generalized IMC–PID method, Lee and Shamsuzzoha in [6], designed a PID controller to achieve better load disturbance rejection and by using set-point filter reference tracking was achieved. In [1], S. R. Rao and A. S. Rao designed direct synthesis-based controller for integrating process with time delay. Tao and Zhang in [a] implemented two-degree-of freedom (2-DoF) control for open loop unstable process and they tried to improve both set-point filter performances as well as load disturbance reduction. Initially they utilized PD controller and they got optimized values simultaneously and separately. Panda and vijayan in [7] implemented double feedback method to draw into stable operation and to achieve better performance of system. A PID controller which works based on IMC (internal model controller) is implemented in the outer loop to made set-point tracking is satisfactory and the inner loop is tuned with the auto tuning Ziegler-Nichols method to stabilize the system. Vilanova and Alfaro in [9] presented 2DoF PI controller for robust tuning and they used model reference optimization procedure with servo and regulatory closed-loop transfer functions targets. It is concluded from their work that there may be always a tradeoff between performance and robustness [8]. Robustness can be maintained with the factor of maximum sensitivity Ms, for desirable robustness of ISOPDT models and IPDT models, the maximum sensitivity range is $1.2 \leq Ms \leq 2.0$ for stable processes and it should be $Ms \geq 2.0$ for unstable process. In [10], Ajmeri M and Ali A altered Parallel Control Structure (PCS) which was derived from 2-DoF Control Structure for some class of unstable processes in isothermal chemical reactors that reflects multi steady state solutions. They proposed new tuning methods using direct synthesis (DS) method.

In [11], Kasi, V. R. and Majhi used DS method to find the approximate ranges of parameters and they implemented a new nature inspired optimization technique with the name called Particle Swarm Optimization algorithm. In that work the authors achieved satisfactory performance during load disturbance rejection and set-point tracking.

Most of the tuning methods described above concentrate on either load disturbance rejection or monitoring of set-points. It can be achieved by selecting proper objective function during the design. Always a tradeoff takes place between these two (controller and set-point filter) performances. So, the objective here is to design a PID controller which can reduce the load disturbance and to design a set-point filter for satisfactory reference tracking. Initially the approximate ranges of PID controller parameters will be obtained by using Direct Synthesis method and the values of parameters are captured from Cuckoo Search Algorithm.

The proposed approach is contrasted with the current algorithm for Particle Swarm Optimization (PSO) and evaluated under the uncertainty parameter.

2. Mathematical relations

The structure of PID controller is taken in the parallel form as

$$G_c(s) = K_p(1 + \frac{1}{T_i s} + T_d s) \tag{1}$$

Where K_p, T_i and T_d are parameters of the controller to be optimized. The transfer function of set-point filter is given in (2) and β is set-point coefficient.

$$G_f(s) = \frac{1}{1+s\beta T_i} \tag{2}$$

Now let $G(s)$ be a transfer function of an SOPDT stable/unstable process by including steady state gain K and time delay θ, then

$$G(s) = \frac{Ke^{-\theta s}}{(sT_1 \pm 1)(sT_2+1)} \tag{3}$$

where T_1 and T_2 ($T_1 > T_2$) are time constants.

The PID controller is primarily designed for the rejection of load disturbance by implementing the direct synthesis process. Then, for reference tracking, a set point filter is designed. In terms of load disturbance, the desired tf can be taken as

$$\left(\frac{Y}{D}\right)_d = \frac{\frac{T_i s}{K_p}e^{-\theta s}}{\left(\frac{\theta}{T}T_{c1}s+1\right)^3} = \frac{G}{1+GG_c} \tag{4}$$

Where T is the time constant and θ is the effective time delay of the SOPDT process model. Here T_{c1} is the desired tuning parameter.

An expression shown below comes out by including transfer functions G and G_c and from the comparison of the denominator parts

$$s^3\left(\frac{T_1T_2T_i}{KK_p} - T_iT_d\theta\right) + s^2\left(\frac{T_{i(T_1 \pm T_2)}}{KK_p} + (T_iT_d - \theta T_i)\right) + s\left(T_i - \theta \pm \frac{T_i}{KK_p}\right) + 1 = 0$$

$$T_{c1}^3 s^3 + 3T_{c1}^2 s^2 + 3T_{c1}s + 1 = 0 \tag{5}$$

For stable system, the controller parameters from above expression (5) are given in (6), (7) and (8)

$$K_p = \frac{-T_{c1}^3\theta^2 - 3T_{c1}^2T\theta^2 + T_{c1}(3T_1T_2T^2 + 3T_1T^2\theta + 3T_2T^2\theta) + T_1T_2T^3 + T_1T^3\theta + T_2T^3\theta}{\theta^2K(T_{c1}+T)^3} \tag{6}$$

$$T_d = \frac{\theta\left(T_1T_2T^3 + 3T_1T_2T^2T_{c1} + 3T_1T_2T_{c1}^2T - T_{c1}^3(T_1\theta + T_2\theta + \theta^2)\right)}{-T_{c1}^3\theta^2 - 3T_{c1}^2T\theta^2 + T_{c1}(3T_1T_2T^2 + 3T_1T^2\theta + 3T_1T^2\theta) + T_1T_2T^3 + T_1T^3\theta + T_2T^3\theta} \tag{7}$$

$$T_i = \frac{\theta(-T_{c1}^3\theta^2 - 3T_{c1}^2T\theta^2 + T_{c1}(3T_1T_2T^2 + 3T_1T^2\theta + 3T_1T^2\theta) + T_1T_2T^3 + T_1T^3\theta + T_2T^3\theta)}{(T_1\theta + T_2\theta + \theta^2 + T_1T_2)T^3} \tag{8}$$

Similarly, for unstable system the controller parameters from (5) are given in (9), (10) and (11)

$$K_p = \frac{T_{c1}^3\theta^2 + 3T_{c1}^2T\theta^2 + T_{c1}(3T_1T_2T^2 + 3T_1T^2\theta - 3T_2T_{c1}T^2\theta) + T_1T_2T^3 + T_1T^3\theta - T_2T^3\theta}{\theta^2K(2T_{c1}+T)^3} \tag{9}$$

$$T_d = \frac{\theta\left(T_1T_2T^3+3T_1T_2T^2T_{c1}+3T_1T_2T_{c1}^2T+T_{c1}^3(-T_1\theta+T_2\theta+\theta^2)\right)}{T_{c1}^3\theta^2+3T_{c1}^2T\theta^2+T_{c1}(3T_1T_2T^2+3T_1T^2\theta-3T_1T_{c1}T^2\theta)+T_1T_2T^3+T_1T^3\theta-T_2T^3\theta} \quad (10)$$

$$T_i = \frac{\theta(T_{c1}^3\theta^2+3T_{c1}^2T\theta^2+T_{c1}(3T_1T_2T^2+3T_1T^2\theta-3T_1T_{c1}T^2\theta)+T_1T_2T^3+T_1T^3\theta-T_2T^3\theta)}{(T_1\theta-T_2\theta-\theta^2+T_1T_2)T^3} \quad (11)$$

Now to track the reference input, the filter coefficient β (0 to 1) used and below expression is the desired transfer function of closed loop system.

$$\left(\frac{Y}{R}\right)_d = \frac{(1+T_is+s^2T_iT_d)e^{-\theta s}}{\left(\frac{\theta}{T}T_{c1}s+1\right)^4} \quad (12)$$

The actual expression for set-point response is

$$\left(\frac{Y}{R}\right) = \frac{GG_cG_f}{1+GG_c} \quad (13)$$

which can also be written as

$$\left(\frac{Y}{R}\right) = \left(\frac{Y}{D}\right)G_cG_f \quad (14)$$

By rearranging the function in terms of desired transfer function

$$\left(\frac{Y}{R}\right)_d = \left(\frac{Y}{D}\right)_d G_cG_f \quad (15)$$

Substituting the desired transfer function in above expression and by comparing can get

$$\beta = \frac{\theta T_{c1}}{TT_i} \quad (16)$$

After verifying so many outcomes it is observed that during set-point response, β in (16) causes more peak overshoot. So
it is desired to adopt the following set-point filter coefficient.

$$\beta = 1 - \frac{\theta T_{c1}}{TT_i} \quad (17)$$

Where T_{c1} is the desired time constant and can be obtained by applying Cuckoo search algorithm.
The goal here is to reduce Integral Square Error (ISE)[18].
The objective function *ISE* is

$$ISE = \int_0^\infty e_r(t)^2dt \quad (18)$$

where, e_r is the error in system.

3. The cuckoo search algorithm

The Cuckoo search algorithm is an algorithm inspired by nature that belongs to the meta-heuristic community that imitates the behaviour of a family of birds called cuckoo brood parasitism.[12]. These bird types used to lay their own eggs in other host bird nests by raising the chance of hatching by choosing newly spawned nests and eliminating established eggs. By assuming that the eggs are their own, the host birds will take care of those eggs. Some of the host birds might, however, notice that the eggs are not their own and then they can either create new nests in some other new places through the eggs or through them. The similar analogy is used to solve several optimization problems. The steps related to the algorithm are as it follows:

Step1: Specify the parameters for Cuckoo hunt such as number of nests (n), step size (α), discovering probability (pa) and maximum number of generations for conclusion.

Step2: Generate the initial locations of nests (Eggs of host birds) by assigning set of random values to each variable as shown below:

$$nest_{i,j}^0 = Round\left(x_{j,\min} + rand[x_{j,\max} - x_{j,\min}]\right) \quad (19)$$

Where $nest_{i,j}^{(0)}$ the initial value of the j_{th} variable for the i_{th} nest; and $x_{j,min}$, $x_{j,max}$ are the minimum and the maximum permissible values for the j_{th} variable; *rand* is a random number in the interval [0, 1].Because of the discrete nature of the problem, the round function is carried out.

Step3: Based on the quality of new cuckoo eggs produced with Levy flights from their positions, all nests except for the best one is replaced as:

$$nest_i^{(t+1)} = nest_i^{(t)} + \alpha.S(nest_i^{(t)} - nest_{best}^{(t)}).r \tag{20}$$

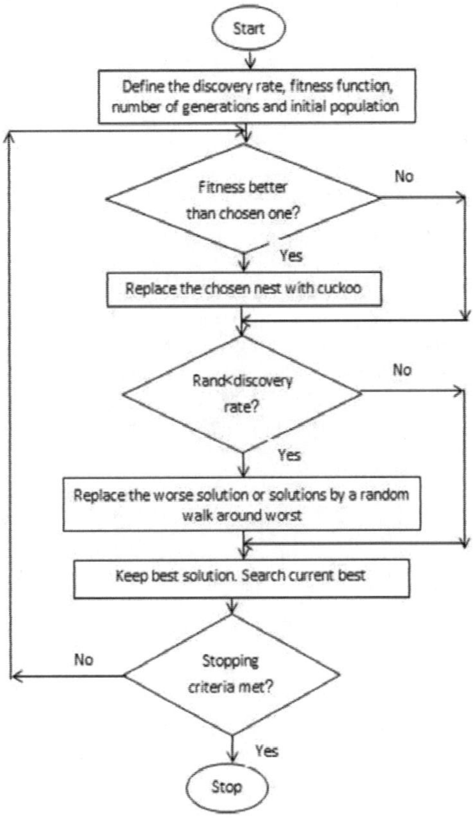

Figure 1. The Cuckoo Search Algorithm Flow Map

Where $nest_i^{(t)}$ is the current position of i_{th} nest, α is step size; *r* is a random number from a regular normal distribution and $nest_{best}^{(t)}$ is the best nest position so far and *S* is a random walk based on the Levy flights and the step length *S* can be determined using following formulas [13].

$$S = \frac{u}{|v|^{1/\beta}} \tag{21}$$

The values of u and v are drawn from normal distribution as:

$$u \sim N(0, \sigma_u^2),\ v \sim N(0, \sigma_v^2) \tag{22}$$

$$\sigma_u = \left\{\frac{\Gamma(1+\beta).\sin(\frac{\pi\beta}{2})}{\Gamma[(1+\beta)/2].\beta.2^{(\beta-1)/2}}\right\}^{1/\beta} \quad , \quad \sigma_v = 1 \tag{23}$$

Step4: The probability matrix used for each solution to discover the aliened eggs given as

$$P_{ij} = \begin{cases} 1 & ; rand < P_a \\ 0 & ; rand \geq P_a \end{cases} \tag{24}$$

Existing eggs will be replaced by random phase walks with a good quality of new produced eggs from their current positions as follows:

$$S = rand.\big(nests(randperm1(n),:) - nests(randperm2(n),:)\big)$$

And $nest^{(t+1)} = nest^{(t)} + S.P$ (25)

Where randperm1 and randperm2 are random permutation functions used for various permutation rows added to the matrix of nests, P is the matrix of probability.

Step5: ***Termination Criterion***

Steps are repeated until a termination condition is reached by producing new cuckoos and finding alien eggs. [14].

4. Results and discussions

Example 1: Consider a stable SOPDT process $G_1(s) = \frac{2e^{-s}}{(10s+1)(5s+1)}$

The parameter values of controller which suit for the process $G_1(s)$ are mentioned in Table 1. For this process M_s value is given as 2.0. The output characteristics of system with the PID-PSO controller [7] along with proposed controller are plotted in Fig 2. Here the load disturbance is given in the form of a step function with the magnitude of − 0.5 exactly at time t = 40sec. The algorithm given by Venkata Ramana gives a better set-point output, but settling time is very high and so system is slow and during disturbance also system takes more time to reach steady state. The suggested method has strong rejection of load disturbance and also takes less rise time and settles time during rejection of load disturbance compared to the PID-PSO method. This method is also undergone parameters variation with the time constants (T_1 and T_2) as −10% and K and θ as + 10% and the corresponding output is shown in Fig 3.

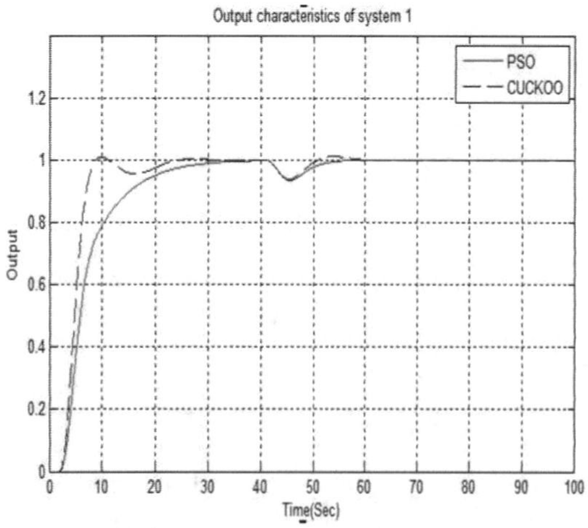

.**Figure 2.** Output characteristics of $G_1(s)$

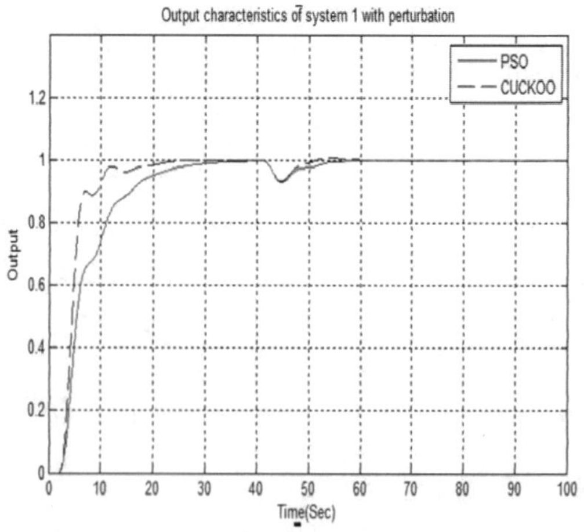

Figure 3. Output characteristics of $G_1(s)$ with perturbation

Example-2: Consider an unstable SOPDT process $G_2(s) = \frac{e^{-0.5s}}{(2s-1)(0.5s+1)}$

The parameter values of controller which suit for the process $G_2(s)$ are mentioned in Table 1. For this process M_s value is given as 2.98. The output characteristics of system with the PID-PSO controller [7] along with proposed PID-CSA controller are plotted in Fig 4. Here the load disturbance is given in the form of a step function with the magnitude of − 0.5 exactly at time t = 40*sec*. The algorithm given by Venkata Ramana gives a better set-point output, but settling time is very high and so system is slow and during disturbance also system takes more time to reach steady state. The proposed method has better load disturbance rejection and also requires less rise time and settling time during load disturbance rejection when compared with the PID-PSO method. This method is also undergone parameters variation with the time constants ($T1$ and $T2$) as −10% and K and θ as + 10% and the corresponding output is shown in Fig 5

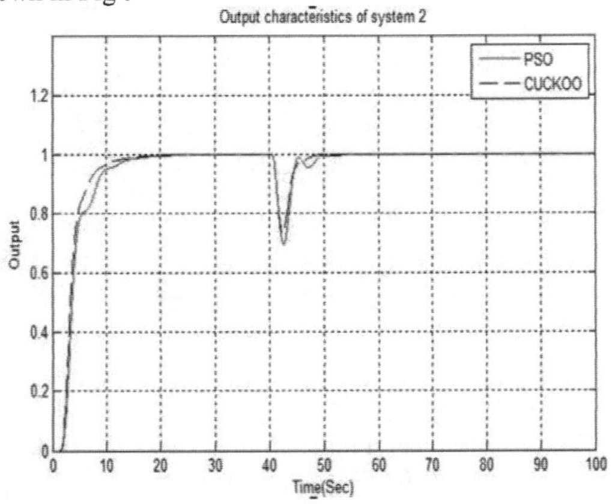

Figure 4. Output characteristics of $G_2(s)$

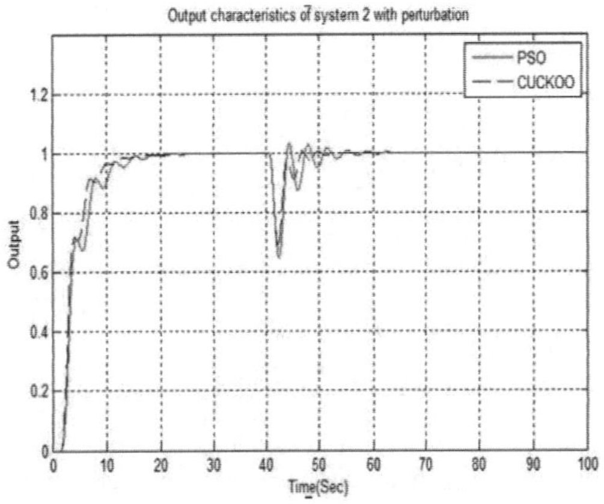

Figure 5. Output characteristics of $G_2\ (s)$ with perturbation

Example 3: Consider another unstable SOPDT process $G_3(s) = \frac{e^{-0.939s}}{(5s-1)(2.07s+1)}$

The parameter values of controller which suit for the process $G_3(s)$ are mentioned in Table 1. For this process M_s value is given as 2.95. The output characteristics of system with the PID-PSO controller [7] along with proposed PID-CSA controller are plotted in Fig 6. Here the load disturbance is given in the form of a step function with the magnitude of − 0.5 exactly at time $t = 40sec$. The algorithm given by Venkata Ramana gives less settling time and less rise time, but the proposed method reduces peak overshoots and also reduces oscillations of response while reaching steady state when comparing with the PID-PSO method. This method is also undergone parameters variation with the time constants (T1 and T2) as −10% and K and θ as + 10% and the corresponding output is shown in Fig 7.

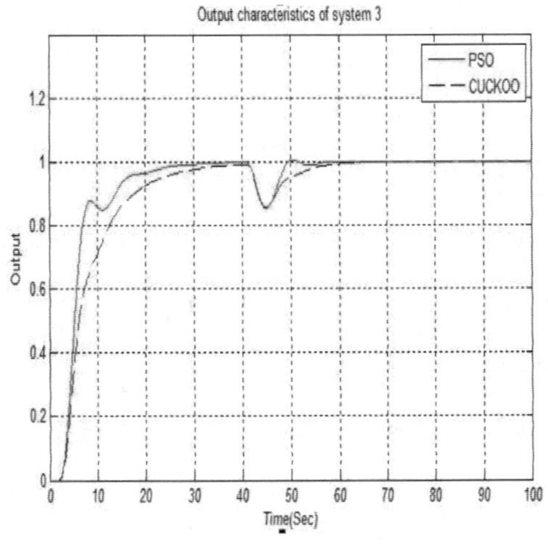

Figure 6. Output characteristics of $G_3\ (s)$

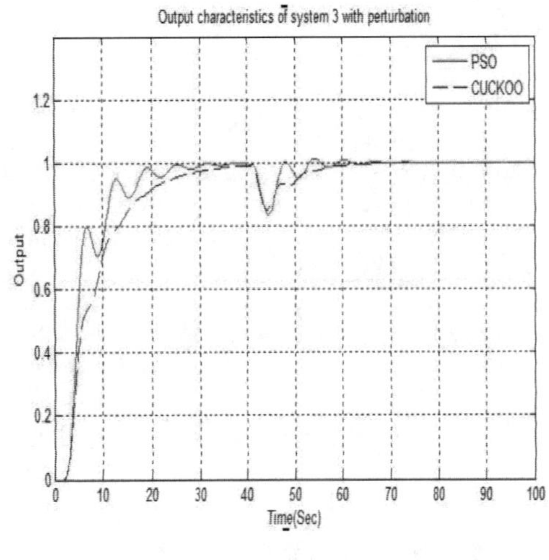

Figure 7. Output characteristics of $G_3\ (s)$ with perturbation

Table 1: PID controller parameters for stable and unstable process

System/process	Algorithm	K_p	K_i	K_d
$G_1(s) = \frac{2e^{-s}}{(10s+1)(5s+1)}$	PSO	7.7213	1.1962	15.0023
	CSA	7.3978	1.8434	15.9410
$G_2(s) = \frac{e^{-0.5s}}{(2s-1)(0.5s+1)}$	PSO	3.0748	0.5804	1.0929
	CSA	3.1770	0.7105	1.4837
$G_3(s) = \frac{e^{-0.939s}}{(5s-1)(2.07s+1)}$	PSO	5.1481	0.7305	7.0039
	CSA	5.0313	0.5085	7.8654

5. Conclusion

In this work the design of PID controller to control the stable and unstable SOPDT processes is performed. The ranges of controller parameters are obtained by applying Direct Synthesis method. The narrowed values of parameters obtained by applying Cuckoo Search optimization algorithm. The proposed algorithm performance compared with the existed PSO algorithm and the algorithm generated optimal values to fulfill the control objective. The Set-Point philtre architecture, however, does not require any additional parameters.

References

[1] A. S. Rao, V.S. R. Rao, and M. Chidambaram. **Direct synthesis-based controller design for integrating processes with time delay**, *J Franklin Inst*, Vol. 346, No.1, pp. 38–56, 2009.

[2] D. Chen, and D.E. Seborg. **PI/PID controller design based on direct synthesis and disturbance rejection**, *Ind Eng Chem Res,* Vol. 41, pp. 4807– 4822, 2002.

[3] S. Nema, and P.K. Padhy. **Identification and cuckoo PI–PD controller design for stable and unstable processes**, *Trans Inst Meas Control*, Vol. 37, No. 6, pp.708–720, 2015

[4] S. Nema, and P.K. Padhy. **PI–PD controller for stable and unstable processes,** *Int J Syst Control Commun*, Vol. 5, No. 2, pp.156–165, 2013

[5] M. Ajmeri, and A. Ali. **Direct synthesis-based tuning of the parallel control structure for integrating processes**, *Int J Syst Sci*, Vol. 46, No.13, pp. 2461–2473, 2015

[6] M. Shamsuzzoha, and M. Lee. **IMC–PID controller design for improved disturbance rejection of time-delayed processes**, *Ind Eng Chem Res*, Vol. 46, No.7, pp.2077–2091, 2007

[7] V.Vijayan, and R.C. Panda. **Design of PID controllers in double feedback loops for SISO systems with set-point filters**, *ISA Trans*, Vol. 51, pp. 514–521, 2012

[8] R. Vilanova, V.M. Alfaro, and O. Arrietia. **Simple robust auto tuning rules for 2-DoF PI controllers**, ISA Trans, Vol.51, No.2, pp. 30–4, 2012

[9] V.M. Alfaro, and R. Vilanova R. **Model-reference robust tuning of 2-DoF PI controllers for first-and second-order plus dead-time controlled processes**, *J Process Control*, Vol. 22, pp. 359–374, 2012

[10] M.Ajmeri, and A. Ali A. **Two-degree freedom control scheme for unstable processes with small time delay**, *ISA Trans*, Vol. 56, pp. 308–326, 2015

[11] V. R. Kasi, S. Majhi, and A. K. Gogoi. **Identification and PID tuning techniques for stable and unstable SOPDT processes**, *International Journal of Dynamics and Control*, Vol. 6, No.3, pp.1194-1206, 2018.

[12] X.S. Yang, and S. Deb. **Cuckoo search via Lévy flights, in Proceedings of world congress on nature & biologically inspired computing (NaBIC2009)**, *IEEE Publications*, USA, pp 210–214, 2009.

[13] X.S. Yang, and S. Deb. **Engineering Optimization by Cuckoo Search,** *Int J Math Model Numer Optim*, Vol. 1, pp 330-343, 2010.

[14] W. Buaklee, and K. Hongesombut. (2013, May). **Optimal DG allocation in a smart distribution grid using Cuckoo Search algorithm**, In 2013 10th International Conference on Electrical Engineering/Electronics, Computer, Telecommunications and Information Technology, pp. 1-6, May,2010

[15] S. Vivek, and M. Chidambaram. **An improved relay auto tuning of PID controllers for critically damped SOPDT systems**. *Chem Eng Commun*, Vol.199, No.11, pp.1437–1462, 2012

[16] H.P. Huang, J.C. Jeng, and K.Y.Luo. **Auto-tune system using single-run relay feedback test and model-based controller design**, *J Process Control,* Vol. 15, No.6, pp. 713–727, 2005

[17] P.K. Padhy, and S. Majhi. **Relay based PI–PD design for stable and unstable FOPDT processes**, *Comput Chem Eng*, Vol. 30, No.5, pp. 790–796, 2006

[18] S. Majhi, and D.P. Atherton. **Autotuning and controller design for processes with small time delays**, *IEE Proc Control Theory Appl*, Vol. 146, No.5, pp. 415–425, 1999

[19] T.Liu, W. Zhang, and D. Gu. **Analytical design of two-degree-of-freedom control scheme for open-loop unstable processes with time delay**, *Journal of Process Control, Vol.* 15.5, pp. 559-572, 2005.

[20] G.T.Chandra Sekhar, D.Vijay Kumar, B. Manmad Kumar, and P.Ramana. **Design and Analysis of BFOA optimized PID controller with derivative filter for frequency regulation in distributed generation system**, *International Journal of Automation and Control*, Vol. 12, No.2, pp.291-323, July, 2018.

[21] Gorripotu T.S., Pilla R. **Golf-Worm Swarm Optimized 2DOF-PIDN Controller for Frequency Regulation of Hybrid Power System**. In: Das A., Nayak J., Naik B., Dutta S., Pelusi D. (eds) Computational Intelligence in Pattern Recognition. Advances in Intelligent Systems and Computing, Vol 1120. Springer, Singapore, Feb. 2020.

Lead-free ferroelectric barium titanate -based thin film for tunable microwave device application

Thu-Hien Vu [1,*], Nguyen T. M. Phuong [1] and Tai Nguyen[2]

[1]International Training Institute for Materials Science, Hanoi University of Science and Technology (HUST)
[2]Materials Research and Technology Department, Luxembourg Institute of Science and Technology, 41, rue du Brill, L-4422 Belvaux, Luxembourg
* Corresponding e-mail address: hien.vuthu@hust.edu.vn

Abstract. Lead-free ferroelectric barium titanate-based ($Ba_{0.85}Sr_{0.15}Zr_{0.1}Ti_{0.9}O_3$, BSZT) thin films have been successfully deposited on Pt- coated silicon substrates by a spin-coating sol-gel method. Microstructure and dielectric respones of the Pt/BSZT/Pt thin film capacitors were investigated in detail. X-ray diffraction analysis results indicated that BSZT thin films were well crystallized in tetragonal perovskite structure with a random crystal orientation at rather low temperature of 650 °C. Dielectric constant and dielectric loss of the thin film capacitors were performed at various bias voltage ranges and frequency from 1kHz up to 5MHz. Hysteresis C-V curves with two maxima confirmed the ferroelectric nature in the film. The BSZT capacitors provide excellent dielectric tunability ~ 42– 69% and high figure-of-merit (FOM) ~ 4–17 at different bias voltages.

1. Introduction

Ferroelectric materials possess many outstanding properties including large dielectric constant, high piezoelectric & pyroelectric coefficients with a wide Curie temperature range. Thus, they have been the subject of intense research for various applications including dynamic random access memories (DRAMs), non-volatile random access memories (NVRAMs), high-frequency (GHz) bypass capacitors and microwave tunable devices [1–3]. For many years, lead-based ferroelectric family (commonly abbreviated as $PbZr_{1-x}Ti_xO_3$ or PZT) of both ceramic and thin film has always been the leading commercial material for above-mentioned applications owing to their superior ferroelectric properties. Nevertheless, the rather high Pb content (~ 60 wt%) raises environment and human health concerns during their utilization. The US and European countries have restricted and even banned electronic devices containing lead element since 2002 [4]. As a consequence, research and development of friendly environmental ferroelectric materials which completely do not contain lead (Pb) with properties that closely match those of PZT has been becoming an urgent issue until nowadays [5,6].

Among the various lead-free ferroelectric materials, such as $BaTiO_3$ (BT), $BiFeO_3$ (BFO), $(Na_{0.5},K_{0.5})Bi_{0.5}TiO_3$ (BNKT), $Na_{0.5}K_{0.5}NbO_3$ (KNN), recently, the complex system barium titanate-based $[Ba_{1-x}Sr_xZr_yTi_{1-y}O_3]$ has received much attention as promising candidates for application in electrically tunable microwave devices at room temperature because of the compatibility between dielectric constant as well as the dissipation factor of these composition systems [7]. Additionally, the advance in recent development of thin and thick ferroelectric film technology also aids to integrate with microelectronic circuits, then leads to substantial miniaturization of microwave device and reduction in producing cost. Therefore, intensive research efforts have been concentrated on this direction during the past decade to obtain desired material properties with high tunability and large figure of merit [8]. However, it is well documented that a small change in composition, porosity, grain size, grain boundaries *etc.*, will lead to remarkable modification the physical properties of ferroelectrics. These effects are of the most important issues for practical applications of ferroelectric materials. In general, the Curie temperature (T_C) of barium titanate-based thin films, which intimately involves to the working temperature of the microelectronic devices, are still low, while the dielectric loss is rather high (up to few ten percent) [9]. Many experimental efforts are still underway in order to clarify the underlying physics for interplays between the microstructure and ferroelectric properties [10–12]. For instance, Chong *et al.* reported that 40% aluminum oxide doped $Ba_{0.5}Sr_{0.5}TiO_3$ thin film on (100)-LaAlO3 substrates by the pulsed laser deposition technique reduced the loss tangent from 3% to 1.1% and enhanced the FOM ~ 7.33 up to 15.45 in comparison to the pure one [13]. For $Ba_{1-x}Sr_xTiO_3$ (BST) ferroelectric materials, Ba/Sr ratio was attributed to strongly effect on their dielectric properties. Recently, Kuppusami and co-workers noted that decreasing Sr content from x = 0.5 to 0.3 resulted to shift of Curie point near to room temperature and reduced the oxygen vacancy concentration approximately one magnitude order, however, the better dielectric properties (larger dielectric constant and lowest loss tangent) were observed for x=0.4 [14]. There still exists considerable debate about the "doping effect". Hence, it is necessary to put more effort into understanding such role of dopant on both microstructure and the overall properties of ferroelectric materials based on barium titanate.

In this study, we focused on tailoring composition *via* simultaneously substituting Ba and Ti elements in barium titanate (BT) with 15% mol of Sr and 10% mol of Zr to investigate its influence on the microstructure and dielectric properties of the resulting BSZT thin films. The thin film samples were fabricated on plantinum electrode (Pt- coated Si substrates) *via* the simple sol-gel techniques. The dielectric response as well as the tenability of the BSZT films were studied as a function of frequency and different external applied electric field ranges.

2. Experimental

2.1. BSZT thin film fabrication

$Ba_{0.85}Sr_{0.15}Zr_{0.1}Ti_{0.9}O_3$ (BSZT) thin films were fabricated on $Pt/Ti/SiO_2/Si(100)$ substrate via sol-gel techniques. Starting materials for preparing BSZT sol are salts $Ba(CH_3COO)_2.3H_2O$, $Sr(CH_3COO)_2$, and alkoxides $Zr(OC_3H_7)_4$ (Merck, 70% in propanol) and $Ti(OC_3H_7)_4$ (Merck, 98%). Aetic acid (CH_3COOH, glacial) was used as solvent to dissolve barium and strontinum axetate salts. While 2-methoxyethanol ($CH_3OCH_2CH_2OH$, 2-MOE) was used for diluting the two alkoxides. After dissolving $Ba(CH_3COO)_2$ and $Sr(CH_3COO)_2$ in glacial acetic acid, then the so-called mixture 1 was heated up at 120 oC for five hours with a constant stirring. The mixture 2 included $Zr(OC_3H_7)_4$ and $Ti(OC_3H_7)_4$ with the desired molar ratio was diluted using 2-MOE solvent at 120 oC during stirring for at least three hours. Mixutere 1 and 2 were then added together under continuous stirring at 80 oC to get a homogeneous and transparent solution. To adjust viscosity and stabilize the sol, formamide (NH_2CHO) and acetylacetone ($CH_3COCH_2COCH_3$) were added to the resulting solution. The whole solution was then refluxed for at least 1 hour to obtain homogeneous mixing among the components. In final, BSZT sol was filtered through the 2 μm diameter filter paper to get stock solution.

The above-mentioned stock sol was coated onto $Pt/Ti/SiO_2/Si$ substrates at 4000 rpm for 30 s by a Spinner 150i. The as-deposited BSZT thin films were followed by a heating procedure: first dried at a low temperature of 150 °C for 12 min to get rid of the water vapour and residual solvent; then pyrolyzed at 450 °C in air for 10 min. In order to achieve a desired film thickness (~ 400 nm), the coating and deposition process were usually doned from 6 to ten times. The films finally were crystallized at about 650 °C in the air by rapid annealing (flowing rate was ~25 °C/min) for 30 min to obtain polycrystalline perovskite phase.

2.2. Microstructure and electrical characterizations

The crystal structure of the BSZT powder samples was examined by Synchrotron X–ray diffraction (SXRD) experiments with a Beamline SAXS of the Synchrotron Light Research Institute (Thailand). The corresponding crystal structure for BSZT thin films were examined by X-ray diffractometer using CuKα radiation (XRD; Philips X'pert). The surface morphology of the films was performed by scanning electron microscopy (SEM) using a Nova NanoLab 600 microscope at 15 kV. For electrical measurements, series of capacitors with structure metal–insulator–metal (MIM) were patterned. Pt top electrodes (100 nm thick) were sputtered through a shadow mask ~100×100; 150×150 and 200×200 μm^2 in area. The capacitance–voltage (C–V) and capacitance– frequency (C–f) characterizations were measured via a four probes-station equipped with Keithley SCS-4200 Semiconductor characterization system. The voltage limitation of this equipment is ± 30V.

3. Results and discussion

Figure 1 shows the X-ray diffraction (XRD) results of BSZT thin films deposited on Pt electrode, and SXRD pattern of the BSZT powders prepared with the same precursor solution. The film and the powder exhibited single phase perovskite structure with no detectable impurity or secondary phase such as pyroclore phase. The XRD patterns of BSZT thin films are indexed in tetragonal perovskite structure with space group *Pmm4* according to the analysis of SXRD data of BSZT powder [15]. Four well-defined peaks, identified as the (100), (110), (200), and (210) diffraction planes of BSZT tetragonal perovskite structure, are clearly observed in the XRD patterns.

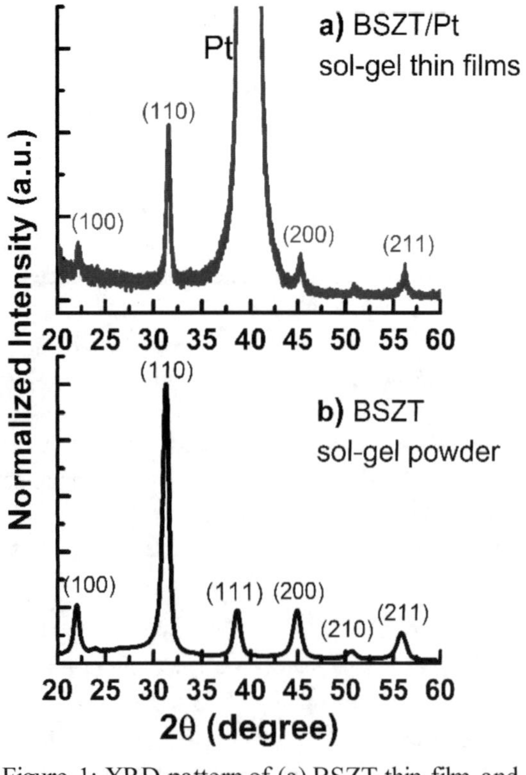

Figure 1: XRD pattern of (a) BSZT thin film and (b) corresponding nano powder.

In order to obtain the information concerning the crystallite size, D, and preferential film orientation from XRD data, the Scherrer's formula and texture coefficient TC(hkl) according to the Harris's expression as given by equation 1 and 2 [16]:

$$D = \frac{K\lambda}{\beta cos\theta} \quad (1)$$

$$\mathrm{TC(hkl)} = \frac{\frac{I(\mathrm{hkl})}{I_o(\mathrm{hkl})}}{n\sum\frac{I(\mathrm{hkl})}{I_o(\mathrm{hkl})}} \quad (2)$$

where K≈ 0.9, β is the full width at half-maximum (FWHM) of highest (110) diffraction peaks in radian, θ is the Bragg angle, and λ is the CuKα wavelength (1.54439 Å). For the equation 2, I(hkl) is the integrated intensity of (hkl) plane of the BSZT thin films, I_o(hkl) is the relative intensity taken from JCPDS card no.: 34-0129 and n is the number of diffraction peaks. The calculated data for the crystallite size and the preferred orientation of the BSZT thin films is listed in table 1.

It is well known that X-ray intensities are function of atomic structure. Any certain deviation in the texture coefficient from unity for a particular Miller plane corresponds to change in atomic structures in respect to the plane. The result of texture analysis shows that the BSZT thin films on Pt substrate exhibit randomly oriented crystallites. And the crystallite sizes in the BSZT thin films as estimated from XRD data is around 25.18 nm.

Table 1. Parameters on crystallite size, *D*, and texture coefficients TC(hkl) of the BSZT thin films and nano powder.

Planes (hkl)	2θ		TC(hkl)
	BSZT powder	BSZT/Pt	BSZT/Pt
(100)	22.00	22.13	1.15
(110)	31.28	31.56	1.03
(111)	38.62	-	-
(200)	44.91	45.23	1.02
(210)	50.64	50.95	1.27
(211)	55.88	56.20	1.28
Crystallite size, D (nm)	25.18		

It is well known that X-ray intensities are function of atomic structure. Any certain deviation in the texture coefficient from unity for a particular Miller plane corresponds to change in atomic structures in respect to the plane. The result of texture analysis shows that the BSZT thin films on Pt substrate exhibit randomly oriented crystallites. And the crystallite sizes in the BSZT thin films as estimated from XRD data is around 25.18 nm.

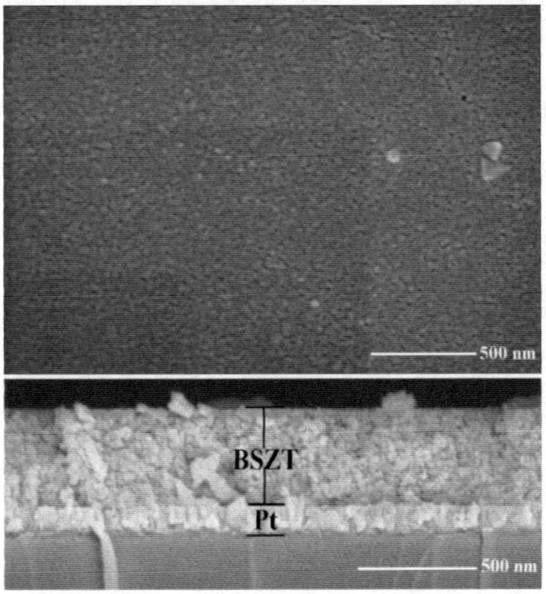

Figure 2: Surface morphology and cross-sectional SEM images of BSZT films on Pt electrode.

The surface morphology of the BSZT thin films were examined by SEM over a 1×1 μm2 scan area. Figure 2 presents the typical SEM images of BSZT films on Si substrate.

The BSZT thin film possesses a smooth and dense surface with grain sizes ranging from 27 to 49 nm. The lower parts of figure 2 show the cross-section profiles of the BSZT film. As seen in the image, the thickness of the BSZT is approximately 355 nm for 10 layers, which turns out approximately 35.5 nm for each coating layer. The BSZT film on Pt-coated Si substrate exhibits coherently interfaces between the films and the substrates.

The dependence of capacitance measurements on the external applied dc voltages for Pt/BSZT/Pt thin film capacitor structures were performed with a small ac signal of 30 mV amplitude and *dc* bias voltage up to ± 30V. The corresponding dielectric

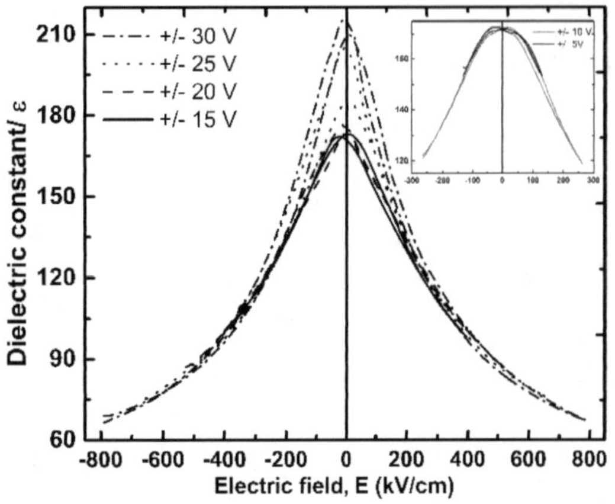

Figure 3: Dielectric constant *vs* electric field of Pt/BSZT/Pt capacitors. Insert shows the butterfly shape at lower sweeping bias voltage ranges.

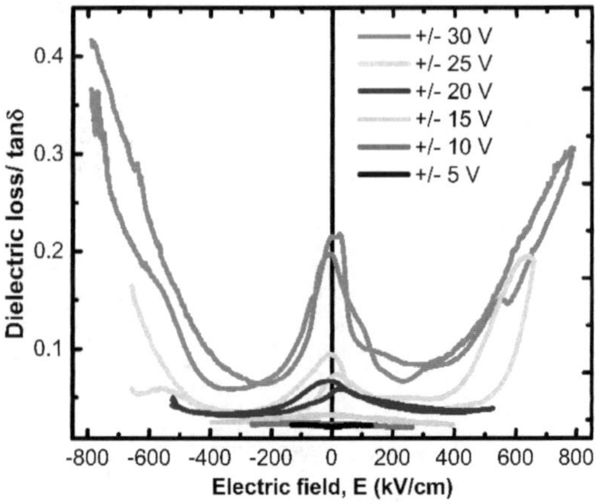

Figure 4: Dependence of dielectric losses of BSZT films on the applied *dc* electric field.

constant−electric field (ε−E) and dielectric loss−electric field (tanδ−E) curves as shown in figure 3 and figure 4 were calculated from the capacitance datas using the paralleled plate capacitor formula and assuming the homogeneous film thickness.

As observed, the dielectric constant curves exhibited apparent hysteresis, the so-called butterfly shape with two typical maxima in the lower electric field region (-400 ≤ E ≤ 400 kV/cm, see insert in figure 3). As the occurrence of butterfly hysteresis in the C−V curves is usually described to the process of ferroelectric polarization switching, the films are ferroelectric in nature at low applied voltages. Increasing the *dc* bias voltage, the butterfly-shape trails off gradually decrease and disappears completely in higher electric field. These observation was also noted by Bao and co-worker for BZT-xBCT thin film [17], and Chan *et al.* for (Ba,Sr)(Zr,Ti)O_3 samples [18]. This phenomenon is often attributed to the presence of local polar nanoregions (PNRs) or micropolar clusters, which are associated with the typical ferroelectric relaxor behaviour due to the heterogeneity of polarization domain structure [19,20]. In addition to the reduced ferroelectricity of the sample, the dielectric constants of the capacitors are found to increase with increasing sweeping voltage range. The maximum capacitance densities at zero bias voltage of Pt/BSZT/Pt measured at the maximum applied voltage (30 V) is ~ 0.50 μF/cm^2. It corresponds to the dielectric constant, ε of ~ 214.58 and dielectric loss, tanδ ~ 0.186 (figure 4 and table 2). In contrast to the capacitance density trend, tanδ varies considerably with applied electric field. Specifically, the tanδ increases and decreases suddenly when the capacitor was switched to the higher electric field range (E ≥ ± 600 kV/cm). It was stated that dielectric loss of parallel-plate ferroelectric thin-film capacitor contributes from many sources. In sol-gel BSZT thin films, it cannot rule out the presence of defects such as oxygen vacancies. Beside, residual stress due to lattice mismatch or different thermal coefficient between thin film and substrate, free carriers, and interfacial effects between film material and top/bottom electrodes are all possible sources of higher loss. Experimental studies aimed to elucidate the mechanisms of dielectric loss as well as minimize them are currently underway on the BSZT thin films deposited on different types of electrode (noble metal Pt, Au and conductive oxide $LaNiO_3$ and $SrRuO_3$) and different multi-valent dopants such as Mn and La.

In order to estimate the possibility of these materials for tunable capacitor applications, the overall voltage dependence of capacitance as well as the dielectric loss for Pt/BSZT/Pt capacitor are evaluated. Interestingly, figure 5 indicates that the Pt/BSZT/Pt capacitor provides rather excellent tenability with applied dc electric field, E. The tunablity of capacitance is defined as $\{\delta C = \left[1 - \frac{C(E)}{C(o)}\right]\}$, where $C(E)$ and $C(o)$ are the capacitance at nonzero and zero electric field, respectively. For E = 800 kV/cm (i.e. at a bias voltage of ± 30 V), the δC reaches to maximum value ~ 69 %. The figure of merit $[FOM = \frac{\delta C}{tan\delta}]$ is ~ 4 – 17 for BSZT capacitor in the higher electric field region (table 2). These obtained values are comparable to those values (δC ~ 40–50% and FOM~ 6–16) observed for SRO/(Ba,Sr)TiO_3/SRO capacitor [21].

Table 2. Dielectric parameters of BSZT thin films obtained at different bias voltages.

Bias Voltage (V)	Dielectric constant, ε (E=0V)	Dielectric loss, tanδ (E=0V)	Tunability, δC (%)	FOM
± 15	172.87	0.024	42.08	17.03
± 20	175.72	0.054	53.13	9.84
± 25	204.51	0.071	63.12	8.89
± 30	214.58	0.186	69.09	3.83

Figure 6 displays the dielectric constant and dielectric loss of BSZT thin films as a function of frequency in the range from 1 kHz – 5 MHz. The thin-film capacitor exhibited a noticeable dielectric dispersion and dielectric relaxations in the measured frequency range with a peak in tanδ–f curve

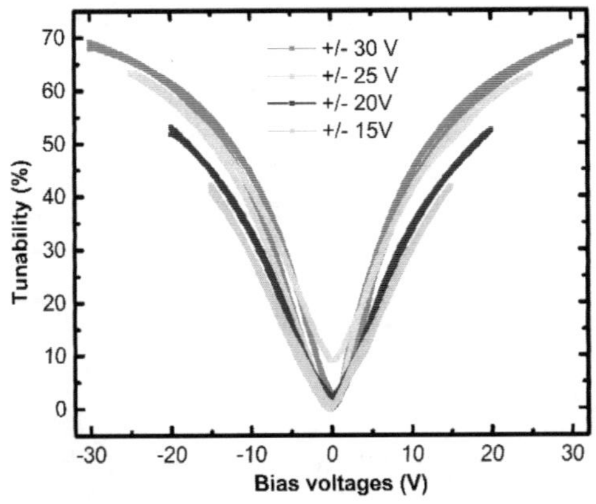

Figure 5: Voltage dependence of tenability of BSZT thin films performed at 100 kHz.

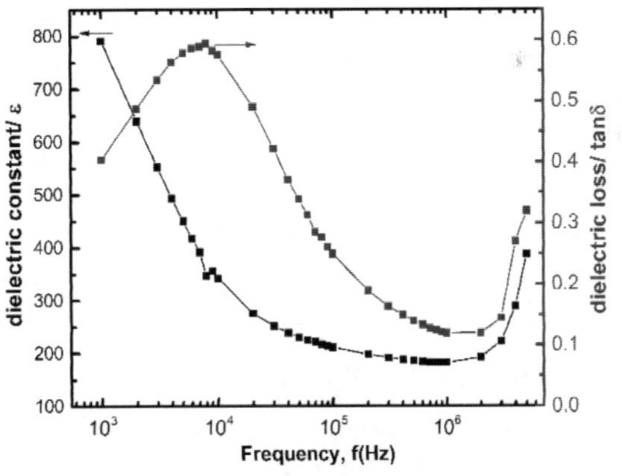

Figure 6: Frequency dependence of dielectric constant and loss tangent of Pt/BSZT/Pt capacitor measured at room temperature.

around 10 kHz.

In dielectric materials, several mechanisms are responsible for polarization such as electronic, ionic, dipole reorientation-related, and space charge or interfacial polarization [22,23]. The degree to which each mechanism contributing to the overall polarization of the material depends on the applied frequency range. Generally, electronic and ionic relaxation are often related to very fast switching of weak dipoles. These oscillations can be measured at very high frequencies from gigahertz (GHz) to petahertz (PHz) (10^9–10^{15} cycle/s). Permanent dipole reorientation can be tuned only frequencies from megahertz (MHz) to GHz (10^6–10^9 cycle/s) [2]. However, the dielectric properties, in fact, are strongly influenced by the complexity of porosity, the grain boundary densities, domain wall motion, crystallographic structure, and ionic space charge carriers. Therefore, it is speculated that the relaxation observed at several ten kHz spectrum in figure 6 might relate to ionic space charge carriers, such as the oxygen vacancies and defects generated during film growth, interface polarization located at grain boundary or the interface between the film and electrode. Waser and co-authors reported that oxygen vacancy and related defects were the main reasons for higher dc conductivity and increasing loss tangent in oxide ferroelectric materials in the low frequency region [24]. Because of their positively charged with respect to the regular lattice, thus oxygen vacancies experience an electromigration toward the cathode under a *dc* electric field. The oxygen vacancies pile up at the cathode and are supposed to be compensated by electrons injected from external applied bias voltage. This could be the reason why the tanδ observed in figure 4 in the negative electric field region is often greater than that of the dielectric loss in the positive electric field band. In addition, during the degradation, coloration effects near the electrodes is also observed quite clearly. Specifically, the BSZT thin film area under measured capacitor structure was blistered, which resulted in the top electrode peeling off the thin film surface.

4. Conclusions

Lead-free ferroelectric barium titanate-based (BSZT) thin films were fabricated on metal $Pt/Ti/SiO_2/Si(100)$ substrates by the effective sol-gel method. The films are of polycrystalline perovskite phase with the grain sizes ranging from 27 to 49 nm and coherent grain boundary. Butterfly-shaped (C−V)/or (ε−E) characteristics revealed the ferroelectricity of thin films. Capacitance of Pt/BSZT/Pt can be easily tuned in the wide voltage range without broken structure. The best parameters to compare the quality of the films are obtained with a tenability of 42% and figure of merit, FOM ~ 17 at 15 V. This result demonstrates that the as-deposited BSZT thin film on $Pt/Ti/SiO_2/Si(100)$ substrate are highly potential candidate to apply for electrically tunable devices.

Acknowledgements

This research is funded by the Hanoi University of Science and Technology (HUST) under project number T2020-SAHEP-020.

References

[1] J.F. Scott, Applications of Modern Ferroelectrics, Science 315 (2007) 954–959.
[2] K. Uchino, Ferroelectrics Devices, 2nd Ed., CRC Press, Inc., FL, United States, 2019.
[3] L.W. Martin, A.M. Rappe, Thin-film ferroelectric materials and their applications, Nat. Rev. Mater. 2 (2017) 16087.
[4] 'EU-Directive 2002/95/EC: Restriction of the Use of Certain Hazardous Substances in Electrical and Electronic Equipment (RoHS), Off. J. Eur. Union. 46 (2003) [L37] 19–23.
[5] W.W. Wolny, European approach to development of new environmentally sustainable electroceramics, Ceram. Int. 30 (2004) 1079–1083.

[6] T.R. Shrout, S.J. Zhang, Lead-free piezoelectric ceramics: Alternatives for PZT?, J Electroceram. 19 (2007) 111–124.
[7] C.L. Sengupta, S. Sengupta, Breakthrough advances in low loss, tunable dielectric materials, Mat Res Innov. 2 (1999) 278–282.
[8] A.K. Tagantsev, V.O. Sherman, K.F. Astafiev, J. Venkatesh, N. Setter, Ferroelectric Materials for Microwave Tunable Applications, J. Electroceramics. 11 (2003) 5–66.
[9] S. Ezhilvalavan, T. Tseng, Progress in the developments of (Ba,Sr)TiO_3 (BST) thin films for Gigabit era DRAMs, Mater. Chem. Phys. 65 (2000) 227–248.
[10] X.G. Tang, K. Chew, H.L.W. Chan, Diffuse phase transition and dielectric tunability of $Ba(Zr_yTi_{1-y})O_3$ relaxor ferroelectric ceramics, Acta Mater. 52 (2004) 5177–5183.
[11] J. Chen, Y. Zhang, C. Deng, X. Dai, L. Li, Effect of the Ba/Ti Ratio on the Microstructures and Dielectric Properties of Barium Titanate-Based Glass–Ceramics, J. Am. Ceram. Soc. 90 (2009) 1350–1353.
[12] S. Tio, V. Ramana, L. Feng, W. Lin, B.S. Murty, Effect of grain size on dielectric and ferroelectric properties of, J. Adv. Ceram. 4 (2015) 46–53.
[13] K. Chong, L.B. Kong, L. Chen, L. Yan, C.Y. Tan, T. Yang, et al., Improvement of dielectric loss of doped $Ba_{0.5}Sr_{0.5}TiO_3$ thin films for tunable microwave devices, J. Appl. Phys. 95 (2004) 1416.
[14] B. Vigneshwaran, P. Kuppusami, S. Ajithkumar, H. Sreemoolanadhan, Study of low-temperature dependent structural, dielectric, and ferroelectric properties of $Ba_xSr_{1-x}TiO_3$ (x = 0.5, 0.6, 0.7) ceramics, J. Mater. Sci. Mater. Electron. 31 (2020) 10446–10459.
[15] P.T. Nguyet, N. Thi, M. Phuong, V.T. Hien, V.N. Hung, Preparation and Microstructure of Acetate-based Lead-free BSZT Ferroelectric Thin Films Using Sol-gel Technique, VNU J. Sci. Math. - Phys. 35 (2019) 55–61.
[16] G.B. Harris, X. Quantitative measurement of preferred orientation in rolled uranium bars, Philos. Mag. Ser. 7. 43 (1952) 113–123.
[17] Y. Lin, G. Wu, N. Qin, D. Bao, Structure, dielectric, ferroelectric, and optical properties of $(1-x)Ba(Zr_{0.2}Ti_{0.8})O_3-x(Ba_{0.7}Ca_{0.3})TiO_3$ thin films prepared by sol–gel method, Thin Solid Films. 520 (2012) 2800–2804.
[18] N.Y. Chan, D.Y. Wang, Y. Wang, J.Y. Dai, H.L.W. Chan, The structural and in-plane dielectric/ferroelectric properties of the epitaxial (Ba, Sr)(Zr, Ti)O_3 thin films, J. Appl. Phys. 115 (2014) 234102.
[19] P. Taylor, G.H. Haertling, Ferroelectrics PLZT electrooptic materials and applications — a review, Ferroelectrics. 75 (1987) 25–55.
[20] I. Stolichnov, A. Tagantsev, N. Setter, S. Okhonin, P. Fazan, J.S. Cross, et al., Dielectric breakdown in (Pb,La)(Zr,Ti)O_3 ferroelectric thin films with Pt and oxide electrodes, J. Appl. Phys. 87 (2000) 1925.
[21] Y.A. Boikov, T. Claeson, Dielectric Response of $Ba_{0.05}$ $Sr_{0.95}TiO_3$ (110) Films to Variations in Temperature and Electric Field, Phys. Solid State. 57 (2015) 957–961.
[22] A.A. Bokov, Z.-G. Ye, Dielectric Relaxation in Relaxor Ferroelectrics, J. Adv. Dielectr. 2 (2012) 1241010.
[23] L. Dissado, Chap. 10: Dielectric Response, in: S. Kasap, P. Capper (Eds.), Handb. Electron. Photonic Mater., 2nd Ed., Springer, Cham, 2017: pp. 219–245.
[24] R. Waser, T. Baiatu, K.-H. Hardt, dc Electrical Degradation of Perovskite-Type Titanates: I, Ceramics, J Am Ceram Soc. 73 (1990) 1645–1653.

Refining Concrete's micro-structure by enzymatically-induced carbonate precipitation through urease activity of bacteria

[1]V Srinivasa Reddy, [2]Ch Sasikala, [3]S Shrihari

[1]Professor of Civil Engineering, GRIET, Hyderabad, Telangana, India
[2]Professor of Environmental Science and Technology, JNTUH, Hyderabad, Telangana, India
[3]Professor of Civil Engineering, VJIT, Hyderabad, Telangana, India.

Corresponding Author: vempada@gmail.com

Abstract. This paper presents the mechanism of improving the microstructure of concrete by enzymatically-induced carbonate precipitation through urease activity of bacteria *Sporosarcina pasteurii.* This calcium carbonate or calcite crystal precipitated during the biological activity fills the pores in the concrete imparting the concrete the dense microstructure. The presence of 10^5 per milliliter of mixing water yields high compressive strength due to precipitation of highest amount of calcium crystals in the concrete pores thereby plugging the micro and macro pores in the concrete refining the micro-structure of the concrete. SEM images, XRD plots and thermo-gravimetric analysis confirmed the presence of calcite mineral in the bacterial mortar specimens. The presence of spores of *Sporosarcina pasteurii* in the bacterial mortar specimens of 365 days old are found through phase contrast microscopic images. This confirms the sporulation ability of bacteria *Sporosarcina pasteurii.* Concrete specimens induced with bacteria showed considerably less water absorption capacity compared to normal specimens. This decrease in water absorption capacity of bacteria induced concretes is credited to the reduction of pores in the concrete due to calcite mineral precipitation filling the pores reducing the typical pore radius of concrete. Average pore diameter in bacteria incorporated concrete decreases drastically due to calcite mineral precipitation in the pores because of microbial metabolic activity. The total pore volume also decreased in case of bacteria incorporated concrete samples.

Keywords: Bacterial concrete, MICP, durability, Sporosarcina pasteurii, bio-mineralization

1. Introduction

Bio-mineralization is a chemical process during bacterial metabolic activity yielding minerals in the crystalline form. Microbial induced calcite precipitation (MICP) deposits calcium carbonate in the concrete pores and cavities by urease enzyme in bacteria by hydrolyzing the urea which is a calcium based system. Different species of bacteria producing urease enzyme are used for calcite precipitation such as B. subtilis, Sporosarcina pasteurii (formerly Bacillus pasteurii), Bacillus sphaericus, Bacillus cophnii, Aerobacter aerogenes, Shewanella , B. megaterium, B. thuringiensis, Bacillus sp. CR2, Deleya halophila, Kocuria flava CR1, Methylocystis parvum, Myxococcus xanthus, Proteus mirabilis, Halmonas eurihalina, Pseudomonas denitrificans, Sporosarcina ginsengisoli, Bacillus pseudifirmus, L. sphaericus CH5, Bacillus cereus, Sporolactobacillus sp., and Microbacterium sp. Certain pathogenic bacteria such as Helicobacter pylori, Proteus vulgaris, Staphylococcus aureus, Proteus vulgaris and Pseudomonas aeruginosa also produce urease and are involved in the formation of calcium carbonate.

2. Objectives

The objective of the study is to refine the concrete's micro-structure by enzymatically-induced carbonate precipitation through urease activity of bacteria. To demonstrate the efficacy of the enhanced micro-structure various experiments are conducted such as determination of optimum cell count to be used for effective production of calcium carbonate crystals, water absorbtion capacity and porosity studies. High quantities of calcite mineral presence is observed using SEM, XRD and thermo-gravimetric images.

3. Methodology

3.1 Cell Concentration

This section of study identifies the most optimum cell count required to produce maximum amount of calcium carbonate crystals. So various cell counts of *Sporosarcina pasteurii* solutions are prepared and used as mixing water to prepare concrete mortar cubes and assessed for their compressive strength to determine the cell concentration of bacteria *Sporosarcina pasteurii.* The presence of 10^5 per milliliter of mixing water yields high compressive strength due to precipitation of highest amount of calcium crystals in the concrete pores thereby plugging the micro and macro pores in the concrete refining the micro-structure of the concrete.

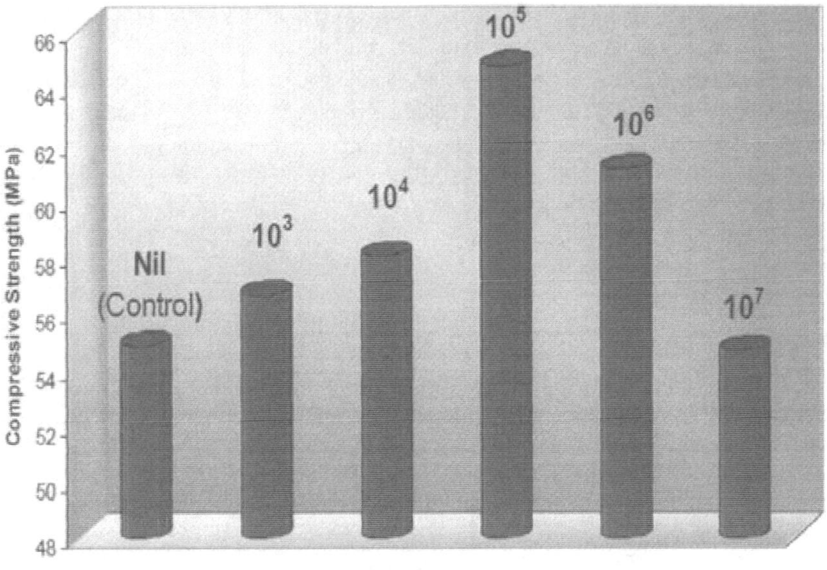

Figure 1. Optimum cell count for maximum calcite crystals precipitation

3.2 Factors affecting the properties of bacterial concrete

The investigations are carried out to understand the role of each constituent of bacterial concrete and their synergic effect.

Table 1. *Factors affecting the properties of bacterial concrete*

Cement mortar cube sample made with	Cement Mortar Cube Compressive strength (MPa)	
	@ 28 days	% increase
1. Distilled water	54.3	-
2. *Sporosarcina pasteurii* + Nutrient broth solution	56.2	3.5
3. Nutrient broth solution only	55.1	1.5
4. *Sporosarcina pasteurii* + Nutrient broth + $CaCO_3$ supplement solution	65.3	20.3
5. *Sporosarcina pasteurii* + Nutrient broth + 5% bwc Calcium lactate additive	68.3	25.8
6. *Sporosarcina pasteurii* + Nutrient broth + $CaCO_3$ supplement solution + 5% bwc Calcium lactate additive	68.4	25.9
7. Distilled water and 5% bwc Calcium lactate additive	60.3	11.1

*bwc – by weight of cement

It can be observed that the concrete cube strength made with normal distilled water (sample 1) is 54.3 MPa at 28 days age of curing. When bacteria and nutrients suspension is used as mixing water instead of normal distilled water there is not significant improvement in compressive strength but when calcium nutrient supplement in added to bacteria and nutrients broth, the improvement of strength is very substantial due to abundant availability of calcium source for calcite mineral precipitation which will eventually fills the pores in the cementitous materials leading to dense microstructure. Since calcium supplement in the form of nutrient is expensive so an inexpensive calcium source such as calcium lactate is added to mortar dry mixture (5% bwc) to facilitate the mechanism of bio-mineralization in bacterial concrete. This study also proves that the expensive component calcium nutrient supplement can be replaced with calcium lactate to obtain similar strength development. Table 1 shows the effect of each constituent of bacterial concrete on the strength development and it proves that addition of calcium complement will facilitate favorable environment for the bacterial activity.

3.3 Scanning Electron Microscopic studies

Scanning electron microscopic studies on various samples made with various cell counts of *Sporosarcina pasteurii* showed that 10^5 cell count of *Sporosarcina pasteurii* addition to concrete precipitates maximum amount of calcite crystals giving the concrete optimal dense micro-structure as shown in figure 2.

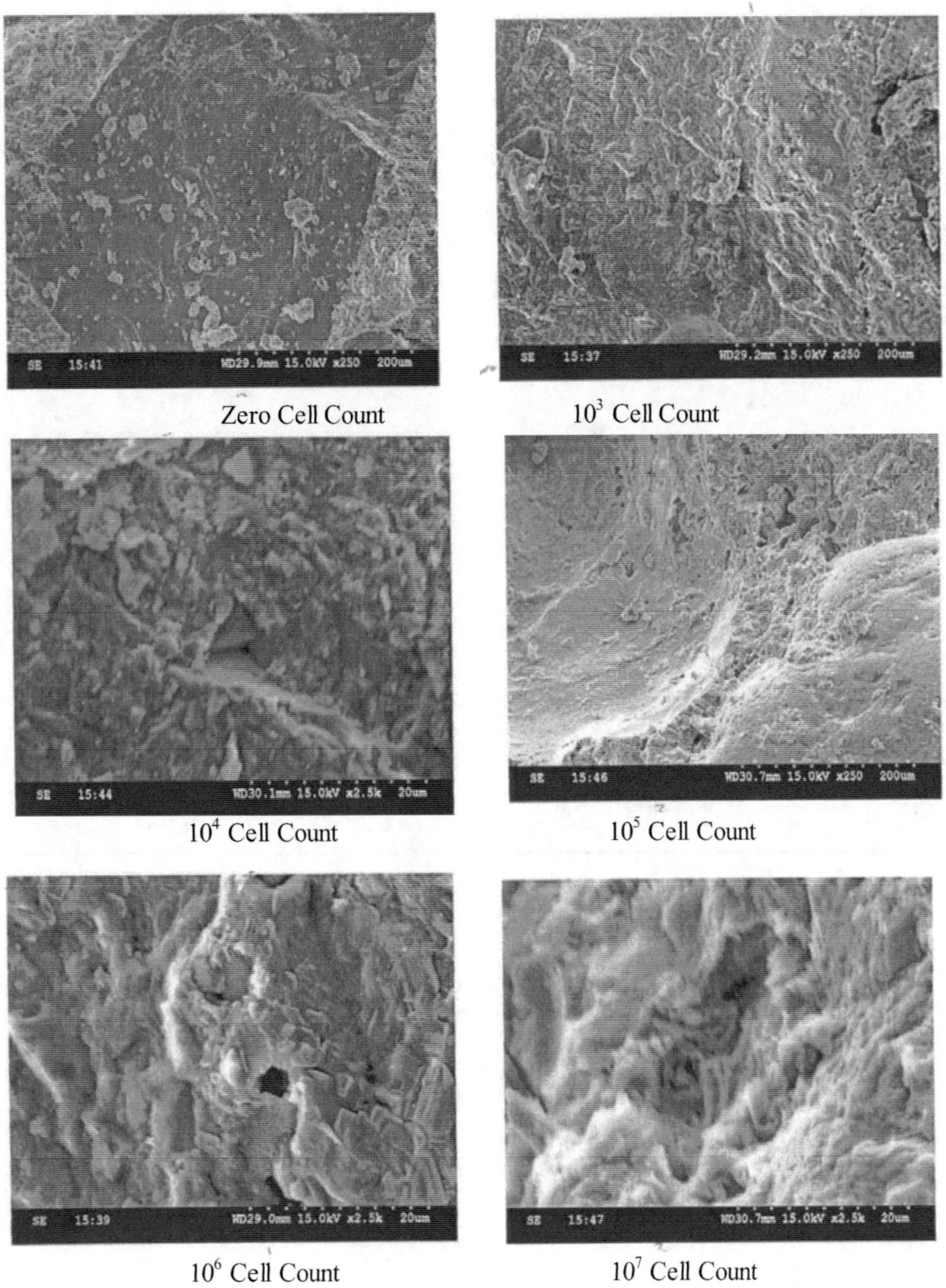

Figure 2. SEM images of cement mortar specimens made with various counts of *Sporosarcina pasteurii* cells

3.4 X-ray diffraction studies

X-ray diffraction studies shows that *Sporosarcina pasteurii* incorporated concrete has high amount of calcite mineral in its microstructure which can be observed by the spikes of a high peaks indicating presence of calcium mineral as shown in figure 3b.

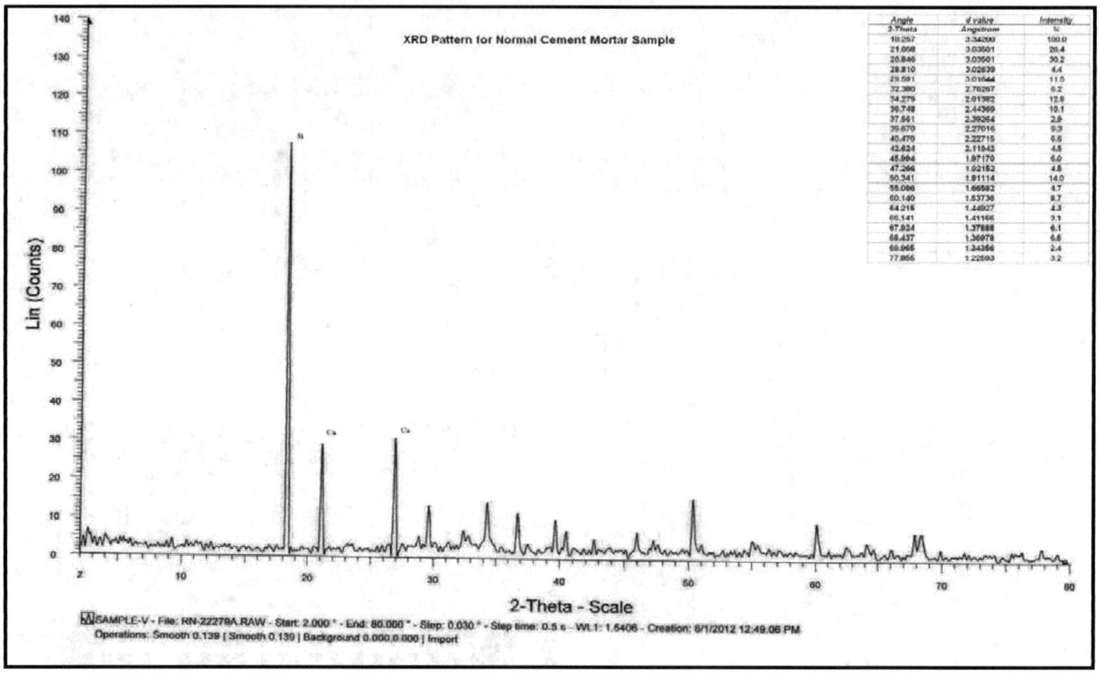

Figure 3a. Diffractogram of normal cement mortar specimens

IOP Conf. Series: Materials Science and Engineering 1091 (2021) 012061 doi:10.1088/1757-899X/1091/1/012061

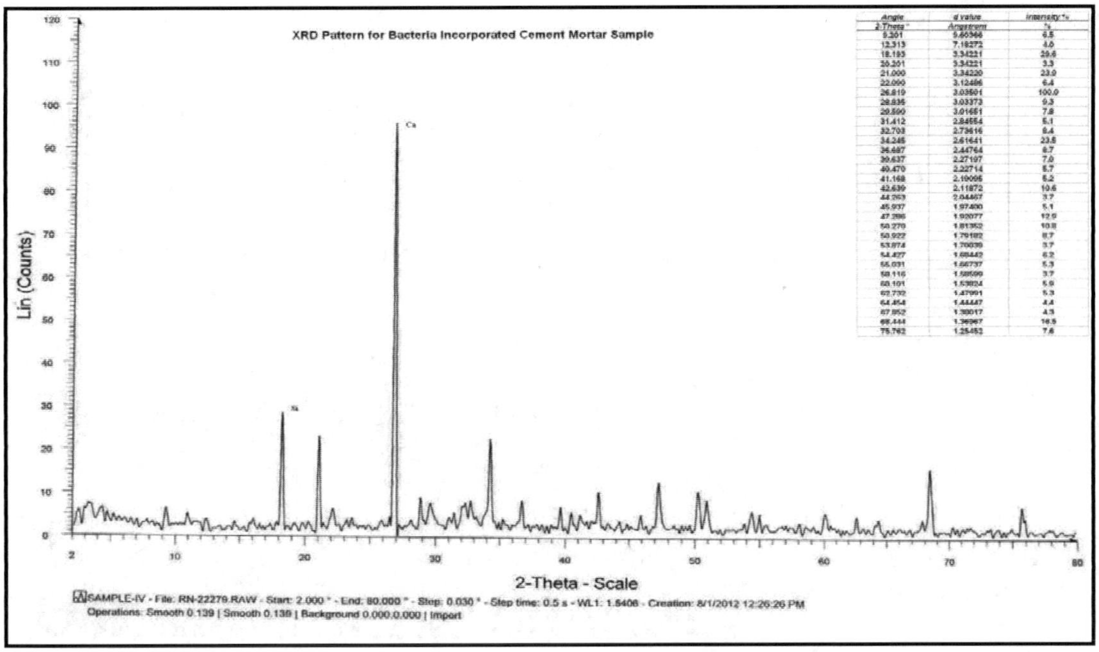

Figure 3b. Diffractogram of bacteria incorporated mortar specimen's shows the abundant presence of Ca^{2+} and precipitation was inferred to as calcite ($CaCO_3$) crystals

3.5 Thermo-gravimetric Studies

Thermo-gravimetric analysis shows that *Sporosarcina pasteurii* incorporated concrete has high amount of calcite mineral so on heating beyond 591°C makes calcite mineral to decompose into lime and carbon dioxide so there is sudden weight loss in the concrete as shown in figure 4.

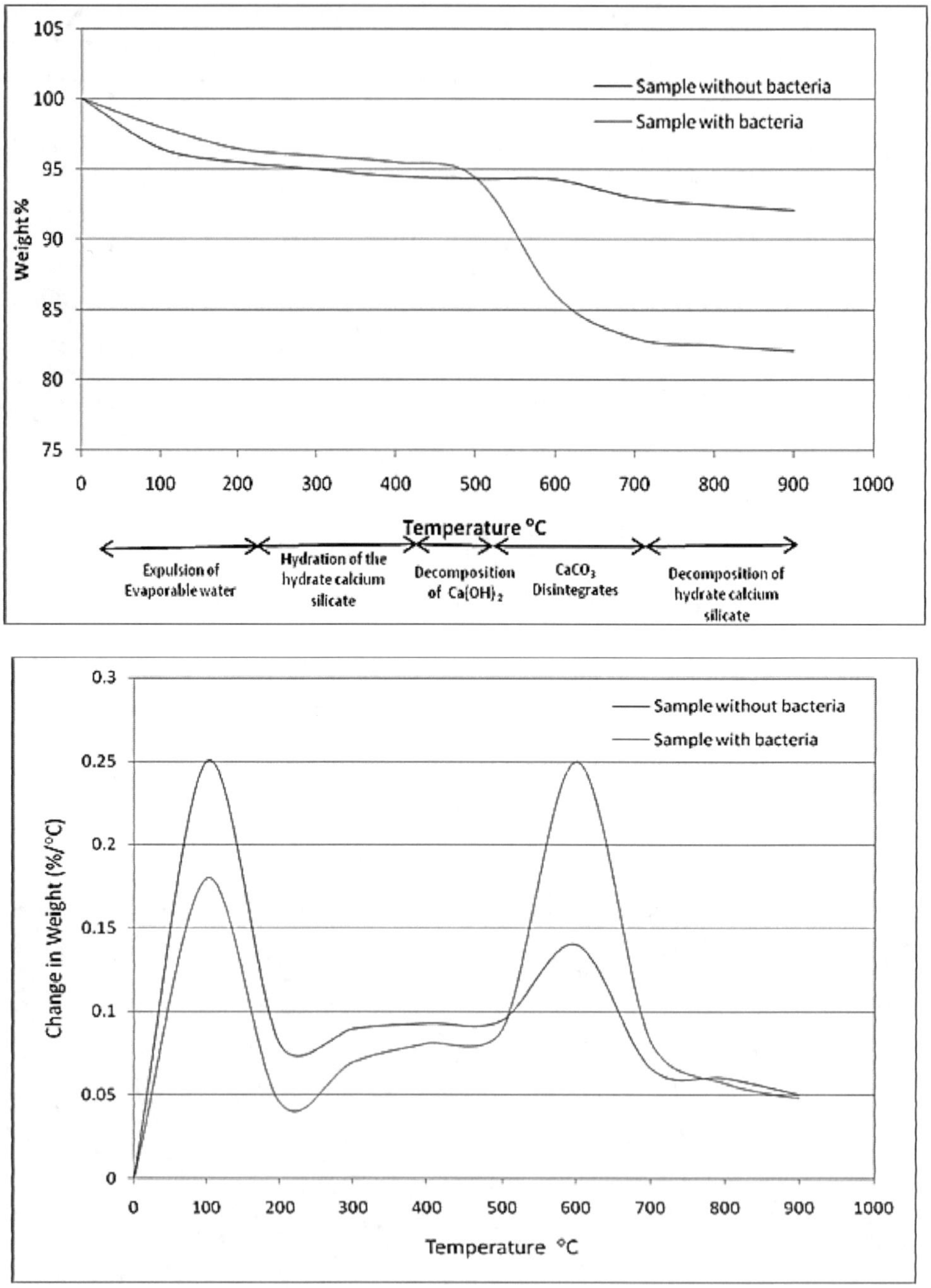

Figure 4. TGA Results for cement mortar specimens with and without bacteria showing weight loss and Change in weight loss per ^{o}C

3.6 Sporulation Test

The presence of spores of *Sporosarcina pasteurii* in the samples of bacteria of 365 days old are found through phase contrast microscopic images. This confirms the sporulation ability of bacteria *Sporosarcina pasteurii.*

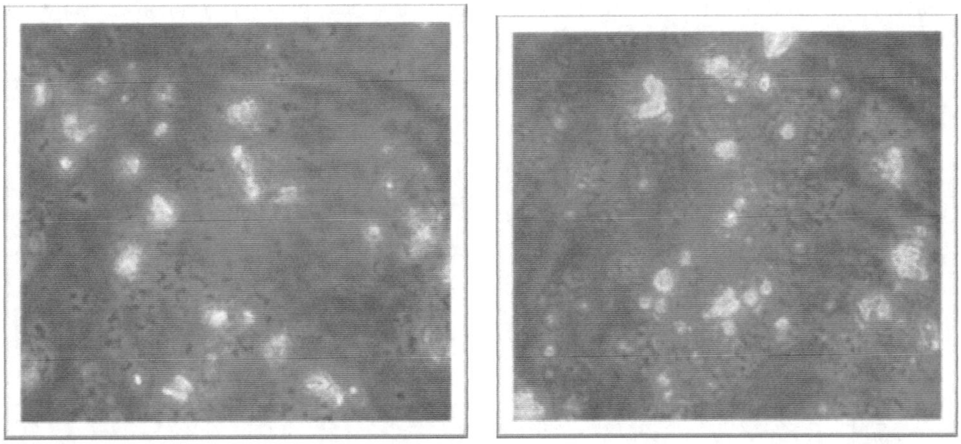

Figure 5. Phase contrast microscopic images shows white calcium carbonate crystals formation

3.7 Permeation properties

This section investigates the water absorbtion capacity and porosity of bacteria induced concrete samples conforming to standard ASTM C 642-13.

Table 2. Water Absorption at regular time intervals of normal and bacteria induced concrete

Measurement Intervals ti (min)	Normal Concrete				Bacteria induced concrete			
	M20		M40		M20		M40	
	mo= 2.49 kg		mo= 2.51 kg		mo= 2.51 kg		mo= 2.53 kg	
	mi(kg)	Mi(%)	mi(kg)	Mi(%)	mi(kg)	Mi(%)	mi(kg)	Mi(%)
0	2.49	0.00	2.51	0.00	2.51	0.00	2.53	0.00
15	2.51	0.80	2.55	1.59	2.52	0.40	2.54	0.40
30	2.59	4.02	2.56	1.99	2.54	1.20	2.55	0.79
60	2.60	4.42	2.58	2.79	2.56	1.99	2.55	0.79
90	2.61	4.82	2.58	2.79	2.58	2.79	2.56	1.19
180	2.62	5.22	2.58	2.79	2.58	2.79	2.56	1.19
480	2.62	5.22	2.58	2.79	2.58	2.79	2.56	1.19
1440	2.63	5.62	2.58	2.79	2.58	2.79	2.56	1.19
2880	2.63	5.62	2.58	2.79	2.58	2.79	2.56	1.19

Table 3. Permeation properties of controlled and bacteria incorporated concrete specimens for different grades

	Controlled Concrete		Bacteria incorporated concrete	
	M20	M40	M20	M40
Water Absorption Capacity (WAC) (%)	5.62	2.79	2.79	1.19
Porosity, P	0.14	0.07	0.04	0.03
Decrease in Porosity	-	-	72%	57%

Concrete specimens induced with bacteria showed considerably less water absorption capacity compared to normal specimens. This decrease in water absorption capacity of bacteria induced concretes is credited to the reduction of pores in the concrete due to calcite mineral precipitation filling the pores reducing the typical pore radius of concrete.

3.8 Pore Structure Analysis

Brenauer- Emmett-Teller method) (BET) Nitrogen (N2) nitrogen adsorption method is adopted to analyze the pore structure of the bacteria induced concrete in terms of specific surface area, pore size distribution, and pore volume conforming to DIN 66131.

Table 4. Pore diameter, Total pore surface area, Total pore volume and porosity of bacteria incorporated concretes

	M20 Controlled	M20 Bacterial	M40 Controlled	M40 Bacterial
Average Pore Diameter (nm)	19.121	3.386	5.084	3.097
Total Pore Surface Area (m^2/g)	1.747	5.047	3.346	5.088
Total pore volume (cc/g)	0.0161	0.0071	0.0137	0.0057
Porosity (%)	1.6	0.7	1.37	0.57

Average pore diameter in bacteria incorporated concrete decreases drastically due to calcite mineral precipitation in the pores because of microbial metabolic activity. The total pore volume also decreased in case of bacteria incorporated concrete samples. Porosity in bacteria incorporated concretes is reduced by 20 to 60%. Total Pore volume in bacteria incorporated concretes is reduced by 20 to 60%.

4.0 Conclusions

The following conclusions are derived from the above experimental investigations-

1. The presence of 10^5 per milliliter of mixing water yields high compressive strength due to precipitation of highest amount of calcium crystals in the concrete pores thereby plugging the micro and macro pores in the concrete refining the micro-structure of the concrete.
2. SEM images, XRD plots and thermo-gravimetric analysis confirmed the presence of calcite mineral in the bacterial mortar specimens
3. The presence of spores of *Sporosarcina pasteurii* in the bacterial mortar specimens of 365 days old are found through phase contrast microscopic images. This confirms the sporulation ability of bacteria *Sporosarcina pasteurii.*
4. Concrete specimens induced with bacteria showed considerably less water absorption capacity compared to normal specimens. This decrease in water absorption capacity of bacteria induced concretes is credited to the reduction of pores in the concrete due to calcite mineral precipitation filling the pores reducing the typical pore radius of concrete.
5. Average pore diameter in bacteria incorporated concrete decreases drastically due to calcite mineral precipitation in the pores because of microbial metabolic activity. The total pore volume also decreased in case of bacteria incorporated concrete samples. Porosity in bacteria incorporated concretes is reduced by 20 to 60%. Total Pore volume in bacteria incorporated concretes is reduced by 20 to 60%.

5. 0 Acknowledgements

This work is funded by JNTUH Hyderabad through TEQIP-III Collaborative Research Scheme (Project No: JNTUH/TEQIP-III/CRS/2019/Civil/03). Work carried by Dr. V.Srinivasa Reddy, GRIET Hyderabad as the Principal investigator along with Prof. Ch. Sasikala, Centre for environment, IST, JNTUH Hyderabad and Dr. S Shrihari, VJIT Hyderabad as Co-investigators. We sincerely express our appreciation to institutions JNTUH Hyderabad and GRIET Hyderabad.

6.0 References

[1] Stocks-Fischer, S., Galinat, J.K., and Bang, S.S., "Microbiological precipitation of $CaCO_3$", Soil Biology and Biochemistry, v. 31, pp. 1563-1571, 1999.

[2] Ramakrishnan, V., Bang, S.S., Ramesh, K.P., and Mulukutla, S., "Performance of Bacterial Cement", Proceedings of 24th International Conference of Cement Microscopy, San Diego, California, pp. 336-354, April 2002.

[3] Ramakrishnan, V., Bang, S.S., Srinivasan Neelamegam., and Ramesh, K.P., "Durability of Cement Mortar Made With Different Concentrations of Bacteria", Proceedings of 25th International Conference of Cement Microscopy, Richmond, Virginia, April 2003.

[4] Ramachandran, S.K., Ramakrishnan, V., and Bang, S.S., "Remediation of Concrete using Microorganisms", ACI Materials Journal, v.98, No.1, pp. 3-9, Jan-Feb 2001.

[5] Kantzas, A., Ferris, F.G., Jha, K.N., and Mourits, F.M., "A Novel Method of Sand Consolidation through Bacteriogenic Mineral Plugging", paper presented at the CIM Annual Technical Conference, Calgary, June 7-10, 1992.

[6] Gollapudi, U.K., Knutson, C.L., Bang, S.S., and Islam, M.R., "A New method for Controlling Leaching through Permeable Channels", Chemosphere, v. 30, No. 4, pp. 695-705, 1995.

[7] Zhong, L., And Islam, M.R., "A New Microbial Process and its Impact on Fracture Remediation", paper presented at the 70th Annual Technical Conference and Exhibition of the Society of Petroleum Engineers, Dallas, Texas, Oct 22-25, 1995.

[8] Ramakrishnan, V., Ramesh Panchalan., and Bang, S. S., "Bacterial Concrete" Proceedings of SPIE-International Society for Optical Engineering on Smart Materials, Australia, Vol. 4234 (2001), pp. 168-176, 13-15 December 2000.

[9] Ramakrishnan, V., Ramesh Panchalan., and Bang, S.S., "Bacterial Concrete- A Self Remediating Biomaterial" Proceedings of 10th International Congress on the Polymers in Concrete, Hawaii, May 2001.

Exploring nanotechnology for diagnostic, therapy and medicine

T. Chithrakumar[1], M Thangamani[1], and S.Pon miruthu vasini[3]
[1]Assistant professor, Computer Science and Engineering, Sri Ramakrishna engineering college, Coimbatore
[2] Associate Professor, Information Technology, Kongu Engineering College, Tamilnadu, India
[3]Student of Computer and Technology, Kongu a Engineering College, Tamilnadu
E-mail: manithangamani2@gmail.com

Abstract. In current days, knowledge enhancement is growing more. Nanotechnology can applied to self-repeating machines and mechanical technology, compound designing, mechanical designing, science, natural designing, and electrical designing. Nanotechnology has the great preference to improves produce new and superior product. Nanomaterials to perceive avoid and eliminate pollution. Nanotechnologies are blueprint, invention and applied in medicine, therapy, diagnostic, sequencing and environmental protection. The usage of nanotechnology is rapidly increases in entire future market. This article explore the nanotechnology for medicine, diagnostic and therapy.

Keywords: Nanotechnology, drug delivery, Medical diagnosis and Therapy and Medicine using nanomaterials.

2. Introduction

Nanotechnology is wide as to be unhelpful as a show to tackling challenges of risk governance, risk management and insurance. The main properties of nanotechnologies are nanoparticles, carbon nanotubes and quantum dots. For the most part, when individuals talk about nanotechnology, they mean structures of the size 100 nanometers or more modest. There are 1,000,000 nanometers in a millimeter. Nanotechnology attempts to make materials or machines of that size. Individuals are doing various kinds of work in the field of nanotechnology. Most flow work takes a gander at making nanoparticles that have exceptional properties, for example, the manner in which they dissipate light, ingest X-beams, transport electrical flows or warmth, and so on At the more "sci-fi" end of the field are endeavors to make little duplicates of greater machines or truly novel thoughts for structures that make themselves. New materials are conceivable with nano size structures. It is even conceivable to work with single particles.

There has been a ton of conversation about the eventual fate of nanotechnology and its threats. Nanotechnology might have the option to develop new materials and instruments which would be extremely valuable, for example, in medication, PCs, and making clean power is helping plan the up and coming age of sun based boards and proficient low-energy lighting. Then again, nanotechnology is new and there could be obscure issues. For instance, if the materials are awful for individuals' well being or for nature. They may badly affect the economy or even huge regular frameworks like the Earth itself. A few gatherings contend that there ought to be rules about the utilization of nanotechnology.

3. Related Works

Nanotechnology today [1] has spread out in countless various territories of sciences in light of its huge advantages as far as improving the exhibition in numerous fields. Various types of nano-sized structures are nanowires, nanorods, nanoribbons, nanotubes, nanobelts, nanofibers, quantum spots and nanoparticles. Nanotechnology manages [2] materials and collaborations at sub-atomic and nuclear level, measured somewhere in the range of 0.1 and 100 nm. The little size of the nanoparticles and the progressions of their physical and substance properties contrasted with that of their macromolecular analogs offer numerous preferences for contemporary medication, including improved medication conveyance, open doors for a superior and noninvasive diagnostics, and focused on treatment with decreased unfavorable and fundamental impacts. Improved biomedical collections are used [3]. Directed medication conveyance is a technique for conveying drug in a remarkable manner so that the grouping of the medication at the objective site is enhanced, weight of the medication to different tissues is diminished and poisonousness limited. There are different novel ways to deal with convey medications to the objective locales as nanocapsules, nanocrystals, nanoemulsions, dendrimers, polymeric micelles, nanotubes and monoclonal antibodies. Undoubtedly, development of nanotechnology [5] has set exclusive standards in natural science and translational medication, which is perhaps because of the way that the extents of useful components in science are in the nanometer run and consequently advance the association among nanomaterials and organic frameworks.

Nanotechnology [6] has improved medication conveyance to those regions which be negative for macromolecules. It is suggesting new implantable detecting innovations along these lines giving exact clinical data. The mix of nanotechnology and medication has made another field nanomedicine to upgrade human medical services. In this investigation, a portion of the nanotechnology is used for diabetes mellitus treatment than pancreas transplantation. Electrospinning create [7] an empowering nanotechnology level to producing a rich nonmaterial in biosensing, drug conveyance, regenerative medication and tissue designing. Another angle enthralls [8] their utilization in food bundling materials that have very high gas boundaries and antimicrobial properties and nanosensors which can distinguish microorganisms. There are enormous possibilities of nanotechnology in farming wherein the majority of the examination projects are in their beginning stage, and it will doubtlessly blast all entryways of agri-food area with solid plans and purposes. Ebb and flow modalities [9] of assurance and therapy of various sicknesses, particularly malignant growth have critical obstructions, for model, helpless affectability or particularity and Medicine poison levels independently. Nonoparticles are created for malignancy identification and medicines are focused for antibodies.

In clinical sciences [10] they are utilized for symptomatic or helpful purposes. They can likewise be applied in the readiness of nanovaccines and nano adjuvants. Their utilization in the therapy of malignant growth and in quality treatment opened the entryway for another period in medication. As of late, different utilizations of nanotechnology began to discover their way in the veterinary area. They progressively attack animal therapeutics, diagnostics, creation of veterinary immunizations, ranch disinfectants, for animal rearing and generation, and even the field of animal nourishment.

Pomegranate tree [11] is one of the most established organic product trees known to people (4000 and 3000 BCE). Pomegranate blossoms as an image of life, lastingness, prosperity, femaleness, ripeness, information, everlasting status, and blessedness. In endeavors to sort out the best wellspring of phenolic mixes in human eating routine, pomegranate gets a lot of fame attributable to the natural impacts that applies through free extremist rummaging abilities by its phenolic mixes. Pomegranate has ascended to notoriety for its clinical applications since antiquated occasions. Advances in computerized advances [12] have release new open doors for making additional dependable, cost and time successful, more secure and versatile techniques for diagnosing, overseeing and curing illnesses.

A couple of instances of cutting edge nano and computerized innovations are approved for identifying and curing illnesses. Plasma therapy is as yet arising as another medical services innovation, however antimicrobial resistant diseases and tumors with practically zero unfavorable results. Nanotechnology [13] is applied for cardiovascular infections treatments.

The fate of nanomedicine [14] is energizing! It includes the designing of materials for researching basic organic issues and identifying and curing infection. Notwithstanding unprecedented [15] advances that have been made in disease treatment, the quantity of malignant growth cases keep on flooding, making it the main source of death across the world. Therefore, early recognition is one of the vital viewpoints in the fight against the illness. Screening and early conclusion assume a urgent part for compelling therapy and to bring down the malignancy death rate. Malignancy nanotechnology is another branch in science that gives a connection among nanotechnology and clinical disease research.

Surendiran, A., et al.[16] Treatment of various diseases and diagnosis of current modalities, particularly cancer have key constraints such as specificity, drug toxicities, lack of sensitivity respectively. Nanoparticles are being developed for detecting cancer. For disease identifying uses paramagnetic nanoparticles, nanoshells, quantum dots and nanoparticles. In diagnostics and therapeutics the nanotechnology is being used. Patil, M., et al.[17] to medicine there is increasing optimism in nanotechnology and significant advances in the identify, cure and anticipation of sickness is brought by dentistry. Nanomedicine is called as producing attention in the nanotechnology and future medical field. The core technology is destined by Molecular technology and future impact of nanotechnology in medicine and dentistry is predicted. In science and technology [18] nanomedicine is a virtually latest field.

Singh, M., et al.[19] suggested a nanotechnology atomic scale tailoring of materials to battle and prevent the diseases. In the 21st century nanomedicine [20] is in infancy and Nanotechnology in modern medicine are the potential to change medical research. Procedures, such as gene delivery systems, drug delivery etc and therapeutic purposes, imaging, detection, etc are applied by Nanomedical devices. In many aspects, for cancer therapy [21] Nanotechnology has revolutionized. The treatment pattern is changing radically. The main role of nanotechnology is rectifying the lack of the conventional chemotherapies and on discriminating identifying of the cancerous cells. The traditional chemotherapies side effects are removed effectively. Hofmann-Amtenbrink et. al.[22] discussed about pharmaceutics (nanopharmaceutics) and nanoparticles. The key of this the examining the industrial applications, legal and regulatory frameworks and influence of research funding.

Martis, E. A., et al [23] with the application of nanotechnology Many devices are developed at a lightning speed. Nanotechnology is minimizing the toxicities and it is boon in making the treatment. This paper describes the nanotechnology in medicine and their devices and their applications. Lavenus, S., et al[24]For surface modifications Nanotechnologies are used increasingly in dental implants. Improve their clinical success rate and to direct the nature of peri-implant tissues nanometer-controlled surfaces are directed. At the betterment of human life nanomedicine [25] and Nanotechnology are complementary disciplines. Devices of size 1–100 nm and for designing tools, the emerging branch of science are Nanotechnology. Nanomedicine is defined as study of using clinical practice and nanotechnology in medical research.

Saini, R., et al. [26] described the definition of nanotechnology is involved in application of materials, characterization, synthesis and design. In medicine there are many for nanotechnology applications and some are discussed in this paper. The science and engineering consists of application of materials, synthesis, devices and design, which has lowest amount functional association in at least 1 dimension on the nanometer scale or 1 billionth of a meter is known as nanotechnology. Degree of specificity can probably [27] be converted into tissue explicit medical fields and targeted cellular designed to achieve maximal therapeutic affected with nominal side effects and great promise to biomedicine is nanotechnology [28]. In the previous decade the nanotech [29] gave the chance for molecular disease imaging, therapeutic intervention tools, sensing clinically relevant markers, which have the tendency to convert the pasture of medicine.

The components of a nano toolbox are quantum dots, nanoparticles, nanowires, cantilevers and nanopores for sensing. The latest field of technology improvement in medicine and human biology is nanotechnology. The study of toxicity of nanoparticles [30] on human health is nanotoxicology. Nano-particles are developed by devices and medical field. In the area of tissue culture, genetic side effect, bio materials and molecular biology the nanomedicines are used. Dolomatov, M. Y., et al. [31], monitor the real time pollutants nanoparticles are used. In water treatment the use of nanomaterials is cost effective. The medical imaging and the controlled drug delivery is represented by nanostructures. The nanomedicine application was done by Ayurvedic Bhasma which throws the light on the living beings. In therapeutics, the usage of nanoparticulate metal is common in Ayurveda. Organometallic ethno nanomedicine in nanomedicine is the concept of Ayurvedic Bhasma.

The frontier of medicine [33] is nanotechnology which plays an significant position in reaching the drug and in diversity expansion. The protected transport vehicle for hydrophobic drugs such as siRNA, DNA and mRNA are not bioavailable which are called nanoparticles and suggested about the challenges for effective delivery and various forms of disease treatment. Owen, A., et al [34] using novel CLB, this paper is concentrated on biodegradable nanoparticles and biocompatible with the tunable biodegradation kinetics and hydrophobicity synthesized. In the organic nanoparticles in medicine, change of gold salt to the communicating with gold nanoparticles is achieved by the simple mixing of EGCG with the gold precursor because the development is different. Mishra, S.,et al [35] human attention is concentrated on meter level and fraction. Range of RBC is 10^{-6} m, the range of protein molecules is 10^{-9} m and range of atom is 10^{-11}. The RBC is 10 times smaller than drug particles. So that majority of drugs are wasted with the decrease in the efficiency.

4. Conclusion

Advances in nanotechnology assure latest materials and structures that are the root of solutions for humanizing strength and optimizing accessible power and resources, sustaining a lively wealth, increasing the standard of living and enhancing national security. Life sciences and health care applications use emerging medical devices and nano-enabled pharmaceuticals. The diagnostics and nanoagents are intelligences to identify human cancer at earliest level. In future investigation can be extends by applying the nano technology to drug discovery from protein to protein interaction network

References

[1] Kargozar, S. and Mozafari, M., 2018. Nanotechnology and Nanomedicine: Start small, think big. *Materials Today: Proceedings*, *5*(7), pp.15492-15500. Nikolova, M., et al. (2020). "Nanotechnology in medicine." Drug Discovery and Evaluation: Methods in Clinical Pharmacology: 533-546.

[2] Hui, E., Di Carlo, D. and Pennathur, S., 2019. Guest editorial special section on IEEE EMBS Conference on Micro and Nanotechnology in Medicine. *IEEE Transactions on NanoBioscience*, *18*(2), pp.214-215.

[3] Mostafavi, E., Soltantabar, P. and Webster, T.J., 2019. Nanotechnology and picotechnology: A new arena for translational medicine. In *Biomaterials in translational medicine* (pp. 191-212). Gupta, R. (2017).

[4] Chen, S., Li, R., Li, X. and Xie, J., 2018. Electrospinning: An enabling nanotechnology platform for drug delivery and regenerative medicine. *Advanced drug delivery reviews*, *132*, pp.188-213.

[5] Saxena, S.K. and Singh, Y.K., A review of the emerging nanotechnology applications in medicine treatment.

[6] Liu, D., Szili, E.J. and Ostrikov, K., "Plasma medicine: Opportunities for nanotechnology in a digital age." Plasma Processes and Polymers 17(10), 2020.

[7] Shwetha, K. and Deveswaran, R., "The use of nanotechnology in cardiovascular disease." Applied Nanoscience 8(7): 1607-1619, 2018.

[8] Cheon, J., Chan, W. and Zuhorn, I., "The Future of nanotechnology: cross-disciplined progress to improve health and medicine." Accounts of chemical research 52(9). 2019 2405-2405.

[9] . Kargozar, S. and Mozafari, M., "Nanotechnology and Nanomedicine: Start small, think big." Materials Today: Proceedings 5(7), 15492-15500, 2018.
[10] Surendiran, A., Sandhiya, S., Pradhan, S.C. and Adithan, C., "Novel applications of nanotechnology in medicine." Indian Journal of Medical Research 130(6), 2009.
[11] Patil, M., Mehta, D.S. and Guvva, S., "Future impact of nanotechnology on medicine and dentistry." Journal of Indian society of periodontology 12(2), 2008..
[12] Boisseau, P. and B. Loubaton, "Nanomedicine, nanotechnology in medicine." Comptes Rendus Physique 12(7), 2011, 620-636.
[13] Singh, M., Singh, S., Prasad, S. and Gambhir, I.S., "Nanotechnology in medicine and antibacterial effect of silver nanoparticles." Digest Journal of Nanomaterials and Biostructures 3(3), 2012, 115-122.
[14] Logothetidis, S. "Nanotechnology in medicine: the medicine of tomorrow and nanomedicine." Hippokratia 10(1), 2006, 7-21.
[15] Sutradhar, K. B. and M. Amin , "Nanotechnology in cancer drug delivery and selective targeting." International Scholarly Research Notices 2014.
[16] Hofmann-Amtenbrink, M., Hofmann, H., Hool, A. and Roubert, F., "Nanotechnology in medicine: European research and its implications." Swiss medical weekly, 2014.
[17] . Martis, E.A., Badve, R.R. and Degwekar, M.D., "Nanotechnology based devices and applications in medicine: An overview." Chronicles of young Scientists, 3(1), 2012
[18] Lavenus, S., Louarn, G. and Layrolle, P., "Nanotechnology and dental implants." International journal of biomaterials 2010.
[19] Basavaraj, K, "Nanotechnology in medicine and relevance to dermatology: Present concepts." Indian journal of dermatology 57(3), 2012.
[20] Saini, R., Saini, S. and Sharma, S., "Nanotechnology: the future medicine." Journal of cutaneous and aesthetic surgery 3(1), 2010.
[21] Silva, G. A., "Introduction to nanotechnology and its applications to medicine." Surgical neurology 61(3): 2004, 216-220.
[22] Van Tassel, P. R. "Nanotechnology in medicine: nanofilm biomaterials." The Yale journal of biology and medicine 86(4), 2013.
[23] Lin, H. and R. H. Datar,. "Medical applications of nanotechnology." The National Medical Journal of India 19(1): 27-32, 2006.
[24] Abdussalam-Mohammed, W., "Review of Therapeutic Applications of Nanotechnology in Medicine Field and its Side Effects." Journal of Chemical Reviews, 243-251, 2019.
[25] Dolomatov, M. "Electron phenomenological spectroscopy and its application in investigating complex substances in chemistry, nanotechnology and medicine." Journal of Materials Science and Engineering, 2013.
[26] Sharma, R. and P. Prajapati "Nanotechnology in medicine: Leads from Ayurveda." J Pharm Bioallied Sci 8(1), 2016, 80-81.
[27] Mukherjee, A. and S. Bhattacharyya Nanotechnology in medicine. Biotechnology Business-Concept to Delivery, Springer, 2020, 57-64.

IOP Conf. Series: Materials Science and Engineering 1091 (2021) 012063 doi:10.1088/1757-899X/1091/1/012063

Utilization of industrial wastes in the production of sulfate-resistant cement

Z A Mukhamedbaeva[1,4], I A Chorieva[1], E I Kurbanov[2], Kh A Adinaev[3]

[1]Department of silicate materials and rare noble metals, Tashkent Chemical-Technological Institute, 32 Navoi street, Tashkent 10011, Uzbekistan
[2]Namangan cement factory, 74 Madaniyat street, Chust district, Namangan region, Uzbekistan
[3] Termez branch of the Tashkent State Technical University named after I. Karimov, 288 a Islam Karimov street, Termez city 732000, Uzbekistan

[4]Corresponding author e-mail: mzamira1946@gmail.com

Abstract. In connection with the possibility of creating, on the basis of secondary raw materials and wastes of industrial enterprises, the production of new building materials of increased resistance in aggressive environments, there is a further need for systematic research and their use in the production of Portland cement. In this regard, the tailings of concentration of ferrous and non-ferrous metals are promising in terms of the content and reserves of useful components in comparison with the dump deposits of mining enterprises. At the concentration plants of Almalyk Mining And Metallurgical Complex, tens of millions of tons of "tailings", formed during the enrichment of lead and copper-bearing ores and rich in Al_2O_3, Fe_2O_3, etc. are thrown into dumps annually, which are valuable components for the production of cement. The article presents the results of research on topical problems of the use of tailings from the lead concentration and copper smelting factories of the AMMC in the production of sulfate-resistant Portland cement as an active mineral additive. The influence of mineralogical compositions of clinkers of cement plants in Uzbekistan on the phase composition and properties of the obtained sulfate-resistant cement in the presence of lead-copper active mineral additives is analyzed. It is shown that additives play an important role in forming the structure of a cement stone and increasing its strength. Acceleration of the process of hydration and hardening of sulfate-resistant cement is observed from a change in the mineralogical composition towards a decrease in the content of tricalcium aluminate and a decrease in the basicity of clinker due to high-silica waste, which meets the requirements for sulfate-resistant cements.

1. Introduction

Currently, at the world level, scientific research is being carried out to develop ingredients and compositions that improve the construction and technical properties of cement, create highly efficient energy and resource-saving technologies, and improve their performance indicators. The existing scientific foundations for the production of highly efficient sulfate-resistant cements using secondary resources of various industries have been developed. The influence of the mineralogical composition of clinker on the phase composition and properties of products in the presence of technogenic additives is analyzed.

The question of the influence of the mineralogical composition of Portland cement clinker on the corrosion resistance of cement stone in an aggressive environment was studied by many scientists from abroad and the CIS: Isakhodzhaev B.A. [1], Kansepolskiy I. S. [2], Moskvin V.M. [3], Kuznetsova T.V. and Atakuziev T A [4], Davila J. [5], Lokher F.B. [6], Akula P. [7], Reddy B. [8], Sycheva L. [9], Pashchenko A. [10], Klassen V.K. [11], Taylor H. [12], Ezziane K. [13], Choriva I.A. and Mukhamedbaeva Z.A. [14].

The high mineralization of surface and subsoil waters in a number of regions of Uzbekistan require corrosion-resistant Portland cement for the construction of hydro technical and irrigation facilities [1-4]. Using local raw materials many types of sulfate-resistant Portland cements for the needs of Uzbekistan industry was developed by Kansepolskiy I.S. [2]. The sulfate-containing cement from low-aluminate and technogenic raw materials was obtained by Sycheva L.I. and Bakeev D.V. [9].

Recently mechanism of formation of cement and effect of concentrated solutions of salts (brines) on blended cement hydration using calorimetry methods was studied by Nocun-Wczelik W. and Meedala M. [15]. Connection between processes of deterioration of concrete structures and formation of thaumasite in concretes and mortars containing fly ash was investigated by Mulenga D.M. and others [16].

For chemical and mineralogical composition of sulfate-resistant cement, special attention is paid to study of the corrosion-resistance of cements during hardening in aggressive environments. To increase the resistance of cement under sulfate solutions attack, the mineralogical composition of the initial clinker is very important. Studies show that the sulfate resistance of Portland cement is achieved with a reduced content of tricalcium aluminate and a moderate amount of alite [17-18]. Especially high salinity of groundwater takes place in the Aral Sea region of Uzbekistan, where construction structures are subjected to complex sulfate-magnesia-chloride aggression [19-21].

In this regard, the purpose of this study was the scientific justification and development of technology for the production of sulfate-resistant cements using flotation waste from metallurgical industries.

2. Research methods

As the objects of research the wastes from the enrichment of lead and copper-containing ores of the Almalyk Mining And Metallurgical Complex (AMMC) and clinker from cement plants of Uzbekistan was chosen.

For the research, modern methods of analyzing raw mixtures and reaction products were used. Chemical analysis of raw materials was determined according to GOST 5382-91. The dependence of the properties of cements on individual factors was studied in accordance with GOST 310-89. The most common and aggressive in practice 3% $MgSO_4$ solution and 5% Na_2SO_4 solution, corresponding to subsoil and ground waters, were used as aggressive media. The samples were tested after 3, 7, 14, 28, 90, 180, 360 days of hardening. The assessment of the sulfate resistance of cements was carried out according to the resistance coefficient, which after 6 months is at least 0.8. The test for the hydraulic activity of lead production waste (***SOF***) and flotation waste from copper production (***MOF***) wastes was determined according to the standard method of absorbing the amount of lime from a saturated solution by grams of additive. XRD-method and Bruker AXS D8 Advance diffractometer, Bruker, Germany was used (Cu-Kα - cathode, step - 0.05, shooting speed 2 sec) for identification of crystal phases. The hydrated samples were dehydrated with acetone at $t \approx 20^\circ C$. The survey parameters are as follows: 2θ angle interval from 6 to 65°, CuKα radiation, tube voltage 24 kV, anode tube current 14 mA.

3. Results and its discussion

Non-additive Portland cement grade 400 of “Akhangarantsement” JSC, “Bekabadcement” JSC and “Kuvasaycement” JSC (Uzbekistan), tailings of lead-production and flotation waste from copper-enrichment plants was chosen as initial components, the chemical composition is given in table 1.

Phase compositions of matrix material for the production of cement mortars based on clinkers of Akhangaran, Bekabad and Quvasay cement plant, sand and additives SOF and MOF were revealed using X-ray phase analysis. Clinkers of cement plants are mainly represented: C_3S - d = 0.303; 0.296;

0.260; 0.218; 0.192; 0.176 nm, C_2S - d = 0.385; 0.277; 0.272; 0.260; 0.208 nm., C_3A - d = 0.273; 0.269; 0.216; 0.202; 0.192; 0.154 nm., C_4AF - d = 0.269; 0.264; 0.218; 0.204; 0.192; 0.182 nm.

Table 1. Chemical composition of initial materials, mass.%

Initial materials	SiO_2	Al_2O_3	Fe_2O_3	CaO	MgO	SO_3	CaO_{fr}	R_2O	P.p.p.	Other	∑
Akhangaranport-landcement	22,68	4,55	3,65	65,48	2,45	0,30	0,12	-	-	0,87	100
Bekabadportland-cement	20,63	4,52	4,05	65,92	1,73	2,15	0,43	-	-	0,70	100
Quvasayportland-cement	21,78	4,81	4,08	64,14	1,89	0,64	0,37	-	-	0,60	98,31
Tails, SOF	45,75	8,72	7,19	14,59	7,10	2,98	-	2,98	8,83	-	98,14
Flotation waste, MOF	60,68	14,02	9,54	1,37	0,11	5,69	-	4,13	4,11	-	99,65

The chemical compositions of Portland cements grade 400 of AOO "Akhangarantsement", AOO "Bekabadcement" and "Kuvasaycement" (Uzbekistan) are given in table 2.

Table 2. Calculated mineral composition of clinkers (production of Uzbekistan cement industry)

Cement production plant	KN	n	c	Minerals content, mass. %				
				C_3S	C_2S	C_2A	C_4AF	∑
Akhangaran	0,91	2,29	1,33	64,0	18,0	7,2	11,3	100,5
Bekabad	0,91	2,33	1,19	65,3	10,8	7,5	12,1	95,7
Quvasay	0,95	2,45	1,18	57,5	19,1	5,8	12,4	99,4

Table 3 shows the results of changes in the salt resistance of the Akhangaran, Kuvasay, Bekabad no-additive Portland cements, depending on the action of the additives MOF and SOF.

Phase compositions were revealed using X-ray phase analysis of matrix material for the production of cement mortars based on clinkers of Akhangaran, Bekabad and Quvasay cement plant sand additives SOF and MOF. Clinkers of cement plants are mainly represented: C_3S - d = 0.303; 0.296; 0.260; 0.218; 0.192; 0.176 nm, C_2S - d = 0.385; 0.277; 0.272; 0.260; 0.208 nm., C_3A - d = 0.273; 0.269; 0.216; 0.202; 0.192; 0.154 nm., C_4AF - d = 0.269; 0.264; 0.218; 0.204; 0.192; 0.182 nm.

Table 3 shows the results of changes in the salt resistance of the Akhangaran, Kuvasay, Bekabad no-additive Portland cements, depending on the action of the additives MOF and SOF.

Based on the above data, we can conclude that the addition of SOF and MOF additives to the compositions of Portland cements of different mineralogical composition sharply increases its sulfate resistance. Increasing the sulfate resistance of cements due to the introduction of SOF and MOF additives can be explained by a change in the structure of the cement stone. The introduction of SOF and MOF additives leads to a decrease in the proportion of C_3A in cement and to the binding of calcium hydroxide released during hydration, which leads to a significant increase in the resistance of cements against sulfate corrosion. An increase in salt resistance is associated with a change in the mineralogical composition of cements towards a decrease in the content of C_3A and a decrease in the basicity of clinker. The introduction of additives MOF and SOF reduces the C_3A content from 9-10% to 2-3%, the basicity of cement to 0.84-0.86, which meets the requirements for sulfate-resistant cements.

Table 3. Strength for sulfate resistance of cements with the addition of metallurgical waste MOF and SOF

№	Introduced additive	Aggressive environment	Compressive strength of the sample, *MPa, days*						
			3	7	14	28	90	180	360
			Akhangaran Portland cement						
1.	MOF	3% $MgSO_4$	30	38	40	46	66	80	94
2.	SOF	3% $MgSO_4$	38	42	48	59	71	87	103
3.	No additives	3% $MgSO_4$	14	19	20	29	33	36	50
4.	MOF	5% Na_2SO_4	32	38	41	50	73	75	80
5.	SOF	5% Na_2SO_4	38	41	47	58	64	77	89
6.	No additives	5% Na_2SO_4	15	19	25	33	39	38	38
			Quvasay Portland cement						
1.	MOF	3% $MgSO_4$	38	42	43	49	77	82	96
2.	SOF	3% $MgSO_4$	30	48	75	87	88	90	90
3.	No additives	3% $MgSO_4$	24	26	27	30	33	35	37
4.	MOF	5% Na_2SO_4	35	38	40	47	75	81	97
5.	SOF	5% Na_2SO_4	30	40	60	74	75	91	95
6.	No additives	5% Na_2SO_4	18	18	20	23	25	37	45
			Bekabad Portland cement						
1.	MOF	3% $MgSO_4$	43	48	54	63	79	98	106
2.	SOF	3% $MgSO_4$	33	36	40	54	78	96	102
3.	No additives	3% $MgSO_4$	33	41	50	53	50	51	48
4.	MOF	5% Na_2SO_4	41	50	54	62	85	106	110
5.	SOF	5% Na_2SO_4	35	40	48	66	84	102	109
6.	No additives	5% Na_2SO_4	26	28	35	41	45	46	47

The data obtained are consistent with the results of determining the coefficient of resistance of samples in aggressive media, depending on the time of their hardening. The dependence of the coefficient of resistance of the samples and aggressive media on the hardening period are shown in figure 1.

It is generally known that the interaction of cement with water begins almost instantly, while hydration particles are gradually deposited on the cement grains, water flows inside, new surfaces of non-hydrated cement particles are exposed, which leads to accelerated dissolution of minerals and crystallization of new formations. While exposure of unhydrated surfaces prevails over the process crystallization of neoplasm's, a decrease in the strength of the cement stone is not observed. Over time, within the setting time, the number of neoplasms increases, few new non-hydrated surfaces are found, and the strength increases and does not decrease.

As a result of our research, physicochemical basis for a new progressive energy-saving technology for the production of highly efficient sulfate-resistant cements using waste from metallurgical industries developed, considered the issues of the formation of hydrate phases in silicate material, such as low-basic calcium hydrosilicates, tobermorite, hydrocarbosilicates, as well as gel-like CSH structures, the regularities of the reaction were established when mineral additives MOF and SOF are introduced into the composition of Portland cements, amorphous silica sorts out $Ca(OH)_2$ and other impurities contained in additives, fixing them on the surface of silica particles.

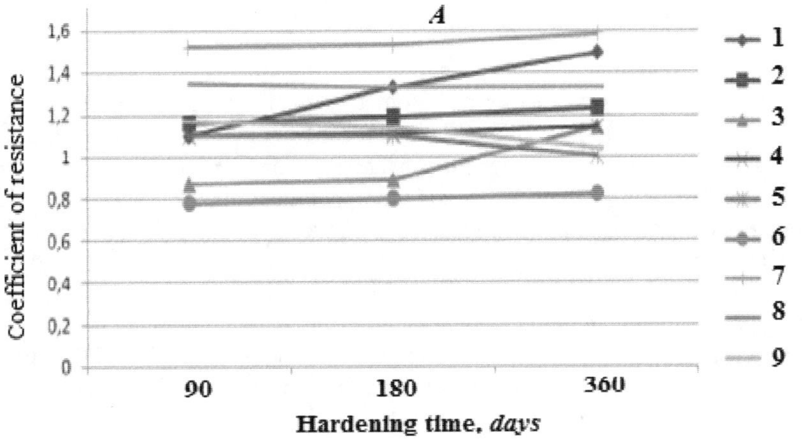

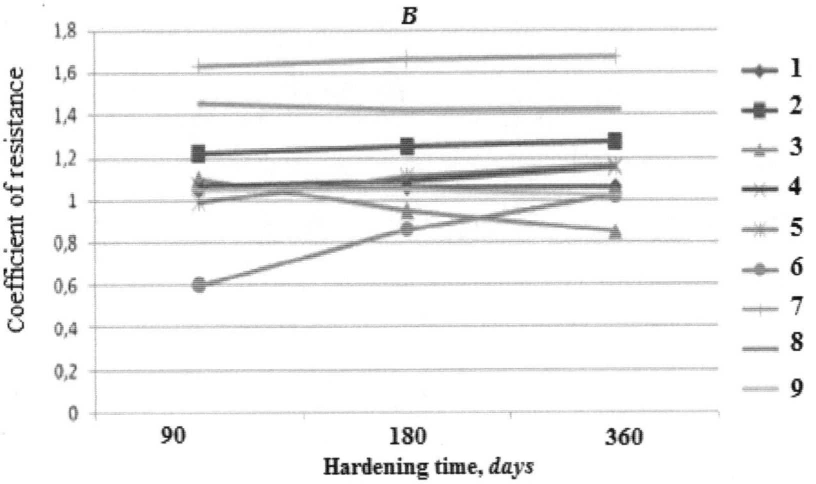

Figure 1. Coefficients of resistance of samples of Akhangaran, Kuvasay and Bekabad Portland cements in a solution of 3 % $MgSO_4$ (A) and 5 % $MgSO_4$ (B):

1-Akhangaran Portland cement with the addition of MOF;
2-Akhangaran Portland cement with SOF additive;
3-Akhangaran Portland cement without additive;
4-Kuvasay Portland cement with the addition of MOF;
5-Kuvasay Portland cement with SOF additive;
6-Kuvasay Portland cement without additive;
7-Bekabad Portland cement with addition of MOF;
8-Bekabad Portland cement with SOF additive;
9-Bekabad Portland cement without additive.

As a result, the activating ability of the additives increases and, consequently, due to the acceleration of the removal of Ca^{2+} ions from the liquid phase, the formation of hydrated minerals and the formation of cement stone are intensified.

4.Conclusion

A complex of studies of lead and copper production wastes was carried out, the results of which indicate the constancy of the content of the main components in them and the possibility of using the production wastes as a mineral additive in the production of cements. The physical and mechanical properties of the studied cements fully meet the requirements for sulfate-resistant cement, the most optimal amount of additive is 5-10 %. Based on the experiments carried out, the large role of the mineralogical composition and activity of the matrix Portland cement clinker in the production of cements with SOF and MOF additives is confirmed. The results obtained state that for the manufacture of cements with SOF and MOP additives, it is advisable to use alumina clinker as a matrix clinker. A comparative assessment of the $Ca(OH)_2$ content in the composition of the cement stone containing SOF and MOF allows us to conclude that the introduction of 5% SOF and 3% MOF favorably affects the process of $Ca(OH)_2$ binding into hydro silicates and calcium hydroaluminates, and also in carbonate containing hydrated products.

References

[1] Isakhodzhaev B A and Khodzhaev N T 2003 Mineral resource base of the Republic of Uzbekistan for obtaining composite materials *Composite materials* **3** 58-60.
[2] Kansepolskiy I S, Pulatov Z P and Dyatlov I P 1974 *Gliezh Portland cement for hydraulic structures* (Tashkent: Fan) p. 405.
[3] Moskvin V M, Ivanov F M etc. 1980 *Corrosion of concrete and reinforced concrete, methods of their protection* (M.: Stroyizdat) p. 384.
[4] Kuznetsova T V, Atakuziev T A and Iskandarova M. 1984 *Use of high-iron waste from non-ferrous metallurgy as an additive to non-ferrous sulfate cements* (Tashkent: Fan) p. 150.
[5] Davila J M, Sarmiento A M , Fortes J C, Santisteban M, Leiva M, Cordoba F, Cabello J E and Grande J A 2021 *Construction and Building Materials* **269** 121817 DOI: 10.1016/j.conbuildmat.2020.121817
[6] Lokher F V 1973 *Fifth Inter.Cong. on Cement Chemistry* (Moscow: Stroyizdat) pp 101-105
[7] Akula P, Little D N 2021 Mineralogical characterization and thermodynamic modeling of synthesized ettringite from Ca-Al-SO_4 suspensions *Construction and Building Materials* **269** 121304 DOI: 10.1016/j.conbuildmat.2020.121304
[8] Reddy B M 2020 *Journal of Physics: Conf.Series* 1706 (1) 012137 DOI: 10.1088/1742-6596/1706/1/012137
[9] Sycheva L V and Bakeev D V 2009 Preparation and properties of sulfate-containing cements based on low-aluminate raw materials *Cement* **6** 23-29
[10] *Theory of cement* 1991 Ed. A.A. Pashchenko (Kiev: Budivelnik) p 168
[11] Klassen V K etc 2008 *Man-made materials in the production of cement* Monograph (Belgorod: BSTU Publishing House) p 126
[12] Taylor H 1996 *Chemistry of cement* (Moscow: Mead) p 560
[13] Ezziane K, Kadri E, Hallal A and Duval R 2010 Effect of mineral additives on setting of blended cement by the maturity method *Mater And Struct* **43 p** 393-401
[14] Chorieva I A, Mukhamedbaeva Z A, Mukhamedbaev AA, Tokhtakhunova G A, Aripova B Kh and Abdukadirov A A *Sulphate-resistant cement* Patent IAP 0581 Date of receipt 11/20/2015 2016 0445.
[15] Nocun-Wczelik W and Meedala M 2003 Studies of cement hydration kinetics in concentrated

solutions of salts *Ann. chim. Sci. mater*. **28** 19-26
[16] Mulenga D M, Stark J, Nobst P 2003 *Cement and Concrete Composites* **25** 8 907-912 DOI: 10.1016/S0958-9465(03)00136-7
[17] Taylor F W and Richardson H 2015 *Cement Chemistry* (F.W. Taylor ICE Publishing) p 552
[18] Shlyakhova E A and Martemyanova Yu N 2012 *Science of Science* **4** 1-5
[19] Kabulova L B, Atakuziev T A and Dzhandullaeva M S 2015 *Austrian Journ.of Technical and Natural Sciences* **9-10** 111-113
[20] Chorieva I A, Mukhamedbaeva Z A, Atakuziev T A and Adinaev Kh A 2018 *Chemical Industry* **1** 43-50.
[21] Babakhanova Z, Aripova M, Khodjaev N and Khamidov R 2016 *Chemistry & Chemical Technology* **10** 3 343-348

Comparative dielectric study of binary mixtures of clove and cinnamon oil

Mohammad Shafi Khan[1], Vishal Singh Chandel[2*] and Satyendra Pratap Singh[3]
[1]Dr. ZHIIC, Nidholi Kalan, Etah, UP, India
[2]Rajkiya Engineering College, Ambedkar Nagar, UP 224122, India
[3]Department of Physics, AIAS, Amity University, Noida, UP 201301, India
chandel.integral@gmail.com

Abstract: The present paper deals with the dielectric study (dielectric constant, dielectric loss) of two medicinal oils, cinnamon and clove oil and their binary mixtures at different temperatures and frequencies. HP 4194A impedance gain/phase analyzer and temperature controller (Julabo, model number F-25, microprocessor controlled) were used for determination of dielectric parameters and maintaining the temperature of pure oils their binary mixtures.

Key Words: Clove Oil, Cinnamon Oil, Dielectric constant, Dielectric loss, Binary Mixture.

1. Introduction

Food is fuel for body and keeps our mind fit and working. Adulterated food products are responsible for an abundant loss both physically and mentally. With the development in the methods of identifying the adulteration, it becomes imperative that food we consume must be pure. For example, the mustard oil is used as the major cooking oil in India and it is highly recommended in food because it contains two essential fatty acids namely linolenic acid and α- linolenic acid which our body cannot make itself.

Edible oils are vital component of foods content either extracted from plant or from other sources. The edible oils give their characteristic flavors and textures to food. They are important in cosmetics, pharmaceuticals, lubricants and in various industries. They are made edible by purification and undergo variety of chemical and physical processes such as heating, distillation etc. before consumption.

Many semi-empirical equations have been developed to relate the physio-chemical properties of oils such as change in their density, viscosity, surface tension, etc. With the help of these semi-empirical equations, we can predict change in the properties of oils under various physio-chemical conditions.

The physio-chemical properties of oils depend on constituents of the oil, frequency, temperature, age, adulteration etc. In the present paper, we have reported dielectric properties of two medicinal oils (clove and cinnamon) and their binary mixtures. Various studies have already been done by researchers, the dielectric properties of eleven fats and cooking oils at 300 MHz and at different temperature have been reported by Pace et al [1]. The Dielectric behaviour of edible unsaturated oils and their binary mixtures was reported by Shilpi Agarwal et al [2]. The dielectric study of many edibles oils and fatty acids were measured by Lizhi et al. over the frequency range of 100 Hz-1MHz [3]. The dielectric properties of oil used in engines and bone fatty oil were reported by other researchers [4]. The optical study of various oils has also been reported by our group [5]. Šegatin et al recently reported dielectric and dipole Moment study of edible oils after thermal treatment [6].

2. Material and Methods

The oils under investigation, clove and cinnamon, have been purchased from local market.

2.1. Clove Oil

Clove oil is extracted from cloves. It is rich in various minerals mainly iron, calcium, phosphorus, sodium, potassium, hydrochloric acid, and vitamin A and vitamin C. The main constituent of clove oil is Eugenol. This oil is used in aroma industry and as a local antiseptic and anesthetic medicine, for diarrhea, hernia, bad breath, vomiting, nausea, etc. intestinal gas [7]. It is also used in the manufacturing toothpaste, soaps, creams and lotions, perfumes etc. It is used as home remedy for dental care tooth ache, sore gums and mouth ulcers and provides relief from pain and stress.

The chemical structure of Eugenol is given in figure 1 below-

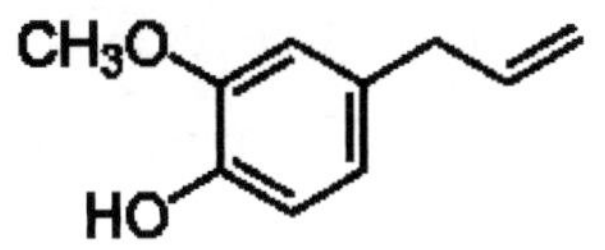

Eugenol (4-Allyl-2-methoxyphenol)

Fig.1:Chemical structure of Eugenol

2.2. Cinnamon Oil

This oil is extracted from Cinnamomum verum and Cinnamomum cassia tree of Lauraceae family. The major constituents of this oil are eugenol, eugenol acetate, cinnamic aldehyde and benzyl benzoate. This oil is used to cure rheumatism, infection of the respiratory tract, arthritis and general pains and period pains. The chemical structures of the few constituents of cinnamon oil is given in figure 2 below-

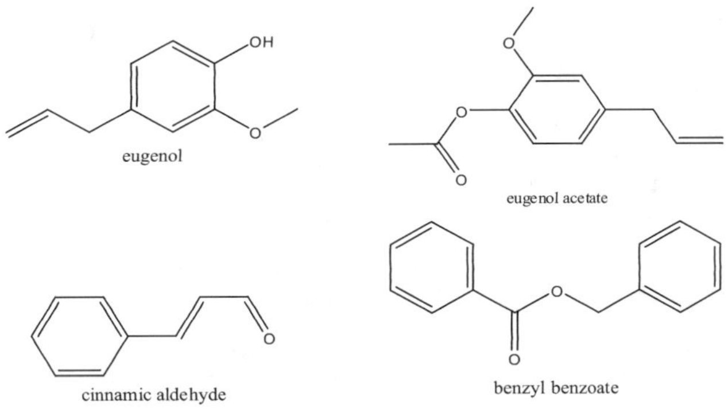

Fig.2:constituents of Cinnamon Oil

3. Experimental Details

The dielectric parameters of two medicinal oils, cinnamon and clove oils and their binary mixtures, dielectric parameters (ε′constant, ε″ loss and tan δ, loss tangent) between the temperature range 30^0C to 50^0C, and between the frequency 10 kHz to 10 MHz have been reported. The dielectric parameters were determined with the help of an impedance gain/phase analyzer HP 4194A and the temperature of oils was maintained with the help of temperature controller Julabo F-25 (microprocessor controlled) and details of these devices have also been published in our previous papers [8-9]. All the samples investigated that is pure and binary mixtures have been designated as 1, 2, 3, 4 and 5, where sample 1 is pure oil (A), sample 2 is (75% oil A+ 25% oil B), sample 3 is (50% oil A+ 50% oil B), sample 4 is (25% oil A+ 75% oil B), and sample 5 is 100% pure oil (B).

4. Results and Discussion

Figure (3) is representing the variation of ε′ and ε″ with $\log_{10}$ frequency (Hz) at indicated percentage of impurity (the second oil is treated as impurity in first oil) and at constant temperature 30^0C for the pure and its binary mixtures (clove oil and cinnamon oil).

Figure (4) is showing the variation of ε′ and ε″ with percentage impurity contents at indicated frequencies and at constant temperature of 30^0C while figure (5) shows the variation of ε′ and ε″ with temperature at indicated percentage of impurity and at constant frequency 50Hz for the pure and binary mixtures of clove oil and cinnamon oil.

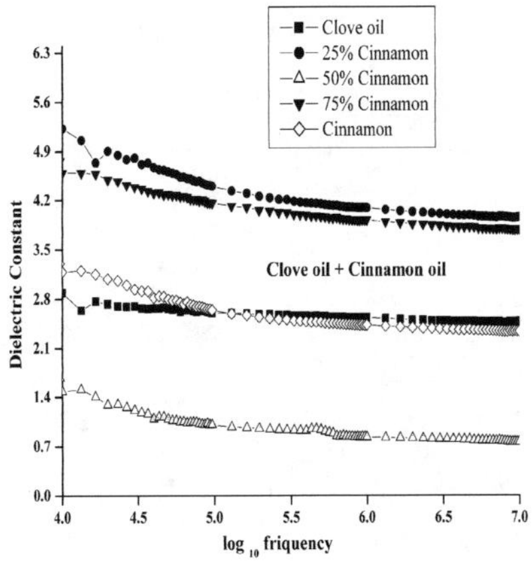

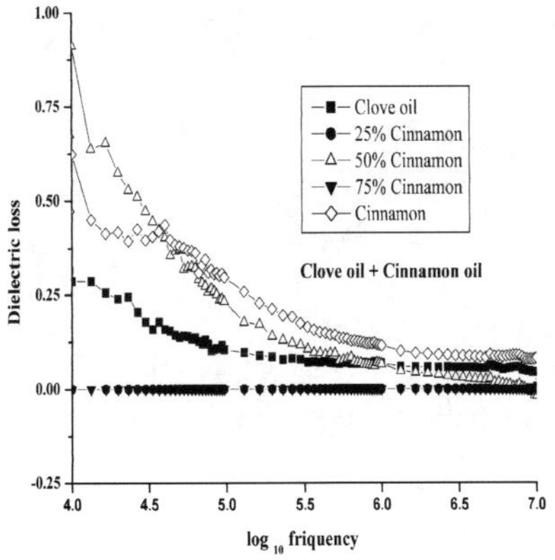

(a) (b)

Fig.3: Frequency dependence of dielectric constant ε′ and dielectric loss ε″of clove oil and cinnamon oil at indicated impurity (in percentage) and at temperature 30^0C.

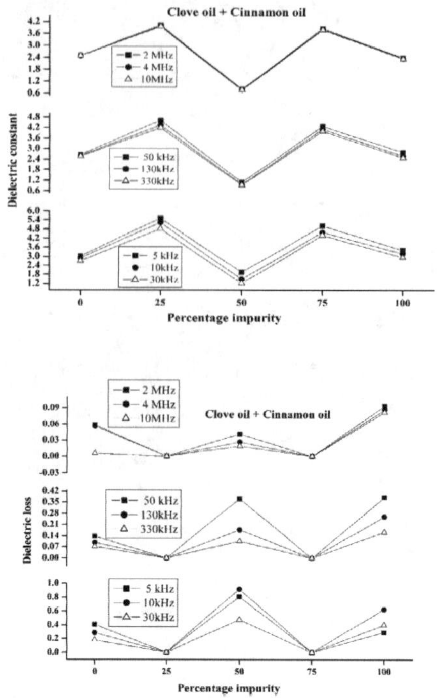

(a) (b)

Fig.4: Variation of dielectric constant ε′ and dielectric loss ε″ with percentage impurity of clove oil and cinnamon oil at indicated frequency and 30^0C.

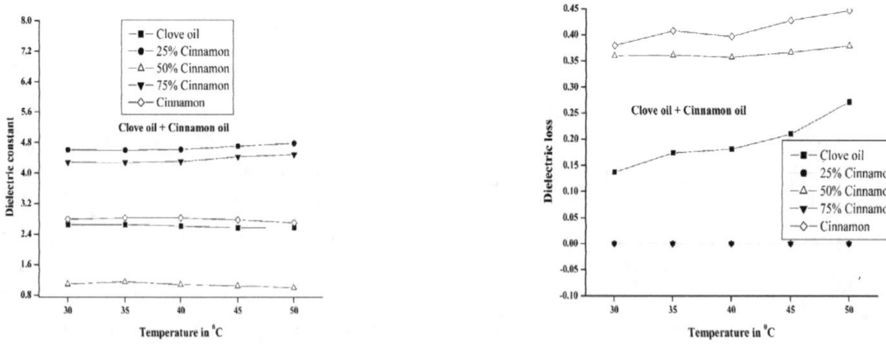

(a) (b)

Fig.5: Temperature dependence of dielectric constant ε′ and dielectric loss ε″of clove oil and cinnamon oil at indicated impurity (in percentage) and at frequency 50 kHz.

4.1. Dielectric Properties of Pure Oils and their Binary Mixtures

4.1.1 Frequency Dependence

Figure (3) is showing the variation of dielectric constant and dielectric loss with $\log_{10}$ frequency at a constant temperature 30^0 C for clove oil and cinnamon oil. It can be seen from the figure (3), that the dielectric constant value of pure clove oil is less than the cinnamon up to certain frequency then reverse effect is seen. For all the samples, pure and binary mixtures, the nature of the dielectric constant curve is almost same i.e. it decreases with increase in frequency. For the samples 2, & 4 the value of dielectric constant is more than the dielectric constant of pure oils while for the sample 3 it is less than the dielectric constant of pure oils. This type of behaviour was reported by Sorichetti et al [10] in 2005.

Figure (3b) is showing the variation of dielectric loss ε'' with natural log of frequency at constant temperature 30^0C for the clove oil and cinnamon oil. The dielectric loss value for the samples 2, 3 and 4 are in between the dielectric loss value of sample 1 and sample 5. The dielectric loss value for the samples 4 is almost zero. The dielectric loss value for sample 1 decreases sharply up to 50 kHz frequency then it decreases slows down. Similar type of nature has also been reported by Saraev et al [11].

4.1.2 Composition Dependence

Figure (4a) is presenting the variation of dielectric constant with percentage impurity at a constant temperature 30^0C for frequency (5kHz, 10kHz, 30kHz, 50kHz, 130kHz, 330kHz, 2MHz, 4MHz, and 10MHz) for the said oils and their binary mixtures. The dielectric constant decreases sharply initially with small impurity addition. When percentage of second oil increases up to 50% then the rate of decrease for dielectric constant is very slow with impurity addition. Therefore, we can say that the dielectric constant values decrease with increase in impurity with very slow rate. The similar nature of the curve for binary mixture of edible unsaturated oils at 300 kHz has been reported by other workers [2]. Figure (4b) is presenting the variation of dielectric loss with percentage impurity at a constant temperature 30^0C for frequency (5kHz, 10kHz, 30kHz, 50kHz, 130kHz, 330kHz, 2MHz, 4MHz, and 10MHz) for the selected oils and their binary mixtures. The dielectric loss values are also decreasing with increase in percentage impurity. The value of dielectric loss up to 10% decreases sharply at all the shown frequencies and then its decrease slows down with increase in percentage impurity.

4.1.3 Temperature Dependence

The variation of ε' with temperature at a frequency of 50 kHz is presented in figure (5a) for the said oils and their binary mixtures. It can be seen that the dielectric constant decreases with increase in temperature for all the samples. From the figure (5a) it is also observed that the dielectric constant values for sample 1 are quite high as compared to the sample 5 and rest of the samples, the dielectric constant values are in between sample 1 and sample 5. This type of behaviour has also been reported by Tasic et al [12].

The variation of dielectric loss with temperature at a frequency of 50 kHz are presented in figure (5b) for the said oils and their binary mixtures, it is observed that the dielectric loss values for all the samples decreases with increase in temperature. Again, it can be seen that in figure (5b) the dielectric loss values for samples 1 are moderately high as compared to the sample 5. The changes can be attributed to constituents of the oils which consist of mixtures of esters of the trihydric alcohol i.e. glycerol and fatty acids.

5. Conclusions

The dielectric constant and dielectric loss of all the oils are showing decreasing tendency with increase in applied frequency, while the dielectric constant is found to decrease with increase in temperature and the dielectric loss is found to increase with increase in temperature. It can also be concluded that the dielectric constant and dielectric loss of binary mixtures of oils is showing unpredictable behavior. The type of study can also be applied to explore the purity of any liquid sample for example, fluoride content in water.

Acknowledgement

Authors are thankful to TEQIP-III scheme of REC Ambedkar Nagar for providing the financial assistant of the project.

References

[1] Pace P E, Westphal W B and Goldblith S A 1968. J. Food Sci. 33 30
[2] Agrawal S and Bhatnagar D 2005. Indian journal of pure Applied Physics 43 624
[3] Hu Lizhi K Toyoda I. Ihara 2008. Journal of Food Engineering 88 151
[4] Ramana CH V, Shoaib Md. and Malakondaiah K 2009. International journal of chemical Engineering Research 1 (2) 161
[5] Singh S P, Chandel V S and Manohar R 2017. European Journal of Advances in Engineering and Technology 4(10) 744
[6] Šegatin N, Žontar T P and Ulrih N P 2020. Foods (9) 900 1
[7] Jadhav B K, Kishanchandra R Khandelwal, Ketkar Anant R and Pisal Sambhaji S 2004. Drug Dev Ind Pharm. 30 (2) 195
[8] Khan Mohammad Shafi and Chandel V S 2010. JPAS. 16 69
[9] Singh S P, Chandel V S and Manohar R 2017. Journal of Ayurveda and Integrative Medicine 9(1) 53
[10] Sorichetti P A and Romano S D 2005. Physics and Chemistry of Liquids 43(1) 37
[11] Saraev, D. B et al 2005. Oil and Gas Business, 1-9.
[12] Tasic R D and Klofutar 1999. Acta Chim. Slov. 46(4) 511

Damage tolerance design of transport aircraft

H Vishnu[1], R Ajith Kumar[1,3], R Sujith[1], R Suraj[1] and K E Girish[2]

[1] Department of Mechanical Engineering, SCMS School of Engineering and Technology, Kerala, India
[2] Director, Bangalore Aircraft Industries Ltd., Bangalore, India

Abstract. In a transport aircraft, there are normally two spars to withstand bending loads. A large part of this bending moment is drawn from the main spar. The main spar in large transport wings is an integrally machined part that is physically connected to the skin and ribs. At some fastener openings, the mechanical fastening contributes to extreme tension concentration where a fatigue crack can originate. This crack will form first in the flange under service loading and then expand into the spar network. If not detected during service and repaired, this crack growth can lead to catastrophic failure. This investigation aims at examining the main spar's alternate structural design to make it damage tolerant. With one more intermediate flange and at a height of one-third from the bottom flange, the spar construction will be in two separate parts. The bottom flange and web can fail in the case of fatigue cracking, but the top flange, web, and intermediate flange will remain intact and can be designed to bear the requisite load limit design. An approach to finite element modelling and analysis is also used to analyze all forms of spar constriction and test the principle of harm tolerance nature.

1. Introduction

Damage tolerance is a feature of a system that refers to its ability to safely retain faults before they can be damaged by repair. The approach to engineering design to allow for risk tolerance is based on the premise that in any system there will be defects and these defects spread with use. In aerospace engineering, this technique is also used to handle the expansion of structural cracks by applying the concepts of fracture mechanics.

The analysis of risk tolerance of aerial structural components [1] provides descriptions of the types of damage that are likely to occur during production and damage caused in operation. Harm resistance is the ability to withstand fracture for a given period of time from the pre-existing cracks and is an integral feature of materials whose failure could result in loss of life or property. The second line of protection against premature fracture is then provided by the integration of structural configurations and materials that are resistant to sub-critical crack growth and final fracture. “Operational survivability” was introduced [4] in 1976 similar to previous work. It contained the principles of fail-safe and hazard tolerance. Aviation regulations for transport aircraft were adopted in 1994, where the definition of operating survivability was assigned as the primary one. In C-130 aircraft [5], due to the consequences of hard use, various cracks are occurred. Damage tolerance design greatly lessens the possibility of such cracks causing a serious accident or requiring excessive down time for emergency repair. Cracks such as these were contained sufficiently by damage tolerance deign to permit safe operation until the using command was able to ferry the aircraft to a repair base.

Under conditions of severe usage, damage can develop and propagate with unexpected rapidity. A damage tolerant structure, however, greatly increases the chances of damage being detected before it reaches critical proportions affecting safe operation. Cracking of a spar web is a typical wing damage mode that may be treated analytically. Damage tolerance is achieved by ascertaining that un-failed wing structure is capable of redistributing the damaged spar web load without causing propagation of the damage. Due to fatigue break, numerous catastrophic failures of aircraft systems occurred in the 20th century [2]. In 1948, a Martin-202 aircraft with 40 passengers crashed because of wing failure caused by a fatigue fracture in a 7075-T6 wing spar joint. In 1948, a Martin-202 aircraft with 40 passengers crashed because of wing failure caused by a fatigue fracture in a 7075-T6 wing spar joint.

2. Problem Definition

The primary aim is to examine the main spar's alternative structural architecture to make it tolerant of damage. At a height of $1/3^{rd}$ from the bottom flange, the spar construction would be in two different components with one or more intermediate flange. The bottom flange and web can fail in the case of fatigue cracking, but the top flange, web, and intermediate flange will remain intact and can be designed to bear the requisite load limit design. In order to analyze all forms of spar construction and verify the principle of damage tolerance architecture, a finite element modelling and analysis approach will be employed.

2.1. Geometrical configuration

The required part is modeled in CATIA and the same was imported into NASTRAN for the analysis. The 3D sketch of the model used for the analysis is shown in the figure 1.

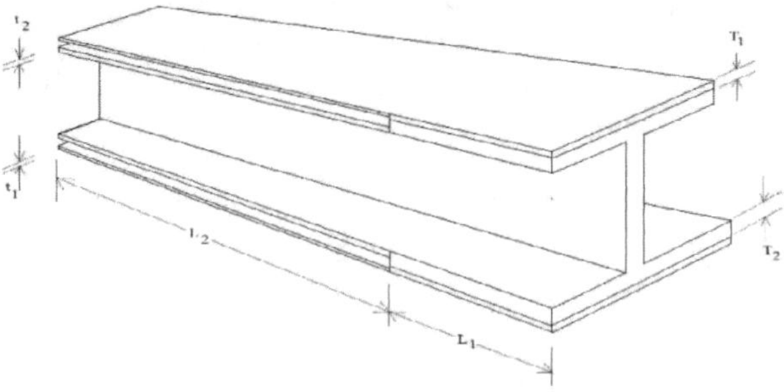

Figure 1. Isometric view of the spar

The material used for the model is an aluminum alloy. The name of the material is AL 2024 T351 alloy. The geometrical specifications of the model is taken as L_1 = 500 mm, L_2 = 4000 mm, T_1 = 2.5 mm, T_2 = 5 mm, t_1 = 1.5 mm and t_2 = 3 mm.

3. Stress analysis of spar

The finite element mesh model consists of quad, tria and bar elements. The total number of elements is shown in table 1. Also the finite element analysis of spar is shown in figure 2.

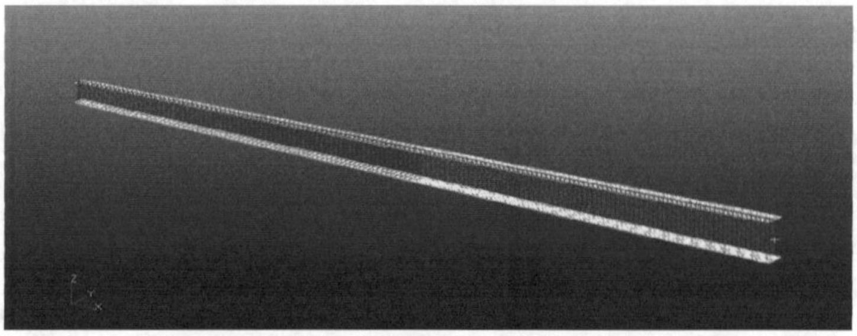

Figure 2. Finite element form of spar

Table 1. Model summary of wing spar

Parameter	Value
Grid points	39265
Bar elements	900
Quad elements	10592
Tria elements	56

3.1. Load calculations

The various steps used for calculating the load in spar in discussed in this section. The all up weight is 2500 kgf while the load factor is taken as 3. Therefore the total design load will be 7500 kgf. This load is distributed on the wings and fuselage as 80% and 20% respectively. Therefore the load on both the wings of aircraft will be 80% of the total design load (7500 kgf) and that will be equal to 6000 kgf. In this study, we are concerned with the single wing. Hence the load on a single wing will be half of the load in both wings and that will be equal to 3000 kgf. Further, the load on front spar will be 75% of load in each wing and it will be equal to 2250 kgf. This load on the spar is given as distributed load on the model as shown in figure 3. The load at tip must be found out as shown in figure 4. It was calculated by momentum balance and was found to be 802.875 kgf.

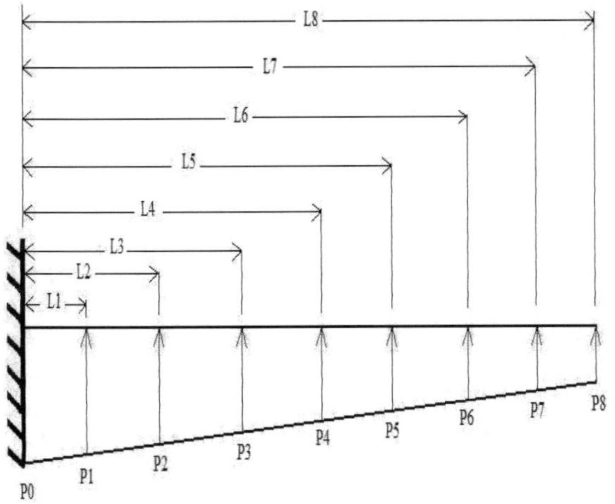

Figure 3. Load distribution of spar

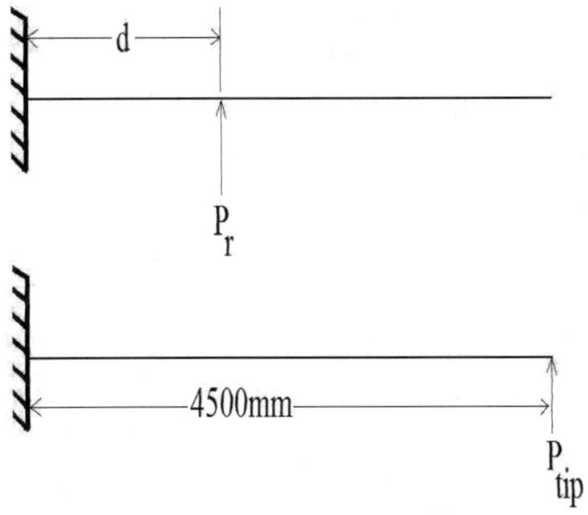

Figure 4. Load calculation at tip

Table 2. Value of load along length

Parameter	Load value along length								
Length (mm)	L_0=0	L_1=562.5	L_2=1125	L_3=1687.5	L_4=2250	L_5=2812.5	L_6=3375	L_7=3937.5	L_8=4500
Load (kgf)	P_0= 438	P_1=394	P_2=328	P_3=272	P_4=227	P_5=206	P_6=175	P_7=119	P_8=92

4. Evaluation of damage tolerance

For the damage tolerance analysis of the spar, stress intensity factor is to be calculated. The stress intensity factor is calculated using Modified virtual crack closure integral method (MVCCI). In order to validate this method, a classical problem was chosen. A rectangular plate with a crack at the center was taken for the validation purpose. The schematic diagram of this plate is shown in figure 4.

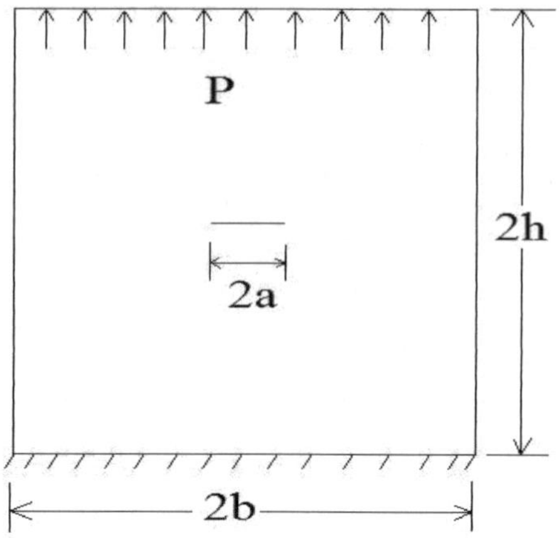

Figure 5. Schematic diagram of rectangular plate with crack at center

The input data for the given problem are P = 2000 kg, 2h = 800 mm, 2b = 400mm and thickness = 2mm. The rectangular plate as shown in the above figure was modelled analysed in the Patran. From the data obtained from the analysis, using MVCCI method, stress intensity factor was calculated. These values are compared with experimental proved value. It was found that SIF values obtained through FEM method is same as the experimentally proved values. Hence this FEM method is validated and it can be used for SIF calculation of any geometry.

4.1. Procedure for the theoretical calculation of SIF

At right angles, in an infinite plane, to a uniform stress field σ, the stress pressure factor for a through crack of length 2a is given by

$$K_1 = \sigma\sqrt{\pi a} \tag{1}$$

If the crack is centrally positioned on a finite plate of 2b width and 2h height, the stress strength element has an estimated relationship given by

$$K_1 = \sigma\sqrt{\pi a} \times \frac{\left[1 - \frac{a}{2b} + 0.326\left(\frac{a}{b}\right)^2\right]}{\sqrt{1 - \frac{a}{b}}} \tag{2}$$

where σ is remotely applied stress, a is half of the crack length and b is half of plate width

4.2. MVCCI method for calculating SIF

In this method, SIF is calculated with the help of data obtained from FEM method. As per this method, stress intensity factor, K is given by

$$K = \sqrt{E \times G} \tag{3}$$

where G is the strain energy release rate which is given by

$$G = \frac{1}{2\Delta a} \times \Delta v \times \frac{F}{t} \tag{4}$$

where E is the Young's Modulus, Δa is length of element near the crack tip, Δv is opening node displacement, F is the force acting at crack tip and t is the thickness of the plate.
The energy dissipated during fracturing per unit of newly generated fracturing surface area is the strain energy release rate (or energy release rate). For fracture mechanics, this quantity is important because the energy that must be given to a crack tip for it to expand must be balanced by the amount of energy dissipated due to the creation of new surfaces and other dissipative processes such as plasticity. Data needed for calculation of K which is to be obtained from the FEM analysis is shown in the figure 5.

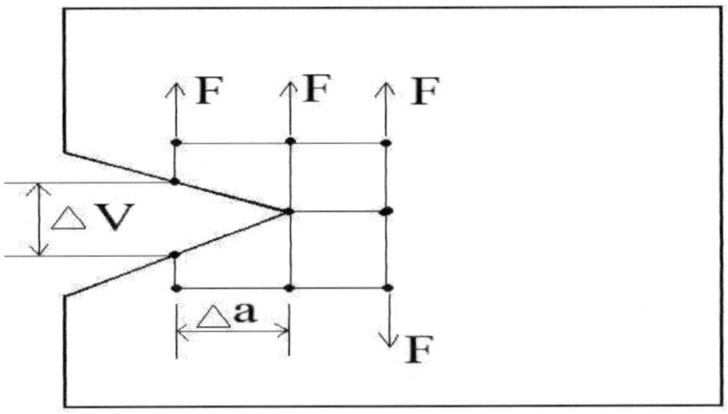

Figure 6. Parameters required for SIF calculation

5. Results and discussions

In the initial sage, the spar was analyzed and the maximum stress occurring region was found out. The stress values occurred at the corresponding regions is given in the table below. In this case, stress occurred due to tension is taken into consideration.

Table 3. Maximum shear stress values at each part

Part	Stress (kg/mm^2)
Top skin	-25.8
Top flange	-31.4
Bottom skin	7.33
Bottom Flange	19.9

After finding out the maximum shear stress region, using MVCCI method, Stress Intensity Factor was calculated. The SIF values for each crack length was found out and tabulated in table 4.

Table 4. Values of K for different crack lengths

Parameter	Crack length and SIF values					
Crack length	5	8	9.4	11	12	46
SIF (K)	11.661	14.454	15.871	17.203	18.443	48.208

The variation values of SIF along the crack length is shown in figure 7. It is clear from the graph that the SIF values are getting higher as the crack length is increased and hence the flange will get failed. After the complete failure of flange, the crack get initiated through the web and propagate throughout the spar which result in the failure of the aircraft structure.

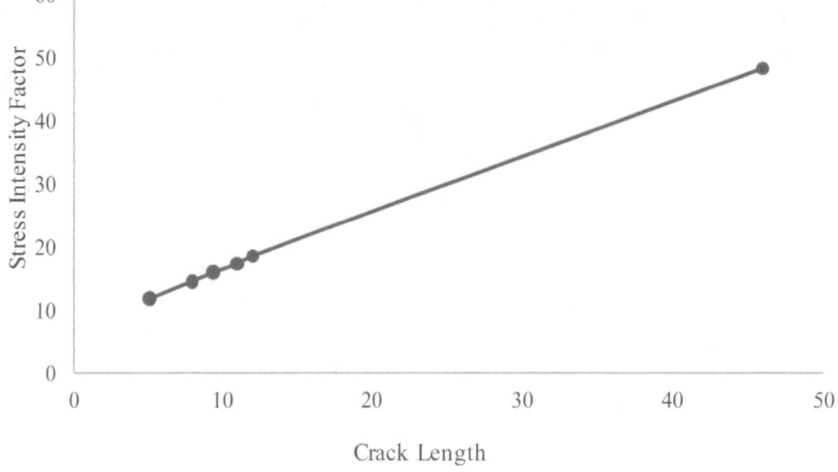

Figure 7. Variation of SIF along crack length

In the further step, crack was simulated through the web and SIF was calculated for the different crack lengths. The variation of SIF for different crack lengths in web along crack length is plotted in figure 8. From the graphical plot it is clear that SIF value keep on increasing which finally results in the failure of the structure. So the design was further modified.

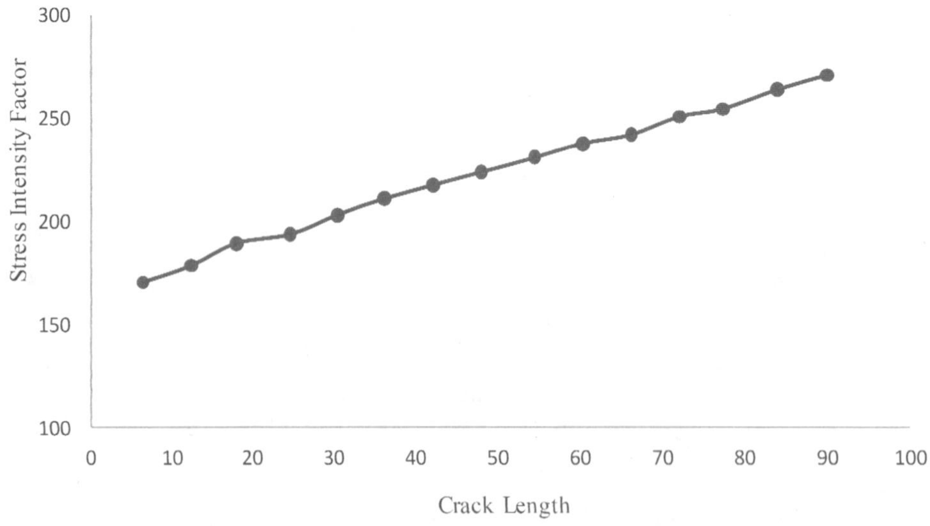

Figure 8. Variation of SIF along crack length without flange

So in order to modify the model, an intermediate flange is introduced at a height one-third from the bottom flange and the same procedure was followed. Here also a graph is plotted to depict the variation of SIF along the crack length with intermediate flange in figure 9. It can be seen from the plot that the SIF value got reduced when the crack is approaching near the intermediate flange. So it is clear that by introducing an intermediate flange, the crack propagation can be arrested to an extent, hence the wing spar can be made as damage tolerant.

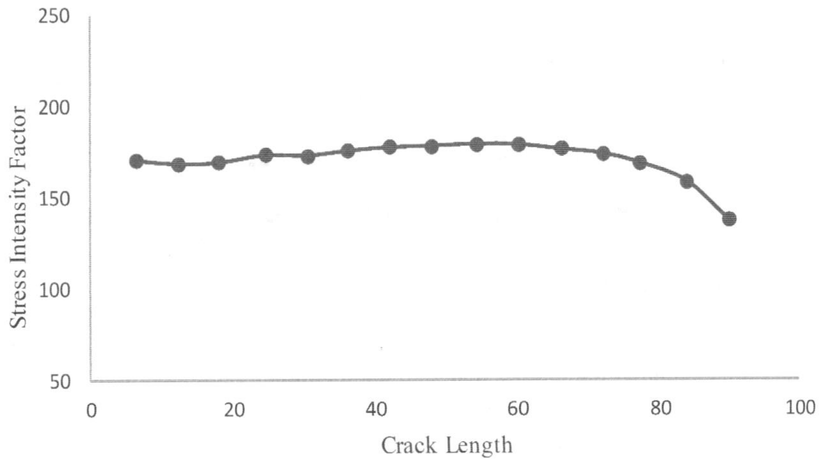

Figure 8. Variation of SIF along crack length with intermediate flange

6. Conclusions

In this study the damage tolerance design of transport aircraft is studied. The stress analysis is carried out using finite element method (FEM). The stress on the bottom skin and bottom flange are obtained and it was found that stress on bottom flange is greater than that of bottom skin. After doing the finite element analysis, the Stress Intensity Factor (SIF) was calculated verified using MVCCI method. The damage tolerance calculation is carried out by finding out the SIF at different crack lengths. Modifications was done on the model to minimize the damage tolerance and it was found that the structure will fail in all cases until the use of intermediate flange. The stress analysis and SIF calculation has been carried out in the presence of intermediate flange. The results with the presence of intermediate flange show that SIF decreases when the crack get nearer to intermediate flange.

7. References

[1] Kiran R, Sachin R,Hemanth U, Ashok S, Hakeem S A and Narayana K B 2007 Damage tolerance analysis of aero structural components *Tata Consultancy Services*

[2] Jaap Schijve 2009 Fatigue damage in aircraft structures, not wanted, but tolerated?' *International Journal of Fatigue* vol 31 pp 998–1011

[3] Friedrich G B, Azzedine C and Hans A R 1999 Computational fracture analysis by means of Virtual Crack Closure Integral Method *MECOM99 6th Argentine Congress on Computational Mechanics, Mendoza*

[4] Nesterenko G I, Nesterenko B G 2009 Ensuring structural damage tolerance of Russian aircraft *International Journal of Fatigue* vol 31 pp.1054–1061

[5] Shuler W T and Morcock D S 2012 Damage tolerance and logistic transport design *J.Aircraft* vol 6 no 5 pp 416–424

[6] Molent L 2010 Fatigue crack growth from flaws in combat aircraft *International Journal of Fatigue* vol 32 pp 639-649

[7] Barter S, Molent L, Goldsmith N, Jones R 2005 An experimental evaluation of fatigue crack growth *Engineering Failure Analysis* vol 12 pp 99–128

[8] Shuler W T and Morcockj D S 2012 Damage tolerance and logistic transport design *Journal of Aircraft* vol 6 no 5 pp 416–424

[9] Castro P M, Tavares S M O, Richter T V , Matos P F P, Moreira P M G P and Silva L F M 2010 Damage tolerance of aircraft panels *Revista da Associação Portuguesa de Análise Experimental de Tensões*

[10] Grandt F 2011 Damage tolerant design and nondestructive inspection - Keys to aircraft airworthiness *Procedia Engineering* vol 17 pp 236–246
[11] Adeel M 2008 Study on damage tolerance behavior of integrally stiffened panel and conventional stiffened panel *World Academy of Science, Engineering and Technology* vol 45 pp 1026–1030
[12] David B 1982 Elementary engineering fracture mechanics *Martinus Nijhoff Publishers*, pp 408–433
[13] Silva L F M, Gonçalves J P M, Oliveira F M F and de Castro P M S T 2000 Multiple-site damage in riveted lap-joints: Experimental simulation and finite element prediction, *International Journal of Fatigue,* vol 22 pp 319–338
[14] Rans C D, Alderliesten R C and Straznicky P V 2009 Assessing the effects of riveting induced residual stresses on fatigue crack behaviour in lap joints by means of fractography, *International Journal of Fatigue,* vol 31 pp 300–308
[15] Horst P 2006 Widespread fatigue damage–qqqqqqqAn issue for aging and new aircraft *Key Engineering Materials,* vol 324-325 pp 1–8

Analysis on impact of 'switching off light' event on indian grid frequency

Dr.T.Logeswaran[1], N.Loganathan[2] Dr.S.Senthil kumar[3], Dr.R.Uthirasamy[4] Ravivarman Shanmugasundaram[5]

1 Associate Professor, Department of Electrical and Electronics Engineering, Kongu Engineering College, Erode, Tamil Nadu
2 Assistant Professor, Department of Electrical and Electronics Engineering, Sri Krishna College of Engineering and Technology, Coimbatore, Tamil Nadu
3 Assistant Professor, Department of Electrical and Electronics Engineering, University College of Engineering Ariyalur, Ariyalur, Tamilnadu
4 Associate Professor, Department of Electrical and Electronics Engineering, KPR Institute of Engineering and Technology, Arasur, Coimbatore, Tamilnadu
5 Professor, Department of Electrical and Electronics Engineering, Vardhaman College of Engineering, Hyderabad.
Email: logeskongu@gmail.com, logukirsh@gmail.com, senthil21575@gmail.com, rusamy83@gmail.com, ravivarman@vardhaman.org
Corresponding author's e-mail: logeskongu@gmail.com

Abstract. Electricity places a vital role in the development of economy of a country. Electrical power grids are the most complex of all human engineered systems. This paper presents the progress made in the energy grid and outlines the power scenario in India. It reports the impact created by the event "Switching off lights" on grid frequency. Role of POSOCO in the management of the event has been vividly described. The paper provides us the most important input for prioritizing measures or steps to be taken to cope up with the situation and possible action plan have been summarized.

1. Introduction

Of all the inventions made by human, electrical power grids are said to be most complex engineering systems. Electricity generation is one among the eight core industries of India. According to a report from international energy agency, India stands in third position globally as the largest producer of electricity and it also holds the third place as largest consumer of electricity. National Electricity grid of India has an installed capacity of 370.106 GW as of March 2020. Other highlight is that India has moved from power deficit to power surplus.

Prime minister asks the people of India to light the candle to drive out the darkness created by Corona virus and to switch-off the lights for nine minutes. The paper describes the impact created by the event "Switching off light" over the grid frequency. Section I deals with the introduction. Section II describes the evolution of Indian grid. Section III describes the "light off" event. Section IV portrays the results and waveforms obtained during the event. Section V highlights the lessons learnt from the event and conclusion is brought out in sectionVII.

2. Indian Grid

In the country of India, electricity production began in the year 1897. It initiated with setting a hydroelectric power plant near Darjeling. Later in the year 1899, Calcutta Electric Supply Corporation built a thermal power station. Initially during these years, the power sector growth is very slow. Till the end of the year 1920, the total installed capacity is about 130 MW. Though they saw a slow growth at the initial stage, they got a real boost during the period from 1920 to 1940 because of increase in industrialisation. [1],[2]

Ganga –Canal grid in UP is said to be India's first grid connected system. It consists of 59 transformer substation and it was developed in the year 1930. In the year 1937, Mettur Thermal Power Plant and Pykara Power Plant were interconnected and Papanasam hydroelectric station was also attached to this system in the year 1945[3].The importance of interconnection of grid power systems was realized well just after the independence. The interconnection of systems came into existence but the important feature, i.e., exchange of power, of any interconnected system is missing [4].

In 1960's, to enhance the progress of power sector, it is planned to have regional grids. In 1981, Indian government had a policy decision of developing national grids, which leads a way to the development of integrated operations of transmission systems that includes central and regional one [5].

The objectives of national grid are to ensure reliable operation, transfer of power from a surplus area to deficit area and so on. Indian grid is divided into five regional grids:1) Northern 2) Eastern 3) Western 4) North- Eastern and 5) Southern [5]. This motto "One Nation - One Grid - One Frequency" has been achieved on December 2013 when southern grid is connected to central grid. Grid control is performed at different levels of hierarchy known as

- National Load Despatch Center (NLDC)
- Regional Load Despatch Center (RLDC)
- State Load Despatch Center (SLDC) and
- Area Load Despatch Center (ALDC) [6]

Power grid is functioning in a way that it make possible the transfer of power from one area where there is deficit to another area where power demand is more.

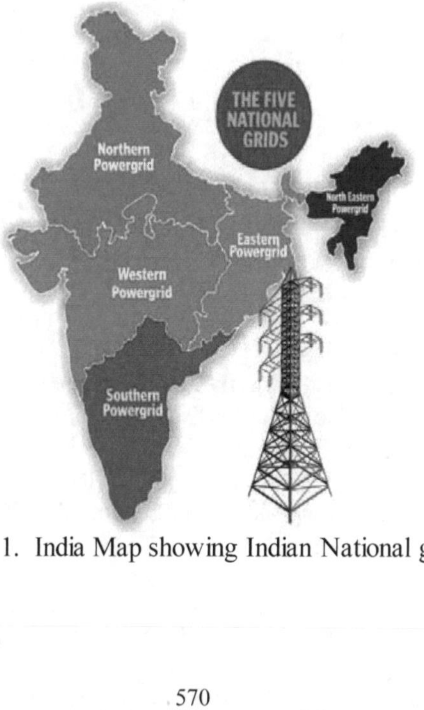

Figure 1. India Map showing Indian National grids

Total installed capacity as on March 2020 is shown in the table 1 and table 2.

Table 1. Installed Capacity of India (Sector Wise)

Sector	MW	% of Total
Central Sector	93,477	25.2%
State Sector	103,322	27.9%
Private Sector	173,308	46.8%
Total	3,70,106	

Source: Central Electricity Authority (CEA)

Table 2. Installed Capacity of India (Fuel Wise)

Fuel	MW	% of Total
Total Thermal	2,30,600	62.8%
Coal	1,98,525	54.2%
Lignite	6,610	1.7%
Gas	24,937	6.9%
Diesel	510	0.1%
Hydro (Renewable)	45,699	12.4%
Nuclear	6,780	1.9%
RES* (MNRE)	87,028	23.5%
Total	370,106	

Source : Central Electricity Authority (CEA)

3. Switch Off Light Event - Case Study

Switching off light event is elucidated in different stages. It starts from the call by PM, then the fear spread among the public, the plan by the POSOCO and also the execution by them and the lessons learnt from the event.

3.1. The Call

In order to dispel the darkness created by the corona virus, Prime Minister of India urges the people of India to light a candle and to switch off lights for about 9 minutes by 9.00 pm on April 5th 2020.

Figure 2. The Call for the event Source: The Economic Times (April 4th 2020)

3.2. The Fear

The ideas floating on the social media is that they have to switch off all the household equipments and gadgets for safety. By considering the messages, the Ministry of Power issues a notice which clarify the necessity of switching off lights only and to keep the other home appliances in ON condition.

3.3. The plan

The National power grid observes this event as a unique event. It anticipates a 12.9 GW dip in demand considering only the lights to be turned off. POSOCO had a meeting with officials of all the State Load despatch centre. Persons from major hydro power plants also participated in the meeting. POSOCO issued an advisory note. On April 4th night mock exercise of hydro-ramping was made. The same is repeated on April 5th morning.

THE ECONOMIC TIMES

Business News › Industry › Energy › Power

Search for News

Benchmarks › Nifty CLOSED 9,859.90 306.55

Stock Screener› Top Growth Stocks | Stocks with Regular Payout | Mid-Cap Growth Stocks

Search, Select & Invest in Top Stocks

FEATURED FUNDS Axis Long Term Equity Direct Plan-Growth ★★★★★ 5Y RETURN 8.39 % INVEST NOW

National power grid terms the 9 PM light-off event unprecedented, estimates 12.9 GW dip in demand

BY DEBJOY SENGUPTA, ET BUREAU | UPDATED: APR 04, 2020, 07.27 PM IST

Figure 3. The POSOCO's view over the event Source: The Economic Times (April 4th 2020)

3.4. The Execution

The actions taken are as shown in the table III. The hydro generation was ramped accordingly, so that it meets the decrease or increase in demand. Some other advance actions such as switching off a transmission line were taken to keep the voltage within limit. Frequency will be brought to 49.9 Hz at 20:30 Hrs and 50.15 Hz at 21:09 Hrs. The event was managed in a marvellous way and there were no unpleasant happenings and also the power systems are well maintained within tolerable limits. POSOCO thanks all the stakeholders for their cooperation to make the event successful.

Table 3. Actions taken on the day of Light OFF Event

S.No	Time	Actions taken
1	20:45 Hrs	Hydro generation is maximised
2	20:45 Hrs to 21:10 Hrs	Generation reduction of 17543 MW (to match the demand reduction)
3	20:45 Hrs to 21:10 Hrs	Thermal, Gas and Wind generation reduced
4	21:10 Hrs to 21:27 Hrs	Hydro generation was ramped up (to match the increase in demand)

4. Results and Discussions

4.1. Change in frequency and Voltage:

The Power Quality analyser instrument [Fluke Model 1748] is used for recording the voltage and other related electrical parameters at LT side. The change in frequency during the event is noted. Variation of the voltage during the event is recorded and corresponding waveforms are shown below.

Inferences from the waveforms are as follows. There is a large dip in frequency during the event i.e., between 20:50 Hrs to 21:15 Hrs.

4.2. Variation of frequency

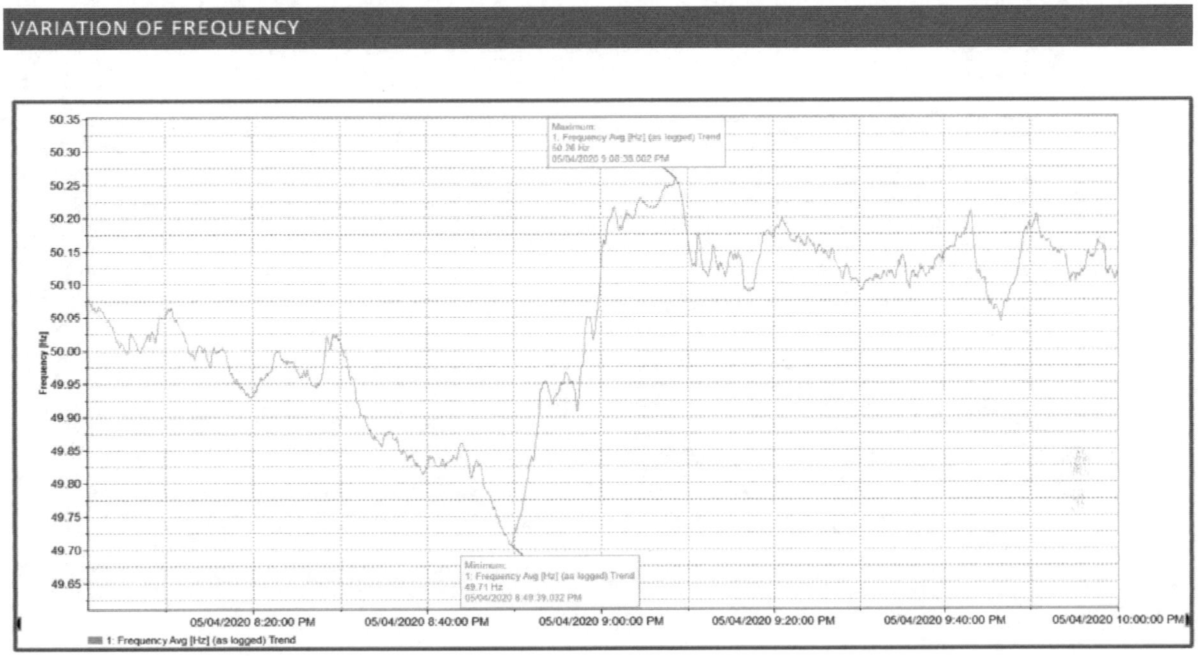

Figure 4. Frequency Waveform

4.3. Variation of voltage

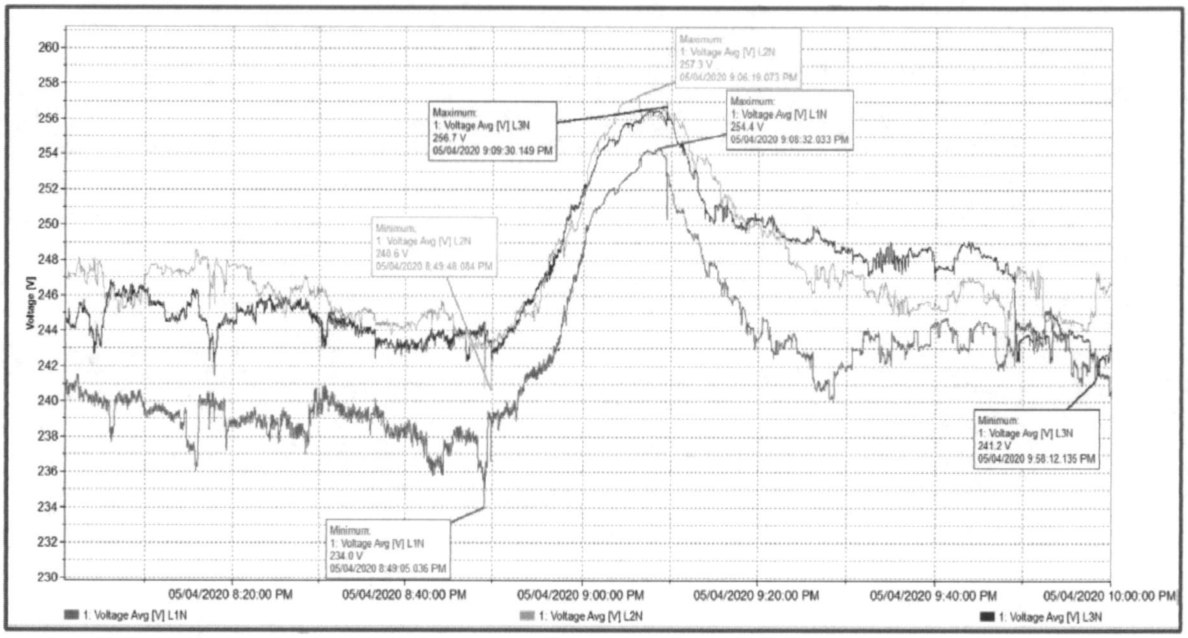

Figure 5. Voltage Waveform

4.4. Overview of frequency and voltage

ES.016 Pinnac 9 Min 5 Apr 2020			
Voltage [V]	L1N	L2N	L3N
Max	**254.5 V** 05/04/2020 9:07:38 PM	**257.6 V** 05/04/2020 9:06:20 PM	**256.8 V** 05/04/2020 9:07:17 PM
linear Avg	**242.5 V**	**247.5 V**	**247.1 V**
Min	**231.4 V** 05/04/2020 8:48:21 PM	**239.4 V** 05/04/2020 8:49:48 PM	**238.7 V** 05/04/2020 9:53:05 PM
Current [mA]	L1	L2	L3
Frequency [Hz]	L1N		
Max	**50.26 Hz** 05/04/2020 9:08:38 PM		
linear Avg	**50.04 Hz**		
Min	**49.71 Hz** 05/04/2020 8:49:37 PM		
V THD [%]	L1N	L2N	L3N
Max	**3.0 %** 05/04/2020 8:06:20 PM	**2.7 %** 05/04/2020 8:06:07 PM	**3.1 %** 05/04/2020 8:18:09 PM
linear Avg	**2.7 %**	**2.4 %**	**2.7 %**
Min	**1.9 %** 05/04/2020 9:03:33 PM	**1.8 %** 05/04/2020 9:03:33 PM	**1.9 %** 05/04/2020 9:06:27 PM

Figure 6. Overview of frequency and Voltage

4.5. *Variation of frequency and voltage*

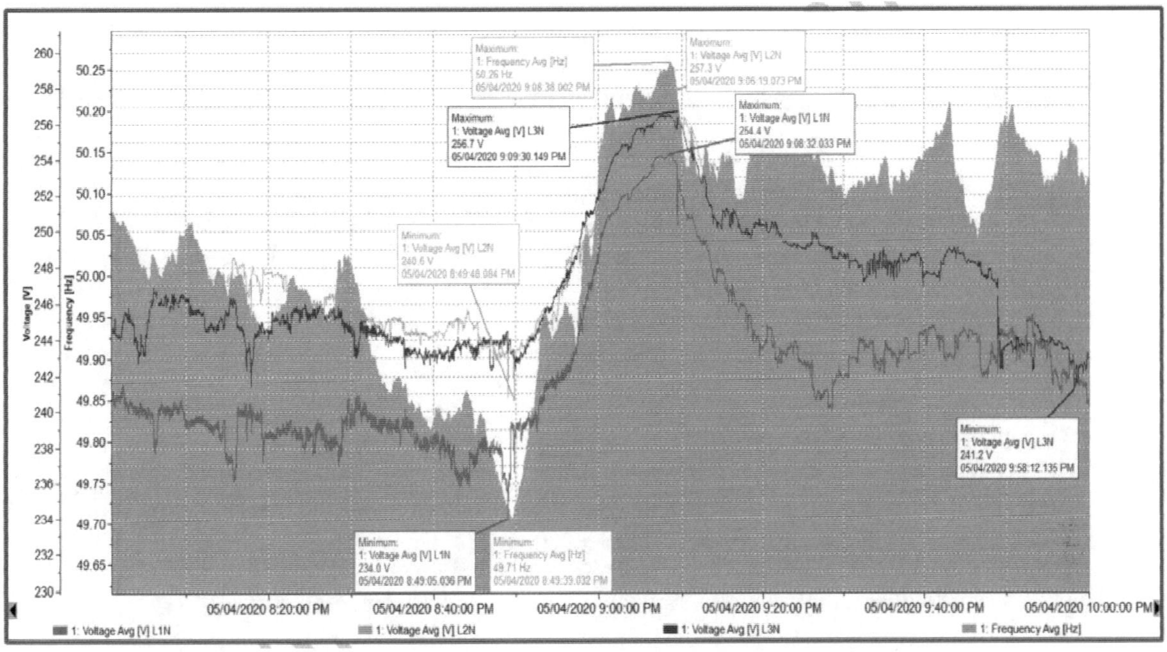

Figure 7. Voltage and frequency Waveform

5. Lessons Learnt

The lessons learnt from the light off event are as follows.

- The power grid has flexibility which is proven by the event on April 5th 2020.
- The flexion is well executed by the dedicated utility personnel.
- The POSOCO plans to drop the grid frequency before the starting the event has worked well.
- In real situations, flexion cannot be done manual and therefore the automation is necessary and AI becomes essential.
- This study is relevant to higher penetration of renewable energy into grid.

6. Conclusions

Utilization of renewable energy is highly increasing and higher renewable energy penetration into the grid is seen and need to focus this area for research. The world economic forum's framework on "future of electricity" recommends these terms: electrification, decentralization and digitization for the transformation of future electricity. Progress towards Smart Grid is essential and Smart engineers have to play their role to reap the full benefits of smart grid and also to make the Indian power sector more prosperous.

Acknowledgment

Author wishes to acknowledge, Sachin Shelar. Assistant Professor, Department of Electrical Engineering, AISSMS Institute of Information Technology, Pinnac Sumanshree, Karvenagar, Pune for providing recordings and data for this paper. Authors wishes to acknowledge the POSOCO for their preliminary report on the event. This paper is dedicated to all smart Engineers of the future.

References

[1] Mohd. Rihan, Mukhtar Ahmad, M. Salim Beg, "Developing Smart Grid in India: Background and Progress" IEEE 2012.
[2] K. K. Aggarwal, "Problems of future power development in India", Power Engineer, July 1966, pp. 88-92
[3] K. P. P. Menon,"Power System Interconnections in India", Power Engineer, October 1951, vol. 1, No. 4 pp. 172-185.
[4] K. M. Sankaran, "Establishment of Super–grids and Organizational Procedure for Their Control", Power Engineer, October 1960, vol. 10, N. 4. pp. 179-188.
[5] M. Hayath, S. Swayambhu, "Interconnection between the Hirakud Hydro-electric System and the Rourkela Thermal Power Station", Power Engineer, October 1960, vol. 10, N. 4 pp. 176-178.
[6] S.K.Soonee, P.K.Agarwal, and R.K.Porwal, "A Novel Settlement/Arbitrage Scheme-Use of Asynchronous Inter-Regional links in India", in proc. ICPSODR-2006, Dept. of Electrical Engineering, ITBHU, Varanasi.
[7] Vivek Pandey, "Electricity Grid Management in India- An Overview", Electrical India, Nov, 2007, Vol. 47, No. 11.
[8] www.cea.nic.in
[9] Preliminary Report on Pan India Lights Switch Off Event on 5th April 2020 by Power System Operation COrporation Limited (POSOCO), 2020
[10] www.posoco.in
[11] (2020) https://powermin.nic.in/en/content/power-sector-glance-all-india

Impact of dimensionless length scale parameter on material dependent thermoelastic attenuation and study of frequency shifts of rectangular microplate resonators

Resmi R. [1,a)], V. Suresh Babu[2] and M.R. Baiju [3]

[1] *University of Kerala , LBS Institute of Technology for Women, Thiruvananthapuram, Kerala-695012, India.*
2 APJ Abdul Kalam Technological University, Govt. Engineering College, Wayanad, Kerala-670644, India.
3 University of Kerala; Kerala Public Service Commission, India; Kerala-695004, India.

[a)]Corresponding author: resmilbs@gmail.com, resmi@lbsitw.ac.in

Abstract. Quality factor is one of the decisive performance parameters of micro/nano plate resonators based sensors and filters .Various energy dissipation mechanisms limit the maximum attainable quality factor, and in micro/nano scales thermoelastic damping is a critical energy loss mechanism. When the devices are scaled down, in order to accurately model micro/nano scale resonators, non-classical elasticity theories like Modified Couple Stress Theory (MCST) are valid. In this paper, by including a length scale parameter (l) the size effects were incorporated in the analysis and thermoelastic energy dissipation and related attenuation were found to be diminished by selecting a dimensionless length scale parameter i.e. l/h. Thermoelastic attenuation and frequency shifts related to thermoelastic damping in microplate rectangular resonators with five structural materials (SiC, polySi, diamond, Si and GaAs) were investigated with $l/h = 0$.The attenuation related to thermoelastic energy dissipation and frequency shifts were found to be minimum for polySi material. With the inclusion of size effect, the impact of dimensionless length scale parameter on thermoelastic attenuation of polySi based microplate resonators was also explored. The conventional thermoelastic damping analysis in rectangular plate was modified by applying Modified Couple Stress Theory and the impact of dimensionless length scale (l/h) on thermoelastic energy attenuation was simulated numerically using MATLAB 2015.

Keywords: Micro/nanoplate rectangular resonators, thermoelastic attenuation, frequency shifts, size effects, dimensionless length scale parameter.

Introduction

Micro and nanoelectromechanical systems (MEMS and NEMS) based thin rectangular plates are one of the major structures emerged as the essential components for sensors [1] and resonators [2]. Most of these applications exploited the physical and thermal properties of thin film materials for focused applications [2] . Micro/nano structures are universally used due to their unique advantages of high sensitivity, mass and cost, high surface-to-volume ratio etc. [3]

When resonators are used Quality Factor (QF) is one of the essential features to ensure high sensitivity and resolution and gives a representation of dissipated energy due to both energy loss mechanisms- extrinsic and intrinsic [3]. Thermoelastic damping (TED) is a significant intrinsic energy loss mechanism at micro scales which decides the maximum attainable QF in the resonator denoted by Q_{TED} [4] .Thermoelastic damping limited energy attenuation arises due to the coupling between temperature and elasticity of the materials which declines QF. Presence of TED in different structures was identified in various research works [5]. The existence of TED as a prominent energy loss mechanism in homogeneous, isotropic, Euler–Bernoulli micro-beams was first identified by Zener in 1937 [6]. Lifshitz and Roukes studied the impact of different geometrical parameters on TED and derived an exact closed-form expression in slender beams [7]. Nayfeh and Younis expressed Q_{TED} of microplates analytically in terms of their structural mode shapes by utilizing the perturbation method [8] .

Classical theories are inadequate to explore the size effects due to the lack of the material length scale parameter; hence, higher order continuum theories must be used to predict the size dependencies. The size dependence effects of materials have been experimentally observed by scaling down the microstructure size [9]. Indeed, the size effect is essential in mechanical analyses of various structures [10]. To include the size effects in the vibration analysis, non-classical continuum theories are used, the most common is the Modified Couple Stress Theory (MCST) developed by Yang et al [11]. Tsiatas et al. analysed isotropic micro-plates based on the Kirchhoff plate model and suggested using MCST for static analyses that need only one material length scale parameter, which is enough to capture the size effect [9]. Zhong et al. analysed thermoelastic damping and the factors affecting TED in the size-dependent microplate resonators based on modified couple stress theory [12]. Vahid Borjalilou et al. analysed TED in microplates based on MCST and the dual-phase-lag heat conduction model [13]. Resmi R. et al. explored thermoelastic damping dependent quality factor, thermoelastic frequency and figure of merit of rectangular plates applying modified couple stress theory [14].

In this paper, Section 2 gives the analytic equations of thermoelastic attenuation related to energy dissipation and frequencies of a rectangular microplate applying size effect based on MCST. Section 3. presents the results and discussions, where section 3.1. elaborates material dependent thermoelastic attenuation and frequency shifts. Section 3.2. discusses the impact of dimensionless length scale parameter on attenuation. The plots of the attenuation and frequency shifts for different materials are also illustrated in Section 3.1. The conclusions of the work are given in Section 4.

2.Expression for Thermoelastic Energy Attenuation and Frequency Shifts of a Rectangular Microplate

In classical elasticity theory, stress and stain energy is related to strain tensor and independent of the rotation vector . According to MCST, presented by Yang et al. [11] the strain energy density is a function of both strain tensor (conjugated with stress tensor) and curvature tensor (conjugated with couple stress tensor).

In our analysis, a rectangular microplate of length L, width W, and constant thickness h under plane stress was analysed using five different materials at a temperature of T_0=293K.In the study, the coupled thermoelastic fields are assumed and the equations to have the simple-harmonic vibrations as [12]

$$W(x,y,t) = w_0(x,y)e^{iwt}, \qquad \theta(x,y,z,t) = \theta_0(x,y,z)e^{i\omega t} \tag{1}$$

where W indicates the mid plane displacement of the microplate , θ is the temperature distribution in the microplate and t is the time.

The motion equation of the microplate can be given by

$$\left[Ƥ_0 + Ƥ_1 + \frac{Ƥ_0}{2}\acute{r}\,(1 + f(\omega))\right]\nabla^4 W - \rho h\ddot{W} = 0 \tag{2}$$

where f (ω) is the complex function; $\acute{r} = E\alpha^2 T_0/C_p$ (E-Young's modulus,α-thermal expansion coefficient, T_0-equilibrium temperature, C_p -specific heat capacity at constant pressure ; Ƥ is a parameter depends on material properties such as E (Young's Modulus), υ (Poisson's ratio) and thickness of the microplate and the variables are $Ƥ_0 = Eh^3(1-\upsilon)/\ 12(1+\upsilon)(1-2\upsilon)$ and

$Ꝑ_1 = El^2h/2(1 + \upsilon)$.

In Eq.(2), $Ꝑ_0\ \nabla^4 W$ and $\rho h\ddot{W}$ are associated with the classical thermoelastic model ; $Ꝑ_1\ \nabla^4 W$ represents the size effect and $\frac{Ꝑ_0}{2}\acute{r}\ (1 + f(\omega)\ )$ denotes the effect of thermoelastic coupling.

When the microplate is simply supported, the modes of vibration (normal) of the microplate are

$$w_0(x, y) = \sum_{m=1}^{\infty}\sum_{n=1}^{\infty} B sin\frac{m\pi x}{\mathrm{L}} sin\frac{n\pi y}{\mathrm{W}} \tag{3}$$

where B is a constant, m and n are modal numbers.

For a simply supported plate [12][12],

$$\left[\pi^4\left(\frac{m^2}{\mathrm{L}^2}+\frac{n^2}{\mathrm{W}^2}\right)^2 - \frac{\rho h\omega^2}{Ꝑ'}\right] Be^{iwt} sin\frac{m\pi x}{\mathrm{L}} sin\frac{n\pi y}{\mathrm{W}} = 0 \tag{4}$$

where $Ꝑ' = Ꝑ_0 + Ꝑ_1 + Ꝑ_0/2.\acute{r}\ [1 + f(\omega)]$.

For a microplate resonator, $\pi^4\left(\frac{m^2}{\mathrm{L}^2}+\frac{n^2}{\mathrm{W}^2}\right)^2 - \frac{\rho h\omega^2}{Ꝑ'} = 0$ (5)

The thermoelastic frequency dispersion relation between ω_i and mode numbers (m, n) for the thermoelastic plate can be obtained as

$$\omega_i = \pi^2\sqrt{\frac{Ꝑ'}{\rho h}\left(\frac{m^2}{\mathrm{L}^2}+\frac{n^2}{\mathrm{W}^2}\right)} = \omega_0\sqrt{1+\frac{Ꝑ_1}{Ꝑ_0}+\frac{\Delta_E}{2}[1+f(\omega)]} \tag{6}$$

where ω_0 is the isothermal value of the eigenfrequency of the microplate,

and $\omega_0 = \pi^2\sqrt{Ꝑ_0/\rho h}(m^2/\mathrm{L}^2 + n^2/\mathrm{W}^2)$.

From Eq. (6), frequency is complex; the real part Re(ω) indicates the new eigenfrequencies of the rectangular microplate resonators in the presence of thermoelastic coupling, and the imaginary part Im(ω) gives the attenuation of the vibration.

The real and imaginary parts of the frequency can be extracted from Eq. (6)

$$Re(\omega_i) = \omega_0\left[\sqrt{1+Ꝑ_1/Ꝑ_0} + \frac{\acute{r}}{4\sqrt{1+Ꝑ_1/Ꝑ_0}} \times \left(1 - \frac{6}{\xi^3}\frac{sinh\xi - sin\xi}{cosh\xi + cos\xi}\right)\right] \tag{7}$$

$$Im(w) = \omega_0\frac{\acute{r}}{4\sqrt{1+Ꝑ_1/Ꝑ_0}}\left(\frac{6}{\xi^3}\frac{sinh\xi - sin\xi}{cosh\xi + cos\xi} - \frac{6}{\xi^2}\right) \tag{8}$$

where $\xi = b\sqrt{\omega \mathrm{u} t_0/2\chi}$.

The amount of thermoelastic damping, expressed in terms of the inverse of the quality factor, is given by [12]

$$Q^{-1} = 2\left|\frac{|Im(\omega)|}{Re(\omega)}\right| \tag{9}$$

3. Results and Discussions

To validate the analytical study, rectangular plate resonators with simply supported boundary condition, length L= 200 μm, width W= 200 μm, and thickness h=10μm using five different materials were examined at a temperature of T_0=298 K. The mechanical and thermodynamic properties of all the materials are taken from [14]. Numerical simulations of the analytical expressions were done with MATLAB 2015.

3.1. Thermoelastic Attenuation and Frequency Analysis

The universal behaviour of the normalized frequency shift $[Re(\omega) - \omega_0]/\omega_0 \acute{r}$ and the normalized attenuation $Im(\omega)/\omega_0\acute{r}$ as a function of the dimensionless variable ξ of a polySi based rectangular microplate is shown in Figure 1($\xi = h\sqrt{\omega_0/2\chi}$ where h - thickness of the plate; ω_0 - isothermal value of the eigenfrequency of the microplate and χ -thermal diffusivity of the material used). The normalized attenuation which is a measure of thermoelastic energy dissipation in microplate resonators, increases first, peaks at ξ=2.2246, and then decreases again with the dimensionless variable ξ, as depicted in Figure1. Thermoelastic energy dissipation shows Lorentzian behaviour and is found to be tightly bound between two Lorentzians in the variable $\frac{\xi^2}{\sqrt{24}}$ and varies as ξ^2 for small values of ξ and as $\frac{1}{\xi^2}$ for large values of ξ [7].

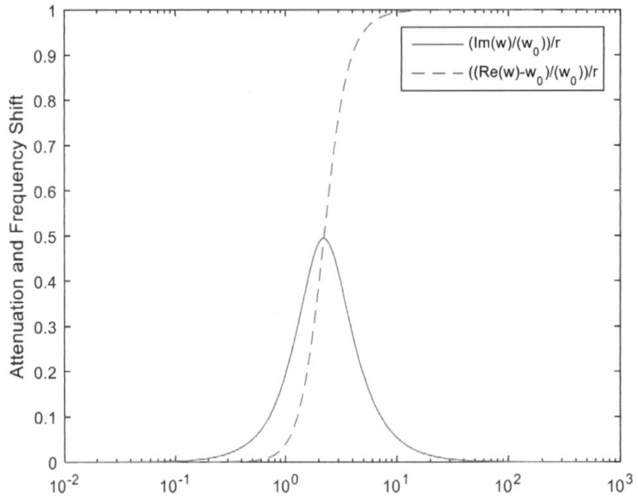

Figure 1. Normalized attenuation and frequency shift with ξ for a polySi rectangular microplate resonator with simply supported boundary condition; T_0=298 K; mode (m ,n) with m=1 and n=1; L= 200 μm, width W= 200 μm, and thickness h=10μm.

To study the effect of material parameters on attenuation and frequency shift, as a measure of the energy dissipation, $Im(\omega)/\omega_0 \acute{r}$ is plotted with the dimensionless variable ξ, and the variation of TED for different materials is plotted in Figure 2. From the analysis SiC has the largest attenuation, while polySi has the lowest .

Figure 2. (a), (b), (c), (d), and (e) shows the variations of thermoelastic attenuation associated with thermoelastic energy dissipation and the frequency shifts with ξ for (a) polySi, (b) diamond, (c) Si, (d) GaAs, and (e) SiC. All five selected structural materials show a step for the normalized frequency shift, $[Re(\omega) - \omega_0]/\omega_0 \acute{r}$, which is a universal behaviour at ξ=2.2246. The rectangular plate model analysed in this work can be converted to the classical thermoelastic plate model by setting l = 0, where the size effect is not considered. When rectangular microplate resonators are considered, the energy dissipation peak occurs when the resonance frequency is in the order of its effective relaxation rate. Moreover, the attenuation increases with ξ for small values and decays for large ξ values as shown in Figure 2. which is applicable to all five structural materials selected.

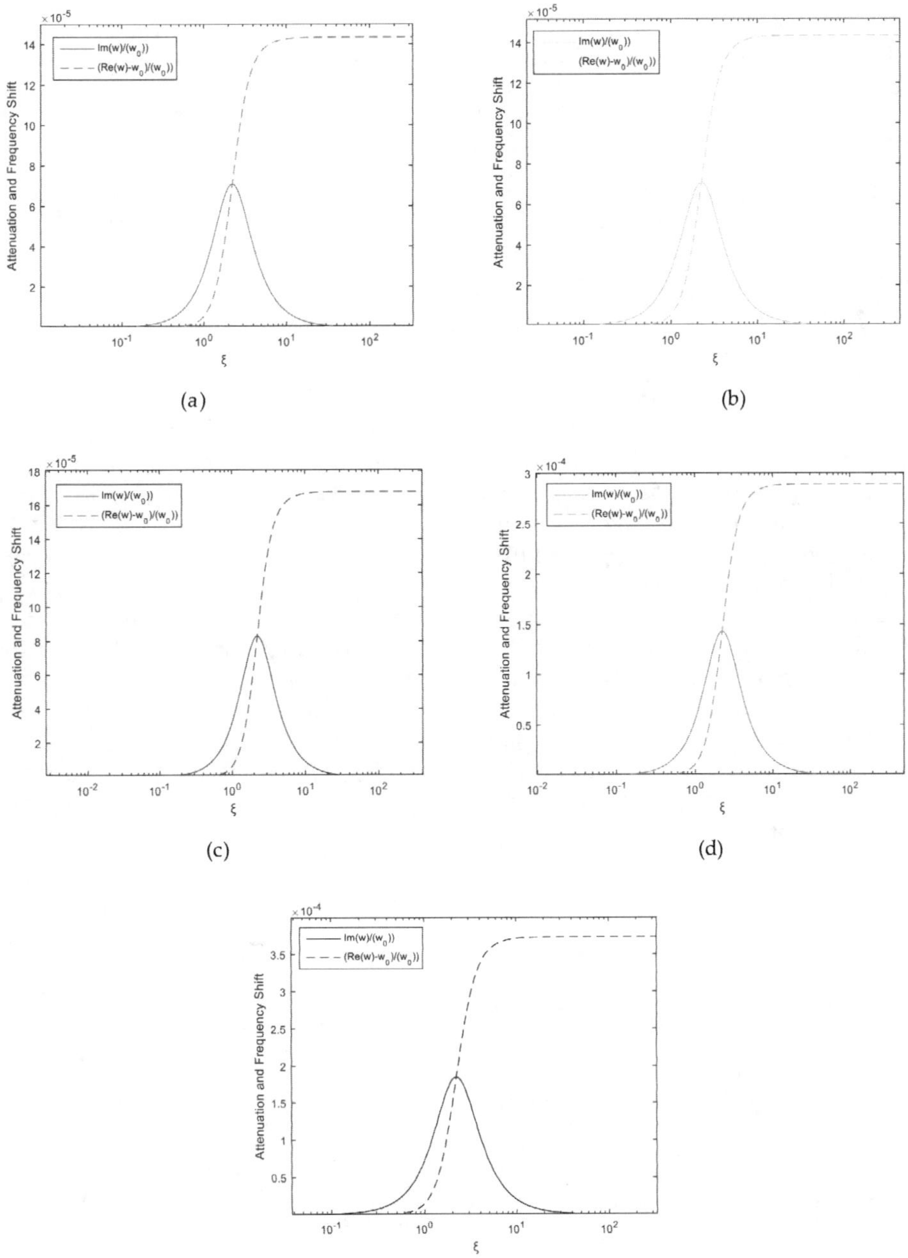

(e)

Figure 2. Variation of attenuation and frequency shift with ξ for rectangular microplate resonators with simply supported boundary condition; T_0=298 K; mode (m ,n) with m=1 and n=1; L= 200 μm, width W= 200 μm, and thickness h=10μm using five different structural materials (a) polySi, (b) diamond, (c) Si, (d) GaAs, and (e) SiC.

The material order in which attenuation and frequency shift diminishes with ξ in our work is found to be in the order of SiC> GaAs >Si > diamond >polySi as shown in Table 1.As the frequency shift decreases, attenuation also declines and the quality factor gets enhanced.

Table 1. Attenuation and frequency shift of rectangular microplate resonators with simply supported boundary condition; mode (m ,n) with m=1 and n=1; L= 200 μm, width W= 200 μm, and thickness h=10μm using five different structural materials

Material	Attenuation	Frequency Shift
PolySi	6.71E-05	1.31E-04
Diamond	7.23E-05	1.43E-04
Si	8.29E-05	1.68E-04
GaAs	1.43E-04	2.89E-04
SiC	1.84E-04	3.74E-04

3.2. *Impact of Dimensionless Length Scale Parameter (l/h) on Thermoelastic Attenuation*

As size effects are included in the study, the size-dependence of the material properties enhances the bending rigidity and hence the values of thermoelastic frequencies. By considering the material length scale parameter l, the normalized frequency shift $[\mathrm{Re}(\omega) - \omega_0]/\omega_0\, r^*$ is elevated, and the normalized attenuation $\mathrm{Im}\,(\omega)/\,\omega_0\, r^*$ is diminished more than that in the classical model. This demonstrates that the energy dissipation in the microplate resonators may have a large reduction due to the size effect. Figure 3.depicts the variation of attenuation without size effect (l/h=0) for different structural materials. The material order in which attenuation reduces with ξ is in the order of SiC.> GaAs > Si > diamond >polySi. When size effects are included, the peak damping diminishes with the material length scale parameter. It is thus confirmed that energy dissipation decreases with the dimensionless material length scale l/h, and, as a result, the size-dependent quality factor increases. Figure 4. illustrates the size effects i.e. l/h =0,0.2,0.5and 1 when polySi is applied as the structural material. The thermoelastic damping limited quality factor increases with the dimensionless material length scale parameter (l/h) for microplate resonators with all boundary conditions and in all vibrating modes.

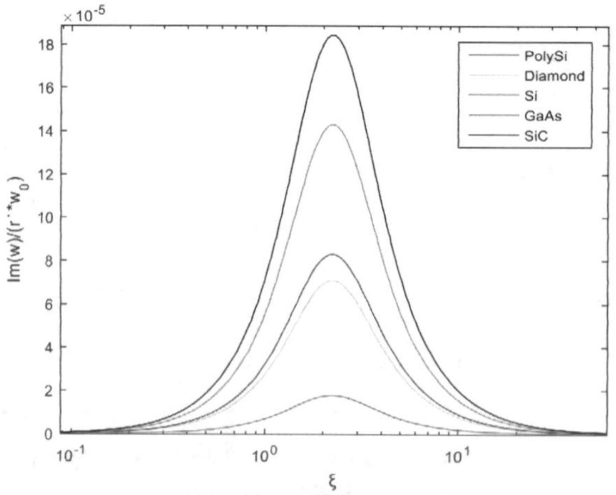

Figure 3. Variation of attenuation and frequency shift with ξ for rectangular microplate resonators with simply supported boundary condition; T0=298 K; mode (m ,n) with m=1 and n=1; L= 200 μm, width W= 200 μm, and thickness h=10μm ; l/h=0 using five different structural materials .

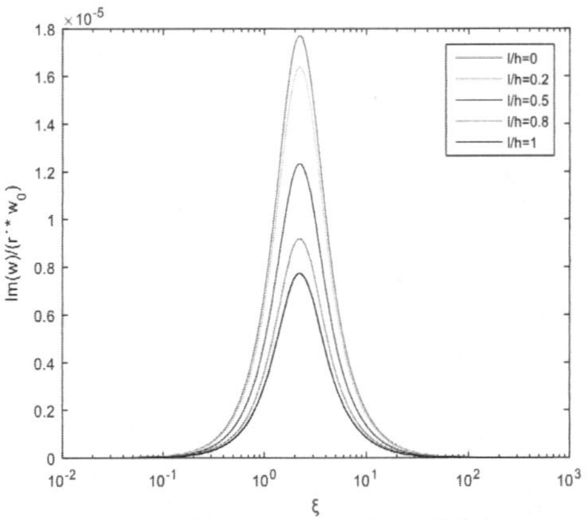

Figure 4. Variation of attenuation and frequency shift with ξ for a polySi based rectangular microplate resonator with simply supported boundary condition; T_0=298 K; mode (m ,n) with m=1 and n=1; L= 200 µm, width W= 200 µm, and thickness h=10µm ; *l/h*=0,0.2,0.5,0.8,1 .

4.Conclusion

Various energy dissipation mechanisms limit the maximum attainable quality factor and in micro/nano scales, thermoelastic energy dissipation is a critical energy loss mechanism. When the devices are scaled down, in order to accurately model micro/nano scale resonators, non-classical elasticity theories like Modified Couple Stress Theory (MCST) are valid. By including a dimensionless length scale parameter *(l/h),* the size effects are incorporated in the analysis and thermoelastic energy dissipation and related attenuation were found to be diminished. In this paper, thermoelastic attenuation and frequency shift related to thermoelastic damping in microplate resonators were investigated without applying size effects (*l/h* =0).The material order in which thermoelastic attenuation diminishes is SiC.> GaAs > Si diamond > >polySi. The impact of dimensionless length scale parameter on thermoelastic attenuation was also explored with polySi as the structural material. Thin beams with properly selected structural material with length scale parameters vibrating in higher modes provide large Q_{TED} and helps engineers to design microplate resonators with less energy dissipation and high quality factors.

REFERENCES

[1] Chen J.Y., Hsu Y.C., Lee S.S., Mukherjee T., Fedder G.K., 2008, Modeling and simulation of a condenser microphone, Sensors and Actuators A 145–146: 224-230.
[2] M. Gologanu, C.G. Bostan, V. Avramescu, O. Buiu, Damping Effects in MEMS Resonators, Sensors and Wireless Lab Bucharest (SWLB), IEEE Conference,pp:67-74,2012.
[3] V. T. Srikar and S. D. Senturia, *Thermoelastic damping in fine grained polysilicon flexural beam resonators*, J. Microelectromech. Syst., vol. 11, no. 5, pp. 499–504, 2002.
[4] Duwel, R. N. Candler, T. W. Kenny, and M. Varghese, *Engineering MEMS resonators with low thermoelastic damping*, J. Microelectromech.Syst., vol. 15, no. 6, pp. 1437–1445, 2006.
[5] R. E. Mihailovich and N. C. MasDonald, *Dissipation measurements of vacuum-operated single-crystal silicon micro resonators*, Sens. Actuat. A. Phys., vol. 50, no. 3, pp. 199–207, 1995.
[6] Zener C.: Internal friction in solids I. *Theory of internal friction in reeds*. Phys. Rev. 52(3), 230 (1937).
[7] Lifshitz, R., Roukes, M.L.: *Thermoelastic damping in micro-and nanomechanical systems*. Phys. Rev. B 61(8), 5600 (2000)

[8] Nayfe,A.H., and Younis,M.I., Modeling and simulations of thermoelastic damping in microplates. Journal of Micromechanics and Microengineering, 2004, 14(12): 1711-1717.]
[9] Tsiatas, G.C., Yiotis, A.J.: Size effect on the static, dynamic and buckling analysis of orthotropic Kirchhoff-type skew micro-plates based on a modified couple stress theory: comparison with the nonlocal elasticity theory. Acta Mech. 226(4), 1267–1281 (2015)
[10] Park, S.K., Gao, X.-L.: Bernoulli–Euler beam model based on a modified couple stress theory. J. Micromech. Microeng. 16, 2355–2359 (2006)
[11] Yang, F., Chong, A., Lam, D.C.C., Tong, P.: Couple stress based strain gradient theory for elasticity. Int. J. Solids Struct.39(10), 2731–2743 (2002).
[12] Zhong, Z.Y., Zhang, W.M., Meng, G., Wang, M.Y.: Thermoelastic damping in the size-dependent microplate resonators based on modified couple stress theory. J. Microelectromech. Syst. 24(2), 431–445 (2015).
[13] Vahid Borjalilou, Mohsen Asghari, Small-scale analysis of plates with thermoelastic damping based on the modified couple stress theory and the dual-phase-lag heat conduction model, Springer-Verlag GmbH Austria, part of Springer Nature 2018, https://doi.org/10.1007/s00707-018-2197-0
[14] Resmi R, M. R. Baiju, V. Suresh Babu "Thermoelastic damping dependent quality factor analysis of rectangular plates applying modified coupled stress theory" . AIP Conference Proceedings 2166, 020029 (2019); https://doi.org/10.1063/1.5131616.

BIOGRAPHY

Resmi R. received B.Tech Degree in Electronics Engineering from Cochin University of Science and Technology, India in 2000 and M.Tech. Degree from College of Engineering Trivandrum, University of Kerala. She is currently working as an Assistant Professor in LBS Institute of Technology for Women, Kerala, India. Her research interests include RF MEMS Resonators and Sensors Design, Energy Dissipation analysis in MEMS/NEMS structures, Biomedical Signal Processing and Instrumentation, Image Processing, VLSI Systems Design etc. She is pursuing her Ph.D. in the Department of Electronics and Communication, College of Engineering Trivandrum, under University of Kerala. Orcid Id: https://orcid.org/0000-0002-0849-5509

V. Suresh Babu received B.Tech. Degree in Electronics and Communication Engineering from Kerala University, Trivandrum, India, in 1989, M. Tech. Degree in Integrated Electronic Devices and Circuits from IIT Madras in 1996 and Ph.D. Degree in Electronics and Communication in 2013 from Kerala University.
From 1990 to 1991, he was with the Kerala State Electronics Development Corporation Ltd., Trivandrum, India. Since 1991, he has been a member of the faculty of the Department of Electronics and Communication Engineering, in Government Engineering Colleges in Kerala. Presently he is Professor in Electronics and communication Engineering, Government Engineering College Wayanad, Kerala, India. His areas of interest are solar photovoltaics and nano-electronic devices. Orcid Id : https://orcid.org/0000-0001-8024-2621

M. R. Baiju received B.Tech. Degree in Electronics and Communication Engineering from College of Engineering, Trivandrum, India, in 1988, and M. Tech. Degree in Electronics Design and Technology and the Ph.D. Degree in Power Electronics from the Centre for Electronics Design and Technology, Indian Institute of Science, Bangalore, India, in1997 and 2004. From 1988 to 1991, he was with the National Thermal Power Corporation Ltd., New Delhi, India. Since 1991, he has been a member of the Department of Electronics and Communication Engineering Faculty, College of Engineering, Trivandrum, Kerala. His areas of interest are Multilevel Inverter Control strategies, MEMS and VLSI systems.

Studies on strength behaviour of green self cured concrete with ecosand with impact of fly ash and ground granulated blast furnace slag

Vivek Kumar C[1], K Vamsi Krishna[2], Suresh B[3], S Venkatacharyulu[4],

[1,4]Assistant Professor, Civil Engineering Department, GRIET, Hyderabad, India.
Corresponding Author : vivekkumarcr@gmail.com.
[2]Assistant Professor, Civil Engineering, Malla Reddy Engineering College, Hyderabad, India.
[3]Assistant Professor, Civil Engineering, Institute of Aeronautical Engg, Hyderabad, India.

Abstract. Construction firms now use artificial sand or else manufactured sand in concrete for a day instead of natural river sand, where the excavation prohibition applies to river sand along with the implementation law. Natural sand or River sand are one of the constituents of concrete taken from a river. The problems of deterioration of the ecosystem and costly and the condition of the natural sand of river is rising day by day. About the Global Natural sand consumption has increased because of excessive use of concrete to ensure a very strong demand for river san, where Eco Sand is necessary to replace natural silica. Potable Water, meanwhile, is also the extremely ubiquitous feature for the concrete production and curing activity, and it is now insufficient for a limited period due to a superior growth in human population growth. Self-curing concrete helps to reduce the use of potable water, in contrast to the traditional curing system. In supplement to chemical admixture with the concrete mixture, this paper deals with the various studies of fresh and hardened concrete specimens with various replacement level of Eco sand and findings with the self-curing agent mixed with PEG 400 with Fly Ash and GGBFS proportioning. The results are correlated with the replacement of Eco sand with river sand with both ordinary self-curing concrete and blended self-curing concrete with M30 grade for the different combinations of blended proportions. Attempts have been made to research with Eco Sand to identify the properties of blended self-curing concrete.

1. Introduction

Portland cement is typically used to make concrete and other materials having cement characteristics such as fly ash and slag, where both cement acts as a binder of the aggregates. Consequently, In concrete, cement reacts with water and forms a hardened cement mass. In addition, the additives that are mineral admixtures such as fly ash, slag, are added to the production of cement for property enhancement. Curing, particularly in the early ages of concrete strength, is a very important component that remains important for the hydration process. However, by traditional curing methods, appropriate curing is not continuously probable, and it is suggested to make the entrenched water available for curing in order to solve this challenge on curing process.

Self-curing agents should be applied to reduce water evaporation from the concrete surface and improve the water retention potential of the concrete compared to conventional concrete. The self-curing agents apply to surface moisture in the atmospheric state. The hydrogen bond between molecules of polymerized fresh concrete water reduces surface water evaporation as time leads to concrete reinforcement. Considerable autogenous deformation and early-age cracking can result due to hydration and other processes after drinking water, which is not frequently available.

Because of its chemical shrinkage striking during cement hydration, empty cavity formed within the cement paste, prominent to a reduction in its internal relative humidity and shrinkage influencing the early age cracking. Eco sand is a very fine grain, a bi-product of cement production, that can be used in concrete to boost performance. Its micro-filling effect decreases concrete pores and provides additional resistivity to moisture and thus durability. It has more consistent grading than several
extracted aggregates. Efficient and cost-effective use of waste material and thus performs as well as natural sand. The main extent of the paper is experimentally to examine the effect of polyethylene glycol (PEG 400) on strength characteristics of Self-curing concrete with replacements of Eco Sand. With use of eco sand instead of collected or extracted silt natural sand can help designers and engineers tackle environmental concerns. Finally, the compressive strength, tensile strength and bending of concrete are being evaluated in this analysis by utilizing eco sand, cement and super plasticizer in the given moulded specimens.
P. Magudeaswaran et al.[3] are investigating green high-performance concrete with the materials such as silica fume, eco sand, fly ash and coarse aggregates. The research to be carried out assess the physical and mechanical properties of replacement of cement with silica fumes between 7.5-15 % and about 15-40 % of fly ash. Shubham Gupta et al.,[5] performed the analysis on Eco sand replacements for fine aggregates varies from 5 to 20 % and the mechanical analysis is analyzed. Compressive strength spilt tensile strength and flexural strength are included. Grade of the M40 is unique for the analysis and the rheological tests are also examined to evaluate the study on utilization of water.
The concrete is investigated with eco-sand and steel sag by Nandini Suri et al.,[7] through replacing coarse aggregate with steel slag and 30 percent substitution of eco sand as fine aggregate, the goal is to improve the strength of the concrete. The blend consisted of 30% to 45 &% of slag. It is observable that as compared to conventional blends, 50 % substitution and above tends to decrease the strength found. Nagraj T.S et.al [2] (1996) stated that rock dust consumes more cement compared to sand because of its higher surface area, which enhances workability and also studied the impact of rock dust and pebbles as aggregates in cement and concrete and discovered that crushed stone dust could be used to substitute concrete with natural sand.

The purpose of the experimental study was to understand the strength characteristics of internal cured blended concrete and to acquire the optimal percentage of PEG 400 and replacement level of Flyash and GGBFS which are self-curing. Attempts have thus been made to facilitate internal curing in concrete through the addition of self-curing admixtures, which may produce concretes that are equal or superior to traditional forms of curing. The aim of this research is to vary the mechanical characteristics of concrete such as compressive strength, split tensile strength and by varying the % PEG proportion by cement weight for both ordinary self-curing and blended self-curing concrete with partial replacement of Eco Sand in Fine Aggregate for grade concrete M30.

2. Materials

All the constituents of concrete making materials are properly tested under laboratory condition. The materials are listed as follows.

Polyethylene Glycol is an ethylene oxide group, general formula H(OCH2CH2)nOH. Where n represents the average number molecular weight in ethylene oxide group ranges from 400 to 4000. In many medical applications, PEG is non-toxic, odourless, lubricating, neutral pH benefit & non-irritating, and even used. The effect of admixture (PEG 400) on compressive, split tensile strength and rupture modulus was analyzed by Jagannadha Kumar et al. The proportion of Polyethylene Glycol 400 by cement weight ranges from 0 % to 2 % for M20 and M40 blends.

It was found that 1% of PEG 400 by weight of cement was optimum for M20, while 0.5 % for M40 grade concrete for achieving maximum strength without compromising workability. Joseph et al., experimented on self-cured concrete with PEG400 for M20 grade of concrete also suggested 1% of PEG400 by weight of cement gives optimum strength for M20 grade of concrete and gets increased the workability. Shikha Tyagi et al., (5) made M25 and M40 grade of concrete and found the optimum dosage of PEG-400 to be 1% for M25 and 0.5% for M40 grades of concrete.

As per previous investigations, it is founded that, Polyethylene Glycol (PEG) is widely used self-curing compound, which is taken to be chemical admixture in this experimental work having molecular weight as 400, pH value as 6.5 and specific gravity of PEG 400 as 1.12 respectively.

Ordinary Portland (OPC 53) cement and a particle density of 3.05 have been used to comply with both the IS 1489-1991 specification. Locally sourced aggregates of 4.75mm and 20.0mm sizes were used for sieve analysis, in conjunction with IS: 383-1970, these values of 2.74, 0,25 % and 3.05 respectively, the specific gravity, water content and fineness modulus are determined.

Eco sand is a very fine grain, a bi-product of cement manufacturing, that can be used in concrete to boost performance. Its micro-filling effect decreases concrete spaces and gives greater resistivity to moisture and thus resilience. It has more robust grading than many aggregates that have been extracted. Efficient use of waste products and thus cheap and cost-effective. It works just as well as fine aggregate. Eco sand contains numerous advantages, including such efficient energy, fire proof, dead load mitigation, environmentally sustainable, reliable, light weight, relatively inexpensive, and lower construction costs.

The Green Sand or Eco sand used in this experimental work should be conforming to the 4.75 mm sieve passage. The main properties were also been examined and it is having the specific gravity as 2.48, bulk density as 1500 kg/m^3 and fineness modulus as 4.33, whereas specific gravity as 2.62, bulk density as 1542 kg/m^3 and fineness modulus as 3.09 for naturally occurring sand, respectively.

Normally employed quality drinking water conforming to IS:456-2000. In this study, it was acquired from the National Thermal Power Plant, Dadri, where fly ash Class F was used. The characteristics of this fly ash are light grey and the specific gravity ranges from 1.90-2.55 if the uniformity is 3.1-10.7. Apart from diminished permeability, GGBFS blended concrete is considerably more resistant to the ingress of chloride ions through concrete.

GGBFS was procured in Jindal South West Steel Ltd, Karnataka mainly having SiO_2 as 35.20%, Al_2O_3 as 19 % and CaO as 34.90 %. GGBFS is an aluminosilicate with major components such as SiO2, Al2O3, CaO, and MgO, and it has been discovered that the incorporation of GGBFS promotes flowability and extended compressive power.

The addition of Flyash and ground granulated blast furnace slag has also been found to provide coral aggregate-concrete with lower early strength capacity but superior early age strength due to late development of pozzolanic reaction. As a super plasticizer, a polycarboxylate-type, new-generation high-range water reducing admixture confirming ASTM C494 was used to enhance the flow or workability of blends with a decreased water-cement ratio.

Table 1. Chemical Composition of Class F Fly Ash

Chemical properties	SiO_2	Al_2O_3	Fe_2O_3	CaO	MgO	Na_2O	K_2O	SO_3	*Loss of Ignition*	*Cl*
Compounds %	45.98	23.55	4.91	18.67	1.54	0.24	1.80	1.47	2.31	0.0053

3. Mix Proportioning of M30 Mix

The mix proportioning for M30 of binary and ternary mixed concretes developed with optimal mixtures of Fly ash (FA) and Ground Granulated Slag furnace mixtures was based on the Indian Code for Mix design approach in IS 10262: 2019. Numerous experimental mixes had to be done on mixed concrete mixes made of varying combinations of Fly Ash (FA) and Ground Granulated Blast furnace slag (GGBFS). Slump Cone experiments were performed out from the above literatures and studies for workability and dosage variations of 0.5 percent, 1 percent and 1.5 percent were taken for both natural sand and Eco sand as seen in Table 2.

Table 1. Proportions of Blended Self Cured Concrete with M30 grade mix

Material	Quantity (kg/m^3)
Cement	350
Fine aggregate (River Sand)	744.25
Fine aggregate (Eco Sand)	744.25
Coarse aggregate	1314.18
Polyethylene Glycol (PEG 400)	2.28
Super Plasticizer	3.5
Water	175

As per IS 456-2000, clause 8.2.4.2, the maximum cement content is limited to 450 kilograms per cubic meter of concrete. Revised quantities in kg per cubic metre for grade (M30) mixed concrete mix are reached in subsequent trail mixes without compromising the necessary strength property. The estimated quantity of OPC is 350 kg and pozzolanic materials such as Flyash (FA) and Ground Granulated Blast Furnace Slag (GGBFS) have been segregated from the literature studies on different dosages formulated in Table 3.In the desired design mix, as per code provision IS:1199-1959, the percentage of PEG 400 kept constant as 1 % for the slump value also increases consequentially in different replacements of Flyash and Ground Granulated Blast Furnace Slag with the blended mixing parameters of River sand and Eco-Sand as shown in Table 2.

Table 2. Slump values for various PEG 400 percentages for Natural & M-Sand

Percentage of PEG	Slump in Natural Sand	Slump in Eco Sand
0.0%	74	62
0.5%	92	89
1.0%	**112**	**123**
1.5%	140	131

Table 3. Mix Proportion of M30 grade Blended mix for various percentages for Flyash & GGBFS

Mix (%)	Cement	Flyash	GGBFS	C.A	Sand %	River Sand	Eco sand	S.P	Water
M0 (100)	350	--	--	1314.18	100	744.3	--	3.5	140
M1 (100)	350	--	--		100	--	744.3		
M2 (60+25+15)	227.5	87.5	35.0		70+30	520.0	223.3		
M3 (60+15+25)		35.0	87.5		70+30	520.0	223.3		
M4 (60+25+15)		87.5	35.0		80+20	595.5	148.9		
M5 (60+15+25)		35.0	87.5		80+20	595.5	148.9		
M6 (65+20+15)		70.0	52.5		70+30	520.0	223.3		
M7 (65+15+20)		52.5	70.0		70+30	520.0	223.3		
M8 (65+20+15)		70.0	52.5		80+20	595.5	148.9		
M9 (65+15+20)		52.5	70.0		80+20	595.5	148.9		
M10 (65+10+25)		35.0	87.5		70+30	520.0	223.3		
M11 (65+25+10)		87.5	35.0		80+20	595.5	148.9		
M12 (65+10+25)		35.0	87.5		70+30	520.0	223.3		
M13 (60+20+20)		87.5	35.0		80+20	595.5	148.9		
M14 (60+20+20)	210.0	70.0	70.0		70+30	520.0	223.3		
M15 (70+10+20)		70.0	70.0		80+20	595.5	148.9		
M16 (70+20+10)	245.0	70.0	35.0		70+30	520.0	223.3		
M17 (70+10+20)		70.0	35.0		80+20	595.5	148.9		
M18 (70+20+10)		35.0	70.0		70+30	520.0	223.3		
M19 (70+20+10)		35.0	70.0		80+20	595.5	148.9		

IOP Conf. Series: Materials Science and Engineering 1091 (2021) 012068 doi:10.1088/1757-899X/1091/1/012068

** All of the materials given in this table have a kg/m³ unit.*
*** Slump figures for different SCM mixtures of cement from M1 to M18 with 1 % addition of PEG 400 .*

4. Mechanical Properties of Blended SCC mix with M-Sand replacements

Within that blended mixture of self-curing concrete together along with SCMs, the mixing among all materials was controlled for a period of about two minutes. The amounts of polycarboxylate-type superplasticizer was adjusted with each suitable mix with Natural sand and Eco Sand replacement quantities to meet the optimum workability without any kind of segregation resistance over fresh condition. Using these compressive cubic moulds including split-tensile cylindrical moulds preceding the mixture of concrete with the replacement percentages of eco sand of 20 percent and 30 percent with natural sand based on previous literature review, different specimens were produced with accurate blending of Flyash and Ground Granulated Blast Furnace Slag in various proportions.

After the castings were poured with a very well consolidated mixture of self-curing concrete, the fresh concrete surface was levelled and the specimens were placed in the moulds for 24 hours or one day under laboratory conditions, whilst the mould surface was covered by plastic wrap and then the demoulded specimens were left at room temperature during experimental investigation.

Self Curing Concrete mixtures with their strength parameters blended with Supplementary Cementitious components in various proportions of different percentages of Class F Flyash and Ground Granulated Blast Furnace Slag of various mixed percentages of M0 (Conventional mix), M1 (Ecosand Mix) and M2 to M18 has been tabularized in Table no.4.

In accordance with Table 3, the cube and cylinder specimens of replacements of Eco sand with natural sand with 20% and 30% with respect to previous studies, are casted for 7 days and 28 days, curing traditional full curing and self-curing concrete samples held at a temperature shadow of 300c ± 30c. The specific curing was carried out till the corresponding concrete specimens was examined after 7 & 28 days for compressive and split tensile strength. Cubes with dimensions of 100 mm x 100 mm x 100 mm were formed. The experiment is having adequate time will be provided for 24 hours of hardening of concrete and then it should be cured for 28 days also the specimen should be loaded to its failure load after 7 days and 28 days in the compressive testing machine. Split tensile strength test was carried out through cylindrical samples of sizes 150 mm diameter and 300 mm height at the 7 days and 28 days curing employing compression testing machine.
It is an indirect measure to assess the cylindrical samples' tensile strength.

Table 3. Strength parameters of M30 grade Blended SCC for various ratios for Flyash & GGBFS with various replacement levels of natural sand and Eco sand

Mix (%)	Cement	Flyash	GGBFS	Sand %	River Sand	Eco sand	Compressive Strength (MPa)		Split Tensile Strength (MPa)	
							7 Days	28 Days	7 Days	28 Days
M0 (100)	350	--	--	100	744.3	--	10.24	34.49	1.89	5.14
M1 (100)	350	--	--	100	--	744.3	11.12	35.6	2.14	4.98
M2 (60+25+15)	227.5	87.5	35.0	70+30	520.0	223.3	11.29	36.9	2.19	5.23
M3 (60+15+25)		35.0	87.5	70+30	520.0	223.3	10.49	35.88	2.23	4.98
M4 (60+25+15)		87.5	35.0	80+20	595.5	148.9	11.55	36.7	2.34	5.14
M5 (60+15+25)		35.0	87.5	80+20	595.5	148.9	11.24	35.88	2.19	5.11
M6 (65+20+15)		70.0	52.5	70+30	520.0	223.3	10.66	33.4	1.98	4.68
M7 (65+15+20)		52.5	70.0	70+30	520.0	223.3	10.74	34.58	1.91	4.71
M8 (65+20+15)		70.0	52.5	80+20	595.5	148.9	11.49	35.48	2.01	4.5
M9 (65+15+20)		52.5	70.0	80+20	595.5	148.9	11.89	36.8	2.16	4.69
M10 (65+10+25)		87.5	35.0	70+30	520.0	223.3	11.29	37.4	2.49	4.7
M11 (65+25+10)		87.5	35.0	80+20	595.5	148.9	12.3	38.7	2.7	4.98
M12 (65+10+25)		87.5	35.0	70+30	520.0	223.3	11.4	38.49	2.38	4.88

M13 (60+20+20)	210.0	35.0	87.5	80+20	595.5	148.9	12.4	38.8	2.5	5.34
M14 (60+20+20)		70.0	70.0	70+30	520.0	223.3	10.4	34.2	2.11	5.12
M15 (70+10+20)		70.0	70.0	80+20	595.5	148.9	11.14	34.8	2.14	5.21
M16 (70+20+10)	245.0	35.0	70.0	70+30	520.0	223.3	12.33	37.8	2.22	5.01
M17 (70+10+20)		70.0	35	80+20	595.5	148.9	12.48	38.48	2.49	5.3
M18 (70+20+10)		35.0	70.0	70+30	520.0	223.3	11.49	35.8	2.44	5.29
M19 (70+20+10)		70.0	35	80+20	595.5	148.9	11.98	36.21	2.31	5.35

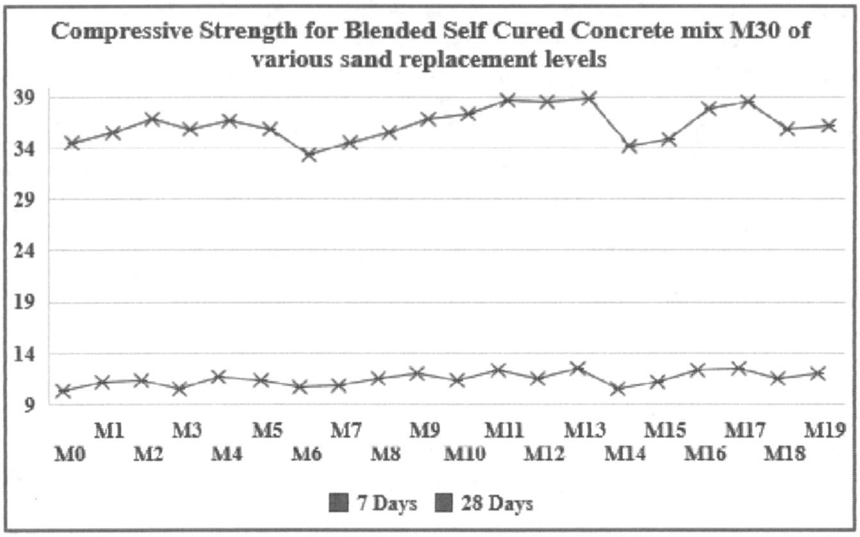

Fig 1. Compressive Strength of M30 grade Blended SCC for various ratios for Flyash & GGBFS with Eco sand replacement levels of 20% & 30 %

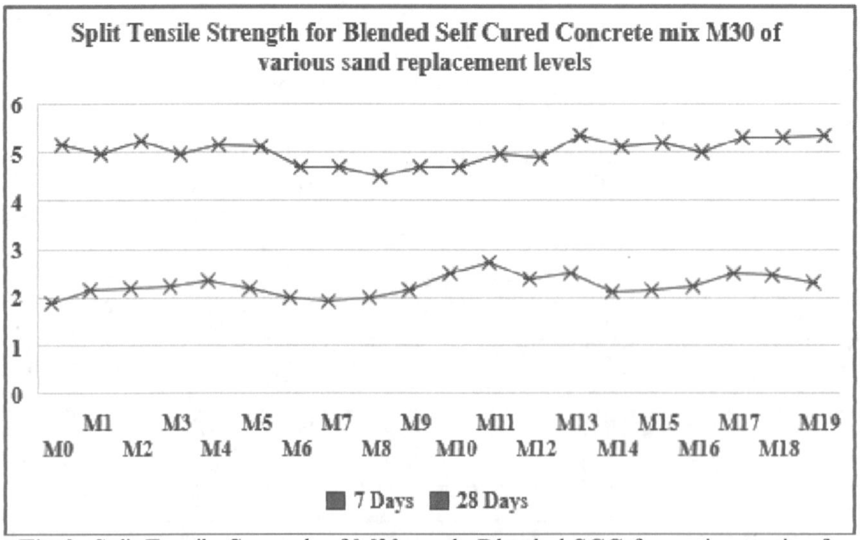

Fig 2. Split Tensile Strength of M30 grade Blended SCC for various ratios for Flyash & GGBFS with Eco sand replacement levels of 20% & 30 %

5. Results and Discussions

The mechanical properties like workability, compressive strength, split tensile strength and acid resistance which been improved with binary and ternary blended with the Flyash and GGBFS in the concrete mix of grade M30 with several mixes of River Sand with Eco Sand of 20% and 30 %

replacement. After this percentage of replacement, the mechanical properties having slight increase and decrease in the compressive strength parameters and very similar for eco sand to the naturally occurring river sand. Therefore, an optimum value of 1% of Polyethylene glycol based upon the workability is maintained throughout the blended mix and added to the mix.

In the slump test, a workable slump of 112 mm for natural sand and 123 mm for Polyethylene Glycol 400 for Eco sand was obtained for a water-cement ratio of 0.45, which is taken in mixed design in accordance with IS 10262:2009.

For all replacements of Flyash and GGBFS when eco sand from 20 % to 30 % have been incorporated such that there was an increase in mechanical properties in the strength parameter in mixes of M2 (60% OPC +25% FA+15% GGBFS with replacement of 70 % natural sand and 30 % eco sand having 36.9 MPa, M4 60% OPC +25% FA+15% GGBFS with replacement of 80 % natural sand and 20 % eco sand with 36.7 MPa, M8 65% OPC +15% FA+20% GGBFS with replacement of 80 % natural sand and 20 % eco sand with 35.48 MPa , M11 having the mix proportion of 65% OPC +15% FA+10% GGBFS with replacement of 80 % natural sand and 20 % eco sand with 38.7 MPa, , M12 with 65% OPC +10 % FA+25 % GGBFS with replacement of 70 % natural sand and 30 % eco sand with 38.49 MPa, M13 having blended mix of 60% OPC +20% FA+20% GGBFS with replacement of 70 % natural sand and 30 % eco sand having strength of 38.8 MPa and M17 with mix of 70% OPC +20% FA+10% with replacement of 80 % natural sand and 20 % eco sand having 38.48 MPa compressive strength respectively for their blended mix pattern as shown in fig 1.

The split tensile strength of mix around 15 to 20 % Flyash and GGBFS with eco sand was found to increase to a maximum of 10 % when compared to the control mix. While for the mixes M2 (60% OPC +25% FA+15% GGBFS with replacement of 70 % natural sand and 30 % eco sand, M4 60% OPC +25% FA+15% GGBFS with replacement of 80 % natural sand and 20 % eco sand, M8 65% OPC +15% FA+20% GGBFS with replacement of 80 % natural sand and 20 % eco sand with 35.48 MPa , M13 having the mix proportion of 60 % OPC +20 % FA+20 % GGBFS with replacement of 80 % natural sand and 20 % eco sand, , M15 with 70 % OPC +10 % FA+20 % GGBFS with replacement of 70 % natural sand and 30 % eco sand, M17 with mix of 70% OPC +20% FA+10% with replacement of 80 % natural sand and 20 % eco sand and at last with M19 with mix of 70% OPC +15 % FA+15 % GGBFS with replacement of 80 % natural sand and 20 % eco sand, shown in fig 2, incorporating with Eco sand increase in tensile strength by varying GGBFS percentages from 15% to 25 % and decrease in split tensile strength at the age of 28 days have been noticed down from mixes ranging from M6 to M10.

6. Conclusions

Based on the results obtained from this experimental work for blended self cured concrete with M30 mix incorporated with Eco Sand with replacements with Natural sand, the following conclusions have been drawn:

It is found that there is a reduction in workability characteristics as the percentage substitute of natural sand by 20 % and 30 % of Eco sand, which have been is enhanced with a PEG 400 as 1 %. The shape and surface texture of eco sand, that imparts more internal friction having slight reduction the flow characteristics of concrete, are the factors for reducing workability.

The use of eco sand rather than natural river sand doesn't always affect the compressive strength of concrete throughout this experimental analysis. The replacement of cement with high volume fly ash approximately 10 to 25 percent raises the compressive strength marginally, while the rise in fly ash volume steadily reduces the compressive strength and breaks the tensile strength. Whereas the blending of GGBFS provides the initial early strength and increased workability of M30 grade self cured mix with PEG 400 as 1% ranging percentage varies from 10 % to 20 % of blended mix.

In this experimental analysis, it is noted that the use of Eco sand rather than River sand doesn't really affect the compressive strength of the composite substitution concrete with high volume fly ash approximately 10 percent, marginally increasing the compressive strength, and the rise in fly ash volume will decreases gradually the compressive strength.

GGBFS provides greater improvement in -the compressive strength while increasing the variation from 15 % to 25 %. Moreover, with the blended mix of 20 % of Ground Granulated Blast Furnace Slag provides greater strength in all the mixes of Self Cured Concrete, which did not affects the other parameters of concrete strength.

The blended concrete mixture of 12% to 20 % of Class F Flyash with eco sand have achieved an appropriate increase in compressive strength when compared to control concrete mix at the age of 28 days, when compared to river sand, there is no significant difference in dry shrinkage in concrete made with eco sand.

7. References

[1] Plunge W.H (2001) Vision 2030,"A vision for the U.S. Concrete Industry" *Concrete International,* **23** 25-34

[2] Kumar Mehta P (2007) Cements and Concrete Mixtures for Sustainability *Proceedings of Structural Engineering* World Congress (Bangalore, India,)

[3] Ruhal Pervez Memon, Abdul Rahman Bin Mohd. Sam, et.al, (2018) A Review: Mechanism, Materials and properties of Self-Curing Concrete, ARPN Journal of Engineering and Applied Sciences, **13** (24) ISSN 1819-6608.

[4] Specification and Guidelines for Self-Compacting Concrete, EFNARC Association House, 99 West Street, Farnham, Surrey GU9 7 EN, UK

[5] Nimitha Vijayaraghavan and Vayal A S 2013 Effect of manufactured sand on durability properties of concrete *American journal of Engineering Research (AJER)* **2** 437-440

[6] A. Elahi, P. A. M. Basheer, S. V. Nanukuttan and Q. U. Z. Khan,(2012), Mechanical and Durability Properties of High Performance Concrete Containing Supplementary Cementitious Materials, pp 292-299 .

[7] M. Radinski, Janolek, (2012).Investigation into the Synergistic Effects in Ternary Cementitious Systems Containing Portland Cement, Fly ash and Silica Fume, pp 451-459

[8] A. R. Bagheri, H. Zanganeh and M. M. Moalemi, (2012), Mechanical and Durability Properties of Ternary Concretes Containing Silica Fume and Low Reactivity Blast Furnace Slag, 663-670.

[9] K. Turk, (2012), Viscosity and Hardened Properties of Self-Compacting Mortars with Binary and Tenary Cementitious Blends of Fly Ash and Silica Fume, pp 326-334.

[10] T. S. Nagaraj and Zahida Banu, (1996), Efficient Utilization of Rock Dust and Pebbles as Aggregates in Portland Cement Concrete, *The Indian Concrete* Journal. pp. 53-56.

[11] T. S. Nagaraj, (2000), Proportioning Concrete Mix with Rock Dust as Fine Aggregate, *CE and CR J.* pp. 27-31.

[12] Donza H., Cabrera O. and Irassar E.F. 2002 High-strength concrete with different fine aggregates *Journal of Cement and Concrete Research* **32** 755-1761

[13] Magudeaswaran P, P. Eswaramoorthi.et.al, (2015), Green high performance concrete using eco sand and industrial wastes Experimental Study on Self Compacting Concrete Contains Partially Manufactured Sand and Recycled Clay Roof Tile, *International Journal of Chemical Sciences · January 2015,* **13** (2), 661–671.

[14] IS 383:1970 Indian Standard Specification for coarse and fine aggregates from natural sources for concrete (New Delhi, India: Bureau of Indian Standards)

[15] Ganeshwaran P A and Suji P (2013) Behavioral studies on potential aspects of self compacting concrete with M-sand and Fly Ash *International Journal of Emerging Technology and Advanced Engineering* **3** 192-197 2250-2459.

[16] Dr.M.Palanisamy P.Sudheer kumar, (2019), Aggregate replacing with Marbles in M30 grade concrete, *International Journal of Management, Technology and Engineering*, 9 583-588

[17] Bilodeau A and Malhotra V. 2008 High- Volume Fly Ash System: Concrete Solution for Sustainable Development ACI Materials Journal 97-116 January-February
[18] Pillai Anil Kumar (2013) Sustainability and Durability in Concrete, *Bulletin of ACCE(India)* **12** 40-43
[19] Deo Shirish V. and Pofale Arun D. (2015) Parametric Study for Replacement of Sand by Fly Ash
for Better Packing and Internal Curing *Open Journal of Civil Engineering* **5** 118-130
[20] IS:12269-2013 Ordinary Portland cement,53 grade-specification (New Delhi, India: Bureau of Indian standards)
[21] IS 456-2000 Code of practice for plain and reinforced cement concrete (New Delhi, India: Bureau
of Indian standards)
[22] IS 9103-1999 Concrete admixtures- specification (New Delhi, India: Bureau of Indian standards)
[23] BS: 5075- 3:1985," Concrete admixtures. Specification for super plasticizing admixtures"
[24] IS 516:1959 Reaffirmed in 2004,"Methods of tests for strength of concrete" (New Delhi, India: Bureau of Indian standards).

Durability studies on Geopolymer Mortar Cubes and their Impact Strength characteristics of Geopolymer Ferrocement Panels

Magudeaswaran P[1], M.Palanisamy[2], Akula Prakash[3], Srikanth T[4]

[1]Professor, Civil Engineering Department, Aditya Institute of Technology, Coimbatore, India.
[2]Professor, Department of Civil Engineering, BITS, Warangal, India.
Corresponding Author : mpalanisamym@gmail.com
[3]Assistant Professor, Department of Civil Engineering, GRIET, Hyderabad, India.
[4]Associate Professor, Department of Civil Engineering, GRIET, Hyderabad, India.

Abstract. Ferrocement, a thin element, is cast-off as a structural construction as well as a repair material. In this investigation, geopolymer ferrocement panels was made by using ingredients such as fine aggregate, fly ash, water, and the binder solution. The binder solution consists of a combination of Sodium hydroxide and Sodium silicate solution in the ratio of 1:2.5. In this experiment setup, the size of the ferrocement panel adopted was 250mm x 250mm x 30mm. The optimum water/ binder ratio of 0.416 was selected in this experiment. The water/binder ratio is the ratio of solution (NaOH, Na2SiO and water) to fly ash. Two techniques of curing such as heat curing and ambient curing were adopted for this study. The slabs were cast with different types of meshes with single layer, double layers, and triple layers. The different meshes used were square woven, square welded and expanded metal mesh. The impact of different type of meshes and number of layers, preliminary energy absorption, ultimate energy absorption and ductility index were studied. The purpose of the present study is to examine the performance of fly ash based geopolymer ferrocement panels under impact loading and durability tests are conducted out in geopolymer cubes and this study concludes that ferrocement will certainly be one of the best structural alternatives for RCC in the future.

1. Introduction

Global energy sources are power stations, using coal-like fuel; therefore, significant quantities of fly ash are developed. Actually, only a small amount of the ash is used; the majority is handled, leading to severe complications, potential threats of air pollution, contamination along with leaching, etc. Not just because of the scarcity of fly ash as its raw resources, and also because its development is environmentally clean, and also because the significant focus granted to the modern concept of geopolymer cement has been prompted. The basic philosophy of the production of ordinary Portland cement (OPC) is that emissions of greenhouse gases are around 7% of the overall atmospheric greenhouse gas discharge. The substitute of Portland cement with fly ash is considered by its agents to minimize the 'imprint' of concrete greenhouse gases, as the development of one ton of Portland cement produces around one ton of CO2 correlated with zero CO manufactured using existing fly ash. The new development of fly ash produces around twenty to thirty tons of CO2, 2 per ton of fly ash, i.e. the emission of coal. Since Portland cement generates approximately 2 billion tons worldwide, supplementing any large portion of this cement with fly ash could significantly reduce carbon emissions in combination with construction.

The geopolymer reaction mechanism can be described in two stages (Buchwald and Schulz 2005): In activation, the alkaline pore solution disintegrates the solid network to produce low molecular weight reactive silicate and aluminate organisms. Alkaline solutions containing alkali hydroxides, alkali silicates and/or alkali carbonates are used to activate the solid. Pure alumino-silicate materials, e.g. metakaolin and even certain forms of fly ash, were set in the setting reaction by a condensation reaction which leads to the formation of polymers of alumino-silicate. The latter reaction distinguishes the geopolymer from other varieties of activated alkali materials (such as alkali activated slag) because the result wasn't C-S-H however a polymer. The best-known material for producing geopolymer is metakaolin, however due to the limitation on the quantity available it is not suitable as a concrete binder; this has resulted in the substitution of metakaolin with other aluminosilicate materials such as, fly ash and rice husk ash.

Nath and Sarker (2010) aimed to improve the application of Fly ash-based geopolymer concretes suitable for ambient highly scalable by developing Geopolymer Concrete suitable for curing at ambient temperature to concrete structures. Geopolymer pastes have also been tested for setting time, workability and strength parameters when curing at 20-23oC. Results for the different amounts of slag and alkaline activator in the mix were compared.

Sanjayan et.,al reported that the results of a study on the effect of elevated temperature on geopolymer paste, mortar and concrete made using fly ash as a originator. The geopolymer was integrated with sodium silicate and potassium hydroxide mixtures and with several experimental parameters have been analysed such as specimen and aggregate sizing, aggregate and superplasticizer type. This experiment recognizes the specimen and aggregate size are the most important factor that direct geopolymer behaviour at higher temperatures (800 °C) whereas aggregate sizes greater than 10 mm stemmed in excellent strength performances in both ambient and elevated temperatures. Loss of strength in geopolymer concrete at elevated temperatures is recognized to the thermal incongruity between the geopolymer matrix and aggregates.

Rangan et.,al (2005) described that the review of the performance of fly ash based geopolymer concrete to sulphate attack, where specimens were saturated in sodium sulphate solution and sulfuric acid solution for several stages of time and performance of geopolymer concrete is premeditated by assessing the consequence on the compressive strength, change in length and in mass. The performance of geopolymer concrete under sulphate exposure has been studied by soaking the specimens in sodium sulphate solution and sulfuric acid solution. Later with the 12 weeks of exposure, by examining the difference in compressive strength, change in mass and length of the specimens, the results demonstrated that in type of sodium sulphate, sulphate attack did not have substantial effect on geopolymer concrete.

2. Materials

Low-calcium fly ashes with a carbon content of less than 5% but sometimes as high as 10% are typically class F. Class F fly ash is available in large concentrations and is developed with either anthracite, pozzolanic bituminous properties and a CaO content of < 5%. It contains a higher combination (greater than 70%) of silica, alumina and iron than fly ash class C. Only when cement reacts with water will Class F Flyash react with the by-products developed. During concrete curing, it most efficiently moderates heat gain and is thus considered an ideal cementitious material in concrete weight and high strength blends. For the same cause, Class F Flyash is the solution to a wide variety of problems in the time. Class F Fly ash obtained from the Mettur thermal power plant has been used in this investigation.

Table 1. Chemical Composition of Class F Fly Ash

Chemical properties	SiO_2	Al_2O_3	Fe_2O_3	CaO	MgO	Na_2O	K_2O	SO_3	*Loss of Ignition*	*Cl*
Compounds %	45.98	23.55	4.91	18.67	1.54	0.24	1.80	1.47	2.31	0.0053

Use of clean and dry river sand available locally. For casting all the specimens with special gravity as 2.65 and fineness module as 2.25, sand passing through IS 4.75mm sieve will be used.

The alkali activator used in this assessment was a solution based on sodium silicate, indicating that there was sodium silicate and sodium hydroxide in the activator. The sodium hydroxide solution was prepared by dissolving sodium hydroxide pellets at least 1 day prior to mixing with deionized water.

Aqueous sodium silicate solution is being used in current investigation. And according to manufacturer, silicates were supplied as a bonding agent to the detergent business and the textile industry, although sodium silicate has been used in 16 molar concentrations for the development of geopolymer mortar and sodium hydroxide pellets. In order to obtain the flexural strength of the slab, three types of wire mesh were used with different layers. In single layer, double layer & triple layer, square woven, square welded and expanded metal mesh are used. For the three styles, the diameter of the wire is 1.5 mm and the mesh opening is 15 mm.

Table 2. Wire Mesh Specifications

Sl.No	Property	Wire Mesh Specifications
1.	Mesh shape	Square woven, welded and Expanded metal
2.	Length x Width in mm	760 x150
3.	Size of Mesh opening (mm)	15 x 15
4.	Joint Thickness (mm)	7.22
5.	Wire Diameter (mm)	1.5
6.	Density(g/cm3)	0.95
7.	Ultimate tensile strength (MPa)	34.5

3. Mix Proportioning and Methodology

A mixing ratio of 1:2.5 is used to prepare the Geo-polymer mix. Both for ambient curing and heat curing, the Water-Binder ratio is taken as 0.376, 0.386.0.396 and 0.416. Since fly ash showed a low binding property, in addition to increasing the binding property in the fly ash mortar, a binder solution has to be formulated. The binder solution is prepared up of a 1:2.5 combination of sodium hydroxide and sodium silicate solution. Extra water can be added to this, according to workability, although the fly ash falls from multiple plants which have different water absorption properties. The maximum water content to be added in combination is 0.06 % and 0.02 % of water to fly ash ratio is the minimum water content to be added in addition.

Detailing of single-layer, double-layer & triple-layer slabs. All three slabs have same size of 250 X 250 mm and the thickness of the slab is 30 mm, the mesh is distributed at a distance of 4 mm in the single layer slab (cover) and the double layer slab also has 4 mm cover on both top and bottom then the mesh is kept at a 22 mm interval, and the third layer slab also has a 4 mm cover on both top and bottom therefore the mesh is kept at an 11 mm interval. Nearly 36 slabs were cured by ambient curing, and another 36 slabs were cured by heat curing. The slab specimens underneath ambient curing were left without disturbance at room temperature for 28 days and the samples under heat curing were held for 24 hours in a heat chamber at 60 43 degrees C. The surface of the specimens was then whitewashed to prevent surface shrinkage.

The primary objective of this study is to study the strength and durability characteristics of geopolymer mortar because it is the source material for ferro-geopolymer slab and to evaluate the effect of water absorption and porosity on the durability of geopolymer mortar specimens based on fly ash. The resistance against corrosion of geopolymer mortar specimens based on fly ash is also evaluated using carbonation depth. It also predominantly examines the impact behavior of a 250x250x30 mm geopolymer ferrocement slab with a fly ash to sand ratio of 1:2.5 and investigates the behavior of geopolymer ferrocement panels completely reliant on fly ash under impact loading. The impacts of various mesh types and number of layers, initial energy absorption, final energy absorption, and ductility index were examined in this study.

4. Durability parameters

Durability studies on geopolymer mortar, comprising saturated water absorption, apparent porosity and depth of carbonation, and their results on saturated water absorption at 28, 56 and 90 days on different ratios 1:2.5 & 1:3 are evaluated under laboratory condition. It was noted that the absorption of saturated water declines with the increase in the content of fly ash. At 28 days of curing for 1:2.5 GPM, the saturated water absorption percentage is higher than that of 1:3 GPM. Flyash behaves as a filler material that fills the pores, thereby reducing the absorption of water. The water absorption of concrete also decreases with increasing in the addition of fly ash. The porosity ratio of 1:2.5 for fly ash to sand with a water binder ratio of 0.416 was found to be lower than 1:3 at the age of 28 days. The value of porosity decreases as the w/b ratio increases.

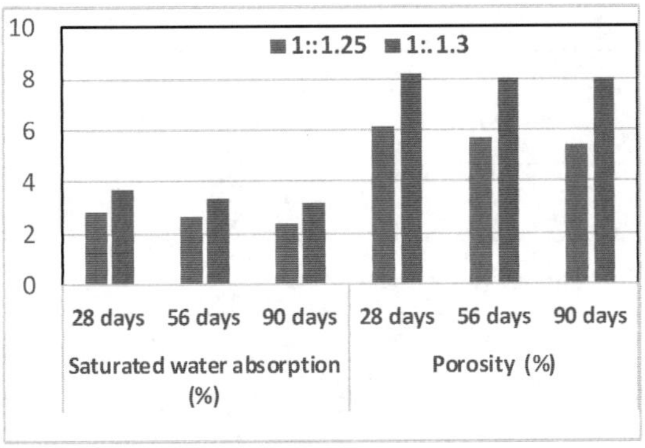

Fig 1 : Saturated Water Absorption & Porosity at 28, 56 & 90 days for 1:3 &1:2.5 ratio

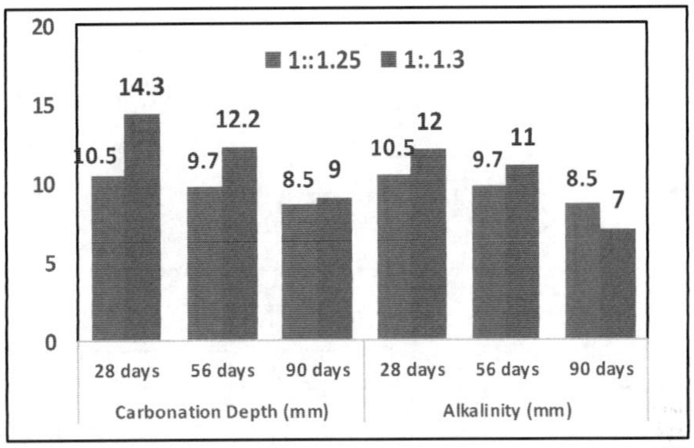

Fig 2 : Carbonation Depth & Alkalanity at 28, 56 & 90 days for 1:3 &1:2.5 ratio

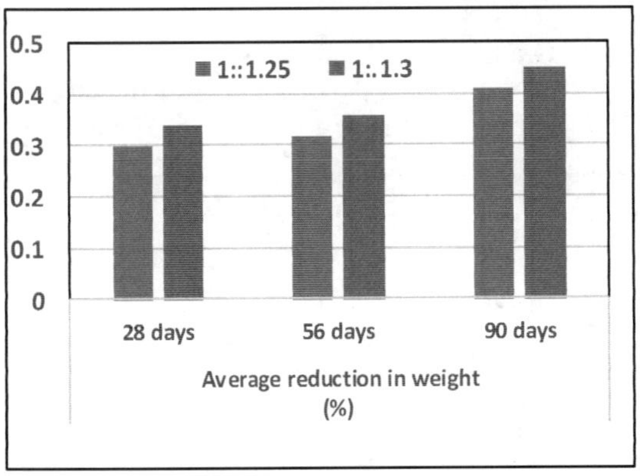

Fig 3 : Average Reduction in Weight at 28, 56 & 90 days for 1:3 &1:2.5 ratio

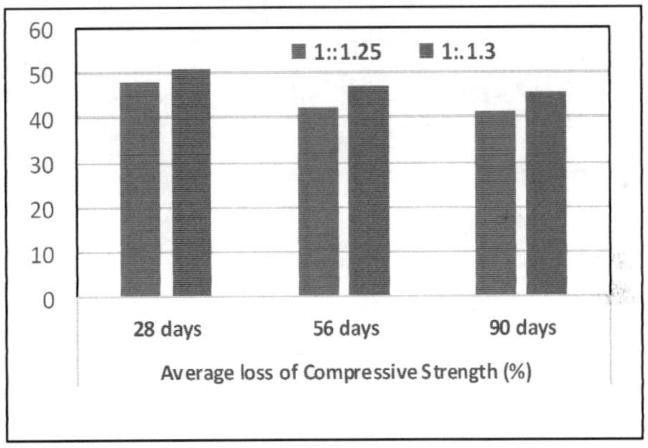

Fig 4 :Average Loss of Compressive Strength at 28, 56 & 90 days for 1:3 &1:2.5 ratio

The carbonation depth of the geopolymer mortar sample indicates that the depth of the 1:3 geopolymer mortar mix is higher than the 1:2.5 geopolymer mortar mix. The alkalinity measurement value of the geopolymer mortar sample is noted to be higher than the 1:2.5 alkalinity measurement value of the 1:3 geopolymer mortar mix.

The compressive strength ratio of 1:2.5 for fly ash to sand is blended with a water binder ratio of 0.416 at 28 days of age, increasing the overall strength. The table illustrates that the weight reduction of the specimens is reduced with an increase in the fly ash content & activator solution and the loss of compressive strength is decreased.

The number of blows corresponding to the first crack and final crack and impact energy absorbed for each crack and their ductility index values are shown in the Initial and Ultimate Energy Absorption in Table 3 to 7. From these observations, the importance of energy absorption increases as the mesh layers increase. Better results are shown among the three welded mesh meshes.

The number of mesh layers and form of mesh plays a major role in the energy absorption of the effect. Ductility Index of the geopolymer ferrocement panel is the ratio of Propagation Energy(Ep) and Initiation Energy(Ei).

Table 3 Energy Absorption and Ductility Index in Ambient Curing 0.416

Sl No	Types of Mesh	Layers	No. of Blows First crack(N)	No. of Blows Final Crack(N)	Height Drop (m)	E in Joules Initial	E in Joules Final	Ductility Index
1	Square Woven	1	5	7	1.18	202.6	283.64	1.4
		2	7	10	1.18	283.64	405.23	1.43
		3	7	11	1.18	283.64	445.7	1.57
2	Expanded	1	8	13	1.18	324.2	526.9	1.62
		2	9	15	1.18	364.7	607.9	1.67
		3	11	19	1.18	445.7	770.1	1.73
3	Square Welded	1	12	21	1.18	486.39	851.19	1.75
		2	13	23	1.18	526.9	932.2	1.76
		3	14	25	1.18	567.4	1013.0	1.78

Table 4 Energy Absorption and Ductility Index in Ambient Curing 0.396

Sl No	Types of Mesh	Layers	No. of Blows First crack(N)	No. of Blows Final Crack(N)	Height of Drop (m)	E in Joules Initial	E in Joules Final	Ductility Index
1	Square Woven	1	6	8	1.18	1.33	324.64	1.33
		2	6	8	1.18	1.33	324.64	1.33
		3	7	11	1.18	1.57	445.7	1.57
2	Expanded	1	8	12	1.18	1.50	486.18	1.50
		2	8	13	1.18	1.62	526.97	1.62
		3	9	15	1.18	1.67	607.9	1.67
3	Square Welded	1	10	17	1.18	1.70	689.04	1.70
		2	11	19	1.18	1.73	770.12	1.73
		1	6	8	1.18	1.33	324.64	1.33

Table 6 Energy Absorption and Ductility Index in Ambient Curing 0.386

Sl No	Types of Mesh	Layers	No. of Blows First crack(N)	No. of Blows Final Crack(N)	Height of Drop (m)	E in Joules Initial	E in Joules Final	Ductility Index
1	Square Woven	1	6	7	1.18	1.16	283.74	1.16
		2	6	8	1.18	1.33	324.26	1.33
		3	7	10	1.18	1.43	445.7	1.43
2	Expanded	1	8	12	1.18	1.50	486.4	1.50
		2	8	13	1.18	1.62	526.9	1.62
		3	8	14	1.18	1.65	567.4	1.65
3	Square Welded	1	9	16	1.18	1.67	648.5	1.67
		2	10	17	1.18	1.69	689.06	1.69
		3	11	18	1.18	1.70	729.6	1.70

Table 7 Energy Absorption and Ductility Index in Ambient Curing 0.376

Sl No	Types of Mesh	Layers	No. of Blows First crack(N)	No. of Blows Final Crack(N)	Height of Drop (m)	E in Joules Initial	E in Joules Final	Ductility Index
1	Square Woven	1	6	7	1.18	253.2	284.74	1.18
		2	6	8	1.18	273.2	334.5	1.37
		3	7	10	1.18	293.7	435.7	1.45
2	Expanded	1	8	12	1.18	327.6	456.4	1.60
		2	8	13	1.18	354.26	526.8	1.71
		3	8	14	1.18	334.2	577.3	1.69
3	Square Welded	1	9	16	1.18	365.8	658.5	1.71
		2	10	17	1.18	445.8	680.06	1.68
		3	11	18	1.18	455.8	789.6	1.78

The number of blows corresponding to first crack and final crack and impact energy absorbed for each crack and their ductility index values in Heat Curing are shown in the Table 8 to 11. The energy absorption and ductility index for the specimens of Square woven, square welded and Expanded metal mesh were used in single layer, double layer & triple layer in Heat curing of 0.416,0.396,0.386 and 0.376 have been tabulated below from table 8 to 11.

Table 8 Energy Absorption and Ductility Index in Heat Curing 0.416

Sl No	Types of Mesh	Layers	No. of Blows First crack(N)	No. of Blows Final Crack(N)	Height of Drop (m)	E in Joules Initial	E in Joules Final	Ductility Index
1	Square Woven	1	5	8	1.18	202.6	324.12	1.6
		2	7	12	1.18	283.64	486.18	1.71
		3	8	14	1.18	324.12	567.14	1.75
2	Expanded	1	8	15	1.18	324.2	607.5	1.875
		2	9	17	1.18	364.7	689.06	1.89
		3	10	19	1.18	405.4	769.9	1.9
3	Square Welded	1	12	23	1.18	486.18	932.42	1.92
		2	13	27	1.18	527.02	1094.5	2.06
		3	13	28	1.18	527.02	1135.1	2.15

Table 9 Energy Absorption and Ductility Index in Heat Curing 0.396

Sl No	Types of Mesh	Layers	No. of Blows First crack(N)	No. of Blows Final Crack(N)	Height of Drop (m)	E in Joules Initial	E in Joules Final	Ductility Index
1	Square Woven	1	7	11	1.18	283.74	445.87	1.57
		2	8	13	1.18	324.2	526.97	1.62
		3	9	15	1.18	364.7	607.9	1.67
2	Expanded	1	10	17	1.18	405.35	689.06	1.7
		2	11	19	1.18	445.87	770.12	1.73
		3	11	19	1.18	445.87	770.12	1.73
3	Square Welded	1	11	20	1.18	445.87	810.66	1.8

		2	12	23	1.18	486.39	932.25	1.92
		3	13	27	1.18	526.92	1094.4	2.06

Table 10 Energy Absorption and Ductility Index in Heat Curing 0.386

Sl No	Types of Mesh	Layers	No. of Blows First crack(N)	No. of Blows Final Crack(N)	Height of Drop (m)	E in Joules Initial	E in Joules Final	Ductility Index
1	Square Woven	1	7	10	1.18	283.74	445.7	1.43
		2	8	12	1.18	324.2	486.4	1.5
		3	8	13	1.18	324.26	526.9	1.62
2	Expanded	1	9	16	1.18	526.9	648.5	1.67
		2	10	17	1.18	445.7	689.06	1.69
		3	11	18	1.18	445.86	729.6	1.7
3	Square Welded	1	11	19	1.18	445.86	770.12	1.73
		2	11	20	1.18	445.86	810.6	1.8
		3	12	22	1.18	486.8	891.7	1.89

Table 11 Energy Absorption and Ductility Index in Heat Curing 0.376

Sl No	Types of Mesh	Layers	No. of Blows First crack(N)	No. of Blows Final Crack(N)	Height of Drop (m)	E in Joules Initial	E in Joules Final	Ductility Index
1	Square Woven	1	6	8	1.18	243.2	324.26	1.33
		2	7	10	1.18	283.74	445.7	1.43
		3	7	11	1.18	283.74	445.86	1.48
2	Expanded	1	8	12	1.18	324.2	486.4	1.5
		2	8	13	1.18	364.7	526.9	1.62
		3	8	14	1.18	445.7	567.4	1.65
3	Square Welded	1	11	18	1.18	445.86	729.6	1.7
		2	11	19	1.18	445.86	770.12	1.73
		3	11	20	1.18	445.86	810.6	1.8

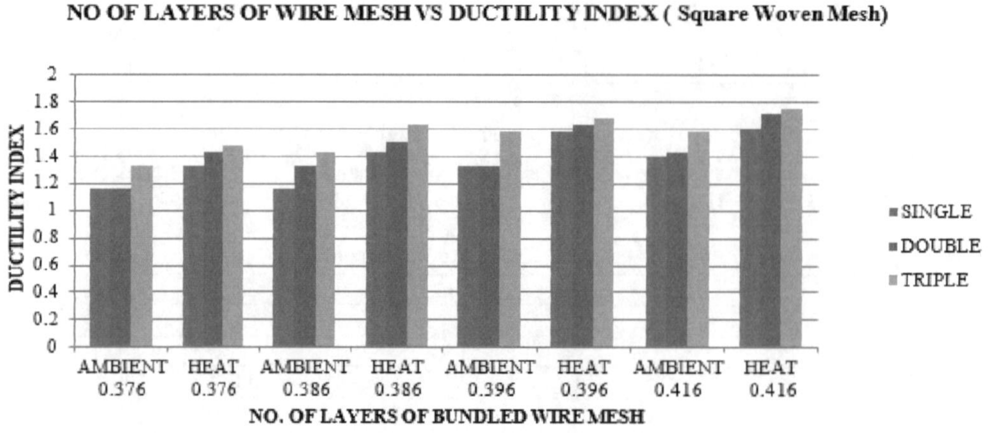

Fig 6 Performance of square woven meshes in each water binder ratios 0.376, 0.386, 0.396 & 0.416

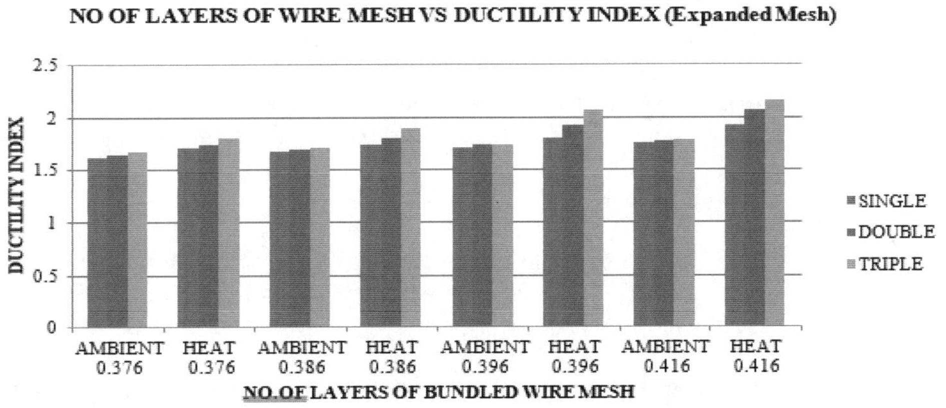

Fig 7 Performance of expanded meshes in each water binder ratios 0.376, 0.386,0.396 & 0.416

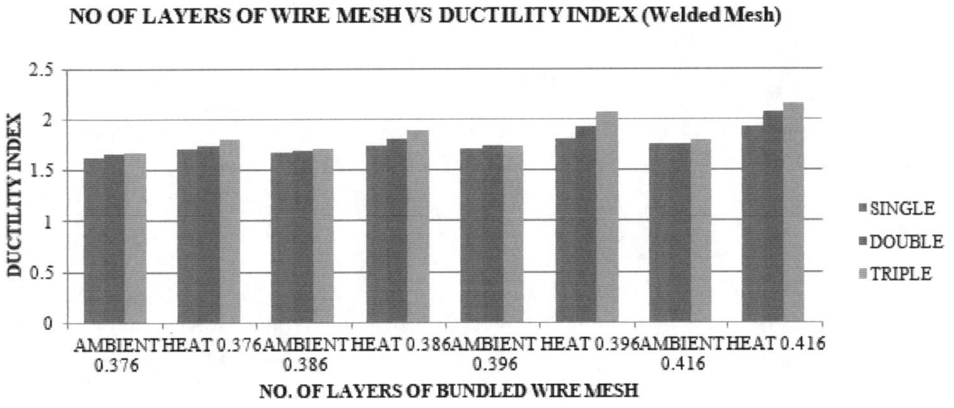

Fig 8 Performance of welded meshes in each water binder ratios 0.376, 0.386,0.396 & 0.416

5. Conclusions

Based on the findings during the present study, following conclusions can be drawn. The saturated water absorption percentage at 28 days curing for 1:2.5 GPM is higher than that of 1:3 GPM. The saturated water absorption percentage decreases with increase in water binder ratio. This low water absorption level is a good indicator of limited open porosity that can inhibit the high flow of water into the mortar.

The loss of weight was observed to be lower in geopolymer mortar specimen when compared with conventional cement mortar. The property of improvement of compressive strength with ages enhances the performances of the fly ash mix hence the reduction in wear. Geopolymer mortar specimens showed greater resistance against corrosion when compared with conventional cement mortar specimens.

The geopolymer mortar specimens showed very little loss in strength and the percentage of strength loss was observed to be higher in 1:3 ratio of flyash to sand. Geopolymer mortar specimens showed greater resistance to acid environments when compared with conventional cement mortar specimens. Due to the presence of number of wire meshes in the path of impact ,it required more number of blows to cause failure. Increase in number of layers of wire meshes increases the energy absorption and ductility index significantly. As the thickness of ferrocement panel increases the energy absorption and ductility index increases . Fluid binder ratio(X4) is main parameter which is influencing the ductility index characteristics. Heat curing have high initial and final energy absorption and ductility index than that of Ambient curing.

6. References

[1] ACI Committee 549-R -97 State of the art report on ferrocement, ACI549, IR93. Guide for the design, Construction, and repair of ferrocement. Reapproved 1999

[2] Apha Sathonsaowaphaka, Prinya Chindaprasirt, Kedsarin Pimraksa, (2009), "Workability and strength of lignite bottom ash geopolymer mortar", *Journal of Hazardous Materials* 168, pp. 44–50.

[3] Anitha .M, R Thenmozhi, S Deepa Shri, (2012) Behaviour of hybrid ferrocement slab subjected to impact", *International Journal of Emerging Trends in Engineering and Development*, 4,

[4] Bakharev T,(2005) "Geopolymeric materials prepared using Class F fly ash and elevated temperature curing", *Cement and Concrete Research*, 35(6), pp.1224-1232

[5] Balasubramanian, Senthil.R, Santhakumar.A.R. R, (2003) "Testing for ferrocement composite properties" *S.R.M NSFCCE*, pp. 4.1 - 4.6.

[6] Boshra Aboul-Anen, Ahmed El-Shafey, and Mostafa El-Shami, (2009), "Experimental and Analytical Model of Ferrocement Slabs", *International Journal of Recent Trends in Engineering*, Vol. 1, No.6, pp.25-29.

[7] Davidovits J, "Geopolymers: Inorganic Polymeric New Materials",(1991), *Journal of Thermal Analysis*, 37, pp.1633-1656.

[8] Davidovits J, "Chemistry of Geopolymeric Systems, Terminology," (1999), Geopolymerè'99, *Geopolymer International Conference, Proceedings*, 30 June – 2 July, pp. 9-39.

[9] Djwantoro Hardjito, Chua Chung Cheak & Carrie Ho Lee Ing, (2008), "Strength and Setting Times of Low Calcium Fly Ash-based Geopolymer Mortar", *Modern Applied Science*, 2(4), pp.3-10,(2008),.

[10] Djwantoro Hardjito, Tsen M.Z, (2008)"Strength and Thermal Stability of Fly Ash-Based Geopolymer Mortar", *The 3rd International Conference-ACF/VCA*, pp.144-150.

[11] C Vivek Kumar, Mohammed Aiman Parvez, T Srinivas, Strength Behaviour of Blended Self Cured Concrete made with Cement by partial replacement of Supplementary Cementitious Materials (SCM's), E3S Web of Conferences, 184 01100 (2020), DOI: 10.1051/e3sconf/202018401100, (2020).

[12] Paramasivam P and R Sri.Ravindra rajah, Effect of arrangement of reinforcement on mechanical properties of ferrocement, *ACI Structural Journal*, pp 03-11, 1988.

[13] Songpiriyakij S "Effect of Temperature on Compressive Strength of Fly Ash based Geopolymer Mortar", Silikaty 48(4), pp.183-197, (2000).

[14] Sudarsana Rao. I,Vaishali G,Ghoropade .N, Ramana V,(2010) Response of SIFCONE two way slab under impact loading, *International Journal of Impact Engineering*, 37,pp 452-458.

[15] Suresh Thokchom, Dr.Partha Ghosh and Dr. Somnath ,(2009) "Acid Resistance of Fly ash based Geopolymer mortars", *International Journal of Recent Trends in Engineering*, 1, pp.36- 40.

[16] Suresh Thokchom, Dr.Partha Ghosh and Dr. Somnath Ghosh, (2009) "Effect of Water Absorption, Porosity and Sorptivity on Durability of Geopolymer Mortars", *ARPN Journal of Engineering and Applied Sciences*, Vol. 4, No.7, pp.28-32.

[17] C Vivek Kumar, Patam Manisha, Pooja Sadula,Behavior of Monotonic Loading for Glass Fibre based High Performance Concrete in External Beam-Column Joint using ANSYS. Analysis, E3S Web of Conferences, 184 01088 (2020), DOI: 10.1051/e3sconf/202018401088, (2020).

[18] Carrasquillo RL (1985) "Production of high strength pastes, mortars, and concrete, very high strength cement-based materials". *Materials Research Society Symposia Proceedings* 42:151–168.

[19] IS: 383-1970 (reaffirmed 1997) ―Specification for Coarse and Fine Aggregates from Natural Source for Concrete‖, New Delhi.

[20] IS: 12269-1987 (reaffirmed 1999) ―Specification for 53 grade Ordinary Portland Cement, New Delhi.

[21] IS: 456-2000 (reaffirmed 2005) ―Plain and Reinforced Concrete – Code of Practice‖, Fourth

Revision, pp.14.

Comparative analysis of deflection in a bimetal

R K Jaya Kumar, T Ashokkumar and P Sivakumar

Department of Aeronautical Engineering, Sri Ramakrishna Engineering College, Coimbatore-22, India.

Corresponding Author : jayakumarrk7@gmail.com

Abstract: A Bimetal is a combination of two different metals, which is employed in electrical appliances as a circuit breaker to prevent the overflow of current. The primary objective of this work is to examine the deflection produced in the bimetal component, due to electro-thermal load induced when the current flows through it. Copper and stainless steel are the most widely used metals having different coefficient of linear thermal expansion were taken as a bimetal composition and the same has been compared with Stainless steel and Aluminum. Modeling and analysis of the bimetal component, with different amperes, has been done using ANSYS and resulted stainless steel - Aluminium combination were shown higher deflection during the Analysis.

Keywords: Bimetal, Thermal load, Deflection

I. INTRODUCTION

Bimetals are employed in various engineering applications, such as thermometers, thermostats, miniature circuit breakers, etc. It consists of two different metal pieces that are bonded together that expands when heated and contract once it cooled. Due to different coefficient of thermal expansion, exposing the strip to temperature induces thermal stress in the strip. The resulting change of curvature or bending, in response to temperature change, is a fundamental property of all thermostatic bimetals. In effect, the temperature change is converted into mechanical displacement. The behavior of bimetals is predictable and repeatable. The components used for bimetals are selected for their temperature characteristics as well as their thermal conductivity, stability, strength, workability, and electrical properties. This effect is mainly used for the fabrication of temperature-controlled, residual-current circuit breakers with over current protection. The triggering function is performed by a bimetallic plate element, which has to deform within a given time to generate a force and trigger the switching mechanism. Recently, thermally activated bimetallic micro beams have been the mainstays of MEMS technology [1]. As electro thermal bimetal actuators have the advantage of a large force and displacement under a low driving voltage [2]. The first model of the bimetallic beam was developed by Timoshenko [3]. It plays a significant role in equipment's to prevent the flow of excessive current. Thermostatic bimetal is a classic element mainly used for detection of small over currents caused by overloads. It is the main structural component of the thermal overload trip unit and thermal overload relays. It scans and evaluates the current flowing through the device (circuit breakers, motor starters) and instructs the device to act directly on its switching mechanism.

IOP Conf. Series: Materials Science and Engineering 1091 (2021) 012070 doi:10.1088/1757-899X/1091/1/012070

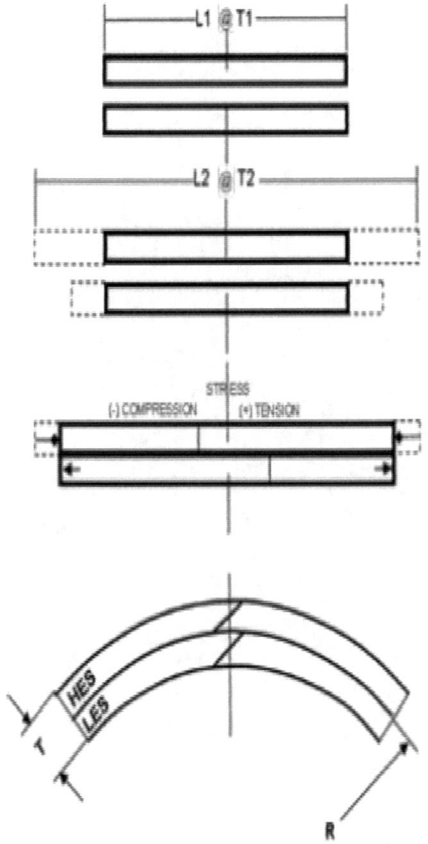

Two metal strips of identical length (at a given temperature) having high and low coefficients of thermal expansion.

When the temperature rises, their relative lengths will change due to different coefficient of thermal Expansion

When the strips are bonded together, and the temperature is raised, the high expansion strip will be under compression and the low expansion strip will be under tension.

These forces (compression and tension) produce a moment which causes the element to bend in a uniform arc. Bimetal bending is directly proportional to the difference in the coefficient of expansion and the temperature change of the component strips, and inversely proportional to the thickness of the combined strips. The amount of bending is also affected by the ratio of the moduli of elasticity of the two strips and by their thickness ratio.

Fig 1. Bimetal strip with different coefficient of Expansion [4]

The fundamental characteristics of a bimetal is flexivity, also known as specific curvature. It is an important property of bimetal that is defined by the curvature change along its longitudinal axis. This is expressed as [4]

$$F = \frac{\left(\frac{1}{R_2} - \frac{1}{R_1}\right)t}{(T_2 - T_1)}$$

Where

F	-	Flexivity in °C
R1&R2	-	Radii of curvature of the active and passive sides in mm
t	-	Thickness of the strip in mm
T_1 & T_2	-	Temperatures in °C

2. METHODOLOGY

The work has been carried out using numerical method using ANSYS Software. ANSYS is a finite element modeling software for numerically solving a wide variety of mechanical problems. ANSYS Workbench was designed to make it easier to simulate multiphysics coupling and used to perform a comprehensive study, integrated Simulations and in addition to that it also used to perform various types of structural, thermal, fluid, and electromagnetic analysis. The progress of this work has divided into three phases namely Electric, Thermal and Structural to execute the process in a sequential manner. It is known that output from the electrical module has taken as an input to thermal module and the corresponding results from the same has been feed to structural module's input to obtain the desired result called mechanical displacement of the bimetal

2.1 Generating a Model

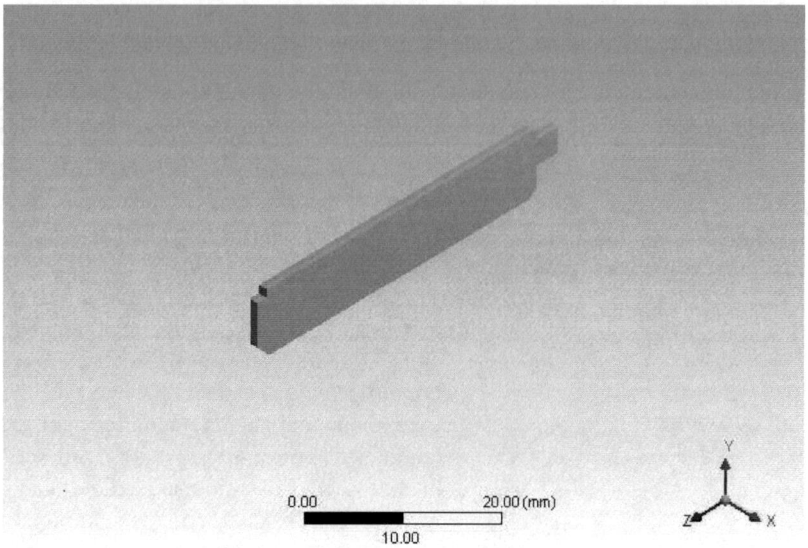

Fig 2. Modeling of Bimetal

Preprocessing is also called as model preparation, which is the first step in Finite Element Analysis. During this step, material properties of the model has defined using Engineering data module where material properties can also be customized as per the real time environment and the modeling of the bimetal had done using design modeler as per the geometry of the component. Design modeler is primarily intended to provide geometry to an analysis environment as per the requirements of the user (i.e) it can be created and modified. The following table represents material properties of the bimetal

Table 1. Material Properties of a bimetal

S.No.	Parameters	Material Properties	
		Copper	Stainless steel
1	Density	8.3e-06kg/mm^3	7.85e-06kg/mm^3
2	Young's Modulus	1.1e5MPa	2e5MPa
3	Poisson's Ratio	0.34	0.3
4	Bulk Modulus	1.1458e5MPa	1.67e5MPa
5	Shear modulus	41045MPa	76923MPa
6	Isotropic Secant Coefficient of Thermal Expansion	1.8e-5 /°C	1.2e-5 1 /°C

2.2 Meshing

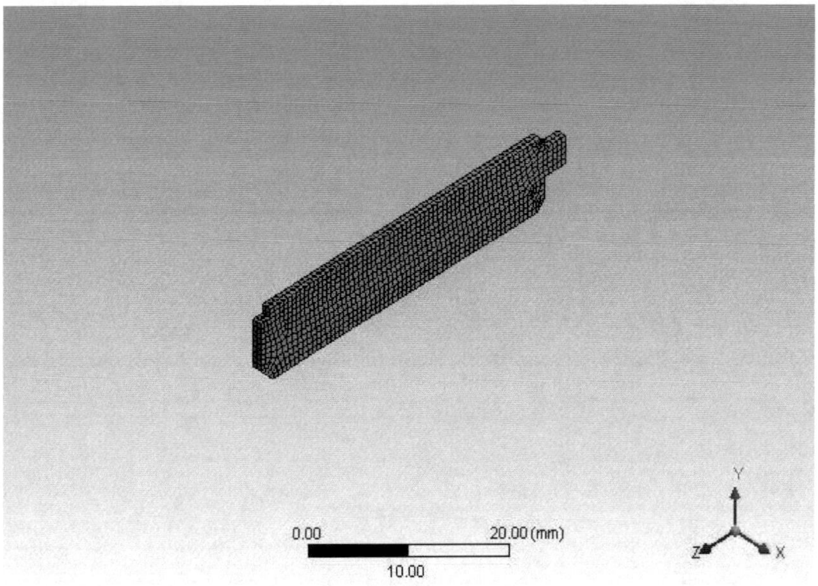

Fig 3. Meshed model of Bimetal

Meshing is nothing but a discretization of continuous body into finite number of elements. Generally, by increasing a greater number of elements leads to accuracy of the results. But at the meantime, increasing the number of elements also increases the processing time of the entire Analysis. The most common way is that allowing it to generate automatically, however more or less controlled. Generating fine mesh on the model has done during this step. The number of nodes generated on the geometry is 12327, which is a coordinate location in space where the degrees of freedom (DOFs) are defined. And the number of elements generated is 1643.Cordinate system employed for this analysis is global Coordinate system which corresponds to define nodes in the entire body

2.3 Boundary Conditions

The following Boundary Conditions have been taken to carry out the analysis in an electric module.

Table 2. Bimetal Boundary Conditions

S. No.	Particulars	Values
1	Initial temperature	22 °C
2	Current	1 to 32 A
3	Convection	All Faces of the component
4	Time	Unit time (s)

The bimetal is subjected to different range of current (1 – 32A) for a unit time at the face having higher thermal coefficient of expansion.

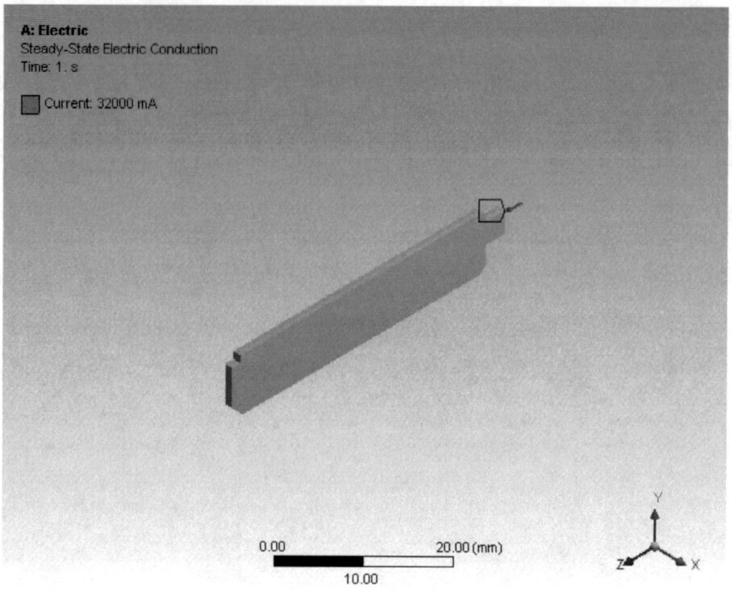

Fig 4. Current to the Bimetal

For the conversion of electrical current into heat energy it is necessary to ensure the Joule heating has been employed, which is defined as the physical effect by which the pass of current through an electrical conductor produces thermal energy or otherwise called as heat. It is evident that Joule heating is directly proportional to the flow of current and the maximum Joule heat generated is 12.83 W/mm for stainless steel – copper combination. The generated output/solution from module1 (electric) has taken as an input for module2.

2.4 Transient Thermal

Transient thermal analysis is the evaluation of how a system responds to fixed and varying boundary conditions over time. For fixed boundary conditions, the time to reach a steady state temperature can be evaluated, as well as how long operating conditions can be sustained before reaching a threshold temperature. For time-varying boundary conditions, transient analysis can show the resulting thermal response of the system. It is used to evaluate transitions in operating conditions before the system returns to steady state. In such cases, it is typical to find the steady-state solution to one set of operating conditions and then perform a transient analysis using a different set of operating conditions.

It was taken that convection on the surface of bimetal on all faces is uniform and temperature of the component is considered to be 22 °C. Convection is the mode of heat transfer between a surface and a fluid moving over it. The energy transfer in convection is predominately due to the bulk motion of the fluid particles, though the molecular conduction within the fluid itself also contributes to some extent. If this motion is mainly due to the density variations associated with temperature gradients within the fluid, the mode of heat transfer is said to be due to Natural Convection [5] and it takes place evenly on the bimetal, when thermal stress has induced on the component at different range of current.

2.5. Thermal Load

Thermal load can be found theoretically by using the formula. It can be used to compare the variation produced among theoretical result, numerical result and experimental result.[5]

Total heat load can be found using the following formula, Q = AU∧t

Q - Total Heat Load
A - Area
U - Coefficient of heat Conductivity
∧t - Difference in temperature

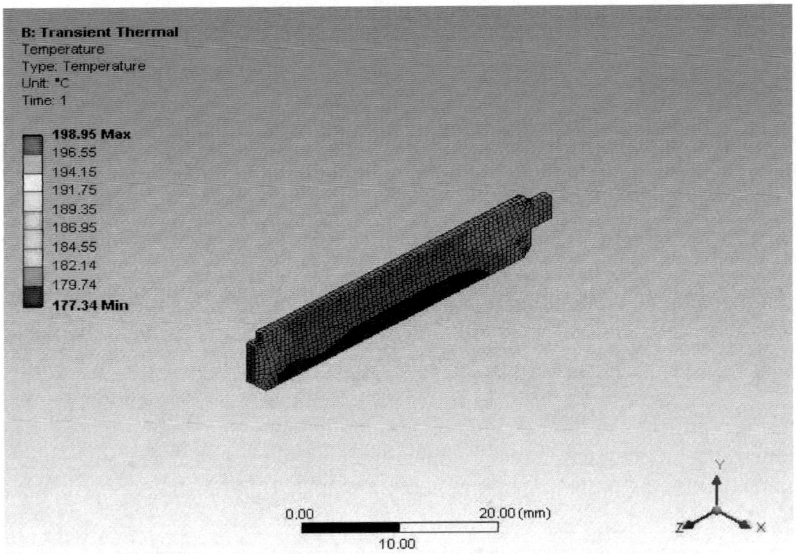

Fig 5. Temperature Distribution on Bimetal

It has been observed from the Fig.5. Temperature distribution on bimetal, the maximum temperature produced in the bimetal (copper and stainless steel) at 32A is 198.95°C for a unit time and minimum temperature at the lower surface is 177.34°C. Depend upon the flow of current to the component, joule heating takes place in the bimetal. This obtained thermal load result, can be fed as an input to Transient Structural module to find the deflection of the component at various levels of electric current.

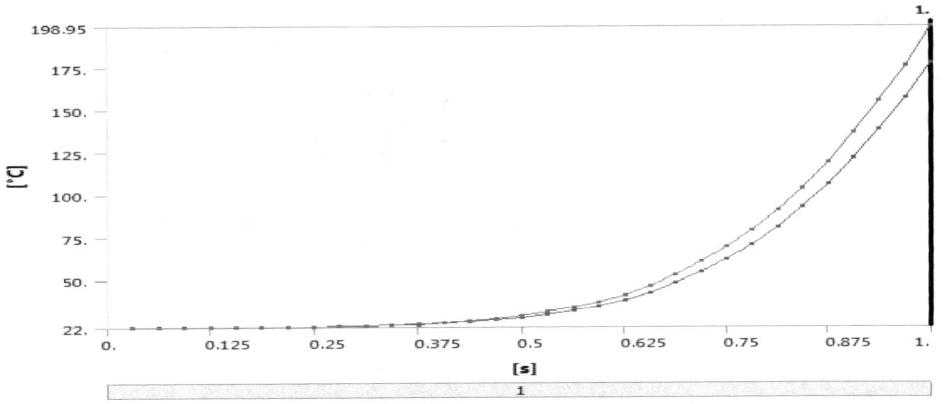

Fig 6. Time Vs Temperature

From the above graph, it was known that from the plots (minimum and maximum) there was no major significant variation in temperature up to 0.5s, then gradual increase in temperature with respect to time at various current level. In other words, it can be simply stated as time directly proportional to temperature when current flows through it for the particular material combination. In order to obtain the deflection at one end of the metal, due to thermal load which is generated by the flow of current and it is required to fix the other end. It acts as a cantilever beam to get desired result.

2.6 Thermo Structural Analysis Result

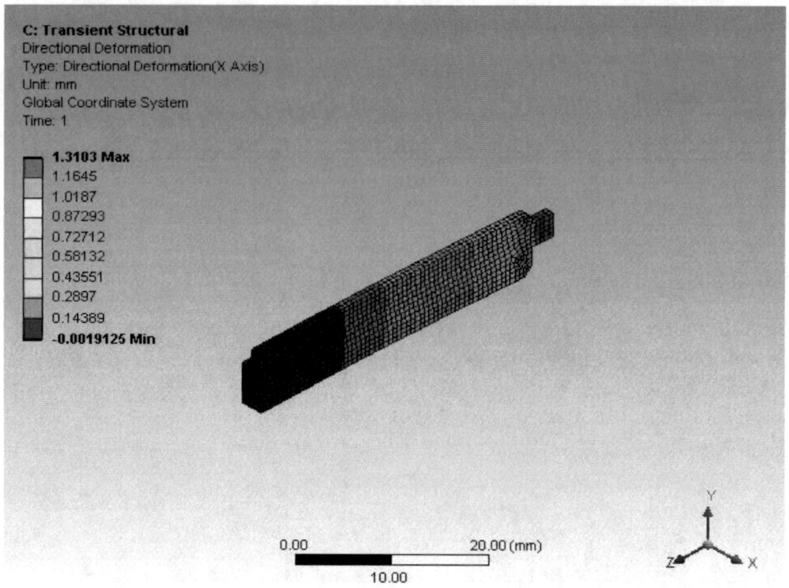

Fig 7. Thermo Structural Deflection on Bimetal

From the analysis result, maximum deflection produced at the tip due to increased temperature of material composition and activate the cutoff switch to break the circuit from operation. Maximum deflection produced in the bimetal is 1.31mm at 198.95 °C at the end, when 32A current flows in a unit time. By changing the material properties to Stainless Steel and Aluminium with the corresponding current, the obtained temperature and deflection produced in the bimetal was also tabulated

3. Results and Discussion

Table 3. Comparison of Bimetal combination Results

S.No.	Stainless Steel – Copper		Stainless Steel – Aluminum	
1	Max. Temperature (°C)	Max. Deflection (mm)	Max. Temperature (°C)	Max. Deflection (mm)
	198.95	1.31	318.97	3.56

It is also inferred from the analysis; structural deformation is proportional to temperature and current flowing through it. From the analysis result, it has been observed that maximum temperature produced in the stainless steel – copper combination is 198.95°C and deflection produced is 1.31mm and the Maximum temperature for the stainless steel- Aluminum Combination is 318.97°C and the deflection produced is 3.56mm for the corresponding current flow through the bimetal. This variation in deflection is mainly due to difference in thermal conductivity, coefficient of thermal expansion, density and specific heat of copper and Aluminum. In detail, the density of the copper is higher than aluminium. Hence it is heavier compared to aluminium conductors it requires strong structures and hardware to bear the weight. More ductile and has high tensile strength. Compared to Aluminium it has good conductivity and low resistivity but high in cost. Compared to copper it is Less in weight and of Low cost but less in conductivity. More resistivity, as it has a chance of oxide surface formation which results in heating. Due to increased price of copper, aluminium was preferred and it is economical. Aluminum wires are coated with special grease and wiring devices to prevent any accident due to heat.

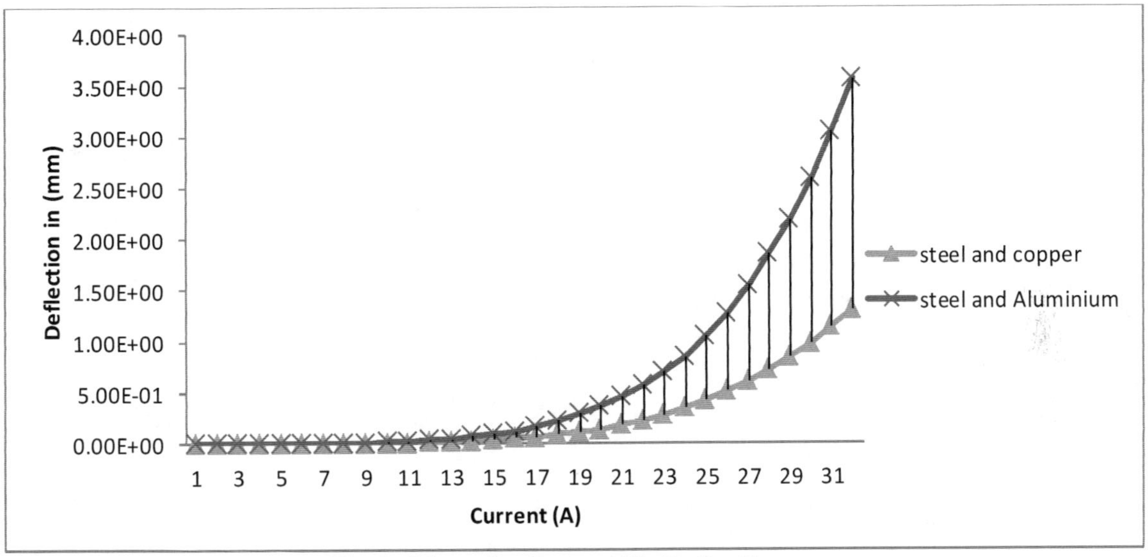

Fig 8. Bimetal - Current Vs Deflection

The above graph has been plotted against Current flow and the obtained deflection from numerical results, for a unit time and the deflection produced for stainless steel and Aluminum combination is comparatively higher than the copper – stainless steel.

4.Conclusion

The following conclusions were made from the analysis results, they are. Difference in temperature distribution and joule heating across the length of the bimetal plays a significant role and directly influences the deflection of the bimetal. Deflection of the bimetal is directly proportional to the current flow through the metal and generation of temperature for copper - stainless-steel and stainless steel – Aluminum combination. Deflection produced in the stainless steel and aluminum combination is 2.25mm higher than the widely used conventional combination. Maximum deflection produced at an elevated temperature will vary depends upon the type of material employed. So appropriate material selection needs to be considered. However, depends upon the applications, material combination of the bimetal can be varied for higher or lower deflection to protect electrical components.

References

[1] Chu W H, Mehregany M and Mullen R.L 1993 *J. Micromech Microeng*. **3** 4
[2] Ross D S, Cabal A, Trauernicht D and Lebens J 2005 *Sens. Actuators A, Phys*. **119** 537
[3] Timoshenko S 1925 *J. Opt. Soc. Am.* **11** 233
[4] Engineering Materials Solutions 1916 *Thermostatic Bimetal designers Guide* (USA:Attleboro)
[5] R.C.Sachdeva 2008 *Fundamentals of Engineering Heat and Mass Transfer* New Age

International Publishers p 223
[6] Zou Q, Sridhar U and Lin R 1999 *Sens. Actuators A, Phys* **78** 212
[7] Suocheng W,Yongping H and Shuangjie L 2015 *J. Semicond* **36** 1
[8] Sedighi M and Dardashti, B.N 2012 *Mater. Phys. Mech* **14** 37
[9] Dufva K.E, Sopanen J.T and Mikkola A.M 2005 *J. Sound Vib.* **280** 719
[10] Vallejo D G, Mikkola A.M and Escalona, J.L 2007 *Nonlinear Dyn.* **50** 249
[11] Abbas L K, Rui X and Marzocca P 2015 *Multibody Syst. Dyn.* **33** 163
[12] Gerstmayr J and Shabana A A 2006 *Nonlinear Dyn.* **45** 109

IOP Conf. Series: Materials Science and Engineering 1091 (2021) 012071 doi:10.1088/1757-899X/1091/1/012071

Numerical investigation of inter-linked parallel micro-channel heat sinks

Munib Qasim Ansari[1,2], **Guobing Zhou**[1]*

[1]School of Energy, Power and Mechanical Engineering, North China Electric Power University, Beijing 102206, PR China.
[2]Department of Mechanical Engineering, COMSATS University Islamabad, Sahiwal 57000, Pakistan.

* Corresponding author. zhougb@ncepu.edu.cn (G. Zhou).

Abstract. Parallel micro-channel heat sinks linked with a joint are investigated in this paper. Channels are analyzed for laminar flow (water) under constant heat flux. The considered channels with same hydraulic diameter (776 μm) are investigated for two different dimensional variations of connecting link with parallel flow and the comparison of parallel and counter flow in the channels. It is found that the varying dimension of joining link has a little influence on the over-all thermal hydraulic performance. Performance index decreases in the cases of either increasing the height (J) or width (W) of the joint at higher Reynolds number (1000 and 1200). Convective heat transfer (Nu) marginally improves with the increasing J. Inter-linked micro-channels with counter flow has given improved convective heat transfer but with an increase in frictional factor. The Nu with counter flow is found almost 15% higher than that with parallel flow at Re = 1200. The design can be further explored based on present calculations for a positive addition of designing micro-channel heat sinks.

1. Introduction

The demand for micro-electronic systems is growing rapidly and provides a tough task in a high expanse of thermal control. In this perspective, several methods are being established to keep the device temperature under the critical limit where high heat flux is generated. Application of micro-channel heat sinks is an advanced temperature reduction methodology for the removal of a large amount of heat from a small area [1, 2]. However, straight smooth channels are not enough to meet the desired needs [3, 4]. Sasaki and Kishimoto [5] reported the effect of channel height and width at various pressure drop conditions and found that optimized dimensions are needed for a better performance of micro-channel. Ghajar et al. [6] investigated micro-tubes and found the significant increase of friction factor with decreasing hydraulic diameter of channel (D < 1372 μm). Lin and Kandlikar [7] experimentally examined the rough micro-channels. The heat transfer enhancement for surface roughness height of 96 μm (D_c = 690 μm) in the laminar fully developed region was 3.78 times higher than that of smooth channels, while the increase of corresponding friction factor was also around 3.7 times. Moradikazerouni et al. [8] studied the flow performance of micro-channel heat sinks with air as the working fluid. Five different configurations (rectangular, circular, triangular, hexagonal and a straight slot) of channels were investigated for D = 2.47 mm to 10.85 mm. The results demonstrated that, among all the channel configurations, the triangular shape provides the maximum thermal performance for cooling a micro-channel heat sink. Fan et al. [9] conducted a heat transfer analysis of heat sinks (mini-channel) for various oblique angles, varying widths of the secondary channels, and for different Reynold numbers under the (laminar) flow condition. The authors stated the improvement in performance (heat transfer) and decrease in the pressure drop penalty only for cylindrical oblique finned heat sinks. The Literature shows that the micro-channel flow and heat transfer characteristics vary with the dimensional and shape variations. It is further noted that the innovative shapes and modifications of micro-channels can improve the heat transfer but also lead to higher pressure drop [10-12]. Therefore, in this research inter-linked parallel micro-channels are considered for the analysis. The size of "joint" is varied to study the effects on the thermal-hydraulic performances. The channels are also further explored for counter flow and characteristics are compared with the parallel flow.

2. Physical model and numerical methods

2.1 Configurations of micro-channel heat sinks

Rectangular inter-linked (Parallel and 3D) micro-channels are considered for the investigation. The upper and lower channels are connected through a joint at the centre of the channels, as shown in Fig. 1. The width, length, height, and hydraulic diameter of the channels are represented as 'W', 'L', 'H' and D, respectively. The working fluid (water) flows along the channel length and along the x-axis (Fig. 1(b)). Where J and G are the height and width of the joint (connecting the channels), respectively. The dimensional parameters of the channels are given in Table 1.

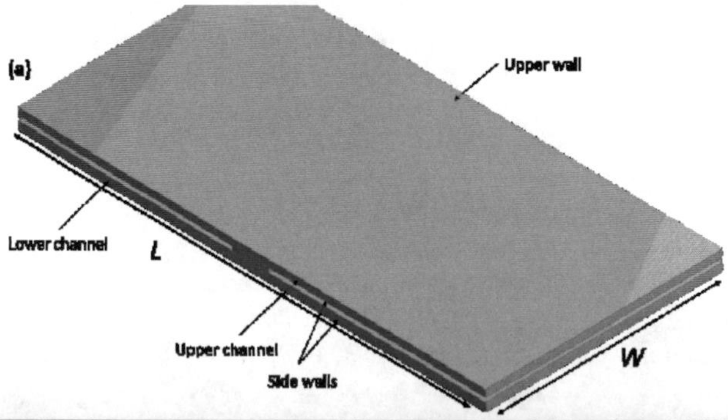

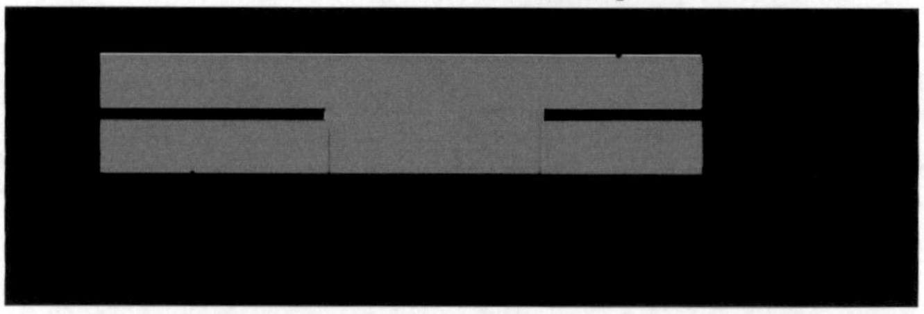

Fig. 1. Schematic of inter-linked micro-channels: (a) isometric view; (b) Front view.

Table 1. Parameters of inter-linked micro-channels.

Parameters	H (μm)	D (μm)	J (μm)	G (mm)	Flow type
Dimensions	400	776	50, **200**	0.5, **1**, 2	**Parallel**, Counter

Note: Bold faced parameters remain constant in other variations.

2.2 Governing equations and evaluation parameters

Parallel micro-channels (3D) are analysed, and the flow is assumed steady and incompressible with constant properties. The gravitational force is not considered in this work. The channel walls are heated, while thermo-physical properties are assumed to be constant. The assumptions of non-slip boundary condition, flow continuum, and Navier-Stokes equation are reasonable in laminar flow regime of the study. The governing equations of continuity, momentum, and energy are written based on the above assumptions:

Continuity equation: $\nabla\vec{V} = 0$ (1)

Momentum equation: $\rho(\vec{V}.\nabla\vec{V}) = -\nabla P + \nabla.(\mu\nabla\vec{V})$ (2)

Energy equation: $\rho C_P(\vec{V}.\nabla T_f) = k_f\,\nabla^2 T_f$ (3)

where P, c_p, μ, T_f, and k_f are fluid pressure, specific heat, dynamic viscosity, fluid temperature, and thermal conductivity, respectively.

For the convenient evaluation of results, Reynolds number (Re), Fanning friction factor (f), and Nusselt number (Nu) are defined as

$$Re = \frac{\rho U D}{\mu} \quad (4)$$

$$f = \frac{\Delta P D}{2\rho U^2 L} \quad (5)$$

$$Nu = \frac{hD}{k_f} \quad (6)$$

$$h = \frac{q}{T_w - T_f} \quad (7)$$

$$\eta = \frac{Nu/Nu_p}{(f/f_p)^{1/3}} \quad (8)$$

where U, ρ, h, q, Nu_p, f_p and η are inlet velocity, fluid density, convective heat transfer coefficient, heat flux, Nusselt number of simple smooth channels, Fanning friction factor of simple smooth channels and overall performance index, respectively. The Nu_p and f_p are also calculated for the evaluation of overall performance.

2.3 Boundary conditions and solution method

The velocity and temperature at the channel inlet: $u = u_{in}$, $v = w = 0$, $T = T_{in}$. The upper wall of upper channel and lower wall of lower channels are kept under constant heat flux (q =1200 W/m^2) (Fig. 2(b)). The remaining walls of the channels are assumed adiabatic, $\partial T/\partial x= \partial T/\partial y=\partial T/\partial z=0$. For the outlet, the pressure is set as P_{out} = 0. The above mentioned governing equations and boundary conditions are solved by the finite volume method with Fluent program [13]. The Second-order upwind scheme for discretization of governing equations along with the SIMPLE algorithm for the velocity-pressure coupling is utilized to perform the numerical analysis. The residual criteria of 10^{-6} are set to verify the convergence of continuity, momentum, and energy equations.

3. Results and Discussion

The inter-linked parallel micro-channel heat sinks are investigated for varying dimensions of linking "joint" and "flow types". The discussion is divided into three different sections. The model is first validated before the analysis which shows good agreement with the reference study [7] and laminar flow theory [11]. The deviations of the values calculated by the present model and reference values are not more than 3%.

3.1 Effect of varying width of the joint of the parallel micro-channel heat sinks

In this section, parallel connected channels (Fig. 1) with parallel type of flow are studied for the varying width of connecting link (Table 1). Fig. 2(a) shows the effect on friction factor in terms of Reynolds number. The trends of f are found consistent within the considered range of variations. The difference of f among the considered cases is not more than 2%. Fig. 2(b) elaborates that the effect of G variations on Nu. It is observed that for lower value of G (0.5mm) and at higher Re, the value of Nu is lesser than the other variations (G = 1 and 2 mm). It seems that the effect of flow mixing is lesser with lower size of G. The difference of Nu between the channels of G = 2 mm and G = 0.5 mm is almost 4% at Re = 1000. However, the Nu of all the channels increases with increasing Re. The performance index is calculated to get the overall performance evaluation and comparison with the respective smooth channels. Fig. 2(c) shows that the overall performance index decreases with the increase of Re. Fig. 2(c) further shows that η is very closer to "1" and additional rise of Nu is compensated by the additional pressure drop.

Velocity streamlines of channels with G = 1 mm and 2 mm are plotted in Figs. 2(d) and (e). It shows with the increasing size of G, the fluid has started stagnated between the channel and not adding anything to the flow modification. Therefore, the overall analysis of Fig. 2 show that the channel with G = 1 seems the better choice in the considered range of dimensions.

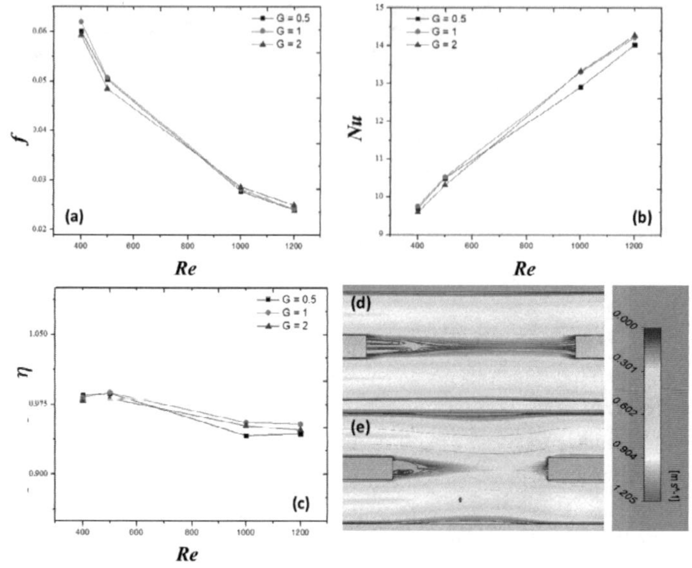

Fig. 2. Effect of varying width of inter-linked channel (J = 200 μm): (a) f vs. Re; (b) Nu vs. Re; (c) η vs. Re. * Units of G are in mm. Velocity streamlines at Re = 500: (d) G = 2 mm; (e) G = 1 mm.

3.2 *Effect of varying height of the link, connecting the parallel channel heat sinks*

In this section, the inter-linked micro-channels are investigated for varying link height (J) and parallel flow type. Fig. 3(a) shows that the effect of varying J on f is almost negligible and the trends are consistent. However, Fig. 3(b) elaborates that the decrease of J (link height) results in lower value of Nu. The difference of Nu among them is around 3% at Re = 1000. It seems that lowering the height of link (J) gives less time to the fluid for mixing and decreases the contact time. The overall performance index is plotted in Fig. 3(c), which shows that at the lower value of Re (400) the channels with J = 200 μm give slightly better performance than J = 50 μm. However, no significant enhancement of performance is noted at any of the Re values and either of the cases.

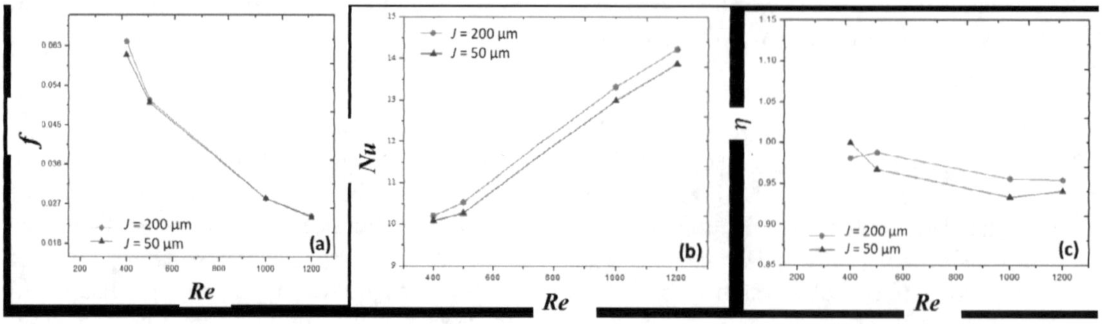

Fig. 3. Effect of varying width of inter-linked channel (G = 1mm): (a) f vs. Re; (b) Nu vs. Re; (c) η vs. Re.

3.3 Comparison of parallel and counter flow in inter-linked micro-channel heat sinks

It is deducted from the above cases that addition of inter-link between the channels with parallel flow is adding only subtle improvement in the performance of micro-channel heat sinks. To further explore the design, considered inter-linked micro-channels are also investigated for counter flow. The comparisons have been made in Fig. 4. Fig. 4(a) indicates that the counter flow in parallel channels is adding to pressure drop. The increase of friction factor is more than 23% at $Re = 500$. Fig. 4(b) shows that counter flow improves the over-all heat transfer of the considered channels. The difference between the Nu values of both the flow type is also increasing with increasing Re. The difference of Nu between them is almost 15% at $Re = 1200$. The rise of convective heat transfer seems due to the more flow mixing, which is also the reason of higher friction factor. Discussion of Figs. 4(a) and (b) lead to the need of performance index calculations. Fig. 4(c) presents the performance comparison of both types of channel. It shows that the overall performance of counter flow is slightly higher than the parallel flow due to the respectively higher friction factor values.

Figs. 4 (d) and (e) show the velocity streamlines of both the flow type. It shows that the flow velocities are higher for counter flow (Fig. 4(d)) with more mixing and generation of the secondary flow can also be seen at the joining point of channels. However, the flow in parallel flow is consistent as in normal smooth channels and the flow is stagnated at the joining point and has less influence on the flow mixing. This shows that the counter flow type can be used where the higher heat transfer is a dare need.

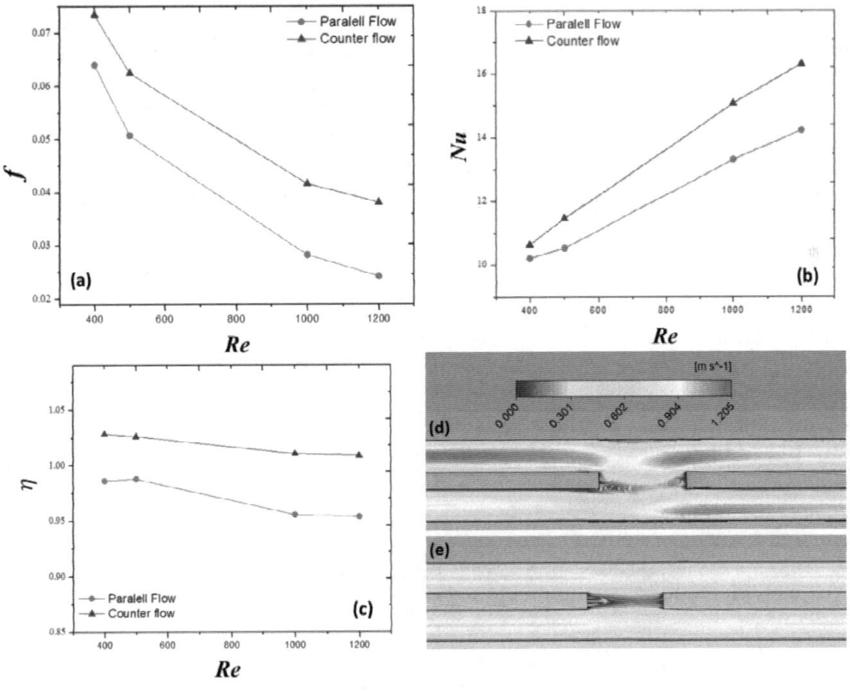

Fig. 4. Comparison of parallel and counter flow (a): f vs. Re; (b) Nu vs. Re; (c) η vs. Re. Velocity streamlines at $Re = 500$. (d) Counter flow; (e) Parallel flow

Conclusions

The flow and heat transfer characteristics of inter-linked paralell micro-channels are investigated in this research. These channels are studied for dimensioanl variations of the joint connecting the channels. For further details, performance comparison of parallel and counter flow types are also made in this study. Following conclusive points are noted:

- The parameteric (dimensional) variation of joining link has a little influence on the over-all thermal hydraulic performance of the inter-linked channels.
- Nu slightly improves with the increasing height of joint (J). However, the variation of J has negligible effect on f.
- Inter-linked micro-channels with counter flow give improved convective heat transfer, though together with higher pressure drops. The Nu is found almost 15% higher than the parallel flow at Re = 1200.
- The design can be further explored on the basis of dimensional parameters and could be a positive addition in the designing of micro-channel heat sinks.

References

[1] Z. Khattak, H.M. Ali, Air cooled heat sink geometries subjected to forced flow: A critical review, International Journal of Heat and Mass Transfer 130 (2019) 141-161.
[2] S.A. Jajja, W. Ali, H.M. Ali, Water cooled minichannel heat sinks for microprocessor cooling: Effect of fin spacing, Applied Thermal Engineering 64 (2014) 76-82.
[3] Y. Li, G. Xia, D. Ma, Y. Jia, H. Wang, Characteristics of laminar flow and heat transfer in microchannel heat sink with triangular cavities and rectangular ribs, International Journal of Heat and Mass Transfer 98 (2016) 17-28.
[4] G. Xie, Y. Li, F. Zhang, B. Sundén, Analysis of micro-channel heat sinks with rectangular-shaped flow obstructions, Numerical Heat Transfer, Part A: Applications 69 (2016) 335-351.
[5] S. Sasaki, T. Kishimoto, Optimal structure for micro grooved cooling fin for high-power LSI devices, Electronics Letters 22 (1986) 1332-1334.
[6] A.J. Ghajar, C.C. Tang W.L. Cook, Experimental investigation of friction factor in the transition region for water flow in minitubes and microtubes, Heat Transfer Engineering 31 (2010) 646-657.
[7] T.Y. Lin, S.G. Kandlikar, An experimental investigation of structured roughness effect on heat transfer during single-phase liquid flow at microscale, Journal of Heat Transfer 134 (2012) 101701.
[8] A. Moradikazerouni, M. Afrand, J. Alsarraf, O. Mahian, S. Wongwises, M.-D. Tran, Comparison of the effect of five different entrance channel shapes of a micro-channel heat sink in forced convection with application to cooling a supercomputer circuit board, Applied Thermal Engineering 150 (2019) 1078-1089.
[9] Y. Fan, P.S. Lee, L.W. Jin, B.W. Chua, Experimental investigation on heat transfer and pressure drop of a novel cylindrical oblique fin heat sink, International Journal of Thermal Sciences 76 (2014) 1-10.
[10] M.Q. Ansari, G. Zhou, Influence of structured surface roughness peaks on flow and heat transfer performances of micro- and mini-channels, International Communications in Heat and Mass Transfer 110 (2020) 104428.
[11] M.Q. Ansari, G. Zhou, Flow and heat transfer analysis of microchannels structured with rectangular surface roughness, Chemical Engineering and Processing-Process Intensification 156 (2020) 108066.
[12] Y. Li, G. Xia, Y. Jia, D. Ma, B. Cai, J. Wang, Effect of geometric configuration on the laminar flow and heat transfer in microchannel heat sinks with cavities and fins, Numerical Heat Transfer, Part A: Applications 71 (2017) 528-546.
[13] A. Fluent, Theory Guide and User's Guide, Ansys Inc, USA (2015).

Friction stir welding and processing-a perspective review

Sanjeev Verma[1, a)] **and Vinod Kumar**[2, b)]

[1]*Reseach Scholar, Mechanical Engg. Department, Punjabi University, Patiala-147002, India*
[2]*Professor, Mechanical Engg. Department, Punjabi University, Patiala-147002, India*
[a)] Corresponding author(s): er.verma.11@gmail.com

[b)] vk_verma5@rediffmail.com

Abstract. The paper reviews the basic principles of friction stir welding, its methodology, and all other technical aspects. As reported by different researchers, the effect of varying welding variables such as tool geometry, tool-tilt angle and tool materials on the microstructure and mechanical properties of the welded materials, and weld quality are reported and discussed. The paper covers the various mechanical and metallurgical properties of similar/dissimilar aluminium alloys and other metal joints welded using friction stir welding (FSW). The attempt has been made to present a comprehensive review of some of the salient works carried out by various researchers pertaining to friction stir processing (FSP) /FSW of aluminium alloys. The paper also identifies the existing research gaps in the area of FSW, which should be explored in future.

Keywords: Friction stir welding, welding parameters, Al alloys, mechanical properties, microstructure, dissimilar alloys.

1. INTRODUCTION

Friction stir welding (FSW), a solid state welding process, was invented in 1999 by the welding institute [1] and is used to join various non ferrous materials like aluminium, copper, magnesium etc. In this welding technique, a non consumable tool pin profile rotates between the parent materials to generate friction heat, which stir the material and the stirred material joined. In FSW (figure 1) a rotating tool is passed through the joint line to produce frictional heat. The rotating tool stirs and mixes the materials around the tool pin and with the translation of the tool the material is also stirred from the front to back of the rotating tool pin to complete the welding process. [2-3]

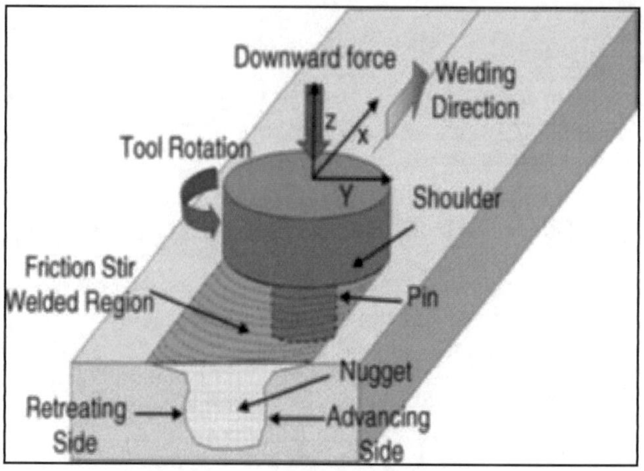

Figure 1. Schematic of FSW. [2,3]

In comparison to conventional welding processes, FSW is an eco-friendly and cleaner welding technique, which doses not produce fumes, smoke or arc glare. The main advantages of FSW are less distortion and no melt-reacted defects. The macrographs shown in figures 2(a) and 2(b) shows the four different welded regions marked as base material zone (BMZ), heat affected zone (HAZ), thermo-mechanically affected zone (TMAZ) and stir zone (SZ)/weld nuggets on both advancing and retreating sides.[4]

Figure 2(a). Macrographic view of FS welded regions for AA6061 alloy on both sides. [4]

HAZ (Zone B) lies closer to weld center and does not exhibit, plastic deformation. The material is plastically deformed in TMAZ (zone C). The highest strain is in the SZ/weld nuggets (zone D) zone, with the evidence of recrystallisation in this region [4].

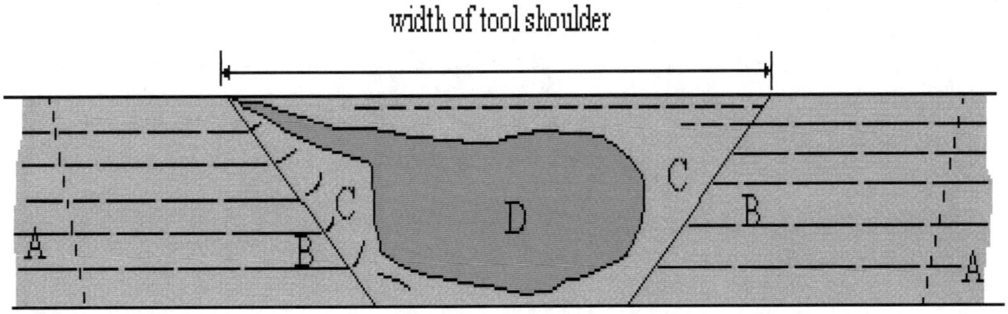

Figure 2(b). Schematic illustration of different zones in friction stir welded AA 6061 alloys [4] (**A:** BMZ, **B:** HAZ, **C**: TMAZ and **D**: SZ).

Mishra and Ma [5] in their review article discussed the developments in the area of FSW and FSP. The review article mainly described the impact of different process parameters of FSW/FSP on the mechanical and metallurgical properties of aluminium alloys as well as the mechanisms involved in forming the weld and refining the microstructure. The salient benefits of FSW in terms of metallurgical, environmental and energy are summarized in table 1 below:

Table 1. Metallurgical, Environmental & Energy benefits of FSW. [5]

Metallurgical benefits of FSW	Environmental benefits of FSW	Energy benefits of FSW
Very low distortion in welded parts. Robust in-phase process. High dimensional stability with repeatability Nil alloying elements loss No cracking Excellent improvement in metallurgical properties Replace various attachments and fasteners.	No requirement of shielding gas Solvent and grinding wastes are eliminated Cleaner surface produced No use consumable material	Material used improve because of very less wastage. Less energy requirement as compared to other welding techniques. Reduced fuel consumption in lightweight aviation, automotive and marine applications.

2. FSW MATERIAL COMBINATION STUDIES

2.1. FSW of Similar Aluminium Alloys

Prasanna et al. [4] analyzed the role of heat treatment processes and using different profiles of tool pin during FSW of AA6061 (aluminium alloy). The study employed the use of various types of tool pin profiles for producing the butt joints while maintaining the tool rpm, welding speed and axial force as 1200, 14mm/min and 7kN, respectively. The effect of various heat treatment processes annealing, normalizing and quenching was also investigated on the metallurgical aspects and mechanical properties of the welded joints. The use of hexagonal shaped pin profile of the tool was noticed to produced the welded joint having good ductility and tensile strength. Rohilla and Kumar [6] investigated the impact of tool geometry on the mechanical properties of FSW butt joint of AA6061 plate having 6mm thickness. The tensile strength, micohardness and percentage elongation during single and double pass were determined by using half square (HS), full square (FS), half cylindrical (HC) and full cylindrical (FC) tool pin profiles. The joint obtained by using single pass weld exhibited lower percentage elongation and tensile strength as compared to joints produced by employing double pass welding. Raf et al. [7] examined the role of various FSW parameters on the tensile behavior rods shaped AA7075-T6. The study indicates that the most critical parameters influencing the joint strength of FS welded AA70755 are spindle speed, friction pressure and burn-off length. The findings also revealed that FSW can be employed to attain a joint efficiency of 89% for AA7075-T6 through proper selection of welding parameters. Bisadi et al. [8] studied the role of various parameters on the microstructural aspects, mechanical properties and weld defects in lap joint composed of AA5083 and produced through FSW. Amongst the various welding zones, the nugget zone was found to have the better grain size and thus exhibiting higher hardness. The best joint properties were noticed corresponding to tool rpm of 825 and a weld speed of 32 mm/min. Cavaliere et al. [9] analyzed the influence of welding parameters on the microstructural aspects mechanical properties FS welded joints made of AA6056. The experiments were performed at three different tool rpm (*i.e.* 500, 800 and 1000) and welding speed (*i.e.* 40, 56 and 80 mm/min.) and their effect was investigated on the micro-hardness, tensile and fatigue (low and high cycle) behavior of the welded joint. During the fatigue tests, the ratio R (= $\sigma_{min}/\sigma_{max}$) was kept 0.1. The specimen corresponding to welding speed of 56 mm/min. was noticed to have the highest fatigue properties in the low cycle fatigue regime.

2.2. FSW of Dissimilar Aluminium Alloys

Park and Kim [10] examined the influence of tool rotation speed and tool traverse speed on the stirring phenomenon and friction heat generated during FSW of AA5052-O and AA6061-T6. The mechanical strength of FS welded joints, obtained for various combinations of welding parameters, was evaluated. The maximum weld strength was noticed for the welded joint corresponding to tool rpm of 1600 and transverse speed of travel speed of 61 mm/min. The examination of plastic flow behavior of the material in the welded joint indicated that with the increase in stirring effect and decrease in traverse speed the defects in the welded joint decrease. Park et al. [11] investigated the role of material locations on the properties of FS welded joints for AA5052-H32 and AA6061-T6 alloys. The study revealed that the mixing of the materials in the welded joints depends on the location of the base materials. The materials were observed to have proper mixing when AA6061-T6 was on the retreating side and AA5052-H32 was on the advancing side, as compared to the case where AA5052-H32 was on the retreating side and AA6061-T6 was on the advancing side during the FSW process. For A5052-H32 alloy the lowest micro-hardness was noticed in the HAZ, and as a consequence the fracture during the tensile test was observed on the AA5052-H32 side for both types of material arrangements. Sundaram and Murugan [13] investigated the tensile behavior of FS welded joints made of two dissimilar aluminium alloys (*i.e.* AA2024-T6 and AA5083-H321). The study used five different types of pin profiles of the tool. A mathematical model using central composite design and response surface methodology (RSM) was used to predict the optimal solution for the experiment. The maximum tensile strength and maximum elongation were obtained for the welded joint corresponding to tool pin having tapered hexagonal profile. However, the lowest tensile strength and the lowest elongation were noticed for the welded joint produced using tool-pin with straight cylindrical profile. The study also reveals that with the increase in tool rpm the tensile strength of the welded joint increases initially, reaches a maximum value, followed by decrease beyond a certain tool speed. Madhusudhan et al. [14] investigated the microstructure and mechanical properties of FS welded joints between plates made of AA 6262-T6 and AA 7075-T6. The study employed tool having square pin profile and different values of welding parameters such as tool rotation speed, welding speed and axial force. The welded joints fabricated corresponding to tool speed of 1200 rpm, welding speed of 0.6 mm/sec. and axial force of 9 kN exhibited the best mechanical properties. Ghaffarpour et al. [15] investigated the role of varying the welding parameters on the microstructure and mechanical properties of FS welded 6061 and 5083 aluminium alloys. To predict the optimal mechanical properties, RSM was used. It has been observed from the optimal solution that the amongst the various welding parameters responsible for deciding the tensile strength of the welded joint, the tool rotation speed is the most dominant. Palanivel et al. [16]

investigated the influence of tool-pin profile and tool rpm on the tensile strength and microstructural properties of FS welded joints between dissimilar AA5083 and AA6351 alloy. The study employed different types (five types) of tool pin profiles and three different tool rotation speeds for obtaining the welded joints. The study demonstrated three different regions in weld zone. The welded joint obtained using tool with straight square pin profile and rotating at 9500 rpm possess the maximum tensile strength of 273 MPa.

2.3. FSW of Aluminium Alloys with other Materials

Dressler et al. [17] made an attempt to join TiAl6V4 and 2024-T3 by using FSW technique. The study focused on the feasibility of the joint apart from investigating the microstructure, tensile strength and hardness of the welded joints. The ultimate joint strength was found to be 73% of the basic material strength AA2024-T3. Pandya and Menghani [18] examined the impact of welding parameters on the ultimate tensile strength of FS welded joints between AA6061-T6 and CU dissimilar by using Sn powder as filler material. The study used different combinations of tool rpm, tool speed, tool-pin offset and gap (for adding Sn filler powder). The study concluded that the tool rotation speed and tool travel speed are the major parameters that affect the strength of the welded joint. The joints prepared were also investigated for microstructural changes, micro-hardness measurement and X-ray diffraction analysis. The study revealed that the optimum tensile properties are attained for the welded joint produced with 0.05 mm gap for Sn filler material.

3. PROCESS PARAMETERS STUDIES

The most significant process parameters in FSW of similar or dissimilar materials are tool geometry and rotation speed, welding speed, axial force and tilt angle of the tool. Mishra and Ma [5] studied various aspects pertaining to FSW and noticed that tool geometry is the most significant parameter in affecting the material flow during FSW process. It has been found that the tool geometry influences the speed of the FS welding process. Amongst various tool-pin profiles, the threaded cylindrical pin profile and concave shoulder pin profile are widely used. The use of higher tool rotational speed has been shown to result in high frictional heat due to higher temperature generated during the course of FSW process. Koilraj et al. [19] analyzed the influence FSW parameters like tool profile, tool rpm, tool transverse speed, and the ratio of tool shoulder to tool-pin diameter on the tensile strength of the joint. The study demonstrated that wide range of FSW parameters can be used to produce highly efficient and defect free welded joints. The study also suggested the recommended range of process parameters for achieving the best tensile properties of the welded joint. The use of Analysis of variance (ANOVA) suggested that the ratio of tool shoulder diameter to tool-pin diameter is the most significant factor in yielding the soundness of the welded joint. Besides, the tool-pin profile and welding speed are also play a significant role in deciding the properties of the welded joint.

4. MECHANICAL PROPERTIES AND MICROSTRUCTURAL CHARACTERIZATION

Fujii al. [20] studied the effect of varying the amount of carbon in IF (interstitial free), S12C (low carbon) and S35C (medium carbon) steels on the FSW. The strength of S12C steel was noticed to increase with the increases in welding speed. Figure 3 shows the ultimate tensile strength of IF, S12C and S35C carbon steels FSW joints for different welding speeds. It is noticed that with the increase in welding speed, the strength of IF steel joint decreases slightly, attributed to decrease in heat input. In case of FSW of S12C steel, the strength of the joint is observed to increase with the increase in welding speed, attributed to decrease in amount of heat input. The maximum strength of the joint for S35C steel has been noticed around a welding speed of 200 mm/min. Figures 4 (a) -(c) depict Vicker's hardness values of FS welded joints corresponding to IF, S12C and S35C steels obtained at different welding speed.

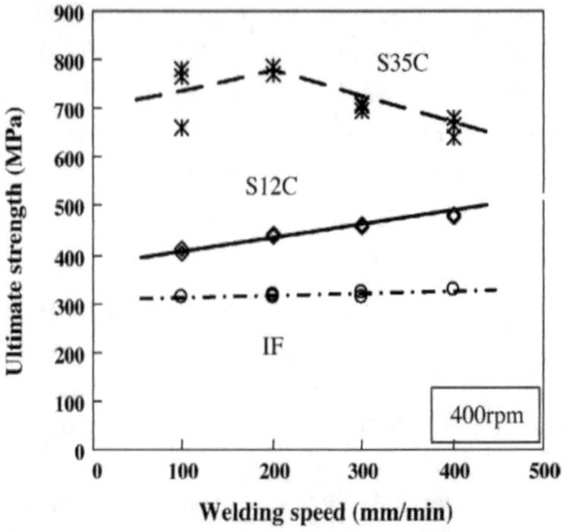

Figure 3. Effect of welding speed on ultimate tensile strength of FS welded carbon steels. [20]

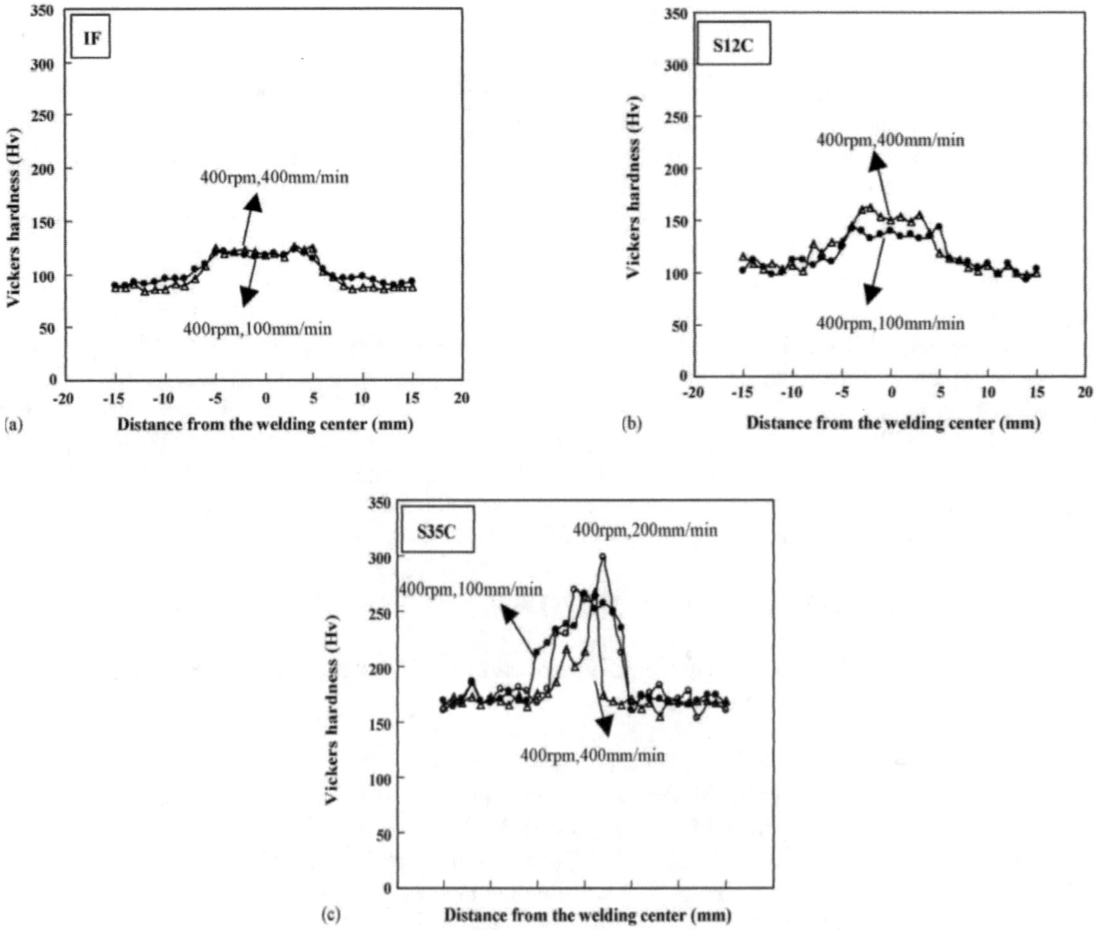

Figure 4. Microhardness profiles for FS welded carbon steels: (a) IF, (b) S12C and (c) S35C steels. [20]

The microstructure examinations of IF, S12C and S35C steels reveal uniform grain structure comprising of equiaxed grains with the sub-structures noticed in the SZ. It has been found that the grain sizes in the welded SZ center were 5 and 6μm for the highest and lowest levels of input conditions, respectively, which are quite small as compared to that noted in BMZ (24μm). It has also been noted that in IF steel single-phase ferrite region is observed no transformation has been observed in the region. S12C steel shows that under all welding conditions ferrite-pearlite micro-structure were formed with a small amount of pearlite. Figure 5 shows microstructure of S12C and S35C steels.

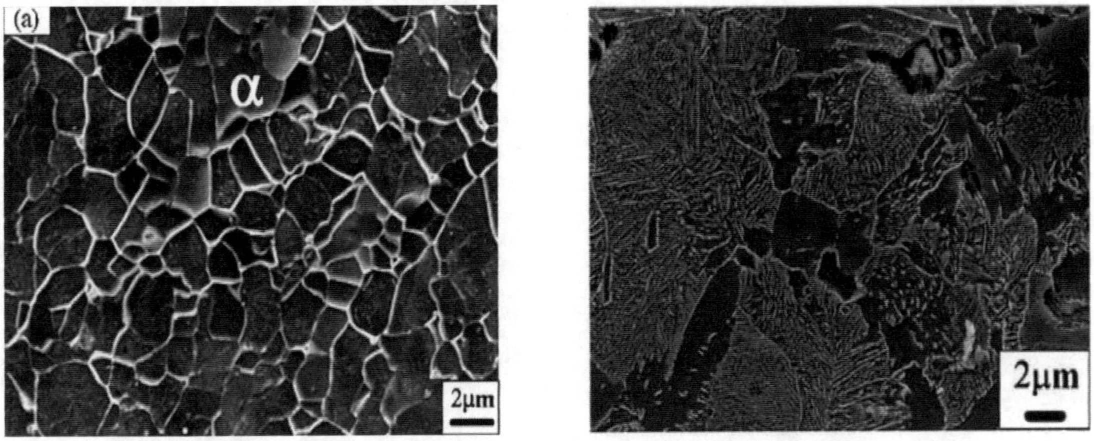

Figure 5. Microstructures of (a) S12C and (b) S35C steels. [20]

Charita et al. [21] examined the various grain growth phenomenon observed during FSW of different alloys. The study reveals that abnormal grain growth (AGG) is not observed in pure aluminium. AGG occurs when the normal grain growth gets suppressed. A few samples of optical macrographs showing AGG, upon treatment at 763K for 1 hour, are shown in figure 6. The study indicates that during FSW process, the primary factors contributing to the presence of AGG in the friction stirred materials are consistent with the inhomogeneous pattern of deformation.

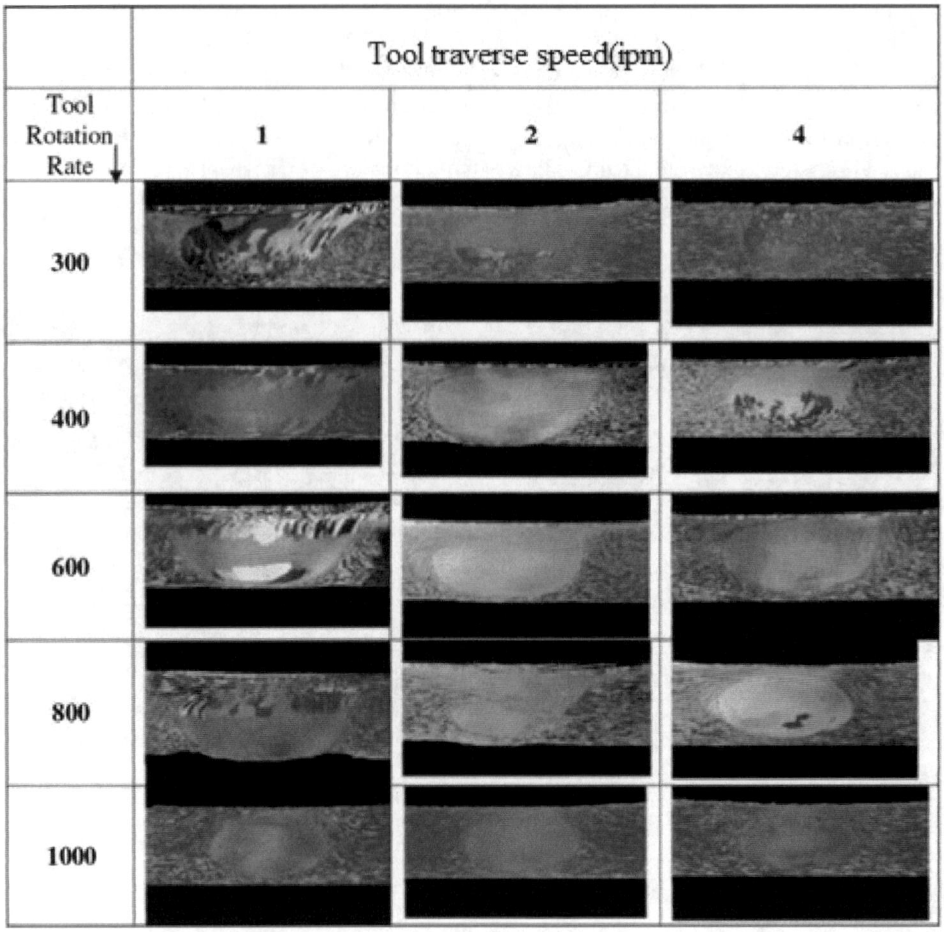

Figure 6. Optical macrographs of FSP 7075 Al alloy exhibiting AGG corresponding to different combinations of process parameters. [21]

5. CONCLUSIONS

Friction stir welding is an effective welding process that can be used to join different types of aluminium alloys and other materials. FSW is a solid-state joining technique that possesses numerous benefits as compared to other welding processes. The most important factor for the development of defect-free welds is tool geometry. The profile of the tool pin must be designed in such a way that the material stain rate in the stir zone increases and the clogging of material with the tool surface decreases. The origin of weld defect is clogging of base material with tool which results to reduce the joint strength. To accomplish sound and defect free weld joints, parameters such as welding speed, tool rpm, tool profiles, tilt angle of the tool, and axial force play an important role. There are three distinct zones in the friction stir welded joints, viz., nuggest zone, HAZ and TMAZ. The optimum heat generated during the course of FSW is dependent on the thermal conductivity of the tool material, initial temperature of the tool and base material, and the amount of heat contributed due to phase deformation in the different zones. Each of these parameters directly or indirectly influences the material flow and heat generated during the FSW process. FSW is observed to exhibit a significant increase in the tensile strength, elasticity, ductility, fatigue strength and toughness of the welded joints, as compared to joints obtained using fusion welding techniques. It has been observed that the welded joint exhibits lower fatigue strength as compared to base materials, but significantly higher fatigue strength than that noticed for laser welded and MIG welded joints. Preheating of the material to be welded is advantageous for improving the consistency of the weld and besides improving the weld quality. FSW can be successfully employed for joining different materials such as copper, steel, titanium, composites and magnesium, apart from joining aluminium alloys. It is revealed that more research work is needed to produce reliable and cost-effective joining of aluminium and other lightweight alloys, in addition to welding of harder materials like steels and composites.

6. REFERENCES

[1] R. Palanivel, P. Koshy Mathews, N. Murugan and I. Dinaharan, Effect of tool rotational speed and pin profile on microstructure and tensile strength of dissimilar friction stir welded AA5083-H111 and AA6351-T6 aluminum alloys, *Materials and Design*, 2012; **40**: pp. 7–16.

[2] Arvin Bagheri, Taher Azdast and Ali Doniavi, An experimental study on mechanical properties of friction stir welded ABS sheets, *Materials and Design*, 2013; **43**: pp. 402–409.

[3] K. Elangovan and V. Balasubramanian, Influences of pin profile and rotational speed of the tool on the formation of friction stir processing zone in AA2219 aluminium alloy, *Materials Science andEngineering A*, 2007; **459**: pp. 7–18.

[4] P.Prasanna, Dr.Ch.Penchalayya and Dr.D.Anandamohana Rao, Effect of tool pin profiles and heat treatment process in the friction stir welding of AA 6061 aluminium alloy, *American Journal of Engineering Research*, 2013; **07**: pp-07-15.

[5] Mishra RS and Ma ZY, Friction stir welding and processing, Mater Sci Eng R Rep, 2005; **50**:1–78.

[6] Rohlla Puneet and Kumar Narinder, Experimental investigation of tool geometry on mechanical properties of friction stir welding of AA6061, *Internationaljournal of innovative technology and exploring engineering(IJITEE)*, 2013; vol. **03.**

[7] H. Khalid Rafi, G.D. Janaki Ram, G. Phanikumar and K. Prasad Rao, Microstructure and tensile properties of friction welded aluminum alloy AA7075-T6, *Materials and Design*, 2010; vol.**31**,pp. 2375–2380.

[8] H. Bisadi , M. Tour and A. Tavakoli, The influence of process parameters on microstructure and mechanical properties of friction stir welded al 5083 alloy lap joint, *American Journal of Materials Science*, 2011; **1(2)**: pp. 93-97.

[9] Effect P. Cavaliere, G. Campanile, F. Panella and A. Squillace, Effect of welding parameters on mechanical and microstructural properties of AA6056 joints produced by friction stir welding, *Journal of Materials Processing Technology*, 2006; vol.**180**, pp. 263–270.

[10] J. C. Park and S. J. Kim, The effect of traveling and rotation speeds on mechanical properties during friction stir welding of dissimilar Al alloys, *Defect and Diffusion Forum*, 2010; vol. **297- 301**, pp. 590–595.

[11] S. K. Park, S. T. Hong, J. H. Park, K. Y. Park, Y. J. Kwon, and H. J. Son, Effect of material locations on properties of friction stir welding joints of dissimilar aluminium alloys, *Science and Technology of Welding and Joining*, 2010; vol. **15**, no. 4, pp. 331–336.

[12] R. Palanivel , P. Koshy Mathews, The tensile behaviour of friction-stir- welded dissimilar aluminium alloys, *Materials and technology*, **45** (2011) 6, pp. 623–626.

[13] S. N. Sundaram and N. Murugan, Tensile behavior of dissimilar friction stir welded joints of aluminium alloys, *Materials and Design*, 2010; pp. 4184–4193.

[14] R. Madhusudhan, M.M.M.Sarcar, N.Ramanaiah, K.PrasadaRao, An experimental study on the effect of weld parameters on mechanical and micro structural properties of dissimilar

aluminium alloy FS welds", *International Journal of Modern Engineering Research*, 2012; vol.**2**, Issue.4, pp. 1459-1463.

[15] M. Ghaffarpour, B. M. Dariani, A. H. Kokabi and N. A Razani, Friction stir welding parameters optimization of heterogeneous tailored welded blank sheets of aluminium alloys 6061 and 5083 using response surface methodology, *ProcIMechE Part B: J Engineering Manufacture* **226(12),**2012; pp. 2013–2022.

[16] R. Palanivel , P. Koshy Mathews, N. Murugan and I. Dinaharan, Effect of tool rotational speed and pin profile on microstructure and tensile strength of dissimilar friction stir welded AA5083-H111 and AA6351-T6 aluminum alloys", *Materials and Design* **40** (2012); pp.7–16.

[17] Ulrike Dressler, Gerhard Biallas and Ulises Alfaro Mercado, Friction stir welding of titanium alloy TiAl6V4 to aluminium alloy AA2024-T3, *Materials Science and Engineering A*, vol. **526** (2009); pp. 113–117.

[18] Shailesh N Pandya and J V Menghani, A parametric investigation on effects of friction stir welding process parameters on mechanical properties of AA6061-T6 to Cu dissimilar joints, *Indian Journal of Engineering and Materials Sciences*, vol. **25**, 2018; pp. 131-146.

[19] M. Koilraj, V. Sundareswaran, S. Vijayan and S.R. Koteswara Rao, Friction stir welding of dissimilar aluminum alloys AA2219 to AA5083-optimization of process parameters using taguchi technique", *Materials and Design* (2012); vol. **42**, pp. 1-7.

[20] H. Fujii , L. Cui , N. Tsuji , M. Maeda, K. Nakata and Kiyoshi Nogi, " Frictions stir welding of carbon steels", *Materials Science and Engineering A*, vol.**429** (2006); pp. 50–57.

[21] I. Charita, and R.S. Mishra, "Abnormal grain growth in friction stir processed alloys", *Scripta Materialia*, vol. **58** (2008); pp. 367–371.

Synthesis and characterzation of graphite based shape memory polymers

Ranganatha Swamy MK[1], U S Mallik[2], Udayakumar V[3]
[1]Assistant Professor, Department of Mechanical Engineering, Faculty of Engineering and Technology, JAIN DEEMED TO BE UNIVERSITY, Bengaluru-562112
[2]Professor, Department of Mechanical Engineering, Siddaganga Institute of Technology, Tumkur-572103
[3]Professor and Head, Department of Chemistry, MVJ College of Engineering, Bengaluru-560067

mk.ranganatha@jainuniversity.ac.in,usm_sit@yahoo.co.in,udayakr@gmail.com

Abstract. Shape memory polymers are the class of materials which regains its shape up on external stimulus like heat, magnetic field, light etc. these polymers have got wide range of applications in biomedical and automobile industries. There are many researches going on about the shape memory polymers and it is better than other smart materials in some characteristics. This paper covers about the strength of the polymer after adding graphite to the polymer. The addition of graphite would gradually increase the characteristics of the polymer. These polymers are flexible and have less strength. The primary idea of improving the characteristics was to increase the strength of the polymer. We have added graphite to peg based polymer which has increased the hardness significantly. Many tests were conducted on the graphite-based shape memory polymer like tensile test, hardness test and dynamic mechanical analysis, it was observed that the characteristics were enhanced by adding graphite to the polymer.

Keywords: Shape memory polymer, Cross-linking, Shape memory effect, thermomechanical,

1. Introduction

The materials which regains its original shape when an external stimulus is applied on it is known as shape memory polymers, the external stimulus might be heat, light, magnetic field etc. these smart materials are classified into shape memory alloys, shape memory ceramics and shape memory polymers [19]. SMPs have lower density, better strain recovery rate, and lightweight and most importantly it is cost effective. These polymers can replace other smart materials in many of the applications. It is mainly used in actuators, textile industries and in aerospace applications. Even though, shape memory alloys are better, there are some obvious reasons for going for SMPs like poor corrosion resistance, high mass density, low strain etc. The polymers which react to the temperature is called thermoresponsive polymers [19].

Literature studies shows that the addition of certain additives or fillers would enhance the properties of the polymer. Sometimes the addition of certain cross-linking agents would increase the stability of the polymer, these cross-linking agents might be functional groups such as alcohol, here in or experiment butane-diol (BDO) is used as the chain extender [18]. The addition of certain materials would increase the bond strength which in turn increases the hardness of the shape memory polymer [1]. The addition of hard segment would decrease the shape recovery speed of the polymer. The glass transition of the polymer will increase by adding more of hard segment [2]. The addition of the graphite in the polymer increases the cross-linking ability of the polymer which in turn increases the hardness of it. Graphite also increases the melting point and electrical conductivity of the polymer which makes the polymer suitable for electrical applications.

Researches are to be carried out on adding different additives to the shape memory polymer as it has got many advantages over other smart materials. Materials like graphene, CNT and other allotropic forms of carbon can also be added to the polymer by checking its solubility to enhance the properties of the polymer.

2. Materials and Methods

2.1. Materials

The chemicals of the soft and the hard segment is reported [3]. PEG-6000 is used as the soft segment. It is required to add 58% of soft segment and 42% of hard segment in order to get optimal result. Graphite was added to this concentration and the polymer was obtained. The hard segment used are 4,4′-diphenylmethane diisocyanate (MDI), isophorone diisocyanate (IPDI), PEG-200 and 1, 4-butanediol (BDO). The concentration of PEG-200 is 10 wt% and the concentration of BDO used is 4.2 wt%.

2.2. Methodology

The set-up consisted of 2- neck round bottom flask of 500 mL, hot water bath along with a magnetic stirrer, temperature controller and a thermometer. The flask was continuously purged with dry nitrogen to prevent the reaction from contaminants. A measured amount of PEG-6000 was added to round bottom flask along with 25mL of DMF up on continuous stirring and purging of nitrogen gas till it reaches 90 degrees. Required amount of graphite was added to the solution and is dissolved completely, then we add suitable molar concentration of IPDI and 2 drops of DBTDI, the reaction is kept for 2 hours. After 2 hours, suitable amount of MDI and PEG-200 with 10mL of DMF solvent was added to the solution and the reaction was continued for 1 hour. Then, the temperature was reduced to 60 degrees and 10 drops of BDO with 5mL of DMF was added and the reaction is continued for another hour. The solution was poured into the glass mould and cured in 3 stages inside a hot air oven. Firstly, the solution is cured at 60 degrees for 12 hours. Then, it is cured at 80 degrees for 24 hours. Lastly, the solution was cured at 100 degrees for 8 hours. Finally, the polymer is taken out from the oven and removed from the glass mould.

3. RESULTS AND DISCUSSION

Tensile test, hardness test and dynamic mechanical analysis was carried out on graphite-based shape memory polymers, and it was found that the addition of graphite increased the hardness of the polymer. With respect to above test the shape memory test was also performed which shows 100% recovery with a span of 20seconds at 30°C.It can be found that the addition of graphite to PEG matrix nowhere affected the shape memory properties.

3.1. Tensile test

Tensile test is carried out to find out the tensile characteristics of the specimen like peak load, breakage points and other parameters to know the strength of the specimen. Graphite based specimen was cut into dumbbell shape and it was fed into the UTM. Uniform load was applied on the polymer and the stress-strain relationship was found out. At one point, the specimen fails to take further load, the brittle materials will have less toughness whereas flexible materials will have more toughness.

Table1: Tensile test results for varying graphite composition of different samples:

Graphite (g) Wt%	Ultimate tensile strength (MPa)	Young's modulus (MPa)	Percent elongation to break
0	36.5 (62)	1957	2.6 (3.5)
0.05	13.4	600	40
0.1	17.3	170	226
0.15	41.8	55	628

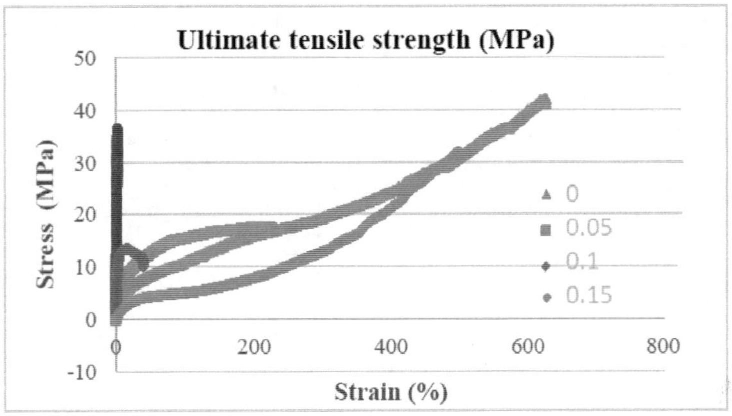

Figure 1. Tensile test graph for various wt% of graphite

It was found that the addition of graphite reduces the tensile strength of the specimen. In other words, the specimen becomes more flexible by reducing the graphite content but in contrast, it is very important to note that the flexibility of the polymer after adding graphite was not reduced to a greater extent. The reason behind reduction of strength is the SMP becomes brittle with the addition of more graphite content, hence and optimum wt% of graphite is added to PEG matrix to obtain good ultimate strength without compromising the shape memory effect. Table 1 shows the tensile test values with respect to graphite composition, the strength is 41.8Mpa for the sample having 0.015wt% of graphite and the elongation is found to be 628, this shows the addition of graphite in higher percentage has not affected the elongation.

3.2. Hardness test

The hardness of the specimen is defined as the resistance to permanent indentation. The hardness of the natural rubber and rubber could be found out through Rockwell hardness test and Shore hardness test. Shore hardness test was carried out on the specimen in which the spring-loaded needle was placed against the sample and different pressures were applied on the sample at different intervals of time.

IOP Conf. Series: Materials Science and Engineering 1091 (2021) 012073 doi:10.1088/1757-899X/1091/1/012073

The indentation caused on the sample gives the respective pressure applied on the specimen. Hardness of the specimen is defined through the hardness number.

Table 2: Hardness test results for different samples:

Graphite wt%	Durometer	Indenting foot	Time (s)	Load (Kgf)	Hardness
0	Shore D	1.1 mm – 1.4 mm diameter	20	60	56.2
0.1	Shore D	1.1 mm – 1.4 mm diameter	20	60	59.4
0.05	Shore D	1.1 mm – 1.4 mm diameter	20	60	65.3
0.15	Shore D	1.1 mm – 1.4 mm diameter	20	60	71.5

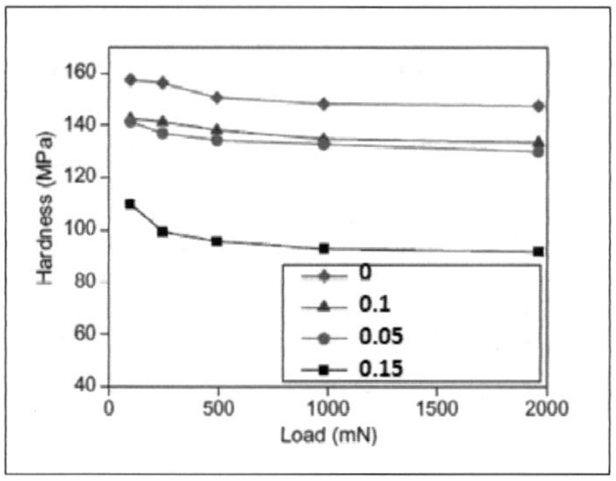

Figure 2. Hardness values with respect to load.

By increasing the graphite content of the polymer, we can increase the hardness of the specimen but, there will even be experimental anomalies which are to be taken into consideration for example, 0.05g of graphite specimen has got more strength than 0.1g of graphite specimen. As shown in figure 2 it is very important to follow every step of the methodology properly to get the desired properties of the polymer. As a whole, we can conclude that the addition of graphite will significantly increase the hardness of the polymer by compromising very less flexibility.

3.3. Dynamic mechanical analysis

DMA is an analysis in which a small stress is applied on a cyclic manner which allows the material to respond to frequency, temperature, stress and other factors. The basic objective of performing DMA on a specimen is to find stiffness and damping which is expressed in terms of tan delta and modulus. The storage modulus is represented as E' and the loss modulus is represented as E". The ratio of loss modulus to storage modulus is tan delta which is nothing but damping.

The value of glass transition could be found out by DMA, glass transition temperature is a temperature at which the polymer becomes pliable. It is very important to know the glass transition temperature of the specimens in order to perform shape memory test. This temperature can be found out by plotting a graph between log scale of E' and temperature. The graph plotted in figure 3 shows a peak, the corresponding value of the temperature gives the glass transition temperature of the specimen.

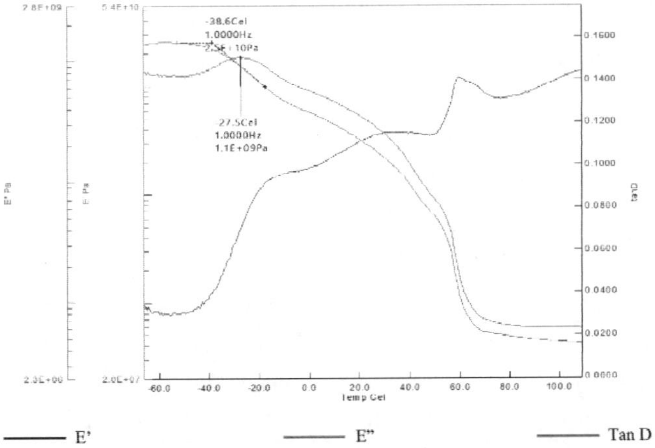

Figure 3. DMA graph for 0.1g of graphite

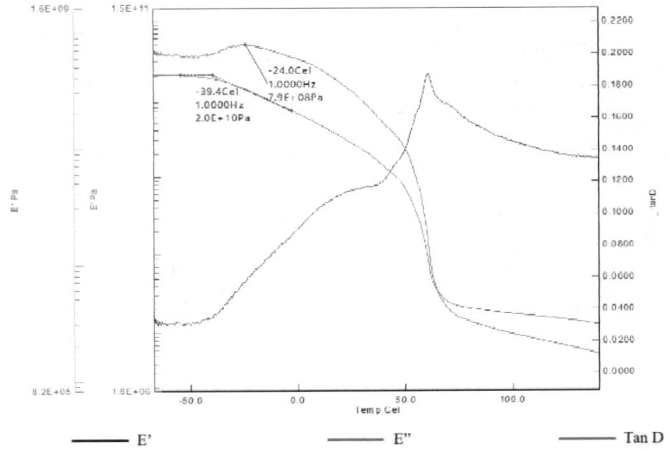

Figure 4. DMA graph for 0.05g of graphite

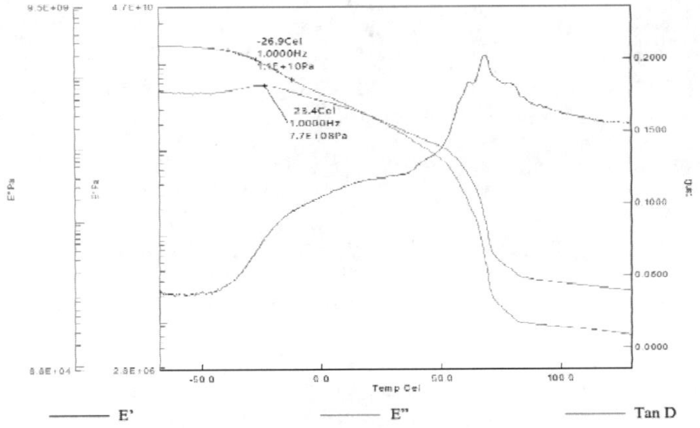

Figure 5. DMA graph for 0.15g of graphite

The graph plotted in figure 4 and figure 5 drawn taking log scale of E' against temperature to find the glass transition temperature of the specimens and it was found that the glass transition temperature gradually increased up on the addition of graphite.

Table 3: DMA analysis results for different composition.

Freq. Hz	Temp. Cel	E'(G') Pa	E"(G") Pa	dL um	tanD	Ft mN	Time min
5	36.94734	8.22E+08	1.6E+08	126.3881	0.194992	1248.163	0.15
2	36.94016	7.35E+08	1.48E+08	127.0825	0.201341	1248.163	0.283333
1	36.93109	6.73E+08	1.39E+08	129.8603	0.206434	1248.163	0.466667
0.5	36.92641	6.16E+08	1.31E+08	131.9436	0.211851	1248.163	0.683333

Table 3 gives the information of storage modulus, loss modulus, and tan delta. In figure 5, the storage shear modulus G' and the loss shear modulus G" are drawn against the temperature from - 5°C to 150°C. Until accomplishing the region of the Tg there is a drop away modulus G' dynamically with increasing temperature. Changes in G' with temperature are incredibly divergent for low and high frequencies starting from that area. For frequencies higher than 2Hz, there is an event of a fast decrease in estimations of the capacity modulus.

3.4. Shape Memory Effect

The SMP sample, in the beginning, rolled/bend to 100% on 60°C (>Tg), and by cooling it to ambient temperature the distorted shape was fixed (~20°C). After adjusting the hot plate to 60°C the sample was placed on top and the changes in shape were recorded after some time.

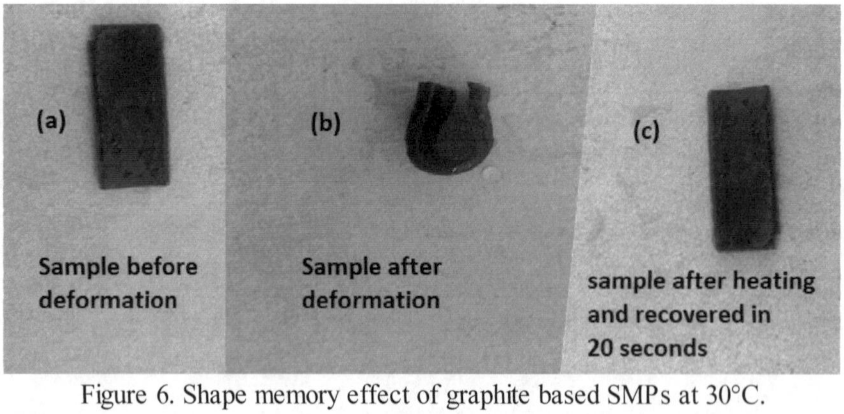

Figure 6. Shape memory effect of graphite based SMPs at 30°C.

Figure 6 shows the SME of blank SMP's and 4.2 demonstrates the SME of SMP's filled with graphite as a filler along with transition temperature and time, in figure 6 (a) the flat rectangular sample appears before deformation, figure 6 (b) demonstrates the sample after deformation (bent to u shape) which has been deformed in cold state, figure 6 (c) demonstrates the sample which has been recovered to flat rectangular i,e. earlier permanent shape upon heating to 30°C for a time of 20second.

4. CONCLUSION

The addition of graphite significantly increases the hardness of the polymer by compromising minimum flexibility. Graphite is much cheaper than many other allotropic forms of carbon, it could be clearly seen that the research on this work is still in progress. All the three test results give the following inference:

1. Graphite based SMP's have better hardness when compared to SMP's but not so flexible after seeing the characterization result. It could be used where hardness is of greater significance.
2. The tensile strength of the graphite-based polymer is not much lesser than SMP's.
3. The glass transition temperature of the polymer increases with the addition of graphite.
4. DMA results shows the graphite based SMP's are good for damping related applications which shows good tan δ values.
5. Shape memory test reveals that graphite based SMP's shows 100% shape memory effect.

References

[1] Zheng, Y. W.; Li, J.; Lee, E.; Yang, S. Light-Induced Shape Recovery of Deformed Shape Memory Polymer Micropillar Arrays with Gold Nanorods. RSC Adv. 2015, 5, 30495−30499.
[2] Christenson EM, Anderson JM, Hiltner A, Baer E. Polymer 2005;46:11744–54.
[3] Raquez J-M, Vanderstappen S, Meyer F, Verge P, Alexandre M, Thomassin J-M, et al. Design of cross-linked semicrystalline poly(epsiloncaprolactone)-based networks with one-way and twoway shape-memory properties through Diels–Alder reactions. Chem Eur J 2011;17(36):10135–43.
[4] Zhao, J.; Shan, J.; Van Assche, G.; Tenhu, H.; Van Mele, B. Demixing and Remixing Kinetics in Aqueous Dispersions of Poly(Nisopropylacrylamide) (PNIPAM) Brushes Bound to Gold Nanoparticles Studied by Means of Modulated Temperature Differential Scanning Calorimetry. Macromolecules 2009, 42, 5317−5327.
[5] Zhao, J.; Hoogenboom, R.; Van Assche, G.; Van Mele, B. Demixing and Remixing Kinetics of Poly(2-isopropyl-2-oxazoline) (PIPOZ) Aqueous Solutions Studied by Modulated Temperature Differential Scanning Calorimetry. Macromolecules 2010, 43, 6853−6860.
[6] Manzoor Ahmad, Bin Xu, Hendra Purnawali, Yongqing Fu, Weimin Huang, Mohsen Miraftab and Jikui Luo: High Performance Shape Memory Polyurethane Synthesized with High Molecular Weight Polyol as the Soft Segment: applied sciences ISSN 2076-3417.
[7] Yayun Liu, Jun Zhao, Lingyu Zhao, Weiwei Li, Hui Zhang Xiang Yu,and Zhong Zhang; High Performance Shape Memory Epoxy/Carbon Nanotube Nanocomposites: ACS applies materials and Interface. 2010, 464, 267−270.
[8] Tobushi, H.; Hashimoto, T.; Ito, N.; Hayashi, S.; Yamada, E. Shape fixity and shape recovery in a film of shape memory polymer of polyurethane series. J. Intell. Mater. Syst. Struct. 1998, 9, 127−136.
[9] Chun BC, Cho TK, Chung YC. Eur Polym J 2006; 42:3367–73.
[10] Lee HS, Wang YK, Hsu SL. Macromolecules 1987;20:2089–95.
[11] Ning L, Ning WD, Kang YS. Macromolecules 1997; 30:4405–9.
[12] Ping P, Wang WS, Chen XS, Jing XB. J Polym Sci Part B: Polym Phys 2007; 45:557-70.
[13] Lendlein, A.; Langer, R. Biodegradable, Elastic shape-memory polymers for potential biomedical applications. Science 2002, 296, 1673−1677.
[14] Tobushi, H.; Haray, H.; Yamaday, E.; Hayashiz, S. Thermomechanical properties in a thin film of shape memory polymer of polyurethane series. Smart Mater. Str. 1996, 5, 483−491.
[15] Yan B, Gu S, Zhang Y. Polylactide-based thermoplastic shape memory polymer nanocomposites. Eur Polym J 2012;49(2):366–78.

[16] Sun L, Huang WM. Thermo/moisture responsive shape-memory polymer for possible surgery/operation inside living cells in future. Mater Des 2010;31(5):2684–9.
[17] Zhu Y, Hu JL, Choi KF, Meng QH, Chen SJ, Yeung KW. Shape memory effect and reversible phase crystallization process in SMPU ionomer. Polym Adv Technol 2008;19(4):328–33.
[18] Tiffany Williams1, Michael Meador1, Sandi Miller1, Daniel Scheiman2 1NASA Glenn Research Center, Polymers Branch, Structures and Materials Division, 21000 Brookpark Rd. Cleveland, OH, 44135. 2ASRC Aerospace, 21000 Brookpark Rd., Cleveland, OH 44135.
[19] Manzoor Ahmad, Bin Xu, Hendra Purnawali, Yongqing Fu, Weimin Huang, Mohsen Miraftab, Jikui Luo. High Performance Shape Memory Polyurethane Synthesized with High Molecular Weight Polyol as the Soft Segment. ISSN 2076-3417. 31 May 2012

Overview on second & third order resonant switch mode power converter for variable input voltage applications

Reshma P Eldho[1], Ragasudha C P[2], Anchal Chhabra[3]
[1]Assistant Professor, Department of Electrical and Electronics Engineering, CMR Institute of Technology, Bengaluru, India, reshma.e@cmrit.ac.in
[2]Assistant Professor, Department of Electrical and Electronics Engineering, Presidency University, Bengaluru, India, ragasudha@presidencyuniversity.in
[3]Assistant Professor, Department of Electrical and Electronics Engineering, Sir M. Visvesvaraya Institute of Technology, Bengaluru, India, anchal_eee@sirmvit.edu

Abstract— DC-DC converters have wide applications in the field of battery charger, distributed generation, power supplies and so on. There are various isolated and non-isolated topologies available which has to be chosen based on the size, cost , weight, application, power range and duty of the converter. Due to increasing applicability of DC-DC converters, it is essential to enhance the efficiency, controllability and output regulation to a step ahead. Resonant converters at high frequency has gained attention in the market to a large scale due to its low size and weight. With the developing enthusiasm for this subject, it is essential to have a detailed review to sum up and update a unique reference about the soft switching configurations for better performance of the converter. The paper provides a brief overview in two and three element soft switching configurations. Moreover, The current technologies evolved in the field of resonant converters are also briefed for wide input voltage applications.

Keywords— DC-DC Converter, Soft switching, ZVS, ZCS, isolated/non isolated converters.

1. Introduction

DC-DC converters are widely employed in the fields related to automation, battery chargers, distributed generation, power supplies, LED Drivers, laptop adaptors, Electric vehicles and so on. The DC-DC converter behaves as an interface between the input supply and the output-load while changing the levels of voltages between them. It can be done by controlling the switching elements with the most suitable modulation schemes and designing the reactive components so as to obtain the desired output. For variable inputs like renewable energy sources, more specifically photovoltaic systems, the output greatly depends on the climatic/operating conditions which are often unpredictable and highly variable. Output of such systems needs to be regulated for a broad input range of voltage. In order to deal with such conditions, the converters ought to be equipped with appropriate control strategies so as to obtain regulated output while keeping up high efficiencies. As a result, DC-DC converters and related switching elements needs to be chosen appropriately as its choice is very crucial in the gross quality determination of the system as a whole [1]. There are numerous DC-DC converters available currently. The isolated converter topologies includes flyback, Dual active full/half bridge, forward, push pull converters and so on whereas buck, boost, buck-boost, cuk and sepic-zeta converters [2]-[3] are the non-isolated topologies [4]-[6]. An efficient DC-DC converter requires high power density, limited losses, less cost and reduced size.

In earlier days, the converter output often being controlled using discrete components like resistive load and diodes. Unlike linear regulators which generates high heat dissipation, the switched mode regulators utilize semiconductor switching devices like transistors, MOSFETS etc to convert input DC supply to pulsating output. Moreover, switched mode regulators consists of inductive and capacitive elements handling rectifiers and regulators for regulating the output to a predefined value. However, the switching losses during the switch-on and switch-off tend to interrupt the performance of the converter. As the operating frequency increases, the number of switch turn on and off transitions per unit time is more.[7]-[8] Consequently, It increases the switching losses at the transition period due to the non-zero values of voltage and current. This happens as the switching device will not turn on immediately when a voltage is applied to the gate. It takes a small delay to reach the steady state. Other limitations of hard switching includes limited frequency, high EMI, bulky structure and weight.

Despite these disadvantages, hard switched converters are widely used, studied and implemented in a wide range of applications [9]. In soft switching, the limitations can be reduced to some extend by forcing the switch to turn on at zero voltage condition, ZVS and zero current condition (ZCS) during the switch turn off. Moreover, Soft switching can help reduce the size of the circuit by allowing lossless switching transitions at high frequencies [10]. Figure 1 shows the DC-DC converter classification based on the switching operation.

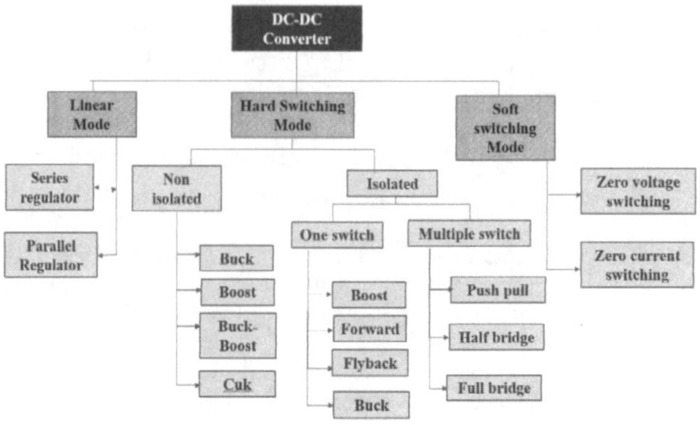

Figure 1: DC-DC Converter topologies

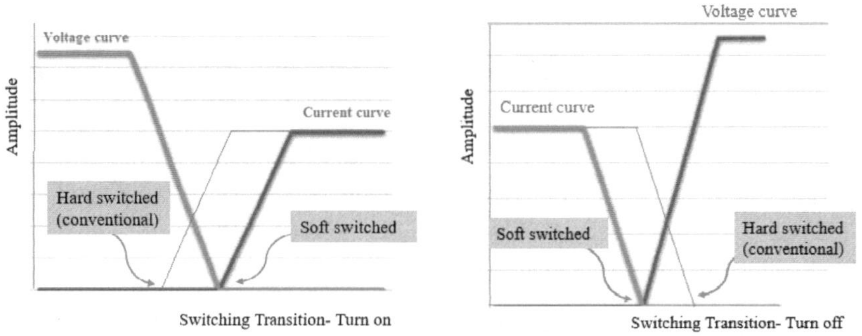

Figure 2. Hard/soft switching transition for (a) turn-on and (b) turn off modes

In this paper, different soft switching DC-DC converter configurations are mentioned and examined based on the recent development in respect of efficiency, steady state oscillation, number of components and voltage gain. The merits and demerits of the recent improvement within this scope are also pointed. This paper is arranged as the following. Section II showcases the classification of soft switching topologies. In Section III and IV, the second order and third order resonant topologies are analysed and the advantages and commonly used topologies are presented. Eventually, section V draws the conclusion.

2. Soft Switching Converter Topologies

Figure 2 shows the switching transition for the conventional hard switching and soft switching during turn on and turn off periods. During turn on, hard switching makes high switching losses as the switch allows the current to flow through it before completely opening the switching element. Soft switched DC-DC Converters have wide applications as they can improve the performance of converter circuits by reducing the switching losses and electro-magnetic interference. The switching losses can be minimized by achieving the Zero Current/Zero voltage switching (ZVS/ZCS). At Resonance, the resonant components inductor and capacitor elements charges and discharges simultaneously. The converters using this as the principle are called as resonant converters [11]. A series-parallel combined resonant converter uses the reactive elements providing the hybridized features of series and parallel resonance [12].Series Parallel Hybrid Resonant Converters are possible by arranging the Resonant Capacitor and Inductor in different orientation. Some of the commonly used topologies includes LLC, LCC, LCLC, LCCL Converters and so on. The switching frequency and the quality factor determines whether buck/boost operation need to be performed. Soft switching converters are classified depending on the operating principles as (a) Resonant power converters(RPC) (b) Multi and Quasi resonant converters(QRC) (c)Resonant transition converters(RTC).

2.1 Resonant Power Converters(RPC)

An efficiently designed resonant converter can attain resonance over the entire load and line range. A classical DC-DC resonant converter be made up of three portions- a control switching circuit, a resonant tank and a rectification network [13]. Figure 3 shows the basic structure of DC-DC resonant power converter. The initial controlled pulsating network switches the input dc to high frequency square waves for generating the current and voltage at the input side of the resonant tank. Commonly used switching topologies includes half bridge and full bridge inverter which is chosen according to the power requirements. The resonant tank network(RTN) consists of inductor and capacitive elements which charges and discharges allowing back and forth energy transfer in the network. The resonant tank is responsible to reduce the stresses and switching losses in the network. The quality of the resonant converter is assessed considering the losses and resonant bandwidth. This can be analysed using Q factor for which larger the value of Q factor can reduce the oscillations and minimise the degree of damping. To achieve the desired degree of damping the soft switched converters can be grouped as series resonant converters(SRC), parallel resonant converter(PRC) and hybrid/ series parallel resonant converters. The simplest resonating network includes SRC or PRC comprises of a resonant inductor or a resonant capacitor connected in series or parallel respectively with the load and the power supply. Proper isolation guarantees the converter usage at high power applications. The resonant tank output is fed to the rectifier network which aims at rectifying the sinusoidal output to pulsating DC and then filter it to required DC output waveform.

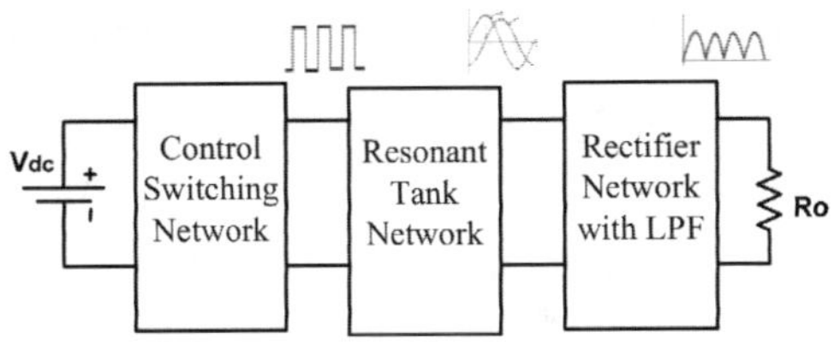

Figure 3. Structure of DC-DC resonant converter

Series parallel resonant converters includes combination of SRC and PRC which includes many popular topologies. It includes LLC, LCL, CLC, CLL where LLC and LCC are most popular.

2.2 Quasi Resonant(QRC) Converter

Noise and heat developed can be suppressed using QRC which combines a resonant switch along with the resonant element. While operated at higher frequencies QRC improves the output power quality. For high voltage applications, to reduce the size of the circuit, switching frequency is increased. However, high switching frequency leads to increased switching stresses which requires a large heat sink. One of the demerits of Quasi resonant converter is that it can operate either at ZVS or at ZCS but not both concurrently by resonating only at a particular region in the cycle. It involves both resonant and non resonating intervals in its operation. By arranging an inductor in series with the switch allows ZCS during turn off and arranging a capacitor in parallel with the resonant switch allows ZVS during turn on thereby reducing switching losses.

For ZVS, the resonant capacitor, C_r is connected side by side with the switching element, S keeping resonant inductor, L_r in cascade with C_r and S . However, for ZCS, C_r is kept in parallel with the series combination of Lr and S. Figure 4 shows examples of Quasi resonant converters which satisfies

ZCS and ZVS condition respectively. A multistage dual direction converter using two half-bridge gallium nitride modules is discussed in [14]. A reduced voltage stresses as a result of ZVS is achieved during turn on. Moreover, GaN switches reduces on-resistance and faster switching speed compared with Si based switches. It minimizes switching losses during turn off as well. In [15], A dual quasi flyback resonant converter is designed and implemented with auto regulated structure for parallel/series input. In order to improve the performance of the converter, the input transformer structure can be varied to either cascade/parallel based on the input voltage.

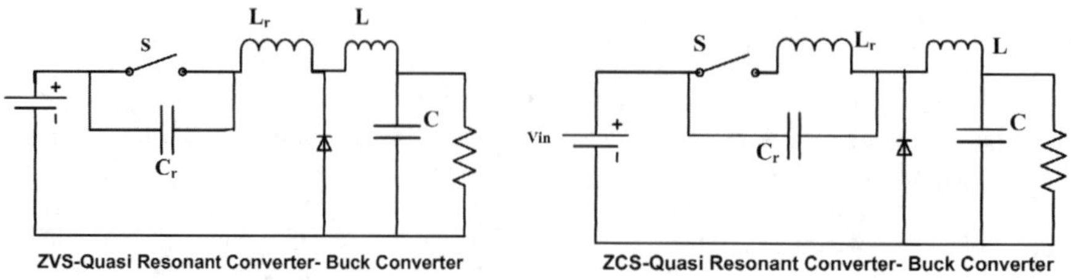

Figure. 4 :DC-DC Buck converter satisfying a) ZVS-QRC and b) ZCS-QRC

2.3 Resonant Transition Converter(RTC)

Resonant transition converter(RTC) is a modified quasi resonant converter with the combination of four elements. Unlike, quasi resonant converter ZCS and ZVS can be obtained from RTC concurrently as shown in figure 5 (b) and 5 (c). In addition to the resonant elements L_r and C_r, RTC includes the auxiliary components switch, S_2 and diode, D_2. However, it increases the current stresses in the main switch. A high performing LCL-T resonant converter is discussed in [17] achieving load independent output current for automotive LED drives. It helps control the brightness of LED by achieving regulated output current using phase shift control modulation strategy at fixed frequency. Moreover, a detailed analysis is done for LCL-T resonant converter and compared with LC^3L and LCLC multi resonant converters which then proved to have better efficiency than the said topologies. In [18], a variable structure multi resonant converter which can operate in FB-CLTCL, FB-LCLCL and HB-LCLCL modes (FB-Full bridge and HB- Half bridge) based on different switching configurations is analyzed and implemented . The resonant converter operates at high conversion efficiency and has wide voltage gain with narrow bandwidth.

2.4 Multi resonant converter(MRC) Converters

MRC's are selected to get rid of the demerits of current and voltage oscillations and thereby to achieve ZVS as shown in figure 5(a). To reduce the effect of leakage inductance of transformer and parasitic components of switching elements, soft switching is implemented using a forth order LLCC-Load resonant converter(LRC) in [16]. It consists of a half bridge inverter isolated with a full bridge rectifier with elevated potentiality in line as well as load regulation with wide variation in the input.

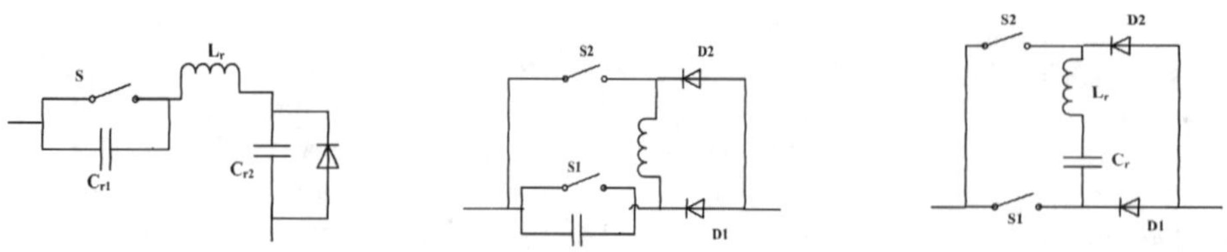

Figure. 5: a) Multi resonant LCC resonant switching b) ZVT- switching c) ZCT switching

3. Second Order/ Two Element Resonant Tanks

Number of resonant elements decide the transfer function order. The smallest resonant network is possible making use of a resonant inductor and capacitor in series or parallel combinations with load and the supply. There are eight possible resonant tank combinations in the two element resonant topologies. However , not all of them can work as a resonant converter with pulsating voltage source inputs due to due to the lack of voltage regulation capability and requirement of current sources.

3.1 Parallel Resonant Converter(PRC)

Figure 6(b) and 6(c) shows parallel resonating tank circuit consisting of reactive elements. PRC consist of two resonating elements, a resonant inductor, L_r and a resonant capacitor, C_r in parallel to the load and the supply. In PRC, the energy stored in inductor and the capacitor is continuously oscillating which makes the resultant current between them to zero. The instantaneous values of inductor and capacitor currents are always equal which is the basic principle behind the resonant oscillation. Parallel resonant converter allows amplification of the input signal and provide high gain nearby the resonant frequency. A combined topology based on a three phase inverter with two-LLC resonant tank interleaved converter topology is explained in [19] with less number of switching elements and other components. Comparing to conventional isolated three phase interleaved topology, cost and size of the equipment is considered lower. The topology achieves ZVS above resonant frequency for all the switches used[20].

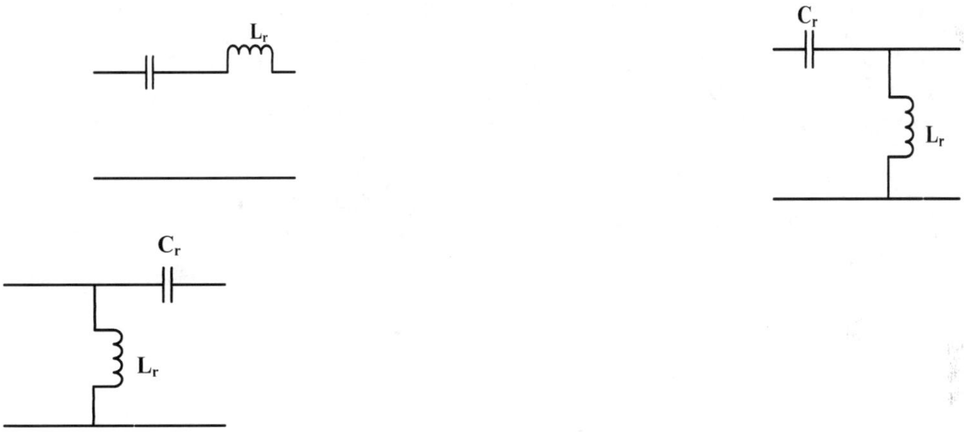

Figure. 6: a) Series resonant tank (b) and (c) Parallel resonant tank

3.2 Series Resonant converter(SRC)

Figure 6 (a) shows a series resonating circuit which helps to regulate the output. Series resonance doesn't help to amplify the input signal. Instead, it can only provide a buck function for which the gain can only be unity or lower. SRC is a promising solution for regulating the input voltage over a wide range of input voltage. A bridgeless isolated serial resonating circuit is proposed on the transformer secondary to improve the power factor with less switching elements over wide input voltage ranges[21]. In [22], an efficient isolated converter for electric vehicle -V2G application is discussed with full bridge inverter primary and voltage doubler secondary. The proposed bidirectional topology allows step up/step down capability in forward and reverse direction respectively providing high conversion ratio and high efficiency over wide battery voltage range. A dual mode rectifier type series resonant DC-DC converter [23] is configured with advantages including efficiency and input voltage range improvement. The proposed dual mode rectifier is configured on the secondary side of the isolation transformer with voltage gain range doubled for PV applications.

4. Three Element Resonant Tank

Three element resonant topology consist of 3 resonating elements which combines the advantages of both SRC and PRC. It includes two inductors and one capacitor or vice versa. For three element resonant topologies, thirty six possible configurations are possible by arranging the inductor and capacitors in different orientations. In these, twenty three can be utilized in combination with a voltage source. In these 23, 15 can be used in voltage and current regulation. Eight of the configurations have notch resonance making it difficult to control. The commonly used third order resonant networks includes LLC, LCL, CLC, CLL were LLC and LCC being the most popular. Voltage gain of LCC is better than LLC which makes LCC being more used in applications were high output voltage is required. Such an LCC and LLC resonating circuit is shown in figure 7(a) and 7(b). The main benefit of resonant converters is that the converter can achieve soft switching over wide load ranges. As a result, it is important to analyse the necessary conditions of resonance. The switch achieves zero voltage by allowing the parasitic capacitor energy of the switch to discharge completely. Enough dead time has to be provided to the switches to discharge the output capacitor of MOSFET to zero. These topologies can be categorized as the combination of a half and full-bridge inverter and a full-bridge or centre-tapped rectifier on the primary and the secondary side respectively, were the centre-tapped rectifier are not in priority as a result of the elevated voltage stress on the diodes. The output voltage or current can be controlled in a broad area by changing the switching frequency for LLC. In [24], a dual output topology with LLC resonant tank is discussed capable of achieving high power density and satisfying the ZVS and ZCS condition. Instead of using two separate DC-DC converters, the proposed topology discussed the converter allowing multiple outputs of 200-400 V for the traction and 48 V for auxiliary supplies using switched frequency and phase angle control respectively.

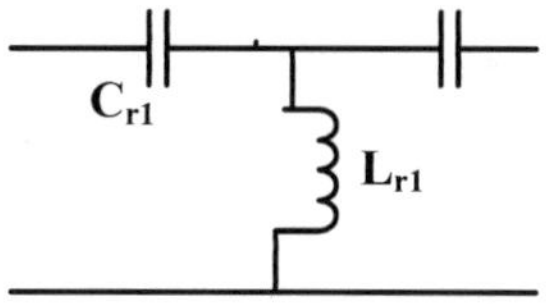

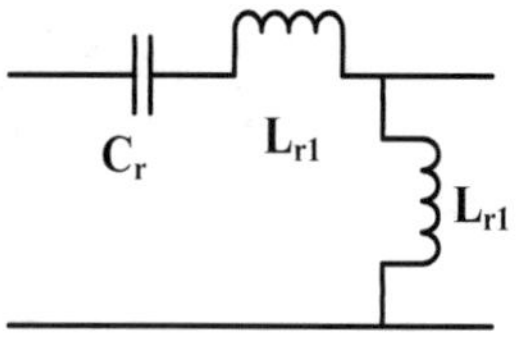

Fig 7 (a): LCC resonant tank (b) LLC resonant tank circuit

5. Conclusion

In the paper, different soft switching DC-DC converter topologies are mentioned and examined based on the recent development in terms of efficiency, steady state oscillation, number of components and voltage gain. Resonant converters have a vast class of topologies on the basis of number of passive storage elements and auxiliary elements. Different second and third order resonant tank configurations are investigated pointing out the merits and demerits of each configuration. Moreover, the most commonly evolved resonant configurations are identified and integrated.

References

[1] Chewale, Mangesh A., et al. 2018 *A Review on Isolated and Non-isolated DC-DC Converter for PV Application*, International Conference on Control, Power, Communication and Computing Technologies (ICCPCCT). (IEEE.)

[2] Nithara, P.V. et al. *V-f Controlled Autonomous Wind Energy Conversion System using Z2 Transformer Connected STATCOM*, (International Journal of Engineering and Advanced Technology) vol. 8 pp 1492-1496

[3] Bhaskar, Mahajan Sagar, et al. 2020 *Survey of DC-DC non-isolated topologies for unidirectional power flow in fuel cell vehicles*. (IEEE Access) vol. 8 pp 178130-178166.

[4] Arulselvi, S., K. Deepa, and G. Uma. 2005 *Design, analysis and control of a new multi-output flyback CF-ZVS-QRC.* (IEEE International Conference on Industrial Technology)
[5] Amir, Asim, et al. 2019 *Comparative analysis of high voltage gain DC-DC converter topologies for photovoltaic systems.* (Renewable energy) vol. 136 pp 1147-1163.
[6] Buvana, D., B. Stalin, and K. Viji. 2020 *Analysis Of Coupled Inductor Based Converter For The Grid-Connected PV System* (Solid State Technology) vol 63 pp 1749-1756.
[7] Shayeghi, Hossein, et al. 2020 *Non-isolated DC–DC converter with a high-voltage conversion ratio.* (IET Power Electronics) vol 13 pp 3797-3806.
[8] Valipour, Hamed, et al. 2020 *Extended Range Bridgeless PFC Converter with High Voltage DC Bus and Small Inductor* (IEEE Transactions on Power Electronics).
[9] Sayed, Khairy, Mohammed G. Gronfula, and Hamdy A. Ziedan. 2020 *Novel soft-switching integrated boost DC-DC converter for PV power system* (Energies) vol. 13 pp 749.
[10] Faraji, Rasoul, and Hosein Farzanehfard. 2020 *Fully Soft-Switched Multiport DC–DC Converter With High Integration.* (IEEE Transactions on Power Electronics) vol. 36 pp 1901-1908.
[11] Arshadi, Sayed Abbas, et al. 2020 *Three-Phase LLC Battery Charger: Wide Regulation and Improved Light-Load Operation.* (IEEE Transactions on Power Electronics)vol 36 pp1519-1531.
[12] McLaughlin, Prescott H., Ziyu Xia, and Jason T. Stauth. 2020 *A Monolithic Resonant Switched-Capacitor Voltage Regulator With Dual-Phase Merged-LC Resonator* (IEEE Journal of Solid State Circuits vol 55 pp 3179-3188.
[13] Salem, Mohamed, et al. *Resonant power converters with respect to passive storage (LC) elements and control techniques–An overview* (Renewable & Sustainable Energy Reviews) vol 91 pp 504-520.
[14] Moradisizkoohi, Hadi, Nour Elsayad, and Osama A. Mohammed 2019 *Experimental Demonstration of a Modular, Quasi-Resonant Bidirectional DC–DC Converter Using GaN Switches for Electric Vehicles.* (IEEE Transactions on Industry Applications) vol 55 pp 7787-7803.
[15] Wang, Jian-Min, et al. 2019 *The Novel Quasi-Resonant Flyback Converter With Autoregulated Structure for Parallel/Serial Input* (IEEE Transactions on Industrial Electronics) vol. 67 pp 992-1004.
[16] Khoshsaadat, Alireza, Arash Khoshooei, and Mohammad Mohammadi 2020 *Analysis and Design of a New Current-Source Output Load Resonant Converter With High Capability in Line and Load Regulation* (IEEE Transactions on Circuits and Systems I: Regular Papers).
[17] Khatua, Mausamjeet, et al. 2020 *High-Performance Megahertz-Frequency Resonant DC-DC Converter for Automotive LED Driver Applications.* (IEEE Transactions on Power Electronics.)
[18] Chen, Mengying, et al. 2018 *A Variable-Structure Multi-Resonant DC–DC Converter with Smooth Switching.* (Energies) vol 11 pp 2240.
[19] Salem, Mohamed, et al 2020 *Three-Phase Series Resonant DC-DC Boost Converter With Double LLC Resonant Tanks and Variable Frequency Control* (IEEE Access) vol 8 pp 22386-22399.
[20] Ramesh, P., Sharmeela, C. 2014 *A coupled inductor based DC-DC converter with the technique of Soft Switching and voltage Clamping* (National Conference on Electrical Information and Communication Engineering NCEICE).
[21] Bakeer, Abualkasim, et al. 2020 *Wide Input Voltage Range Operation of the Series Resonant DC-DC Converter with Bridgeless Boost Rectifier* (Energies) vol 13 pp 4220.
[22] Bai, Changkyu, et al. 2019 *Highly efficient bidirectional series-resonant DC/DC converter over wide range of battery voltages* (IEEE Transactions on Power Electronics) vol 35 pp 3636-3650.
[23] Shen, Yanfeng, et al. 2018 *A 1-MHz series resonant DC–DC converter with a dual-mode rectifier for PV microinverters.*(IEEE Transactions on Power Electronics) vol 34 pp 6544-6564.
[24] Li, Guangdi, et al. 2019 *Hybrid Modulation of Parallel-Series LLC Resonant Converter and*

Phase Shift Full-Bridge Converter for a Dual-Output DC–DC Converter." IEEE Journal of Emerging and Selected Topics in Power Electronics vol 7 pp 833-842.

Experimental studies on mechanical and durability characteristics of lc$_3$ concrete

S Kavya Sree[1,a], *B D V Chandra Mohan Rao*[2,b], *Maganti Janardhan Yadav*[3,c], *P Swamy Naga Ratna Giri*[4,d]

[1]*PG Student, Civil Engineering Department, JNTUH CEH, Hyderabad, Telangana, India.*

[2]*Professor, Civil Engineering Department, VNRVJIET, Hyderabad, Telangana, India.*

[3]*Professor, Civil Engineering Department, JNTUH CEH, Hyderabad, Telangana, India.*

[4]*Assosiate Professor, Civil Engg. Dept., Sree Chaitanya College of Engineering, Karimnagar, Telangana, India.*

a) siddamshettykavyasree@gmail.com
b) bdvcmrao@gmail.com (Corresponding Author)
c) prof.mjyadav@gmail.com
d)psnrgiri24@gmail.com

Abstract: In the present day context utilization of natural resources is increasing day by day, which is a burden for future generations. Limiting the usage of natural raw materials and development of sustainable methodologies is needed in all aspects of construction activities. In this regard, manufacturing of cement requires large amount of limestone deposits, fuel energy and on the other hand cement manufacturing liberates carbon dioxide (CO_2) into atmosphere. Liberation of CO_2 results in global warming which leads to climatic changes. Hence there is a need to develop an alternative material superior to cement without compromising its properties. To address the above problem present study is focused on examining the usage of Limestone Calcined Clay Cement (LC_3) as partial replacement to clinker to an extent of 50% in concretes. M 30 grade of concrete is considered for investigation with three types of binders (i.e. LC_3, Ordinary Portland Cement (OPC) and Pozzolana Portland Cement (PPC). From the studies on strength and durability it is evident that the performance of LC_3 concrete is better than OPC and PPC based concrete.

Keywords: LC_3; Strength; Durability; Absorption; Desorption; RCPT

1. INTRODUCTION

Now a day's adapting to sustainable development is a challenging task i.e. without compromising the quality and durability of structural elements. By manufacturing of cement a lot of CO_2 is released to atmosphere which is responsible for green house gases, as cement is manufactured from local available materials and industrial bye products like fly ash slag etc. [1]. Benefit of blending Portland Pozzolana Cement (PPC) for reduction of the carbon emissions and fuel energy capacity is upto to certain amount only. More quantity of CO_2 emissions can be reduced by making alternative cements through blending highly reactive pozzolana materials like calcined clay with the combination of limestone [2, 3]. Utilization of secondary cementious materials (SCM's) for making concrete is increased to enhance the quality of concrete structures, in line to that by using of LC_3 concrete can achieve desired properties without compromising the quality [4, 5]. LC_3 is treated at low level temperatures about 800°c-900°c when compared to cement for the activation of pozzolanic action, further it changes to metakaolin which is highly reactive pozzolana [6, 7]. The pozzolana activity for calcined clay is materializing mainly due to the formation of metakaolin while treating clay minerals. Further it reacts as pozzolanic material with calcium hydroxide, water and sulphate to form calcium alumina silicate hydrate and alumina ferrite phases [8, 9]. For achieving sufficient workability and early age strength there is no need of high purity of kaolinite in clay deposits, about 40% of kaolinite content can perform well by making some changes in the particle size of cement compounds [10, 11]. The main idea of combining limestone with calcined clay is, limestone helps to promote the clinker hydration, facilitates nucleation of hydrates by providing surface area and responsible for hydration reactions in the presence of aluminates. The quantity of limestone reacted depends on the purity of aluminate sources, presence high kaolinite content responsible for the development of strength in LC_3 concrete [12, 13]. By replacement of clinker with LC_3-50% has shown similar properties as that of OPC. Present study focused to examine the performance of LC_3 based concrete in terms of mechanical and durability characteristics and to compare the results with OPC and PPC based concretes.

2. IMPORTANCE OF THE STUDY

Due to the increase of cement demand there is a need to alter cementious materials without compromising the properties. This can be achieved by blending clinkers into binary and ternary binders with the locally available materials. Several authors had investigated on such type blended cements like limestone calcined clay cement which can be future cementitious material from the sustainable point of view. It is necessitated to find out the characteristics of binder while it is being mixed with several ingredients like fine aggregate, coarse aggregates and water. Present study is focused to know the potential of LC_3 concrete by conducting experiments on M 30 grade concrete and the results of LC_3 concrete are compared with OPC and PPC based concretes.

3.0 MATERIALS AND EXPERIMENTS

Present study consisted of standard grade concrete (M 30), designed according to IS 10262-2019. The details of mix proportions for M 30 is (1:1.42:2.73:0.46) with 0.25% of super plasticizer is used by weight of cement (Ploy carboxylates either based) to maintain slump at 80-100 mm The variables in the study includes type of binder (LC_3, OPC and PPC) and age of curing (i.e. 7, 28 days) for determining the mechanical properties, Compressive strength, Split tensile strength and Flexural strength tests are conducted and for durability performance, Rapid chloride ion penetration test, Absorption and Desorption tests were conducted on concrete specimens after 28 days curing. Typically a concrete specimen labelled Mix MO represents M 30 grade of concrete with 0.46 w/c containing OPC, Mix MP represents M 30 grade of concrete with 0.46 w/c containing PPC and Mix ML represents M 30 grade of concrete with 0.46 w/c ratio containing Limestone Calcined Clay Cement.

3.1 Materials used:

53 grade cement as per IS 12269-2013 is used in present study and the physical characteristics are given in Table 1. PPC conforming to IS 1489-1 (1991) is used for investigation and the properties are shown in Table 1. LC_3 procured from society for Technology and Action for Rural development Authority, New Delhi, India is considered for investigation and the physical properties of LC_3 are given in Table 1. Fine aggregate (FA) of size less 4.36 mm and conforming to zone-II as per IS: 383-2016 is used and Coarse aggregate (CA) consisted of two sizes (20mm and 10mm) mixed in the ratio of 60:40 is used to obtain well graded aggregate as per codal provisions IS: 383-2016. Physical Properties of FA and CA are given in Table 2.

Table 1 : Physical ccharacteristics of OPC, PPC and LC_3

S. No.	Property	OPC	PPC	LC_3
1	Normal consistency	30%	31%	32%
2	Initial setting time (minutes)	73	83	65
3	Final setting time (minutes)	230	295	185
4	Specific gravity	3.11	2.89	2.99
5	Fineness	4%	8%	5%
6	Compressive strength (MPa)	54.6	49.2	53.5

Table 2: General properties of CA and FA

S. No.	Property	CA	FA
1	Bulk Density (g/cc)	1.64	1.73
2	Specific Gravity	2.65	2.57
3	Fineness Modulus	7.32	2.73
4	Void ratio	1.01	0.92
5	Porosity	36%	41%

3.2 Experiments conducted:

For determining the mechanical characteristics of concrete specimens the following tests are conducted, i) Compressive Strength Test (CST) is carried on cube specimen of size 150×150x150 mm and as per IS 516-1959 [13] test is conducted on 3000 kN Compression testing machine. ii) Split Tensile Strength Test (STST) as per IS 5816-1999 [14] is conducted on the cylinder specimen of size 150 mm dia. and 300 mm height and iii) Flexural Strength Test (FST) is conducted on prism of size 500×100×100 mm by two point loading system as per IS codal provisions. To determine the performance of LC_3 following tests are considered: i) Rapid Chloride ion Penetration Test (RCPT) is performed according to ASTM C 1202-97 [15] on specimen of 50 mm thick and 100 mm diameter ii) Absorption Test (AT) is used to measure the rate of absorption of water by the concrete by finding the increase in the mass of a specimen as a function of time when cubes are immersed in water for a period of 72 hr. iii) Desorption Test (DT) is done to know the rate of evaporation of water in the concrete by finding the decrease in the mass of a specimen as a function of time when cubes are kept in hot air oven for 72 hours at 105°c. The percentage weight gain and weight loss of specimens is recorded for further analysis of results.

4. RESULTS AND DISCUSSIONS

4.1 Compressive Strength Test:

Compressive strength test results of M 30 grade concrete with different type binders is presented in Figure 1. It is found that the mixes attain the required strength after curing for 28 days. From the 7, 14 and 28 days results it is evident that the LC_3 concrete had performed better than that of OPC and PPC based concretes. The improvement of strength in LC_3 concrete is mainly due the uniform microstructure and dense morphology in inner core structure. Reason for achieving better strength values is due to the formation of more quantity of alumina in reactive product and the presence of calcium carbonate in limestone allows producing carbon aluminate hydrates. On the other hand PPC is a binary system whereas as LC3 is ternary system, by comparing both it is clear that the early strength gain during first 7 days is more in LC_3 concrete which resolves the problem of low strength gain at early ages.

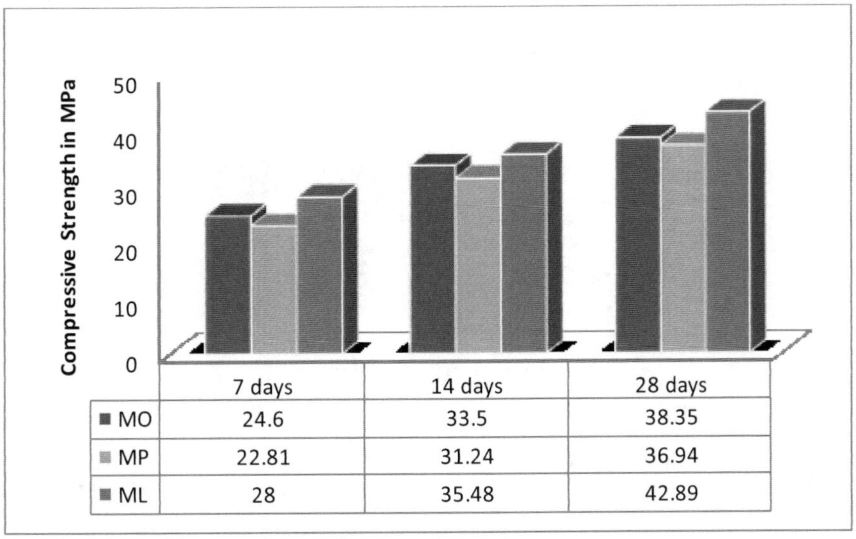

Figure 1 Compressive strength results for M 30 grade concrete with different binders

4.2 Split Tensile Strength and Flexural Strength Test:

Figures 2, 3 shows the results of M 30 grade concrete with different types of binders. It is exhibited that split tensile strength is almost similar in all types of concrete irrespective of type of binder. From Figure 3, it is clear that enhanced performance of LC_3 concrete is noticed when compared with OPC and PPC based concretes. An average of 15 % increase in flexural strength is observed in LC_3 concrete. This is due to the dense compact morphology of LC_3 develops the crack bridge in the inner layer structure by forming silicon chains.

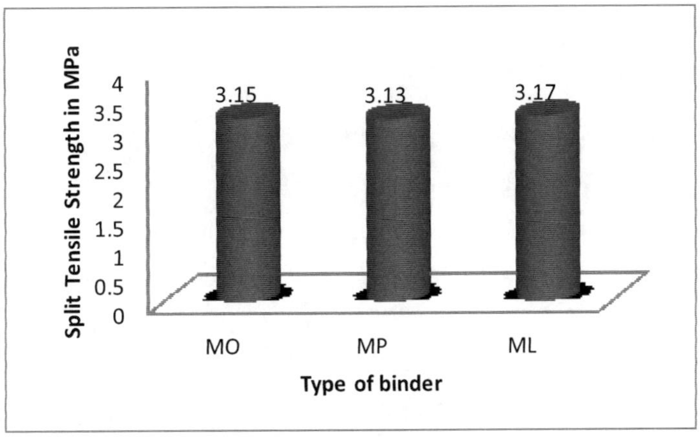

Figure 2 Split Tensile strength results for M 30 grade concrete with different binders

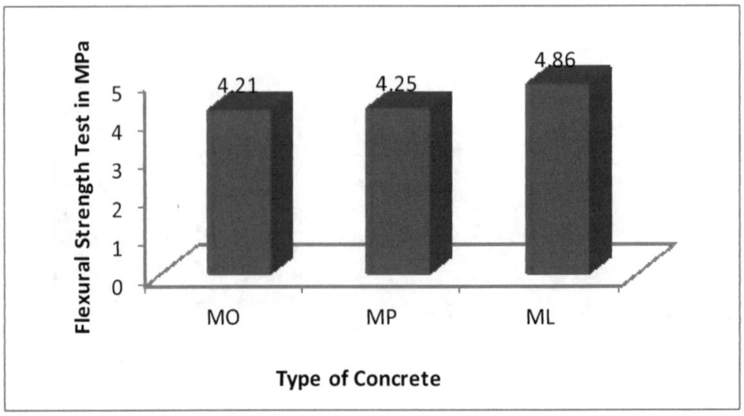

Figure 3 Flexural Strength results for M 30 grade concrete with different binders

4.3 RCPT Results:

Table 3 shows the result for rapid chloride ion penetration test for M 30 grade concrete with different types of binders. From the values of current and charge passed through specimen it is evident the ingress of chloride ions into LC_3 concrete specimens is less compared to other specimens. From this it is clear that the durability of LC_3 concrete specimens are enhanced with the RCPT results, i.e LC_3 concrete specimens are less vulnerable to chloride ingress, which is due to the refinement of pore sizes and having low pore threshold radius value. Due to the presence of fibrillar-like core structure LC_3 based concrete helps in strengthening the concrete structures form environmental effects like chloride attack and carbonation. By using LC_3 binder, the chloride resistance is enhanced by about 85-90% when compared to OPC and PPC based concretes.

Table 3: Rapid Chloride ion Penetration Test Results of M 30 grade concrete with different binders

S.No.	Mixes	Charge passed (Coulombs)	Chloride ion permeability as per ASTM C 1202
1	AO	1632	Low
2	AL	817	Very Low
3	AP	1385	Low

4.4 Absorption and Desorption Test Results:

In addition to RCP test, Absorption and Desorption tests were also conducted to get more qualitative results for M 30 grade concrete with different binders. From Figure 4, it is evident that the rate of absorption all specimens is linear upto 24 hours after that the curve is almost straight for all specimens irrespective of type of binder. From the graph, it is clear that LC_3 concrete specimen shown lesser water absorption when compared to OPC and PPC based concrete specimens. Pore inter connection is less in LC_3 concrete and the inter layer space between the calcium ions and silicate chains is filled with water and calcium ions. The rate of water absorption is less in case of LC_3 concrete, whereas in OPC based concrete specimens the rate of water absorption is more due to less dense microstructure and inter connectivity of pores is more as compared to PPC based concrete. Ternary blended and binary blended cements are more durable than that of OPC due to the effect of fine grained particles and compact core structure of concrete.

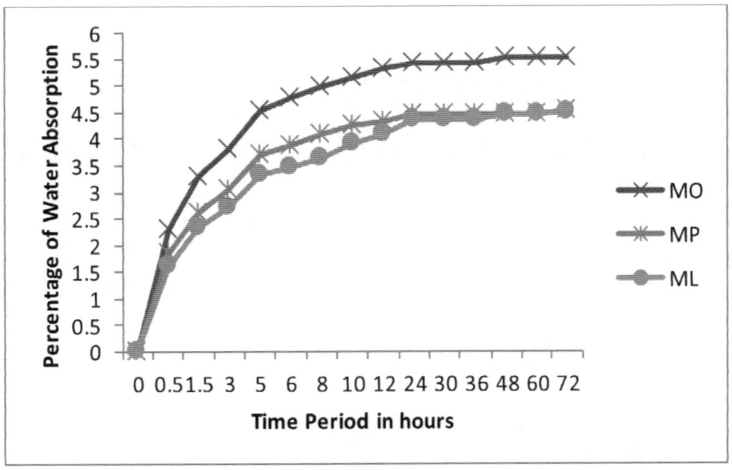

Figure 4 Percentage of water Absorption for M_{30} grade concrete with different binders

Figure 5 shows the percentages of water desorption for M 30 grade concrete with different binders. From figure, it is observed that the loss of water is more in initial hours i.e. up to 24 hours and then the curve is almost straight. At a temperature of 105°C, freely available water molecules are evaporated in concrete specimens, these voids present in the concrete are responsible for the ingress of harmful chemicals into concrete. Concrete with less percentage water desorption is more durable than that of other concrete specimens. From the curves, it is found that the LC_3 concrete specimens prone to less desorption compared to OPC and PPC concrete specimens. It is clear that the LC_3 concrete has dense microstructure and compact core structure which does not allow the water desorption and this is mainly due to the presence of finer particles in LC_3 concrete and more reactivity pozzolana action is taken place which intern requires more amount of water to form secondary gel phase. There is a coincidence in durability and strength results, which helps to strengthen the performance of LC_3 concrete in all aspects.

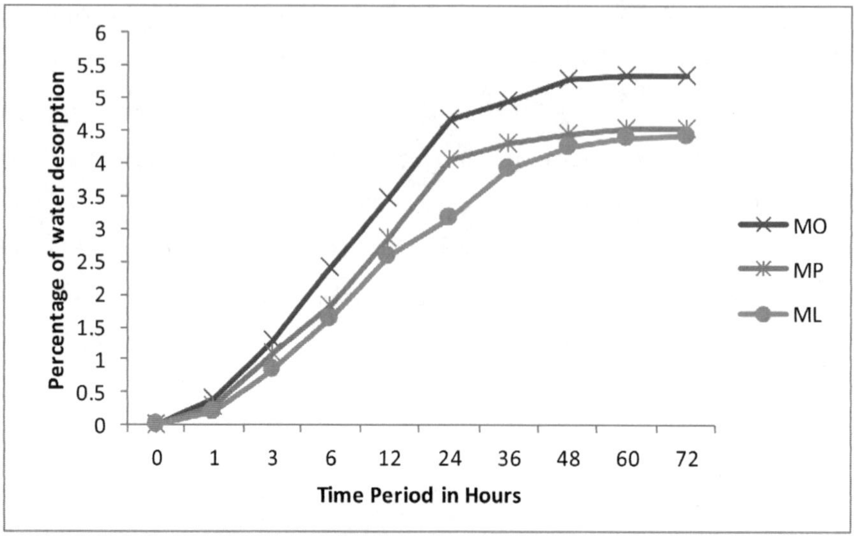

Figure 5 Percentage of water desorption for M 30 grade concrete with different binders

5. CONCLUSIONS

From the detailed experimental studies on mechanical and durability characteristics of LC_3 concrete and comparison with OPC and PPC based concretes, the following are the findings.

- Concrete made with LC_3 shows better strength when compared to OPC and PPC binder concretes. Based on 7 day strength of LC_3, it is clear that early strength can be achieved through this pozzolanic material when compared to PPC.
- The performance of LC_3 concrete is far better than that of OPC concrete especially in chloride ion penetration test. From results of RCPT, it is clear that LC_3 concrete specimen observed less ingress of chloride ions compared to OPC and PPC concrete specimens, which is due to the presence of dense and uniform morphology.
- From absorption and desorption test results, it is evident that LC_3 concrete specimen resulted in low permeability which could be a better in case of achieving good durable concrete. Enhanced properties of LC_3 concrete are mainly due to the pore refinement and the dense microstructure.
- From over all studies, results suggested that the high performance pozzolana activity can be achieved through LC_3 binder system without compromising the strength and durability properties when compared to OPC binder system.

REFERENCES

[1]. Naqi, A. and Jang, J.G., 2019. Recent progress in green cement technology utilizing low-carbon emission fuels and raw materials: A review. *Sustainability*, *11*(2), p.537.
[2]. Rakhimova, N. and Rakhimov, R., 2020, July. Advances in development of calcined clays as supplementary cementitious materials. In *IOP Conference Series: Materials Science and Engineering* (Vol. 890, No. 1, p. 012085). IOP Publishing.
[3]. Khan, M.S., Nguyen, Q.D. and Castel, A., 2020. Performance of limestone calcined clay blended cement-based concrete against carbonation. *Advances in Cement Research*, *32*(11), pp.481-491.
[4]. Parashar, A.K., Gupta, N., Kishore, K. and Nagar, P.A., 2020. An experimental investigation on mechanical properties of calcined clay concrete embedded with bacillus subtilis. *Materials Today: Proceedings*.
[5]. Gobinath, R., Awoyera, P.O., Praveen, N., Babu, V.A., Sai, P.S. and Prathibha, K., 2020. Effects of calcined clay on the engineering properties of cementitious mortars. *Materials Today: Proceedings*.
[6]. Dhandapani, Y., Sakthivel, T., Santhanam, M., Gettu, R. and Pillai, R.G., 2018. Mechanical properties and durability performance of concretes with limestone calcined clay cement (LC3). *Cement and Concrete Research*, *107*, pp.136-151.
[7]. Abdullah, M.E., Jaya, R.P., Shahafuddin, M.N.A., Yaacob, H., Ibrahim, M.W., Nazri, F.M., Ramli, N.I. and Mohammed, A.A., 2018, May. Performance of Kaolin Clay on the Concrete Pavement. In *IOP Conference Series: Materials Science and Engineering* (Vol. 358, No. 1, p. 012049).
[8]. Reddy, A.N. and Meena, T., 2018. A study on compressive behavior of ternary blended concrete incorporating alccofine. *Materials Today: Proceedings*, *5*(5), pp.11356-11363.
[9]. Liu, Y. and Ling, T.C., 2018, October. Potential use of calcined kaolinite-based wastes as cement replacements in concrete–an overview. In *IOP Conference Series: Materials Science and Engineering* (Vol. 431, No. 3, p. 032006). IOP Publishing.
[10]. Rai, S. and Tiwari, S., 2018. Nano silica in cement hydration. *Materials Today: Proceedings*, *5*(3), pp.9196-9202.
[11]. Tsivilis, S., Tsantilas, J., Kakali, G., Chaniotakis, E. and Sakellariou, A., 2003. The permeability of Portland limestone cement concrete. *Cement and concrete research*, *33*(9), pp.1465-1471.
[12]. Tiwari, A.K. and Chowdhury, S., 2016. Relative evaluation of performance of limestone calcined clay cement compared with Portland pozzolana cement. *Journal of Asian Concrete Federation*, *2*(2), pp.110-116.
[13]. INDIAN STANDARDS: 516 -1956 (Reaffirmed 2013) 'Indian Standard Methods of Tests for Strength of Concrete', *Bureau of Indian Standards*, New Delhi, India, 1956.
[14]. IS: 5816. (1999). Indian standard splitting tensile strength of concrete. *Bureau of Indian Standards*.
[15]. Joshi, R.C., Chatterji, S., Achari, G. and Mackie, P., 2000. Reexamination of ASTM C 1202—Standard Test Method for Electrical Indication of Concrete's Ability to Resist Chloride Ion Penetration. *Journal of testing and evaluation*, *28*(1), pp.59-61.

Performance studies on limestone calcined clay based concrete

K Almas[1,a)], B D V Chandra Mohan Rao[2,b)], Maganti Janardhan Yadav[3,c)], P Swamy Naga Ratna Giri[4,d)]

[1]PG Student, Civil Engineering Department, JNTUH CEH, Hyderabad, Telangana, India.
[2]Professor, Civil Engineering Department, VNRVJIET, Hyderabad, Telangana, India.
[3]Professor, Civil Engineering Department, JNTUH CEH, Hyderabad, Telangana, India.
[4]Assosiate Professor, Civil Engg. Dept., Sree Chaitanya College of Engineering, Karimnagar, Telangana, India.
a) almasyasmin111@gmail.com
b) bdvcmrao@gmail.com (Corresponding Author)
c) prof.mjyadav@gmail.com
d) psnrgiri24@gmail.com

Abstract: In the construction field, lot of improvement in sustainable development is going on from last few decades, mainly use of pozzolanic materials are increasing day by day for the production of cement. On the other hand a lot of anthropogenic CO_2 gas is liberating in to atmosphere while manufacturing cement. To alter the emission of CO_2, use of alternative cementious materials is one of the best practices like fly ash, rice husk ash, GGBS and Metakaolin. Due to the use of such materials there is a lag in early strength development in concrete. Hence there is need to develop a substitute material for cement without negotiating in properties. In the present paper Limestone Calcined Clay Cement (LC_3) as an alternative to Ordinary Portland Cement (OPC) is examined for strength and durability and the results are compared with Portland Pozzolana Cement (PPC) for M 50 grade of concrete. From results it is found that the use of LC_3 concrete is beneficial in terms of strength and durability.

Keywords: Calcined clay; Limestone; Compressive Strength; RCPT; Sustainability.

1. INTRODUCTION

In the present days context, for the production of concrete requires large quantity of resources and energy consumption is also more for manufacturing cement, which is also a pollution intensive material. Due to this, cement industry is facing lot of problems like: (i) Anthropogenic CO_2 gas is liberating to atmosphere. (ii) More quantity of fuel is required for clinkering process. (iii) degradation of natural materials and (iv) Cost of manufacturing is increased per unit [1, 2]. To address these problems, somehow benefit of blending Portland Pozzolana Cement (PPC) for reduction of the carbon emissions and fuel energy capacity is upto to certain amount only. More quantity of CO_2 emissions can be reduced by making alternative cements through blending highly reactive pozzolana materials [3, 4]. Utilization of secondary cementitious materials (SCM's) for making concrete is increased to improve the quality and durability of concrete structures, in line to that a new LC_3 concrete can achieve desired properties without compromising the quality [5, 6]. LC_3 is treated at low level temperatures of about 800°c-900°c when compared to cement for the activation of pozzolanic action, further it changes to metakaolin which is highly reactive pozzolana [7]. The pozzolana activity for calcined clay is materializing mainly due to the formation of metakaolin while treating clay minerals. Further metakaolin reacts as pozzolanic material with calcium hydroxide, water and sulphate to form calcium alumina silicate hydrate and alumina ferrite phases [8, 9]. For achieving sufficient workability and early age strength there is no need of high purity of kaolinite. In clay deposits, about 40% of kaolinite content can perform well by making some changes in the distribution of particles of cement compounds [10, 11]. The main idea of combining limestone with calcined clay is, limestone helps to promote the hydration of clinker by facilitating suitable surface area for nucleation of hydrates and responsible for hydration reactions in the presence of aluminates. The quantity of limestone reacted phases depends on the purity of aluminate sources, presence high kaolinite content responsible for the development of strength in LC_3 concrete [12, 13]. Present study focused to examine the performance of LC_3 based concrete in terms of mechanical strength and durability.

2. IMPORTANCE OF THE STUDY

Several investigators has concentrated on developing alternative materials for cement. Due to the increase of cement utilization there is an urgency to alter cementitious materials without compromising the strength and durability, which can be achieved by combining two or more cementitious materials. Few of authors had investigated on blended cements like limestone calcined clay cement which can be future cementitious material in the sustainable point of view. It is a basic thing to find out the characteristics of such type of binder while it is being mixed with several ingredients like fine aggregate, coarse aggregates and water. Present investigation is focused to know the efficiency of limestone calcined clay concrete by conducting tests on M 50 grade of concrete and the values of LC_3 specimens are compared with OPC and PPC based concrete specimens.

3. METHODOLOGY AND EXPERIMENTS

Present study is carried out to investigate the performance of LC_3 based concrete and compared with OPC and PPC based concretes. In the present paper M 50 grade of concrete is considered and designed according to IS 10262-2019. The details of proportions of mix are shown in Table 1. The parameters in the study include grade of concrete (M_{50}), Type of binder (LC_3, OPC and PPC) and age of curing (i.e. 7, 14 and 28 days). To determine the strength properties, compressive, split tensile and flexural strength tests are carried out on specimens and the durability studies are Rapid chloride ion penetration test and absorption and desorption tests. The specimens were designated according to the grade of concrete, type of concrete. Typically a concrete specimen labeled Mix AO represents M 50 grade concrete containing OPC, Mix AP represents M 50 grade concrete containing PPC and Mix AL represents M 50 grade concrete with 0.34 w/c ratio, containing Limestone calcined clay cement.

Table 1- Mix proportions for M 50 grade concrete

Materials	Quantities (kg/m^3)
Binder (OPC, PPC and LC_3)	450
Fine aggregate	627
Coarse aggregate	1212
Water	153
Super plasticizer	7

3.1 Materials used:

[1]. Ordinary Portland Cement (OPC): 53 grade OPC conforming to IS 12269-2013 is used in present study and its physical properties are given in Table 2.

[2]. Portland Pozzolana Cement (PPC): PPC conforming to IS 1489-1 (1991) is used for investigation and the properties are shown in Table 2.

[3]. Limestone Calcined Clay Cement (LC_3): It is procured from society for Technology and Action for Rural development Authority, New Delhi, India. Table 2 shows the physical properties of LC_3.

[4]. Fine aggregate (FA) and Coarse aggregate (CA): As per IS: 383-2016 codal provisions fine and coarse aggregate were used. The properties are shown in Table 3.

[5]. Super Plasticizer: Poly carboxylic ether based water-reducing admixture confirming ASTM C494 [20] is used as super plasticizer for better workability.

Table 2 Properties of OPC, PPC and LC_3

S.NO	PROPERTY	OPC	PPC	LC_3
1	Normal consistency	31 %	32 %	33%
2	Initial setting time	33 minutes	32 minutes	35 minutes
3	Final setting time	460 minutes	480 minutes	270 minutes
4	Specific gravity	3.11	2.98	2.79
5	Fineness	4%	7%	6%
6	Compressive strength in MPa	54.1	50.2	53.5

Table 3: General properties of CA and FA

Properties	CA	FA
Bulk Density (g/cc)	1.62	1.71
Specific gravity	2.78	2.57
Fineness modulus	7.44	2.67
Void ratio	1.01	0.92
Porosity	41%	49%

3.2 Experiments:

(a) Compressive Strength Test (CST): The cube of size 150×150x150 mm is considered and as per IS 516 (1959) [14] test is conducted on 3000 kN compression testing machine.

(b) Split Tensile Strength Test (STST): The test is carried out on the cylinder of 150 mm dia. and 300 mm height as per IS 5816-1999 [15].

(c) Flexural Strength Test (FST): The prism of size 500×100×100 mm is used to determine the flexural strength of concrete specimens and as per IS 516 codal provisions, the test is carried out on specimens.

(d) Rapid Chloride ion Penetration Test (RCPT): This test is performed according to ASTM C 1202-97 [16]. Specimen of 50 mm thick and 100 mm dia. is considered for testing. The charge passed (Total) is found out and is used to rate the quality of the concrete according to ASTM C 1202 rating.

(e) Absorption Test (AT): The absorption test is used to measure the rate of absorption of water by the concrete specimen by finding the increase in the mass of a specimen with function of time when cubes are immersed in water for a period of 72 hrs.

(f) Desorption Test (DT): This test is done to know the rate of evaporation of water in the concrete by finding the decrease in the mass of a specimen with function of time when cubes are kept in hot air oven for 72 hours at 105°c. The cubes which are cured upto 28 days are taken from curing tank and surface dried before they are kept in hot air oven.

4. RESULTS AND DISCUSSIONS

4.1 Compression Test:

Figure 1 shows the results for compressive strength for M 50 grade concrete with different types of binders. From 28 days results it is clearly observed that the mixes attain the target strength. Based on the 7 days results it is evident that the LC_3 concrete has shown better performance in terms of early strength gain when compared to PPC concrete. By seeing the 28 days test results it is evident that LC_3 specimens performed better than that of other binder concretes. The enhancement of strength is mainly due the formation of dense and compact micro structure of LC_3 particles. On the other hand PPC based concrete is shown less strength compared to OPC and LC_3. Reason for achieving better strength values is mainly due to the dense uniform morphology of LC_3 and a large quantity of alumina is available in calcined clay which will react with calcium carbonate present in limestone to produce alumina phases. On the other hand PPC is a binary system and LC3 is ternary system, while comparing both it is found that the early strength gain during first 7 days is more in LC_3 concrete which resolves the problem of low strength gain at early ages.

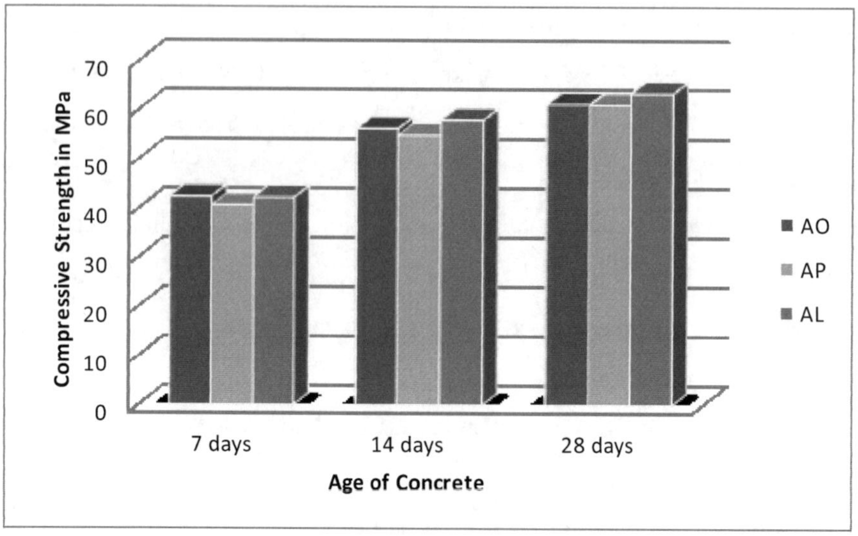

Figure 1 Compressive strength results for M 50 grade concrete with different binders

4.2 Split Tension Test:

Figure 2 shows the split tensile strength results for mixes A with different types of binders. From STST results it is noticed that, similar values are obtained for all types of concrete irrespective of type of binder. All the STST values are on par with marginal differences only, the effect of pozzolana activity is not much but LC_3 concrete specimens attained strength more than that of OPC and PPC based concrete specimens.

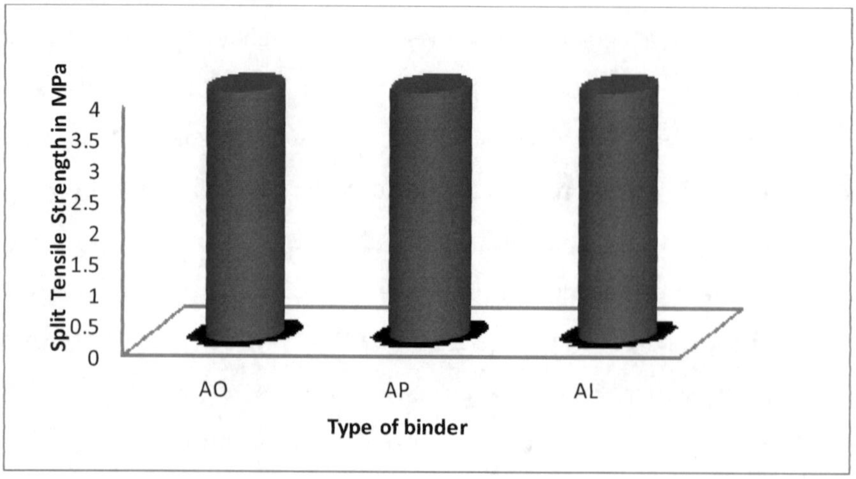

Figure 2 Split Tensile strength Test results for M 50 grade concrete with different binders

4.3 Flexural Strength Test:

Figure 3 shows the result for flexural strength for mixes A with different binders. It is clearly observed that the performance of LC_3 concrete is better when compared with OPC and PPC based concretes. An average of 12 percentage of increase in flexural strength is observed.

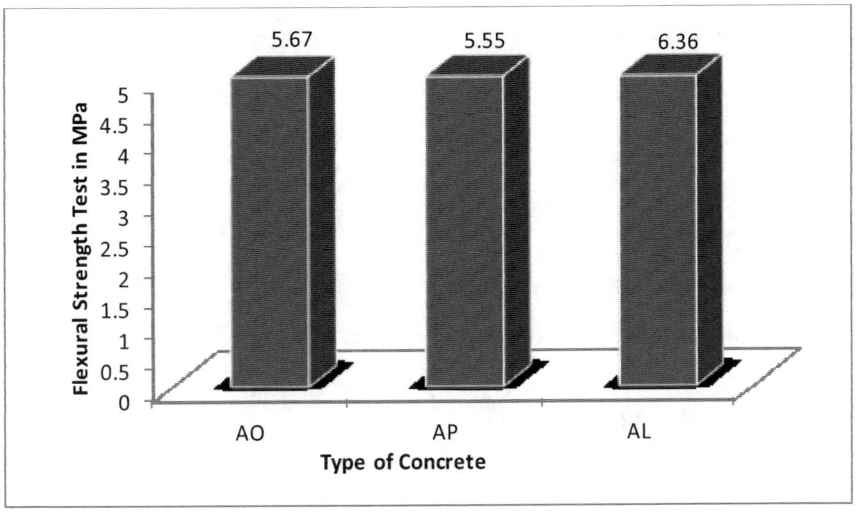

Figure 3 Flexural Strength Test results for M 50 grade concrete with different binders

4.4 RCPT Results:

Table 4 shows the RCPT results for M 50 grade concrete with different types of binders. From the values of charge passed through specimen it is evident that the ingress of chloride ions into LC_3 concrete specimens is less compared to other specimens. From this it is clear that the strength properties are enhanced with the RCPT results. LC_3 concrete specimens are less vulnerable to chloride ingress, which is due to the refinement of pore sizes and having low pore threshold radius value. Due to the presence of fibrillar-like core structure LC_3 based concrete helps in strengthening the concrete structures form environmental effects like chloride attack and carbonation.

Table 4: RCPT Results for M_{50} grade of concrete with different binders

S. No.	Mixes	Charge passed (Coulombs)	Chloride ion permeability as per ASTM C 1202
1	AO	1011	low
2	AP	897	Very low

3	AL	391	Very low

4.5 Absorption and Desorption Test Results:

Table 5 shows the absorption for M_{50} grade of concrete with different binders. From table 4 it is observed that the rate of absorption all specimens is linear upto 6 hours than the curve is almost straight for all specimens irrespective of type of binder and grade of concrete. On other hand it is clear that LC_3 concrete specimen shown lesser water absorption when compared to OPC and PPC based concrete specimens. As pore inter connection is less in LC_3 concrete and the inter layer space between the calcium ions and silicate chains is filled with water and calcium ions. The rate water absorption is less in case of LC_3 concrete. whereas in OPC based concrete specimens the rate water absorption is more due to less dense microstructure and inter connectivity of pores is more as compared to PPC based concrete.

Table 5 Percentage of water Absorption for M 50 grade concrete with different binders

Time (hours)	Water Absorption (%)		
	AO	AP	AL
0	0	0	0
0.5	1.445	1.353	0.998
1.5	2.088	1.91	1.397
3	2.489	2.229	1.756
5	3.052	2.667	2.235
6	3.253	2.786	2.355
8	3.413	2.945	2.514
10	3.694	3.184	2.794
12	3.895	3.383	2.914
24	4.037	3.702	3.353
30	4.037	3.742	3.353
36	4.037	3.742	3.353
48	4.037	3.861	3.512
60	4.057	3.861	3.512
72	4.128	3.901	3.512

4.6 Desorption Test Results:

Table 6 shows the percentages of water desorption for M 50 grade concrete with different binders. From the values of desorption it is found that the loss of water is more in initial hours i.e. up to 24 hours and then the curve is almost straight. At a temperature of 105° C freely available water molecules are evaporated in concrete specimens. These voids present in the concrete are responsible for the ingress of harmful chemicals into concrete. Concrete with less percentage water desorption is more durable than that of other concrete specimens. From curves it is found that the LC_3 concrete specimens prone to less desorption compared to OPC and PPC concrete specimens. It is evident that the LC_3 concrete has dense microstructure and compact core structure which does not allow the water desorption and this is due to the presence of finer particles in LC_3 concrete and more reactivity pozzolana action is taken place which in turn requires more amount of water to form secondary gel phase. There is a clear coincidence in durability and strength results, which helps to strengthen the performance of LC_3 concrete in all aspects. Ternary blended concretes are more durable than that of OPC due to the effect of fine grain particles and compact core structure of concrete. Enhanced properties of LC_3 concrete is mainly due to the high reactivity pozzolana action which fills the space and contribute to the better performance in terms strength and durability.

Table 6 Percentage of water desorption for M 50 grade of concrete with different binders

Time (hours)	Water Desorption (%)		
	AO	AP	AL
0	0	0	0
1	0.269	0.154	0.153
3	0.732	0.54	0.344
6	1.425	0.965	0.575
12	2.003	1.389	1.264
24	2.851	2.239	2.108
36	3.352	2.702	2.529
48	3.815	3.104	2.973
60	3.93	3.411	3.127
72	4.046	3.679	3.243

5. CONCLUSIONS

Based on the strength and durability tests on the use of LC_3 in concrete and comparison with OPC and PPC based concretes, the following are the findings.

- Based on the strength properties, LC_3 concrete performed better in terms of compressive strength and flexural strength.
- From RCPT results it is observed that the LC_3 concrete cubes are subjected to low ingress of chloride ions compared to OPC and PPC concrete cubes.
- By water absorption and desorption test results it is evident that the limestone calcined clay concrete cubes resulted to low percentage of water absorption and desorption compared to other concrete cubes. Due to the fewer pores interconnection between the particles and the forming fibril structure morphology helps to decrease the ingress of harmful agents into the concrete further enhances the durability.
- From over all studies, it is concluded that the LC_3 concrete specimens shown enhanced performance than that of concrete made with OPC and PPC.

REFERENCES

[1]. Schmidt, W., Alexander, M. and John, V., 2018. Education for sustainable use of cement based materials. *Cement and Concrete Research*, *114*, pp.103-114.

[2]. Rakhimova, N. and Rakhimov, R., 2020, July. Advances in development of calcined clays as supplementary cementitious materials. In *IOP Conference Series: Materials Science and Engineering* (Vol. 890, No. 1, p. 012085). IOP Publishing.

[3]. Scrivener, K., Martirena, F., Bishnoi, S. and Maity, S., 2018. Calcined clay limestone cements (LC3). *Cement and Concrete Research*, *114*, pp.49-56.

[4]. Dhandapani, Y., Sakthivel, T., Santhanam, M., Gettu, R. and Pillai, R.G., 2018. Mechanical properties and durability performance of concretes with limestone calcined clay cement (LC3). *Cement and Concrete Research*, *107*, pp.136-151.

[5]. Abdullah, M.E., Jaya, R.P., Shahafuddin, M.N.A., Yaacob, H., Ibrahim, M.W., Nazri, F.M., Ramli, N.I. and Mohammed, A.A., 2018, May. Performance of Kaolin Clay on the Concrete Pavement. In *IOP Conference Series: Materials Science and Engineering* (Vol. 358, No. 1, p. 012049).

[6]. Avet, F. and Scrivener, K., 2018. Investigation of the calcined kaolinite content on the hydration of Limestone Calcined Clay Cement (LC3). *Cement and Concrete Research*, *107*, pp.124-135.

[7]. Scrivener, K. and Favier, A., 2015. *Calcined clays for sustainable concrete*. RILEM.

[8]. Vizcaíno Andrés, L.M., Antoni, M.G., Alujas Diaz, A., Martirena Hernández, J.F. and Scrivener, K.L., 2015. Effect of fineness in clinker-calcined clays-limestone cements. *Advances in Cement Research*, *27*(9), pp.546-556.

[9]. Bishnoi, S., Maity, S., Mallik, A., Joseph, S. and Krishnan, S., 2014. Pilot scale manufacture of limestone calcined clay cement: the Indian experience. *Indian Concr. J*, *88*(6), pp.22-28.

[10]. Ferreiro, S., Canut, M.M.C., Lund, J. and Herfort, D., 2019. Influence of fineness of raw clay and calcination temperature on the performance of calcined clay-limestone blended cements. *Applied Clay Science*, *169*, pp.81-90.

[11]. Krishnan, S., Emmanuel, A.C. and Bishnoi, S., 2019. Hydration and phase assemblage of ternary cements with calcined clay and limestone. *Construction and Building Materials*, *222*, pp.64-72.

[12]. Liu, Y. and Ling, T.C., 2018, October. Potential use of calcined kaolinite-based wastes as cement replacements in concrete–an overview. In *IOP Conference Series: Materials Science and Engineering* (Vol. 431, No. 3, p. 032006). IOP Publishing.

[13]. Gettu, R., Pillai, R.G., Santhanam, M., Rathnarajan, S., Basavaraj, A., Raju, S. and Dhandapani, Y., 2019. Service life and life-cycle assessment of reinforced concrete with fly ash and limestone calcined clay cement.

[14]. IS: 516-1956 (Reaffirmed 2013) 'Indian Standard Methods of Tests for Strength of Concrete', *Bureau of Indian Standards*, New Delhi, India, 1956.

[15]. IS: 5816. (1999). Indian standard splitting tensile strength of concrete. *Bureau of Indian Standards*.

[16]. Joshi, R.C., Chatterji, S., Achari, G. and Mackie, P., 2000. Reexamination of ASTM C 1202—Standard Test Method for Electrical Indication of Concrete's Ability to Resist Chloride Ion Penetration. *Journal of testing and evaluation*, *28*(1), pp.59-61.

Water absorption and soil burial degradation of Poly (lactic) acid (PLA)/Oil palm fibres (OPF) composite.

Norsuriati Hapizi[1], Rahmah Mohamed[1]

Green Polymer Research Group, Faculty of Applied Science, Universiti Teknologi MARA, UiTM Shah Alam, 40450 Selangor DE, Malaysia

Abstract. : This research investigated the biodegradation of cellulose based PLA with oil palm composites on two sizes of fibre (0.25mm and 1.0mm). Poly (lactic acid) (PLA) was blended with oil palm fibre using twin screw extrusion technique.The use of oil palm can enhance the biodegradability of PLA since it is compostable under soil burial and easier absorb by water before disintegration for a few weeks. PLA was found to be biodegradable but rather slow in compostability unless at high temperature and humidity. In this study,the biodegradation rate was tested using soil burial and water absorption test. The samples were buried in soil and tested in aqueous water for specific duration at different pH. The higher the fibre content of 25%, the higher the weight loss and water absorption through soil burial and water absorption testing respectively. For soil burial testing, tsample PLA OPF 25% higher weight loss of 3.7% were found for sample fiber size 1.0mm as compared to 1.6% (0.25mm). While for water absorption testing, sample has different behaviour at pH 5.5, 7 and 8. The higher the fibre content, the higher water and soil burial weight loss. For bigger fibre sizes of1.0 mm, higher water absorption and mass loss reading were observed.

1. Introduction

Petroleum-based plastic production has risen drastically, reaching around 350 million tonnes per year (Garside, 2019). There are also billions tons of plastic waste accumulated due to inappropriate usage and disposal of plastic. This will lead to terrestrial and marine ecosystem pollution due to most of the plastic waste goes to river and oceans. In order to overcome this problem, biodegradable polymer were introduced besides to replace conventional petroleum-based polymer and to reduce environmental pollution. Biodegradable is the tendency of objects to disintegrate (decompose) by the action of micro-organisms such as bacteria or biological fungi (with or without oxygen) while being assimilated into the natural environment. The biodegradable polymers that synthesised from bio-based polymer have been studied was Poly (lactic Acid) (PLA).

PLA has good mechanical properties and biodegradability compared to polystyrene (Martin & Averous, 2001). Semi-crystalline polymer is chosen for properties that require strong mechanical properties and toughness. PLA is, however, much more costly than conventional plastics, and the rate of degradation is still slow (Ramsay, Langlade, Carreau, & Ramsay, 1993).

To address the low degradation rate of PLA, blending with other natural polymers that have a higher degradation rate is recommended. Natural fibres that are completely degradable when combined with thermoplastic material are used to minimise manufacturing costs and adverse environmental effects. Today, the use of natural fibre namely oil palm fibers available abundant in Malaysia are of great importance in composite research and industry. Composites using natural fibre act as replacement for synthetic fibre in composites such as recyclability, renewability, biodegradability, abundant, resistance to corrosion, non-toxicity, high flexibility, competitive mechanical properties, low density, less energy consumption, minimum abrasive resistance to processing equipment and waste disposal problems[6]. Properties of composites were extensively studied [10, 14, 15].

Testa et al done an experiment on biodegradability of oil palm with poly (lactic acid (PLA) using garbage-processing machine shows that after composting for 30 days, PLA/Oil Palm's biodegradability decreases by 38%. The tensile strength decreases rapidly during the first 5 and 10 days. After 30 days, due to the degradation of cellulose in the oil palm fibres, the tensile strength still showed a significant decrease of 10%. At the beginning of the 14 days, the percentage of weight loss result are same but after 30 days the percentage of weight loss increase drastically about 40%. These indicates that the enzymatic degradation in the cellulosic polymer chains degrades fibre strength. Decreasing strength of fibre and polymer shows that the polymer is degrade into carbon dioxide and water. Before composting, surface of the specimen is been cover by the biodegradable resin. According to Ochi, S. 2008, the exposure of the fibre is obtained due to the resin on the sample surface is degrade.

In studies on effects of Kenaf and Rice Husk on Water Absorption and Flexural Properties of Kenaf/CaCO3/HDPE and Rice Husk/CaCO3/HDPE Hybrid Composites By NAZuhaira, RM etal (2015), the effects of natural fiber for both rice husk and kenaf had been investigated [13] ; water absorption increased with the presence of kenaf and rice husk fillers. Natural fibers in composite significantly influence water absorption properties due to natural characters of fibers which contain cellulose, hemicellulose and lignin structures.

2. Experimental Analysis

2.1 Raw materials

Materials that were used in this study is Poly (lactic Acid) (PLA) of extrusion grade resin with Melt Index MI of 2, while Oil Palm Fibres (OPF) as reinforced fibre were incorporated to modify the mechanical properties which also acts as fillers.

2.2 Methodology

2.2.1 Chemical treatment of Oil Palm Fibre. Before every process, the sample need to be dried to eliminate moisture in the oven or vacuum. The cleaned oil palm sample need to be treated with 6% Sodium Hydroxide (NaOH) solution for 24 hours and then washed and dried in an oven at 40°C for 24 hours.

2.2.2 Mixing and Compounding. Oil palm fibre and PLA were blended and the mixture extruded into twin screw extruder of length/diameter ratio of 16:1 before palletized. The temperature of the extruder of the machine were set to different temperature zones ranging from 160-175°C and the rotation speed is 330 r/min. Next, the extruded will be trim using trimming machine before. Then the sample were compressed using Hot Press Machine.

3. Results and Discussions

3.1 Water Absorption

Water absorption test was conducted to determine the absorption of the sample when immersed in the different pH system The duration of immersion was taken for 6 weeks.

3.1.1 Water absoption in different pH. PLA Virgin showed minimum water absorption value. PLA Virgin sample absorb less moisture than sample that compounded with oil palm fibres. The water absorption value for all samples that immersed in acidic (pH 5.5) water solution have been reported in Figure1. Results clearly indicated that sample of PLA Virgin showed lowest water absorption value which is 1.06% over 16 weeks. Figure 1 showed the increasing water absorption values with addition of fibre content. According to Huner (2015), Zahari et al. (2015), and Chen et al. (2016), water absorption increased as the fibre content increase due to the hydrophilic nature of fibre that they are cellulose fibre. Upon exposure to the water absorption, the fibre initially swells. Besides, as results of swelling, micro cracks can appear in a brittle matrix and in turn can lead to larger water transport through the fibre matrix interface. Higher water absorption were exhibited by fibre size 1.0mm, this is followed by 0.25mm. This is was due to the larger fibre sizes impose less surface interaction between matrix and fibres.

The water absorption value for all samples that immersed in neutral (pH 7) were exhibited as in Figure 2. Results clearly indicated that sample of PLA Virgin showed minimum water absorption value which is 1.1% over 6 weeks. Figure 2 shows increasing in water absorption values with addition of fibre content. It can be interpret that from the fibre size 0.25mm, values of sample of PLA/3% OPF, PLA/7% OPF,PLA/15%OPF and PLA/25%OPF showed 1.6%,2.34%,2.3% and 6. 48% increased water absorption respectively at room temperature. According to Huner (2015), Zahari et al. (2015), and Chen et al. (2016), water absorption increased as the fibre content increase understandably due to the hydrophilic nature of cellulose fibres. Fibre matrix interface breaksdwon and water are diffused into the interface and microcracks region and ingression of water could initiate degradation hotspots. For composite of fibre size 1.0mm, water absorption of PLA/3% OPF, PLA/7% OPF, PLA/15%OPF and PLA/25%OPF showed 2.59%,3.07%,3.35% and 8.46% respectively. Higher water absorption for higher loading, was imparted by the larger size of fibres having less surface interaction and for bonding between matrix and fibre.

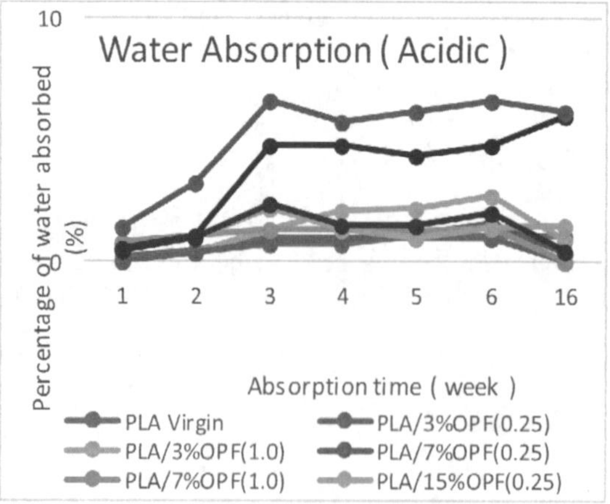

Figure 1: Graph of water absorption test in acidic condition PH 5.5.

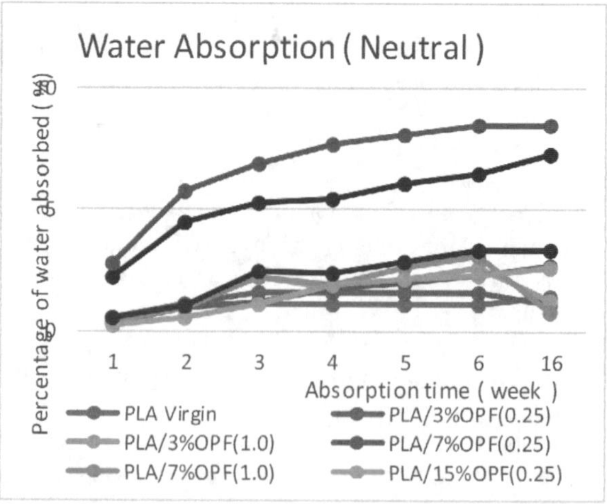

Figure 2 :Graph of water absorption test in neutral condition PH 7.

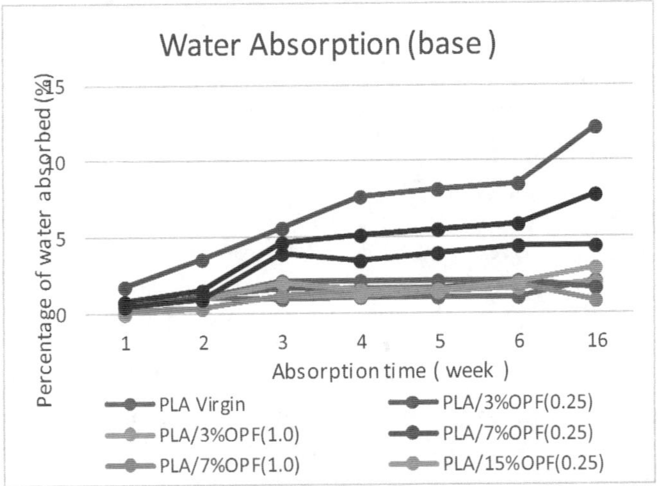

Figure 3 : Graph of water absorption test in base condition at PH 8.

Figure 3 showed increasing water absorption values with addition of fiber content. For 0.25mm , PLA/25%OPF showed highest water ~ 5.58% whilst for fiber size 1.0mm, absorption value can go up to8.56% for highest fiber composition of 25% This is was due to the larger size of fiber which cause higher water uptake.

3.2 Soil Burial Test

Degradation of sample was measured based on dry weight loss before and after buried time. While a dry weight loss can be used to indicate biodegradation rate of sample during soil burial condition. It can be observe that, as the soil burial duration increased, the percentage of weight loss also increased and more significant increase was found when incorporate with oil palm fibre. It also can be observe that, the higher the percentage of fibre, the higher the percentage of weight loss over 10 days. And the percentage of weight loss of sample PLA/3% OPF, PLA/7% OPF, PLA/15% OPF and PLA/15% OPF are more higher compared to PLA Virgin. These results indicate that the introduction of OPF had accelerate the degradation of PLA. This is because ligno-cellulosic materials are more easily attacked by microorganisms. This occur due to the increased in water uptake of the composites which causes major deterioration. Swelling by water uptake also led to the micro cracking of the composites resulting in weight loss. The activities of microorganisms in the soil reduced the sample weights. The voids caused by the presence of moisture also permitted the colonization by microorganisms.

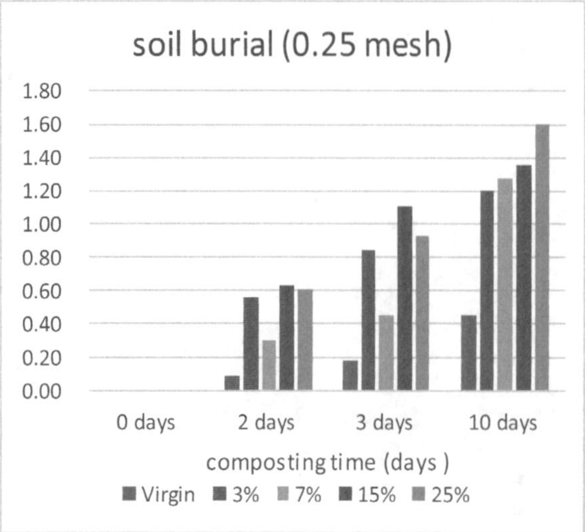

Figure 4: Graph of soil burial result for fiber size 0.25 mm.

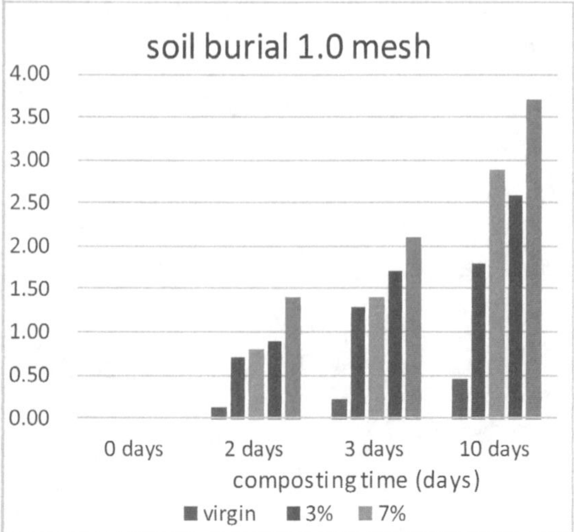

Figure 5: Graph of soil burial result for fiber size 1.0 mm.

Figure 4 and 5 shows two size of oil palm fibre when incorporate with PLA. From this figure, different size of fibre gives different percentage of weight loss. The sample shows a progressive loss of the mechanical properties in relation to the biodegradation time. It can be observed that, the bigger size of fibre leads to higher percentage weight loss. For specimen PLA/3% OPF which the fibre size 1.0 mm has higher percentage of weight loss across 10 days which 1.8% as compared to fibre size 0.25mm was only 1.27%.The trend of this graph was similar to other percentage of fibre loading whereas the higher size of fibre shows higher percentage of weight loss. For specimen 7% of oil palm fibre loading with fibre size 1.0mm and 0.25mm, the percentage of weight loss was 2.9% and 1.2% respectively. Next, across 10 days, the percentage of weight loss for specimen 15% of oil palm fibre loading 2.6% and 1.35% according to fibre size 1.0 and 0.25mm respectively. Lastly for specimen PLA/25% OPF of 1.0 mm shows the highest percentage of weight loss than 0.25mm. At this stage, hydrophilic group on the surface of the OPF absorbed water and microorganisms leading to degradation. The increased of fibre content, accelerated the degradation of PLA. This indicate that, size of fibre 1.0 mm was the most optimum biodegradability due to the highest loss weight of the sample.

4. Conclusion

PLA composite system has element of biodegradability compared to the conventional plastic polymer. This study had shown biodegradation of PLA/OPF in soil burial and their water absorption behaviour. Oil palm fibre/PLA could be biodegraded by microorganisms in the plantation soil.

The higher the percentage of fibre in composite system, the higher the percentage of water absorbed and weight loss. This experiment were done at room temperature. Hydrolysis of PLA matrix must occur for further degradation of PLA while for fibre, the higher the size of fibre the higher the percentage of water absorbed and weight loss. This had been exhibited via biodegradation rate of the larger 1.0mm as compared to reduced size (0.25mm).

References

[1] R. L. Shogren, W. M. Doane, D. Garlotta, J. W. Lawton, and J. L. Willett, "Biodegradation of starch/polylactic acid/poly(hydroxyester-ether) composite bars in soil," *Polym. Degrad. Stab.*, vol. 79, no. 3, pp. 405–411, 2003.

[2] M. Akrami, I. Ghasemi, H. Azizi, M. Karrabi, and M. Seyedabadi, "A new approach in compatibilization of the poly(lactic acid)/thermoplastic starch (PLA/TPS) blends," *Carbohydr. Polym.*, vol. 144, pp. 254–262, 2016.

[3] S. Ochi, "Mechanical properties of kenaf fibers and kenaf/PLA composites," *Mech. Mater.*, vol. 40, no. 4–5, pp. 446–452, 2008.

[4] Z. Xiong *et al.*, "Preparation and characterization of poly(lactic acid)/starch composites toughened with epoxidized soybean oil," *Carbohydr. Polym.*, vol. 92, no. 1, pp. 810–816, 2013.

[5] M. Babaee, M. Jonoobi, Y. Hamzeh, and A. Ashori, "Biodegradability and mechanical properties of reinforced starch nanocomposites using cellulose nanofibers," *Carbohydr. Polym.*, vol. 132, pp. 1–8, 2015.

[6] A. M. Adole, J. M. Yatim, and S. A. Ramli, "SCIENCE & TECHNOLOGY Kenaf Fibre and Its Bio-Based Composites : A Conspectus," vol. 27, no. 1, pp. 297–329, 2019.

[7] N. Saba, M. T. Paridah, and M. Jawaid, "Mechanical properties of kenaf fibre reinforced polymer composite: A review," *Constr. Build. Mater.*, vol. 76, pp. 87–96, 2015.

[8] V. Fiore, G. Di Bella, and A. Valenza, "The effect of alkaline treatment on mechanical properties of kenaf fibers and their epoxy composites," *Compos. Part B Eng.*, vol. 68, pp. 14–21, 2015.

[9] O. M. L. Asumani, R. G. Reid, and R. Paskaramoorthy, "The effects of alkali-silane treatment on the tensile and flexural properties of short fibre non-woven kenaf reinforced polypropylene composites," *Compos. Part A Appl. Sci. Manuf.*, vol. 43, no. 9, pp. 1431–1440, 2012.

[10] B. F. Yousif, A. Shalwan, C. W. Chin, and K. C. Ming, "Flexural properties of treated and untreated kenaf/epoxy composites," *Mater. Des.*, vol. 40, pp. 378–385, 2012.

[11] S. Lv, Y. Zhang, J. Gu, and H. Tan, "Biodegradation behavior and modelling of soil burial effect on degradation rate of PLA blended with starch and wood flour," *Colloids Surfaces B Biointerfaces*, vol. 159, pp. 800–808, 2017.

[12] Ochi, S. (2008). Mechanical properties of kenaf fibers and kenaf/PLA composite. Mechanics of Materials, 40(4-5), 446-452. https://doi.org/10.1016/j.mechmat.2007.10.006

[13] *Noor Zuhaira Abd Aziz*, *Rahmah Mohamed*, *Mohd Muizz Fahimi M.* World Academy of Science, Engineering and Technology International Journal of Industrial and Manufacturing Engineering Vol:9, No:4, 2015

[14] Mohd Nurazzi, N, Khalina, A.1,2*, Sapuan, S. M. Dayang Laila, A. H. A. .1 , Rahmah, M and Hanafee, Z. A Review: Fibres, Polymer Matrices and Composites Pertanika J. Sci. & Technol. 25 (4): 1085 - 1102 (2017)

[15] R Mohamed. Physical and morphological properties of filled calcium carbonate/kenaf fibre/rice husk polypropylene hybrid composite., AIP Conference Proceedings 1985 (1), 040009

AUTHOR INDEX